8th RILEM International Conference on Mechanisms of Cracking and Debonding in Pavements

RILEM BOOKSERIES

Volume 13

RILEM, The International Union of Laboratories and Experts in Construction Materials, Systems and Structures, founded in 1947, is a non-governmental scientific association whose goal is to contribute to progress in the construction sciences, techniques and industries, essentially by means of the communication it fosters between research and practice. RILEM's focus is on construction materials and their use in building and civil engineering structures, covering all phases of the building process from manufacture to use and recycling of materials. More information on RILEM and its previous publications can be found on www.RILEM.net. Indexed by SCOPUS and Springerlink.

More information about this series at http://www.springer.com/series/8781

Armelle Chabot · William G. Buttlar
Eshan V. Dave · Christophe Petit
Gabriele Tebaldi
Editors

8th RILEM International Conference on Mechanisms of Cracking and Debonding in Pavements

Springer

Editors
Armelle Chabot
LUNAM Université, IFSTTAR
Bouguenais
France

William G. Buttlar
Newmark Civil Engineering Laboratory
University of Illinois at Urbana–Champaign
Urbana, IL
USA

Eshan V. Dave
Department of Civil and Environmental Engineering
University of New Hampshire
Durham, NH
USA

Christophe Petit
IUT, Département Génie Civil
Université de Limoges
Egletons
France

Gabriele Tebaldi
Department of Civil and Environmental Engineering and Architecture
University of Parma
Parma
Italy

ISSN 2211-0844 ISSN 2211-0852 (electronic)
RILEM Bookseries
ISBN 978-94-024-0866-9 ISBN 978-94-024-0867-6 (eBook)
DOI 10.1007/978-94-024-0867-6

Library of Congress Control Number: 2016941090

Printed on acid-free paper

This Springer imprint is published by Springer Nature
The registered company is Springer Science+Business Media B.V. Dordrecht

Preface

A pavement is a complex structure composed of several layers made up of different types of heterogeneous materials. This multilayer system rests on a semi-infinite soil and is subjected to severe climate variations and heavy traffic loads. In their lifetime, pavements undergo degradation due to different mechanisms, of which cracking is among the most important. To appraise this type of degradation in mechanical analyses, boundary conditions applied to the pavement structure must be known. Moreover, the damage and the fracture behavior of all the material layers as well as interfaces must be understood.

Prior to 2000 and at the initiative of the Rilem Technical Committee (TC) 97-GCR (application of Geotextiles to Crack prevention in Roads) led by Louis Francken, a series of four international RILEM conferences on "Reflective Cracking" was held in Liege (RC1989) (RC1993), Maastricht (RC1996) and Ottawa (RC2000). The objective was to present up-to-date information on the rehabilitation of cracked roads with bituminous overlays which was the solution adopted worldwide to delay the cracking propagation in pavements (see the TC 157-PRC document).

To approach cracking problems in pavements in a more general sense, the scope of the conference was extended to other modes of cracking modes such as fatigue, aging, or top-down cracking. Then, a successful series of three conferences on "Cracking in Pavement" was held in Limoges (CP2004), Chicago (CP2008) and Delft (CP2012).

The purpose of the 8th Rilem international conference is to coordinate with the activities of the Rilem TC 241-MCD (2011–2016), which aims at developing a deeper fundamental understanding of the mechanisms responsible for cracking and debonding in asphalt concrete and composite (e.g., asphalt overlays placed on PCC or thin cement concrete overlay placed on asphalt layer) pavement systems. The objective of this event is to present the results of TC 241-MCD as well as the latest advances in research to analyze mechanical damage and its detection in multilayer systems. This will favor discussions between different research communities to help apply these advances to pavement structures. Eventually, the aim is to be able to better detect the initiation and the propagation of cracks in pavement and also to have tools that make possible the search for technical solutions able to prevent (or to limit) cracking in usual and emergeing structures.

Armelle Chabot
Chair of the 8th Rilem International Conference MCD2016

Organization Committee

Steering Committee

Chair
Dr. Armelle Chabot (Secretary of the Rilem TC 241-MCD), LUNAM Université, IFSTTAR, France

Co-Chairs
Prof. Bill Buttlar (Leader of the Rilem TC 241-MCD), University of Illinois at Urbana Champaign, USA
Dr. Eshan V. Dave (TG1 leader of the Rilem TC 241-MCD), University of New-Hampshire, USA
Prof. Christophe Petit (TG2 leader of the Rilem TC 241-MCD), Université de Limoges, France
Dr. Gabriele Tebaldi (TG3 leader of the Rilem TC 241-MCD), Università Degli Studi Di Parma, Italy

International Scientific Committee (ISC)

Dr. A. Apeagyei, University of Nottingham, UK
Dr. A. Beldens, BRRC, Belgium
Prof. R. Blab, Vienna University of Technology, Austria
Dr. D. Bodin, ARRB, Australia
Dr. R. Botella Nieto, Universitat Politécnica de Cataluña, Spain
Prof. D. Breysse, Université Bordeaux 1, France
Ing. Y. Brosseaud, LUNAM Université, IFSTTAR, France
Prof. W.G. Buttlar, University of Illinois, USA
Prof. F. Canestrari, Università Politecnica Delle Marche, Italy
Dr. A. Chabot, LUNAM Université, IFSTTAR, France
Dr. E. Chailleux, LUNAM Université, IFSTTAR, France
Prof. K. Chatti, Michigan State University, USA

Dr. O. Chupin, LUNAM Université, IFSTTAR, France
Dr. E. Dave, University of Minnesota-Duluth, USA
Dr. M. De Beer, CSIR Built environment, South Africa
Dr. X. Dérobert, LUNAM Université, IFSTTAR, France
Dr. A. Destrée, BRRC, Belgium
Prof. H. Di Benedetto, Université de Lyon, ENTPE, France
Dr. A. Duarte, University of Illinois, USA
Prof. F. Dubois, Université de Limoges, France
Prof. A.G. Dumont, EPFL, Switzerland
Prof. A. Ehrlacher, Ecole des Ponts Paris Tech, France
Prof. M. François, Université de Nantes, GEM, France
Dr. T. Gabet, LUNAM Université, IFSTTAR, France
Prof. A. Graziani, Univ. Politecnica Delle Marche, Italy
Dr. J. Grenfell, University of Nottingham, UK
Dr. S. Guessasma, INRA, France
Dr. F. Hammoum, LUNAM Université, IFSTTAR, France
Ing. P. Hornych, LUNAM Université, IFSTTAR, France
Prof. R. Kim, North Carolina University, USA
Dr. M.E. Kutay, Michigan State University, USA
Prof. F. Lebon, Université de Marseille, France
Prof. L. Legi Bariani Bernucci, Escola Politécnica da USP, Brasil
Prof. A. Loizos, National Technical University of Athens, Greece
Ing. P. Marsac, LUNAM Université, IFSTTAR, France
Dr. A. Millien, Université de Limoges, France
Prof. R. Miró, Universidad Politécnica de Cataluña, Spain
Prof. N. Moës, Ecole Centrale de Nantes, GEM, France
Prof. Emeritus AAA. Molenaar, TU Delft, The Netherlands
Dr. M.L. Nguyen, LUNAM Université, IFSTTAR, France
Dr. J. Pais, University of Minho, Portugal
Dr. K. Park, Yonsei University, Korea
Prof. M. Partl, EMPA, Switzerland
Prof. F. Pérez-Jiménez, Universidad Politécnica de Cataluña, Spain
Prof. C. Petit, Université de Limoges, France
Prof. D. Perraton, Ecole de Technologie Supérieure de Montréal, Canada
Ing. J.-M. Piau, LUNAM Université, IFSTTAR, France
Prof. G. Pijaudier-Cabot, Université de Pau et Pays de l'Adour, France
Dr. C. Plati, National Technical University of Athens, Greece
Dr. O. Pop, Université de Limoges, France
Dr. S. Pouget, Eiffage, France
Dr. M. Quiertant, Université Paris Est, IFSTTAR, France
Prof. C. Raab, EMPA, Switzerland
Prof. H. Reis, University of Illinois, USA
Prof. J.R. Roesler, University of Illinois, USA
Prof. R. Roque, University of Florida, USA
Dr. C. Sauzéat, Université de Lyon, ENTPE, France

Dr. T. Sedran, LUNAM Université, IFSTTAR, France
Dr. J.-M. Simonin, LUNAM Université, IFSTTAR, France
Prof. J. Silfwerbrand, KTH Royal Institute of Tech., Sweden
Prof. J.B. Soares, Univ. Federal do Ceara, Brasil
Dr. M. Takarli, Université de Limoges, France
Dr. G. Tebaldi, Univ. of Parma, Italy and University of Florida, USA
Dr. J.-M. Torrenti, Université Paris Est, IFSTTAR, France
Dr. F. Toutlemonde, Université Paris Est, IFSTTAR, France
Dr. A. Toumi, INSA Toulouse, France
Prof. A. Turatsinze, INSA Toulouse, France
Prof. Emeritus J. Uzan, Technion—Israel Institute of Technology, Israel
Dr. J.M. Vanderbossche, University of Pittsburgh, USA
Prof. M.P. Wistuba, Technische Univ. Braunschweig, Germany
Dr. A. Zofka, Road and Bridge Research Institute, Poland

Local Organizing Committee

Brigitte Beilvert, IFSTTAR/MAST/LAMES, France
Dr. Armelle Chabot, IFSTTAR/MAST/LAMES, France
Dr. Emmanuel Chailleux, IFSTTAR/MAST/MIT, France
Dr. Olivier Chupin, IFSTTAR/MAST/LAMES, France
Valérie Fournier, IFSTTAR/MAST, France
Maïssa Gharbi, IFSTTAR/MAST/LAMES, France
Ing. Bernard Jacob, IFSTTAR/DS, France
Jacques Kerveillant, IFSTTAR/MAST/LAMES, France
Joëlle Labarrere, IFSTTAR/COSYS, France
Isabelle Larrue, IFSTTAR/SG-DGN, France
Ing. Vincent Le Cam, IFSTTAR/COSYS/SII, France
Mathieu Le Pen, IFSTTAR/COSYS/SII, France
Hanan Nasser, IFSTTAR/MAST/LAMES, France
Dr. Mai Lan Nguyen, IFSTTAR/MAST/LAMES, France
Dr. Jean-Michel Simonin, IFSTTAR/MAST/LAMES, France
Franck Tartrou, IFSTTAR/MAST/LAMES, France
Dr. Andrea Themeli, IFSTTAR/MAST/MIT, France
Dr. François Toutlemonde, IFSTTAR/MAST, France

Contents

Part I
Cracking in Asphalt Materials: Fundamental Bituminous Material Characteristics

Cracking and Linear Visco Elastic Binder Properties

Geoffrey M. Rowe, Mark J. Sharrock and Sérgio Raposo

Abstract The contribution of an asphalt binder to the cracking performance of a mixture is strongly influenced by information relating to the relaxation properties and stiffness. In the 1980s considerable interest existed in the relationship between the relaxation properties and cracking. The relaxation spectra can be captured by the R-value over a limited range of stiffness. This can be related to bottom up fatigue, durability and thermal cracking when the stiffness information is known. Analysis has been presented that demonstrates the interrelationship between parameters used in France, United Kingdom and the USA through linear visco-elastic modeling of a limited region of the master curve. This has been used in several studies recently and has been shown to be useful for judging material used in airfield, high RAP products and other additives in the asphalt market. The use of the linear visco-elastic methods can also lead to a better understanding of methods used to describe fracture—for example the Vialit pendulum test and the direct tension test. While these tests are conducted often in a non-linear region the data can be better explained when interpreting the stiffness results with a frame work developed by linear visco elastic considerations. The paper discusses some aspects of the techniques for development of master curves, interrelationships, linear visco-elastic properties and the relationships between the various parameters proposed for use in different countries.

Keywords Rheology · Relaxation · Cracking · Interrelationships

G.M. Rowe (✉) · M.J. Sharrock · S. Raposo
Abatech, Blooming Glen, USA
e-mail: growe@abatech.com

M.J. Sharrock
e-mail: msharrock@abatech.com

S. Raposo
e-mail: sraposo@abatech.com

A. Chabot et al. (eds.), *8th RILEM International Conference on Mechanisms of Cracking and Debonding in Pavements*, RILEM Bookseries 13,
DOI 10.1007/978-94-024-0867-6_1

1 Importance of Relaxation

In the late 1960s Jongepier and Kuilman (1969) described the relaxation spectra by using a log-normal distribution. This work was further adopted by Ishai et al. (1988) to describe standard deviation of relaxation spectra, "S". While the concept of relaxation spectra involves the consideration of complex mathematic techniques the parameters of interest can be derived similarly by inspection of the resulting master curve in a Black space as noted by Ishai et al. (1988) since it effectively describes the shape of a master curve of G* versus phase angle or frequency. The importance of this description was recognized by Moutier et al. (1990) in that it provided the best correlation with the fatigue of asphalt mixes when considering a range of binder parameters, as shown in Fig. 1a. This was noted to be related to fatigue cracking of road pavements in France.

In the derivation of the CA model the shape of the relaxation spectra was also considered to be skewed from the distribution assumed in the earlier work (Christensen 1992) and this led to the development of the Christensen-Anderson model which defined the complex dynamic modulus as a function of a cross-over frequency, glassy modulus and a rheological index (R).

The R defined above was noted as being "directly proportional to the width of the relaxation spectra" by Christensen and Anderson (1992) and consequently while it is somewhat differently formulated to the "S" used by Ishai et al. (1988) a unique relationship exists between "R" and "S" (Rowe 2015). Differences in the numerical values of "R" and "S" are chiefly due to (log 2) constant used in the CA model and to a minor extent the shape of assumed for the master curve. Other workers such as Dobson (1972) and Dickinson and Witt (1974) have also defined parameter which describes the shape of the master curve in a similar manner to R. The direct correlation between R and these other parameters allows use of the correlations and

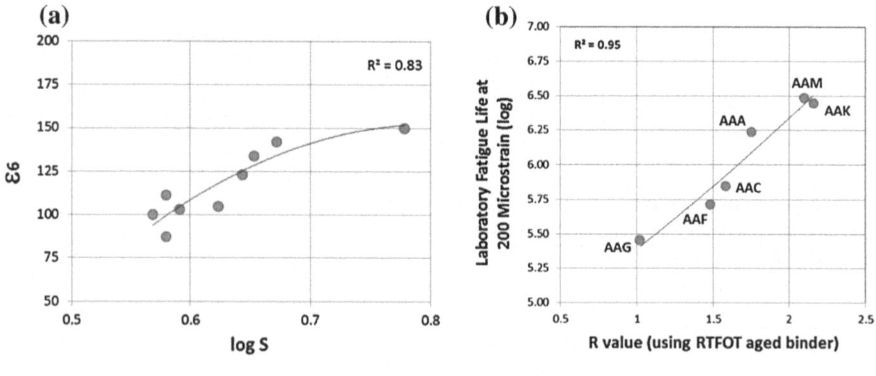

Fig. 1 **a** log S (standard deviation of relaxation spectra) versus strain required to fail a fatigue specimen in 10^6 load applications, **b** rheological index (relaxation parameter from CA model) versus life (log) at 200 micro-strain

data obtained from these experiments to justify the significance of the meaning of the R value used in more recent work.

Some important differences exist in the work developed in this paper compared to the original formulations of the CA model. In this work it has been assumed that temperature dependency is described by variable factors. This is generally a more accepted step in general rheology studies. The fixed values proposed in the SHRP work have not been used. Shift factors move the position of an isotherm with reference to a given temperature to another location on the time or frequency axis. Shifting to produce smooth master curves of stiffness applies when a material exhibits thermo-rheologically simple behavior. That is when isotherms of stiffness (expressed on log scales) are shifted either the log of time or frequency axis then they shall form a smooth curve. The same shift shall apply for all the visco-elastic parameters, e.g. G', G'', G^* and δ. Rowe (2015) has previously detailed the methods used to develop master curves from a combination of DSR and BBR data which involves interconversions from $S(t)$ to G^* and δ. The CA model is used with constraints on stiffness ranging from 10^5 to 10^9 Pa (Christensen 1992; Rowe 2014). Recently the relationship between the relaxation spectra $H(\tau)$ as defined by the CA has been shown in the form of a closed form arithmetic expression (Christensen et al. 1992).

With regard to the importance of this parameter to fatigue and cracking, the correlation of the R value from the recent testing described in this study can be compared to work conducted as part of the validation effort conducted for the mixture performance tests (Deacon et al. 1994). For mixture testing the preferred method adopted by the SHRP A003A team was the bending beam fatigue test. If we use the data within used by Deacon et al. (1994) and compare this with the R value for the RTFOT binder condition we obtain a very high correlation with performance. This example uses a Limestone aggregate which was given the code "RD" during the SHRP program. The performance is shown against the RTFOT ageing condition which has an r^2 value of 0.95, see Fig. 1b. If original or PAV properties are compared the r^2 values are 0.90 and 0.99 respectively. A similar high correlation is not evident with the values of R published at the time of the SHRP study and this may be a result of the determination of R including some different assumptions such as fixed parameters for the temperature susceptibility. This data set is consistent with the earlier work reported by French workers that longer fatigue life associated with traffic loading is obtained with higher values of R.

2 Stiffness Range to Describe Fracture Properties

When evaluating fatigue and fracture one of the major considerations is to consider the time of loading and temperature being used in the test. For many years it has been know that the fracture properties of asphalt binders and mixtures can be related to the binder stiffness (Heukelom 1966). While with newer materials the values of strength may vary (expressed as stress, strain or energy dissipated at an

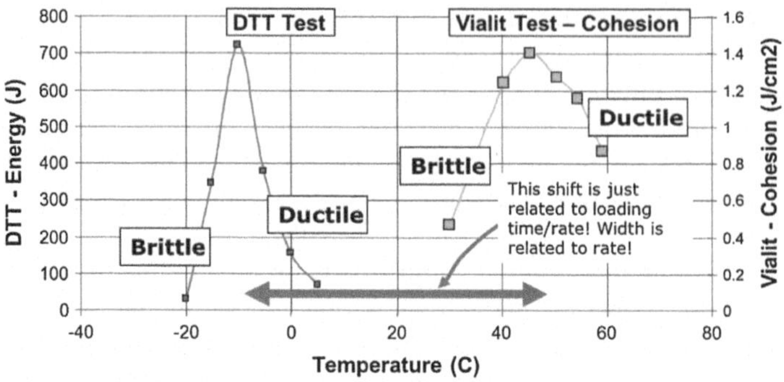

Fig. 2 Comparison of Direct Tension Test (energy under load vs. deformation curve) with Cohesion in Vialit test demonstrating difference in position of peak and width of curve produced

ultimate condition), the range of binder stiffness which typifies the brittle to ductile/flow fracture condition appears to be in the range 1 MPa to 1GPa (Rowe 2016).

The range of stiffness that can be described by the CA model cover this range adequately and thus the CA model provides a useful descriptive tool that can be related to both brittle and ductile fracture properties. The transitions that occur in typical "fracture" or "ultimate" tests can be related to this stiffness range. For example, while the peak temperatures obtained from the Vialit Pendulum Test (VPT) are significantly higher is has been demonstrated that similar ultimate property curves are obtained, which can be better understood if time/rate of loading is included in the calculations (Rowe 2014). For example the data shown in Fig. 2 shows a typical DTT result and a VPT compared to temperature. The differences in position on the temperature scale and the temperature ranges representing brittle to ductile/flow properties for the DTT data (which is much narrower) compared to that from the VPT data, can be directly related to the time of loading differences in the two tests.

If the stiffness of the binder is the controlling factor affecting the temperature spread then it can also be postulated that temperature spread can be considered to be related to the temperature susceptibility of the binder in some manner. This could be defined by an empirical measure or better understood by considering isochronal information which would be dependent upon the rheological fitting parameters. When polymer modified systems are compared to those for conventional binders a much more significant peak and temperature range are obtained. The differences in the results relate to the amount of energy needed to form the fracture surfaces and should be related to a slower rate of crack propagation associated with a tougher material. While this aspect has been traditionally considered for surface treatments with this type of test, conceptually it could be evaluated for asphalt mix performance.

3 Relationships with Cracking

Using stiffness information collected in the LVE region various authors have proposed different measures that relate to cracking (either cold temperature or durability cracking). For single event cold temperature cracking s value of S(60s) and m(60s) of 300 MPa and 0.300 have been used as criteria. These can be related to dynamic properties of G* = 111 MPa and δ = 26.2° (Rowe 2014). The Glove-Rowe parameter (G*cos δ^2/sin δ) has also been shown to relate to cracking (durability). Conceptually the G-R concept can also be used to capture single event cold temperature cracking (Rowe 2014). The visco-elastic transition (VET) concept has been promoted in the UK and it can be shown that these parameters can be calculated directly from the CA-Kaelble model fit parameters. A comparison of USA and UK binders are shown in Fig. 3 along with the values of the G-R parameter. The trend between these parameters is consistent with the various rheological parameters determined.

A further parameter which is currently showing increasing interest is the ΔTc as defined by the difference in critical temperature defined by the S(60s) = 300 MPa and m(60s) = 0.300. This parameter can be directly calculated from the CA-Kaelble model fit or easily deduced from BBR tests. It is strongly dependent upon the R value as are the other parameters discussed.

The parameters discussed above are all related to the stiffness and relaxation properties of the binder in the LVE region.

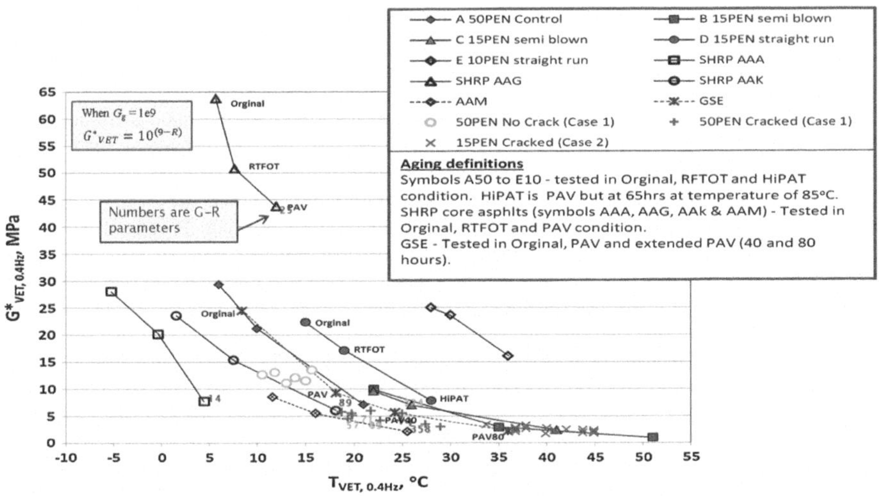

Fig. 3 Linear visco-elastic transition temperature parameters compared to G-R concept for various asphalt binders

4 Conclusions

Based upon the information presented in this paper and referenced work, we can conclude the following:

- Understanding the relaxation properties are critical for developing specifications that relate to cracking.
- The CA model rheological index (R) value is a good shape descriptor of the relaxation spectra as it relates to cracking.
- Fatigue and durability cracking are both related to the relaxation properties as defined by the R-value and the stiffness in the range $1e^5$ to $1e^9$ Pa.

References

Jongepier R, Kuilman B (1969) "Characteristics of the Rheology of Bitumens, Proceedings of the Association of Asphalt Paving Technologists, Volume 38, pp. 98-122.

Ishai I, Brule B, Vaniscote JC, Ramond G (1988) Some Rheological and Physico-Chemical Aspects of Long-Term Asphalt Durability. Proceedings of the Association of Asphalt Paving Technologists, Technical Sessions, Williamsburg, Virginia.

Moutier F, Ramond G, Such C, Bonnot J (1990) Influence of the nature of asphalt cements on the fatigue strength of asphalt-aggregate mixtures under imposed strain. Proceedings of the conference, "The United States Strategic Highway Research Program, Institution of Civil Engineers, Thomas Telford, London, pp. 229 to 240.

Christensen DW (1992) Mathematical Modeling of the Linear Viscoelastic Behavior of Asphalt Cements. PhD Thesis in Civil Engineering, Pennsylvania State University.

Christensen DW, Anderson DA (1992) "Interpretation of Dynamic mechanical Test Data for Paving Grade Asphalt Cements," Journal, Association of Asphalt Paving Technologists, Volume 61, pp. 67-116.

G.R. Dobson, "On the Development of Rational Specifications for the Rheological Properties of Bitumens," Journal of the Institute of Petroleum, Volume 58, No. 559, January 1972.

E.J. Dickinson and H.P. Witt, "The Dynamic Shear Modulus of Paving Asphalts as a Function of Frequency," Transactions of The Society of Rheology, 18:4, 591-606, 1974.

Rowe G (2015) Linear Visco-elastic Binder Properties and Asphalt Pavement Cracking. Proceedings 11th Conference on Asphalt Pavements for Southern Africa.

G.M. Rowe, "Interrelationships in Rheology for Asphalt Binder Specifications," Proceedings, 56th Annual Conference, Canadian Technical Asphalt Association, Winnipeg, MB, Canada. 2014. pp. 457-483.

J.A. Deacon, A.A. Tayebali, G.M. Rowe and C.L. Monismith, "Validation of SHRP A-003A Flexural Beam Fatigue Test," Engineering Properties of Asphalt Mixtures and the Relationship to Performance, ASTM STP 1265, Gerald A. Huber and Dale S. Decker, Eds., American Society for Testing and Materials, Philadelphia, 1994.

Heukelom, W, "Observations on the Rheology and Fracture of Bitumens and Asphalt Mixes", Proceedings, Association of Asphalt Paving Technologists, Volume 35, 1966, pp. 358-399.

Rowe, G.M., "Cohesion, Fracture and Bond – Understanding the Data from the Vialit Cohesion Pendulum Test and Other Fracture Tests from an Analysis of Rheological Properties," The International Journal of Pavement Engineering & Asphalt Technology, Volume 15 Issue 1 (ISSN 1464-8164), May 2014, pp. 20-37.

Evaluation of Fatigue Behavior of Aged Asphalt Mixtures Using the Simplified Viscoelastic Continuum Damage Model

Lucas F. de A.L. Babadopulos, Jorge B. Soares, Jorge Luis S. Ferreira and Luis Alberto H. Do Nascimento

Abstract Aging of asphalt mixtures might play an important role in pavement structural behavior. That should be evaluated using mechanical models. This work aims at identifying the effects of aging on linear viscoelastic and damage behavior of asphalt mixtures, properties that influence the fatigue life of asphalt pavements. That is performed by characterizing a Hot Mix Asphalt (HMA) at four different aging states. The aged materials were obtained by heating loose asphalt mixture in oven at two different temperatures (85 and 135 °C) and aging times (2 and 45 days). Prony series were fitted to complex modulus results. For damage characterization, the Simplified Viscoelastic Continuum Damage (S-VECD) model was adopted. Damage characteristic curves and G^R versus N_f failure envelopes were obtained from controlled crosshead tension-compression test results at 19 °C. It was concluded that aging produces materials that fail for less evolved damage states. However, depending on pavement conditions and layer geometry, and considering the HMA hardening, aging does not necessarily reduce the fatigue life.

Keywords Asphalt mixtures · Aging · Fatigue modeling · S-VECD

L.F.d.A.L. Babadopulos (✉) · J.B. Soares · J.L.S. Ferreira
Universidade Federal do Ceará, Campus do Pici s/n LMP/DET,
Bloco 703, CEP 60455-760 Fortaleza-CE, Brazil
e-mail: lucasbaba@det.ufc.br

J.B. Soares
e-mail: jsoares@det.ufc.br

J.L.S. Ferreira
e-mail: jorgelsf@alu.ufc.br

L.A.H. Do Nascimento
Petrobras, Research and Development Center, Avenida Horácio de Macedo,
950 - Ilha do Fundão, 21915-900 Rio de Janeiro - RJ, Brazil
e-mail: luisnascimento@petrobras.com.br

A. Chabot et al. (eds.), *8th RILEM International Conference on Mechanisms of Cracking and Debonding in Pavements*, RILEM Bookseries 13,
DOI 10.1007/978-94-024-0867-6_2

1 Introduction

Asphalt pavement analysis adopts asphalt mixture properties obtained from specimens fabricated without long-term aging. Nevertheless, asphalt mixtures are known to age as time passes. Although Rolling Thin Film Oven (RTFO) and Pressure Aging Vessel (PAV) tests are currently used for comparing unaged and aged asphalt binders, there is still no recommended method for considering changes in the constitutive behavior of asphalt mixtures due to aging. This work focuses in evaluating the change in fatigue life that asphalt mixtures can present upon aging.

2 Literature Review

Park et al. (1996) presented an application of Schapery's work potential theory that would later produce the S-VECD model, which describes the damage behavior of asphalt mixtures. AASHTO TP 107 (2014) presents test procedures and the calculation process that ends up with the damage characteristic curve (C vs. S curves) for a given material. Such curve relates material integrity (C) to damage accumulation (S), which is a function of the loading history (Underwood et al. 2012). Some of the fatigue failure criteria presented in the literature define a maximum acceptable percentage loss in modulus, usually 50 %. This criterion does not consider differences between the capabilities of materials to undergo damage. Other criteria are based on energy dissipation or on the phase angle trend, which is generally associated to the coalescence of microcracks into macrocracks (failure). In experiments, those criteria are observed to be mode-dependent, thus, not accessing material properties. Contributions by Sabouri and Kim (2014) allowed the combination of the phase angle drop criterion with a pseudo strain energy-based variable, producing a mode-independent failure criterion for fatigue. That variable is the averaged rate of release of the pseudo strain energy (per cycle), G^R, throughout the entire history of the test. Its relation with the number of cycles to failure (G^R vs. N_f curve) for a given asphalt mixture was found to be linear in log-log axis.

In order to simulate asphalt mixtures aging in the laboratory, one can use ovens, either with compacted samples (Walubita 2006; Baek et al. 2012) or with loose samples, prior to compaction (Partl et al. 2012, RILEM TC206 procedure). The main advantage of aging compacted samples is that compaction problems will not be observed after aging. However, this procedure leads to a heterogeneous aging of the sample. The main advantage in loose mixture aging is that a more homogeneous aged mixture is obtained, even though it is noticed that compaction is influenced by the aging state of the material (Babadopulos 2014).

3 Materials and Methods

The Brazilian asphalt mixture investigated in the present research is a dense asphalt concrete with 12.5 mm nominal maximum aggregate size with a PG 64-22 (AC 50/70) binder. For the designed air void content (4.0 %), the required binder content was 6.0 % (by weight of the total mixture). The materials tested are the reference unaged mixture (Age Zero), and that very mixture but subjected to different aging levels. They are named Age X, Y °C, where X denotes the number of days of loose mixture aging in the oven and Y the aging temperature. Specimen fabrication (100 mm diameter) was set to stop at a height of 150 mm with samples of 2630 g of asphalt mixture, targeting a 4 % air void content. The obtained mean air void contents were 4.3 % for Age Zero, 4.5 % for Age 2, 85 °C, 4.7 % for Age 2, 135 °C, and 6.0 % for Age 45, 85 °C. AASHTO T 342 (2011) was the protocol adopted for stiffness characterization. Controlled crosshead harmonic uniaxial fatigue tests were conducted using nine HMA samples per aging condition. The results were used to fit the S-VECD model parameters, after verifying that the damage curves collapsed for the different loading conditions (different strain amplitudes). This procedure was conducted according to AASHTO TP 107 and the G^R based failure criterion was also obtained. Three specimens were tested at each of the three different target strain levels (around 200, 350, and 500 μm/m).

4 Results and Discussion

With respect to stiffness characterization, results showed that the norm of complex modulus ($|E^*|$) gradually increases with aging. Log-log plots of $|E^*|$ master curves (in which the percentage increase in $|E^*|$ with aging is observed) showed that stiffness is mainly affected by aging at low reduced frequencies (or high temperatures), as observed by other authors (Glover et al. 2008). In those frequencies binder properties influence more the bulk response of the mixture. Characterization of E^* allowed the fitting of Prony series (which gives the material relaxation spectra) with 11 elements (Babadopulos 2014). The damage parameter α is directly obtained from the relaxation spectra ($\alpha = 1 + 1/m$, where m denotes the maximum absolute log-log derivative of the relaxation modulus) and it is an input for the S-VECD characterization. Its results for Age 0; Age 2, 85 °C; Age 2, 135 °C; and Age 45, 85 °C were 2.993, 3.089, 3.101, and 3.126, respectively. Figure 1 and Table 1 summarize the obtained damage characteristic curves description (plotted until their respective mean values of the material integrity at failure).

The first trend observed regarding aging is that the damage characteristic curves present higher values of material integrity (C) for the same values of damage accumulation (S). An increase in the material integrity at failure is also observed as aging progresses. This means that the material is failing for less evolved damaged conditions (less damage tolerance). When it comes to failure criteria, good

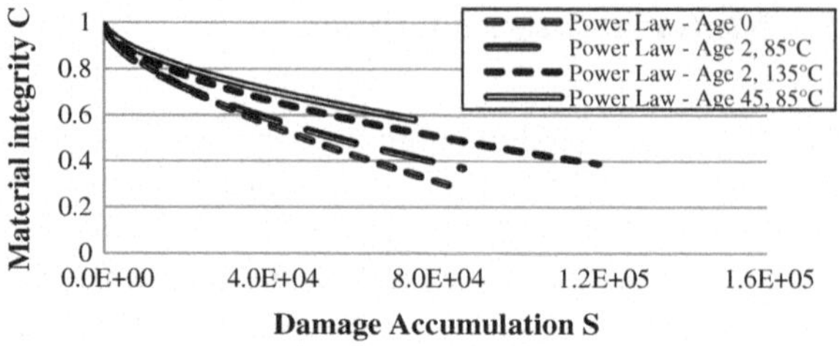

Fig. 1 Damage characteristic curves for the investigated aging states

Table 1 Power law ($C = 1 - C_{11} \cdot S^{C_{12}}$) fitting coefficients for C versus S curves

	Age zero	Age 2, 85 °C	Age 2, 135 °C	Age 45, 85 °C
C_{11}	5.41E−04	1.33E−03	1.04E−03	7.50E−04
C_{12}	6.33E−01	5.42E−01	5.45E−01	5.63E−01

Table 2 Fitting coefficients for G^R versus N_f curves

	Age zero	Age 2, 85 °C	Age 2, 135 °C	Age 45, 85 °C
Multiplier	4.93E+07	1.45E+07	7.60E+07	6.05E+06
Exponent	−1.52	−1.37	−1.51	−1.28
R^2	0.98	0.93	0.98	0.82

agreement between the model proposed by Sabouri and Kim (2014) (linear G^R vs. N_f in log-log axis) to relate averaged rate of release of the pseudo strain energy and fatigue life was observed (R^2 higher than 0.93) for early ages and acceptable agreement for Age 45, 85 °C (R^2 equals to 0.82). Results are summarized in Table 2.

With the calibrated damage models, material level simulations of fatigue behavior for the unaged and for the aged states were performed. The procedure applied by Nascimento et al. (2014) was used, i.e., failure was considered when the simulated G^R at a given cycle matched the one predicted by the fitted G^R versus N_f curve (coefficients in Table 2). A constant strain amplitude solicitation at 10 Hz was simulated at three different temperatures, in order to highlight the model capability of capturing temperature dependency of the fatigue behavior. Consequently, fatigue curves are obtained for three different values of dynamic modulus for each mixture. For the simulations at each temperature, five different strain levels were selected, chosen to achieve 120,200; 60,200; 20,000; 5000; and 1000 loading cycles at failure. The obtained data relate the number of cycles at failure with both the strain level (Whöler curves) and the dynamic modulus, making it possible to fit models for mechanistic-empirical (M-E) design methods (Eq. 1). The obtained parameters from the fitting (Eq. 1) for the tested materials are presented in Table 3. For easier visualization, only the fatigue life simulation results for Age Zero and Age 45, 85 °C were plotted in Figs. 2a–c.

Table 3 Mechanistic-empirical model parameters from fitting using the S-VECD model

	Age Zero	Age 2, 85 °C	Age 2, 135 °C	Age 45, 85 °C
K_1	2.95E+10	1.50E+08	3.96E+13	1.50E+08
K_2	4.44	6.25	5.45	6.25
K_3	−3.14	−3.66	−4.07	−3.66

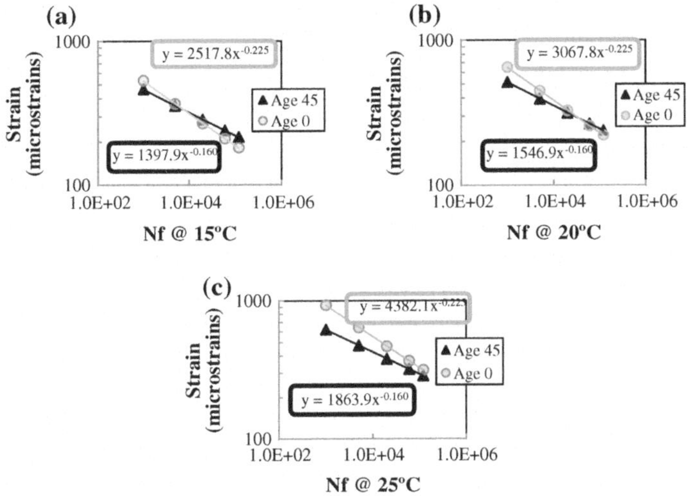

Fig. 2 Constant strain amplitude fatigue simulations for age zero and age 45, 85 °C

From manipulation of Eq. 1 and observation of results presented in Figs. 2a–c, it is noticed that the influence of temperature in fatigue behavior is accounted for by the stiffness term. The powers of the x-variable did not change for the different simulations, i.e., the slopes in the log-log space do not change with temperature, because they depend only on the value of $1/K_2$. It is actually the position of the Whöler curves in the y-axis that is changed by temperature, and this change is driven by dynamic modulus variation with temperature (from Time-Temperature Superposition Principle) and the model parameters K_2 and K_3 in Eq. 1. It can be shown that the change in the position of the Whöler curves in the y-axis is driven by the factor $(K_3/K_2) \cdot \log|E^*|$.

$$N_f = K_1 \cdot \left(\frac{1}{\varepsilon_t}\right)^{K_2} \cdot |E^*|^{K_3} \qquad (1)$$

At 15 °C, within the simulated strain levels (210–470 μm/m), unaged and aged mixtures presented very similar results. Maximum difference between the predicted number of cycles to failure was approximately 30 %, for strains around 210 μm/m (a relatively low difference for fatigue results). At 25°C, within the simulated strain levels (316–620 μm/m), maximum difference between the number of cycles to failure was approximately 80 %, for strain levels around 620 μm/m (from 6000 for Age Zero to 1000 for Age 45, 85 °C). For strain levels around 320 μm/m, the change was of approximately 120,000–60,000, representing a loss of 50 % in fatigue life. One should not forget that the layers thicknesses and stiffness play an important role in the fatigue life of the mixture on the pavement, driving the strain level within the asphalt layers. The aged material could perform better than the unaged one, depending on how the stiffness increase due to aging (inducing a decrease in the strain level in the aged pavement).

5 Conclusions

This work presented the S-VECD characterization of an asphalt mixture at different aging levels and some fatigue simulations. Results showed how the damage characteristic curve (*C* vs. *S* curves) changes with aging. The *C* versus *S* curves were observed to present higher values of material integrity (*C*) for the same values of damage accumulation (*S*) as aging evolves. That fact could mislead to the conclusion that a more resistant material was produced, as *C* is higher. The values of *C* at failure were higher as the aging state has grown, i.e., the material failed for a less evolved damage state. This contributes for a decrease in the fatigue life. However, from the results and simulations, aging does not necessarily reduce fatigue life in asphalt pavements. The consequence of aging can be either positive or negative depending on loading conditions, layers geometry and mixture properties.

References

AASHTO T 342 (2011) Standard method of test for determining dynamic modulus of hot-mix asphalt concrete mixtures. Washington, DC

AASHTO TP 107 (2014) Determining the damage characteristic curve of asphalt concrete from direct tension cyclic fatigue tests (Provisional standard). Washington, DC

Babadopulos LFAL (2014) A contribution to couple aging to hot mix asphalt (HMA) mechanical characterization under load-induced damage. MSc Thesis, UFC, Fortaleza, CE, Brazil

Baek C, Underwood BS and Kim YR (2012) Effects of oxidative aging on asphalt mixture properties. Journal of the Transportation Research Board 2296:77–85

Glover CJ, Martin AE, Chowdhury A, Han R, Prapaitrakul N, Jin X, Lawrence J (2008) Evaluation of binder aging and its influence in aging of hot mix asphalt concrete: literature review and experimental design. Report No. FHWA/TX-08/0-6009-1

Nascimento LAH, Rocha SMN, Nascimento CEH, Kim YR, Chacur M and Martins AT (2014) Uso da mecânica do dano contínuo na caracterização de misturas asfálticas brasileiras. 21st Encontro de Asfalto, IBP, Brazil [In Portuguese]

Park SW, Kim YR and Schapery RA (1996) Viscoelastic continuum damage model and its application to uniaxial behavior of asphalt concrete. Mechanics of Materials 24(4):241-255

Partl MN, Bahia HU, Canestrari F, de la Roche C, Di Benedetto H, Piber H and Sybilski D (2012) Advances in interlaboratory testing and evaluation of bituminous materials. Report STAR 206-ATB. RILEM Technical Committee 206-ATB

Sabouri M and Kim YR (2014) Development of a failure criterion for asphalt mixtures under different modes of fatigue loading. TRB Annual Meeting, Washington, DC

Underwood BS, Baek C and Kim YR (2012) Simplified viscoelastic continuum damage model as platform for asphalt concrete fatigue analysis. Journal of the TRB 2296(1):36-45

Walubita LF (2006) Comparison of fatigue analysis approaches for predicting fatigue lives of hot mix asphalt concrete mixtures (HMA). PhD Dissertation, Texas A&M, College Station, TX

GB5® Mix Design: A New Approach for Aggregate Grading Optimization for Heavy Duty Flexible Pavements

Simon Pouget, François Olard and Ferhat Hammoum

Abstract Since 2008, EIFFAGE has developed an aggregate optimization method (maximum contact and minimum space between aggregate particles) combined with the use of special binders (pure, multigrade or polymer-modified bitumen depending on the traffic concerned), to formulate the high-performance asphalt bituminous mix known as GB5®. During that time, numerous roads designed for heavy traffic have been built using this innovative technique, representing more than 1,000,000 tons at the end of 2014, mainly in France and, more recently, in South Africa. This article presents the result of collaboration between IFSTTAR and EIFFAGE aimed at scientifically relating the specific characteristics of the aggregate skeleton of GB5® to its outstanding mechanical properties, via 2D-image analysis and complete mechanical characterization.

Keywords GB5® · High modulus · Polymer-modified bitumen · Grading optimization

1 Introduction

This study is the result of a long-standing partnership between IFSTTAR and EIFFAGE.

The stakes are high, both for the company and for IFSTTAR: the specific properties of the aggregate structure of GB5® must be related to its outstanding

S. Pouget (✉) · F. Olard
Research and Innovation Department, EIFFAGE Infrastructures,
8 rue du Dauphiné - CS74005, 69964 Corbas Cedex, France
e-mail: Simon.POUGET@eiffage.com

F. Olard
e-mail: francois.olard@eiffage.com

F. Hammoum
MAST/Laboratoire MIT, LUNAM Université, IFSTTAR, CS4,
44341 Bouguenais Cedex, France
e-mail: ferhat.hammoum@ifsttar.fr

A. Chabot et al. (eds.), *8th RILEM International Conference on Mechanisms of Cracking and Debonding in Pavements*, RILEM Bookseries 13,
DOI 10.1007/978-94-024-0867-6_3

mechanical properties (Hammoum et al. 2014; Olard 2012). In addition to the study of GB5®, determining accurately the relationship between the composition, structure and properties of bituminous mix is a major line of research for EIFFAGE and IFSTTAR's researchers.

For this purpose, 2D image analyses and mechanical characterization were performed at IFSTTAR's MIT laboratory and EIFFAGE central laboratory located in Ciry-Salsogne respectively, on three types of mixes: EME2, GB4 and GB5®.

2 Materials and Methods

2.1 *Description of Materials*

To carry out a comparative study, several 0/14 bituminous mixes were used. This material is made of crushed aggregate from the Roche Blain quarry:

- Standard EME2, continuously-graded aggregate skeleton and 5.5 % of 20/30 hard penetration grade binder
- Standard GB4 (so-called "Grave Bitume" class 4), continuously-graded aggregate skeleton and 4.2 % of 35/50 grade binder
- GB5®, 6/10 gap-graded aggregate skeleton and 4.2 or 4.6 % of binder:
    - Multigrade
    - Biprene® 41 IPE (polymer-modified multigrade)

2.2 *Method and Tools*

2.2.1 Image Analysis—Identification of Internal Structure

Bituminous mix consists of a relatively dense assembly of aggregate and binder which gives the mix cohesion and impermeability. The aggregate structure of bituminous mix, which depends on both the components used and the manufacturing and laying conditions, has a direct, decisive influence on the mechanical properties of the mix.

Optical imaging is being increasingly used to obtain digital images with high resolution at micrometric scale. Applications have been developed in numerous fields (biology, biometry, metallurgy and materials science). Although this technique can only be used to observe the surface of a sample but it provides rapid and reliable data.

Based on previous studies (Boselli et al. 1999; Coster and Chermant 2001; Ghosh et al. 1997; Russ and Dehoff 2001), IFSTTAR's MIT laboratory used image processing and analysis methods combining robust mathematical methods able to determine several quantifiers of the microstructure of bituminous materials and, more precisely (Hammoum 2004):

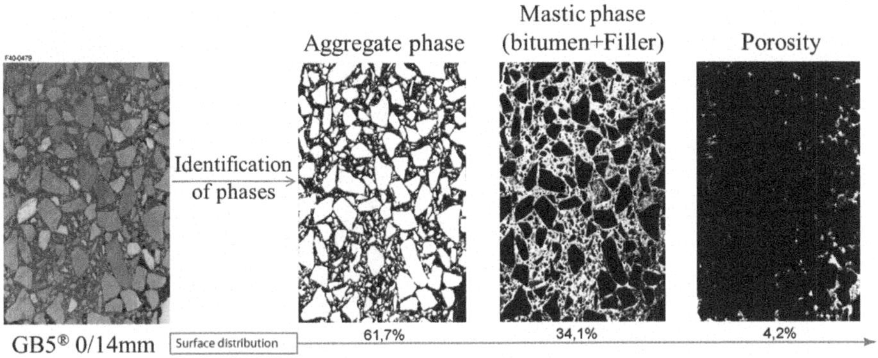

Fig. 1 Identification of the different phases of a material by image analysis of a cross-section

(i) identification of the bituminous material phases (aggregate, mastic and porosity). The contrast of the different phases represents an advantage of the success of this step (Fig. 1).
(ii) description of the arrangement of the aggregates with respect to each other and the distribution of mastic around the aggregate, via a Dirichlet (Voronoi) discretisation.

When applied to bituminous materials for the first time in 2004 (Hammoum 2004), this type of discretisation easily enabled the different phases of the aggregate structure to be identified. Based on a network of straight segments connecting up the centroids of the aggregate particles (Delaunay triangulation), the mediator of each segment can be determined. All mediators form Voronoi tessellations: the image is decomposed of a mosaic of cells. Each cell thus consists of a mastic ring of variable thickness surrounding the aggregate. The mastic surrounding the aggregate is defined as the Voronoi region.

A methodology has been developed to study the optimized aggregate structure of GB5® (dense gap-graded aggregate) and try to relate the aggregate arrangement to the final mechanical properties. This methodology is based on geometric quantifiers that characterize the distribution of the aggregate structure within the mix, in order to compare the mechanical properties of different mixes.

To relate the aggregate structure to the mechanical properties of GB5®, a comparative study was carried out with two standard mixes having relatively similar aggregate characteristics but different mechanical properties (EME2, GB4).

For the three mixes (EME2, GB4 and GB5®), 180 × 500 × 100 mm prismatic specimens were made by EIFFAGE for the purpose of observation by optical imagery. Cuts were then performed by IFSTTAR's MIT laboratory along the three main planes. A high-performance digital camera was used to take images with good resolution (between 20 and 50 μm/pixel) and cover a minimum surface area of 100 cm^2.

With a standard device type of equipment, easily available on the market, a single image pickup is sufficient to cover a representative surface area of the

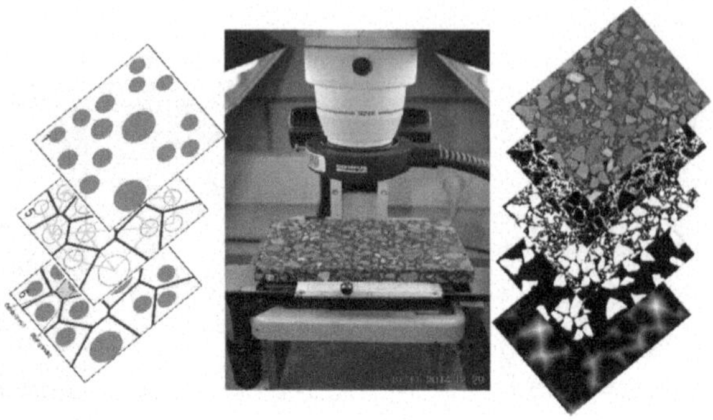

Fig. 2 Photo acquisition system for image analysis

specimen. A study of the representativeness of an image was presented during the Eurasphalt and Eurobitume conference in 2004 (Hammoum 2004).

Using a 150 × 500 mm specimen plate of one of the mixes, a plane section has been removed vertically with a diamond saw. An image is taken of each side of the specimen, under uniform lighting (Fig. 2).

In the following section, we study the distribution of the particles having a surface area of more than 50 mm^2 which form the main skeleton of the mix. This 50 mm^2 minimum limit corresponds to the average surface area of the particles in the area studied.

2.2.2 Mechanical Properties

Four mix design tests were carried out to evaluate the behaviour of the mixes:

- Compactability by using the gyratory compactor test, with measurement of the shear strength to evaluate the mechanical stability of the mix;
- Resistance to rutting;
- Stiffness modulus;
- Resistance to mechanical fatigue (2-point bending).

3 Correlation Between the Internal Structure and the Mechanical Properties

Observation of the aggregate structure of a mix shows that the aggregate forms a network of inter-particle contacts that is quite specific to each mix studied (GB4, EME2, GB5®). Since the mechanical stress supported by the skeleton and transmitted through a network of inter-particle contacts, a correlation must be found between the mechanical properties of the materials tested and the morphological characteristics of the aggregate structure. In this study, the binder differs from one

Table 1 Relationship between the aggregate skeleton and the mechanical properties

Mix	Stiffness modulus (15 °C, 10 Hz)	Rutting depth (%)	$\sum Area/NNDist$ (×1000)
EME2	13,300	3.8	4.5
GB4	14,200	2.7	9.5
GB5® Multigrade	15,500	2.0	14.0
GB5® Biprene IPE	16,200	2.1	18.0

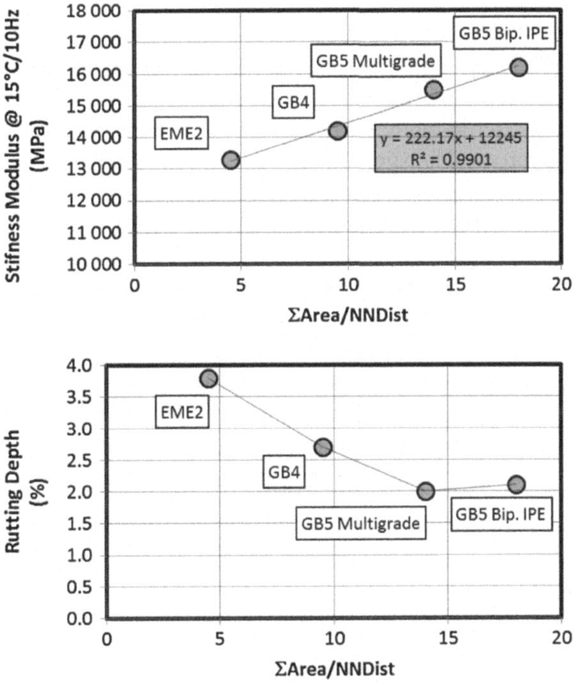

Fig. 3 Correlation between the nearest-neighbour distance of 10/14 aggregate particles and the mechanical properties. *Above* stiffness modulus at 15 °C/10 Hz; *below* rutting depth

material to another. Due to the lack of space, only the rutting resistance and the complex modulus are linked with the aggregate skeleton.

Table 1 gives the results of analysis of the distribution of particles greater than 50 mm^2. The parameter selected is ∑Area/NNDist because it combines both the area of the particle considered and the nearest-neighbour distance. In other words, this parameter considers the interaction zone of a skeleton particle according to its size and distance from its nearest neighbours.

In addition to the characteristics of the binders used in this study, we propose to study the correlation between the distribution of skeleton particles using, as an indicator, the relative distance from particles greater than 50 mm^2 and the mechanical properties of the materials studied.

It can be seen on Fig. 3 that the distribution and the nearest-neighbour distance of the aggregate particles of the skeleton (greater than 50 mm^2) can be correlated with the stiffness modulus. It seems that the stiffness of the material is proportional to the nearest-neighbour distance of the skeleton particles. To a lesser extent, there is also a correlation between the rutting resistance and the nearest-neighbour distance of the skeleton particles.

4 Conclusion

This study, conducted within the framework of a long-standing partnership between IFSTTAR and EIFFAGE, has established correlations between the distribution of the skeleton aggregate particles of bituminous materials and their final mechanical properties. The specific nature of the internal structure of GB5® (maximum contact and packing between aggregate particles) is highlighted by 2D imaging analyses and complete mechanical characterisation on 4 different mixes using the same aggregates nature.

The mechanical properties of GB5® 0/14 mix (with a binder content of 4.6 %) are of the same order (or higher in terms of rutting resistance and stiffness modulus) as those of the standard 0/14 EME2 mix (with a binder content of 5.4 %), thus requiring 15 % less fossil bitumen in the base course. As a result, the GB5® mix clearly presents both a technical and economic advantage. Independently of its carbon content, the re-allocation of bitumen of fossil origin as a subsequent source of energy is beneficial to both the environment and society.

References

Boselli J., Pitcher P., Gregson P., Sinclair I. (1999), Secondary phase distribution analysis via finite body tessellation, Journal of Microscopy, Journal of Microscopy,195, 104-112

Coster M., Chermant J-L. (2001), Image Analysis and Mathematical Morphology for Civil Engineering Materials, Cement and Concrete Composites, 23, p.133-151

Ghosh S., Nowak Z. and Lee K. (1997), Tessellation-based computational methods for the characterization and analysis of heterogeneous microstructures, Composite Science and Technology, 57, pp.1187-1210.

Hammoum F. (2004), Quantitative study of bituminous materials microstructure by digital image analysis, 3rd Eurasphalt & Eurobitume Congress, Vienna, Austria, 12-14 May 2004

Hammoum F., Olard F., Pouget S. (2014), Relations entre les performances mécaniques et la structure interne de l'enrobé GB5® : Analyse d'image basée sur la discrétisation de Voronoï, Journées thématiques -Techniques d'Imagerie pour la Caractérisation des Matériaux et des Structures du Génie Civil – 20 and 21 March 2014 – Clermont-Ferrand.

Olard F. (2012), " GB5 mix design: high-performance & cost-effective asphalt concretes by use of gap-graded curves & SBS modified bitumens ", Road Materials and Pavement Design Volume 13, Supplement 1 (Special Issue: Papers from the 87th Association of Asphalt Paving Technologists' Annual Meeting, April 1-4, 2012), pp. 234-259.

Russ J.C. et Dehoff R.T. (2001), Practical Stereology, Plenum Press, New-York, 2nd edition, pp. 307

Modelling the Hysteresis Loops of Hot Mix Asphalt

Taher M. Ahmed

Abstract This study presents the first attempt to use the Bouc-Wen model to model non-linear hysteresis loops of hot mix asphalt (HMA) samples tested in fatigue in controlled strain mode using a dynamic shear rheometer (DSR). A non-linear least squares optimization was used to identify the Bouc-Wen parameters. In this work, cylindrical samples (12 mm in diameter, 50 mm in height) fabricated from a HMA called dense bitumen macadam (DBM) were tested in fatigue. The outcome of this work confirmed the existence of a good agreement between modelled and experimental hysteresis loops.

Keywords Fatigue · Bouc-Wen · HMA · DSR

1 Introduction

Stress-strain relationships in a fatigue test produce loops which indicate non-linear hysteretic behaviour. The changes in the slope of these loops with the number of cycles demonstrate degradation of the material, i.e. reduction in stiffness. Many studies have succeeded in developing models for predicting this non-linear behaviour, and the Bouc-Wen model (Bouc 1967; Wen 1976) is one of the most popular models used for this purpose. This model has been used extensively to study the degradation of the hysteretic force in structures subjected to extreme dynamic loads such as earthquakes where there are random vibrations (Baber and Wen 1981; Sues et al. 1988); it has also been used in modelling brittle and elastic materials, such as

T.M. Ahmed (✉)
School of Engineering, University of Liverpool, Liverpool, UK
e-mail: taher@liverpool.ac.uk; alanitaher@yahoo.com

T.M. Ahmed
Engineering College, Anbar University, Anbar, Iraq

A. Chabot et al. (eds.), *8th RILEM International Conference on Mechanisms of Cracking and Debonding in Pavements*, RILEM Bookseries 13,
DOI 10.1007/978-94-024-0867-6_4

concrete, steel and soils (Sues et al. 1988; Sasani and Popov 2001; Shih and Sung, 2005). Modelling the degradation in materials can be taken from different perspectives based on the degradation concept; in this work, Bouc-wen model has been used to model the degraded hysteresis loops which reflect the cracks growth in HMA under fatigue test.

2 Objectives of the Study

Due to the simplicity and versatility and practical applications of the Bouc-Wen model (Spanos and Kougioumtzoglou 2011), it was selected to model degraded hysteretic behaviour for asphalt mixtures tested in fatigue using a DSR. The goodness of fit between the experimental and modelled data was also evaluated.

3 The Bouc-Wen Model (Formulation and Identification)

The Bouc-Wen hysteresis model is a first-order non-linear differential equation, as shown in "Eq. 1". This model was originally introduced by Bouc (1967) and later developed by Wen (1976), so it is called the Bouc-Wen model.

$$\dot{z} = A\dot{u} - \beta|\dot{u}||z|^{n-1}z - \psi\dot{u}|z|^n \tag{1}$$

where: A, β, ψ and n are parameters controlling the shape of hysteresis loops, and z(t) and u(t) are hysteresis forces and displacements. "Equation 1" is the Bouc-Wen model for the non-degrading case (Sues et al. 1988). In the degrading case, the model is capable of reproducing the degradation behaviour of material as hysteresis loops, but more parameters were added to "Eq. 1" by Baber and Wen (1981) to produce the final model for the degrading case, as shown in "Eq. 2

$$\dot{z} = A\frac{1}{1+\delta_\eta W} - \delta_A\frac{W}{1+\delta_\eta W} - \delta_v\frac{W\left(\beta\frac{|\dot{u}|}{\dot{u}}|z|^{n-1}z + \psi|z|^n\right)}{1+\delta_\eta W} - \frac{\beta\frac{|\dot{u}|}{\dot{u}}|z|^{n-1}z + \psi|z|^n}{1+\delta_\eta W} \tag{2}$$

where: δ_A, δ_η and δ_v are degradation parameters and W is dissipated energy. In this study, a standard non-linear least squares algorithm (NLSA) has been employed (Nocedal and Wright 2006) to identify the unknown parameters A, β, ψ, n, δ_A, δ_η and δ_v in "Eq. 3". To simplify the solution, the experimental data from the hysteresis loops were used to find the shear stress rate ($\dot{z}$), strain rate ($\mathring{u}$) and energy (W), as shown in "Eq. 3".

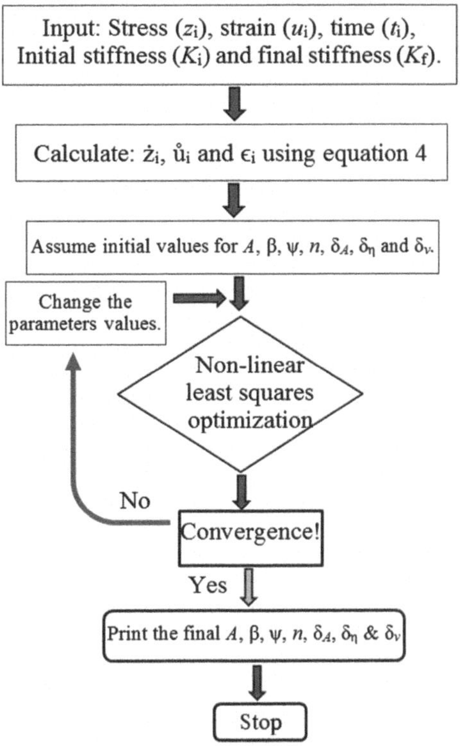

Fig. 1 Flow chart for identifying the Bouc-Wen parameters

$$\dot{z}_i = \frac{dz}{dt} = \frac{\Delta z}{\Delta t} = \frac{z_{i+1} - z_i}{t_{i+1} - t_i}; \quad \dot{u}_i = \frac{du}{dt} = \frac{\Delta u}{\Delta t} = \frac{u_{i+1} - u_i}{t_{i+1} - t_i};$$
$$W_i = (1 - \alpha)k \sum_{i=1}^{i=n} z_i \times \dot{u}_i \times \Delta t_i \tag{3}$$

where: z_i, z_{i+1}, u_i, u_{i+1} are shear stress and strain at time i and $i + 1$, respectively, and W_i is cumulative dissipated energy at time i. The flow chart for the NLSA used to identify the Bouc-Wen parameters is presented in Fig. 1.

4 Materials and Experimental Work

In this work, DSR cylindrical samples (12 mm in diameter and 50 mm in height) were produced from DBM mixes prepared in the lab using two types of aggregate, limestone (L) and granite (G), with one binder (160/220 Pen). Samples were tested in fatigue using the DSR technique in strain according to a previous study (Ahmed and Khalid 2015). Table 1 shows the relevant details and test conditions.

Table 1 Mix ID and material details (Ahmed and Khalid 2015)

Mix ID	Mix properties			Test conditions		
	Content (%)	G_{bulk}	Air voids (%)	Strain (%)	Frequency (Hz)	Temperature (°C)
DBM-L	5.2	2.374	4.9	0.30	2.5	25
DBM-G	5.2	2.290	7.5	0.30	2.5	25

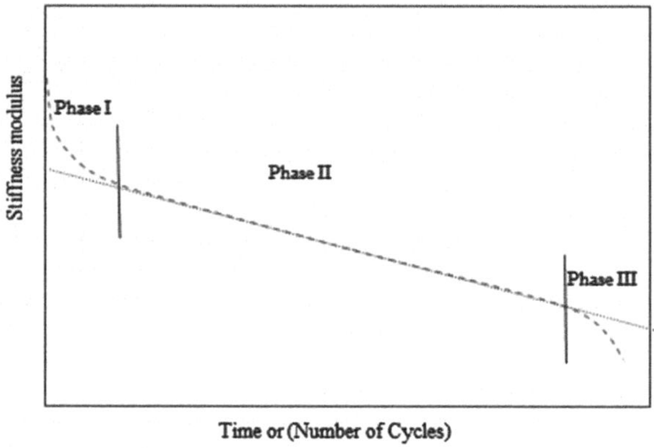

Fig. 2 Shear stiffness modulus against time or number of cycles

5 Results and Discussion

Over the course of repeated loading, materials in fatigue life exhibit three phases: phase I (rapid decrease in stiffness); phase II (a quasi-stationary phase); and phase III (complete failure phase), as shown in Fig. 2 (Di Benedetto et al. 2004). Based on these phases, two strategies were used for modelling the hysteresis loops, as shown in the following sections.

5.1 Hysteresis Loops from Phases I-III (Strategy–1)

The data for the hysteresis loops for one sample from each mix were used to identify the Bouc-Wen parameters using the NLSA as detailed in Fig. 1. A set of Bouc-Wen parameters is listed in Table 2 for all the data from phase I to III. Figure 3 is a representative data set for the experimental and modelled results. It can be seen from the figures that the Bouc-Wen model is incapable of capturing the hysteresis loops during the initial time, but the accuracy of capture of the hysteresis loops improves gradually. However, the degradation in the material is clear through the changes in the hysteresis loops with the progression of time, where the loop size and shear stress decrease as time goes on, and this is a good indication that the model predicts the degradation of the materials reasonably well.

Table 2 Bouc-Wen parameters of DBM-G and DBM-L samples for phases I-III (strategy–1)

Mix ID	$\|G_o\|$	A	β	ψ	n	δ_A	δ_μ	δ_ν
DBM-L	86.01	1.287	25.477	−19.658	0.895	8.374E−03	−4.631E−03	−4.557E−03
DBM-G	98.95	1.171	23.859	−16.034	1.113	2.276E−03	−1.669E−03	−1.456E−03

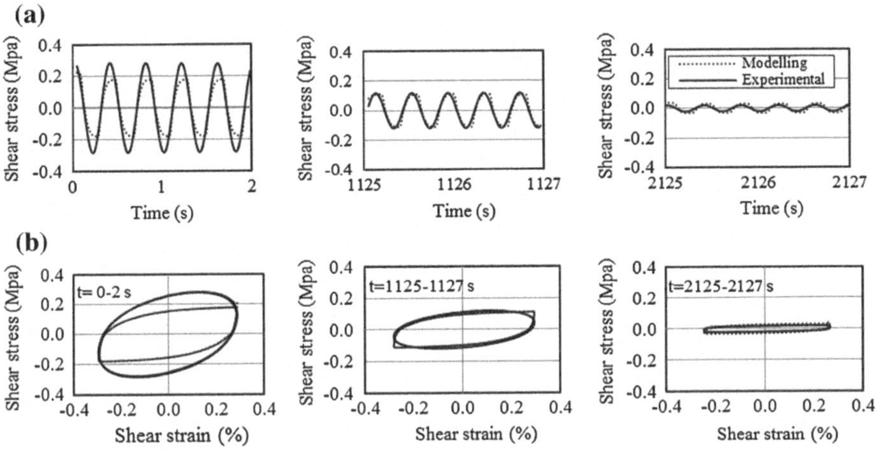

Fig. 3 **a** Experimental and modelled stress amplitude at different times; **b** experimental and modelled hysteresis loops at corresponding time of DBM-G

5.2 *Hysteresis Loops for Separated Phases (Strategy–2)*

The hysteresis loops in strategy-1 were split into two groups: the first group comprised phase I and second group comprised phases II and III together. The data for each group were then used to identify the Bouc-Wen parameters. The same procedure used in strategy–1 was used to identify the parameters in strategy–2 and presented in Table 3 for phase I and phases II-III. The consistency between the experimental and modelled loops is demonstrated in Fig. 4 for phases I and II-III.

Table 3 Bouc-Wen parameters of tested samples for phases I and II–III (strategy–2)

Mix ID	$\|G_o^*\|$	A	β	ψ	n	δ_A	δ_μ	δ_ν
Phase I								
DBM-L	86.01	1.185	21.588	−14.228	1.104	−6.983E−03	3.596E−03	5.607E−03
DBM-G	98.95	1.731	34.752	−27.430	1.053	−1.018E−01	1.141E−01	1.387E−01
Phase II–III								
DBM-L	68.41	0.745	23.454	−15.435	1.093	5.221E−03	−5.083E−03	−5.507E−03
DBM-G	54.77	1.059	40.979	−33.527	1.026	1.342E−02	−1.137E−02	−1.083E−02

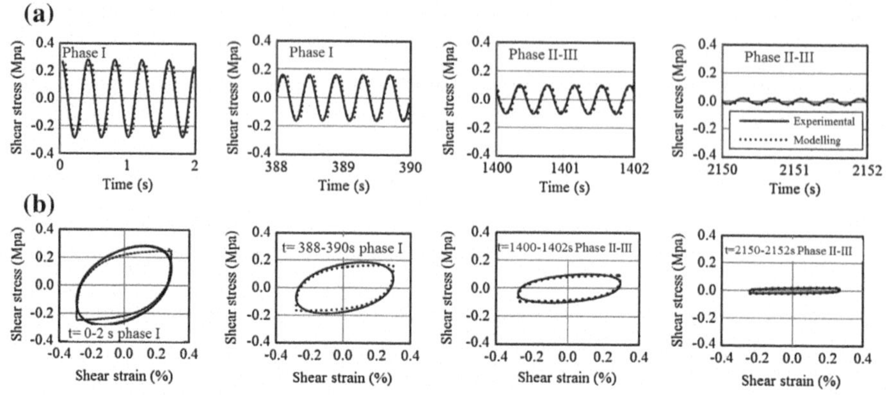

Fig. 4 **a** Experimental and modelled stress amplitude at different times; **b** experimental and modelled hysteresis loops at corresponding time of phase I and II-III of DBM-G

The Bouc-Wen model clearly performs better in capturing the hysteresis loops in strategy–2 than in strategy–1. Furthermore, the degradation in the material is also clear, as shown through the decrease in loop sizes and shear stress response with the progression of time. From Tables 2 and 3, it is noticeable that each sample has different identified parameters; this is because each sample has different properties, i.e. stiffness modulus and volumetric properties. This represents a challenge in the simulation for other samples with different properties.

To verify the solution of Bouc-Wen model, the predicted complex shear modulus ($|G_z^*|$) and differences (D) between the experimental complex shear modulus ($|G^*|$) and $|G_z^*|$ were calculated using "Eq. 4" and plotted in Fig. 5.

$$|G_z^*| = \frac{Z_\tau}{\gamma_0}; \quad D = 100 \times \frac{|G^*| - |G_z^*|}{|G^*|} \tag{4}$$

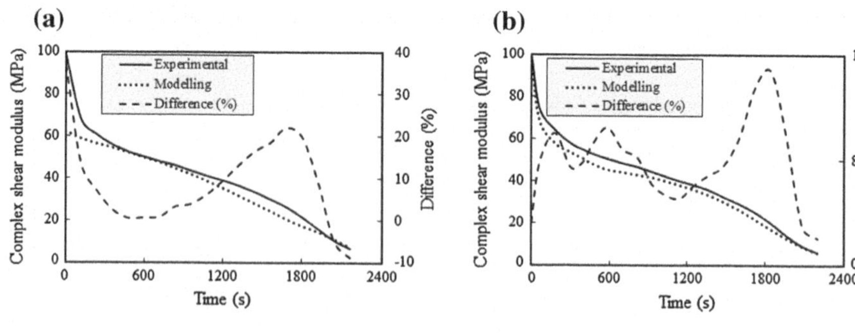

Fig. 5 Experimental and predicted stiffness with differences for: **a** strategy-1; **b** strategy-2

It can be seen that the $|G_z^*|$ is always less than the actual $|G^*|$ in both strategies. On the other hand, the D range between them in strategy-1 is higher than in strategy-2; in strategy-1, the D range was about (40 to −10) % whereas in strategy-2 it was less than 15 %. Overall, the complex shear modulus decreases with time, and this is a good indication that the Bouc-Wen model is able to mimic the degradation in the material. Generally, strategy–2 enhanced the capability of the Bouc-Wen approach to capture and model the hysteresis loops.

6 Conclusions

From the results of this study, the conclusions have been reached.

The Bouc-Wen model was used successfully in modelling the hysteresis loops of HMA samples tested in fatigue using a DSR machine in strain test mode. The Bouc-Wen model's solution is based on seven identified parameters that are related to the shape of the loops and the degradation in materials. Two strategies were used to identify Bouc-Wen parameters, which were based on the failure phases of HMA during fatigue life. Strategy–1 was based on using the hysteresis loops' data for all phases (I-III) at the same time, while strategy–2 was based on dividing the hysteresis loops into two groups: phase I separately and phases II and III together. The results revealed that the capability of the Bouc-Wen model to model the hysteresis loops of strategy–2 was better than that for strategy–1, as better agreement was presented between the experimental and modelled results. Bouc-Wen parameter values differ based on the properties of the samples, and this is a challenge that needs more research in order for it to be used in simulations for different samples.

Acknowledgments The author wish to acknowledge the financial support of the Iraqi government for the award of a research scholarship enabling this work.

References

Ahmed, Taher M. and Khalid, Hussain A. 2015. A New Approach in Fatigue Testing and Evaluation of Hot Mix Asphalt Using a Dynamic Shear Rheometer. *6th International Conference Bituminous Mixtuers and Pavements.* Thessaloniki, Greece, 10-12 June 2015.

Baber, T.T. and Wen, Y.K. 1981. Random Vibration of Hysteretic, Degrading Systems. *Journal of Engineering Mechanics,* 107, 1069-1083.

Bouc, R. Forced vibration of mechanical systems with hysteresis. Preceedings of the 4th International Conference on Nonlinear Oscillations, Prague, Czechoslovakia, 1967.

Di Benedetto, H., De La Roche, C., Baaj, H., Pronk, A. and Lundström, Robert. 2004. Fatigue of bituminous mixtures. *Materials and Structures,* 37, 202-216.

Nocedal, Jorge and Wright, S 2006. Numerical optimization, series in operations research and financial engineering. *Springer, New York.*

Sasani, M. and Popov, E.P. 2001. Seismic energy dissipators for RC panels: analytical studies. *Journal of Engineering Mechanics,* 127, 835-843.

Shih, M.H. and Sung, W.P. 2005. A model for hysteretic behavior of rhombic low yield strength steel added damping and stiffness. *Computers and Structures,* 83, 895-908.

Spanos, PD and Kougioumtzoglou, IA. Harmonic wavelet-based statistical linearization of the Bouc-Wen hysteretic model. Proceedings of the 11th International Conference on Applications of Statistics and Probability in Civil Engineering, ICASP, 2011. 2649-2656.

Sues, R.H., Mau, S.T. and Wen, Y.K. 1988. Systemes I dentification of Degading Hysteretic Restoring Forces. *Journal of Engineering Mechanics,* 114, 833-847.

Wen, Y. K. 1976. Method for random vibration of hysteretic systems. *Journal of Engineering Mechanics,* 102, 249-263.

Prediction of Hot Mix Asphalt Stiffness Behavior by Means of Multiscale Modeling

Lukas Eberhardsteiner, Bernhard Hofko and Ronald Blab

Abstract Cracking behavior and fatigue performance of hot mix asphalt (HMA) are the important factors, when it comes to evaluate the durability and structural lifetime of bituminous pavement constructions. Both effects are strongly affected by resulting stresses in the bituminous layers due to traffic and/or temperature change, whereat HMA stiffness has huge impact on the magnitude of these stresses and, hence, on the durability. It is obvious that a reliable characterization and prediction of HMA stiffness is crucial. A suitable way for the description of the viscoelastic response of a material is continuum micromechanics, where the mechanical, volumetric and morphologic properties of the constituents of the material are considered to predict the homogenized, "overall" material behavior. Thereby, the material is observed on different, reasonably chosen lengths scales allowing for a description of mechanical effects where they occur. These model assumptions were validated extensively showing good accordance between experimental results obtained from 4 PB-PR tests and model predictions with only the mechanical properties of the constituents (bitumen, aggregate) and the volumetric composition as model input. The results of this investigation suggest that applying the presented technique can lead to a significant reduction of experimental efforts in mix design.

Keywords Asphalt · Multiscale · Microstructure · Stiffness prediction

---

L. Eberhardsteiner (✉) · B. Hofko · R. Blab
Institute of Transportation, Vienna University of Technology, 28/230-3,
1040 Gußhausstraße, Vienna, Austria
e-mail: lukas.eberhardsteiner@tuwien.ac.at

B. Hofko
e-mail: bernhard.hofko@tuwien.ac.at

R. Blab
e-mail: Ronald.Blab@tuwien.ac.at

A. Chabot et al. (eds.), *8th RILEM International Conference on Mechanisms of Cracking and Debonding in Pavements*, RILEM Bookseries 13,
DOI 10.1007/978-94-024-0867-6_5

1 Introduction

The cracking behavior of bituminous materials is strongly influenced by stresses induced by traffic loading and temperature changes. Additionally, the fatigue performance of the hot mix asphalt (HMA) used in the bituminous base course, which is considered as relevant for the design of bituminous pavements, is affected by resulting strains. As an indicator for the resistance against deformation, the temperature-dependent stiffness behavior of HMA has a huge impact on the magnitude of these stresses and strains and, hence, the durability.

Naturally, a reliable description of the viscoelastic behavior plays a crucial role. Beside laboratory tests for the characterization of the stiffness (according to EN 12697-26 (ONI 2004)) material models are used to predict the mechanical properties of HMA. Such material models should account for the binder stiffness as well as the volumetric compositions. Two examples of frequently applied semi-empirical models for the estimation of the stiffness modulus of HMA are the Shell model (Van der Poel 1954) and the Hirsch model (Christensen et al. 2003).

To link microstructural characteristics to effective material properties, more sophisticated material models are needed. To observe a macro-homogenous material as a micro-heterogeneous medium (multiscale approach) allows for the physical understanding of material behavior and, thus, enables the description of mechanical effects like effects of cracking, accumulative damage processes (fatigue) or aging impacts.

Continuum micromechanics has proven its ability to provide a good framework for multiscale modeling as multi-step homogenization schemes were introduced for several materials [like e.g. concrete (Scheiner and Hellmich 2009)] in recent years. In terms of hot mix asphalt, Lackner et al. [amongst others in (Lackner et al. 2006)] proposed a model distinguishing 5 levels of observation, which was extended recently by taking the microstructure at the bitumen level into account (Eberhardsteiner et al. 2015b).

While the mentioned HMA multiscale model is described briefly in the following section, details on identification and validation experiments are given in Sect. 3. Before concluding the work, the result of recently performed, extensive validation including over 100 data points is presented.

2 Continuum Micromechanics for HMA Stiffness Characterization

The framework of continuum micromechanics (Hill 1965; Zaoui 2002; Suquet 1997) provides solutions for the description of the mechanical behavior of a material. Regarding bulk material as micro-heterogenous body filling a macro-homogeneous representative volume element (RVE) allows for the consideration of features like material properties, morphology or volume content of constituents at different length scales. Thereby, the microstructure within such an RVE is described by reasonably chosen, quasi-homogenous subdomains, so-called material phases, with known physical properties (volume fractions, elastic/viscoelastic properties, …). The characteristic length of an RVE l has to fulfill the law of separation of scales, $d \ll l \ll L$, where d represents the characteristic length of the inhomogeneities within the RVE, and L denotes the characteristic lengths of geometry or loading of a structure built up by the material defined on the RVE. If a single phase exhibits a heterogeneous microstructure itself, RVEs can be introduced within this phase in order to estimate its mechanical behavior. These RVEs have dimensions, which fulfill $l_2 \leq d$, and imply again smaller inhomogeneities with characteristic lengths $d_2 \ll l_2$, and so on.

By solving matrix-inclusion problems of the Eshelby-Laws type the mechanical behavior of the bulk material can be estimated. The physical properties of the material phases and their interactions are used to correlate homogenous deformations acting on the boundary of the RVE and resulting (average) stresses. As described in (Lackner et al. 2006, Eberhardsteiner et al. 2015a) this leads to a homogenized stiffness tensor $\mathbf{C}^{\text{hom}}$ of the general form

$$\mathbf{C}^{\text{hom}} = \left\{\sum_r f_r \mathbf{c}_r : \left[\mathbf{I} + \mathbf{P}_r^0 : (\mathbf{c}_r - \mathbf{c}^0)\right]^{-1}\right\} : \left\{\sum_s f_s \left[\mathbf{I} + \mathbf{P}_s^0 : (\mathbf{c}_s - \mathbf{c}^0)\right]^{-1}\right\}^{-1} \tag{1}$$

with the volume fraction f_{i}, the stiffness tensor $\mathbf{c}_{\text{i}}$ and the Hill tensor $\mathbf{P}_{\text{i}}^0$ accounting for the morphology of material phase i. $\mathbf{I}$ denotes the forth-order unity tensor.

For hot mix asphalt, 5 scales of observation can be distinguished (see Fig. 1): the bitumen-scale, the mastic-scale, the mortar-scale, the asphalt-scale and the macroscale. To estimate the homogenized stiffness behavior (in terms of $\mathbf{C}^{\text{hom}}$) on the next higher level, Eq. (1) is applied to the mastic-, mortar- and asphalt-scale taking

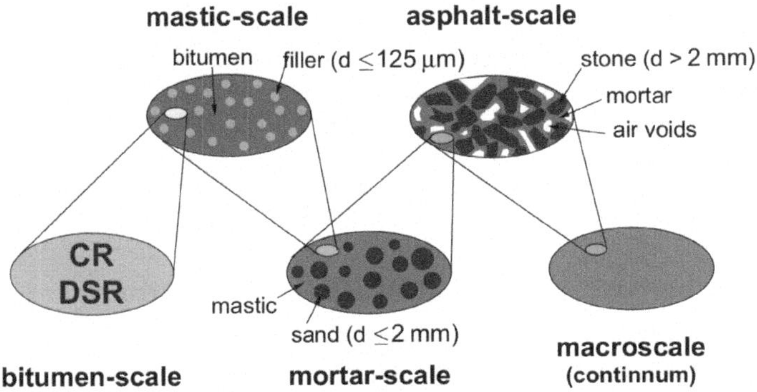

Fig. 1 Micromechanical representation of hot mix asphalt by means of a three-step homogenization scheme (Lackner et al. 2006)

the respective material phases and their properties into account. Finally, the homogenized viscoelastic HMA stiffness behavior can be derived.

3 Experiments

A reliable prediction of HMA stiffness behavior strongly depends on the proper determination of input parameters. While volume fractions can be obtained from mix design (aggregate gradation as well as binder and air void content), the mechanical properties of the constituents have to be identified. Aggregate exhibits elastic behavior, which is usually declared by the manufacturer (through Young's modulus and Poisson ratio). The viscoelastic behavior of bitumen can be characterized through Dynamic Shear Rheometer (DSR) tests, which were performed on 3 different polymer-modified bitumen PmB 45/80-65 and one paving grade bitumen 70/100 at 20 °C and frequencies f of 0.01, 0.1, 1, 3, 5, 8, 10, 20, 30 and 40 Hz. The results were transformed into the time (t) domain using

$$t = \frac{1}{2\pi f} \tag{2}$$

to allow a characterization of the viscoelastic response by a power-law type compliance function according to (Füssl et al. 2014), reading

$$J(t) = J_0 + J_a \cdot \left(\frac{t}{\tau}\right)^k \tag{3}$$

with J_0 as the elastic shear compliance, J_a and k as viscous parameters, and τ as a time variable, which are identified by minimizing the error—using nonlinear least square fitting—between experimentally obtained creep compliances from DSR identification tests and predicted creep compliances.

To show the significance of predictions obtained from the model described in Sect. 2, validation experiments using four point bending tests according to EN12697-26 (ONI 2004) were conducted to determine the stiffness behavior of 4 different HMA mixtures on 3 samples each. These tests were performed at 20 °C and frequencies of 0.1, 1, 3, 5, 8, 10, 20, 30 and 40 Hz. Hence, over 100 data points were used for the validation.

4 Comprehensive Validation of Model Predictions

The multistep homogenization scheme described in Sect. 2 was used to predict the Young's modulus of the investigated HMA mixtures. Thereby, each of the samples was investigated individually by taking the viscoelastic behavior of the bitumen

Fig. 2 Validation of HMA multiscale model in terms of |E*| for mixtures with polymer-modified bitumen PmB 45/80-65 and paving grade bitumen 70/100

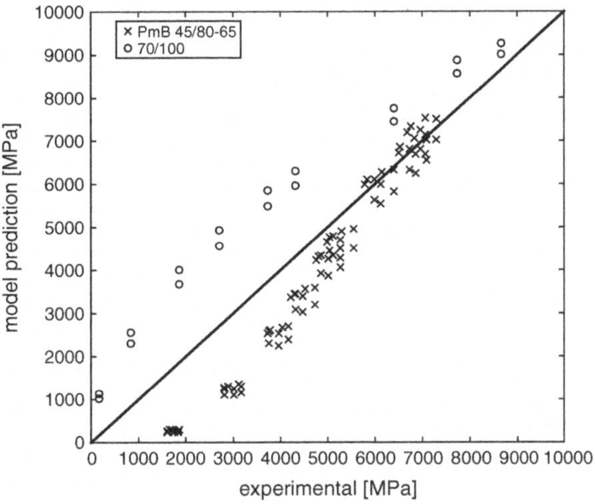

actually used (identified in DSR tests), the elastic properties of the aggregates (as declared by the manufacturer) and the specific volumetric composition (derived from gradation, binder and air void content) into account.

Figure 2 shows the correlation between experimental results and model predictions. While the Young's modulus is overestimated for HMA mixtures containing paving grade bitumen, the model predictions of the stiffness of samples with polymer-modified bitumen fit experimental results well, especially at higher frequencies. This is expressed by a coefficient of determination between $R^2 = 0.64$ and $R^2 = 0.96$.

5 Conclusions

As a parameter for the resistance against deformation, the stiffness behavior of hot mix asphalt (HMA) strongly influences the durability in terms of resistance against cracking and fatigue perfomance. Hence, a reliable prediction of the mechanical behavior is crucial. Continuum micromechanics provide a framework to estimate the viscoelastic behavior of HMA in terms of a multiscale model. Thereby, the volumetric composition of the mixture, the mechanical properties of the constituents and their morphology are considered on five levels of observation (bitumen-, mastic-, mortar-, asphalt- and macroscale).

To validate these model assumptions, the results of four point bending stiffness tests were compared to model predictions for 4 mixtures containing paving grade bitumen 70/100 or polymer-modified bitumen PmB 45/80-65. While the elastic behavior of the aggregates was defined by the manufacturer declaration, the viscoelastic response of the binder was identified in dynamic shear rheometer tests and the volumetric composition was derived from gradation as well as binder and air void content.

This validation confirms that the proposed model is able to predict HMA stiffness well, showing a coefficient of determination between $R^2 = 0.64$ and $R^2 = 0.96$. This approach not only allows for proper pavement design but is also able to reduce the experimental effort in mix design significantly.

References

CHRISTENSEN, D. W., PELLINEN, T. & BONAQUIST, R. F. 2003. Hirsch model for estimating the modulus of asphalt concrete. *Journal of the Association of Asphalt Paving Technologists, Vol 72*, 97-121.

EBERHARDSTEINER, L., FÜSSL, J. & BLAB, R. 2015a. Prediction of Hot-Mix Asphalt Stiffness - A Multiscale Approach. *TRB 94th Annual Meeting Compendium of Papers.*

EBERHARDSTEINER, L., FÜSSL, J., HOFKO, B., BLAB, R., GROTHE, H., HANDLE, F. & HOSPODKA, M. 2015b. Influence of asphaltene content on mechanical bitumen behavior - Experimental invesitgation and micromechanical modeling. *Materials and Structures,* 48, 3099 - 3112.

FÜSSL, J., LACKNER, R. & EBERHARDSTEINER, J. 2014. Creep Response of Bituminous Mixtures - Rheological Model and Microstructural Interpretation. *Meccanica,* 49, 2687-2698.

HILL, R. 1965. Continuum micro-mechanics of elastoplastic polycrystals. *Journal of Mechanics and Physics of Solids,* 13, 89 - 101.

LACKNER, R., BLAB, R., EBERHARDSTEINER, J. & MANG, H. A. 2006. Characterization and multiscale modeling of asphalt - Recent developments in upscaling of viscous and strength properties. *Computational Mechanics: Solids, Structures and Coupled Problems,* 6, 599-622.

ONI 2004. ÖNORM EN 12697-26: Bituminous mixtures - Test methods for hot mix asphalt - Part 26: Stiffness.

SCHEINER, S. & HELLMICH, C. 2009. Continuum microviscoelasticity model for aging basic creep of early-age concrete. *Journal of Engineering Mechanics,* 135, 307 - 323.

SUQUET, P. 1997. Continuum micromechanics, Springer, Wien - New York.

VAN DER POEL, C. 1954. A general system describing the visco-elastic properties of bitumen and its relation to routine test data. *Journal of Applied Chemistry,* 4, 221 - 236.

ZAOUI, A. 2002. Continuum micromechanics: Survey. *Journal of Engineering Mechanics (ASCE),* 128, 808-816.

Simulation of the Asymptotic Behaviour of Bituminous Mixtures Using the Discrete Element Method

M.D. Nguyen, F. Froiio, B. Cambou, H. Di Benedetto and C. Sauzéat

Abstract Use of the Discrete Element Method (DEM) to simulate the asymptotic elastic behaviours of bituminous mixtures (low-temperature-high-frequency or high-temperature-low-frequency) is investigated. Aggregates are represented in the numerical specimen by 10,000 spherical particles, with size distribution according to the grading curve of the reference material (the French GB3 bituminous mixture). Aggregates below 1 mm and bitumen are considered in the mastic phase, represented through interaction laws between particles. The preparation procedure of the numerical specimens ensures isotropy and enables detection of the networks of direct (contact) interactions between aggregates and of interactions via the mastic interface. Normal and tangential elastic interaction laws were used for the simulation of the asymptotic behaviours. Several numerical tests were performed to assess the capability of the discrete model to reproduce the asymptotic behaviours of the reference material in terms of complex Young's modulus and complex Poisson's ratio.

Keywords Bituminous mixtures · Discrete element method · Complex modulus · Multiaxial behaviour

M.D. Nguyen (✉) · H. Di Benedetto · C. Sauzéat
University Lyon, École Nationale des Travaux Publics de l'État, LTDS, 69518 Vaulx-EN-Velin, France
e-mail: minhduc.nguyen@entpe.fr

H. Di Benedetto
e-mail: herve.dibenedetto@entpe.fr

C. Sauzéat
e-mail: cedric.sauzeat@entpe.fr

F. Froiio · B. Cambou
University Lyon, École Centrale de Lyon, LTDS, 69134 Ecully, France
e-mail: francesco.froiio@ec-lyon.fr

B. Cambou
e-mail: bernard.cambou@ec-lyon.fr

A. Chabot et al. (eds.), *8th RILEM International Conference on Mechanisms of Cracking and Debonding in Pavements*, RILEM Bookseries 13,
DOI 10.1007/978-94-024-0867-6_6

1 Introduction

The term Discrete Element Method (DEM) refers to a class of numerical tools for the modeling of large assemblies of quasi-rigid particles. It was originally developed as an adaptation of molecular dynamics algorithms (Cundall and Strack 1979) and is nowadays considered as a standard tool for the micromechanical modeling of granular materials (Cambou et al. 2009). It is being considered with growing interest for the modeling of bituminous mixtures (e.g., Collop et al. 2007, You et al. 2008). The DEM simulations presented in this paper are performed with the commercial software PFC^{3D} (ITASCA 1999) and aim at modelling the asymptotic states of the reference material, as limit elastic states. This study is the first step and can be easily extended introducing local interaction law taking into account viscoelastic, plastic and damage effects. With these local laws, viscous effects, cracking and failure will be modelled.

A few elements concerning the asymptotic behaviour of bituminous mixtures are first recalled in Sect. 2. The numerical simulations described in Sect. 3 refer to the assemblage of the numerical specimen, whose testing is reported in Sect. 4. Conclusions as to the modelling of the asymptotic states of the reference material are finally drawn in the Sect. 5.

2 Asymptotic Behaviour of the Reference Material

Complex modulus tests performed at Univ. Lyon/ENTPE consists in the application of sinusoidal axial stress (σ_1) to cylindrical specimens and the measurement of strains along the axial direction and in the radial plane (ε_1 and ε_2, resp.). If the response of the material is linear viscoelastic and isotropic, the time evolution of these quantities can be represented, at a given temperature, by the sinusoidal functions $\varepsilon_1(t) = \varepsilon_{01}\sin(\omega t)$, $\sigma_1(t) = \sigma_{01}\sin(\omega t + \varphi_E)$, and $\varepsilon_2(t) = \varepsilon_{02}\sin(\omega t + \varphi_\nu + \pi)$, where t is time, ω is the period of the imposed axial strain and the remaining quantities are either amplitudes or phase angles. Under the same assumptions, by use of complex notation, the characterisation of the material consists in the identification of complex Young's modulus E^* and Poisson's ratio ν^* as functions of ω:

$$E^*(\omega) := \sigma_1^*/\varepsilon_1^* = (\sigma_{01}/\varepsilon_{01})e^{j\varphi_E}, \quad \nu^*(\omega) := -\varepsilon_2^*/\varepsilon_1^* = -(\varepsilon_{02}/\varepsilon_{01})e^{j\varphi_\nu} \tag{1}$$

A bituminous mixture called GB3, provided by Eiffage Travaux Publics and used for base course construction in France, was chosen as reference material in this paper. GB3 is produced with pure bitumen (penetration grade 35/50 according to the European standard EN 12591), with bitumen content 4.5 % and aggregates with size distribution following a continuous 0/14 mm grading curve (see Nguyen et al. 2014). The reference material was tested on cylindrical specimens, with diameter

$D \cong 74$ mm and height $H \cong 2D$, cored and sawn from slabs compacted with a French wheel compactor (according to the European standard EN 12697-33:2003 +A1:2007). The air void content of the tested specimens was about 5 %.

For materials which exhibit a "thermorheologically simple" behaviour, measurements of E^* and ν^* can be represented in terms of "master curves" relating their norm or their phase angle to the equivalent frequency $f_{eq} = a_T\,\omega/2\pi$: master curves at different temperatures can be deduced, from a reference temperature, by a simple shift along the frequency axis in log-scale by a shift factor $a_T\,(T)$ (time temperature superposition principle, see Ferry 1980). As an example, Fig. 1 provide the generalised master curve for the complex modulus of the reference materials. In the same figure, the quantities E_0 and E_{00} represent the norm of the complex modulus in the low-frequency-high-temperature and the high-frequency-low-temperature asymptotic limits, respectively. Notice that for the same limits the phase angle φ_E of the complex Young's modulus tends to zero. Analogous plots for the complex Poisson's ratio enable determination of the corresponding asymptotic values of its norm ν (ν_0 and ν_{00}, resp.) and phase angle φ_ν (both null). These quantities can be therefore thought of as the mechanical properties of two linear elastic materials exhibiting the same behaviour as assumed for the linear viscoelastic material in the two asymptotic limits, respectively: they characterise its "asymptotic behaviour". For the reference material it was obtained $E_{00} = 50$ MPa and $\nu_{00} = 0.45$ in the low-frequency-high-temperature limit, as well as $E_0 = 46.4$ GPa and $\nu_0 = 0.25$ in the high-frequency-low-temperature limit (Nguyen et al. 2014). More information on the linear viscoelastic behaviour of the reference material (that was used for a round robin test performed within working group 3 of the RILEM SIB committee) can be found in Perraton et al. (2015).

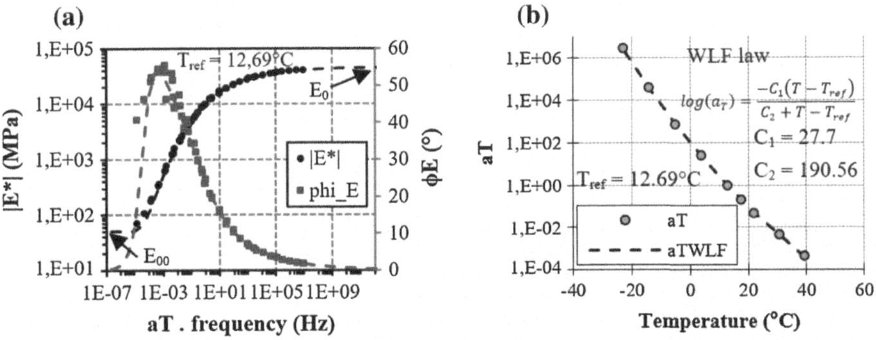

Fig. 1 **a** E^* master curves (norm and phase angle) for the reference material at $T_{ref} = 12.69$ °C (Nguyen et al. 2014), **b** Shift factor a_T for the reference material according to measurements (*green* circular marker) and their fitting with the WLF law (*dashed line*, see Ferry 1980)

3 Assemblage of the Numerical Specimen

The numerical specimen was shaped as the cylinder in Fig. 2a. Aggregates in the range of sieve apertures higher than 1 mm, referred to herein as "bigger aggregates", were represented by 10,000 spherical particles (Fig. 2b) with diameters distributed according to the grading curve of the reference material. Aggregates passing the 1 mm sieve, including fillers, were not represented in the numerical specimens but considered as part of the mastic interface separating bigger aggregates. The role of the mass phase in the deformation process was translated in the form of interaction laws for neighbouring particles in the bigger aggregate range. Gravity was neglected in all the simulations discussed in this paper.

The preparation procedure for the numerical specimen was organised in four steps. *In step 1*, the assigned volume was filled by randomly positioning the spherical particles until their volume fraction attained the value for to bigger aggregates in the reference material (60.7 %). *In step 2*, the assembly underwent quasi-static isotropic compression under "lubricated conditions" up to a confining pressure of 10 kPa: the particles were assumed to interact only by normal contact forces (i.e., repulsive contact forces orthogonal to the contact plane) according to the elastic unilateral contact law in Fig. 2c. The normal contact stiffness K_n (the same for each contact pair) was set to a value endowing the assembly with a Young's modulus in the range assumed for laboratory tests on a granular soil with comparable core material and grain size (i.e., K_n = 720 kN/m for a Young's modulus of 100 MPa ca.). In practice, dashpot elements set at 0.1 of the critical dumping value were used in parallel configurations along with the normal spring elements, essentially to allow the system to dissipate before attaining the final configuration. Notice that great precision on the contact properties was not required at this step, the objective being the development of a "stable" network of force-carrying contacts (network N1). In *step 3*, a uniform "lubricated" expansion of isolated particles allowed them to develop, temporarily, a stable network of

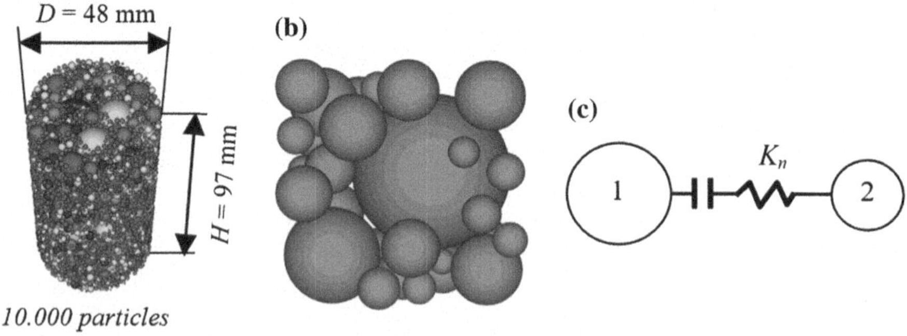

Fig. 2 Numerical specimen: **a** Example initial configuration, **b** Detail of the initial configuration, **c** Unilateral elastic normal interaction law used for the preparation procedure

force-carrying contacts with their closest neighbours (network N2). *In step 4*, the particles expanded in *step 3* were restored to their initial position and size (while keeping track of network N2) and a slight uniform reduction of the particle diameters (by a factor of 0.96) was imposed to restore the volume fraction assigned at step 1 (but affected by the isotropic compression in step 2). The so obtained configuration and interaction network (the union of N1 and N2), both inherently isotropic, are jointly referred to in the remainder of this paper as the "reference configuration".

4 Testing and Measurements

Axial compression tests were performed based on the reference configuration obtained according to Sect. 3, assuming the interaction law in Fig. 3: the mastic interface is sketched to a first approximation as a cylinder with axial length equal to the relative distance a and with the same radius as the smallest particle in the pair. Changes of the relative configuration of the pair are opposed by the force F_n (along the normal direction connecting the centres of the two particles) and by the tangential force F_t (in the tangential plane, orthogonal to the normal direction). F_n is proportional to the variations of a through the normal stiffness K_n, while F_t is proportional to the relative tangential displacement of the two particles via the tangential stiffness K_t. Stiffness K_n, bounded from below due to the limit distance a_{lim}, is computed based on the representative dimensions (a, A) and Young's modulus (e, resp.) of the interface cylinder. It was found anyway that $a_{lim,}$ roughly representing the direct contact limit, didn't significantly affect the results and could be fixed as 5 % of the radius of the smallest particle in the pair. In practice, a weak damping was introduced by dashpot elements in parallel configurations along with both spring elements (cf. Section 3). Stiffness K_t was not computed based on the cylindrical interface, but tuned via the ratio α with the normal stiffness. There are therefore only two adjustable parameters for the contact model: e and α.

The tests were performed under quasi-static conditions (axial strain rate: $5.21 \times 10^{-6}\ s^{-1}$) and without lateral confinement, up to an axial strain of 5×10^{-6} (axial strain in complex modulus tests is of the order of 5×10^{-5}). The values of Young's modulus E and Poisson's ratio ν obtained depending on the parameters (e and α) are shown in Fig. 4. The Young's modulus E increases proportionally to

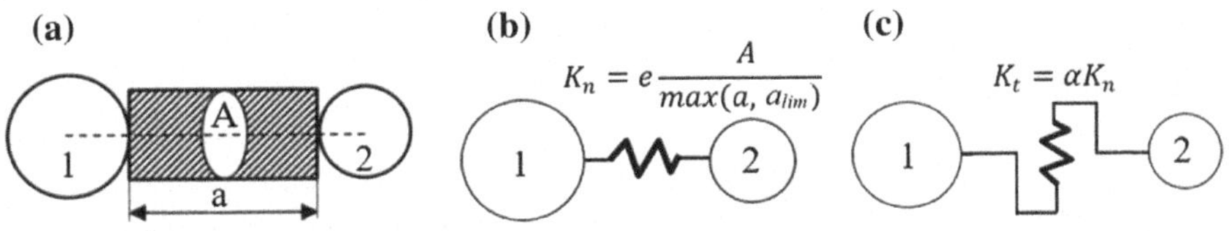

Fig. 3 Interaction models for the axial compression tests: **a** Cylindrical mastic interface between two particles, **b** Normal elastic interaction model, **c** Tangential elastic interaction model

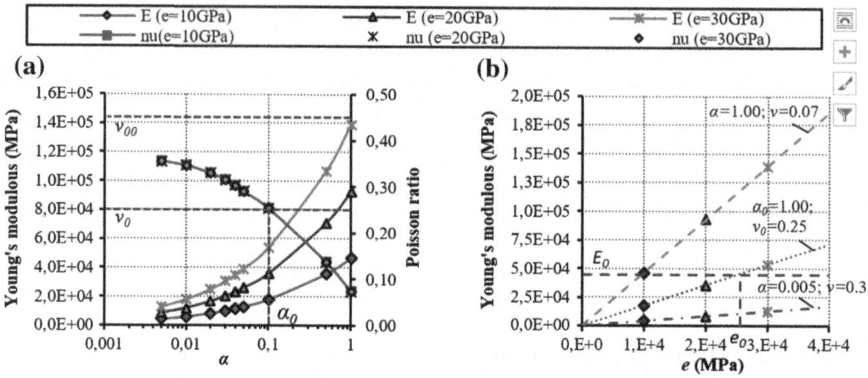

Fig. 4 Evolution of Young's modulus and Poisson's ratio depending on the control parameters: **a** E and ν vs. α, **b** E vs. e

e (at constant α), as expected. The increase of E with α (at constant e) can be understood as expressive of the overall stiffening role played by tangential stiffness K_t. Another result from this figure is the direct dependence of Poisson's ratio on parameter α.

5 Conclusions and Acknowledgments

Concerning the asymptotic states of the reference material, the values of Young's modulus and Poisson's ratio in the high-frequency-low-temperature limit (E_0 and ν_0, see values in Sect. 2), can be referred to the values $e_0 = 26.1$ GPa and $\alpha_0 = 0.1$ of the control parameter for the DEM simulations in Sect. 4. The latter value can be interpolated in Fig. 4a, while that former value can be deduced from the proportionality between E and e (see Fig. 4b). It was not possible, on the other hand, to determine values of the control parameters for the low-frequency-high-temperature limit (i.e., for E_{00} and ν_{00}, see values in Sect. 2) due to the Poisson's ratio not exceeding 0.35, even for negligible values of parameter α. Refinement of the preparation procedure (Sect. 3) or improvement of the elastic interaction laws (Sect. 4) are being considered in order to fill this gap. The authors acknowledge financial support from Univ. of Lyon (*PALSE, ANR-IDEX-0007*).

References

Cambou B, Jean M, Radjai F (2009) Micromechanics of Granular Materials. Wiley-ISTE

Collop AC, McDowell GR, Lee Y (2007) On the use of discrete element modelling to simulate the viscoelastic deformation behaviour of an idealized asphalt mixture. Geomechanics and Geoengineering: An International Journal 2:77-86. doi:10.1080/17486020701243128

Cundall PA, Strack ODL (1979) A discrete numerical model for granular assemblies. Geotechnique 29:47-65. doi:10.1680/geot.1979.29.1.47

Ferry JD (1980) Viscoelastic properties of polymers. John Wiley & Sons, Chiches

Nguyen QT, Di Benedetto H, Sauzéat C (2014) Linear Viscoelastic Domain for Bituminous Mixtures. Pavement Materials, Structures, and Performance, ASCE: 59-68. doi:10.1061/9780784413418.007

Itasca (1999) Particle flow code in three dimensions, Itasca Consulting Group Inc.

Perraton D, Di Benedepetto H, Sauzéat C, Hofko B, Graziani A, Nguyen QT, Pouget S, Poulikakos LD, Tapsoba N, Grenfell J (submitted in 2015) 3Dim experimental investigation of linear viscoelastic properties of bituminous mixtures, Materials & Structures

You Z, Adhikari S, Dai Q (2008) Three-dimensional discrete element models for asphalt mixtures. Journal of Engineering Mechanics 134:1053-1063. doi:10.1061/(ASCE)0733-9399(2008)134:12(1053)

Visco-Plastic Behavior of Bituminous Mixtures: Experiments and Modeling

Pierre GAYTE, Hervé Di Benedetto and Cédric Sauzeat

Abstract Behaviour of bituminous mixtures is very complex. Their rheological properties depend on strain amplitude level and temperature. We can observe linear viscoelasticity properties for very small strain amplitudes and non-linearities for larger strain level. Fatigue and rutting (permanent deformation) can appear for a great number of cycles. Moreover brittle or ductile failure can occur. All of these properties highly depend on temperature. The DBN (Di Benedetto-Neifar) model developed last years at LGCB-LTDS laboratory (University of Lyon/ENTPE "Ecole Nationale des TPE") aims at describing the complex behaviour of bituminous mixtures with a unified formalism. The model is also versatile and may be adapted to take into account or not some specific properties of bituminous materials as failure mechanisms. First, equations of the DBN_{EPP} (Elastic-Perfectly Plastic version of the DBN model) describing the whole three-dimensional DBN model in the linear and viscoplastic domains are recalled. Experiments made at LGCB-LTDS laboratory on bituminous mixtures in the large strain domain are then presented. Finally simulations are proposed based on the DBN_{EPP} model in the viscoelastic and plastic domains.

Keywords Bituminous mixtures · 3D behaviour · DBN model · Viscoplasticity

---

P. GAYTE (✉) · H. Di Benedetto · C. Sauzeat
ENTPE-University of Lyon, LGCB-LTDS (UMR 5513), Rue Maurice Audin, 69518 Vaulx-En-Velin, France
e-mail: pierre.gayte@entpe.fr

H. Di Benedetto
e-mail: herve.dibenedetto@entpe.fr

C. Sauzeat
e-mail: cedric.sauzeat@entpe.fr

A. Chabot et al. (eds.), *8th RILEM International Conference on Mechanisms of Cracking and Debonding in Pavements*, RILEM Bookseries 13,
DOI 10.1007/978-94-024-0867-6_7

1 Introduction

Many external solicitations (mechanical, thermal, chemical, etc.) can occur when considering the behaviour of roadway structures. These effects are often coupled. The main difficulty is then to model these effects and their coupling in order to be able to get realistic performance prediction of bituminous pavements.

The DBN model is an attempt to describe with a unique formulation the different types of behaviour observed for bituminous mixtures developed at Univ. of Lyon-ENTPE. It is based on simple rheological elements and its calibration is made from experimental results on bituminous mixtures (cf. Figure 1). Due to its high versatility this model is able to describe the behaviour in a wide range of strain amplitude, temperature and number of cyclic loadings. As a result this model can be very simple to use (linear viscoelastic domain) or more complicated (when introducing nonlinearity, fatigue) following required refinement.

This article focuses on the linear and non-linear domain. The DBN_{EPP} model which is able to describe the linear viscoelastic and viscoplastic behaviour of bituminous mixture is first recalled. The experimental campaign launched so as to characterise this type of behaviour is then described. Finally results provided by the experimental tests and simulations made using DBN_{EPP} model are presented so as to illustrate the viscoplastic behaviour and the good agreement between experimental data and simulations using the DBN_{EPP} model.

2 The EPP Version of the DBN Model

The EPP version of the DBN model is developed to describe the linear viscoelastic and the viscoplastic behaviour of bituminous mixtures. The goal is to facilitate introduction of the model in finite elements methods calculations considering that each EPi body is elastic-perfectly plastic which means linear elastic (E) behaviour within the yield surface and perfect plasticity (PP) for larger strain amplitudes (cf. Figure 2). Constitutive equations of the DBN_{EPP} model are recalled by Eqs. 1 and 2. Readers are invited to refer to Gayte et al. (2014) and Di Benedetto et al. (2007) for more detailed information about the DBN model and its simplified EPP version.

$$\underline{\varepsilon} = \sum_{i=0}^{n} \underline{\varepsilon}_i^e + \underline{\varepsilon}_i^p \tag{1}$$

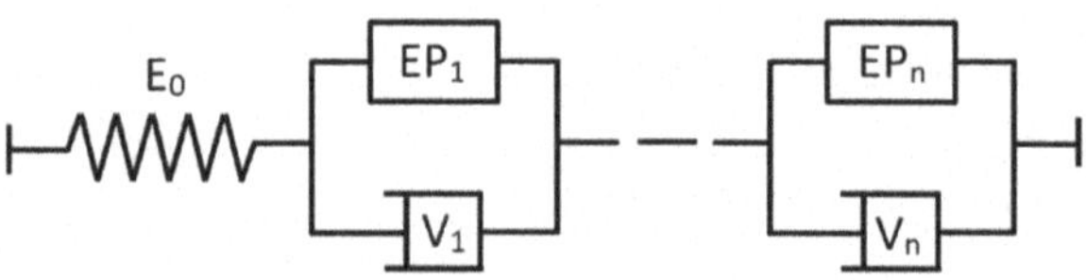

Fig. 1 Analogical 1D representation of the DBN model

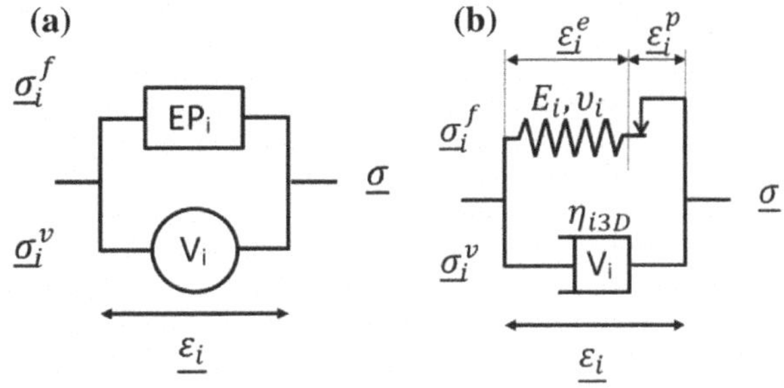

Fig. 2 Representation of stress and strain in body "*i*" (**a**, *left*) and analogical 3D representation of the EPP version of the DBN model (**b**, *right*)

$$\underline{\sigma} = \underline{\sigma}_i^f + \underline{\sigma}_i^v \tag{2}$$

Where $\underline{\varepsilon}$ and $\underline{\sigma}$ are respectively strain and stress tensors of the given material, $\underline{\varepsilon}_i^e$ and $\underline{\varepsilon}_i^p$ are elastic and plastic strains tensors, and $\underline{\sigma}_i^f$ and $\underline{\sigma}_i^v$ are non-viscous and viscous stresses tensors of the body "*i*".

3 Experimental Campaign and Simulations

3.1 Tested Material

One asphalt mix was tested and is stemmed from the experimental campaign developed in Pham et al. (2015). It is a warm mix asphalt with 30 % RAP content (in weight) and 0.4 % of additive ETIMA (in weight of the total bitumen weight). This correspond to a classical mix defined in the French design guide (NF P 98-082) as "Grave Bitume" of class three (GB3) and size of 0/14.

3.2 Test Procedures

Specimen were tested following two steps. At a given constant temperature (0; 15; 30 °C) they were first loaded at 7 frequencies in order to determine the viscoelastic properties (Fig. 3). The amplitude of axial strain during cyclic loading was about 50 μm/m. The number of loading cycles varied from 4 cycles at 0.01 Hz up to 50 cycles at 10 Hz (Gayte et al. 2015). A rest period of 5 min was applied between each frequency change.

Keeping the same initial temperature specimens were then loaded with a constant strain rate (0.01 %/min; 0.1 %/min; 1 %/min) until they reached a strain level of 0.1 % in strain (strain value of 1000 μm/m). Specimens were unloaded until 0 μm/m and reloaded to 0.8 % in strain with the same strain rate. A creep period

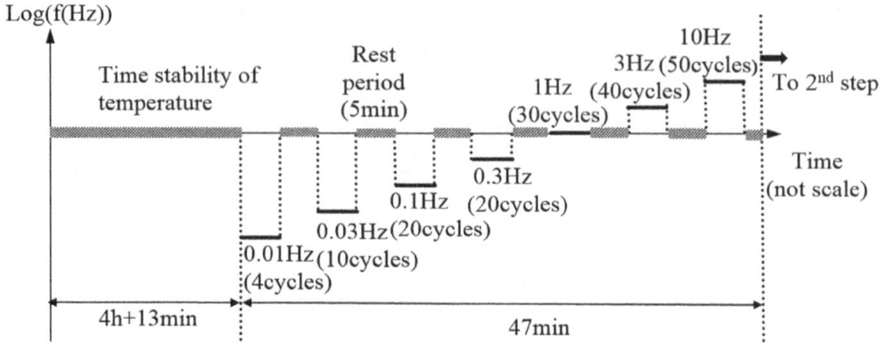

Fig. 3 Chronology of the first step—Cyclic loading at frequencies from 0.01 up to 10 Hz at a given temperature (3 temperatures 0, 15, 30 °C)

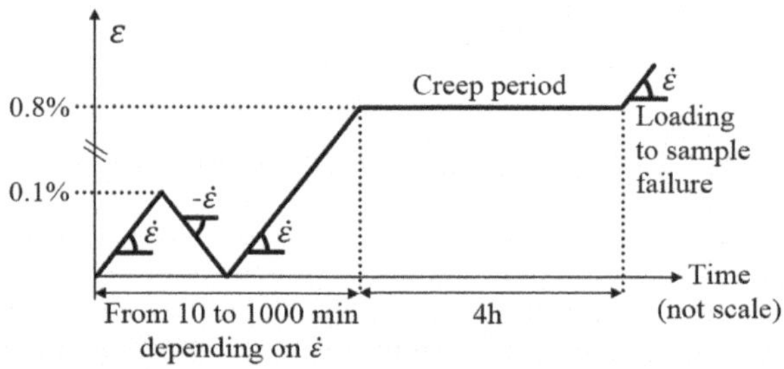

Fig. 4 Chronology of the second step—Loading at a constant strain rate $\dot{\varepsilon}$ followed by unloading, re-loading, creep period and final loading until the specimen fails at the temperature chosen during step 1

was then applied during 4 h before a last loading (still applied with the same strain rate) leading to the failure of the sample (Fig. 4).

3.3 *Experimental Results and Simulations*

Calibration of the DBN_{EPP} model is made from the 2S2P1D model also developed at ENTPE-LTDS laboratory (Olard and Di Benedetto 2003) in the small strain domain (VEL behaviour). In the large strain domain calibration is made from Di Benedetto criterion (Di Benedetto and Yan 1994). Constants of the DBN_{EPP} model are obtained using an optimisation process with $n = 20$ elementary bodies.

Figure 5 ($|E^*|$ and $|\nu^*|$—linear viscoelastic domain) and Fig. 6 (simulation of the 2nd step of experimental tests—viscoplastic domain) are two simulations to illustrate the DBN model and its EPP version are able to model quite correctly the behaviour of bituminous mixture.

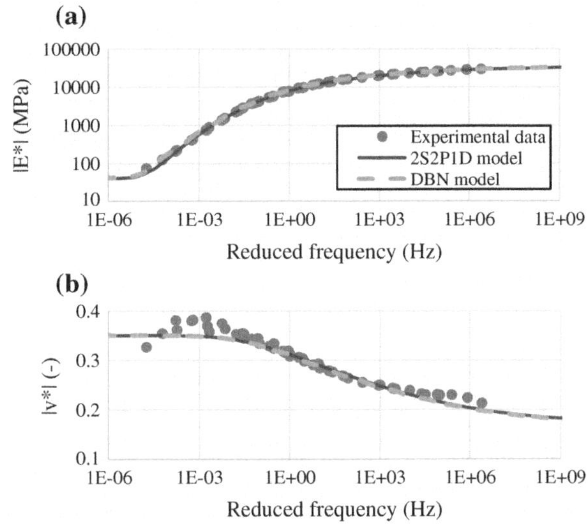

Fig. 5 Experimental (*dots*), simulated with 2S2P1D model (*continuous line*) and with DBN model (*dotted line*) master curves of the complex modulus $|E^*|$ (**a**, *top*) and Poisson's ratio $|\nu^*|$ (**b**, *bottom*) as a function of reduced frequency

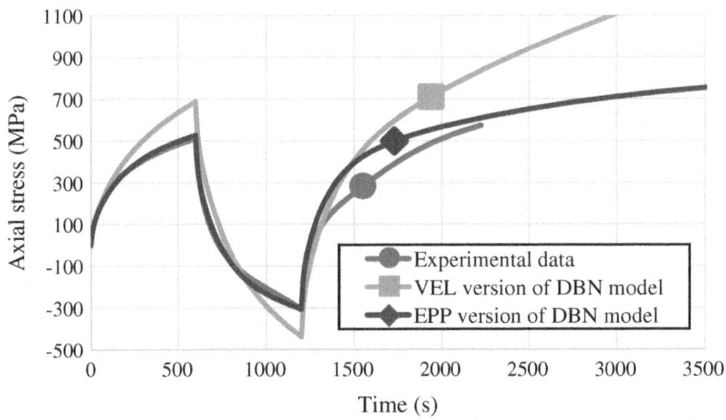

Fig. 6 Axial stress recorded during the second step of the experimental test (*cf.* Fig. 4), modelled using the VEL and the EPP version of the DBN model. Simulations made at $\dot{\varepsilon} = 0.01\ \%/min$ and $T = 14.85°C$

4 Conclusions

From obtained results the following conclusions can be drawn:

- Experimental campaign provided good results so as to calibrate the EPP version of the DBN model and then to model the viscoplastic behaviour of bituminous mixtures.
- In the linear viscoelastic domain master curves of $|E^*|$ and $|\nu^*|$ are well simulated by both the 2S2P1D and the DBN model on the whole range of temperatures and frequencies.
- For larger strain amplitude (up to 0.8 %) EPP version of the DBN model with 20 elements can simulate quite correctly the behaviour of bituminous mixtures.

References

Di Benedetto, H., Mondher, N., Sauzéat, C. and Olard, F. (2007). "Three-dimensional thermo-viscoplastic behaviour of bituminous materials: The dbn model." Road Materials and Pavement Design **8**(2): 285-315.

Di Benedetto, H. and Yan, X. (1994). "Comportement mécanique des enrobés bitumineux et modélisation de la contrainte maximale." Materials and structures **27**(9): 539-547.

Gayte, P., Di Benedetto, H. and Sauzéat, C. (2014). Three dimensional behaviour of bituminous mixtures in the linear viscoelastic and viscoplastic domaines: The dbn model. ISAP (International Society for Asphalt Pavements). Y. R. Kim. Raleigh, NC, USA, Taylor & Francis Group.

Gayte, P., Di Benedetto, H., Sauzéat, C. and Nguyen, Q. (2015). "Influence of transient effects for analysis of complex modulus tests on bituminous mixtures." Road Materials and Pavement Design, 10.1080/14680629.2015.1067246. DOI:10.1080/14680629.2015.1067246.

Olard, F. and Di Benedetto, H. (2003). "General "2s2p1d" model and relation between the linear viscoelastic behaviours of bituminous binders and mixes." Road Materials and Pavement Design **4**(2): 185-224. DOI:10.1080/14680629.2003.9689946.

Pham, N.H., Sauzéat, C., Di Benedetto, H., González-León, J.-A., Barreto, G., Nicolaï, A. and Jakubowski, M. (2015). "Reclaimed asphalt pavement and additives' influence on 3d linear behaviour of warm mix asphalts." Road Materials and Pavement Design, (ahead-of-print): 1-23.

Part II
Cracking in Asphalt Materials: Material Crack Characterization

Effect of Fiber Grid Reinforcement on Crack Initiation and Propagation in Asphalt Concrete

Xiaofeng Gao, Georg Koval and Cyrille Chazallon

Abstract A mechanism to explain the local contribution of fiber grid reinforcements to reduce tensile crack initiation and its propagation is proposed. In the first part, a local model of quasi-brittle rupture based on the fracture mechanics background is presented. This model is then compared to experimental results and predictions of the literature of V-notched samples in tension. In the second part, the effect of fiber grids as boundary reinforcements of the asphalt concrete layer is then analyzed under constant strain conditions. This study shows an improvement of the local performance against the propagation of cracks crossing the grid. The obtained results also indicate that the contribution of the fiber grid reinforcement depends on the stiffness ratio of the reinforcement and the asphalt and increases for higher quantities of fiber (per unity of length). However, this increase is bounded, which may define a limit quantity of fibers to be employed in practice.

Keywords Crack initiation · Crack propagation · Fiber grid · Asphalt concrete

1 Introduction

The mechanical contribution of the fiber grid reinforcement on the inhibition of the initiation and propagation of cracks in asphalt pavements is still not completely understood (Arsenie 2013). Existing models usually combine a linear elastic analysis with a characteristic length in order to deal with crack initiation (Li and Zhang 2006). This length scale depends on material properties like toughness and

X. Gao (✉) · G. Koval (✉) · C. Chazallon
Laboratoire ICUBE—UMR 7357, INSA de Strasbourg, 24 Boulevard de La Victoire, 67084 Strasbourg, France
e-mail: xiaofeng.gao@insa-strasbourg.fr

G. Koval
e-mail: georg.koval@insa-strasbourg.fr

C. Chazallon
e-mail: cyrille.chazallon@insa-strasbourg.fr

A. Chabot et al. (eds.), *8th RILEM International Conference on Mechanisms of Cracking and Debonding in Pavements*, RILEM Bookseries 13,
DOI 10.1007/978-94-024-0867-6_8

strength (Bazant and Pijaudier-Cabot 1989; Maimí et al. 2013) and may reach some centimeters in asphalt concrete (Kim and Hussein 1997). Hence it restricts the definition of smaller cracks by these non-local models.

In this article, a local model for crack initiation (and propagation) is proposed, which allows the analysis of the behavior of the material at the proximity of the reinforcements. Based on finite element simulations, a simple mechanism explaining the interaction between the asphalt concrete and the fiber grid is discussed.

The rupture criterion is described in the first section. In Sect. 3, the numerical results are presented; the model is compared with experimental results of the literature and applied to study the contribution of fiber grids as reinforcement. These results are followed by the conclusions of this work.

2 Local Model for Quasi-Brittle Behavior

Quasi-brittle behavior is characterized by very low levels of deformation before rupture which is usually followed by material cracking. Under monotonic loading, two important material parameters to describe the limit efforts may be identified: strength and toughness.

Tensile strength σ_t is usually defined under homogenous stress conditions, and is more related to crack initiation; while toughness G_{IC} is directly associated to the propagation of large cracks, where the stress and its gradient tend to be very high.

A rectangular pre-cracked sample submitted to a uniform tensile stress σ_0 is shown on Fig. 1a. The ideal value of the maximum tensile stress σ_0 (as a function of the crack length a) for a quasi-brittle material is represented in Fig. 1b. For sufficiently large cracks, the behavior may be defined by the energy variation of the system during crack extensions, which is characterized by energy release rate G and limited by the material toughness G_{IC}. For smaller cracks, the material behavior may be limited by its tensile strength σ_t.

In order to explain this experimental observation, the material is supposed to present micro-defects. Admitting an independent behavior for each small defect, the value of the energy release rate of the cracks orthogonally oriented with respect to the stress can be written in a general shape as:

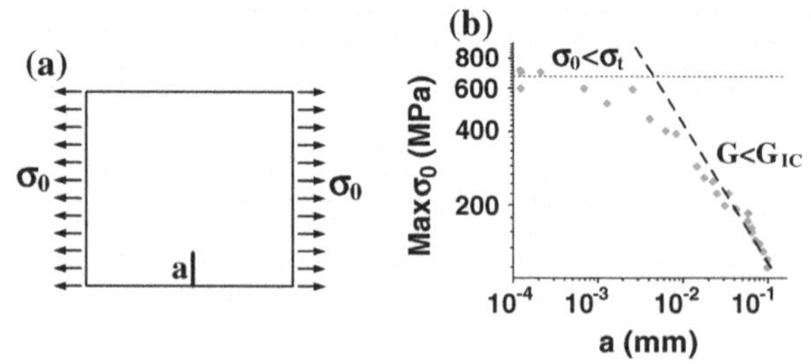

Fig. 1 **a** Cracked sample and **b** rupture stress as a function of the crack length a (Taylor 2010)

$$G = (A\sigma_0)^2 \pi a/E, \tag{1}$$

where A is a value depending on the sample dimensions, and mainly on the crack position. For example, in a very large sample, a crack inside the material presents $G = \sigma_0^2 \pi (a/2)/E$; consequently $A = 1/\sqrt{2} \approx 0.71$. For a single edge crack in a semi-infinite plate subjected to tension (as in Fig. 1a), $G = (1.12\sigma_0)^2 \pi a/E$, meaning that $A = 1.12$ (Sun and Jin 2012). From Eq. 1, (for $G = G_{IC}$) one can get the rupture load $\sigma_{0\,max} = \sqrt{G_{IC}E/(\pi a)}/A$. Considering the influence of A on the rupture load, it would be more likely to observe the propagation of cracks on the boundaries than inside the material. The longest micro-defects (length a_t) dominates the rupture behavior, limiting the maximum load, in absence of pre-cracks or if their size is inferior to a_t, what explains the deviation from the prediction from fracture mechanics. For small pre-cracks ($a < a_t$), the value of G becomes too small, inducing unreal predictions for the maximum load. The value of the transition crack length a_t is then defined, in practice, from the limitation of the maximum load to the tensile strength (for $A = 1/\sqrt{2}$): $a_t = 2G_{IC}E/(\sigma_t^2 \pi)$.

The singular stress at the crack tip turns any criterion based on a limit of the local stress totally useless (the obtained limit loading would tend always to zero). Hence the analysis of the derivative of the energy release rate with respect to the crack size is proposed here:

$$dG/da = (A\sigma_0)^2 \pi/E, \tag{2}$$

which is a local quantity, directly related to the imposed stress σ_0 and independent on the crack size. Considering that tensile rupture in real samples initiate at the boundaries, it implies that the maximum dG/da supported by the material may be

$$\max\, dG/da \approx (1.12\sigma_t)^2 \pi/E, \tag{3}$$

where $A = 1.12$, neglecting any eventual size effect related to the sample dimensions during the evaluation of the tensile strength.

Finally, it is considered that rupture occurs for $G = G_{IC}$ (as usually in fracture mechanics, for large pre-cracks) or $dG/da = \max dG/da$ (for small pre-cracks). This double rupture criterion allows the clear analysis of materials presenting large transition crack lengths a_t, like asphalt concretes ($30\,\text{mm} < a_t < 100\,\text{mm}$) (Kim and Hussein 1997).

3 Numerical Results

The first step is the verification of the robustness of the present approach to deal with crack initiation under complex boundary conditions. Hence the numerical predictions are compared to results of the literature for symmetric plane specimens

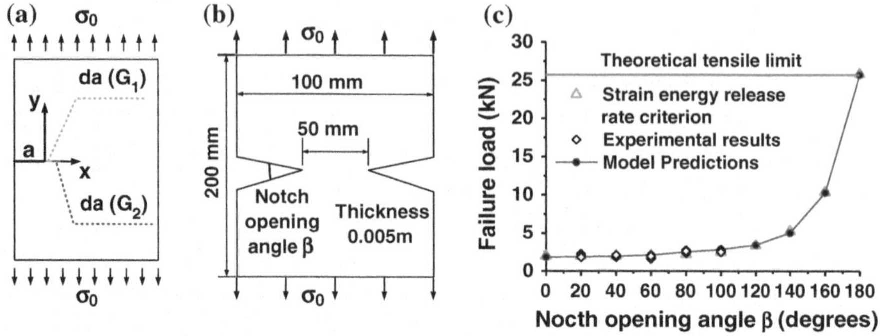

Fig. 2 **a** First and second crack extensions, **b** V-shaped notch sample and **c** the failure load as a function of the notch opening angle β

with V-shaped notches under tension. Thereafter, the behavior of the fiber grid reinforcement of a pavement layer is analyzed.

Plane stress models are calculated with finite element software ABAQUS. The parameters of the material are the same for all the models: Young's modulus $E = 3.3\,\text{GPa}$, Poisson's ratio $v = 0.35$, toughness $G_{IC} = 437.81\ Pa \cdot \text{m}$ and tensile strength $\sigma_t = 102.8\,\text{MPa}$.

The values of energy release rate are obtained by the crack closure method, a local method based on the release of a small segment at the crack tip (Sun and Jin 2012). As shown by Fig. 2a, the first segment provides a first value G_1, while the second release provides a value G_2. The numerical derivative of G is calculated as $dG/da = (G_2 - G_1)/da$, where da is the segment size (or the mesh size, in practice). The rupture is attained if $G_1 = G_{IC}$ or $dG/da = (1.12\sigma_t)^2/E$ (see Eq. 3).

3.1 V-Shaped Notched Samples

V-shaped notches (see example on Fig. 2b) induce singular stresses at the crack tip (like a crack), but the severity of the stress gradient is variable, dependent on the notch opening angle β. These are ideal geometries to study the transition from a crack ($\beta = 0$) to a homogeneous stress field ($\beta = 180°$).

As shown in Fig. 2c, the results of the simulations of the present model are in very good agreement with the failure load obtained by the experiments and numerical predictions of Seweryn and Łukaszewicz (2002), Berto and Lazzarin (2014).

3.2 Fiber Reinforcement

Based on the general results observed for complex boundary conditions shown in Sect. 3.1, the contribution of fiber reinforcement against small cracks ($a < a_t$) on very large samples subjected to tension is analyzed. One supposes that the adhesion between the fiber grid and the asphalt concrete is perfect. The study of two configurations with different crack sizes is shown on Fig. 3a, where the focus is on the improvement induced by a fiber layer on the value of dG/da (Eq. 2), and more precisely on the value of A for a given load σ_0. The effect of the cross section of the fiber $S_f[m^2/m]$ in all analysis can be generalized by $S_f.E_f/E_{ac}$ (see Fig. 3b), where E_f and E_{ac} are the Young's modulus of the fiber and the asphalt concrete, respectively.

According to the simulations, the crack initiation on the boundary (case 1), the value of $A \approx 1.12$ seems independent on the fiber quantity, which suggests a low effect of the fiber reinforcement against crack initiation. On the second case, we study the propagation of a longer crack beyond the fiber grid. The effect of the fiber reinforcement over the variable A as a function of the fiber quantity $S_f E_f/E_{ac}$ is presented on Fig. 3b. Two different contributions of the fiber against the crack opening are presented: as an imposed force (underestimation) and as an imposed displacement (overestimation). The exact behavior is inside these two bounds but cannot be obtained without a tridimensional representation of the crack and the grid. In any case, the increase of the reinforcement quantity $S_f E_f/E_{ac}$ leads to a decrease of the value of A, and a consequent improvement on the local strength against crack initiation (see Eqs. 2 and 3). The value of A is limited at 0.62 for very high quantities of fiber. However, $A = 1/\sqrt{2} \approx 0.71$ corresponds to the value obtained for a crack initiated at any point far from the boundaries; beyond this value, the cracks may initiate (relatively) far from the reinforcement (independently on the fiber quantity).

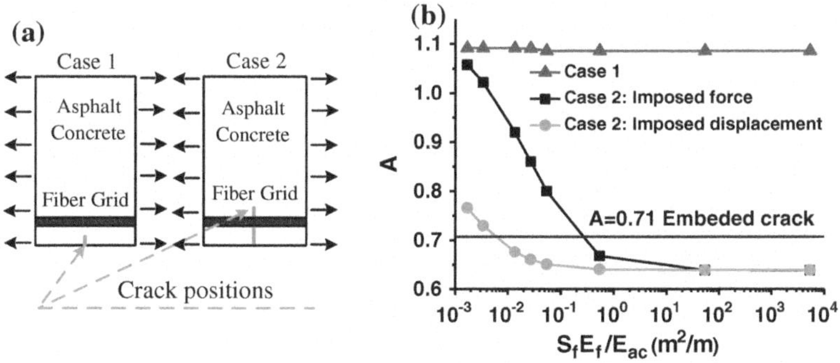

Fig. 3 **a** Two different crack positions and tension direction, and **b** the value of A as function of the fiber quantity

4 Conclusions and Perspectives

The rupture predictions of the proposed model, based on a local analysis of the energy release rate G and its derivative (with respect to the crack length) dG/da, are in good agreement with experimental results and with the predictions of a non-local criterion for V-shaped notched samples of the literature. Furthermore, the formulation can be generally applied to other complex boundary conditions.

A bidimensional analysis is adopted to quantify the contribution of fiber grid reinforcement near a boundary. Low effect of the fibers was observed against crack initiation on the boundary. However, its protection response is clear for cracks crossing the fiber grid. The effectiveness of the grid reinforcement (described by the evolution of the parameter A) seems to depend quasi-logarithmically on the fiber cross section S_f, its Young's modulus E_f and the Young's modulus of the asphalt E_{ac} (trough the relation $S_f E_f / E_{ac}$). This trend stabilizes for high quantities of fiber; nevertheless the fact that cracks may appear elsewhere (for $A < 0.71$), and not at the boundaries of the asphalt sample, imposes a practical limit for the fiber quantity.

Considering these results, the usual quantities of fiber grid reinforcement $0 < S_f E_f / E_{ac} < 0.003$ (Arsenie 2013) are technically inside the effectiveness range identified in these tests.

References

Arsenie IM (2013) Etude et modélisation des renforcements de chaussées à l'aide de grilles en fibre de verre sous sollicitations de fatigue (PhD thesis, Université de Strasbourg).

Bazant, Z. P., Pijaudier-Cabot, G. (1989). Measurement of characteristic length of nonlocal continuum. Journal of Engineering Mechanics, 115(4), 755-767.

Berto, F., Lazzarin, P. (2014). Recent developments in brittle and quasi-brittle failure assessment of engineering materials by means of local approaches. Materials Science and Engineering: R: Reports, 75, 1-48.

Li, J., Zhang, X. B. (2006). Crack initiation prediction for V-notches under mixed-mode loading in brittle materials. Journal of Mechanics of Materials and Structures, 1(8), 1385-1404.

Kim, K. W., El Hussein, M. (1997). Variation of fracture toughness of asphalt concrete under low temperatures. Construction and building Materials, 11(7), 403-411.

Maimí, P., González, E. V., Gascons, N., & Ripoll, L. (2013). Size effect law and critical distance theories to predict the nominal strength of quasibrittle structures. Applied Mechanics Reviews, 65(2), 020803.

Seweryn A, Łukaszewicz A (2002) Verification of brittle fracture criteria for elements with V-shaped notches. Eng. Fract. Mech. 69(13): 1487-1510.

Sun C. T., Jin Z.-H. (2012) Fracture Mechanics, Academic Press.

Taylor D (2010) The theory of critical distances, Elsevier.

Fracture Characterization of Grid-Reinforced Asphalt Pavements

Andrea Graziani, Cesare Sangiorgi and Francesco Canestrari

Abstract The ability of grid-reinforced asphalt systems to improve pavement performance depends on the selection of the appropriate reinforcement product. In particular, the flexural properties of reinforced double-layered asphalt systems can be used to investigate the potential benefits produced by an interlayer reinforcement. This paper presents the results of monotonic three-point bending tests on grid reinforced double-layered asphalt specimens carried out as part of an interlaboratory test organized by the TG4 of RILEM TC 237-SIB focused on the evaluation of pavement grid effects on both flexural strength and interlayer bonding of asphalt pavements. The tests were carried out on beams of different geometry, at different test temperatures and speeds. Results were analyzed in terms of fracture work and by considering energy-based dimensionless indices, which allowed the direct comparison of apparently different bending test procedures. Results showed that grid reinforcement did not noticeably influence the crack-initiation toughness. However, the fracture energy was clearly influenced by the grid presence and by the grid type. This effect was related to the effect on the post-peak deformation phase and to the different failure mechanisms.

Keywords Asphalt pavements · Grid reinforcement · Bending test · Fracture toughness

A. Graziani (✉) · F. Canestrari
Università Politecnica delle Marche, Ancona, Italy
e-mail: a.graziani@univpm.it

F. Canestrari
e-mail: f.canestrari@univpm.it

C. Sangiorgi
Università di Bologna, Bologna, Italy
e-mail: cesare.sangiorgi4@unibo.it

A. Chabot et al. (eds.), *8th RILEM International Conference on Mechanisms of Cracking and Debonding in Pavements*, RILEM Bookseries 13,
DOI 10.1007/978-94-024-0867-6_9

1 Introduction

The use of geosynthetic reinforcement within bituminous bound layers is becoming a common construction and maintenance practice to prevent or delay reflective cracking and rutting and improve fatigue life of pavements (Brown et al. 2001; Montestruque et al. 2004; Nguyen et al. 2013; Sobhan and Tandon 2008; Zielínski 2008). However, performance improvements provided by interlayer reinforcements can be negatively affected and eventually neutralized by the potential de-bonding effect caused by geosynthetics at the interface (Canestrari et al. 2015). Therefore, the successful adoption of reinforcement systems demands for improved research in order to overcome this issue.

The study here described is part of the research project: "Advanced Interface Testing of Geogrids in Asphalt Pavements" promoted by Task Group 4 "Pavement Performance Prediction and Evaluation" of RILEM Technical Committee 237-SIB "Sustainable and Innovative Bituminous Materials". Within this research framework, quasi-static three-point bending (3PB) tests were carried out on double-layered AC specimens, with reinforced and un-reinforced interface, obtained by a full-scale pavement test section Graziani et al. (2014). 3PB results from two different laboratories, using different testing setups, are here analyzed in terms of fracture work and by considering energy-based dimensionless indices, which allow the direct comparison of apparently different bending test procedures.

2 Materials and Methods

The test section used for the production of 3PB samples was a real scale double-layered bituminous pavement. The two AC layers, having a thickness of 50 mm-each, were prepared with the same mixture: a typical Italian dense graded mix with 12 mm maximum aggregate size (AC 12) and 70/100 penetration bitumen dosed at 5.5 % by aggregate weight.

In both reinforced and unreinforced sub-sections, an SBS polymer-modified tack coat emulsion, classified as C 69 BP 3 (EN 13808), was applied on the surface of the lower layer with a rate of 0.25 kg/m^2 of residual binder.

In the reinforced pavements, two different grids were laid. The Carbon Fiber/Glass geogrid (CF) is pre-coated with bitumen and made of carbon fiber rovings in the transversal direction and glass fiber rovings in the longitudinal direction, with a 20 mm^2 mesh. The product is sanded on the upper side, whereas a burn off film is applied on the lower side. The Glass Fiber Reinforced Polymer geogrid (FP) is obtained by weaving continuous alkaline-resistant pre-tensioned glass fibers, covered with a thermosetting epoxy resin (vinylester). The grid has a bi-directional square geometry with flat transversal strands woven into longitudinal twisted strands, with a 33 mm^2 mesh and a thickness of about 3 mm. Apart from the constituent material and mesh size, the main difference among the two geogrids is

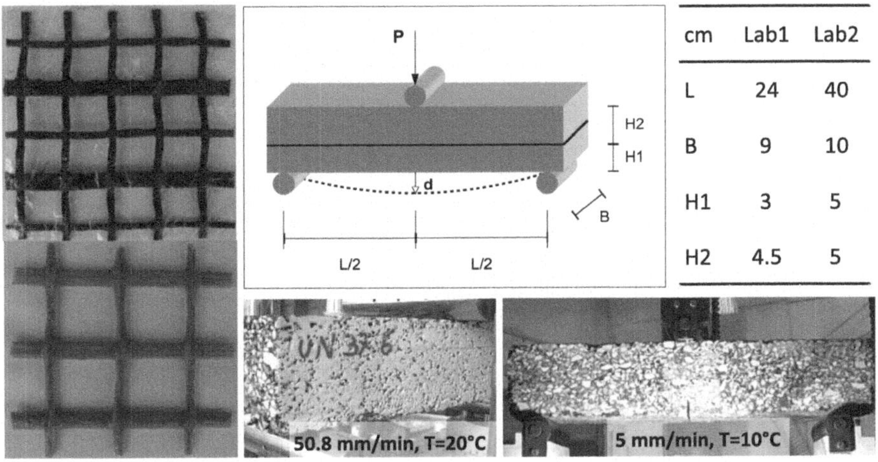

cm	Lab1	Lab2
L	24	40
B	9	10
H1	3	5
H2	4.5	5

Fig. 1 CF (*top*) and FP geogrids, 3PB scheme, double-layered specimens Lab1 and Lab2 (*right*)

their torsional rigidity, also called aperture rigidity. In fact, the FP geogrid is extremely stiff as twisting and distorting its square mesh is very difficult, whereas the CF geogrid mesh is highly flexible and deformable (Fig. 1). Additional details about the characteristics of the grids are reported Canestrari et al. (2015).

Displacement control 3PB tests were performed on prismatic beams of different sizes according to the methodologies adopted by the two Labs. Scheme and dimensions of the 3PB tests are shown in Fig. 1. Load and beam deflection at the mid-span of the specimen are measured through a load cell and a LVDT, respectively. Note that Lab2 samples are notched. 3PB test is suitable to evaluate the crack propagation allowing the comparison between different reinforced systems.

3 Results and Analysis

Typical load-deflection curves obtained by Lab1 and Lab2 are depicted in Fig. 2. The results obtained by the two laboratories were different in magnitude due to the differences in beam geometry, test speed and test temperature. However, the effect of grid reinforcement on the load-deflection behavior was analogous. Unreinforced double-layered specimens showed a rapid strength loss due to upwards crack propagation across the unreinforced interface (Fig. 3a). In CF-reinforced specimens the load-carrying capacity showed a rapid drop after the peak, followed by a phase where the rate of strength loss decreased and then a rapid evolution until collapse (Fig. 3b). Such behavior indicates that CF grid delayed crack propagation across the interface increasing the toughness of the system. Similar to CF-specimens, FP-reinforced specimens showed a drop of load-carrying capacity after the peak. However, this was followed by a post-cracking strength increase of the reinforced

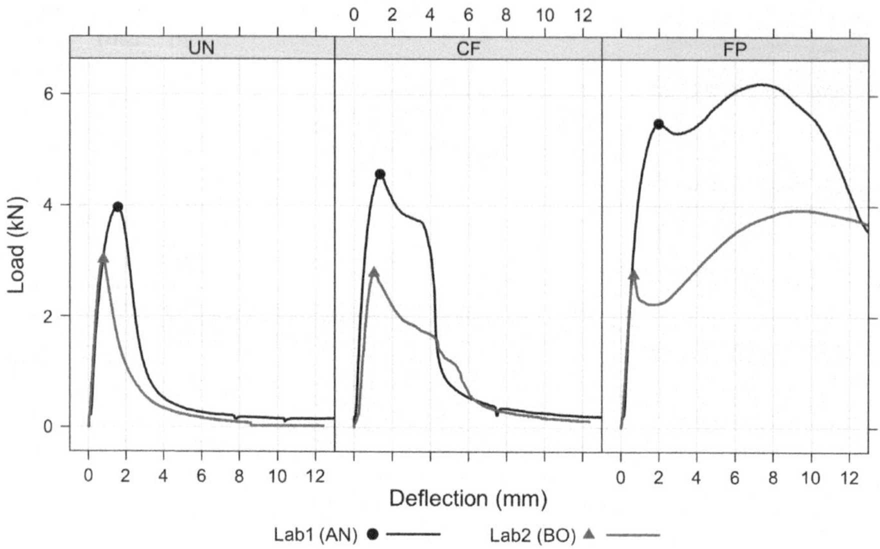

Fig. 2 Typical load deflection curves obtained by Lab1 and Lab2 (the "first-crack" point is highlighted in each curve)

system. Such behavior is due to the fact that the upward crack propagation was stopped by the grid and actually continued along the interface (Fig. 3c). The system failure was due to slippage between the upper and lower layers.

It is important to note that the same fracture behavior was observed by both laboratories, regardless of specimen geometry, presence of notch, test temperature and speed. This suggests that bending tests could be good candidates to become a standard test for evaluating the reinforcing effect of pavement grids.

Fig. 3 Typical failures of unreinforced and reinforced double-layered beams: **a** Lab1 CF system; **b** Lab2 CF system; **c** Lab1 FP system; **d** Lab2 FP system

3.1 Toughness Characterization of Fracture Behavior

The fracture behavior of reinforced double-layered systems in bending tests is similar to that observed for fiber reinforced concrete (FRC). In particular, the post-cracking behavior of CF and FP reinforced systems in bending tests (Fig. 2) can be described as deflection-softening and deflection hardening, respectively (Di Prisco et al. 2009). This suggests that the fracture behavior of grid-reinforced asphalt pavements could be characterized in terms of flexural toughness, i.e., the ability of the reinforced system to absorb energy. Among the various approaches to toughness characterization that have been developed for FRC (Gopalaratnam and Gettu 1995) the calculation of energy-based dimensionless indices was applied to the results of the flexural tests described above.

Dimensionless toughness indices are defined as the ratio of the area under the load deflection curve up to specific deflection limits, to the area of the same curve up to the "first-crack" deflection (d_F). In particular, in the ASTM C 1018 standard the toughness indices I_5, I_{10} and I_{20} are computed at deflection limits of 3 d_F, 5.5 d_F and 10.5 d_F, respectively. The rationale behind these indices is that they depend only on the energy absorption capability of the structural element regardless of other geometrical variables associated with the specimen and testing arrangement. In addition, it can be shown that, for a linear elastic behavior up to the first crack, and perfectly plastic behavior thereafter, the values $I_5 = 5.0$, $I_{10} = 10.0$ and $I_{20} = 20.0$ are obtained.

The values of I_5, I_{10} and I_{20} computed from test results obtained in this study, are summarized in Fig. 4. It is highlighted that the first-crack deflection was defined on the basis of the maximum of the load deflection curve after the initial quasi-linear segment. The three indices are able to highlight the effect of grid reinforcement on the energy absorption capability of the double-layered systems. In particular, the

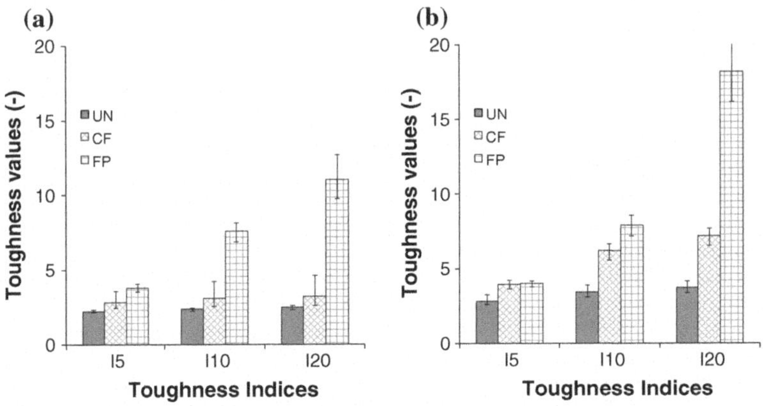

Fig. 4 Dimensionless toughness indices I5, I10 and I20 (average values and range of measured values): **a** Lab1; **b** Lab2

ranking of the three systems (UN, CF and FP) is maintained, regardless of the test setup. The reinforcement effect is particularly evident on I_{20}, i.e., at higher limiting deflections. The hardening behavior of FP systems is particularly evident in the tests carried out by Lab2 (notched beams) where an average $I_{20} = 18.2$ was obtained.

It is finally remarked that the use of dimensionless toughness indices allowed the practical characterization of reinforced systems, regardless of test arrangement, thus confirming that bending tests are suitable for evaluating the reinforcing effect of pavement grids.

4 Conclusions

This paper focuses on the effects of grid reinforcement on the fracture behavior of asphalt pavements. Quasi-static 3PB tests were carried out on double-layered AC specimens, with reinforced and unreinforced interface.

Results showed that the CF grid delayed crack propagation across the interface, whereas FP grid stopped the upward crack propagation, which continued along the interface. The post-cracking behavior of CF and FP reinforced systems was described in terms of flexural toughness. In particular, energy-based dimensionless indices (I_5, I_{10} and I_{20}), originally developed for the analysis of FRC, were computed for both the reinforced and unreinforced systems.

The three indices were able to highlight the effect of grid reinforcement on the energy absorption capability of the double-layered systems. In particular, the reinforcement effect is more evident on I_{20}, i.e., at higher limiting deflections.

It is of special interest to note that the same fracture behavior was observed, regardless of specimen geometry, presence of notch, test temperature and loading speed. This suggests that, bending tests could be good candidates to be standardized for evaluating the reinforcing effect of pavement grids.

References

Brown SF, Thom NH, Sanders, PJ (2001) A study of grid reinforced asphalt to combat reflection cracking. J Assoc Asphalt Pav 70:543-569

Canestrari F, Belogi L, Ferrotti G, Graziani A (2015) Shear and flexural characterization of grid-reinforced asphalt pavements and relation with field distress evolution. Materials and Structures, 48, 959-975 doi:10.1617/s11527-013-0207-1

Di Prisco, M., Plizzari, G., & Vandewalle, L. (2009). Fibre reinforced concrete: new design perspectives. Materials and Structures, 42(9), 1261-1281. doi:10.1617/s11527-009-9529-4

Gopalaratnam, V. S., & Gettu, R. (1995). On the characterization of flexural toughness in fiber reinforced concretes. Cement and concrete composites, 17(3), 239-254. doi:10.1016/0958-9465(95)99506-O

Graziani, A., Pasquini, E., Ferrotti, G., Virgili, A., & Canestrari, F. (2014). Structural response of grid-reinforced bituminous pavements. Materials and Structures, 47(8), 1391-1408. doi:10.1617/s11527-014-0255-1

Montestruque G, Rodrigues R, Nods M, Elsing A (2004) Stop of Reflective Crack Propagation with the Use of PET Geogrid as Asphalt Overlay Reinforcement. Paper presented at the 5th International RILEM Conference on Cracking in Pavements, Limoges, 5-8 May 2004

Nguyen ML, Blanca J, Kerzréhoa JP, Hornych P (2013) Review of glass fibre grid use for pavement reinforcement and APT experiments at IFSTTAR. Road Mater Pavement. 14 (1): 287-308

Sobhan K, Tandon V (2008) Mitigating reflection cracking in asphalt overlay using geosynthetic reinforcements. Road Mater Pavement 9(3):367-387. doi:10.1080/14680629.2008.9690124

Zielínski P (2008) Fatigue investigations of asphalt concrete beams reinforced with geosynthetics interlayer. Paper presented at the 6th RILEM International Conference on Cracking in Pavements, Chicago, 16-18 June 2008: 751-759. doi: 10.1201/9780203882191.ch73

Impact of Loading Rate and Temperature on Tensile Strength of Asphalt Mixtures at Low Temperatures

Daniel Steiner, Bernhard Hofko, Mariyan Dimitrov and Ronald Blab

Abstract Low temperature cracking of an asphalt mix with specific design is primarily influenced by the loading rate and the temperature. For laboratory testing of hot mix asphalt (HMA), several test methods have been developed and standardised in the last decades to assess the resistance to low-temperature cracking. In order to characterise a mixture efficiently and economically, the test has to be kept within acceptable timeframes. Therefore, low temperature test methods apply fixed parameters, as Uniaxial Tensile Strength Test (UTST) and Relaxation Test (RT) according to EN 12697-46. To study impact of time and temperature on the low temperature behaviour, loading rate at UTST and initial stresses at RT are systematically varied at four temperatures (−20 | −10 | 0 | +10 °C). UTST are carried out at 3 strain rates (1.6×10^{-4}–1.6×10^{-2} mm/s). The tensile strength β_t and failure strain $\varepsilon_{failure}$ are determined and evaluated. Furthermore, RTs are performed at 3 different initial stresses, depending on the strength determined by UTST (0.75 | 0.5 | 0.35* βt). This will reveal how the level of initial stress influences the relaxation time and ratio. The results of this study can be used for modelling and simulation of pavement structures employed on bridges for expansions joints with low deformation rates and high absolute deformations.

Keywords Low temperature · Cracking · Tensile strength · Relaxation test

D. Steiner (✉) · B. Hofko · M. Dimitrov · R. Blab
Institute of Transportation Research Center of Road Engineering, Vienna University of Technology, Gusshausstraße, 28/230-3, 1040 Vienna, Austria
e-mail: daniel.steiner@tuwien.ac.at

B. Hofko
e-mail: bernhard.hofko@tuwien.ac.at

M. Dimitrov
e-mail: mariyan.dimitrov@tuwien.ac.at

R. Blab
e-mail: Ronald.Blab@tuwien.ac.at

A. Chabot et al. (eds.), *8th RILEM International Conference on Mechanisms of Cracking and Debonding in Pavements*, RILEM Bookseries 13,
DOI 10.1007/978-94-024-0867-6_10

1 Introduction

The performance of HMA at low temperatures can be characterized by several laboratory test methods. The cracking behaviour at conditions up to −40 °C can be ascribed to two key causes: Thermal cracking and traffic loading. Traffic loading at low temperatures, however, always appears in connection with stresses due to thermal loading. (Hills and Brien 1966) stated that fully restrained pavements build up stresses due to contraction. If these thermal stresses exceed the tensile strength of the asphalt mixture, thermal cracks occur. This point defines the lowest service temperature. This critical cracking temperature, often called T_{Crack}, is usually determined by the Thermal Stress Restrained Specimens Test (TSRST) according to EN 12697-46 (CEN 2012; Monismith et al. 1965; Jung and Vinson 1994; Hofko et al. 2012). At temperatures above T_{Crack}, the thermal stresses are lower than the tensile strength. This means that asphalt pavements can bear additional loads from traffic at these conditions. (Wistuba et al. 2006) Uniaxial tensile stress tests (UTST) (CEN 2012) at different temperatures can describe the evolution of the tensile strength versus temperature. The difference of both stress curves derived from both, TSRST and UTST, is called tensile stress reserve and indicates the stress that can be applied due to traffic loading additional to thermal stresses before a pavement fails. UTST according to EN 12697-46 are carried out at defined constant strain rates. An ongoing research project investigates the technical feasibility of alternative expansions joints on bridges with continuous pavement structures. These structures request asphalt pavements that have the ability to withstand high deformation rates and high absolute deformations. Therefore, tests with amended conditions related to standards were carried out.

In this paper, the results of a laboratory mixed asphalt concrete with maximum aggregate size of 22 mm (AC 22) are shown. In the course of the project additional tests on plant-mixed AC and mastic asphalt (MA) were carried out. Since the low temperature behaviour depends greatly on the binder properties, which are influenced by the harsher processing of mixtures in the asphalt mix plant, a comparison between plant and laboratory mixing conditions is possible. Furthermore a comparison of two fundamentally different mix designs (AC/MA) is possible.

2 Materials and Specimen Preparation

An asphalt concrete with a maximum nominal aggregate size of 22 mm (AC 22) was used for the tests. The coarse aggregate and filler used for the mix is limestone. As a binder an SBS modified PmB 45/80-65 was used. The binder content was set to 4.5 % by mass. The maximum density of the AC 22 was determined to be 2.579 kg/m^3.

The mix was prepared in a laboratory reverse-rotation compulsory mixer, according to EN 12697-35, with a mixing temperature of +185 °C. HMA Slabs (50 × 26 × 4 cm) were compacted in a roller compactor according to EN 12697-33. From the slabs, seven specimens were cut (6 × 6 × 22.5 cm). The air void contents of the specimens range from 2.5 to 3.8 % by volume.

Table 1 Test program for UTST and RT tests

UTST					RT			
Temperature [°C]	−20	−10	0	10	Temperature [°C]	−20	−10	0
$\varepsilon_1 = 0.016$ mm/s	4	2	3	2	$0.75 \times \beta\ (\varepsilon_1)$	2	3	2
$\varepsilon_2 = 0.0016$ mm/s	2	3	3	2	$0.5 \times \beta\ (\varepsilon_1)$	–	2	–
$\varepsilon_3 = 0.00016$ mm/s	2	3	2	2	$0.35 \times \beta\ (\varepsilon_1)$	–	2	–

3 Test Methods and Test Program

Uniaxial tensile stress tests (UTST) were carried out according to EN 12697-46 (CEN 2012). UTST are run at temperatures of +10, 0, −10 and −20 °C with 3 different deformation rates from 0.00016 to 0.016 mm/s. The tensile strength versus temperature βt (T) is obtained from the test.

Relaxation tests (RT) were carried out according to EN 12697-46 (CEN 2012). RT are run at 3 different temperatures of 0, −10 and −20 °C. At −10 °C, the initially applied tensile stress is varied. The applied stress levels are derived from the tensile strength resulted in the UTST tests. The stress levels were set to 0.75* βt, 0.5* βt and 0.35* βt. For the testing temperature 0 and −20 °C the stress level was set to 0.75* βt.

Table 1 gives an overview of the test program. The numbers indicates the number of tested specimens.

4 Results and Discussion

To obtain the effect of the strain rate on the low temperature behaviour of HMA, 16 specimens were tested. Tensile strength and failure strain are influenced by varying the strain rate due to the time dependency of visco-elastic materials. Figure 1 shows the detailed results of the UTST. The correlation between testing strain rate and resulted tensile strength is illustrated. Logarithmic regressions are also presented for each testing temperature in the figure. The coefficients of determination R^2 are equal or above 0.85, which indicates a good correlation of strain rate versus tensile strength. This is, however, not the case for the lowest testing temperature −20 °C. For this temperature the tensile strength seems to be mostly independent from the strain rate. For lower temperatures, the tensile strength is higher, the higher the strain rate is. Figure 2 clearly shows that the tensile strength is higher at lower temperatures. At the highest and lowest temperatures the impact of the strain rate is relatively small, but at a temperature of 0 and −10 °C the courses of the curves are diverse and thus, the tensile strength is affected by the strain rate. In Fig. 3, the failure strains derived from UTST are related to the strain rates for the different test temperatures. Here, once again, the logarithmic regressions are evaluated for all

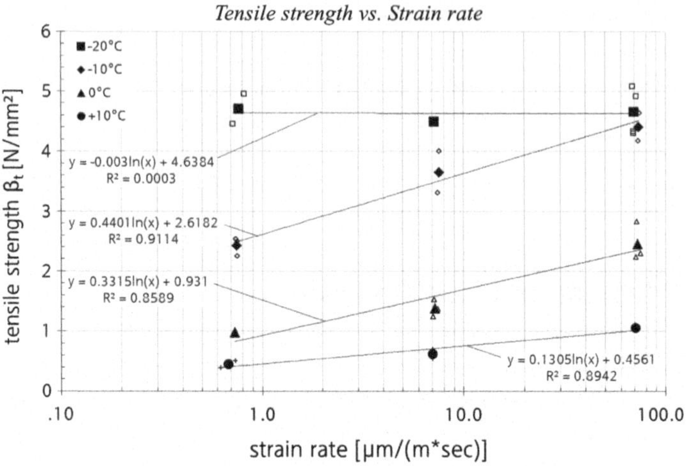

Fig. 1 Tensile strength versus strain rate at different temperatures

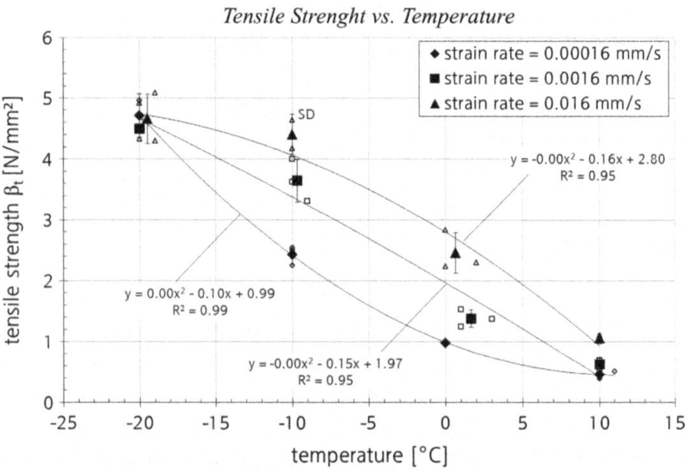

Fig. 2 Tensile strength versus temperature at different strain rates

temperatures. R^2 values between 0.81 and 0.89 indicate an acceptable correlation between failure strain and strain rate. Of particular note is, steeper slopes of the regression lines can be seen at the higher testing temperatures 0 and 10 °C, which indicates of bigger influence of the strain rate at these temperatures.

In Fig. 4, the results of the relaxation test program are shown. The mean values of the relaxation time related to vary initial stress factors are illustrated. The relaxation time is defined as the time, when the stress has dropped to a level, which

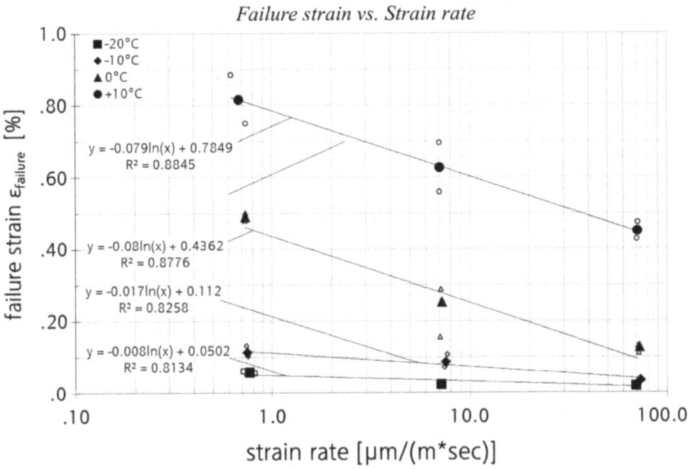

Fig. 3 Failure strain versus strain rate at different temperatures

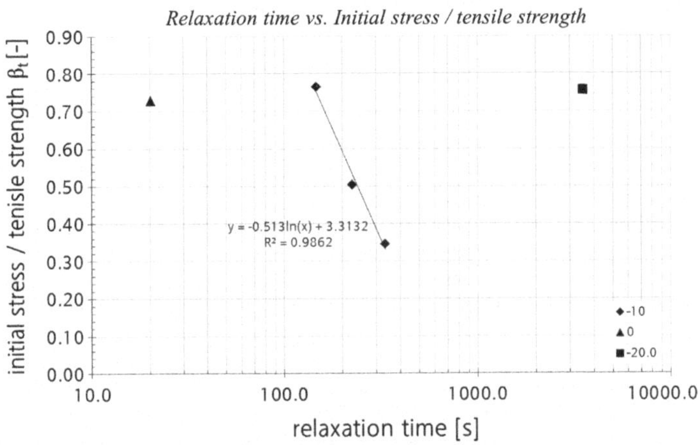

Fig. 4 Relaxation time versus initial stress at different temperatures

is equal to 36.8 % of the initial stress. At a testing temperature of −10 °C, an excellent logarithmic correlation between ratio of initial stress /tensile strength and the relaxation time can be seen. The results of the testing temperatures 0 and −20 °C also points out the roughly logarithmic correlation between temperature and relaxation time at the same initial stresses (0 °C t = 3502 s | −10 °C t = 147 s | −20 °C t = 20 s). This can be explained by the bitumen behaviour at low and moderate temperatures.

5 Summary and Conclusions

The main motivation for the research project presented within this paper is to study the impacts of different testing conditions on the low temperature behaviour, in particular the strain rate at UTST. The results can be employed for modelling and simulation of pavement structures on bridges. Thereby asphalt layers cover expansion joints and have to withstand high absolute deformations at low deformation rates. As both, temperature and loading time influence the HMA behaviour; the material shows brittleness as well as ductility. Variation in testing conditions (temperature or strain rate) can affect the type of cracking. In some cases brittle cracks will occur, in some cases the material will perform ductile, so that under certain circumstances, instead of a discrete crack, micro cracks will distribute over a greater area. This behaviour was also observed by (Pucci et al. 2004) at low strain rates. The results show this in particular at relatively higher testing temperatures 0 and 10 °C. We visually noticed that effect at some tests in our test program. The failure strain is getting greater with decreasing strain rate, although the tensile strength at 10 °C is almost the same. This indicates a ductile behaviour. At the lowest temperature −20 °C tensile strength and strain rate are almost on the same level. At mid testing temperatures, a combination of glassy state behaviour and viscoelastic behaviour is seen.

References

CEN 2012. EN 12697-46: Bituminous mixtures - Test methods for hot mix asphalt - Part 46: Low temperature cracking and properties by uniaxial tension tests. Brussels.

HILLS, J. F. & BRIEN, D. 1966. The cracking of bitumen and asphalt mixes by temperature induced stress. *Proceeding Association of Asphalt Paving Technologists (AAPT),* vol. 35, p. 292-357.

HOFKO, B., BLAB, R. & SPIEGL, M. Enhancing the High Quality Recycling of Asphalt in Austria by Means of Performance Based Testing. *In:* MADELLO, A. & MARCHAND, J.-L., eds. 5th Eurasphalt & Eurobitume Congress, 2012 Istanbul. Foundation Eurasphalt.

JUNG, D. & VINSON, T. S. 1994. *Low-temperature cracking: test selection.*

MONISMITH, C. L., SECOR, G. & SECOR, K. E. Temperature induced stresses and deformations in asphalt concrete. Association of Asphalt Paving Technologists Proceedings, 1965.

PUCCI, T., DUMONT, A.-G. & DI BENEDETTO, H. 2004. Thermomechanical and Mechanical Behaviour of Asphalt Mixtures at Cold Temperature. *Road Materials and Pavement Design,* 5, 45-72.

WISTUBA, M., LACKNER, R., BLAB, R. & SPIEGL, M. 2006. Low-temperature performance prediction of asphalt mixtures used for LLP—new approach based on fundamental test methods and numerical modeling. *International Journal of Pavement Engineering,* 7, 121-132.

Numerical Correlation Between Low Temperature SCB Fracture and IDT Strength of Asphalt Mixture Using FEM Analysis

Augusto Cannone Falchetto, Michael P. Wistuba and Ki Hoon Moon

Abstract Currently, Indirect Tensile (IDT) and Semi Circular Bending (SCB) tests are used for evaluating strength and fracture properties of asphalt mixture at low temperature. While IDT is performed on notchless samples, the SCB peak stress (strength) is obtained on pre-notched specimens which are not representative of initial undamaged pavement conditions. In this paper, the low temperature strength of asphalt mixture is investigated based on IDT and SCB tests, and on 2D finite element simulations. The values of maximum stress at crack tip in SCB simulation are correlated to the experimental IDT strength through non-parametric statistical analysis. The newly proposed relationship can be used to predict the IDT asphalt mixtures strength based SCB fracture tests and, thus, contribute to a better understanding of the low temperature cracking properties in the evolution of asphalt pavements from undamaged to damaged conditions.

Keywords IDT · SCB · Low temperature · Numerical simulation

1 Introduction

Low temperature cracking represents a serious distress for asphalt pavement in cold regions such northern U.S. and Europe. When temperature dramatically drops during winter, tensile stress develops in the restrained pavement layers eventually reaching the material strength and resulting into cracking phenomena (Li and Marasteanu 2010; Moon et al. 2013). The failure properties of asphalt mixture are

A.C. Falchetto (✉) · M.P. Wistuba
Technische Universität Braunschweig, Braunschweig, Germany
e-mail: a.cannone-falchetto@tu-bs.de

M.P. Wistuba
e-mail: m.wistuba@tu-bs.de

K.H. Moon
Korea Expressway Corporation, Gimcheon, South Korea
e-mail: moonx113@umn.edu

A. Chabot et al. (eds.), *8th RILEM International Conference on Mechanisms of Cracking and Debonding in Pavements*, RILEM Bookseries 13,
DOI 10.1007/978-94-024-0867-6_11

crucial for designing asphalt pavement since they are used as input in the Thermal Cracking (TC) model which is part of the current Mechanistic-Empirical Pavement Design Guide (MEPDG) (AASHTO 2008; Cannone Falchetto et al. 2014).

Currently, Indirect Tensile (IDT) creep and strength tests (AASHTO T322 2007) are used to determine thermal stress and corresponding low temperature nominal strength, $\sigma_{N\text{-}IDT}$, of asphalt mixture. However, this test method does not provide information on crack initiation and propagation and, hence, on the fracture mechanisms of asphalt pavement (Li and Marasteanu 2010). Recent research has indicated that a fracture mechanics approach combined with specifically designed tests methods is necessary in order to properly address the resistance of asphalt pavement to thermal cracking (Marasteanu et al. 2002; Li and Marasteanu 2010).

Among the available low temperature test methods, the Semi-Circular Bending (SCB) test has received considerable attention due to experimental simplicity combined to the use of Linear Elastic Fracture Mechanics (LEFM) (Li and Marasteanu 2010). Specimens present a semi-circular shape with 75 mm radius, 30 mm thickness and a 1 mm wide notch of 15 mm in length; parameters such as fracture energy, G_f, and fracture toughness, K_I, can be computed.

Peak stress, $\sigma_{N\text{-}SCB}$, (strength), G_f and K_I are obtained in SCB tests on pre-notched specimens which are not representative of an initial undamaged condition of a pavement. Therefore, if a mathematical/numerical relation between $\sigma_{N\text{-}SCB}$ and $\sigma_{N\text{-}IDT}$ can be developed, both strength and fracture properties can be derived from single SCB low temperature fracture tests without relying on IDT, significantly reducing costs and time required for additional specimens' preparation and testing.

In this paper, the low temperature strength of asphalt mixture is investigated based on IDT and SCB tests and on two-dimensional (2D) numerical simulations. The values of maximum stress at crack tip in SCB simulation are obtained using a commercial FEM code and cohesive zone model (Marasteanu et al. 2002). Then, a correlation between the numerically determined maximum SCB stress and the strength value experimentally measured on IDT specimens is derived through non-parametric statistical analysis (Cannone Falchetto et al. 2014).

2 Materials and Testing

In this paper, three types of asphalt mixtures (Table 1) with different Nominal Maximum Aggregate Size (NMAS), provided from Minnesota Department of Transportation (MN/DOT), were used both for IDT strength and SCB fracture tests. In order to limit the experimental work only one testing temperature, T, was selected, corresponding to lowPG + 10 °C. Both IDT and SCB specimens (3 replicates per test) were obtained from the same set of mixture samples.

IDT tests were performed according to the current AASHTO standard (AASHTO T322 2007) and the nominal strength was computed as:

Table 1 Prepared asphalt mixtures for IDT and SCB testing

Mix ID	Binder (PG)	Air voids [%]	NMAS [mm]	Origin	Test T [°C]
1	PG52-28	7	19.0	MN/DOT	−18
2	PG58-34	4	12.5	MN/DOT	−24
3	PG46-34	4	9.5	MN/DOT	−24

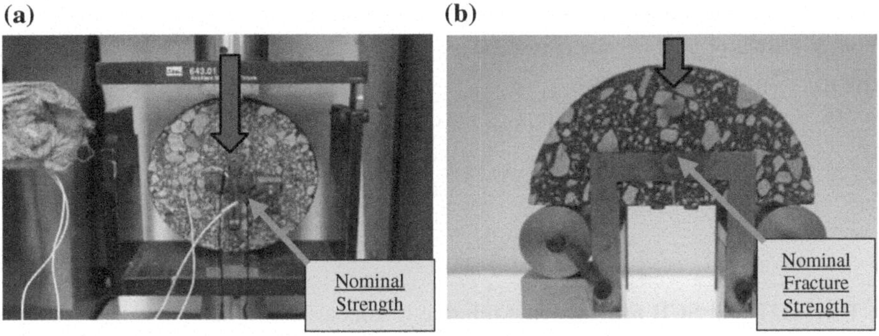

Fig. 1 **a** IDT and **b** SCB testing set up

$$\sigma_{N-IDT}[\text{MPa}] = (2P_{fail})/(\pi \cdot b \cdot D) \tag{1}$$

In Eq. 1, P_{fail} is the peak load; b and D are specimen thickness (40 mm) and diameter (150 mm), respectively. The SCB tests were performed with two displacement sensors: Load Line Displacement (LLD) and Crack Mount Opening Displacement (CMOD). The CMOD signal was used to obtain a stable crack growth during the test with a rate of 0.0005 mm/s. Finally, plots of load versus LLD are generated and K_I and G_f, can be computed as (Li and Marasteanu 2010):

$$K_I[\text{MPa/m}^{0.5}] = \sigma_0 \cdot \sqrt{\pi \cdot a} \cdot \left[Y_{I(S_0/r)} + (\Delta s_0/r) \cdot B\right] \tag{2}$$

$$G_f[\text{J/m}^2] = W_f/A_{lig} \quad \text{and} \quad W_f = \int Pdu \tag{3}$$

where K_I is Mode I stress intensity factor, $\sigma_0 = P/(2rt)$, r is the radius, t is the thickness, Y_I is the normalized stress intensity factor, a is the notch length, Δs_0 and B are fracture parameters and A_{lig} is the area of ligament (Li and Marasteanu 2010). Pictures of IDT and SCB testing set up are presented in Fig. 1.

3 Tests Results and FE Simulation

The experimental results of IDT and SCB tests are shown in Tables 2 and 3.

Table 2 Results of IDT strength tests at PG + 10 °C

Mix ID	Binder (PG)	$\sigma_{N\text{-}IDT}$	$\sigma_{N\text{-}IDT}$	CoV [%]
		Mean [MPa]	St.dev [MPa]	
1	52–28	4.416	0.185	4.01
2	58–34	4.272	0.165	3.86
3	46–34	3.928	0.367	9.34

Table 3 Results of SCB fracture tests at PG + 10 °C

Mix ID	Binder (PG)	G_f	G_f	K_I	K_I
		Mean [kJ/m^2]	CoV [%]	Mean [MPa*m$^{1/2}$]	CoV [%]
1	52–28	0.922	11.67	0.788	1.28
2	58–34	0.789	15.94	0.855	8.53
3	46–34	0.380	14.53	0.771	7.67

Both IDT and SCB tests provide reliable results at low temperature (see the low coefficient of variation CoV). In order to obtain the maximum nominal stress at the crack tip of SCB specimens at peak load, a finite element model (FEM) was developed. ABAQUS (2013) software was used to simulate SCB fracture through a cohesive zone modeling (CZM) of the ligament ahead of crack tip (Fig. 2a).

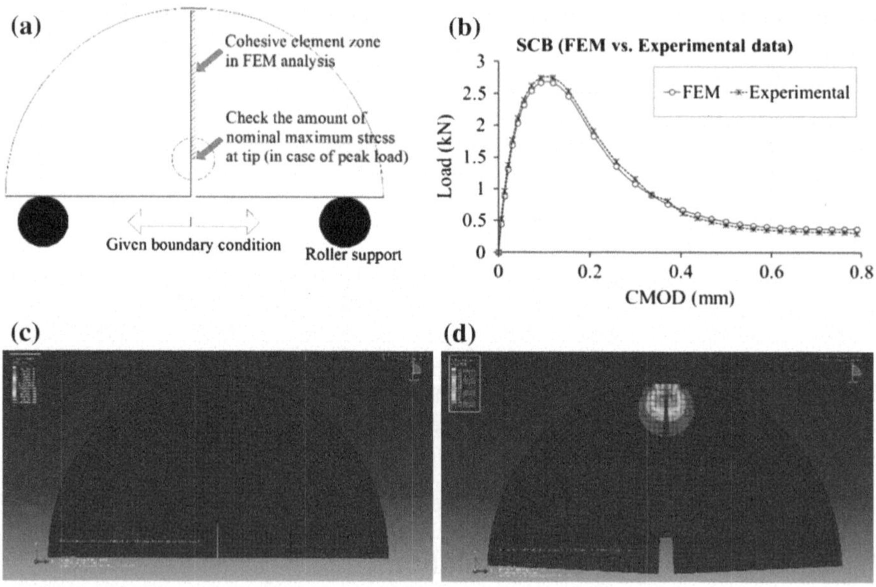

Fig. 2 **a** Schematic of SCB FEM analysis; **b** comparison between FEM and experimental results: Load versus CMOD; configuration **c** before and **d** after cracking

The FEM analysis was calibrated with the experimental load versus CMOD curve so that the CZM parameters could be estimated (i.e., f_t', cohesive strength and G_f, cohesive fracture energy). A schematic of the FEM analysis is shown in Fig. 2. Table 4 presents the material parameters and the SCB nominal maximum stress (SCB strength).

In Table 4, initial values of Modulus E [GPa] and cohesive strength [MPa] were estimated based on Eq. 4 and previous research (Li 2004), respectively.

$$\sigma_{initial} = \frac{\text{Peak Load}_{\text{SCB test}}}{2 \times \text{thickness} \times \text{Radius}} = \frac{K_I}{\sqrt{\pi \times a} \times \cdot \left[Y_{I(S_0/r)} + (\Delta s_0/r) \cdot B \right]} \tag{4}$$

It can be observed that the initial slope of the experimental data is slightly smaller than that of FEM simulation results for each mixture investigated. It is assumed that a refinement of the FEM analysis may improve the accuracy of the simulation. Based on the results in Tables 2 and 3, and on non-parametric statistics (Cannone Falchetto et al. 2014) the relationship between IDT strength and SCB nominal maximum stress is derived and shown in Fig. 3. A good linear relationship can be observed between experimental IDT data and simulation SCB values:

$$\sigma_{\text{N-IDT}}[\text{MPa}] = 0.12 \times \sigma_{\text{N-SCB}}[\text{MPa}] \tag{5}$$

Table 4 Material parameters and maximum stress at crack tip for FEM simulation of SCB test

Mix ID	Modulus E [GPa]	Cohesive strength f_t' [MPa]	Nominal max. stress at crack tip [MPa]			
			No. 1	No. 2	No. 3	Mean
1	18.5	2.58	0.550	0.551	0.554	0.552
2	19.1	2.64	0.530	0.497	0.511	0.513
3	18.6	2.78	0.423	0.523	0.469	0.472

Fig. 3 IDT strength versus nominal maximum stress in SCB test (all mixtures)

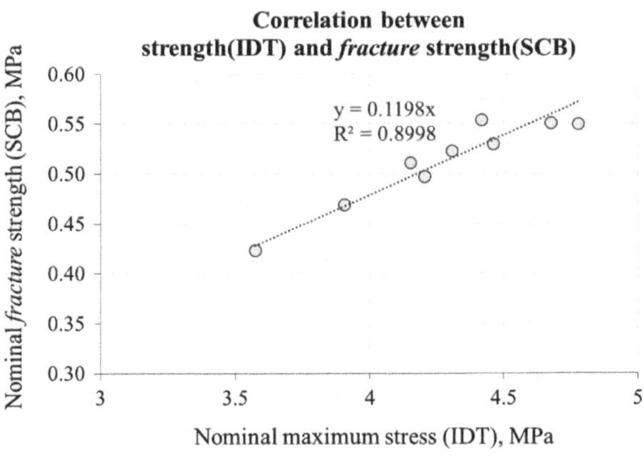

4 Summary and Conclusions

In this paper, a simple FEM code was developed to simulate SCB fracture test with the objective of predicting IDT strength from SCB data. Based on cohesive zone model and SCB experimental data, the FEM simulation were calibrated and a simple linear relationship between experimental IDT strength and nominal maximum stress obtained from the SCB simulation was derived. This finding suggests that strength properties of asphalt mixture can be successfully predicted from SCB fracture testing without performing IDT tests at low temperature.

Obtaining strength and fracture properties from the same test and from the same specimen can reduce the errors potentially associated with the need of relying on different experimental methods. FEM requires some effort to be used, especially for practitioners in department of transportations or in road authorities; however, as long as the model is implemented the use of the software becomes much simpler and faster. In addition, the use of a FEM solution would significantly reduce the experimental time demand required to prepare IDT specimens and to perform the tests.

It should be noted that, only one temperature and a limited number of mixtures were considered in the present research. Therefore, additional experimentation and refinement of the FEM simulation are needed to provide further support to the present findings; this is part of an ongoing research.

Acknowledgements The authors would like to gratefully acknowledge the Minnesota Department of Transportation for providing the materials used in this study

References

AASHTO T322 (2007) Standard method of test for determining the creep compliance and strength of hot mix asphalt (HMA) using the indirect tensile test device. Am Assoc State Highw Transp Off

AASHTO (2008) Mechanistic-empirical pavement design guide, interim edition: a manual of practice. Am Assoc State Highw Transp Off

ABAQUS 6.13 (2013) User manual, Simulia

Cannone Falchetto A, Marasteanu M, Balmurugan S, Negulescu (2014) Investigation of asphalt mixture strength at low temperatures with the bending beam rheometer. Road Mater Pavement Des 15(1):28-44. doi:10.1080/14680629.2014.926618

Marasteanu M, Dai S, Labuz J, Li X (2002) Determining the low-temperature fracture toughness of asphalt mixtures. Transp Res Rec 1789:191-199. doi:10.3141/1789-21

Li X, Marasteanu M (2004) Evaluation of the low temperature fracture resistance of asphalt mixtures using the Semi Circular Bend (SCB) test, Journal of Association of Asphalt Paving Technologies, 73, 401-426. http://worldcat.org/issn/0270932.

Li X, Marasteanu M (2010) The fracture process zone in asphalt mixture at low temperature. J Eng Fract Mech, 77(7):1175-1190. doi:10.1016/j.engfracmech.2010.02.018

Moon KH, Marasteanu, Turos I (2013), Comparison of thermal stresses calculated from asphalt binder and asphalt mixture creep tests. J Mater Civ Eng 25(8):1059-1067. doi:10.1061/(ASCE)MT.1943-5533.0000651

A Study of Longitudinal Crack Which Occurs to the Surface of Asphalt Pavement by Wheel Tracking Test

Toshiaki Hirato, Kenji Himeno and Masato Murayama

Abstract It is known that asphalt pavement is subject to longitudinal cracking that begin at the surface and propagate downward. The main mechanism involved is considered to be bending-induced tensile strain away from the tire or shear-induced near-surface tensile strain at the tire edge. It is difficult to specify the location where such cracks occur because vehicles do not follow precisely the same path. Some cracks might form beneath the tire of a vehicle. However, no reliable method has yet been established. The purpose of the present study was to identify the mechanism involved in longitudinal cracking in asphalt mixtures using an improved wheel tracking test. By varying the temperature and loading conditions, it was determined that cracking occurred beneath the tire. The cracks had similar shapes to the longitudinal cracks that occur in road surfaces. The bottom surface of the asphalt showed no evidence of cracking. To determine the reason for this, stress relaxation in an asphalt mixture was investigated using compression tests. As the results, the stress-relaxation performance of an asphalt mixture becomes higher as temperature increases. Moreover the compressive stress was reduced immediately to about half of its maximum value. This suggests that if the compressive stress in the surface layer is released, the residual strain in the binder course layer would act as a tensile strain, which gives rise to the formation of cracks.

Keywords Longitudinal cracking · Top-down cracking · Wheel tracking · Stress relaxation

---

T. Hirato (✉)
Graduate School of Science and Engineering, Chuo University, Tokyo, Japan
e-mail: a14.gc7k@g.chuo-u.ac.jp

K. Himeno
Department of Civil Engineering, Chuo University, Tokyo, Japan

M. Murayama
Research Laboratory, Toa Road Corporation, Tokyo, Japan

A. Chabot et al. (eds.), *8th RILEM International Conference on Mechanisms of Cracking and Debonding in Pavements*, RILEM Bookseries 13,
DOI 10.1007/978-94-024-0867-6_12

1 Introduction

Top-down cracking is known as longitudinal cracking initiating at the top of the asphalt pavement layer. Top-down cracking occurs not only in thin asphalt pavement layers but also in the thick pavement of expressways (Matsumoto and Nishizawa 1984). In Japan, asphalt pavement is designed under the assumption that cracking is due to tensile strain and is initiated at the bottom face of the asphalt mix; top-down cracking is not reflected anywhere in the design specifications (Japan Road Association 2006). Previous studies have indicated that longitudinal cracking is caused by the tensile strain that arises at the edges of tires or the bending that occurs under passing vehicles (Indiana Department of Transportation 2004; NCHRP 2010). Methods for reproducing longitudinal cracking in the laboratory have been investigated (Himeno et al. 1987; Komoriya et al. 2001), but a suitable approach has still not been identified. This paper describes an investigation into a method for reproducing longitudinal cracking in the laboratory and presents some observations on the mechanisms involved in such cracking.

2 Simulating Longitudinal Cracking Using Wheel Tracking Tests

Longitudinal cracking was reproduced by installing a rubber plate under an asphalt mixture on a wheel tracking device and conducting a tracking test.

2.1 Test Conditions

Table 1 lists the test conditions and Fig. 1 shows a diagram of the test device and a photograph of an actual test. The wheel tracking test was run for up to 8 h to initiate cracking. The rubber plate had a hardness of 30 at 20 °C, as measured using a rubber hardness tester. Steel bars with dimensions of 300 mm × 50 mm × 8 mm were placed along the edges of the rubber plate parallel to the travel direction to prevent the entire test specimen from sinking under the load on the tire. Supporting

Table 1 Conditions during the tests using the wheel tracking device

	Test conditions
Mixture type	Dense-graded asphalt mixture (AC: 5.6 %)
Test temperature (°C)	30, 40, 50
Velocity	42 passes/min
Pressure	0.63 MPa at 60 °C
Specimen	300 × 300 × 50 mm

brackets were also attached to the ends of the test specimen parallel to the travel direction in order to prevent it from deflecting upward under the tire loads.

2.2 Test Results

For a test temperature of 30 °C, no cracks were formed for 8 h. In contrast, for a test temperature of 40 °C, vertical cracks were present on the specimen surface extending along the rolling direction of the tire following 5000 wheel passes. The location of the cracking is shown in Fig. 2a. This was present at the ends of the test specimen where the tire speed was the lowest. As the number of passes increased, these cracks extended towards the center of the tire track. Examination of the behavior of the asphalt mixture showed that its surface was deformed downwards under the tire load, but it recovered to its original height after the load was released.

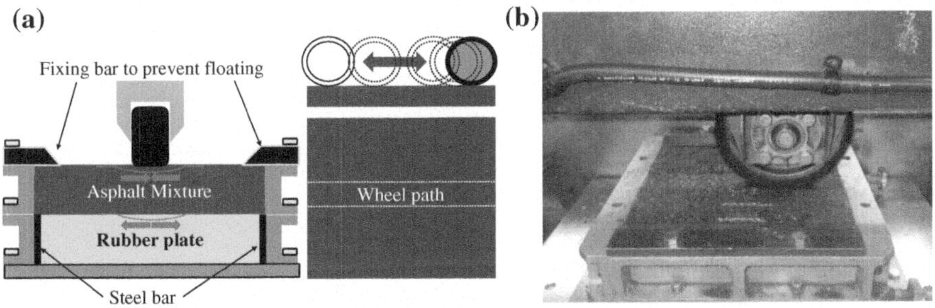

Fig. 1 **a** Schematic diagram of wheel tracking fatigue test, **b** Photograph of actual test

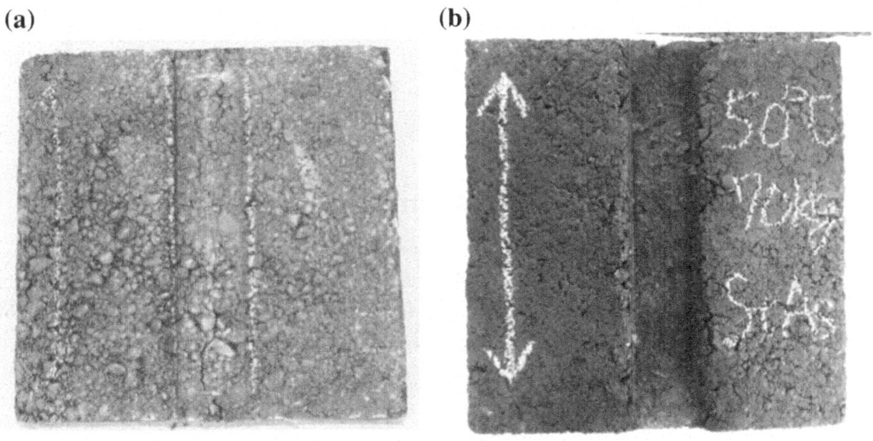

Fig. 2 **a** Vertical cracking near the center of the tire path (test temperature 40 °C, 7560 passes), **b** cracking at the edge of tire path (test temperature 50 °C, 5120 passes)

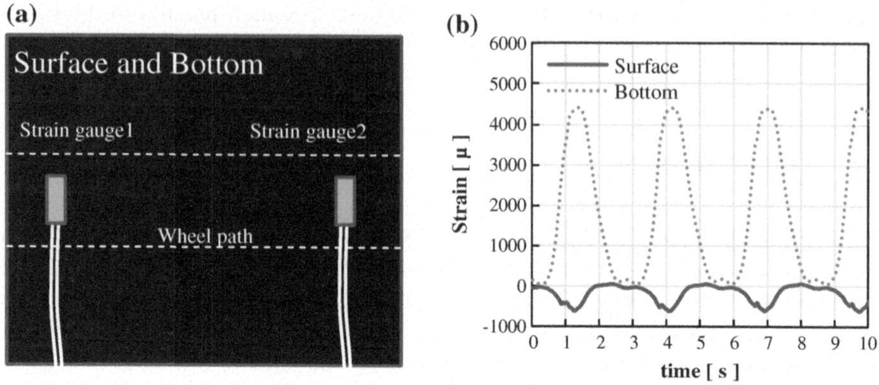

Fig. 3 **a** Locations of installed strain gauges, **b** measured strains at test temperature of 40 °C

For a test temperature of 50 °C and about 5000 wheel passes, vertical cracking occurred at the edge of the tire path, and these cracks extended in a direction transverse to the line of the tread (Fig. 2b). It is thought that these crack occurred by shear forces produced under the tire load.

2.3 Measurement of Strain in Specimens

Strain gauges were attached to the specimens in order to measure the strain, which is the cause of cracking. They were installed on the top and bottom surfaces of the specimens, in the middle of the tire path as shown in Fig. 3a. The strain data in Fig. 3b were obtained during 10 s of measurement.

These results indicate that a compressive strain is present at the upper surface and a larger tensile strain at the bottom surface. When cracking occurred, the compressive strain at the top surface reached about 600 μ. It is quite plausible that similar strains occur at the surface of actual pavements.

3 Evaluation of Stress Relaxation by Compaction Testing

It is assumed that the cracking occurs when the specimen returns to its original shape after the tire load is released. Therefore, we investigated the relaxation of stress in the asphalt mixture after compression. This evaluation consisted of loading a cylindrical specimen until it reached a certain strain and then observing the change of the load as the strain was held constant.

3.1 Test Method

Table 2 lists the test conditions used to investigate stress relaxation. Figure 4a shows a schematic illustration of pavement behavior under a traffic load, Fig. 4b shows specimens cut from asphalt mixture, and Fig. 4c shows the situation of the compaction test. The specimens were cylindrical with a diameter of 100 mm and a length of 100 mm, and were cut from the mixture with their cylindrical specimens perpendicular to the direction of compaction. They were fixed to the test jig with epoxy resin. The mixture type was the same as that described above. The fixed strain was 600 μ, which is the value at which cracking had occurred.

3.2 Test Results

Figure 5 shows the variation in the load during the compression test. The stress in the asphalt mixture was found to relax with time. For a test temperature of 20 °C, the residual stress was still around 70 % of the maximum stress at 0.5 s. At 40 °C, however, the residual stress had relaxed to about 40 % of the maximum at 0.5 s. It is possible that most compressive stress is relaxed as the tire load was removed during the wheel tracking test at 40 °C. After that, strain occurred when the asphalt mixture returned to its original shape, due to the presence of the rubber plate. The strain acts as a tensile strain. This strain may well have been the cause of the cracking along the tire path. The tensile stress was also relaxed at the bottom surface; it is inferred that cracking did not occur here, because the residual strain was compressive.

Table 2 Compression test conditions

	Test condition
Test temperature (°C)	20, 40
Velocity	10 mm/min
Compaction strain	600 μ

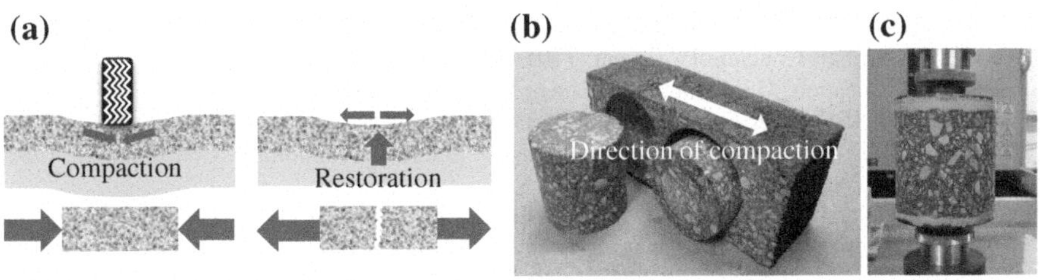

Fig. 4 **a** Schematic of pavement behavior during loading by traffic, **b** view of specimens cut from asphalt mixture **c** the situation of the compaction test

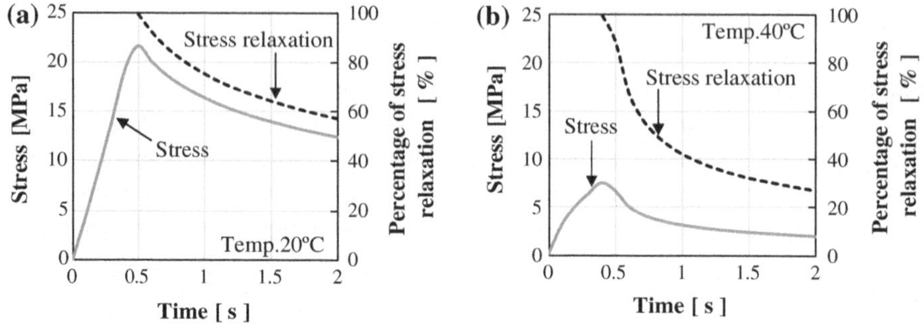

Fig. 5 Variation of stress and degree of stress relaxation with time for tests carried out at **a** 20 °C, and **b** 40 °C

4 Conclusions

The following findings were obtained in this study.

- A method was established to cause cracking in an asphalt mixture directly beneath the wheel path by wheel tracking tests.
- The compressive strain in specimens that had cracked surfaces was about 600 μ; it is entirely plausible that a similar strain occurs in actual roadways.
- The results of compression tests showed that the degree of stress relaxation increased with the test temperature. Wheel tracking tests indicated that at 40 °C, the stress associated with the tire load may relax immediately. When the specimens began to recover to their original shapes due to the pressure from the rubber plate, the acting strain was tensile, which was the cause for the cracking in the wheel path.
- When there are differences in the stress relaxation properties of the layers of an actual pavement due to a temperature difference between the surface course layer and binder course layer, it is possible that cracking similar to that in the wheel tracking tests will occur in the surface of asphalt pavement.

References

Kazushi Komoriya, Takeshi Yoshida, Hiroyuki Nitta: WA-DA-CHI-WA-RE Surface Longitudinal Cracks on Asphalt Concrete Pavement, TRB, 2001

Japan Road Association: Guideline of Pavement Design (In Japanese), 2006.2

Matsuno, S. and Nishizawa, T. Longitudinal Surface Cracking of Flexible Pavement, Proc. Paving in Cold Areas Mini-Workshop, Canada/Japan Science Technology Consultation, 1984, pp.779-796

Himeno, K., Watanabe, T. and Maruyama, T.: Estimation of Fatigue Life of Asphalt Pavement, Procs. 6th International Conference on Structural Design of Asphalt Pavements, Ann Arbor, Vol.1, pp.272-289, 1987.7

Indiana Department of Transportation, FHWA: EVALUATION OF SURFACE (TOP DOWN) LONGITUDINAL WHEEL PATH CRACKING, 2004.9

NCHRP : Top-Down Cracking of Hot-Mix Asphalt Layers: Models for Initiation and Propagation, web-only Document 162, 2010.2

Effect of Ageing and Water Action on the Cracking Resistance of Asphalt Mixtures

Teresa López-Montero, Rodrigo Miró and Félix Pérez-Jiménez

Abstract Cracking in asphalt layers is one of the most frequent distress mechanisms in flexible pavements due to thermal stresses and traffic loads. Furthermore, the ageing phenomenon along with moisture damage are the key factors for evaluating the asphalt mixture cracking resistance. The effect of ageing and water action on the cracking resistance of an asphalt mixture is studied in this paper. To this end, an asphalt concrete mixture, manufactured with two different bitumens (conventional and polymer modified binders), is subjected to ageing in the laboratory. Then, specimens are prepared with the aged mixture and subjected to the action of water. Unaged, aged and water-aged specimens are subjected to a direct tensile test (Fénix test) at different temperatures to obtain the variation of the fracture energy and toughness under these conditions.

Keywords Fracture energy · Fénix test · Ageing · Moisture damage

1 Introduction

An important factor affecting cracking resistance is ageing (Arega et al. 2013). In particular, this phenomenon is characterized by a hardening of the bitumen which affects the rheological and mechanical properties of asphalt mixtures (Menapace et al. 2015). From a temporal standpoint, ageing occurs in two stages (Das et al. 2015). The first one, named short-term ageing, occurs at the moment of

T. López-Montero (✉) · R. Miró (✉) · F. Pérez-Jiménez
Universitat Politècnica de Catalunya – BarcelonaTech, Jordi Girona, 1-3, Edificio B1, 08034 Barcelona, Spain
e-mail: teresa.lopez@upc.edu

R. Miró
e-mail: r.miro@upc.edu

F. Pérez-Jiménez
e-mail: edmundo.perez@upc.edu

A. Chabot et al. (eds.), *8th RILEM International Conference on Mechanisms of Cracking and Debonding in Pavements*, RILEM Bookseries 13,
DOI 10.1007/978-94-024-0867-6_13

manufacturing and laying the hot mix. During long-term ageing, degradation is associated exclusively to the environment during the service life of the mixtures.

The ageing that asphalt mixtures manifest at different stages of their lives can be simulated in the laboratory. Among the laboratory methods to age the mixtures, the procedures established by the SHRP: STOA (Short Term Ageing Oven) and LTOA (Long Term Oven Ageing) are highlighted. The STOA consists of ageing the loose mixture in an oven at 135 °C for four hours. In the case of LTOA, the mixture aged with STOA and compacted is kept in the oven for five days at 85 °C. Recently, the RILEM technical committee has established a procedure for ageing consisting of maintaining the loose mixture for 4 h at 135 °C during short-term ageing and at 85 °C for 9 days during long term ageing (De la Roche et al. 2009).

Moisture damage is another major environmental factor that affects the durability of asphalt mixtures (Airey 2003). It is considered an important factor in the emergence of diverse ways of damage such as cracking, rutting and ravelling (Varveri et al. 2015).

Several tests in compacted mixtures have been developed in an attempt to assess the water susceptibility of bituminous mixtures. Some procedures, such as the one included in the old Spanish standard, immerse the specimens in a water bath at 60 ° C for 24 h or 49 °C for 4 days. Currently, the UNE-EN 12697-12:2009 establishes placing the specimens in a bath at 40 °C for a period of time between 68 and 72 h. Previously, they will have been subjected to a vacuum.

Temperature can have a strong effect on the durability of asphalt mixtures (Behiry 2013). Bituminous mixes show significant changes in their mechanical properties due to differences in thermal susceptibility and the viscoelastic behaviour of the bitumen (Pérez-Jiménez et al. 2013).

The paper presented here studies the effect of ageing and its effect combined with water action. Furthermore, the effect of test temperature in cracking resistance is analysed. To undertake this study, the Fénix test is performed.

2 Methodology

The Fénix test (Pérez-Jiménez et al. 2010) is a monotonic tensile test at a constant displacement velocity. This test consists of subjecting a half cylindrical specimen (Marshall type with a diameter of 101.6 mm) to a tensile stress. The height of the specimen is about 60 mm and a 6 mm length notch is made in it. Two plates are attached to the specimen in its diametral plane by using an epoxy resin. The gluing length on the plate is approximately 50 mm in length and 60 mm in height. These plates are fixed to the loading platen of the press. The notch to induce cracking is situated between the two plates. The test was conducted at a constant displacement rate of 1 mm/min and can be done at different temperatures. In this case, the test was carried out at temperatures of 20, 5 and −5 °C. During the test, the force applied depending on the imposed displacement was recorded. From the resulting load-displacement curve of the Fénix test, a number of parameters can be defined.

These parameters are related to mechanical characteristics of the mixture: the tensile stiffness index, the fracture energy and the toughness index.

A semi-dense asphaltic mix has been selected (AC16S). Its maximum aggregate size is 16 mm and it is centred within the grading envelope. The mix was manufactured with two types of binder, a conventional one (50/70) and a polymer-modified one (PMB 45/80-65). The binder content for the manufacture of the mixture was 4.5 % by weight of mix. The air voids content was approximately 3 %.

To simulate long term ageing, the mixture was kept loose in an oven at 85 °C for 7 days. During the ageing period, the mixture was stirred three times (days 2, 4 and 5), the time interval between agitation being over 24 h. Once the mixture had been aged, the specimens were moulded and compacted. To submit the aged mixture to moisture damage, aged specimens were conditioned according to the standard UNE-EN 12697-12:2009, which involves placing the set of specimens (in this case once they were cut) in a water bath at 40 °C for a period of 72 h. These samples will have been previously subjected to vacuum at an absolute pressure of 6.7 kPa for a period of 30 min. After the conditioning period, the specimens were glued once their surface was dried. Three days later, they were tested.

3 Results and Discussion

The Fénix test allows various parameters related to the crack resistance of the mixture to be obtained. In Fig. 1, the stress against displacement caused during the test for the different conditions is shown. A significant increase in stress is observed when the mixture has been aged or when the test temperature decreases. The effect of moisture damage is hardly appreciable. Test temperature has a greater effect on the stress than ageing, for the degree of ageing considered in this study.

Figure 2 shows the relation between the tensile stiffness index, IRT, and the dissipated energy, GD. There is a big difference between the stiffness of the unconditioned mixture (unaged) and the aged mixture. However, the effect of water action on the rigidity of the mixture is minimal.

In the case of the unconditioned mixture, a maximum in the fracture energy values for a test temperature of 5 °C is observed. This is due to the fact that the

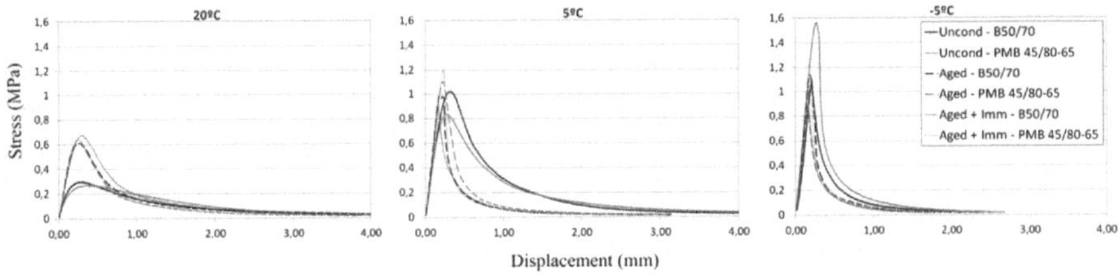

Fig. 1 Fénix results for the different conditions at the test temperatures of 20, 5 and −5 °C (from *right* to *left*)

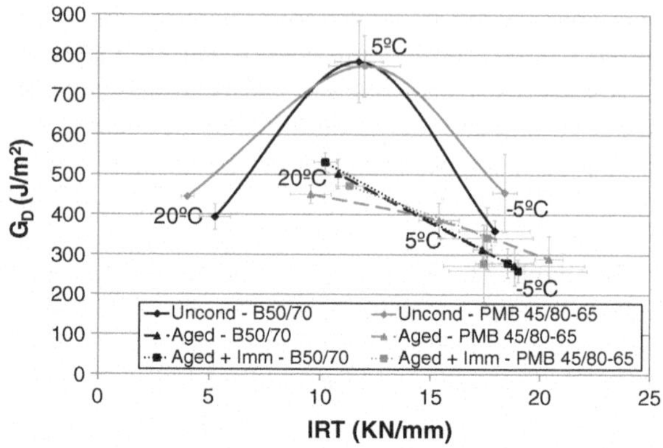

Fig. 2 Fracture energy (GF) versus tensile stiffness index (IRT)

stiffness of the mixture increases with the decrease of test temperature (Fig. 1). The stress that the mixture can withstand increases, the area under the curve of the post-peak stress is greater at 5 °C than the curves described by the specimens tested at 20 °C. At the temperature of −5 °C, the fast drop of the stress after cracking causes a decrease in the area of the curve, despite the increase of the stress.

However, in the case of aged mixtures, there is not a maximum in the fracture energy values. These values decrease as the test temperature decreases. This fact is more apparent for the conventional bitumen (B50/70), although values for the temperature test at −5 °C tend to be close for the aged and unaged samples. This is likely due to the effect of temperature on the stiffness of the mixture. This effect is so pronounced that the ageing effect hardly affects stiffness.

On the other hand, on analysing the effect of water after ageing for a test temperature of 20 °C, it can be seen to cause a slight increase both of the stiffness and the energy for the modified bitumen. For the conventional bitumen, stiffness decreases slightly (and energy increases). These results indicate that the mixture subjected to moisture damage has greater resistance to cracking than the unconditioned one. The water action may cause a slight ageing of bitumen as a result of the temperature of the water bath. When the stiffness increases, energy also increases (even a slight reduction of stiffness occurs in the case of the conventional bitumen). For this type of mixture (dense, made of limestone aggregate, with a good adhesiveness, and a small percentage of air voids) the effect of water results in a slight ageing that is not enough to cause severe damage. Hernández et al. (2014) studied the influence of water on the oxidation of mixtures. These authors concluded that ageing and oxidation are caused not only by the air-asphalt or air-water-asphalt interaction but also by the oxygen in the water (water-asphalt interaction). For this reason, the mixture seems to have a better behaviour. At a test temperature of 5 °C, water tends to increase the stiffness of the mixture and decrease the fracture energy. At −5 °C, the degree of stiffness is so high that the water has no effect on the mixture stiffness.

Figure 3 shows the higher results of the toughness index for the mixture manufactured with modified bitumen (unaged) against those of the mixture made with

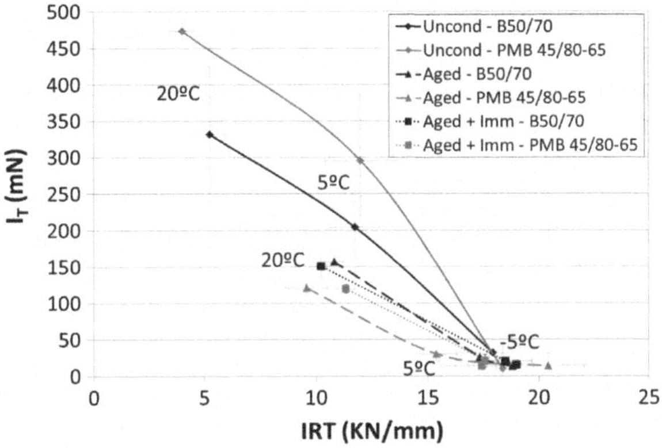

Fig. 3 Toughness index (IT) versus tensile stiffness index (IRT)

conventional bitumen. This is consistent with the lower values of the stiffness index. Both ageing of the mixture and reduction of temperature lead to an increase in stiffness and a decrease in the toughness index, for the two bitumens studied. These results show the brittle behaviour of the aged mixtures. This fact indicates that, as ageing, test temperature has an effect on the brittle behaviour of the mixture. At low temperatures (−5 °C for unaged mixtures and 5 and −5 °C for aged mixtures), values of the toughness index and tensile stiffness index tend to be similar for all the cases studied. The effect of moisture damage hardly changes the values of tenacity.

The results for the mixtures subjected to moisture damage were compared to those for mixtures not subjected to it. They are very similar for this kind of mixture. It could be related to the water conditioning. This kind of conditioning would be scarcely aggressive for this type of mixture (dense, made with limestone aggregate, which has good adhesiveness and a small percentage of voids). Under these conditions, the effect of water action is less than the ageing to which the loose mixture (uncompacted) is subjected. These results also show that the effect of the test temperature is greater than the ageing effect for the degree of ageing considered in this study. This means that mix design properties can become decisive factors in the mixtures behaviour to cracking against external factors, water, ageing or temperature. Furthermore, the Fénix test is sensitive enough to evaluate the effect of each one.

4 Conclusions

This research demonstrates a new method (Fénix test) that is able to study cracking resistance of asphalt mixtures under different conditions (test temperature, ageing and moisture), and which shows its differences in behaviour from the evaluation of fracture energy and toughness of the mixture. This research clearly demonstrates that the Fénix test is a useful tool for studying the effect of ageing and moisture damage under different test temperatures on cracking resistance.

Ageing causes an increase in the stiffness of the mixture and a decrease in toughness, making the mixture more brittle. However, the effect of water action is minimal. This may be because water immersion conditions are poor for this type of mixture (dense, made with limestone aggregate, with good adhesiveness and a small percentage of voids). The water conditioning only produces a slight ageing of the mixture. This shows that both the composition and properties of mix design can become decisive factors on the cracking behaviour against external factors.

Furthermore, temperature can be a determining factor for both the mixture design stage and for the assessment of the mixture behaviour in service. Low temperatures cause an increase in stiffness and a reduction of toughness. Mixtures become more fragile and, therefore, more susceptible to cracking. The effect of temperature, depending on the degree of ageing, may be greater than the ageing effect.

The use of different kinds of bitumens, for the same type of mixture, can influence the mixture response. The mixture made of polymer modified bitumen (PMB 45/80-65) has a better performance against ageing and moisture damage. This shows less fragility than the mixture made of conventional bitumen (B50/70).

Acknowledgements The authors would like to thank the Ministerio de Economía y Competitividad (Spain) for its assistance in the project PROFIS (BIA2012-36508), established within the framework of the VI Plan Nacional de Investigación Científica, Desarrollo e Innovación Tecnológica, co-financed with funds from the European Regional Development Fund (ERDF) of the European Union. The work of the first author was partially funded by the Ministerio de Economía y Competitividad, Spain (Research Grant BES-2013-065678). The authors would also like to acknowledge the company CEPSA for supplying and characterizing the bitumen used in this study.

References

Airey, G. (2003) State of the art report on ageing test methods for bituminous pavement materials. Int J Pavement Eng 4:3 156-176.

Arega, Z., Bhasin, A., De Kesel, T. (2013) Influence of extended aging on the properties of asphalt composites produced using hot and warm mix methods. Cons Build Mat 44:168-174.

Behiry, A. (2013) Laboratory evaluation of resistance to moisture damage in asphalt mixtures. Ain Shams Engineering Journal 4:351-363.

Das, P., Baaj, H., Kringos, N., Tighe, S. (2015) Coupling of oxidative ageing and moisture damage in asphalt mixtures. Road Mat Pav Design 16:265-279.

De la Roche, C., Van de Ven, M., Van den berg, W., Gabet, T., Dubois, V., Granfell, J., Porot, L. (2009) Development of a laboratory bituminous mixtures ageing protocol. In: Loizos, Partl, Scarpas & Al-Qadi (ed) Advanced Testing and Characterization of Bituminous Materials.

Hernández, J., Rondón, H., Fernández, W. (2014) The influence of water on the oxidation of asphalt cements. Cons build mat 71:451-455.

Menapace, I., Masad, E., Bhasin, A., Little, D. (2015) Microstructural properties of warm mix asphalt before and after laboratory-simulated long-term ageing. Road Mat Pav Design.

Pérez-Jiménez, F., Botella, R., Martínez, A., Miró, R. (2013) Analysis of the mechanical behaviour of bituminous mixtures at low temperatures. Cons Build Mat 46:193-202.

Pérez-Jiménez, F., Valdés, G., Miró, R., Martínez, A., Botella, R. (2010) Fénix. Test Development of a new test procedure for evaluating cracking resistance in bituminous mixtures. Transport Res Rec J Transport Res Board 2181:36-43.

Varveri, A., Avgerinopoulos, S., Scarpas, A. (2015) Experimental evaluation of long- and short-term moisture damage characteristics of asphalt mixtures. Road Mat Pav Design.

Effect of Stress Singularity on the Scaling Law of Asphalt Mixture Strength at Low Temperature

Augusto Cannone Falchetto and Michael P. Wistuba

Abstract Asphalt mixtures behave in a quasibrittle manner at low temperatures and, consequently, its nominal strength strongly depends on the structure size. Most of the research performed in the past has addressed this phenomenon either on notched or unnotched specimens; however, the evolution of the pavement conditions can lead to the formation of weaker stress singularities than the conventional $-1/2$ crack-tip singularity. In this paper an experimental study is performed to evaluate the effect of different stress singularities of notchless, V-notch and straight-notch beams of asphalt mixture on the strength size effect. The analysis is based on a generalized weakest link model that combines the energetic scaling of fracture with the strength statistics. The model captures the transition from the statistical to the energetic scaling as the magnitude of the stress singularity increases.

Keywords Asphalt mixture · Strength · Size effect · Stress singularity

1 Introduction

Cracking due to low temperature stresses is the dominant failure mode in asphalt pavements built in cold climates. Therefore, good understanding of strength and fracture properties of paving materials is fundamental for developing durable solutions which can reduce or mitigate the occurrence of thermal cracking.

Recent studies have shown that asphalt mixtures present quasibrittle behavior at temperatures close to the glass transition of the binder constituent (Cannone Falchetto et al. 2014). Quasibrittle materials are brittle heterogeneous materials, such as concrete, and rocks. For structures made of these materials the size of the

A.C. Falchetto (✉) · M.P. Wistuba
Technische Universität Braunschweig, Braunschweig, Germany
e-mail: a.cannone-falchetto@tu-bs.de

M.P. Wistuba
e-mail: m.wistuba@tu-bs.de

A. Chabot et al. (eds.), *8th RILEM International Conference on Mechanisms of Cracking and Debonding in Pavements*, RILEM Bookseries 13,
DOI 10.1007/978-94-024-0867-6_14

material inhomogeneity is not negligible compared to the structure size (Cannone Falchetto et al. 2014) and, hence, the failure behavior depends on structure size.

Two district mechanisms can be identified to describe the size effect on nominal structural strength (Bažant 2005; Bažant and Pang 2007): one is based on the energetic approach to fracture and the other is derived from the statistics of material random strength. The energetic scaling theory is used for structures containing a notch or a preexisting stress-free crack formed prior to the peak load. This type of size effect is usually referred as the energetic (or deterministic) scaling.

The statistical scaling is typical of notchless structures where the peak load is reached when a macro-crack initiates from one material representative volume element (RVE). Through the finite weakest link model (WLM) it is then possible to predict the cumulative distribution function (cdf) of structural strength characterized by an intricate distribution with a far left power-law tail following the Weibull distribution for large-size structures (Bažant and Pang 2007).

These two limiting cases indicate that the type of scaling depends on the order of the stress singularity in the structure. As the stress singularity increases, the statistical scaling should evolve into the deterministic scaling. Understanding such a transitional mechanism is very important since the evolution of the pavement conditions can lead to the formation of stress singularities that are weaker than the conventional "−1/2" crack-tip singularity, as for wider or sharper V-notches. In addition, the experimental results obtained on small specimens are not necessarily representative of the material behavior in the field and, hence, a solid modeling tool for extrapolation purposes is very much needed.

In this paper the evolution of the strength size effect of asphalt mixture at low temperature as function of different stress singularities is experimentally evaluated and modeled based on a generalized weakest link model that combines the energetic scaling of fracture with the finite weakest link model.

2 Generalized Size Effect Law

Let's consider a structure containing a V-notch of angle ϑ, stress singularity λ (Williams 1952) and notch length $a = 0.2D$, where D is the scaling dimension. For strong stress singularities, the scaling mechanisms follow the energetic scaling theory and the peak load is reached when the stress at a certain distance from the notch tip reaches the material tensile strength. For this specific case the following mean strength ($\bar{\sigma}_N$) scaling law can be written (Le 2011):

$$\bar{\sigma}_N = \sigma_s \left[1 + (D/D_{0\vartheta})^{1/\beta_\vartheta}\right]^{\lambda\beta_\vartheta} = \sigma_s \mu(D) \tag{1}$$

where σ_s is the nominal strength at the small-size limit, β_ϑ is a model parameter and $D_{0\vartheta}$ transitional size. When $\lambda = -0.5$ and $\beta_\vartheta = 1$, Eq. 1 becomes the classical

Type II Size effect which is typical of structures with pre-existing crack (Bažant 2005).

The limiting case of zero stress singularity, also known as Type I size effect, is typically observed in unnotched structures, for which a chain of RVEs scheme and the WLM can be used to compute the failure probability and mean strength as:

$$P_f(\sigma_N) = 1 - \prod_{i=1}^{N} [1 - P_1(s_i \sigma_N)] \quad \text{and} \quad \bar{\sigma}_N = \left[C_1/D + (C_2/D)^{rn/m}\right]^{1/r} \tag{2}$$

where P_1 is the cdf of the RVE strength, N is the number of RVEs in the structure, and s_i is the dimensionless stress field, m is the Weibull modulus; n is the order of scaling (n = 1, 2 and 3). C_1, C_2 and r are parameters that can be determined by asymptotic matching (Bažant 2005).

In the case of V-notch structures, a transition from Type I to Type II size effect has to be expected as the stress singularity becomes dominant. In the recent past a generalized WLM was proposed to solve this transitional behavior (Le et al. 2014) where the singular stress zone of the V-notch is separated from the remaining part of the structure. Therefore, the failure probability the entire structure is given by two contributions and can be written in the following form:

$$P_f(\sigma_N) = 1 - (1 - P_{f,V}) \cdot (1 - P_{f,(1-V)}) \tag{3}$$

where the symbols V and $(1 - V)$ indicate the contribution of the V-notch and of the remaining part of the structures. At the large-size limit, the failure probability of the structure is governed by the tail of the strength cdf of one RVE, and, therefore, it must follow the WLM and can be expressed as:

$$\bar{\sigma}_N = s_0\left[\mu^m(D)\Psi_V + \Psi_{(1-V)}\right]^{-1/m} \cdot \Gamma(1 + 1/m)(l_0/D)^{2/m} \tag{4}$$

where s_0 and m are the scaling and the shape parameter (Weibull modulus) of the Weibull distribution (Bažant 2005), Γ is the gamma function, and Ψ_V and $\Psi_{(1-V)}$ are the volume integrals of the stress fields of the two structure contributions. This set of integrals can be obtained by rescaling the actual volume and the integral coordinates by the RVE size, l_0. At the small-size limit, the entire structure consists of few RVEs which in the singular stress zone govern the failure of the entire structure where the mean strength is proportionally related to $\mu(D)$: $\bar{\sigma}_N \propto \mu^{-1}(D)(D/D_0)^{-1/r}$. Since the proposed approach is based on the WLM, by combining and manipulating Eqs. 1–4, it possible to obtain the following generalized strength size effect law (Le et al. 2014):

$$\bar{\sigma}_N = \sigma_0 \left\{ C_a \left[\mu^m(D)\Psi_V + \Psi_{(1-V)}\right]^{-r/m} \left(\frac{D+l_s}{l_0}\right)^{-2/m} \cdot \exp\left[-(\lambda/\lambda_1)^2\right] + \frac{\mu^r(D)D_b}{\exp\left[-(\lambda/\lambda_1)^2\right]D+l_p} \right\}^{1/r} \quad (5)$$

where σ_0 is the reference stress and C_a, r, λ_1, λ_2, l_s, l_p and D_b are constants.

3 Materials and Testing

A conventional asphalt mixture for wearing course was selected for the experimental phase. The mixture consisted of granodiorite aggregate having nominal maximum aggregate size (NMAS) of 11 mm, and a plain PG58-28 asphalt binder (5.6 % by weight). Based on recent study (Cannone Falchetto et al. 2014) the RVE size, l_0, was estimated as double of the average aggregate size which return a RVE volume, V_0, of 480.9 mm^3.

Three-point bending (3 PB) mean strength tests were performed on asphalt mixture beams of four different sizes (1X, 2X, 3X and 4X) with three types of stress singularities: notchless (ϑ = 180°), V-notch (ϑ = 120°) and straight-notch (ϑ = 0°) with a = 0.2D (Fig. 1). The beams presented a thickness, D, to span, L, ratio of 1:6 with constant depth b = 50 mm and were obtained from asphalt mixture slabs. All 3PB tests were performed in crack mouth opening displacement (CMOD) control and the same time to failure was set for all the specimens sizes and types so to achieve an equivalent propagation of the fracture process zone (FPZ) (Cannone Falchetto et al. 2014). Table 1 summarizes the specimens' characteristics while Fig. 1 shows the specimens schematic and the actual tests setup.

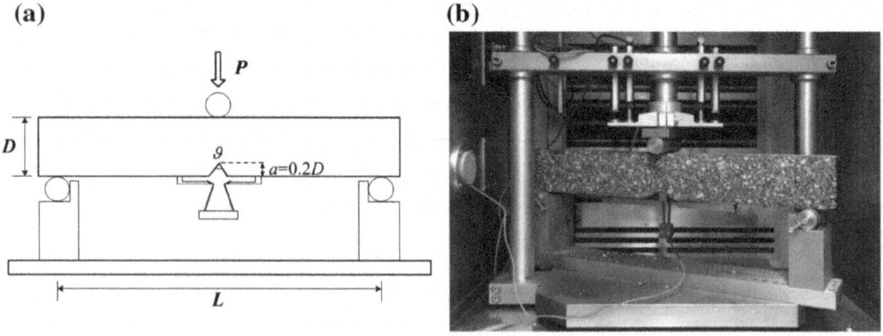

Fig. 1 **a** Schematic and **b** testing setup for three-point bending test on 4X beam with ϑ = 120°

Table 1 Geometrical details and replicates of the beam specimens

Relative size	$D \times L$ (mm)	Notch type (ϑ) − replicates			Total replicates
		180°	120°	0°	
1X	25 × 150	5	4	4	13
2X	50 × 300	5	4	4	13
3X	75 × 450	5	4	4	13
4X	100 × 600	5	4	4	13

4 Results and Size Effect Analysis

The 3PB nominal strength of asphalt mixture was computed as:

$$\sigma_N = c_N P_N/(bD) = 3P_N L/\left(2bD^2\right) \quad \text{and} \quad c_N = 3L/2D \tag{6}$$

where P_N is the peak load, c_N is a constant dimensionless coefficient, L is the length of the beam; D is the thickness of the beam, commonly assumed as scaling dimension in the size effect law; and b is the depth of the beam. Table 2 provides the experimental mean strength values and relative λ for each beam type, while Fig. 2 presents the plots of the nominal stress, versus the relative CMOD obtained by normalizing the CMOD measurements with the beam thickness, D (scaling dimension) for all beams. A progressive reduction of the softening curve as the size increases clearly indicates the presence of a strong strength size effect.

Equation 6 was then applied to the experimental data (Table 2). The size effect evolution is reported in Fig. 3a for the three beam types and the prediction is then extended to different notch angles in Fig. 3b where a three dimensional representation of the nominal strength is proposed.

The statistical parameters m and s_0 were found to be equal to 26 and 8.71, respectively. The model fit reasonably well the experimental strength. A typical Type I and Type II size effect is observed for notchless and straight-notch beams, respectively, with a behavior closer to Type II for the beam with notch angle

Table 2 Mean strength (MPa) and coefficient of variation CoV (%) between parentheses

Notch type	λ	Relative size			
ϑ		1X	2X	3X	4X
180°	0	10.40 (14.5)	9.14 (12.3)	8.53 (11.6)	8.28 (10.5)
120°	−0.3843	7.39 (8.2)	6.51 (7.8)	5.93 (8.1)	5.49 (7.6)
0°	−0.5	6.30 (7.2)	5.58 (7.4)	5.03 (6.9)	4.57 (6.8)

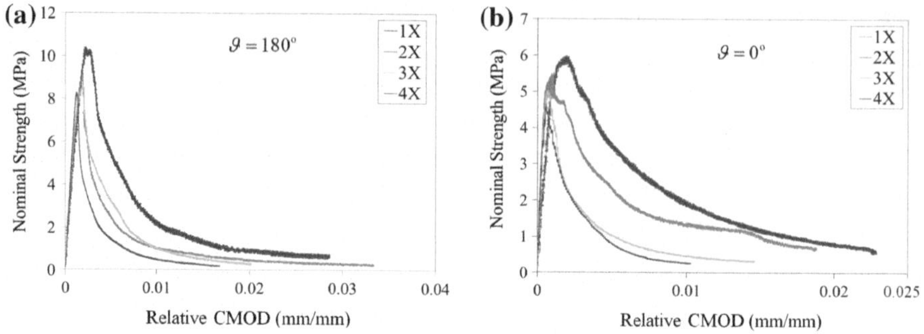

Fig. 2 Nominal stress versus relative CMOD **a** ϑ = 180° and **b** ϑ = 0°

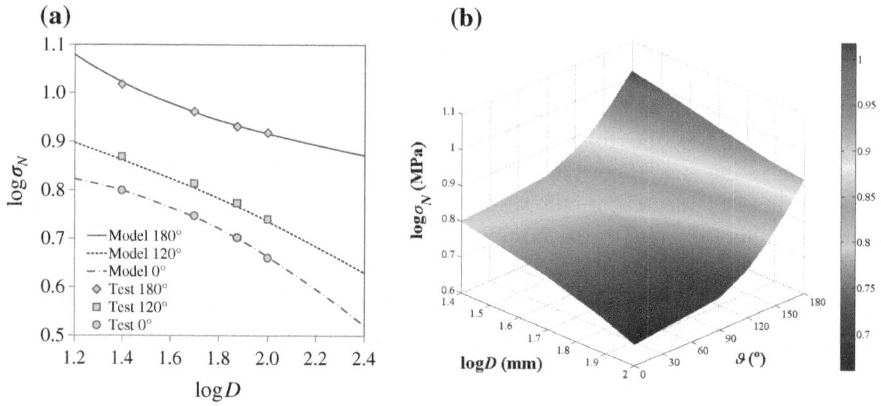

Fig. 3 **a** Generalized size effect model and **b** three-dimensional representation

ϑ = 120°. The 3D plot provides evidence that the transition between the statistical to energetic size effect occurs within a narrow angle interval which is in the range 140°–160°.

5 Summary and Conclusions

The dependence of strength size effect on the stress singularity was investigated by mean of three-point bending tests on asphalt mixture at low temperature. Notchless, straight-notch and V-notch mixture beams were used for the experimental phase. The size effect law is governed by a generalized weakest link model, where the energetic component of scaling is nested into the finite weakest link model. The scaling of structural strength follows the Type II energetic size effect for strong singularities (straight-notch) while for zero singularity it is dominated by the statistical failure. A mixed energetic-statistical behavior is observed for V-notch specimens which represents the case of weak stress singularity.

References

Bažant Z (2005) Scaling of structural strength. Elsevier, London.

Bažant Z, Pang S-D (2007) Activation energy based extreme value statistics and size effect in brittle and quasibrittle fracture. J Mech Phys Sol 55:91-131. doi:10.1016/j.jmps.2006.05.007

Cannone Falchetto A, Le J-L, Turos M, Marasteanu M (2014) Indirect determination of size effect on strength of asphalt mixtures at low temperatures. Mater Struct 47:157–169. doi:10.1617/s11527-013-0052-2

Le J-L (2011) General size effect on strength of bi-material quasibrittle structures. Int J Fract 172:151–160. doi:10.1007/s10704-011-9653-3

Le J-L, Pieuchot M, Ballarini R (2014) Effect of stress singularity magnitude on scaling of strength of quasi-brittle structures. J Eng Mech 140(5):04014011. doi:10.1061/(ASCE)EM.1943-7889.0000693

Williams M (1952) Stress singularities resulting from various boundary conditions in angular corners of plates in extension. J Appl Mech 74:526-528.

Analysis of Cracking Resistance of Bituminous Mixtures Under Monotonic and Cyclic Loads

Félix Pérez-Jiménez, Rodrigo Miró, Adriana Martínez and Ramón Botella

Abstract Although cracking resistance is one of the most important properties affecting the durability of asphalt mixtures, this property is often not considered as a requirement in the mixture design stage, and the bitumen content is determined from other properties (voids content, permanent deformation or water sensitivity, among others). The effect of bitumen content on the cracking resistance of an asphalt concrete mixture (type AC) is analysed in this paper by means of different types of tests. Firstly, a cyclic strain sweep test (EBADE test), to obtain the evolution of complex modulus and dissipated energy density with the number of cycles as well as the failure strain; the EBADE test can provided a first approximation to the fatigue law of the mixtures with different binder contents. Secondly, a direct tensile monotonic test (Fénix test) was used to determine strength and fracture energy of the mixture. The results from both tests were consistent between them.

Keywords Cracking resistance · Fracture energy · Fénix test · EBADE test

1 Introduction

Cracking is among the most common failure mechanisms in bituminous pavements. With time, cracks appear in the surface layer of the pavement. Some of them originate in the lower layer (bottom-up cracking) and some in the upper layer

F. Pérez-Jiménez · R. Miró (✉) · A. Martínez · R. Botella (✉)
Universitat Politècnica de Catalunya – BarcelonaTech, Jordi Girona, 1-3, Edificio B1, 08034 Barcelona, Spain
e-mail: r.miro@upc.edu

R. Botella
e-mail: ramon.botella@upc.edu

F. Pérez-Jiménez
e-mail: edmundo.perez@upc.edu

A. Martínez
e-mail: adriana.martinez@upc.edu

A. Chabot et al. (eds.), *8th RILEM International Conference on Mechanisms of Cracking and Debonding in Pavements*, RILEM Bookseries 13,
DOI 10.1007/978-94-024-0867-6_15

(top-down cracking). Typically the bottom-up cracking is associated with the loads repetition due to traffic, i.e. fatigue cracking, while the top-down cracking is related with the characteristics of the loads in the surface layer (type of load and tires pressure) and the environmental conditions (temperature and humidity) (Yo and Al-Qadi 2008).

Resistance to cracking of the bituminous mixtures is very sensitive to temperature and asphalt binder content of the mixtures, independently of the cause of the crack initiation (Pérez-Jiménez et al. 2011). However, this behavior is not normally taken into account in the design of these materials because of the expensive and time consuming of the tests that characterize it. In general, mixture design is based on other properties of the mixture that are easier to obtain.

This paper presents two methods that provide useful information about the cracking resistance of bituminous mixtures, both under cyclic and monotonic loading. In this case, those tests have been applied to a bituminous mixture manufactured with four different binder contents at 20 °C, showing that both tests deliver the influence of this variable on the cracking resistance of mixtures.

2 Test Methods

The EBADE test is a uniaxial tension-compression strain sweep test (Pérez-Jiménez et al. 2011; Botella et al. 2012; Clopotel et al. 2012). EBADE stands for the Spanish words for strain sweep test. Johnson (2010) proposed that the fatigue properties of an asphalt binder can be obtained by increasing the strain applied in a shear cyclic test using the DSR (Hintz et al. 2011). The EBADE test can apply this concept to asphalt binders, mastics and mixtures using a tension-compression load configuration. However, in this case it was only applied to mixtures. The mixture modality of the EBADE test consists of increasing the strain amplitude every 5000 cycles in 25E-6 mm/mm, starting at 25E-6 mm/mm strain amplitude.

The Fénix test (Pérez-Jiménez et al. 2010; Miró et al. 2014) is a direct tension test that uses similar specimen geometry than that of the SCB test (Molenaar et al. 2002) and applies a loading configuration similar to that of the DC(T) test (Wagoner et al. 2005). The test is named after the research project FENIX that funded its development. The test is performed at a constant displacement rate of 1 mm/min.

3 Materials

A continuously graded mixture was employed, following the standardized AC16S gradation, with a maximum aggregate size of 16 mm. The aggregate used was of limestone origin and the asphalt binder had a penetration of 65 dm at 25 °C and its softening point was at 50.2 °C.

The cylindrical specimens (100 mm in diameter) were manufactured with different binder contents (3, 4, 5 and 6 % in mixture mass) and compacted using the gyratory compactor during 100 cycles at 0.6 MPa and 0.82°. EBADE and Fénix specimens were carved from those cylindrical samples using an special saw. All tests were performed at 20 °C.

4 Results

Regarding the EBADE test results, Fig. 1 shows the evolution of the dissipated energy density with the number of cycles for the four binder contents. Table 1 presents the average values obtained for the initial complex modulus and the failure strain.

As expected, the increase of the binder content of the mixture reduces the stiffness of the mixture and increases its ductility. The specimens with 3 % binder content had a similar initial complex modulus than the 4 % samples, but their failure strain (strain amplitude reached during the strain sweep test) was almost 40 % higher for the latter. The 5 and 6 % showed the same failure strain, however

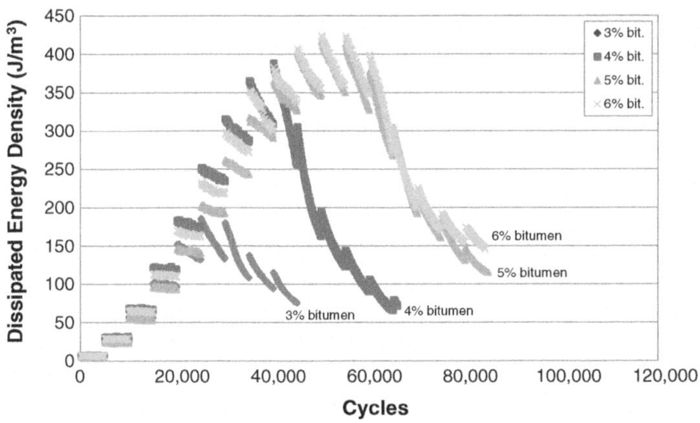

Fig. 1 Variation of dissipated energy density with the number of cycles for the different bitumen contents

Table 1 EBADE test results

Binder content in mixture mass (%)	Initial complex modulus (MPa)	Failure strain (1E-6)
3	6048	150
4	6199	225
5	5372	300
6	4794	300

the initial complex modulus of the 5 % mixture was higher. That would indicate that increasing over 5 % does not improve the ductility of the mixture but reduces its stiffness. Therefore the optimum binder content for this mixture should be around 5 % in mixture mass.

In addition, the EBADE test can provide information about how an eventual fatigue law of the material would look like (Pérez-Jiménez et al. 2015). Since, in one test, several strain amplitudes are applied during 5000 cycles, the test delivers short views of several time sweep tests. By the shape of the dissipated energy density vs. cycles plot at each strain amplitude it can be estimated the evolution of a theoretical time sweep test at that strain amplitude. There are two extreme cases that help estimate the duration of an eventual time sweep test.

In one extreme we have the higher strain levels. It is safe to assume that a time sweep test performed at the highest strain amplitude reached during the EBADE test should last a small number of cycles. Previous research has shown that the number of cycles that a time sweep test at the failure strain lasts is close to 10,000 (Pérez-Jiménez et al. 2013).

On the other extreme case are the lower strain levels. Those strain levels at which the dissipated energy density remains constant do not cause any kind of damage to the material, and therefore a time sweep test carried out at those strain amplitudes should last an infinite number of cycles. For practical purposes, authors have established that 10 million to be considered infinite life for the material. By determining the highest strain level at which the dissipated energy density remains constant in the EBADE test, the strain associated with 10 million cycles is obtained. These values have been determined after comparing data from EBADE test with time sweep tests performed at different strain levels using the same loading configuration (Pérez-Jiménez et al. 2013).

This procedure allows finding two theoretical data points that can be used to sketch a fatigue law of the material, using one simple test. Figure 2 shows the predicted fatigue laws for each binder content obtained using this method.

Using this methodology it was found that the ε_6 of the 3 % mixture was 95 microstrains, the 4 % mixture was 132 microstrains and the 5 and 6 % mixtures both obtained the same ε_6 of 167 microstrains. Therefore, the lowest binder content mixture showed the lower predicted number of cycles to failure for the same strain amplitudes, while the 5 and 6 % presented the highest. However, their initial complex modulus has to be taken into account as well. While the 4 % mixture initial complex modulus was a 15 % higher than the 5 % mixture, its failure strain was 25 % lower and its predicted ε_6 is 21 % smaller. It should be noted that the fatigue laws have been estimated from few data points and they validity is limited to the geometry of the EBADE test.

Regarding the Fenix test results, Figs. 3 and 4 show the stress-displacement curves for each binder content and the variation of the tensile stiffness index and the fracture energy with the binder content.

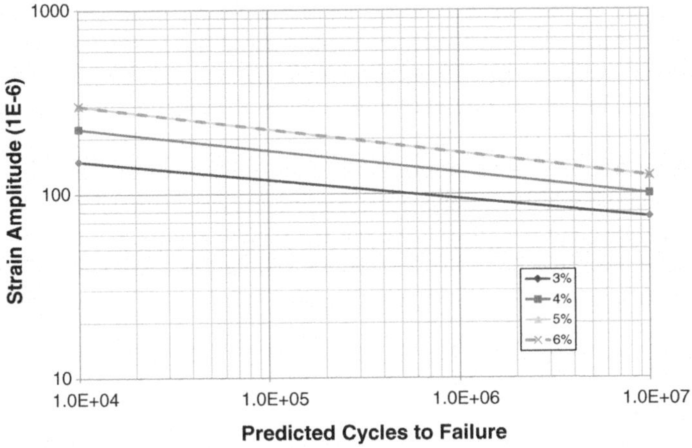

Fig. 2 Predicted fatigue laws for the mixtures with different binder content

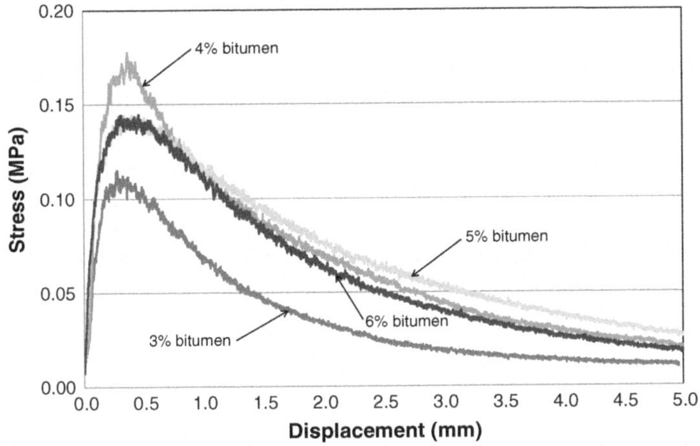

Fig. 3 Results of Fénix test for different bitumen content

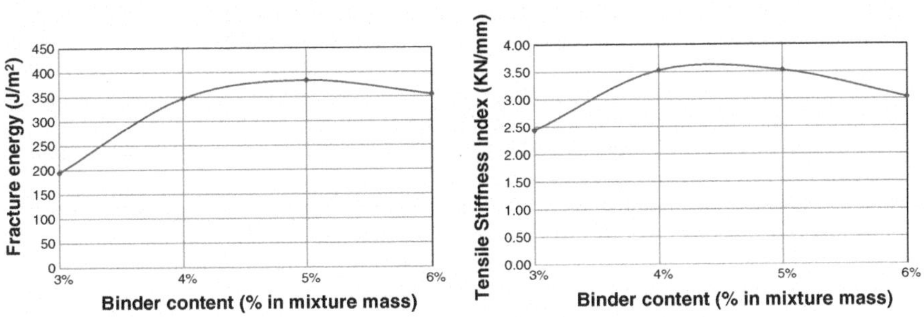

Fig. 4 Results of fracture energy and tensile stiffness index from Fénix test for different bitumen content

The influence of the binder content manifests clearly in the Fénix test results. The mixture with the lowest binder content showed the highest stiffness and the lowest fracture energy. The 4 % content mixture was the one with the highest maximum stress, but its fracture energy was lower than the mixtures with the 5 and 6 % content.

As with the EBADE results, the Fénix test results indicate that the optimum bitumen content should be around 5 % in mixture mass. The stiffness of the 5 % mixture is very close to that of the 4 % but its fracture energy is clearly higher. The 6 % mixture showed too high ductility, which resulted in a reduction of its fracture energy, while the 3 % was too stiff resulting on a reduction of this variable, as well.

5 Conclusions

The two procedures employed during this research project (EBADE and Fénix test) were able to provide meaningful results regarding the characterization of the cracking resistance of bituminous mixtures under cyclic and monotonic loading.

In both cases, the increase of binder content resulted on increase ductility of the mixture and decrease of stiffness.

The results from both tests were consistent between them and indicate that the optimum binder content for the AC16S mixture employed was around 5 % in mixture mass.

In addition, using previous results from time and strain sweep tests, the EBADE test provided a first approximation to the fatigue law of the mixtures with different binder contents that allowed analyzing the influence of the binder content on the fatigue behavior of the mixture.

References

Botella R, Pérez-Jiménez FE, Miró R (2012) Application of a strain sweep test to assess fatigue behavior of asphalt binders. Const Build Mater 36: 906-912.

Clopotel C, Velasquez R, Bahia H, Pérez-Jiménez FE, Miró R, Botella R (2012) Relationship between binder and mixture damage resistance at intermediate and low temperatures. Trans Res Rec 2293: 39-47.

Hintz C, Velasquez R, Johnson C, Bahia H (2011) Modification and validation of linear amplitude sweep test for binder fatigue specification. Trans Res Rec 2207: 99-106.

Johnson C (2010) Estimating asphalt binder fatigue resistance using an accelerated test method. Dissertation, University of Wisconsin-Madison.

Miro R, Martínez A; Pérez-Jiménez F, Botella R (2014) Analysis of cracking resistance of bituminous mixtures using Fenix test. Const Build Mater 59: 32-38.

Molenaar AAA, Scarpas A, Liu X, Erkens SMJG (2002) Semi-circular bending test; simple but useful?. Ass Asph Pav Tech-Proc Tech Ses 71: 794-815

Pérez-Jiménez F, Valdés G, Miró R, Martínez A, Botella R (2010) Fénix. Test Development of a new test procedure for evaluating cracking resistance in bituminous mixtures, Transport Res Rec 2181: 36-43.

Pérez-Jiménez FE, Valdés G, Miró R, Botella R, Campana JM (2011) Effect of thermal stresses on fatigue behavior in bituminous mixes. Trans Res Rec 2210: 90-96.

Pérez-Jiménez FE, Botella R, Martínez AH, Miró R (2013) Estimating the fatigue law of asphalt mixtures using a strain sweep test (EBADE test). Paper presented at EATA Conference 2013, Braunschweig, Germany.

Pérez-Jiménez FE, Botella R, Miró R, Martínez AH (2015) Analysis of the thixotropic behavior and the deterioration process of bitumen in fatigue tests. Const Build Mater *in press.*

Wagoner MP, Buttlar WG, Paulino GH (2005) Disk-shaped compact tension test for asphalt concrete fracture. Exp Mec 45(3): 270-277.

Yoo PJ, Al-Qadi IL (2008) The truth and myth of fatigue cracking potential in hot-mix asphalt: Numerical analysis and validation. Ass Asph Pav Tech-Proc Tech Ses 77: 549-590.

Effects of Ageing on Warm Mix Asphalts with High Rates of Reclaimed Asphalt Pavement

Miguel Perez-Martinez, Paul Marsac, Thomas Gabet, Ferhat Hammoum, Manuela de Mesquita Lopes and Simon Pouget

Abstract Within the framework of the European Project SUP&R ITN a Ph.D. thesis is carried out to study the durability of asphalt mixtures made with the combination of high rates of reclaimed asphalt pavement (RAP) and warm mix asphalt technologies. For this purpose the complex modulus and fatigue resistance of three different asphalt mixtures, including surfactant modified and foamed warm mix asphalts, combined with RAP has been studied. The extra value is given by the application of an ageing procedure based on the oxidation of compacted materials in laboratory. It follows the recommendations of the RILEM TC-ATB TG5, which distinguishes between short and long term ageing. Fourier Transform InfraRed (FTIR) tests were carried out on the extracted bitumens to quantify the oxidation levels. An increase of the norm and a decrease of the phase angle of $|E^*|$ at 15 °C 10 Hz with ageing and RAP addition are experienced for all the mixtures. Similarly the slopes of the fatigue laws tend to increase with ageing. A consistent correlation is observed between these evolutions and the evolution of the carbonyl index. In general, the tendency is similar for all procedures, so the use of warm technologies combined with high RAP amounts may need to be considered.

M. Perez-Martinez (✉) · P. Marsac · T. Gabet · F. Hammoum
MAST, MIT, LUNAM Université, IFSTTAR, 44340 Bouguenais, France
e-mail: miguel.perez-martinez@ifsttar.fr

P. Marsac
e-mail: paul.marsac@ifsttar.fr

T. Gabet
e-mail: thomas.gabet@ifsttar.fr

F. Hammoum
e-mail: ferhat.hammoum@ifsttar.fr

M. de Mesquita Lopes
Polytechnic School University of São Paulo, São Paulo, Brazil
e-mail: mmlengenharia@gmail.com

S. Pouget
EIFFAGE Infrastructures, Vaulx-en-Velin Cedex, France
e-mail: simon.pouget@eiffage.com

A. Chabot et al. (eds.), *8th RILEM International Conference on Mechanisms of Cracking and Debonding in Pavements*, RILEM Bookseries 13,
DOI 10.1007/978-94-024-0867-6_16

Keywords Ageing · Warm asphalt · Reclaimed pavement

1 Introduction

The combination of warm mix asphalt (WMA) and reclaimed asphalt pavements (RAP) is not a new concept, but this coupling seems challenging in order to favour a greener and more sustainable production of asphalt mixtures.

Separately, both procedures have their own advantages and drawbacks, but it is durability what concerns performance during service life. However, these types of asphalt mixes have not been commonly used yet. Thus, few is known about the properties changes induced by their chemical ageing on-site.

In order to improve the knowledge about this evolution, the present study is based on an artificial ageing of two different warm mix asphalt techniques, surfactant additive and foamed procedure, and a reference hot mix asphalt (HMA), with high rates of RAP (50 %). The ageing protocol followed was adapted from the RILEM TC-ATB TG5, consisting in two separate ageing steps, short term and long term ageing. The first one simulates the manufacturing, transport and layout of the mix during construction while the second is expected to reproduce service life evolution.

This research work is focused on the comparison of the mechanical performances through complex modulus and fatigue testing, coupled with the bitumen ageing evolution through Fourier Transform InfraRed testing (FTIR).

2 Methodology

Four different phases can define the followed methodology. The first three are related to laboratory works, materials characterization, mixtures production and performance of the ageing protocol, while the last phase focuses on testing.

The mixture chosen for the study is an asphalt concrete AC10 commonly used in road construction. This selection was made with the aim of reproducing the same formula of an AC10 originally placed in the fatigue carousel at IFSTTAR Nantes, France (Lopes et al. 2014). Virgin gneiss aggregates were employed both for the coarse and fine fractions (0/2, 2/6 and 6/10 mm) in all mixtures.

Moreover, RAP was previously prepared by splitting it in 4 granular fractions for an easier recomposition, and in order to assure the same grading in all mixtures, with and without RAP (Lopes et al. 2014).

The bitumen 35/50 is the same as the original employed in the original AC10 mixture. It has a penetration grade and softening point of 37 (1/10 mm) and 52.8 °C respectively. For the WMA, the first one was modified by adding 0.5 % of surfactant to the virgin 35/50 bitumen, while the foaming process consisted in injecting

1.5 % of water under pressure into the hot bitumen pipe before its introduction into the mixer.

All mixtures were manufactured using a MLPC BBMAX 80 mixer during 75 s of mixing, 15 s of aggregates and fines homogenization (and RAP when used) plus 60 s of mixing when bitumen was added. RAP aggregates were preheated at 110 °C during 2 h before mixing and virgin aggregates were preheated at different temperatures depending on the mixture, as it is summarized in Table 1. Figure 1 shows the methodology followed during the study.

Once the mixtures were manufactured, the ageing phase on the selected ones could take place. The procedure was adapted from the ageing RILEM TC-ATB TG5 protocol in laboratory (De La Roche et al. 2013), consisting in the oxidative thermal ageing of the asphalt mixture during two differentiated phases, a short term and a long term ageing.

The short term ageing consists in maintaining the loose mix at 135 °C during 4 h after manufacture. The material is then compacted and set to 85 °C during 9 days. Both steps are carried out in a ventilated oven.

Then, considering that the work is focused on durability aspects, the dynamic test for complex modulus determination (EN 12697-26, Annex A) and the dynamic test for fatigue resistance (EN 12697-24, annex A) were carried out to assess the mechanical performance of the mixtures. Also, on the recovered bitumen from all mixtures [by the rotary evaporator for binder recovery (EN 12697-3)], the Fourier transform infra-red spectroscopy (FTIR) was applied for ageing characterization.

From the Fourier Transform InfraRed (FTIR) test, only the carbonyl groups (esters, ketones etc.) represented by C=O double bond (ICo), were measured and calculated by the deconvolution method (Marsac et al. 2014) from the raw FTIR spectra. For the dynamic test for complex modulus four trapezoidal specimens were

Table 1 Denomination and manufacture temperature of the studied mixtures

Mixture	Denomination	% RAP	Ageing	Aggregates (°C)	RAP (°)	Manufacture (°C)
Hot mix asphalt	HMA0	0		160	–	160
	HMA0a	0	✗			
	HMA50	50		210	110	160
	HMA50a	50	✗			
Warm mix asphalt with surfactant	WMA0	0		130	–	130
	WMA0a	0	✗			
	WMA50	50		150	110	130
	WMA50a	50	✗			
Foamed warm mix asphalt	FWMA0	0		130	–	130
	FWMA0a	0	✗			
	FWMA50	50		150	110	130
	FWMA50a	50	✗			
				Bitumen always at 160 °C		

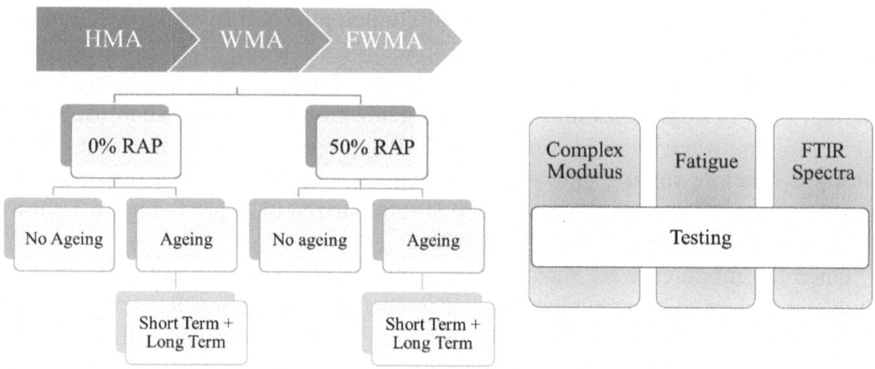

Fig. 1 Mixtures manufactured and tests performed

tested at temperatures from −10 to 40 °C and frequencies from 1 to 40 Hz for the correct study of rheological and stiffness evolution of the samples.

Finally, fatigue testing was carried out by 2 point bending at 10 °C, in strain control mode and at a frequency of 25 Hz. Eighteen trapezoidal samples were tested at three different strain amplitude levels. The fatigue law $\varepsilon = \varepsilon_6\ (N/10^6)^{-1/b}$ was calculated with the parameter ε_6 representing the slope −1/b, and the strain at 1,000,000 cycles of the fatigue law. In order to compare all the mixtures, the value of ε_6 was corrected in relation to the sample air void content (4.5 % as reference) based on the formula $\Delta\varepsilon_6(\mu\text{def}) = 3.3\,\Delta C(\%)$ (Moutier 1991).

3 Analysis of Results

The results obtained from the different mixtures are summarized in Table 2, where the norm of the complex modulus and phase angle at 15 °C 10 Hz of the mixtures, and the fatigue parameters are related to the ICo values from the FTIR spectra.

First of all, it should be noted that ageing induces an increase on the complex modulus results of $|E^*|$ and a decrease of the phase angle for every modality. This even happens when RAP is added and aged (Olard et al. 2012; Poirier et al. 2013). Regardless the technique used on manufacturing, all mixtures performs over the 7000 MPa requested on the standards.

A regression analysis leaded to show that an increment of modulus, $\Delta|E^*| = 460\,\text{MPa}$, is observed for an increment of carbonyl index of $\Delta\text{ICo} = 1$. But when removing from the calculus the values for the FWMA50 mixtures, which are out of the tendency, this increment raises to $\Delta|E^*| = 580\,\text{MPa}$. On the other hand, the phase angle tends observed for all the samples is $\Delta\phi = -0.50$, rising to $\Delta\phi = -0.60$ when FWMA50 is not taken into account. These trends roughly reveal the increase on modulus experienced by the mixtures when oxidation of the binders increases (Mangiafico et al. 2013).

Table 2 Results from FTIR, complex modulus and fatigue testing

Mixture	ICo (%)	Complex modulus |E*| (MPa) 15 °C 10 Hz	Phase angle ϕ (°)	ε_6 Corrected (10^{-6} mm)	Fatigue slope (1/b)
HMA0	1.80	12,497	15.0	115	−5.99
HMA0a	4.96	14,156	10.8	123	−8.35
HMA50	4.08	14,590	12.2	127	−7.08
HMA50a	8.27	15,847	11.3	122	−6.99
WMA0	1.46	12,498	14.6	100	−5.69
WMA0a	5.40	15,492	10.7	123	−7.68
WMA50	3.74	13,721	12.7	119	−5.55
WMA50a	8.39	16,035	11.0	108	−16.70
FWMA0	0.48	10,921	16.1	91	−2.79
FWMA0a	5.40	14,448	12.2	111	−9.00
FWMA50	5.47	13,122	14.0	105	−5.06
FWMA50a	10.35	15,460	11.5	113	−7.08

According to the results shown in Fig. 2 an overall increase in the slope of the fatigue law is observed when the ageing protocol is applied, as well as when RAP is added, i.e. that the fatigue life is getting more sensitive to the strain level.

Particularly, a drastic increase of the slope is observed for the WMA50a mixture. If fatigue laws are plotted in number of cycles by ε, the overall effect of ageing and RAP addition correspond to a clockwise rotation. For the HMA mixtures the mix is going to perform better when aged for strain levels $\varepsilon < 136$ µdef. In the case of WMA mixtures after ageing the surfactant mix develops higher resistance (in terms of ε_6) than the unaged mix. But for the HMA and WMA 50 % RAP mixes, ε_6 decreases with ageing. For these two materials, high values of ε_6 were already obtained before ageing.

For the foam based mixes, ε_6 values increase after ageing. This could be related to a comparatively moderate ageing during the manufacturing process as reflected by the lower ICo level measured. The FWMA mixes perform better after ageing for strain levels $\varepsilon < 123$ µdef for 0 % RAP and $\varepsilon < 157$ µdef for 50 % RAP.

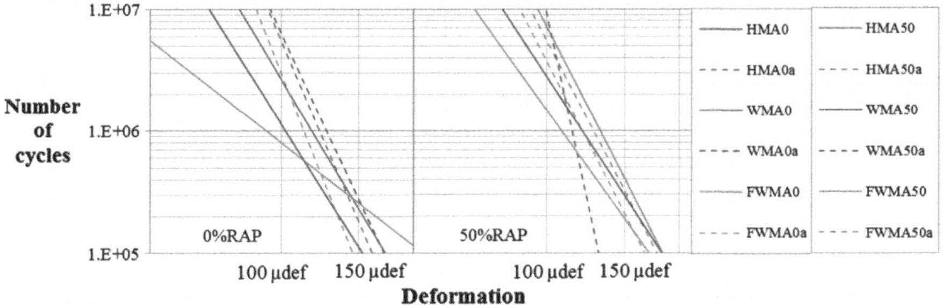

Fig. 2 Fatigue results for mixtures with 0 % RAP and 50 % RAP before and after ageing

4 Conclusions

The research presented in this paper was carried out as part of the Marie Curie Initial Training Network (ITN) action, FP7-PEOPLE-2013-ITN, with the fundings from the European Union's 7th Framework Programme for research, technological development and demonstration under grant agreement number 607524.

The aim of this work was to compare the evolution induced by ageing on the mechanical performance of surfactant and foaming WMA technologies for temperature reduction, coupled with high rates of reclaimed materials. The durability study carried out led to the following conclusions:

- Ageing leads to increase the stiffness of the material, whatever the manufacturing technique used.
- In terms of fatigue resistance, ageing seems to mostly improve ε_6 of mixes with relatively low initial fatigue resistance. According to the fatigue slope, all materials tend to become more sensitive to stress level variations after ageing, following the same trend when RAP is added.
- Generally, the mechanical characteristics of the material commonly used as input for pavement design seems to improve with ageing and RAP addition independently of the manufacturing technique employed.

References

De La Roche, C. et al., 2013. Hot Recycling of Bituminous Mixtures. In M. N. Partl et al., eds. Advances in Interlaboratory Testing and Evaluation of Bituminous Materials - State-of-the-Art. Springer, 361–429

Lopes, M. et al., 2014. Durability of hot and warm asphalt mixtures containing high rates of reclaimed asphalt at laboratory scale. Mat and Struct 48(12): 3937-3948

Mangiafico, S. et al., 2013. Influence of reclaimed asphalt pavement content on complex modulus of asphalt binder blends and corresponding mixes: experimental results and modelling. Road Mat and Pavement, 14(Supplement 1):132–148

Marsac, P. et al., 2014. Potential and limits of FTIR methods for reclaimed asphalt characterisation. Mat and Struct 47(8):1–14

Moutier, F., 1991. Etude statistique de l'effet de la composition des enrobés bitumineux sur leur comportement en fatigue et leur module complexe. Bulletin de Liaison du Laboratoire des Ponts et Chaussées,(172: 71–79

Olard, F. et al., 2012. Formulation et durabilité des enrobés à fort taux de recyclés (jusqu'à 70 %). RGRA 902: 58–62

Poirier, J.-E., Pouget, S. & Leroy, C., 2013. The multirecycling of asphalt mixes (MURE) project. European Roads 23: 13–16

Fracture Energy Evaluation of "Interstitial Asphalt Mixtures"

Simone Musetti, Marco Isola, Gabriele Tebaldi, Elena Romeo and Reynaldo Roque

Abstract Having an adequate coarse aggregate structure in any asphalt mixture may not be enough to accurately distinguish the cracking performance of an asphalt mixture: indeed, it has been proved that the interstitial volume (IV) can affect asphalt mixtures cracking performance. The IV is defined as the volume with in the coarse aggregate structure filled with finer material, binder and air voids (interstitial components). Several surveying activities on pavement sections made with mixtures designed with Dominant Aggregate Size Range (DASR), which is the coarse aggregate that forms the structural interactive network of aggregate, have been developed in the past. This paper presents an experimental study, recently developed at the University of Florida, aimed at investigating how interstitial components are influenced by different variables, as types of aggregates, aggregate

S. Musetti (✉) · E. Romeo
Università degli studi di Parma, Parma, Italy
e-mail: simone.musetti1@studenti.unipr.it

E. Romeo
e-mail: elena.romeo@unipr.it

R. Roque
E.S.S.I.E., Department of Civil and Coastal Engineering,
University of Florida, Gainesville, USA
e-mail: rroqu@ce.ufl.edu

M. Isola
Maccaferri, Inc., Williamsport, USA
e-mail: misola@maccaferri-usa.com

G. Tebaldi
Department of Civil and Environmental Engineering and Architecture,
University of Parma, Parco Area delle Scienze, 181/A, I-43124 Parma, Italy
e-mail: gtebaldi@unipr.it

G. Tebaldi
E.S.S.I.E., Department of Civil and Coastal Engineering,
University of Florida, Gainesville, USA

A. Chabot et al. (eds.), *8th RILEM International Conference on Mechanisms of Cracking and Debonding in Pavements*, RILEM Bookseries 13,
DOI 10.1007/978-94-024-0867-6_17

gradation and by binders. The DASR-IC model was used to identify a range of mixtures to be tested by first designing the coarse aggregate structure with adequate inter-locking and then varying the fine portion of the gradation. Laboratory test results from Superpave Indirect Tension Test (IDT) clearly showed that the IV characteristics have a significant effect on asphalt mixture fracture performance.

Keywords Fracture energy · Asphalt · Interstitial mixtures · Cracking performance

1 Introduction

In the mix design process of asphalt mixtures, the prevention of permanent deformation is commonly the main issue related to gradation's selection. Cracking, water damage and other distresses are considered in the following steps, generally considering the whole mixture. Recently a surveying project performed at the University of Florida to validate a new theoretical approach for aggregate structure (Dominant Aggregate Size Ratio—Interstitial Component, DASR-IC, model) showed different levels of cracking damage for mixtures with apparently similar aggregate distribution. The DASR-IC model is an extension of Bailey method and it is based on packing theory that considers the mixture's behavior dominated by interactive particles that form the structural interactive network of aggregate (the dominant aggregates, DA) and by the volumes within the interstices of the coarse aggregates (the interstitial volume). The Interstitial Components (IC) are considered the combination of the finer part of the aggregates in the asphalt mixture (retained at sieve #30 and below), binder, and air voids, which fill the Interstitial Volume (IV) in-between coarser aggregates. Considering how the DASR-IC model works, the hypothesis made to explain the observed differences in damage is that the interstitial aggregates gradation influences the cracking behavior of asphalt mixtures. This paper shows a research work that proves the previous hypothesis showing how the fracture energy changes for asphalt mixtures with same coarse aggregates gradation and different interstitial aggregates gradations.

2 DASR-IC Model

The DASR-IC model provides a framework for the design and modification of gradations to ensure that mixtures will have sufficient aggregate interlock to resist permanent deformation, as well as adequate durability and fracture resistance (Kim et al. 2006; Guarin 2009; Greene et al. 2011). DASR-IC model defines preliminary acceptable ranges for each parameter based on both laboratory and field data for optimal mixture performance therefore a range of mixture designs is identified to encompass appropriate ranges for each one of the DASR-IC parameters: two DASR

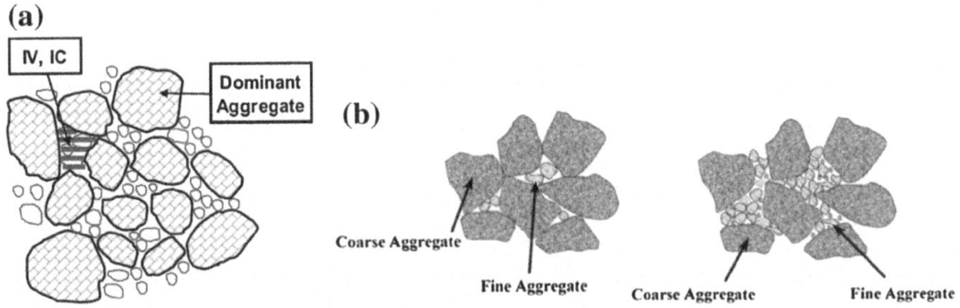

Fig. 1 DASR-IC model. **a** Schematic representation of the DASR-IC model. **b** Schematic representation of the potentially disruptive IC fine aggregate

porosity levels within acceptable range (DASR I & DASR II) and four IC coarseness levels. A new parameter, the Disruption Factor (DF), has been developed (Guarin 2009) to characterize the volumetric distribution of the IC and to determine the potential of fine aggregates to disrupt the DASR structure. The DF is defined as the ratio between the volume of potentially disruptive IC particles and the volume of DASR voids (Fig. 1).

2.1 *Research Approach*

The research's goal is to verify the hypothesis that the gradation of interstitial component influences the cracking behavior of asphalt mixtures. To achieve this goal, two grading curves were designed using DASR-IC model keeping constant the gradation of aggregates retained by #10. Four different gradations were prepared starting from each one of the designed grading curves and varying the gradation of aggregates passing at #10, always remaining inside the acceptance's limits of DASR-IC model. The cracking behavior of HMA mixtures obtained with the above mentioned aggregate gradations was evaluated determining the fracture energy, the creep compliance and the tensile strength in agreement with Superpave IDT and HMA Fracture Mechanics theory developed by Roque. During the testing campaign two aggregate type, Georgia granite and Florida limestone, and one type of unmodified asphalt binder PG 67-22 were used. To evaluate the effects of aging and of damage induced by water, three conditioning methods were used: STOA—short term oven aging, LTOA—long term oven aging and CPPC—cyclic pore pressure conditioning (Isola 2014). All the tests were performed at intermediate temperature (10 °C) in order to meet the basic hypotheses of HMA Fracture Mechanics. Considering that FAR (the ratio between the coarse and fine portions of the IC) is an indicator of the relative coarseness of the IC particle distribution and is defined

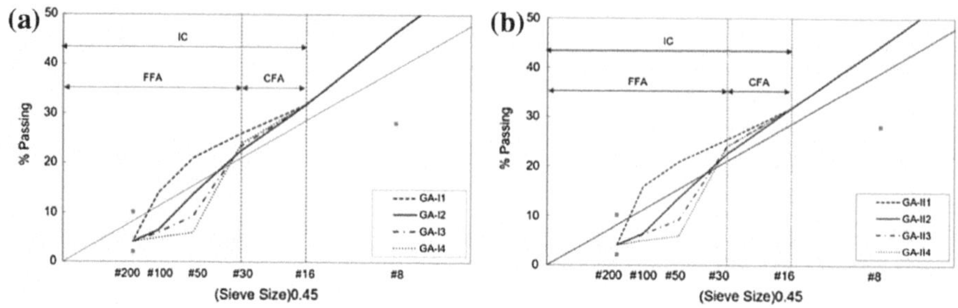

Fig. 2 **a** Designed mixture gradations for Georgia granite (GA) DASR I. Porosity level ~42.0 %. **b** Designed mixture gradations for Georgia granite (GA) DASR II. Porosity level ~45.0 %

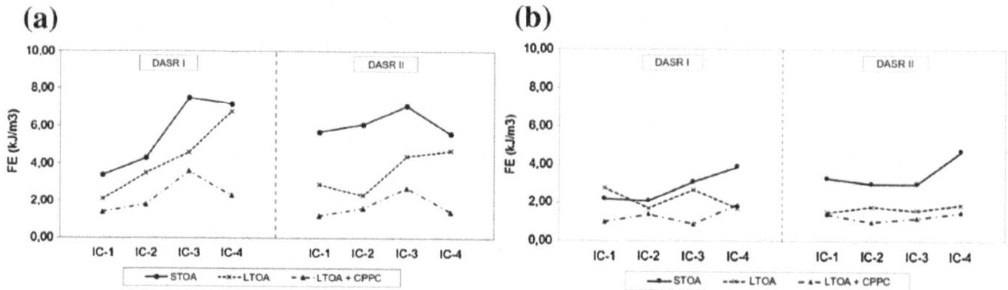

Fig. 3 **a** Georgia Granite FE test results for different conditioning methods. **b** Florida limestone FE test results for different conditioning methods

as the ratio of the coarse portion of the fine aggregate (CFA) and the fine portion of the fine aggregate (FFA). Specifically, CFA includes only the largest particle size of the IC while the FFA is the remaining finer portion of IC particles. Examples of the designed mixtures are shown in Fig. 2 for Georgia granite.

2.1.1 Test Results

Superpave IDT tests were performed at 10 °C to obtain HMA fracture properties for each mixture at each conditioning level. The Superpave IDT is composed of a sequence of three tests (resilient modulus, creep compliance and strength tests) as reported on NCHRP Report 530. According to HMA Fracture Mechanics fracture energy (FE), which is the total energy applied to the specimen until material failure, is determined as the area underneath the stress–strain curve until the first crack appears and it is defined as the sum of dissipated creep strain energy (DCSE) and elastic energy (E.E.).

Where elastic energy is the energy stored and released by the material and dissipated creep strain energy to failure is the absorbed energy that damages the specimen ($DCSE_f$). Figure 3 shows the results of the test carried on the different mixtures (Isola 2014). Even if a clear relation cannot be observed, it is clear that the changes in the interstitial component gradation make not negligible variation in the material's cracking performance and in their evolution related to aging and water induced damage.

3 Future Implementations: The IC Dog-Bone Model

The results clearly show that changes in IV characteristics significantly affected fracture limits (i.e., FE, failure strain) thus the performance for both granite and limestone mixtures: therefore, this kind of analyses is needed to better understand the role of interstitial volumes on mixtures cracking behavior. To further investigate these phenomena a specific test should be developed. In particular, a test capable of characterizing the material inside the interstitial volume, which means the mixture composed by asphalt binder and interstitial components. This material is not mastic so it cannot be tested with the traditional test for mastics. What must be considered is an "interstitial mixture" composed by aggregates with different sizes and a defined gradation (portion of aggregates contained in the interstitial volume) and binder. A specific test will allow to better define the effects of changes in IV characteristics from the DASR structure effects. Several tests used in previous researches were considered as potential bases for the new test. In particular, in order to develop a simple stress-strain distribution test setup, a direct tension test was considered. A research performed by Niu (2011) clearly showed that the new binder Direct Tension test and interpretation system was able to consistently test fracture energy of binder. However, this is not suitable to represent the effect of the IC. Among the different test considered, the most promising specimen's geometry is the one designed by Koh (2009). As shown in Fig. 4 the Dog-bone specimens used for this study (Musetti 2014) have been scaled from the ones used by Koh (2009) reducing the IDT 150 mm (diameter) cylinder gyratory specimen to a 100 mm diameter one, and keeping only the core (internal part of the cylinder) of 80 mm diameter which is not afflicted by the tensile stress due to the coring process (80 mm $\sim$ 3 in. which is the length of the DB specimen). For this specific geometry, failure plane is known a priori and stress concentration is less critical near the ends of the sample as shown by FEA results described below (Musetti 2014).

Finite element analysis of the DB specimen was based on the binder direct tension test model and the DB model developed at the UF (Niu 2011; Koh 2009). In terms of material model, a linear elastic material with stiffness of 1 MPa and Poisson's ratio of 0.4 was considered as a reference and, due to symmetry, only one

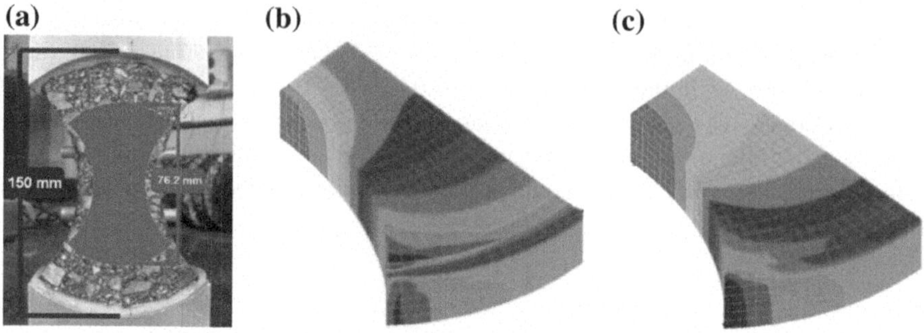

Fig. 4 **a** Dog-Bone geometry. Dog-Bone Finite Elements Model **b** showing the stress and **c** the strain calculated on the y-axis for a 0.4 mm displacement for one eight of the specimen (mid-section, in *red*, presents a major concentration of the displacement)

eighth of the specimen was modeled. The performed finite element simulations showed that the selected geometry is able to concentrate the stress and to induce the maximum stress at the center of the specimen far from the edge, the weakest parts. Therefore, this geometry seems really promising for further investigating the effects of IV characteristics on asphalt mixture performance.

4 Conclusions

The results of the experimental study clearly show that the gradation of interstitial components strongly affect the pavement cracking behavior. For this reason, it's possible to define as bulk material, the mixture composed by interstitial aggregates and asphalt binder, named "interstitial mixture". A specific test to further investigate the effects of the IC on the asphalt mixture performance and to design the IC's gradation (considering primarily the coarse aggregates interlocking) will be implemented (Musetti 2014). The preliminary investigation made with a finite element analysis, shows that a direct tension test performed on a dog-bone specimen may be the proper setup to study IC's performance.

References

Greene J, Kim S, Choubane B (2011) Accelerated Pavement Testing and Gradation-Based Performance Evaluation Method, Transportation Research Record 2225, Transportation Research Board, Washington, D.C., p 119–127

Guarin A (2009) Interstitial Component Characterization to Evaluate Asphalt Mixture Performance, PhD Dissertation, University of Florida, Gainesville

Isola M (2014) Effect of Interstitial Volume Characteristics on Asphalt Mixture Durability and Cracking Performance, PhD Dissertation, University of Florida, Gainesville

Kim S, Roque R, Birgisson B (2006) Identification and Assessment of the Dominant Aggregate Size Range (DASR) of Asphalt Mixture, Journal of the Association of Asphalt Paving Technologists, Vol. 75, p 789–814
Koh C (2009) Tensile Properties of Open Graded Friction Course (OGFC) Mixture to Evaluate Top-Down Cracking Performance, Ph.D. Dissertation, University of Florida, Gainesville
Musetti S (2014) Development of an Interstitial Asphalt Mixture Direct Tension Test for Frac- ture Energy Evaluation, MSc Dissertation, University of Parma, Parma
Niu T (2011) Development of a Binder Fracture Test to Determine Fracture Energy, PhD Dissertation, University of Florida, Gainesville

Significance of Oxidative Aging on the Thermal Cracking Predictions in Asphalt Concrete Pavements

Mohammad Zia Alavi, Elie Y. Hajj and Peter E. Sebaaly

Abstract The resistance of asphalt concrete pavements to thermal cracking is fundamentally related to the asphalt mix stiffness, strength, and contraction properties. These properties continuously change over time with the oxidative aging of the asphalt binder in the mix. Oxidation reaction in long-term mainly results in formation of carbonyl functional group that improves the association among polar components, increasing the asphalt binder stiffness and brittleness. Hence, thermal cracking is generally more prevalent in old pavements. In this paper, a methodology is proposed to refine the mechanistic model for prediction of thermal cracking events over pavement service life in order to account for continuous changes of asphalt mixture critical properties with oxidative aging. Particularly, the changes in asphalt mix relaxation modulus, crack initiation stress, and coefficient of thermal contraction (CTC) were related to the growth in carbonyl groups in the respective binder used in the mix. The approach was examined on few types of asphalt mixes having different air void levels, and with unmodified and polymer-modified binders to be used as a surface course in a specific asphalt pavement. Based on preliminary modeling results, significant differences in thermal cracking performances of mixes were observed when effect of aging was considered.

Keywords Thermal cracking · Oxidative aging · Relaxation modulus · Crack initiation

M.Z. Alavi (✉)
University of California, Davis, CA 95616, USA
e-mail: alavi@ucdavis.edu

E.Y. Hajj · P.E. Sebaaly
University of Nevada, Reno, NV 89557, USA
e-mail: elieh@unr.edu

P.E. Sebaaly
e-mail: psebaaly@unr.edu

A. Chabot et al. (eds.), *8th RILEM International Conference on Mechanisms of Cracking and Debonding in Pavements*, RILEM Bookseries 13,
DOI 10.1007/978-94-024-0867-6_18

1 Introduction

Oxidative aging of asphalt binder results in generation of carbonyl functional group, which increases the association among polar component. This phenomenon alters the physical properties of asphalt binder to turn into harder and brittle material. The aging of binder in a mix over time decreases the durability and increases cracking potential of mix. Therefore, the occurrence of thermal cracking generally becomes more prevalent with the age of pavement.

A comprehensive mechanistic model has been recently developed at the University of Nevada, Reno that accounts for changes in asphalt mix critical properties with aging when predicting thermal cracking events over service life of pavement. The main intention of this model is to predict possible events of thermal cracking over analysis period. This paper summarizes the model components and applied methodologies. The model was examined to evaluate the thermal cracking performances of asphalt mixes having different air void and with unmodified and polymer modified binders when used at surface of pavement at a specific location.

2 Comprehensive Thermal Cracking Model

The developed comprehensive thermal cracking model includes four main components for: (I) temperature prediction, (II) oxidative aging prediction, (III) thermal stress calculation, and (IV) thermal crack event prediction.

Temperature prediction: Hourly asphalt pavement temperature data is estimated using an enhanced heat transfer model which requires hourly air temperature, wind speed, and solar radiation at the pavement location as well as pavement layers' diffusivities and pavement surface albedo and emissivity. At least one years of climatic and meteorological data is required to complete pavement temperature estimation. More information regarding this model and calculation methodology are readily available in the respective literature (Alavi et al. 2013a).

Oxidative aging prediction: The kinetic based oxidative aging model originally developed at Texas A&M University (Han et al. 2013) is implemented to predict the CA of asphalt binder at any depth of the asphalt concrete layer over the analysis period. The inputs of this model include hourly pavement temperature data at the selected depth of asphalt layer, the oxidation kinetics of asphalt binder in the mix (activation energy, hardening susceptibility, and initial CA), respective diameter of air void in the mix, and effective zone of aging in asphalt-aggregate matrix. Having the predicted values of CA over the analysis period, the changes in asphalt mix mechanical properties (i.e., stiffness, strength, and thermal contraction) can be estimated. More information about this model can be found in the respective literature (Alavi 2014).

Thermal stress calculation: Thermal stress at any depth of asphalt layer is calculated from the modified one dimensional viscoelastic constitutive equation

(Boltzmann equation) which accounts for aging effect. Overall, age dependent relaxation modulus and temperature and age dependent CTC are material properties required to perform thermal stress calculation. For this purpose, asphalt mix specimens must be tested at the minimum of two, preferred three, levels of aging.

To obtain age dependent relaxation modulus, asphalt mix dynamic modulus and CA for recovered binder of the mix must be obtained for mixes subjected to different durations of long-term laboratory aging. At each aging level, dynamic modulus data is used to obtain 2S2P1D (two springs, two parabolic elements, and one dashpot) complex modulus function (Olard and Di Benedettto 2003). Continuous relaxation spectrum of asphalt mix is then directly derived 2S2P1D model function using inverse Fourier Laplace transformation. The parameters of Prony series of relaxation modulus is eventually obtained by discretization of continuous spectrum. The age dependency of relaxation modulus is taken into account by finding correlations between 2S2P1D model parameters (6 variables) with increases in CA of the binder used in the mix. As expected, relaxation modulus increased with aging for mixes evaluated in the model development. More information regarding this proposed approach can be found in respective literatures (Alavi et al. 2013b; Alavi 2014).

The temperature dependent CTC function for an asphalt mix is defined by the CTC values (CTC_l and CTC_g) before and after glass transition temperature (T_g). Asphalt mix CTC function is determined from the thermal strain measured by the uniaxial thermal stress and strain test (UTSST) (Alavi et al. 2013b). In the UTSST, thermal stress and strain of an asphalt mix were measured, respectively, from the restrained and unrestrained specimens while cooling from an initial equilibrium temperature (usually 20 °C) under a constant rate (usually 10 °C/hr). The effect of aging on CTC is considered by correlating the changes in CTC function parameters with the increased CA. Details of approach are available in number of literatures by authors (Alavi et al. 2013b, 2015; Alavi 2014; Morian et al. 2014).

Thermal crack event prediction: A thermal crack event is assumed to occur when thermal stress reaches crack initiation stress (CIS) of asphalt mix. The accumulation of these events over design life of pavement is directly associated to the mix thermal cracking performance. The CIS is considered to be age-dependent. The CIS of asphalt mix determined from the UTSST results. Using thermal stress and strain measurements, the modulus of asphalt mix over temperatures can be calculated based on the linear viscoelasticity concept. The crack initiation stage is defined as the point at which the modulus of mix dropped instantaneously. CIS is obtained from measured thermal stress corresponding to the crack initiation stage. CIS was found to decrease with the increased CA for the mixes evaluated during model development. Different percentages of CIS can be defined as the cracking criterion. More information regarding evolution of UTSST results with oxidative aging can be found in respective literatures (Alavi et al. 2013a, b, 2015, Morian et al. 2014).

3 Thermal Cracking Analysis for Selective Asphalt Mixes

Comprehensive thermal cracking analyses (with and without considering oxidative aging effect) were completed for few asphalt mixes. Three laboratory-produced mixes made with PG64-28 polymer-modified binder and aggregates from a local source in Nevada, but compacted at three air voids levels (4, 7, and 11 %) and two field-produced laboratory compacted mixes (at 7 % air void) obtained from local projects in Reno. The filed produced mixes named Moana-PG64-22 (unmodified binder) and Spark-PG58-28 (polymer modified binder). These mixes were supposed to be used at the top lift of asphalt surface layer of a pavement located in Reno, Nevada. The properties of pavement layer materials including those required for prediction of pavement temperature, the oxidation kinetics of asphalt binders used in the mixes, the age dependent characteristics of asphalt mixes (i.e., relaxation modulus, CTC, and CIS) are readily available in the respective literatures by authors (Alavi et al. 2015 and Alavi 2014). The analyses were performed for a period of 20 years started from August 1st 1996 to July 31st 2016. The following observations were made from these analyses:

- For PG64-28 mixes with different air void levels, the developed thermal stress did not reach even 50 % CIS when the oxidative aging effect was not included in crack event prediction.
- When aging is considered in crack prediction, the differences in performance of NV-PG64-28 mixes at different air void levels were observed. As shown in Fig. 1, the accumulative cracking events for these mixes were found to be similar up to 12 years from the construction and became distinguishably different afterwards with more cracking events for mixes with higher air void content. There was no thermal cracking event detected for NV-PG64-28 mix with 4 % air void even at 50 % CIS criterion.

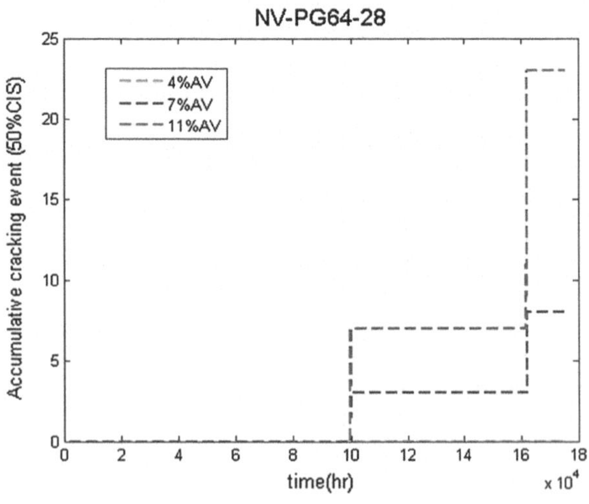

Fig. 1 **a** Accumulative cracking event for NV-PG64-28 asphalt mixes at 4, 7, and 11 % air void

- No thermal cracking event was detected for the Spark-PG64-28 mix even when aging effect is considered. These results are consistent with the current field performance of spark pavement where thermal cracking distress has not been observed yet.
- As shown in Fig. 2, significantly higher thermal cracking events were detected for Moana22 mix when aging effect was included in the analysis. For this mix, the number of accumulative cracking event after 20 years was found to be at least 60 times higher when aging effect was considered in the analysis.
- Moana-PG64-22 mix might show an acceptable thermal cracking performance up to approximately 9 years from the construction date due to the low number of accumulative cracking events. However, after this time the number of cracking events increases drastically due to the effect of oxidative aging on both stiffening the mix and decreasing the stress tolerance of the asphalt mix.

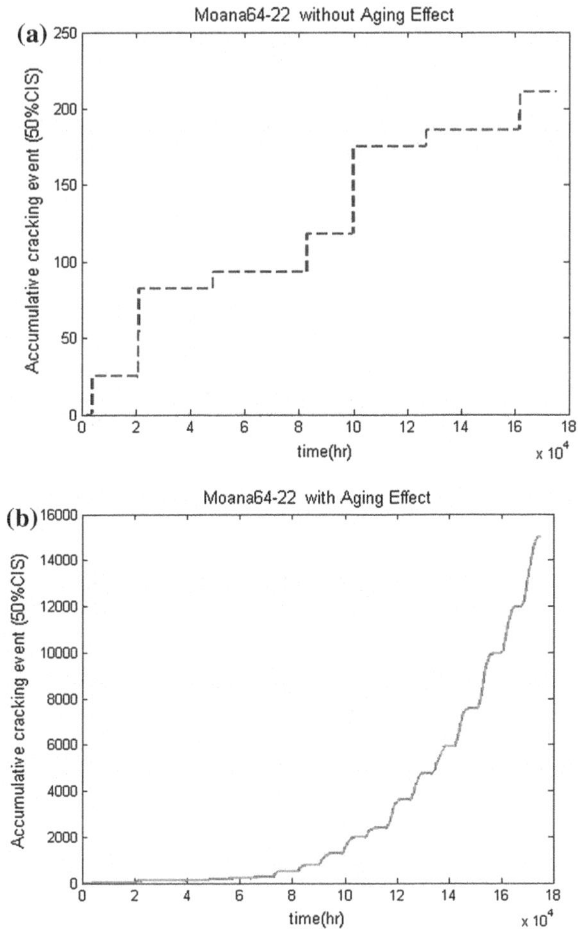

Fig. 2 **a** Accumulative racking event for Moana64-22 mix without considering aging effect, **b** Accumulative cracking event for Moana64-22 mix with considering aging effect

Overall, these observations from the thermal cracking simulations are in line with the field performance of the corresponding asphalt mix especially that the PG64-28 polymer-modified asphalt binder has historically been outperforming the performance of the PG64-22 unmodified asphalt binder in terms of thermal cracking in northern Nevada.

4 Summary and Conclusion

This paper summarizes the components of a proposed comprehensive model for the prediction of thermal cracking event over service life for asphalt concrete pavements. The model offers incorporation of influence of asphalt binder oxidative aging on changing the critical properties asphalt mix (i.e., relaxation modulus, thermal contraction, and CTC) when predicting thermal cracking events. For this purpose, the mechanical properties of a mix over time can be predicted from the increase in CA of the asphalt binder used in the mix. Overall, the model showed promising results in terms of realistic prediction of cracking events for mixes with different air void levels and with unmodified and polymer modified binders. More studies are required to validate the findings of this model for other mixes and to calibrate the model based on the field performance of different asphalt mixes.

References

Alavi MZ, Morian NE, Hajj EY, and Sebaaly PE (2015) Influence of asphalt binder oxidative aging on critical thermal cracking characteristics of asphalt mixtures. Proceeding of the Association of Asphalt Paving Technologies, *In Press*

Alavi MZ, Pouranian MR, and Hajj EY (2013a) Prediction of asphalt pavement temperature profile with finite control volume method. Transportation Research Record, No. 2456, pp. 96-106.

Alavi MZ, Hajj EY, Morian NE, and Sebaaly PE (2013b) Low temperature characterization of asphalt mixtures by measuring visco-elastic properties under thermal loading. ISCORD 2013, 10th ASCE International Symposium on Cold Regions Development, pp. 404-415.

Alavi MZ (2014) Comprehensive methodologies for analysis of thermal cracking in asphalt concrete pavements, Ph.D. Dissertation, University of Nevada, Reno.

Han R, Jin X, and Glover CJ (2013) Oxygen diffusivity in asphalts and mastics. Petroleum Science and Technology, Vol. 31, Issue 15, 1563-1573.

Morian N, Alavi MZ, Hajj EY, and Sebaaly, PE (2014) Evolution of thermo-viscoelastic properties of asphalt mixtures with oxidative aging. Transportation Research Record, No. pp.

Olard F, and Di Benedetto H (2003) General 2S2P1D model and relation between the linear viscoelastic behavior of bituminous binders and mixes. Road Materials and Pavements Design, Vol. 4, No. 2. 185-224.

Viscoelastic Properties of Bituminous Composites Using Multiscale Heterogeneous Numerical Simulation and Micromechanical Analytical Self-consistent Model

F. Fakhari Tehrani, J. Absi, F. Allou and C. Petit

Abstract In this work, detailed study is carried out to develop heterogeneous micromechanical modelling based on multiscale technique to investigate complex response of asphalt mixture. The mechanical behavior of this material is very complex and one must take into account the geometry of the microstructure and the mechanical behavior of the different phases. The properties of such heterogeneous material are highly dependent on the volumetric fraction of bitumen, aggregate structure, air void distributions and interfacial bonding strength between bitumen and aggregates. In this study, the considered bituminous composites are mastic, mortar and asphalt mixes. These composites are treated as biphasic composites, composed of a viscoelastic bituminous binder and an aggregate skeleton with linear elastic properties. In the developed multiscale approach, the homogeneous equivalent properties of biphasic composite are transferred from one scale of observation to the next higher scale of observation. The dynamic modulus of the matrix and elastic properties of aggregates were used as input parameters into micromechanical finite element models qualified by heterogeneous micromechanical models. The

F. Fakhari Tehrani · F. Allou · C. Petit
Groupe d'Étude des Matériaux Hétérogènes—Équipe Génie Civil et Durabilité. Université de Limoges, boulevard Jacques Derche, 19300 Egletons, France
e-mail: fatehftehrani@yahoo.com

F. Allou
e-mail: fatima.allou@unilim.fr

C. Petit
e-mail: christophe.petit@unilim.fr

F. Fakhari Tehrani
Conservatoire national des Arts et métiers, Paris, France

J. Absi (✉)
Université de Limoges, SPCTS, CNRS,
CEC–12 Rue Atlantis, 87068 Limoges Cedex, France
e-mail: joseph.absi@unilim.fr

A. Chabot et al. (eds.), *8th RILEM International Conference on Mechanisms of Cracking and Debonding in Pavements*, RILEM Bookseries 13,
DOI 10.1007/978-94-024-0867-6_19

complex modulus and phase angle were compared to experimental measurements and analytical values obtained by Generalized Self Consistent Scheme (GSCS).

Keywords Micromechanical modelling · Multiscale approach · Self-consistent scheme · Numerical calculation

1 Introduction

Granular skeleton of an asphalt mixture is composed of different combination of aggregates distributed in a size from coarse aggregate to the fraction of fine aggregate (Yin et al. 2011). In this gradation the amount of fine particles is thousand times or even more that of coarse aggregates. This distribution requires a huge computational capacity to construct a numerical model of asphalt concrete with the complete gradation. Therefore, in order to realize the balance between the model fidelity and the computational efficiency, it should be districted several homogenization scale for granular gradation.

In this study, from mastic to asphalt mixes, three scale of observation have been distinguished. At the lowest scale, mastic is composed of filler, i.e., aggregates with a diameter <100 μm are embedded in a bitumen matrix. At the next higher scale, commonly called mortar scale, the mastic is considered as the matrix whereas sand, with a diameter ranging between 100 μm < D < 2 mm, is used for the inclusions. Finally, at the last scale, asphalt can be treated as a biphasic material, with a matrix of mortar and inclusions of aggregates having a size between 2 mm < D < 10 mm. The primary objective of this work is to present a numerical micromechanical modeling based on multiscale technique using digital biphasic models for predicting the dynamic modulus of bituminous materials such as mastic, mortar and asphalt mixes. Then the obtained values will be compared to those obtained with the micromechanical model called Generalized Self Consistent Scheme and with experimental results. (Fig. 1). This model (GSCS) is recognized among the high sophisticated micromechanical model (Christensen and Lo 1979, 1986) which is based on a three-phase model: A spherical inclusion is embedded by a matrix,

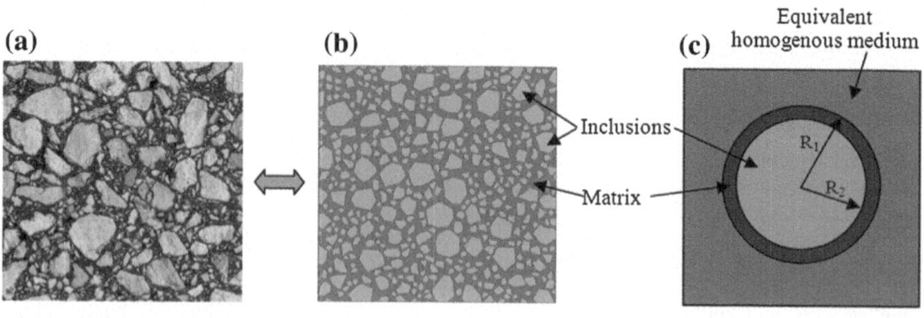

Fig. 1 Schematic representation of micromechanical modeling of bituminous materials

which in turn is embedded in an infinite equivalent medium of the composites. In the last part, stress and strain distribution, within the heterogeneous model, will be presented to reveal the potential region of cracking development.

2 Numerical and Analytical Upscaling Method

In this study, four components are used to construct the coming numerical models. The used mechanical characteristics are issue from Delaporte experimental investigations (Delaporte et al. 2005).

Digital models are obtained from a random generation by using M.O.A. software which is developed in our laboratory in C++ language. The efficiency of this software was studied previously (Fakhari Tehrani et al. 2013). As hypothesis, separated particles are generated to represent a perfect embedding of aggregates by a thin layer of matrix.

The numerical predicted dynamic modulus of bituminous composites were compared to analytical results issues from GSCS micromechanical model. (Christensen and Lo 1979, 1986). The corresponding constitutive equation is:

$$A\left(\frac{G_c}{G_m}\right)^2 + B\left(\frac{G_c}{G_m}\right) + C = 0 \tag{1}$$

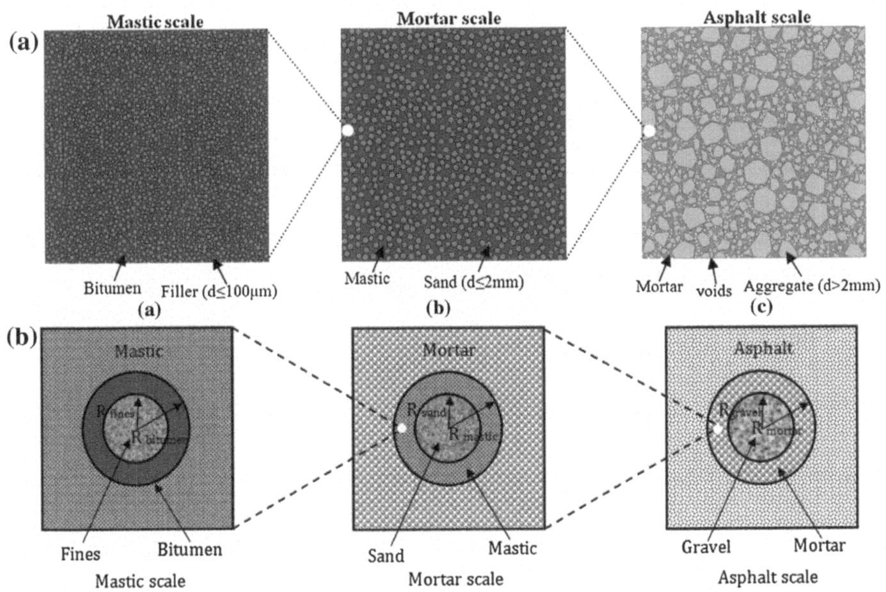

Fig. 2 Multi scale modeling of asphalt mixes using heterogeneous numerical method (**a**) and the generalized self-consistent model (**b**)

where A, B and C are the model parameters which depend on the Poisson ratio of matrix and the volume fraction of inclusions. G_c and G_m are the shear modulus of composite and matrix respectively. By taking into account the four phases for asphalt mixes (bitumen, fines, sand and gravel), the self-consistent model applied is considered in three scales (see Fig. 2).

3 Results and Discussions

Figure 3a, b show respectively the comparison between numerical, analytical and experimental measured dynamic modulus (Delaporte et al. 2005) and phase angle of mastic at T = −10 °C and T = 0 °C. It is observed that the predicted numerical and analytical results followed the same trend as experimental measurements. However the 3D numerical complex moduli are situated blow the original experimental curve. Therefore, to avoid having any cumulative uncertainty of simulation, the experimental measurement of mastic is used as the input for mortar scale.

Figure 4a, b represent respectively the numerical and analytical dynamic modulus and phase angle obtained for mortar scale at different temperatures. As expected, it is observed that the dynamic modulus values are depending on

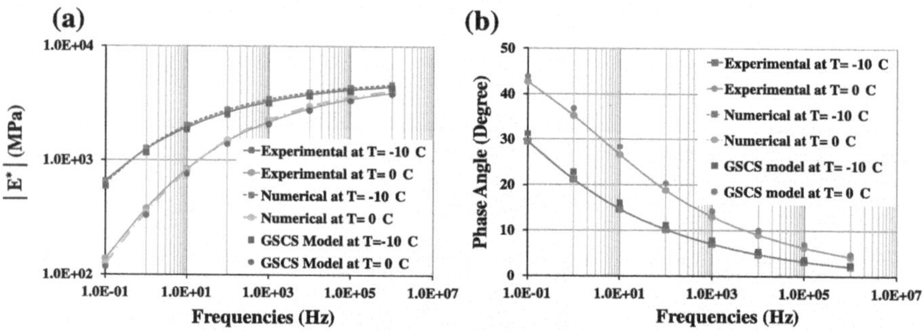

Fig. 3 Numerical and analytical phase angle and master curve of mastic 30 % compared to experimental one at different temperature

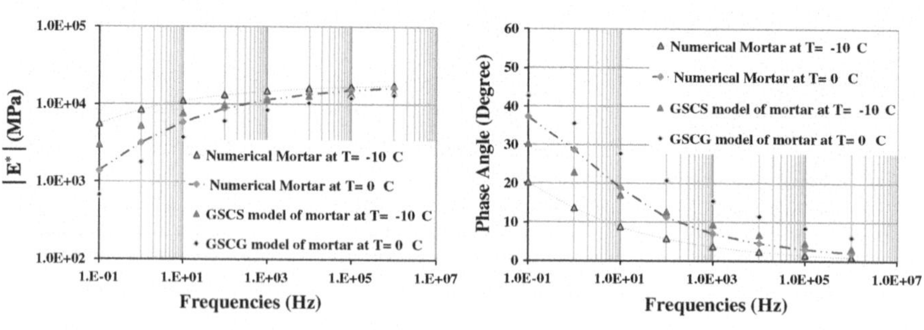

Fig. 4 Numerical and analytical phase angle and master curves of mortar at T = −10 °C and T = 0 °C

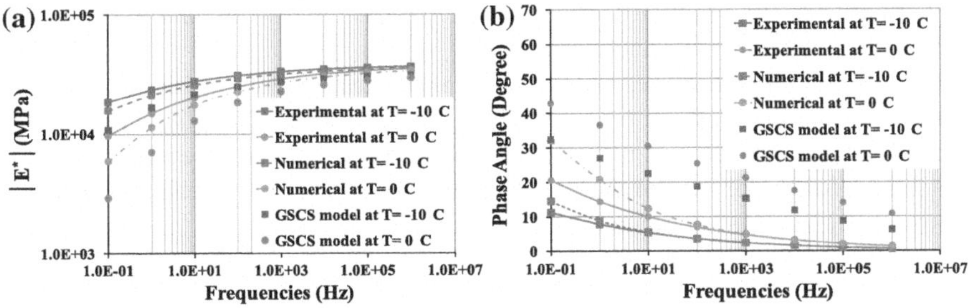

Fig. 5 Numerical and analytical phase angle and master curves of asphalt mixture compared to experimental one at T = −10 °C and T = 0 °C

temperature. When temperature increases, the dynamic modulus values decrease over the entire studied frequency range. However these values become progressively closer when the frequency values increase.

To evaluate the accuracy of our heterogeneous numerical modeling, we compared the dynamic modulus and phase angle master curves obtained by the numerical procedure and analytical model with the homogeneous asphalt mixture laboratory parameters. This comparison is shown respectively in Fig. 5a, b.

It is observed that for high frequencies, the numerical and analytical results give good prediction values of complex modulus of asphalt mix scale in comparison to the experimental ones. However, analytical results are systematically situated below the finite element results. This observation can be explained by the weakness of the analytical models in the case of a high filling rate (almost higher than 40 %) (Buttlar et al. 1999, Yin et al. 2008) for the numerical results issue from T = 0 °C under frequencies loading almost lower than 100 Hz, there is an important gap between the predicted numerical results and experimental measurements. An underlying parameter behind this gap, may be related to the moduli contrast ratio between the mortar matrix phase and the granular phase. According to this assumption, it can be estimated that an increasing of the contrast between the aggregate and mortar moduli from 12.7 ($E_{inclusions}/E^*_{mortar} = 12.7$) at T = −10 °C and f = 0.1–38.9 Hz at T = 0 °C under frequency loading of 0.1 Hz, can influence

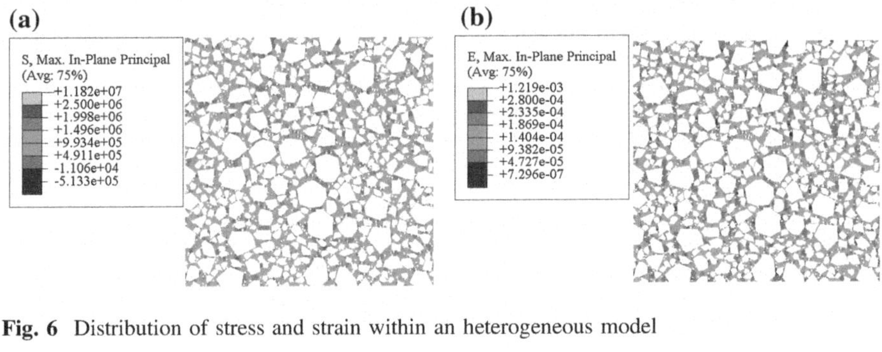

Fig. 6 Distribution of stress and strain within an heterogeneous model

the numerical results obtained at T = 0 °C. Another reason of this gap between predicted numerical results and experimental measurements can be attributed to the limitation of aggregate-aggregate contact in numerical models and lack of considering of the interlock effects between the aggregates, which can play an important role in load transfer between the particles. Figure 6a, b show respectively the stress and strain distribution within the heterogeneous model which reveals the potential region of cracking development.

4 Conclusion

In this work, an attempt has been made to develop a finite elements model to predict the mechanical behavior of bituminous composites. Our approach is based on a homogenization of characteristics obtained from intermediate scales. These intermediate steps represent a progressive technique used to construct the final asphalt mixture model. The major advantage of this technique is that, the larger scale structure, that exhibits complex geometry and constitutive behavior, can be simulated based on material properties on a much smaller scale.

The predicted dynamic modulus of asphalt mixture was compared with that of the experimental test data for different temperatures. It was observed that at low temperature (here −10 °C) the numerical values are slightly less than the experimental measurements values. With increasing in temperature the difference between predicted numerical results and experimental data increases. This observation can be related to the increase of the contrast modulus between matrix and inclusions at high temperature. Another reason can be related to the limitation of aggregate-aggregate contact in the numerical models and lack of interlock effects consideration between aggregates.

References

Buttlar W. G, Bozkurt D, Al-khateeb G. G, & Waldhoff A. S. (1999) Understanding asphalt mastic through micromechanics. Transportation Research Record, 1681, 157-169.

Christensen RM, Lo KH. (1979) Solutions for effective shear properties of three phase sphere and cylinder models. J Mech Physics Solids; 27:315-330.

Christensen RM, Lo KH. (1986) Erratum: Solution for effective shear properties of three phase sphere and cylinder models. J Mech and Physics Solids; 34:639-639.

Delaporte B, Di Benedetto H, Sauzéat C, Chaverot P(2005). Linear viscoelastic properties of mastics: results from a new annular shearing rheometer, and modelling.Trondheim.

FakhariTehrani F, Absi J, Allou F, Petit C (2013) Heterogeneous numerical modeling of asphalt concrete through use of a biphasic approach: porous matrix/inclusions. Computational Materials Science; 69:186-196.

Yin A, Yang X, Yang S, Jiang W. (2011) Multiscale fracture simulation of three-point bending asphalt mixture beam considering material heterogeneity. Eng Fracture Mech; 78:2414-2428.

Yin H.M, Buttlar W.G, Paulino G.H, Di Benedetto H (2008) Assessment of existing micro-mechanical models for asphalt mastics considering viscoelastic effects. Road Materials and Pavement Design, 9: 31-57.

Low Temperature Cracking Problem for Asphalt Pavements in Kazakhstan

Bagdat Teltayev and Boris Radovskiy

Abstract Low temperature cracking is one of the major causes of failure of asphalt pavements in Kazakhstan where minimum pavement design temperature is between −28 and −46 °C. Thermal stresses in the asphalt pavement were calculated as a function of its relaxation modulus and temperature changes. The parameters of modified CAM model for relaxation modulus of bitumen were related with its penetration index and softening point by formulas based on Van Der Poel's nomograph. Hot mix asphalt relaxation modulus has been described by Christensen–Bonaquist model. Hot mix asphalt tensile strength as a function was related to the bitumen stiffness by equation based on Heukelom's data and Molenaar-Li formula. The strength of mix has been also studied by performing tensile tests. The results were analyzed considering the calculated stress increase with temperature drop, and critical temperature was estimated.

Keywords Asphalt pavement · Low temperature · Cracking · Tensile strength

1 Introduction

Kazakhstan is the ninth largest territory in the world. Climate is continental with warm summers and very cold winters. Around a half of the territory requires the bitumen of SUPERPAVE grade PG 58-40. The oxidized bitumen from Western Siberian oil obtained in five oil refineries has been used for road construction. Typical thickness of new asphalt pavement is currently around 15 cm. Low-temperature cracking is one of the major types of failure of asphalt pavement. After the first winter, 20–25 transverse cracks per kilometer appear, after the second or the third

B. Teltayev (✉) · B. Radovskiy
Kazakhstan Highway Research Institute, (KazdorNII) 2A, Nurpeissov Street, Almaty, Kazakhstan
e-mail: bagdatbt@yahoo.com

B. Radovskiy
e-mail: b.radovskiy@att.net

A. Chabot et al. (eds.), *8th RILEM International Conference on Mechanisms of Cracking and Debonding in Pavements*, RILEM Bookseries 13,
DOI 10.1007/978-94-024-0867-6_20

winter the number of cracks is 45–50. To better understand how to reduce the low temperature cracking, we believe we need a comprehensive mechanistic model that includes rheological and fracture properties of asphalt mixture.

Monismith et al. (1965) developed a theoretical calculation method for the thermally induced stress in asphalt pavement as in an infinite viscoelastic beam. This method is currently used for the estimation of critical cracking temperature. Christison et al. (1972) employed five different methods of stress computation and concluded that a potential of low temperature cracking can be evaluated if the mix stiffness and strength characteristics are known. Bouldin et al. (2000) considered the thermally induced stress in asphalt binder as in viscoelastic bar and reported that the midpoint of the binder's glass transition is close to the pavement critical cracking temperature. They also noted that mix with blown bitumen formed severe, wide cracks. The main objective of our study was to develop relationships for prediction of the rheological properties and tensile strength of asphalt concrete with plain bitumen to address the low temperature cracking analysis.

2 Stiffness and Relaxation Modulus of Bitumen

Stiffness modulus $S(t)$ introduced by Van der Poel (1954) is the ratio of the constant applied uniaxial stress to the resulting strain. Based on the test results of 47 bitumens, Van der Poel developed a nomograph to determine the stiffness of bitumen $S(t)$ as a function of temperature T and loading time t if the softening temperature T_s and the penetration index PI are known. To calculate the stiffness, it is convenient to use the BitProps software based on the scanned nomograph and developed by Rowe and Sharrock (Abatech, Inc.). As a model for the stiffness of bitumen, we used Christensen and Anderson (1992) expression:

$$S(t) = E_g \left[1 + \left(\frac{E_g t}{3\eta} \right)^{\beta} \right]^{-\frac{1}{\beta}} \tag{1}$$

where E_g is the uniaxial glassy modulus [MPa], η is the steady-state viscosity [MPa s], β is constant. For binders with penetration index from −3 to +2, the glassy modulus was set to $E_g = 2460$ MPa. After fitting Eq. 1 to Van der Poel's data, the following equations were obtained:

$$\beta = \frac{0.1794}{1 + 0.2084PI - 0.00524PI^2} \tag{2}$$

$$\eta = a_{T\,Ahrr}(T) \cdot \eta(T_r) \quad (at\,T \leq T_s - 10), \eta = a_{T\,WLF}(T) \cdot \eta(T_r) \quad (at\,T > T_s - 10) \tag{3}$$

$$\eta(T_r) = 0.00124\left[1 + 71\exp\left[-\frac{12(20 - PI)}{5(10 + PI)}\right]\right] \cdot \exp\left(\frac{0.2011}{0.11 + 0.0077PI}\right) \quad (4)$$

where T_r is reference temperature, $T_r = T_s - 10$ (°C), $a_T(T)$ is time-temperature superposition function:

$$a_{T\,Ahrr}(\mathrm{T}) = \exp\left[11720 \cdot \frac{3(30 + PI)}{5(10 + PI)}\left[\frac{1}{(T + 273)} - \frac{1}{(T_s + 263)}\right]\right] \quad (5)$$

$$a_{T\,WLF}(\mathrm{T}) = \exp\left[-\frac{2.303(T - T_s + 10)}{(0.11 + 0.0077PI)(114.5 + T - T_s)}\right] \quad (6)$$

For 1910 points in the range of PI from −3 to +2, $(T - T_s)$ from −45 to 10 °C, and t from 10^{-4} to 10^4 s, the standard deviation of $S(t)$ calculated using Eq. 1 from Van der Poel's data was 14.6 % and $R^2 = 0.978$. Rheologically speaking, $S(t)$ is the inverse of uniaxial creep compliance. It is related to the shear creep compliance:

$$J(t) = \frac{1}{G_g}\left[1 + \left(\frac{G_g t}{\eta}\right)^{\beta}\right]^{\frac{1}{\beta}} \quad (7)$$

where G_g is the glassy modulus in shear, $G_g \approx E_g/3 = 820\,\mathrm{MPa}$. There are different methods for converting the shear creep compliance $J(t)$ to relaxation modulus $G(t)$. We used modified Hopkins-Hamming algorithm and approximated $G(t)$ by CAM model (Marasteanu and Anderson 1999):

$$G(t) = G_g\left[1 + \left(\frac{G_g t}{\eta}\right)^{b}\right]^{-\frac{k}{b}} \quad (8)$$

where b and k are constants that can be related using the equation $\int_0^\infty G(t)dt = \eta$. Substituting Eq. 8, we find k = 1 + b. This leads to the formula:

$$G(t) = G_g\left[1 + \left(\frac{G_g t}{\eta}\right)^{b}\right]^{-(1 + 1/b)} \quad (9)$$

where $b = (1/\beta + \ln(\pi)/\ln(2) - 2)^{-1}$.

One can see that the important rheological properties of bitumen can be estimated from the proposed equations using its simplest standard properties: penetration and softening temperature. The uniaxial relaxation modulus of binder can be found as $E_b(t) = 3G(t)$, where its modulus $G(t)$ is given by Eq. 9.

3 Stiffness and Tensile Strength of Asphalt Concrete

To analyze the temperature induced stresses, we need the uniaxial relaxation modulus of asphalt concrete $E_{mix}(t)$. Christensen and Bonaquist (2015) recently improved their model:

$$\begin{aligned} E_{mix}(t) &= P_c\left[E_{agg}\cdot(1-VMA)+E_b(t)\cdot VFA\cdot VMA\right] \\ P_c &= 0.006+0.994[1+\exp(-(0.663+0.586\cdot\ln(VFA\cdot E_b(t)/3) \\ &\quad -12.9VMA-0.171\cdot\ln(\varepsilon_s\cdot 10^6)]^{-1} \end{aligned} \tag{10}$$

where $E_b(t)$ is modulus of binder [MPa], E_{agg} is modulus of aggregate, VMA is voids in mineral aggregate [volume fraction], VFA is voids filled with asphalt, $\varepsilon_s = 0.0001$ is the standard target strain. The same model can be used for the mix stiffness $S_{mix}(t)$ if we replace $E_b(t)$ by stiffness of binder $S(t)$ according to Eq. 1.

To estimate a critical temperature, we need the tensile strength of asphalt concrete f_t as a function of T and t. We combined the Heukelom's experimental data for relative strength (Heukelom 1966) with empirical equation for the maximal strength of mix (Molenaar and Li 2014) and arrived to the following expression for tensile strength:

$$f_t = 0.505\cdot S_{mix}^{0.308}\cdot VFA^{0.849}\cdot\frac{0.774+0.039r+0.141r^{4.547}}{1+0.026r^{3.608}\cdot\exp(1.245r)} \tag{11}$$

where $r = \log(S_g/S)$, S_{mix} is mix stiffness at T = 20 °C and t = 0.08 s.

Example Asphalt mix with granite aggregate (specific gravity $G_{sb} = 2.670$) and 4.8 % bitumen of grade BND 90/130 (by weight of aggregate) from Pavlodar plant in Kazakhstan has VMA = 0.144, and VFA = 0.75. After short-time aging, the bitumen penetration is 70 dmm (at 25 °C), softening temperature T_s = 48 °C, PI = −0.91. Estimate the tensile strength of mix at T = 0 °C and t = 40 s. Using correlation from (Christensen and Bonaquist 2015), $E_{agg} = 7650\ G_{sb}^{1.59} = 36000$ MPa. Stiffness of binder at T = 20 °C and t = 0.08 s according to Eq. 1 is S = 8.163 MPa. Stiffness of mix at T = 20 °C and t = 0.08 s from Eq. 10 equals to $S_{mix} = 5512$ MPa. In Eq. 11 $r = \log(2460/S(T = 0°C,\ t = 40s)) = 2.666$. From Eq. 11, the tensile strength $f_t = 2.85$ MPa. Uniaxial tensile tests of that mixture were performed on testing system TRAVIS (InfraTest GmbH). Specimens of size 40 × 40 × 160 mm were tested with deformation rate 1 mm/min. Tests were performed at temperatures 20, 10, 0, −10, −20, and −30 °C. Test results are given in Fig. 1 together with tensile strength calculated from Eq. 11 for t = 40 s and they are all fairly close to each other.

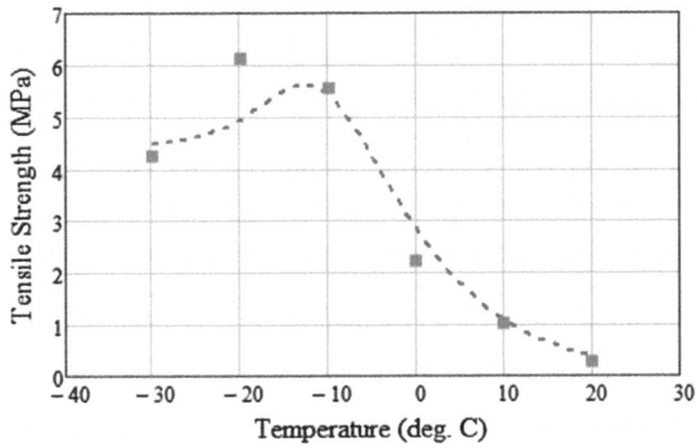

Fig. 1 Measured strength values and those predicted from the material properties by Eq. 11

4 Thermal Stresses and Critical Temperature

Temperature induced stresses in asphalt pavement were calculated as:

$$\sigma(t) = -\int_{0}^{t} E_{mix}(\xi(t) - \xi(\tau))\left(\frac{d}{d\tau}\alpha_{mix}(T(\tau))T(\tau)\right)d\tau \qquad (12)$$

where t is present time [s], τ is passed time [s], $T(\tau)$ is the temperature variation with time [°C], E_{mix} is relaxation modulus [MPa], $\xi(t)$ is reduced time. Coefficient of thermal contraction was assumed constant $\alpha_{mix} = 2.510^{-5} 1/°C$. The properties of binder and mix were taken as before: Ts = 48 °C, PI = −0.91, $E_{agg} = 36000$ MPa, VMA = 0.144, VFA = 0.75. In the first example (Fig. 2), the temperature drops from 5 to −35 °C and returns to 5 °C over 24-h period. In the second example, the cooling rate was 2 °C/h. Strength of mix at t = 1 h is shown on the same graph. The critical temperature is −31.5 °C for the first example and—33 °C for the second one.

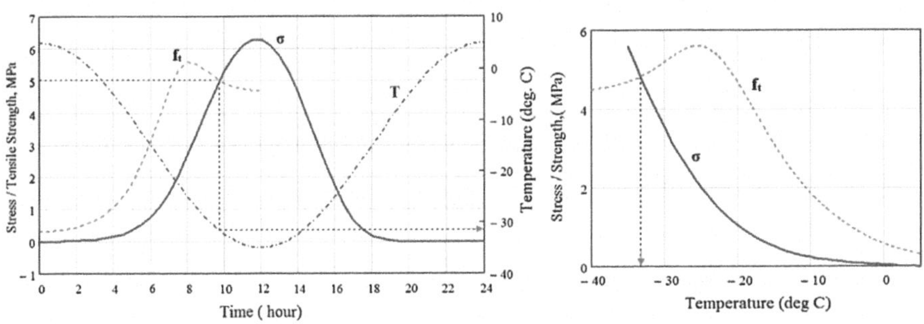

Fig. 2 Thermal stress and estimated fracture temperature: *solid line*—stress σ in asphalt concrete, *dashed line*—its 1 h tensile strength f_t

Probably, the transverse cracking cannot be prevented under the low temperature in Kazakhstan but it can be delayed and its frequency can be reduced. Ability to predict low temperature stresses in asphalt pavement and its strength starting from the properties of binder and mix should be helpful. Damage accumulation induced in asphalt pavement by the subsequent cooling cycles can be also predicted using the proposed formulas. Recently, the modified binders with Elvaloy, Butonal, Calprene, Kraton, and DST were tested, and Elvaloy and Butonal were recommended for road construction. As an alternative solution, softer asphalt grades can be used for low-volume roads in Kazakhstan.

5 Conclusions

A low temperature behavior of asphalt pavement was considered. Rheological properties of plain bitumen such as stiffness, relaxation modulus and complex modulus can be predicted from the proposed formulas. The tensile strength of asphalt concrete has been related with the properties of binder and mix composition. This enables to analyze the thermal stresses in asphalt pavement and the critical temperature. The proposed formulas can be used for mixes with modified binder as well. In that case, the binder stiffness should be determined experimentally.

References

Bouldin M, Dongre R, Rowe G, Sharrock MJ, Anderson D (2000) Predicting thermal cracking of pavements from binder properties - theoretical basis and field validation. J Asph Tech 69:455-488

Christensen DW, Anderson DA (1992) Interpretation of dynamic mechanical test data for paving grade asphalt cements. J Asph Tech 61:67-116

Christensen DW, Bonaquist R (2015) Improved Hirsch model for estimating the modulus of hot mix asphalt, Proceedings of the Association of Asphalt Paving Technologists, Meeting Pre-prints

Christison JT, Murray DW, Anderson KO (1972) Stress prediction and low temperature fracture susceptibility of asphaltic concrete pavements. J Asph Tech 41:494-523

Heukelom W (1966) Observations on the rheology and fracture of bitumens and asphalt mixes. J Asph Tech 35:358– 399

Marasteanu MO, Anderson DA (1999) Improved model for bitumen rheological characterization. Paper presented at Eurobitume Workshop on Performance Related Properties for Bituminous Binders, Luxembourg, May 1999

Molenaar AAA, Li N (2014) Prediction of compressive and tensile strength of asphalt concrete. Int J Pave Res Tech 7:324-331

Monismith CL, Secor GA, Secor KE (1965) Temperature induced stresses and deformations in asphalt concrete. J Asph Tech 34:248-285

Van der Poel C (1954) A general system describing the visco-elastic properties of bitumens and its relation to routine test data. J Appl Chem 4:221-236

Research of Cracks Formation at Transport Facilities of Azerbaijan

Valerii Vyrozhemskyi, Iryna Voloshyna, Kanan Aliyev and Viktor Shumchyk

Abstract The results of asphalt pavement investigation regarding cracks formation on the road in the vicinity of the settlement Gabala and at the ramps and runway of the airport Gabala (Azerbaijan) are presented. Following the research of composition and properties of asphalt mixes, the impact of source materials on the mixes was determined. Mineral component of asphalt concrete is represented mainly by pebbles from mountain streams. Heterogeneity of mineral composition of pebbles causes heterogeneity of their adhesion to bitumen. The properties of original bitumen manufactured in Azerbaijan and of polymer modified bitumen were studied. The effect of viscosity of original bitumen, its susceptibility to cracks formation was studied. High viscosity of modified binder was noted that significantly affects the process of cracks formation in asphalt pavements. Type content before and after aging of original bitumen was studied. Significant decrease in aromatic hydrocarbons in the aged samples and high paraffin content was registered. Parameters that can impair crack resistance of pavement were defined. The properties of asphalt cores taken on the road in specific areas of cracking were studied. The composition of asphalt mixes placed in pavement, debonding in the materials of asphalt pavement, runway and ramps of the airport Gabala were studied. Difference in the nature and in the shape of cracks at these sites was noted. Basing on the research results, the proposals on preventing further cracking on the

V. Vyrozhemskyi (✉) · I. Voloshyna
M.P. Shulgin State Road Research Institute—DerzhdorNDI SE (DNDI), Kiev, Ukraine
e-mail: vv@dorndi.org.ua

I. Voloshyna
e-mail: iv@dorndi.org.ua

K. Aliyev
AzVirt JSC, Baku, Azerbaijan
e-mail: kaliyev@azvirt.com

V. Shumchyk
Belarusian Road Research Institute (BeldorNII), Minsk, Belarus
e-mail: vkshum@tut.by

A. Chabot et al. (eds.), *8th RILEM International Conference on Mechanisms of Cracking and Debonding in Pavements*, RILEM Bookseries 13,
DOI 10.1007/978-94-024-0867-6_21

pavements that will be arranged and the countermeasures to fight cracks in pavements that are in operation were developed.

Keywords Cracks formation · Type content · Aging of bitumen

1 Introduction

Quality level of asphalt pavement is a quantification relative characteristic based on a comparison of aggregate indicators of its quality with the corresponding set of basic indicators. Given level of pavement quality is achieved by production control and is divided into three categories: initial, operational and acceptance level. The objective of the study is to assess the quality of obtained samples of binders and determination of their suitability for the preparation of polymer asphalt that prevent cracks formation on pavement.

2 Field Research

A survey of asphalt pavement was conducted on the road Gabala—airport Gabala and runway of the airport Gabala in 2014. The material of pavements arranged in 2013 corresponds to fine grain dense asphalt type.

The pavement's state is satisfactory except for the following defects:

- Cracking—transversal cracks occurred on the surface of asphalt pavement with a length from 2 to 10 cm, depth from 1 to 1.5 cm and spaced from each other by 2–7 cm (Fig. 1a);
- Segregation of mix—segregation of coarse and fine grains of the mixture's aggregate (Fig. 1b). Segregation may occur at various stages of preparation, transportation and laying the mixture. It can be observed during reloading of mix from asphalt plant into the storage tank; when loading the truck from the tank; when unloading mixture into receiving bin of the paver.

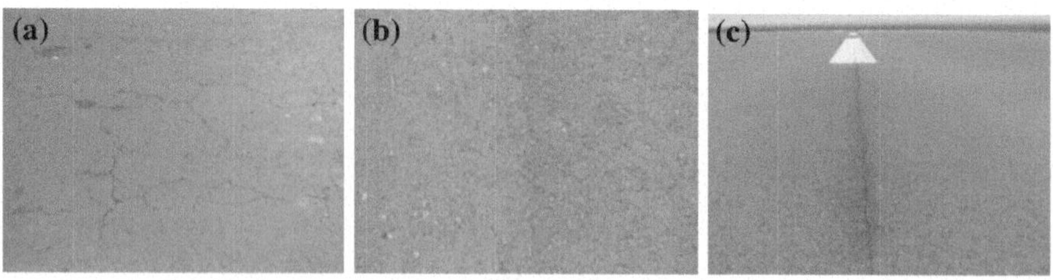

Fig. 1 Photo of defects: **a** Cracking of pavement, **b** Mix segregation, **c** Low quality transversal joints

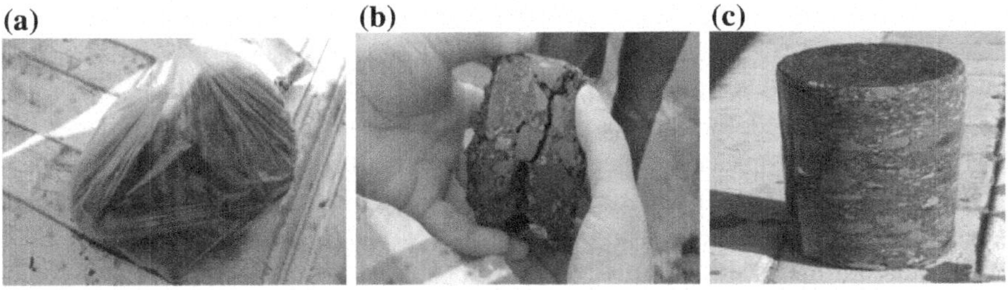

Fig. 2 Photo of cores sample: **a** Disintegrated, **b** Extraction, **c** Damage absence

- Pavement unevenness;
- Low quality transversal joints on the runway of the airport Gabala (Fig. 1c)—formation of a mound and (or) a pothole on the surface of the joint due to incorrect transversal joining of the lanes formed by paver at asphalt laying.

It should be noted that all photos of defects presented in the article were made during the examination of the highway Gabala—Airport and a runway of the airport Gabala.

According to the results of visual inspection, specific locations were chosen and the cores were sampled: in the areas with the net of cracks, in the areas with single cracks and in road surface areas without damage. Depending on the damage, the cores of a different state were sampled. The cores sampled from the area with the net of cracks on the surface were disintegrated (Fig. 2a). In case of single cracks the samples were segregated at the extraction (Figure 2b). Monolithic cores were sampled in case of damage absence (Fig. 2c).

Due to the fact that according to the authors' view, a fundamental factor causing the destruction of asphalt pavement is the bitumen quality, the focus was made on the research of binders based on the Azerbaijani bitumen.

3 Laboratory Research of Bitumen

Laboratory studies of original and modified bitumen samples produced by AzVirt Company were conducted.

Elvaloy AM terpolymer of Azerbaijani production in the amount of from 1.25 to 1.4 % of the binder by weight was used for bitumen modification.

Original bitumen has a high viscosity. Its penetration at 25 °C is 57 × 0.1 mm that allows referring the sample to BND 40/60 grade according to the standards classification of Ukraine and Azerbaijan. According to the European standards this bitumen refers to the grade 50/70.

Original bitumen has a softening point of 51 °C indicating insufficiently high thermal stability of binder (the lower boundary of the requirements of Ukrainian State Standard 4044). Tensile strength of bitumen is 91 cm testifying to a certain cohesive strength. Despite high viscosity, bitumen has sufficiently low brittleness temperature (−24 °C). At the same time, very low tensile strength viscosity for a given viscosity at 0 °C (although it is not specified for the BND 40/60 grade) indicates about low deformability of bitumen at low temperatures. The binder has poor adhesion with macadam (pebbles) but it is relatively high with glass (45 %).

Bitumen has a high dynamic viscosity that indicates to its enhanced shear strength and rutting. The research of bitumen aging in accordance with GOST 18180 and European RTFOT (EN 12607-1) has shown that for such viscosity bitumen is quite intensively ageing (although by residual penetration and softening point changes values it is within the requirements of the standards).

Tested sample of modified bitumen has a penetration of 49 × 0.1 mm, i.e. it refers to the highest viscosity brand according to Ukrainian and Azerbaijani standards. When modifying, the softening point of bitumen raised to 60 °C. However, for such viscosity and polymer type it is not high. Introduction of polymer results in reduced tensile strength of binder at 25 °C to 17 cm. However, this does not indicate deterioration in quality but it shows higher structuredness of bitumen.

Brittleness temperature remains stable. Modification using terpolymer leads to increasing tensile strength at 0 °C (usually in several times) testifying to the improvement of low temperature crack resistance of bitumen. In this case it is not observed: original and modified bitumen have tensile strength at 0 °C of 1 cm.

A sample of modified bitumen has uncharacteristically low (in case of Elvaloy AM) elasticity (64 %). Typically, it ranges from 75 to 85 %.

Bitumen with Elvaloy AM is characterized by high dynamic viscosity. A similar sample has a tendency to increased aging. Changes in softening point and residual penetration are close to the limits of the standards. Tensile strength is reduced dramatically after heating.

The research results of the type content of bitumen by method of thin layer chromatography using the device IATROSCAN MK 6S (Fig. 3) are shown in Table 1.

Insignificant amount of oils and the content of resins and saturated hydrocarbons are contained in the studied bitumen. It should be emphasized that the raw materials

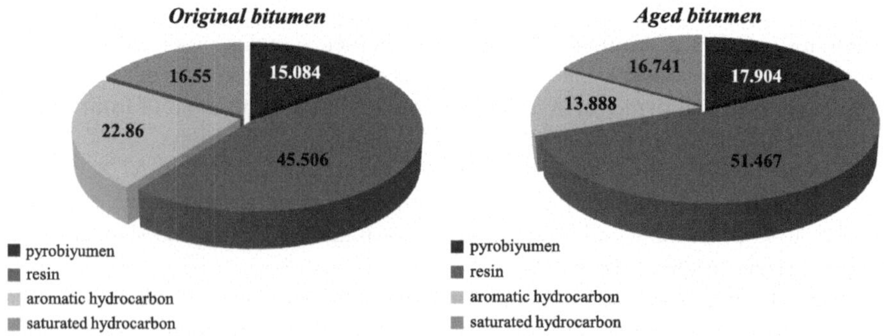

Fig. 3 Type content of bitumen on the device IATROSCAN MK 6S

Table 1 Bitumen type content by weight fraction %

Bitumen type content	Chromatographic column		IATROSCAN MK 6S	
	Original	Aged	Original	Aged
Pyrobitumen	15.2	16.9	15.084	17.904
Resins	40.4	43.6	45.506	51.467
Aromatic hydrocarbons	23.8	15.2	22.860	13.888
Saturated hydrocarbons	20.6	24.3	16.550	16.741

Note Paraffins > 4 %

contain paraffin and high content of resins of II structure type that requires mandatory modification by thermoelastolayers.

Bitumen sample modified by Elvaloy AM has a very high viscosity. For a given viscosity of the binder and considering rather sufficient efficacy of Elvaloy AM, modified bitumen has low softening point and much lower elasticity values.

The modification led to a slump in tensile strength at 25 °C and did not lead to increasing in tensile strength at 0 °C, which is a characteristic of using thermoelastolayers, latex and terpolymers (to which Elvaloy AM refers). Tensile strength of modified bitumen both at 25 and 0 °C does not meet the requirements of Ukrainian and Azerbaijani standards. Increase in dynamic viscosity, as well as slight increase in temperature or stability of brittleness temperature is typical when using Elvaloy AM. A feature of modified bitumen is a substantial change in the properties after heating Radovskyi and Teltayev (2013).

The main reasons for such properties of modified bitumen are high viscosity of original bitumen, and its chemical (type) content which does not provide an effective combination with the given modifier.

Typically high viscosity of modified bitumen sample requires high temperatures of preparation and laying of asphalt mixes thereby promoting further aging of bitumen, the loss of its binding properties and, as a consequence, leads to a rapid deterioration and cracking of asphalt pavement.

4 Investigation of Asphalt Cores

Laboratory studies of samples (cores) were conducted which were selected from asphalt pavement on the road Gabala—Airport and physical and mechanical properties of the asphalt were defined.

These samples were divided into layers and tested. Firstly, average density of the samples was determined. Then the grading of the mineral part and the approximate amount of bitumen was determined.

Average density of the layers of destroyed asphalt samples was from 2.32 to 2.33 g/cm^3.

Average density of the upper and middle layers of asphalt of a monolithic sample was from 2.43 g/cm^3 and of the lower layer—2.42 g/cm^3.

Bitumen amount was from 4.5 to 4.8 % on the road section with the net of cracks. On the sections with single cracks the amount of bitumen was from 5.4 to 5.6 %. And on the sections without pavement damaging the amount of bitumen in the upper layer was from 6.2 to 6.8 %, and in the middle layer—from 5.8 to 6.2 %; in the lower layer—from 5.7 to 6.3 %.

After water flushing of damaged parts of samples, over 50 % of the grains that were not treated with binder were revealed.

In addition, quality of adhesion of bituminous binder to the surface of crushed rock taken from selected sample cores was conducted. The studies were carried involving crushed rock that had been in the mix and were covered with bitumen. The test results showed a good adhesion. This confirms again that the bitumen used for mixture was highly viscous and had the so-called "pseudo adhesion".

The research results of the cores samples that have been selected from asphalt pavement on the road Gabala—Airport have shown that:

- The grading of the upper layer of asphalt corresponds to a fine grain type, while the middle and the lower to porous fine grain asphalt.
- After water flushing of the destroyed part of the samples, more than 50 % of the grains were revealed that were not treated with the binder. Maybe it happened due to the violations of the process technological parameters of asphalt mix preparation or due to low quality materials.

5 Suggestions to Avoid Cracks Formation and Propagation on Asphalt Pavement on the Azerbaijani Studied Facilities

Ways of solving the issue may include the following:

- The use of less viscous original bitumen with penetration at 25 °C from (80 to 90) × 0.1 mm;
- Increasing the content of Elvaloy AM to (1.8–2.0) % by weight of the binder in case of using less viscous bitumen;
- The use of other types of polymers: thermoelastolayer or latex;
- The use of adhesive additives jointly with polymers.

To prevent the propagation of cracks it is necessary to apply protective and recovering compositions Kischynskyi et al. (2012). The use of these materials is a new approach to the recovery of the properties of aged asphalt.

The main purpose of protective impregnating layers is to protect the asphalt pavement from the penetration of external moisture, i.e. pavement waterproofing.

The materials for the arrangement of protective layers also prevent aging of asphalt and the occurrence of erosion.

Materials for protective layers provide:

1. Reduction of water saturation of asphalt.
2. Reduction of the penetrating power.
3. Slowing down the aging process of bitumen during operation of pavement.

References

Kischynskyi SV, Kyrychenko LF, Kopynets IV, Goncharenko YF (2012) Repair of fractured asphalt pavements using membrane technologies. Scientific Production Journal “Road-building materials” № 2, p. 30-34

Radovskyi BS, Teltayev BB (2013) Viscous elastic characteristics of bitumen and their assessment according to standard indicators. Almaty. “Bilim”, p. 152

Part III
Cracking in Asphalt Materials: Material Crack Propagation

Crack Evolution of Asphalt Mixtures Under Compressive Monotonic and Repeated Loads

Yuqing Zhang, Bjorn Birgisson, Fan Gu and Robert L. Lytton

Abstract Field observations and mechanical analyses have shown that cracks accompany rutting in asphalt mixtures under external compressive loads. This study aims to model crack growth in asphalt mixtures under compressive monotonic and repeated loads. With hypothesizing energy equilibrium and viscoelastic Griffith fracture criterion, a damage density characterizing the cracking in a mixture is derived as a function of stress, nonlinear viscofracture strain, asphalt film thickness and bond energy. Crack evolution is modelled by pseudo J-integral Paris' law. Six types of asphalt mixture were tested by monotonic compressive strength tests at 40 °C. Two were further tested at 5 temperatures and 5 loading rates, respectively. Repeated load test results for the same mixtures were obtained in previous studies. It is found that the damage density shows an S-shape curve under a strain-rate controlled monotonic load and an exponential curve under a stress controlled repeated load. Pseudo J-integral Paris' law can capture the overstress softening behavior and model the crack growth in mixtures under a monotonic load. The Paris' law coefficients (*A* and *n*) are independent of loading mode (monotonic or repeated), rate or temperature. Thus they are fundamental material properties and can be used to predict crack growth under varying loading and temperature conditions.

Keywords Asphalt mixture · Crack evolution · Damage density · Paris' law

Y. Zhang (✉) · B. Birgisson
School of Engineering and Applied Science, Aston University,
Aston Triangle, Birmingham B4 7ET, UK
e-mail: y.zhang10@aston.ac.uk

B. Birgisson
e-mail: bjorn.birgisson@aston.ac.uk

F. Gu · R.L. Lytton
Zachry Department of Civil Engineering, Texas A&M University,
3136 TAMU, College Station, Texas 77843, USA
e-mail: tracygufan@tamu.edu

R.L. Lytton
e-mail: r-lytton@civil.tamu.edu

A. Chabot et al. (eds.), *8th RILEM International Conference on Mechanisms of Cracking and Debonding in Pavements*, RILEM Bookseries 13,
DOI 10.1007/978-94-024-0867-6_22

1 Introduction

Field and laboratory observations have shown that cracks accompany rutting in asphalt layers when subjected to external compressive loads (Wang et al. 2003; Underwood et al. 2011). Mechanical analyses demonstrate that the overstress softening behavior under a monotonic compressive load and the tertiary flow of rutting under a repeated compressive load result from cracking alongside plastic deformation (Zhang et al. 2012, 2013; Ramsamooj and Ramadan 1999). Crack initiation criteria were proposed using viscoelastic Griffith theories in compression and in tension, respectively (Zhang et al. 2014a, b; Luo et al. 2015). It was found that cracks start to grow from the peak stress under a monotonic loads and from the flow number under a repeated load. Bond energy and tensile/compressive strength of asphalt mixtures were predicted using the crack initiation criteria.

Crack evolution in the tertiary stage of rutting under the repeated load has been studied in previous studies (Zhang et al. 2013, 2014a, has addressed the temperature). A damage density (ξ in Eq. 1) was employed to quantify the damage caused by cracks and air voids based on continuum damage mechanics (CDM) (Kachanov 1986; Lemaitre and Desmorat 2005). A pseudo-J integral Paris' law (Eq. 1) was used to predict the crack growth in mixtures based on Schapery's viscoelastic damage theory (Schapery 1984). The Paris' law coefficients (A and n in Eq. 1) were obtained from repeated tests for different mixtures. However, the repeated load tests are time-consuming and the analyses are complicated to obtain the coefficients. This study asks if the Paris' law can characterize the overstress softening behavior under a monotonic compressive load, and if the Paris' law coefficients (A and n) are fundamental material properties (independent of loading mode, rate, and temperature) so that they can be used to predict the crack growth in different loading and environmental conditions (e.g., using A and n obtained from monotonic tests at one temperature to predict crack growth at a different loading and/or temperature).

2 Crack Evolution Models for Asphalt Mixtures

The pseudo J-integral Paris' law based on damage density is used to characterize the crack evolution of asphalt mixtures and defined in Eq. 1:

$$\frac{d\xi}{dN} \quad or \quad \frac{d\xi}{dT} = A(\Delta J_R)^n \;\; where: \xi = \frac{A_L}{A_0} = \frac{m\pi c^2}{A_0} = 1 - \frac{\sigma_A}{\sigma_T} \tag{1}$$

where ξ = damage density; $A_L = m\pi c^2$ is lost area due to cracks in a cross section, and m = number of cracks, c = mean crack radius; A_0 = total area of the cross section. A and n = Paris' law coefficients; N = number of load cycles; T = loading time; ΔJ_R = pseudo J-integral; σ^A = apparent (measured) stress acting on A_0 and σ^T = effective (true) stress acting on intact material area ($A_0 - A_L$).

Principles of strain decomposition, redistribution and equilibrium of stored energy are hypothesized to determine ξ and ΔJ_R so that A and n can be computed from a monotonic compressive test. The strain decomposition was developed by the authors (Zhang et al. 2012) based on extended elastic-viscoelastic correspondence principle (Schapery 1984). It has been implemented for repeated load tests and monotonic load tests (Zhang et al. 2014a, b). As shown in Fig. 1, using a pseudo-strain (ε^R), the total strain (ε) is decomposed into viscous strain (ε^{vi}), viscoplastic strain (ε^{vp}), viscofracture strain (ε^{vf}) and nonlinear elastic strain (ε^{ne}). It is demonstrated that $\varepsilon^R = \varepsilon - \varepsilon^{vi} = \varepsilon^{vp} + \varepsilon^{vf} + \varepsilon^{ne}$ and the $\sigma \sim \varepsilon^R$ relation is a linear line before yielding with a slope of Young's modulus (E_0). A viscoplastic yielding analysis indicates that, under a monotonic load, the dissipated pseudo strain energy (DPSE) is used for permanent deformation before the peak stress, whereas for crack growth after the peak. Thus if a nonlinear viscofracture strain is defined as $\varepsilon' = \varepsilon^R - \varepsilon^{vp} = \varepsilon^{vf} + \varepsilon^{ne}$, the overstress softening is represented as a nonlinear elastic fracture process. In a $\sigma \sim \varepsilon'$ coordinate, E' and E_r are defined as secant modulus and recovery modulus, shown in Fig. 1. Note that $E_r < E_0$ due to the softening effect caused by cracks.

The redistribution principle of the stored energy is formulated in $\sigma \sim \varepsilon'$ based on Griffith fracture theory and shown in Eq. 2. It indicates that the total stored energy for a cracked material = the total stored energy for the un-cracked material − released energy due to crack growth + stored energy on the crack surfaces.

$$\Pi = \frac{\sigma_A^2}{2E'} V_0 = \frac{\sigma_A^2}{2E_0} V_0 - \frac{\sigma_A^2}{2E_0} \cdot V_R \cdot 2m + \Delta G \cdot 2m \cdot S_w \tag{2}$$

where Π = total stored energy of a representative volume element (RVE). $V_0 = A_0 t$ = volume of RVE; t = thickness of RVE that can be interpreted as asphalt film thickness. V_R = volume of the material that releases energy; ΔG = bond energy of an asphalt mixture that is the work of adhesion or cohesion per unit of crack surface area. S_w = crack surface area. Under a compressive load, new wing cracks are

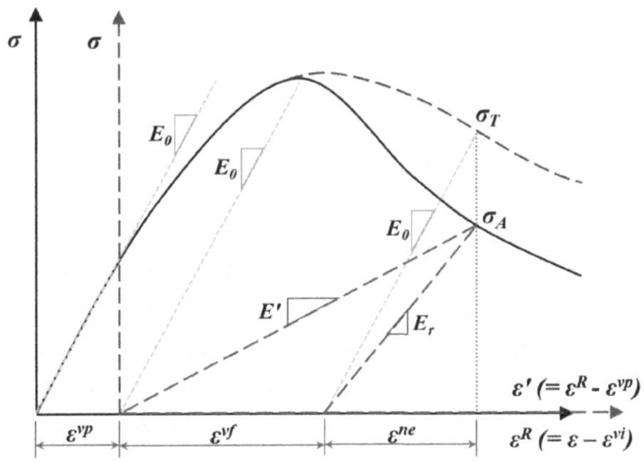

Fig. 1 Stress versus pseudo-strain relation

generated along the direction of external load and it was obtained that $V_R = \pi c^3/12$ and $S_w = 7\pi^2 c^2/48$. Applying Griffith criterion, i.e., $\partial\Pi/\partial c = 0$, to Eq. 2 obtains:

$$c = \frac{7\pi E_0 \cdot \Delta G}{3\sigma_A^2} \tag{3}$$

Substituting $\sigma_A = E'\varepsilon'$ and Eq. 3 in Eq. 2 yields the damage density as:

$$\xi = \frac{72t}{7\pi \cdot \Delta G}\left(\frac{1}{2}\sigma_A \varepsilon' - \frac{\sigma_A^2}{2E_0}\right) \tag{4}$$

The balance principle is originated from CDM theory and hypothesizes that the stored energy in a damaged (apparent) material equals to an idealized undamaged (true) configuration ($\sigma_A^2/2E' = \sigma_T^2/2E_0$). No crack exists in the true configuration, thus the material behaves elastically leading to an elastic unloading with a slope of E_0 and an elastic strain, as shown in Fig. 1. Thus, $\varepsilon^{ne} = \sigma_T/E_0 = \sigma_A/E_r$. Therefore,

$$E' = E_0(1-\xi)^2 \quad and \quad E_r = E_0(1-\xi) \tag{5}$$

Pseudo J-integral is equivalent to the dissipated work per unit crack surface created. Using this definition and Eq. 5 in Eq. 1 yields the crack evolution for a laboratory asphalt mixture sample under a monotonic compressive load:

$$\frac{d\xi}{dT} = A\left[\frac{\partial\left(DPSE^{vf} \cdot V\right)}{\partial(2m\pi c^2)}\right]^n \approx A\left[\frac{H}{2}\frac{\partial}{\partial\xi}\left(\frac{\sigma_A^2}{E'} - \frac{\sigma_A^2}{2E_r}\right)\right]^n = A\left[\frac{H}{2}\frac{\sigma_A^2}{2E_0}\frac{(3+\xi)}{(1-\xi)^3}\right]^n \tag{6}$$

where $DPSE^{vf}$ = dissipated pseudo-strain energy density for crack growth; $V = A_0 H$ is volume of the lab sample and H = sample height. Note that H is used in Eq. 6 rather than t (Eq. 4) as the crack model (Eq. 1) is now applied to a real lab sample and ξ quantifies the overall damage caused by all cracks in the sample. Note that Eq. 6 can be extended to multi-axle loads by replacing $\sigma_A^2/2E_0$ with $(\sigma_{11}^2 + \sigma_{22}^2 + \sigma_{33}^2)/(2E_0) + 2(1 + \upsilon_0)(\sigma_{12}^2 + \sigma_{23}^2 + \sigma_{13}^2)/E_0$, where υ_0 is an elastic Poisson's ratio.

3 Laboratory Tests and Results

Six types of asphalt mixture were fabricated using one binder (PG67-22) at two air void contents (4 and 7 %) and conditioned for three aging periods (unaged, 3-month and 6-month aged at 60 °C). Uniaxial compressive creep (UCC) tests were conducted for all samples at 40 kPa and 40 °C to obtain the viscoelastic properties used for strain decomposition. Uniaxial compressive strength (UCS) tests were performed for all samples at a constant strain rate of 311 με/s (a monotonic load) and at 40 °C. To evaluate temperature and rate effect, UCC and UCS tests were conducted on one mix (4 %-6 month) at 4 more temperatures (45, 50, 55, and 60 °C) and on another mix (7 %-6 month) at 4 more loading rates (18, 65, 622, and 1074 με/s). Note that the repeated load test results on the same types of mixtures were reported in Zhang et al. (2014a) and are employed here for comparison.

Figure 2 shows a typical curve of the damage density calculated by Eq. 4 using ΔG results from Zhang et al. (2014b) and assuming $t = 0.005 \sim 0.01$ mm. The damage density shows an S-shape curve in a strain rate controlled monotonic load test. Its slope (damage density rate) increases and then decreases with loading time, as shown in Fig. 3. In comparison, the damage density has an exponential shape with an increasing slope in stress controlled repeated load tests, as shown in Figs. 5 and 14 in Zhang et al. (2014a). Figure 2 also shows that σ_T (calculated by Eq. 1 using ξ) is greater than σ_A (measured) due to crack growth during the overstress softening process. Figure 3 shows that the mean crack radius (determined by Eq. 3) increases from 0.85 to 2.1 mm with loading time. The match of damage density rate in Fig. 3 between calculated values by Eq. 4 and predicted values from Eq. 6 proves that pseudo-J integral Paris' law can capture the overstress softening behavior and crack growth in asphalt mixtures under a monotonic load and can be used to obtain the Paris' law coefficients (A and n) that are shown in Fig. 4.

Figure 4 demonstrates that the Paris' law coefficients (A and n) for the six types of asphalt mixture determined from monotonic load tests are comparable to that from repeated load tests at the same temperatures. Note that A and n for repeated load tests were presented in a previous study (Zhang et al. 2014a). Figure 5 shows that the Paris' law coefficients of a specific asphalt mixture do not vary significantly with temperatures or loading rates.

One can conclude from Fig. 5 that Paris' law coefficients are fundamental material properties and independent of temperature, loading mode or rate. Note that this conclusion is valid only when pseudo-J integral (J_R) is used in Paris' law. It is

Fig. 2 σ_A, σ_T, ξ versus ε^R

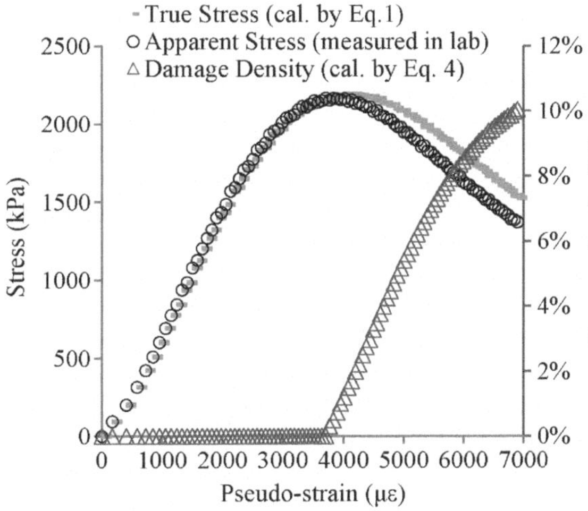

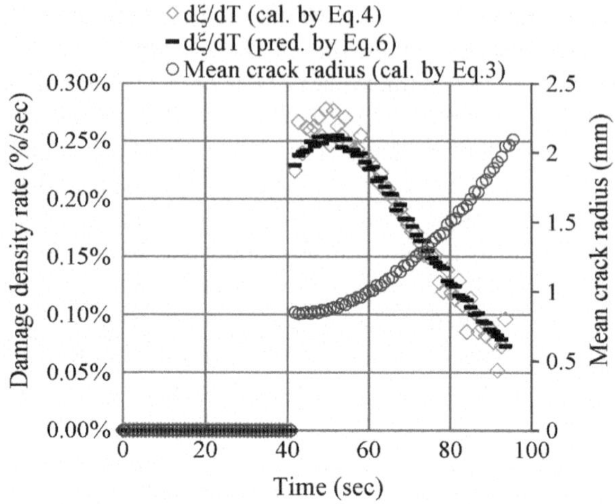

Fig. 3 *dξ/dT* and *c* versus loading time

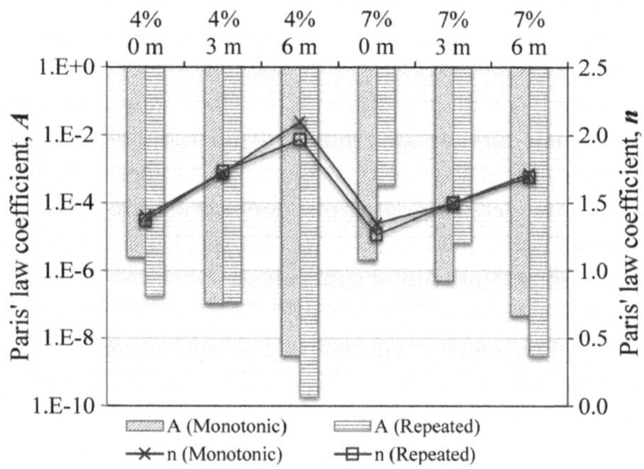

Fig. 4 Paris' law coefficients in monotonic/repeated loads

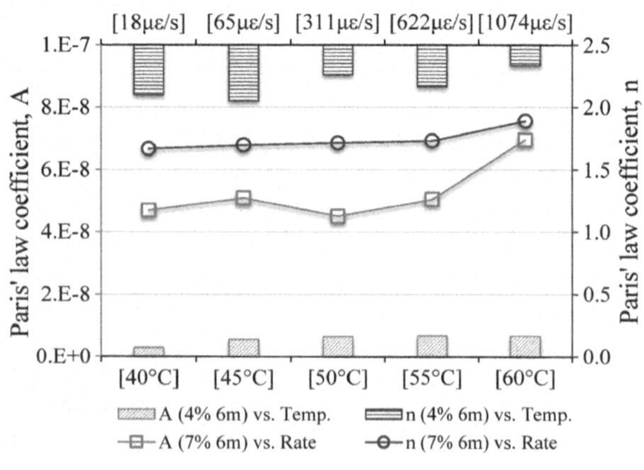

Fig. 5 Paris' law coefficients versus temperature/loading rate

because J_R ($J_R = \int_0^t D(t-s)\frac{dK^2}{ds}ds$) has addressed the temperature effect by creep compliance ($D(t)$) and accounted for loading (mode or rate) effects by stress intensity factor (K). Since repeated load tests are costly and time consuming, monotonic load tests (e.g., compressive strength tests) are recommended to determine the Paris' law coefficients for an asphalt mixture. Then the coefficients can be implemented in Paris' law using an accurate pseudo J-integral to predict the crack growth of the asphalt mixture under different loading and temperature conditions. However, it is noted that the Paris' law coefficients are different for asphalt mixtures in tension and in compression as cracks grow differently in the two cases. Nevertheless, the same methods can be used to model the crack growth in tension.

4 Conclusions

1. Damage density shows an S-shape curve under a strain-rate controlled mono-tonic load and an exponential curve under a stress controlled repeated load.
2. Pseudo J-integral based Paris' law can capture the overstress softening behavior and model the crack growth in asphalt mixtures under a monotonic load.
3. Paris' law coefficients (A and n) are fundamental material properties and independent of temperature, loading mode (monotonic or repeated) or loading rate.
4. Prediction of rutting and cracking in the field requires a coupled 3D modelling of viscoplasticity and viscofracture including their initiation and evolution in a mixture, which is being done and will be presented in a future study.

References

Kachanov LM. (1986) Introduction to continuum damage mechanics, New York, NY: Springer, LLC.

Lemaitre J and Desmorat R. (2005) Engineering damage mechanics: Ductile, creep, fatigue and brittle failures, New York, NY: Springer, LLC.

Luo X, Luo R and Lytton RL. (2015) Energy-based crack initiation criterion for viscoelastoplastic materials with distributed cracks. J of Eng Mech 141: in press. doi: 10.1061/(ASCE)EM.1943-7889.0000830

Ramsamooj DV and Ramadan J. (1999) Prediction of permanent deformation of asphalt concrete in cyclic and monotonic loading. J Test Eval 27: 320-326. doi: 10.1520/JTE12232J

Schapery RA. (1984) Correspondence principles and a generalized j-integral for large deformation and fracture analysis of viscoelastic media. Int J Fracture 25: 195-223. doi: 10.1007/BF01140837

Underwood B, Yun T and Kim Y. (2011) Experimental investigations of the viscoelastic and damage behaviors of hot-mix asphalt in compression. J Mater Civil Eng 23: 459-466. doi: 10.1061/(ASCE)MT.1943-5533.0000197

Wang LB, Myers LA, Mohammad LN, et al. (2003) Micromechanics study on top-down cracking. Transport Res Rec: J Transport Res Board: 121-133. doi: 10.3141/1853-14

Zhang Y, Luo R and Lytton RL. (2012) Characterizing permanent deformation and fracture of asphalt mixtures by using compressive dynamic modulus tests. J Mater Civil Eng 24: 898–906. doi: 10.1061/(ASCE)MT.1943-5533.0000471

Zhang Y, Luo R and Lytton RL. (2013) Mechanistic modeling of fracture in asphalt mixtures under compressive loading. J Mater Civil Eng 25: 1189-1197. doi: 10.1061/(asce)mt.1943-5533.0000667

Zhang Y, Luo R and Lytton RL. (2014a) Anisotropic characterization of crack growth in tertiary flow of asphalt mixtures in compression. J of Eng Mech 140: in press. doi: 10.1061/(ASCE)EM.1943-7889.0000745

Zhang Y, Luo X, Luo R, et al. (2014b) Crack initiation in asphalt mixtures under external compressive loads. Constr Build Mater 72: 94-103. doi: 10.1016/j.conbuildmat.2014.09.009

A Simple Fracture Model for Hot Mix Asphalt Based on Fundamental Fatigue Parameters

Taher M. Ahmed and Hussain Al-Khalid

Abstract This work reports the development of a simple mechanistic fracture model based on a modified Paris' law using the *J*-integral. An internal damage parameter, namely the fracture index (FI_c), was calculated from this model and was used as an index for ranking the fracture performance of different hot mix asphalt materials tested in fatigue. The relaxation test coefficient (m) and the dissipated pseudo strain energy for the fine aggregates matrix (FAM) volume were used as fundamental parameters for deriving this model. The study revealed there is compatibility between the FI_c and phenomenological approach using the number of load cycles at failure. In this work, limestone and granite aggregates were used with two binder grades: 40/60 and 160/220 to prepare four mixtures with two different gradations: gap-graded hot rolled asphalt and continuously graded dense bitumen macadam. The study showed that limestone mixes perform better in fracture than their granite counterparts.

Keywords Paris' law · Fracture · Pseudo-strain and relaxation

1 Introduction

Crack growth due to fatigue brought on by repeated loading has been recognized as an important problem in asphalt pavements. Several models have used different approaches to describe the performance of asphalt concrete. Fracture mechanics is one of approaches used for this purpose, this approach is based on a well-known power function called Paris' law ($dc/dN = AK^n$) (Paris and Erdogan 1963), which

T.M. Ahmed (✉) · H. Al-Khalid
School of Engineering, University of Liverpool, Liverpool, UK
e-mail: taher@liverpool.ac.uk; alanitaher@yahoo.com

H. Al-Khalid
e-mail: khalid@liverpool.ac.uk

T.M. Ahmed
Engineering College, Anbar University Ramadi, Anbar, Iraq

A. Chabot et al. (eds.), *8th RILEM International Conference on Mechanisms of Cracking and Debonding in Pavements*, RILEM Bookseries 13,
DOI 10.1007/978-94-024-0867-6_23

correlates the crack growth per cycle (*dc*/*dN*) with the stress intensity factor (*K*). This model is an approach to investigate the fracture behaviour of brittle or quasi-brittle material within linear elastic fracture (Pirmohammad and Ayatollahi 2014). Based on Schapery's work, a modification was made using the J_R-integral instead of K for fracture analysis of viscoelastic materials "Eq. 1" (Schapery 1984). In literature, modified Paris' law "Eq. 1" was used to develop models. For example, Si et al. (2002) and Masad et al. (2008) derived models to evaluate the fracture performance of asphalt mixes.

$$\frac{dc}{dN} = AJ^n \tag{1}$$

These models were based on an assumption that the mean micro crack length is circular in shape (Si et al. 2002; Masad et al. 2008). This represents a weak point because the crack always creates new surfaces with increasing the crack mouth opening displacement. A concept is called damage density was introduced as method of evaluation in continuum damage mechanics based on effective stress (Darabi et al. 2011). This parameter was used to evaluate fracture damage as damage density in HMA the crack length or crack radius (Xue Luo et al. 2013). Bearing this in mind, an attempt has made to develop a simple fracture model using modified Paris' law and based on the fundamental parameters of fatigue testing, and the damage density concept away from crack radius and crack surface concepts.

2 Objective of Study

This study aims to develop a simple fracture model based on damage density using fatigue parameters represented by the DPSE and the relaxation coefficient (m).

3 Dissipated Energy Fracture Model Calculations

During repetitive loading tests, two kinds of energy are created: applied energy (W_{AE}) and recovered energy (W_{RE}). The difference between these energies indicates the damage in material, this difference in energy is called dissipated energy (Ghuzlan and Carpenter 2006). The formula in "Eq. 2" is used to calculate W_{AE} and W_{RE} in viscoelastic materials (Larson 1999); as shown in "Eqs. 3 and 4"; for more details see (Ahmed and Khalid 2015).

$$W = \int_{t_d}^{t_u} \tau(t) \frac{d\gamma^R(t)}{dt} dt \tag{2}$$

$$W_A^R = \frac{G_{lve}\tau_o\gamma_o}{G_R}\left[\begin{array}{c}\frac{\cos(\delta-\delta_{lve})}{2}\left(\sin^2\left(\frac{\pi+2\delta}{2}\right)+\sin^2\left(\frac{3\pi+2\delta}{2}\right)\right)\\ +\frac{\sin(\delta-\delta_{lve})}{4}(2\pi+4\delta-\sin(\pi+2\delta)-\sin(3\pi+2\delta))\end{array}\right] \quad (3)$$

$$W_R^R = \frac{G_{lve}\tau_o\gamma_o}{G_R}\left[\begin{array}{c}\frac{\sin(\delta-\delta_{lve})}{4}(2\pi+4\delta-\sin(\pi+2\delta)-\sin(3\pi+2\delta))\\ -\frac{\cos(\delta-\delta_{lve})}{2}\left(\sin^2\left(\frac{\pi}{2}+\delta\right)+\sin^2\left(\frac{3\pi+2\delta}{2}\right)\right)\end{array}\right] \quad (4)$$

where W is the DPSE; t_i and t_n are the integration intervals; γ^R is pseudo-strain; τ_o and γ_o are shear stress and strain amplitude respectively; G_{lve} and δ_{lve} are complex modulus and phase angle at linear viscoelastic; G_R is the reference modulus and δ_N is phase angle. Dissipated pseudo-strain energy (W_{DE}) was calculated using ($W_{DE} = W_{AE} - W_{RE}$). In this study, modified Paris' law and damage density (Ø) were used for simulating the internal damage as crack growth in materials. "Equation 5" accounts for the damage density in materials under repeated loading where the damage starts from $Ø_o$ initial damage represented as air voids and increases in rate ($\dot{Ø} = \frac{dø}{dN}$)

$$Ø = Ø_o\dot{Ø}N \Rightarrow \dot{Ø} = \frac{Ø - Ø_o}{N} \quad (5)$$

Damage accumulates as cracks propagation occurs within the matrix because it is weaker than aggregates and can be calculated with damage rate "Eq. 6".

$$D = Ø \times \alpha_v \Rightarrow \frac{dD}{dN} = \frac{dØ}{dN}\alpha_v \quad (6)$$

where D is accumulative damage in a material and α_v is the matrix volume of the sample. The pseudo J_R-integral in modified Paris' law represents the dissipated energy rate to damage rate in materials ($J_R = \frac{dW_D^R}{dN} / \frac{dD}{dN}$). The power law ($W_{DEC} = bN^c$) formula was used to describe the relationship between cumulative DPSE and number of cycles; and the derivative of this equation ($dW_{DEC}/dN = bcN^{c-1}$) represents the DPSE per cyclic load. Substituting "Eqs. 5 and 6" and J_R formula in derivative formula (dW_{DEC}/dN) and combining with the modified Paris' law yields:

$$\frac{Ø - Ø_o}{N} = \frac{A}{(\alpha_v)^n}\frac{(bc)^n N^{cn-n}}{\left(\frac{Ø-Ø_o}{N}\right)^n} \quad (7)$$

Simplifying and rewrite "Eq. 7" with making some changes yields "Eq. 8" which represents a fracture index (FI_c) as a final model.

Table 1 Mix ID, material details and test conditions (Ahmed and Khalid 2015)

Mix ID	Mixes properties				Test Conditions	
	Grade mm	Content %	G_{bulk} Mg/m^3	Air voids %	Strain (%)	Stress KPa
DBM-L	160/220	5.2	2.374	4.9	0.30	150
DBM-G	160/220	5.2	2.290	7.5	0.30	150
HRA-L	40/60	7.8	2.343	2.2	0.25	250
HRA-G	40/60	7.8	2.298	4.0	0.25	400

$$\frac{\O_N - \O_o}{A^{\frac{1}{n+1}}} = \left(\frac{bc}{\alpha_v}\right)^{\frac{n}{n+1}} (N_i)^{\frac{cn+1}{n+1}} \Rightarrow FI_C = \left(\frac{bc}{\alpha_v}\right)^{\frac{n}{n+1}} (N_i)^{\frac{cn+1}{n+1}} \tag{8}$$

where n are $1/m$ and $1 + 1/m$ for controlled strain and stress respectively; m is the exponent of time in the power law of the relaxation modulus (Lee and Kim 1998).

4 Materials and Experimental Works

In this work, two kinds of mixes: a hot rolled asphalt (HRA) and dense bitumen macadam (DBM) were prepared in the laboratory using two types of aggregate: limestone (L) and granite (G), with two binders: 40/60 and 160/220 pen grades. The experimental work included performing relaxation and fatigue tests using the DSR technique in stress and strain modes using an approach presented in a previous study (Ahmed and Khalid 2015). Mixes and test conditions are in Table 1.

5 Results and Discussion

5.1 Relaxation Test and Dissipated Pseudo-strain Energy

The relaxation modulus is the ratio of stress response to constant strain input. This test was carried out to determine the relaxation moduli as a function of time using power law formula ($G(t) = G_\infty + G_1 t^{-m}$) to determine the linear viscoelastic propeties: G_1 and m. Nine DSR-samples were selected to be used in relaxation test to be performed at low shear strain amplitude within the linear region, i.e. 0.002 % for 140 s. Relaxation moduli against time for a typical set of data with the regression coefficients of the relaxation test are demonstrated in Fig. 1. It is clear from the high R^2 values that there is a good correlation between the experimental and fitted data using power law. The same nine samples that were tested for relaxation were also used for fatigue testing according to the conditions in Table 1. Fatigue testing provides the necessary variable values, e.g. phase angle, shear stress and strain, to be used for calculating: W_{AE}, W_{RE} and W_{DE}. The cumulative DPSE

(a)

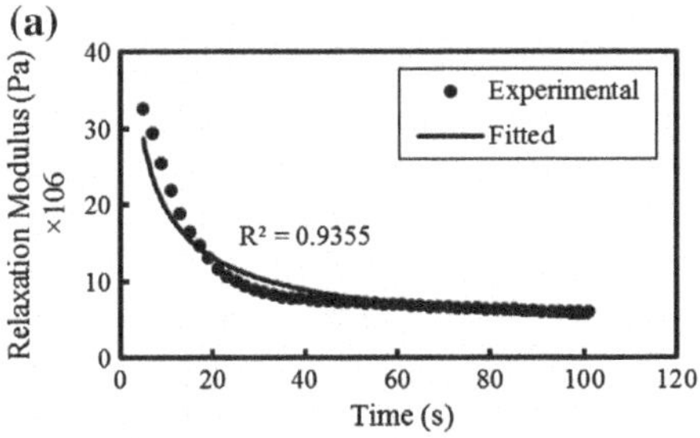

(b)

Mix ID	Relaxation coefficients		R^2
	G_1	m	
DBM-G	51235812	−0.513	0.935
DBM-L	31923580	−0.618	0.908
HRA-G	773280068	−0.797	0.976
HRA-L	327954621	−0.636	0.959

Fig. 1 **a** Typical relaxation modulus against time of DBM-G, **b** regression coefficients of mixes

(a)

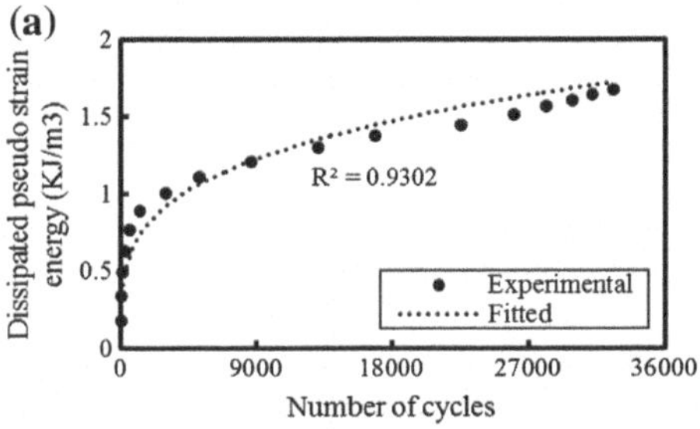

(b)

Mix ID	Controlled strain		
	b	c	R^2
DBM-G	0.0817	0.307	0.961
DBM-L	0.109	0.265	0.930
HRA-G	0.117	0.308	0.949
HRA-L	0.0791	0.288	0.976
Mix ID	**Controlled stress**		
	b	c	R^2
DBM-G	0.0125	0.453	0.968
DBM-L	0.0347	0.422	0.967
HRA-G	0.00924	0.414	0.989
HRA-L	0.00562	0.381	0.979

Fig. 2 **a** Typical DPSE against cycles for DBM-L, **b** Regression coefficients of all mixes

was divided by the matrix volume to calculate the DPSE per unit volume of the damaged part of the mix, because cracks occur in the matrix which is weaker than aggregates. The matrix represents the volume of fine aggregates passing through a 2.36 mm sieve, filler and binder; these components were calculated depending on the volumetric properties of HMA. Then, DPSE was fitted using power law formula. A typical relationship between cumulative DPSE and number of cycles is presented in Fig. 2 with coefficients of regression; a high correlation is confirmed by the high R^2.

5.2 *Fracture Damage*

In order to demonstrate the changes in FI_c at different normalised stiffness modulus, FI_c was calculated at a number of cycle equivalents to 50 and 15 % of the initial stiffness modulus in strain test mode and 50 and 10 % in stress test mode. On the other hand, the fatigue performance was evaluated as a number of cycles at the same normalised stiffness modulus to present a comparison. Figure 3a shows the FI_c results for the mixes tested in strain mode; FI_c increases as decreeing the normalised stiffness modulus for each mix; this indicates that the fracture damage

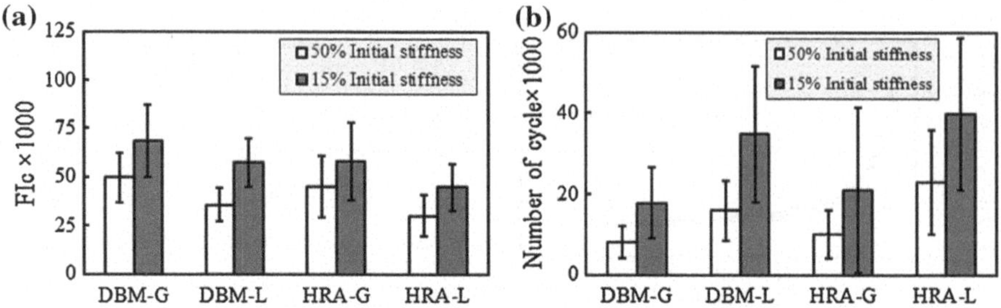

Fig. 3 Performance of mixes tested in strain mode and evaluated in: **a** FI_c at different normalised stiffness modulus, **b** number of cycles at different normalised stiffness modulus

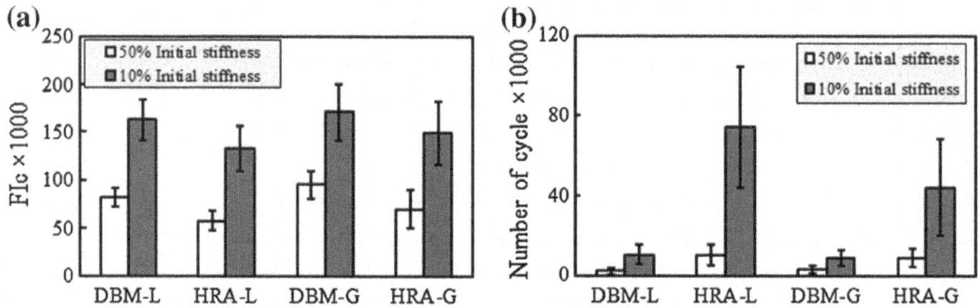

Fig. 4 Performance of mixes tested in stress mode and evaluated in: **a** FI_c at different normalised stiffness modulus, **b** number of cycles at different normalised stiffness modulus

increases as progressing of fatigue testing. It can be seen that an agreement is available in the ranking of order for the mixes between the FI_c and number of cycles as presented in Fig. 3, where the higher FI_c values display lower fatigue performance as a number of cycles. Figure 4 presents the outcomes in stress test mode; also, it is clear the trend of FI_c was as in the strain mode, where FI_c increases as decreasing normalised stiffness in all mixes. On the other hand, the compatibility between the FI_c and fatigue life as a number of cycles is existed; where high FI_c lower fatigue life as shown in Figs. 3 and 4. In both test modes, limestone mixes display better performance than granite mixes, and this is clear in both: FI_c and number of cycles; this is maybe due to that air voids in granite mixes is higher than limestone mixes as shown in Table 1. Remarkable, FI_c revealed lower variation than number of cycles as shown from the error bars, i.e. standard deviation, where there are significant differences between them. So, FI_c can be used in evaluating the fatigue performance based on fracture mechanics with reliability and lower variation as well.

6 Conclusions

Based on this work; a simple fracture model based on modified Paris' law was introduced and used successfully in evaluating the fracture performance of different asphalt mixes. This model was derived depending on the fundamental fatigue parameters represented by the relaxation coefficient (*m*) and the cumulative DPSE for the matrix. The results showed an agreement between the fracture analyses as FI_c and results presented by number of cycles in both modes; where FI_c showed the same behaviour as a number of cycles but at lower significant variation. Results showed that limestone mixes have better performance than granite mixes.

Acknowledgements The authors acknowledge to the Iraqi government for funding this work.

References

Ahmed, Taher M. and Khalid, Hussain A. 2015. A New Approach in Fatigue Testing and Evaluation of Hot Mix Asphalt Using a Dynamic Shear Rheometer. *6th International Conference Bituminous Mixtuers and Pavements*. Thessaloniki, Greece, 10-12 June 2015.

Darabi, M.K., Abu Al-Rub, R.K., Masad, E.A., Huang, C.W. and Little, D.N. 2011. A thermo-viscoelastic–viscoplastic–viscodamage constitutive model for asphaltic materials. *International Journal of Solids and Structures*, 48, 191-207.

Ghuzlan, K.A. and Carpenter, S.H. 2006. Fatigue damage analysis in asphalt concrete mixtures using the dissipated energy approach. *Canadian Journal of Civil Engineering*, 33, 890-901.

Larson, Ronald G. 1999. *The structure and rheology of complex fluids*, Oxford university press New York.

Lee, H.J. and Kim, Richared 1998. Viscoelastic constitutive model for asphalt concrete under cyclic loading. *Journal of Engineering Mechanics*, 124, 32-40.

Masad, E., Castelo Branco, VTF., Little, Dallas N. and Lytton, R. 2008. A unified method for the analysis of controlled-strain and controlled-stress fatigue testing. *International Journal of Pavement Engineering*, 9, 233-246.

Paris, PC and Erdogan, F 1963. A critical analysis of crack propagation laws. *Journal of Basic Engineering*, 85, 528.

Pirmohammad, S and Ayatollahi, MR 2014. Fracture resistance of asphalt concrete under different loading modes and temperature conditions. *Construction and Building Materials*, 53, 235-242.

Schapery, RA 1984. Correspondence principles and a generalizedJ integral for large deformation and fracture analysis of viscoelastic media. *International Journal of Fracture*, 25, 195-223.

Si, Z., Little, DN and Lytton, RL 2002. Characterization of microdamage and healing of asphalt concrete mixtures. *Journal of Materials in Civil Engineering*, 14, 461-470.

Xue Luo, Rong Luo and Lytton., Robert l. 2013. A modified Paris's law to predict entire crack growth in asphalt mixtures. *TRB 13-2800*.

Low Temperature Cracking Properties of Asphalt Mixtures Containing Rubber Bitumen Pellets

Ignacio Artamendi, Bob Allen and Paul Phillips

Abstract This paper evaluates the effect of ground tyre rubber (GTR) bitumen pellets on the low temperature cracking properties of asphalt mixtures. Low temperature properties were evaluated at 0 °C using monotonic and cyclic three-point bending tests. Flexural strength was determined using beam specimens and fracture toughness was determined using two specimen geometries: single edge notched beam and semi-circular bending. Cyclic bending tests under controlled load conditions were also carried out using notched beams to determine crack propagation parameters. Dense surface course mixtures were produced with a paving grade binder, two different polymer modified binders and two pellets concentrations. It was found that the flexural strength increased when GTR bitumen pellets and polymer modified binders were used. Differences in fracture toughness between the mixtures were, however, relatively small. Only one of the mixtures with a polymer modified binder showed a significant higher fracture toughness value. Furthermore, good correlation was found between the fracture toughness values determined from the two specimen geometries. Cyclic test, on the other hand, showed that the mixtures produced with the bitumen pellets and with the polymer modified binders were able to sustain more load applications than the unmodified mixture. Furthermore, crack propagation parameters were affected by both the type of binder and the concentration of pellets in the mixture.

Keywords Rubber-bitumen pellet · Flexural strength · Fracture toughness · Crack propagation

I. Artamendi (✉) · B. Allen · P. Phillips
R&D Department, Aggregate Industries, Coalville, UK
e-mail: ignacio.artamendi@aggregate.com

B. Allen
e-mail: bob.allen@aggregate.com

P. Phillips
e-mail: paul.phillips@aggregate.com

A. Chabot et al. (eds.), *8th RILEM International Conference on Mechanisms of Cracking and Debonding in Pavements*, RILEM Bookseries 13,
DOI 10.1007/978-94-024-0867-6_24

1 Introduction

The use of polymer modified binders (PMBs) in asphalt has increased steadily in Europe and it is now estimated at over 1 Mt a year (Eurobitume 2013). The benefits of polymer modification have been widely reported and include improved performance in terms of rutting, fatigue and thermal cracking. Elastomeric polymers are, however, expensive and are not always readily available. As a result the use of more economical recycled polymers like ground tyre rubber (GTR) has gained considerable attention particularly in Europe.

Ground tyre rubber (GTR) has been used extensively in road pavement for more than half a century. Different processes have been developed for incorporating tyre rubber into asphalt mixtures. These processes can be broadly classified as wet processes where the rubber particles are blended with bitumen at elevated temperatures to produce a rubber modified binder and dry processes where the rubber particles are added directly into the aggregates or into the mixture.

GTR bitumen pellets is a relatively new technology for incorporating tyre rubber into asphalt mixtures. The pellets consist of an asphalt-rubber binder which is pelletized with the aid of a filler. They are free-flowing solid particles that can be transported in sacks or in bulk and stored at the asphalt plant ready for mixing with the aggregates. Furthermore, the amount of pellets added to the mixture will dictate the level of polymer (i.e. rubber) modification and therefore the performance of the mixture.

Although there are many studies dealing with the use of tyre rubber in asphalt, there is no information regarding the effect of GTR bitumen pellets on the performance of asphalt mixtures, particularly the low temperature cracking properties. In this work, two different pellet concentrations were added to a dense asphalt concrete surface course mixture. The same mixture was also produced with a paving grade binder and with two different PMBs. Low temperature fracture properties, flexural strength and fracture toughness, were then determined at 0 °C using different specimen geometries. Crack propagation parameters were also determined from cyclic bending tests on notched specimens.

2 Materials and Specimen Preparation

GTR bitumen paving pellets were supplied by Billian UK Ltd. The pellets were solid particles relatively hard of similar size as the aggregate (typically around 10 mm maximum nominal size) consisting of a rubber-bitumen core coated with a thin layer of filler. Details of the manufacturing process and the composition of the pellets are given somewhere else (Artamendi et al. 2014). The pellets can be stored in bags without the need of heated storage tanks and can be incorporated directly into the mixture.

Table 1 Composition of the mixtures

Component	Mix 1 (%)	Mix 2 (%)	Mix 3 (%)	Mix 4 (%)	Mix 5 (%)
4/10 mm	39.6	39.6	39.6	39.6	39.6
2/6.3 mm	11.3	11.3	11.3	11.3	11.3
0/4 mm	39.5	39.5	39.5	39.3	39.1
LS filler	4.0	4.0	4.0	3.8	3.6
Binder	5.6 (125p)	5.6 (PMB1)	5.6 (PMB2)	4.6 (125p)	3.6 (125p)
GTR pellets	–	–	–	1.4	2.7

An asphalt concrete surface course mixture (AC 10 surf) was used in the study. Granite aggregates and limestone filler were employed. A 100/150 pen (125p) paving grade binder and two polymer modified binders, a 65/95-45 (PMB1) and a 45/80-50 (PMB2), were used to produce the mixtures. Binder content was 5.6 %. GTR pellets were also incorporated into the mixture at different proportions, 1.4 % (GTR1) and 2.7 % (GTR2) by weight of the total mixture. Bitumen 100/150 pen was added at two proportions, 4.6 and 3.6 %, respectively, to give an equivalent total binder content of 5.6 %. Composition of the mixtures is presented in Table 1.

The aggregates and binders were heated in an oven at 170 °C and mixed in a mechanical mixer. GTR pellets, on the other hand, were incorporated directly to the hot aggregates before the bitumen was added. The mixtures were then compacted to $300 \times 300 \times 50\ \text{mm}^3$ slabs using a laboratory roller-compactor. Un-notched beams (UB) were obtained by cutting 300 mm long (*L*), 50 mm thick (*W*) and 50 mm wide (*B*) beams from the slabs. Single edge notched beams (SENB) were obtained by cutting 300 mm long un-notched beams in half. These beams were then notched at the mid-point using a diamond-tip saw. Notch length was 10 mm approximately which gave notch to beam depth ratios (a/W) of 0.2. Semi-circular bending (SCB) specimens, on the other hand, were obtained by coring cylindrical specimens of 150 mm diameter (*D*) and 50 mm thickness (*t*) from slabs manufactured as before. These cylinders were cut perpendicular to the axis, and then notched at the mid-point along their diameter. Notch length was 10 mm approximately which gave notch to radius ratio (a/r) of 0.3.

3 Experimental Work

Flexural strength was determined using un-notched beam (UB) specimens loaded in a three-point bending configuration at a loading rate of 5 mm/min. The span, S, used to test the beams was 220 mm. A 20 kN servo-pneumatic test machine was used to perform the tests. The flexural strength was then determined using the following equation:

$$\sigma_0 = \frac{3PS}{2W^2B} \tag{1}$$

where P is the failure load, S is the span, and W and B are the specimen thickness and width, respectively. Three replicates were used.

Fracture toughness was determined using SENB and SCB specimens loaded in a three-point bending configuration at a loading rate of 5 mm/min. The span, S, used to test the specimens was 120 mm which gave span to length ratio (S/L) and span to diameter ratio ($2s/2r$) of 0.8. Two specimens of each geometry type were employed.

Fracture toughness for SENB specimens was calculated using the following equation:

$$K_I = \frac{3PS}{2W^2B} f\left(\frac{a}{W}\right) \tag{2}$$

where $f(a/W)$ is the geometrical parameter for SENB specimens that depends on the notch depth (a) and the thickness of the specimen (W) (Murakami 1987).

Fracture toughness for SCB specimens was calculated using the following equation:

$$K_I = \frac{4.263P}{2st} f\left(\frac{a}{r}\right) \tag{3}$$

where $f(a/r)$ is the geometrical parameter for SCB specimens that depends on the notch depth (a) and the specimen radius (r) (CEN 2010).

Crack propagation was evaluated by means of cyclic three-point bending tests using SENB specimens. A sinusoidal load of 10 Hz was applied to the specimens. The load—line displacement was measured with the actuator transducer and two external transducers. The vertical displacement with number of cycles was recorded until the specimen failed. Load amplitudes selected varied from 1.25 to 2.5 kN. Test were carried out at 0 °C.

4 Results and Discussions

4.1 Fracture Properties

Flexural strength mean values and standard deviations (error bars) are presented in Fig. 1a. It can be seen that higher flexural strength values were obtained for PMB and GTR modified mixtures compared to the unmodified mixture (Mix 1). Also, the flexural strength of Mix 3 was much higher than the values for the other mixtures. Flexural strength values for the mixtures produced with the GTR pellets (Mix 4 and Mix 5) were similar to that with PMB1 (Mix 2). Statistically, no differences were found between the mixtures with different GTR concentrations.

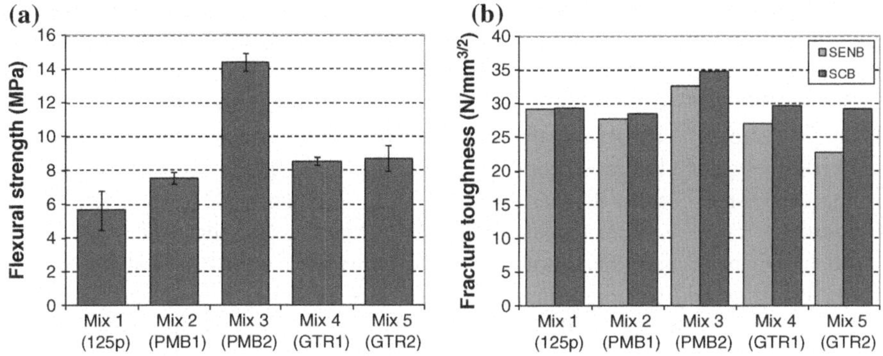

Fig. 1 Fracture properties: **a** flexural strength values, **b** fracture toughness values

Mean fracture toughness values from SENB and SCB tests are presented in Fig. 1b. It can be seen that the fracture toughness value for the mixture modified with PMB2 (Mix 3) was markedly higher than the rest. For SCB tests, differences between the values for these mixtures were relatively small. Regarding specimen geometry, it can be seen that, with the exception of Mix 5, good correlation was found between the fracture toughness from SENB and SCB tests. This suggest that fracture toughness is independent of the specimen geometry whereas fracture energy it is not.

4.2 *Crack Propagation*

Figure 2a shows the evolution of the load-line displacement with number of load cycles for Mix 1 at different load levels. It can be seen that first displacement increased steadily (crack initiation stage) and then rapidly until the specimen failed

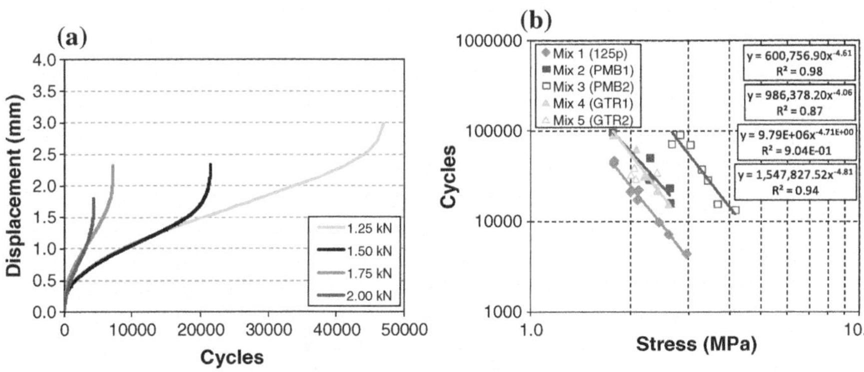

Fig. 2 Crack propagation test: **a** evolution of vertical displacement with number of load cycles (Mix 1), **b** relationship between applied stress and cycles to failure

(crack propagation stage). Furthermore, the duration of the crack initiation and crack propagation stages depended on both, the magnitude of the load applied and the type of mixture.

Figure 2b presents the relationship between the number of cycles to failure and the applied stress. It can be seen that mixtures modified with PMBs and GTR pellets had better resistance to crack propagation that the unmodified mixture. Also, based on the regression curves, the performance of the mixture with low GTR concentration (Mix 4) was similar to that with PMB1 (Mix 2). No clear differences were found between crack propagation of the two mixtures with different GTR concentration. It should be noted, however, that large scatter was observed for Mix 5 (see Fig. 2b). The mixture with PMB2 (Mix 3) was, on the other, the best in terms of crack propagation resistance.

5 Conclusions

Based on the laboratory work the following conclusions can be drawn:

- Mixtures modified with PMBs and GTR pellets had higher flexural strength than the unmodified mixture. Flexural strength of the mixture with PMB2 was the highest.
- Fracture toughness values of the unmodified and modified mixtures were similar with the exception of that with PMB2 which was considerably higher.
- In most cases, good correlation was found between the fracture toughness values determined from the two tests geometries, SENB and SCB.
- Mixtures containing PMBs and GTR pellets had enhanced resistance to crack propagation. The mixture with PMB2 was the most resistant to crack propagation.
- Crack propagation resistance of the mixture with the lower GTR concentration was equivalent to that with a standard PMB. No significant differences were found between the mixtures with different GTR concentration.

References

Artamendi I, Phillips P, Allen B, Evans G (2014) Estimation of the composition of rubber bitumen pellets using thermogravimetric analysis. Paper presented at the 14th International Society of Asphalt Pavements (ISAP) Conference, Raleigh, North Caroline, 1-5 June 2014

Comité Européen de Normalisation CEN (2010) EN 12697-44 Bituminous mixtures – Test methods for hot mix asphalt – Crack propagation by semi-circular bending test

Eurobitume (2013) European Bitumen Consumption 2013 In:Eurobitume, Brussels, Belgium. http://www.eurobitume.eu/system/files/EuropeanBitumenConsumption2013.pdf. Accessed 7 Sep 2015.

Murakami Y (1987) Stress intensity factors handbook. Pergamon, Oxford

Reinforcement of a Cracked Steel Plate Using CFRP Bonding

Emilie Leprêtre, Sylvain Chataigner, Lamine Dieng and Laurent Gaillet

Abstract This paper reports on a series of experiments performed to investigate the effectiveness of different composite bonding configuration on reducing fatigue crack growth propagation. Fatigue crack growth behavior of steel plates with rivet hole in mode-I loading condition, which have been repaired with single and double-side composite patches was investigated. During fatigue tests the crack detection method called 'beach-marking' is adopted to follow the crack propagation shape developing during applied fatigue cycles. Furthermore, effect of patch modulus on the crack growth life of repaired specimens and differences between the crack-front shapes obtained for single and double side repaired are investigated. The experimental results demonstrated that the CFRP patches could reduce the crack growth and extend the fatigue life. Moreover, the CFRP plates with high modulus were found to be much more efficient. This study extends the understanding of CFRP repair and provides some interesting results on crack propagation in CFRP repaired elements.

Keywords Fatigue · Crack-growth · Composite repair · CFRP

E. Leprêtre (✉) · S. Chataigner · L. Dieng · L. Gaillet
Laboratoire Structures Métalliques et à Câbles (SMC), Département Matériaux et Structures (MAST), IFSTTAR (Institut Français des Sciences et Technologies des Transports, de l'Aménagement et des Réseaux), Nantes, France
e-mail: emilie.lepretre@ifsttar.fr

S. Chataigner
e-mail: sylvain.chataigner@ifsttar.fr

L. Dieng
e-mail: lamine.dieng@ifsttar.fr

L. Gaillet
e-mail: laurent.gaillet@ifsttar.fr

A. Chabot et al. (eds.), *8th RILEM International Conference on Mechanisms of Cracking and Debonding in Pavements*, RILEM Bookseries 13,
DOI 10.1007/978-94-024-0867-6_25

1 Introduction

In old steel structures, many failures are due to fatigue. Fatigue cracks propagate indeed at an increasing rate leading to catastrophic failure. Repair of cracked components at an early stage is therefore momentous. Composite repair using adhesively bonded connection has already shown its efficiency in reducing crack growth rate and extending life for several cracked steel components (Bassetti et al. 2000; Jones and Civjan 2003; Liu et al. 2009; Wu et al. 2012).

Most previous studies focused on the fatigue behavior of steel plates notched with a very short initial crack from the edge or from a small center hole. This configuration of notched steel plates is not representative of the real cracks, observed in riveted members, which usually initiate from rivet holes. Typical initial cracks include corner cracks from rivet hole edge and surface cracks from rivet hole bore. Nevertheless, as fatigue crack grows, corner crack or surface crack propagates through the thickness of the plate and the crack front becomes relatively straight (O'Donoghue et al. 1995).

This paper presents the fatigue test results of steel plate with typical dimensions encountered in riveted structures and reinforced by Carbon Fiber Reinforced Plates (CFRP). S235 steel plate are used due to their mechanical properties equivalent to mild steel found in old steel riveted structures (Maraveas et al. 2013). Crack propagation was monitored by the "beach marking" technique and failure modes and fatigue life were recorded. Different lengths are considered: fatigue crack under rivet head when reinforcing or fatigue crack visible beyond the rivet head when reinforcing. Moreover, Normal Modulus (NM) and Ultra High Modulus (UHM) CFRP plates are used with single side and double side bonding configuration.

2 Testing Details

The tested specimens consist of CFRP plates bonded on cracked steel plates with an epoxy adhesive. 18 specimens were designed, amongst which 3 specimens are un-repaired (reference specimens), 12 specimens are single-side repaired and 3 specimens are double side repaired. The effect of plate modulus and initial crack length before reinforcing were studied for both cases.

S235 carbon steel plates of 510 mm long, 90 mm wide and 10 mm thick with center hole are used in the present experiment. The center hole represents rivet hole with classical dimensions on old steel bridges components. In order to initiate one single edge crack from the hole, firstly a small slot was realized using the wire erosion technique, and then a real fatigue crack was propagated applying fatigue loading. The yield stress, tensile strength and Young's modulus of the steel plates are 250 MPa, 510 MPa and 200 GPa, respectively.

All the cracked steel plates are initially divided into two groups depending on the initial length of the crack from the edge of the hole before bonding patch composite repair. The first one has an initial crack length of approximatively 6 mm from the edge of the hole (including the slot length) and the second one has an initial length of approximatively 13 mm.

The plates were reinforced by uni-directional carbon fiber plates with two types of modulus: Normal Modulus (NM) and Ultra-High Modulus (UHM). Based on the manufacture data, NM plate has an elastic modulus of 165 GPa, and a thickness of 1.2 mm. UHM plate has an elastic modulus of 460 GPa and a thickness of 2.3 mm. For each type of plate, bi-component epoxy adhesive associated with the composite plate is used.

The bonding process for all the repaired specimens is as follows: first, the steel bonded area is sandblasted, and then degreased; CFRP plates are also degreased; finally CFRP plates are bonded on steel plates in front of the crack tip with fiber direction perpendicular to the crack propagation direction. The specimens were cured in laboratory conditions at least one week before testing.

For all specimens, the composite plate is bonded at a certain distance from the edge of the hole, taking into account the presence of the rivet head. For this reason, all CFRP plates were bonded at a distance of 10 mm from the edge of the hole. Thus, for a crack length of 6 mm from the edge of the hole, the composite plate is placed outside the crack whereas for a crack length of 13 mm, the composite plate covers a part of the crack. Geometry of repaired steel plates, initial crack length and repaired configuration are shown in Fig. 1.

All specimens were subjected to uniform cyclic loading with a constant frequency of 10 Hz and stress ratio of 0.1 until complete failure of the specimens.

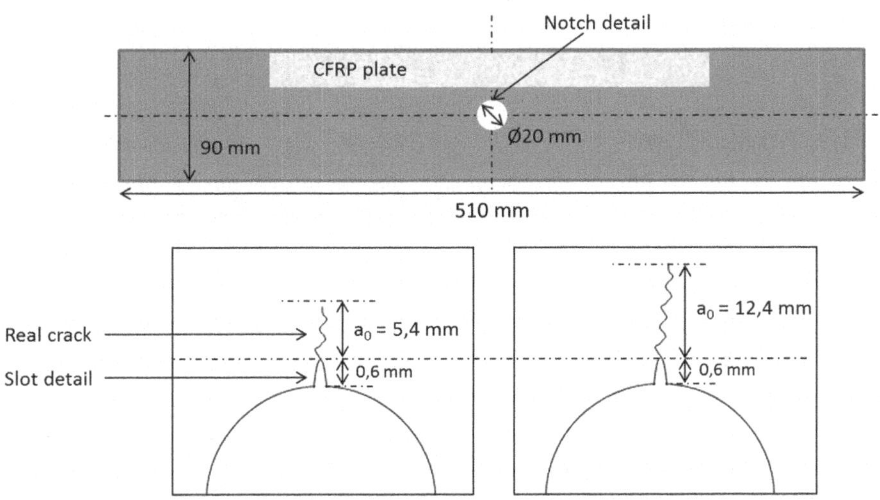

Fig. 1 Configuration of CFRP plates reinforcement for cracked steel plates with center hole (not to scale): plan view

The maximum applied stress is 100 MPa (about 20 % of the ultimate strength of the steel and 40 % of its yield strength), and the minimum applied stress is 10 MPa.

Crack propagation in the repaired and un-repaired specimens was followed by 'beach marking' technique. The beach marking method creates visible marks on the fracture surface by varying the applied stress ranges. Hence, during propagation stage, the applied stress range is reduced cyclically for a short number of cycles, causing change of crack propagation rate and thus creating visible marks on the crack surface. With this technique, observation of real crack size and shape in the thickness of the specimens is possible after complete failure of the plates.

3 Results and Discussions

Details of the specimens and their experimental results are listed in Table 1. The fatigue lifecycle indicated in Table 1 is an average value obtained from three specimens. Moreover, the crack length is measured at the mid-thickness of the plate. Fatigue cycle number to failure for repaired specimens is compared with reference specimens to get the fatigue life increase ratio.

Figure 2a compared the crack length trend for un-repaired specimens and for single-side repaired specimens with initial crack length of 6 mm. It shows that for UHM repairing, fatigue life of the specimens is clearly extended and that the reinforcement action takes effect immediately after the UHM plate is bonding. For NM plate, the fatigue life increase ratio is smaller and due to the low modulus of the plate, the reinforcement action takes effect only when the crack reaches the middle of the composite plate.

Figure 2b shows the crack length versus total number of fatigue cycles for un-repaired specimens and for NM single-side and NM double side repaired specimens with initial crack length of 6 mm. With double-side repair configuration,

Table 1 Experimental results

	CFRP plate type	Initial crack length (mm)	Failure mode	Fatigue cycle number	Fatigue life increase ratio
Reference	–	–	–	403,105	–
MN_NT_NS_A1	NM	6	CF	478,804	1.19
UHM_NT_NS_A1	UHM	6	CF + CD	706,602	1.75
MN_NT_NS_A2	NM	13	CF	520,186	1.29
UHM_NT_NS_A2	UHM	13	CF + CD	672,448	1.67
MN_NT_S_A1	NM	6	CF	966,952	2.40

CF cohesive failure
CD CFRP delamination (separation of carbon fibers from the matrix)

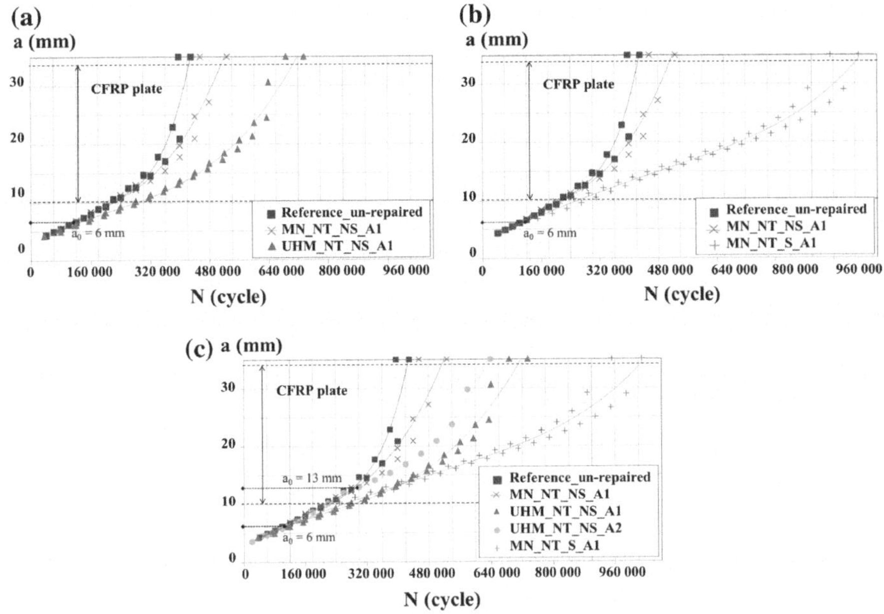

Fig. 2 Crack length versus total number of fatigue cycles: **a** influence of CFRP plate modulus (a_0 = 6 mm, single-side repair), **b** influence of double-side repair (a_0 = 6 mm, NM CFRP plate), **c** all results

the fatigue life increases by a factor of approximately 2.4 over un-repaired specimens. Moreover, NM plate action takes effects earlier than for single-side repair.

Figure 2c shows all results obtained with the different reinforcement configuration and emphasizes that the composite patch method can extend fatigue life of cracked steel specimens. It is interesting to note that the fatigue life increase is bigger for NM single-side reinforcement with initial crack of 13 mm than for initial crack of 6 mm. This phenomenon can be explained by the crack bridging due to presence of adhesive between the crack's lips when composite plate is bonded directly on the surface crack (case a_0 = 13 mm). This phenomenon is less significant for UHM reinforcement configuration because the action of the UHM composite plate takes place immediately after reinforcement.

Two kinds of crack shapes, symmetric and un-symmetric about the mid-thickness axis, are observed for the tested specimens. Some examples are shown in Fig. 3.

For un-repaired specimens (reference specimens) and for double-side repaired specimens, the crack shape tends to be symmetric about the mid-thickness of the plate. Nonetheless, for single-side repaired specimens, crack grew faster on the un-reinforced side, leading to un-symmetric crack shape.

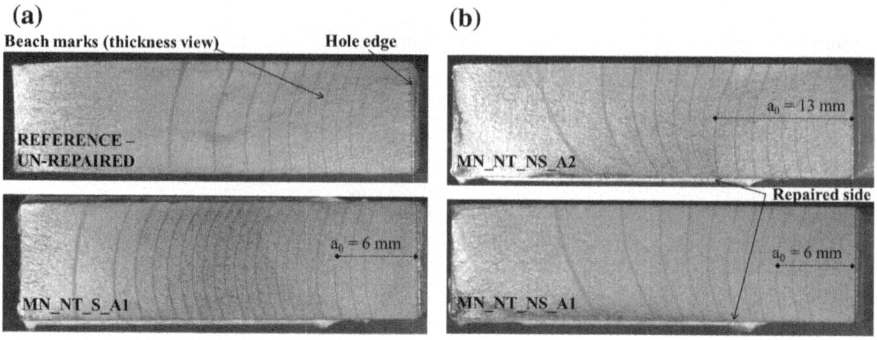

Fig. 3 Fracture surfaces with beach marks: **a** symmetrical beach marks, **b** un-symmetrical beach marks (normal modulus reinforcement)

4 Conclusions

The experimental study of the fatigue behavior of CFRP strengthened steel plates with different degrees of damage and different reinforcement configuration has been carried out.

Significant fatigue life extension was obtained with a range from 20 to 140 % for strengthened specimens compared with un-repaired specimens. Thus, bonding composite patches is effective for retarding fatigue crack propagation at a rivet hole and therefore extending the fatigue life.

Beach marks on the surface cracked of the specimens clearly shows that the CFRP plate restrains the crack propagation. For NM reinforcement, it seems that a late reinforcement at a larger damage level led to bigger extension of the remaining fatigue life. This is due to the low modulus of the NM CFRP plate for which the action of the reinforcement takes effect at a later stage of the propagation of the fatigue crack, and also as a result of the crack bridging phenomenon. For UHM reinforcement, the reinforcement action is immediate and an early repair is more effective in extending the total fatigue life. Future studies will consider the effect of hot driven rivet on the crack propagation for CFRP repaired riveted joint.

References

Bassetti A, Nussbaumer A, Hirt MA (2000) Crack repair and fatigue life extension of riveted bridge members using composite materials. Paper presented at the Bridge engineering conference ESE-IABSE-FIB, March 2000, p.227-38

Jones SC, Civjan SA (2003) Application of fibre reinforced polymer overlays to extend steel fatigue life. J Compos Constr 2003:7:331-8. doi: 10.1061/(ASCE)1090-0268(2003)7:4(331)

Liu HB, Al-Mahaidi R, Zhao XL (2009) Experimental study of fatigue crack growth behavior in adhesively reinforced steel structures. Composite Structures 2009:90(1):12-20. doi: 10.1016/j.compstruct.2009.02.016

Maraveas C, Wang YC, Swailes T (2013) Thermal and mechanical properties of 19th century fireproof flooring systems at elevated temperatures. Construction and Building Materials 2013:48:248-264. doi: 10.1016/j.conbuilmat.2013.06.084

O'Donoghue PE, Atluri SN, Pipkins DS (1995) Computational strategies for fatigue crack growth in three dimensions with application to aircraft components. In: Engineering Fracture Mechanics, vol.52, No.1, pp 51-64

Wu C, Zhao XL, Al-Mahaidi R, Emdad M, Duan WH (2012) Fatigue tests of cracked steel plates strengthened with UHM CFRP plates. Advanced in Structural Engineering 2012:1801-1816. doi: 10.1260/1369-4332.15.10.1801

Characterizing Non-linear Fatigue Crack Growth and Size Effect in Plain Concrete Beams with a Hybrid Effective Crack and Cohesive Zone Model

Nicholas Andres Brake and Karim Chatti

Abstract The mechanical response of concrete is largely influenced by the interlocking coarse aggregate, which supplies the cement matrix with a resisting cohesive crack bridging force. Under low and high cycle fatigue loading, the cohesive stresses can influence the crack growth rate and the structural load capacity. A new method used to quantify the cyclic cohesive zone properties and effective crack resistance in three point bending single edge notch plain concrete specimens of different sizes under both low and high cycle fatigue loading is presented here. For validation, three point bend concrete specimens of two different sizes were tested under crack mouth opening displacement controlled low cycle quasi-static loading and force controlled high cycle fatigue using constant, variable, and random amplitude loading sequences. The results indicate that the cohesive stress-dependent cyclic crack resistance can be quantified and used to effectively characterize the high cycle fatigue non-linear crack growth and size effect.

Keywords Concrete fatigue · R-curve · Size effect · Random loading

1 Introduction

Concrete is a quasi-brittle material that exhibits 'small crack' behavior; the cohesive stresses acting along the concrete crack face strongly influence the fracture characteristics of the material, e.g., the fatigue crack propagation rate and fracture toughness. Plain concrete is made of relatively large aggregates which engender a

---

N.A. Brake
Department of Civil and Environmental Engineering, Lamar University, Beaumont, USA
e-mail: nicholas.brake@lamar.edu

K. Chatti (✉)
Department of Civil and Environmental Engineering, Michigan State University, East Lansing, USA
e-mail: chatti@egr.msu.edu

A. Chabot et al. (eds.), *8th RILEM International Conference on Mechanisms of Cracking and Debonding in Pavements*, RILEM Bookseries 13,
DOI 10.1007/978-94-024-0867-6_26

proportionally large cohesive stress zone near the crack tip (Cox and Marshall 1991) and is responsible, in large part, for the size effect and R-curve behavior (Fett et al. 2000), crack deceleration region in fatigue (Li and Matsumoto 1998) and influencing the endurance limit (Kruzic et al. 2005). Several researchers investigated the relationship between cohesive stresses and size effect (Bazant and Kazemi 1990; Morel 2007; Roesler et al. 2007), and its relationship to the R-curve (Foote et al. 1986; Mai 2002; Bazant and Cedolin 1984; Gallops et al. 2011). In fatigue loading concrete experiences non-linear crack growth, where it initially decelerates and gradually accelerates until failure (Brake and Chatti 2012; Subramaniam et al. 2000).

In pavement engineering, a new mechanistic empirical design protocol was recently launched to predict damage caused by millions of traffic and environmental loadings. The high cycle fatigue damage model is based on an empirical SN curve and a linear damage rule (Miner 1945). Fatigue SN models, although computationally efficient, are limited in that they cannot capture non-linear damage accumulation, size, and the R-ratio effect. A simple but yet powerful alternative approach could be to use a modified Paris law that can capture non-linear crack growth, load history, and size effects. Most importantly, the Paris law is still very computationally efficient in that the number of cycles to failure can be predicted by evaluating a simple integral (Bazant and Planas 1998). However, there is a need for a computationally efficient model that can predict plain concrete non-linear crack growth, size effect, and load capacity under various loading sequences. This paper presents such a model.

2 Model Framework

The model proposed in this paper uses an LEFM framework that considers constant and variable amplitude loading and non-linear crack growth. The model uses a modified Paris law that incorporates the use of a fatigue resistance curve to simulate the effects of cohesive stresses on crack growth. The inclusion of the cohesive stresses can significantly reduce the size dependency observed in Paris constant: C and n, in Portland cement concrete (PCC). The cohesive stresses are described with a fatigue R-curve for computational efficiency; the relationship between the cohesive stresses and fatigue R-curve are shown in Table 1 from Brake and Chatti (2013). Equation 1 shows the modified Paris equation. The variables a and

Table 1 Dimensionally consistent regression equations

$K_{Rc}^{f} = 0.228K_{IC}^{T} + 0.1677, \mathrm{R}^2 = 0.95$	$\Delta a_c = 6.377 \times 10^{-5}\left(\frac{w_c E}{K_{IC}^{T}}\right)^2 + 0.603, \mathrm{R}^2 = 0.97$
$\frac{\lambda_1}{\sqrt{D}} = 3.074\frac{f_t'}{D} + 0.025, \mathrm{R}^2 = 0.89$	$\lambda_2 = -1.053\frac{D}{l_{ch}} + 0.025, \mathrm{R}^2 = 0.89$
$K_{IC}^{T} = \sqrt{G_F^T E}, \boldsymbol{G_F^T} = \int_0^{w_c} \boldsymbol{\sigma_{coh}(\delta)d\delta}$	$l_{ch} = \left(\frac{K_{IC}^{T}}{f_t'}\right)^2$

N represents the crack length and cycle number, respectively. The variable K_I is the mode-I stress intensity given in Eqs. 2 and 3, K_R^F is the fatigue crack resistance curve, σ_N is the nominal stress at mid-span, D is the beam depth, b is the beam width, $k(\alpha)$ is the stress intensity geometric correction factor given in Eq. 4, α is the non-dimensional crack length (a/D), P is the applied load at mid-span, and S is the beam span

$$\frac{da}{dN} = C\left(K_I - K_R^f\right)^n \tag{1}$$

$$K_I = \sigma_N\sqrt{D}k(\alpha) \tag{2}$$

$$\sigma_N = \frac{3PS}{2bD^2} \tag{3}$$

The stress intensity geometric correction factor for a three–point bending single end notch (TPBSEN) specimen with a span to depth ratio (S/D) of 4 is shown in Eq. 4.

$$k(\alpha) = \frac{\sqrt{\alpha}[1.99 - \alpha(1-\alpha)(2.15 - 3.93\alpha + 2.7\alpha^2)]}{(1+2\alpha)(1-\alpha)^{3/2}} \tag{4}$$

The resisting curve function is governed by five variables: K_{Rc}^f, Δa_c, m, λ_1, and λ_2 as shown in Eq. 5. The first three parameters govern the shape of the pre-peak region and the last two govern the post-peak slope. To maintain first order continuity, the functions must be connected at the equivalent crack length, which can be calculated with Eqs. 6 and 7.

$$K_R^f = \left\{ \begin{array}{c} K_{RC}^f\left[1 - \left(1 - \frac{\Delta a}{\Delta a_c}\right)^m\right], \ 0 \le \Delta a < \Delta a_\beta \\ K_{\beta c}^f + \beta(\Delta a - \Delta a_\beta), \ \Delta a \ge \Delta a_\beta \end{array} \right\} \tag{5}$$

where

$$\Delta a_\beta = \left(1 - \left(\frac{\beta \Delta a_c}{m K_{Rc}^f}\right)^{\frac{1}{m-1}}\right)\Delta a_c, m > 1, \frac{\beta\Delta a_c}{mK_{Rc}^f} < 1 \tag{6}$$

$$\beta = \lambda_1 \exp\left[\lambda_2 \frac{\Delta P}{P\text{max}}\right] \tag{7}$$

These parameters are related to the cohesive stress function shown in Eq. 8 using the empirical regression equations shown in Table 1 from (Brake and Chatti 2013), where f'_t is the concrete tensile strength, w_c is the crack opening displacement at zero stress, and c_1 and c_2 are shape parameters. K_{IC}^T is the total fracture toughness, G_F^T is the total fracture energy, and l_{ch}, is the characteristic length.

$$\sigma_{coh}(\delta(x)) = f_t' \left\{ \left[1 + \left(\frac{c_1 \delta(x)}{w_c} \right)^3 \right] \exp\left(-\frac{c_2 \delta(x)}{w_c} \right) - \left(\frac{\delta(x)}{w_c} \right) (1 + c_1^3) \exp(-c_2) \right\} \tag{8}$$

A series of fatigue experiments were conducted to determine the Paris constants from Eq. 1 and to determine their statistical significance with regards to specimen size and loading sequence.

3 Experimental Program

3.1 Materials

The concrete was made of ASTMC-150 Type I cement, natural sand, and a limestone coarse aggregate (nominal maximum size of 25 mm). The water to cement ratio was 0.45 and the air content was 6.5 %. The 28 day compressive strength was 25 MPa. The beam specimens were allowed to cure for one year inside of a humidity room, and were removed one month prior to testing. The average split tensile strength and compressive strength at the time of testing was 3.79 and 40.62 MPa, respectively.

3.2 Notched Three Point Bending Tests

The beams were subjected to (i) quasi-static loading, (ii) constant amplitude fatigue loading, (iii) variable amplitude fatigue loading, and (iv) random amplitude fatigue loading. The span to depth ratio was 4. Beams with two different depths were tested: 50 and 100 mm; where the width, b, was equal to the depth, d, in all the beams. The beams were tested under three different notch-depth ratios: 0.15, 0.33, and 0.5 (Fig. 1).

A total of 48 plain PCC three point bend specimens were tested; six were tested under low cycle quasi-static loading and 34 tested under high cycle fatigue loading. The specimens tested under quasi-static loading were loaded and un-loaded at a COD controlled rate of 0.0005 mm/s until ultimate failure. Under high cycle fatigue loading, the specimens were subjected to constant, variable, and random amplitude loading sequences. Each load was applied at a frequency of 2 Hz. For the constant amplitude tests, stress ratios (SR) between 0.95 and 0.8 and an R-ratio of 0.05 were applied. The compliance and equivalent crack was measured at each loading cycle

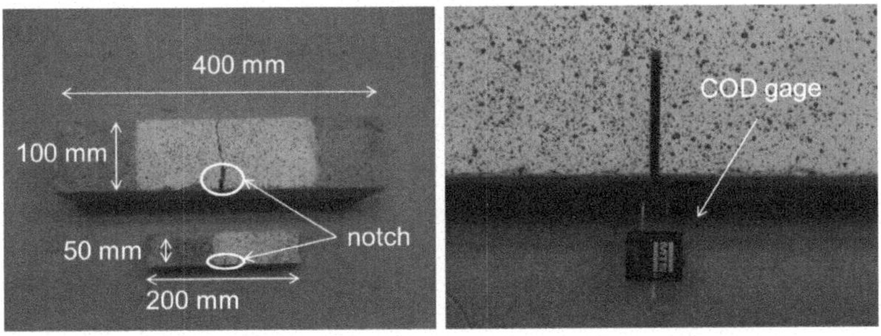

Fig. 1 Photograph of the two different beam sizes, notch, and COD gage

using peak-to-peak crack mouth opening displacement (CMOD) measurements (Jenq and Shah 1985). Equations 9 and 10 show the CMOD and the geometric correction factor, respectively.

$$CMOD = \frac{6PSaV(\alpha)}{EbD^2} \tag{9}$$

$$V(\alpha) = 0.76 - 2.26\alpha + 3.87\alpha^2 - 2.04\alpha^3 + \frac{0.66}{(1-\alpha)^2} \tag{10}$$

For the variable amplitude tests, the stress ratio and R-ratio were changed at different times; the degree of change was commensurate to the degree of damage observed in the specimen. The stress ratios varied between 0.95 and 0.75 and the R-ratios between 0.05 and 0.5. Under random loading, the SR ranged from 0.5 to 0.9 and R-ratio ranged from 0.5 to 0.05.

4 Experimental Results

The results of the fatigue tests are summarized here. Figure 2 shows the log of the crack growth (*da*/*dN*) with respect to the crack extension, Δa, for multiple specimens under the three different loading sequences. The crack resistance curve, K_R^f, was obtained by determining the tensile strength f_t', and critical crack opening displacement, w_c, with a similar procedure used for quasi-static loading which involves inverse analysis at the last recorded cycle. The stress parameters c_1 and c_2 were assumed to be the average values obtained under quasi-static loading, 4.6 and 25.3, respectively. The shape parameter *m*, was also taken to be equal to that obtained under quasi-static loading: 2. The remaining parameters: K_{Rc}^f, Δa_c, λ_1, *and* λ_2 were determined using the cohesive tensile strength, f_t', and crack opening displacement

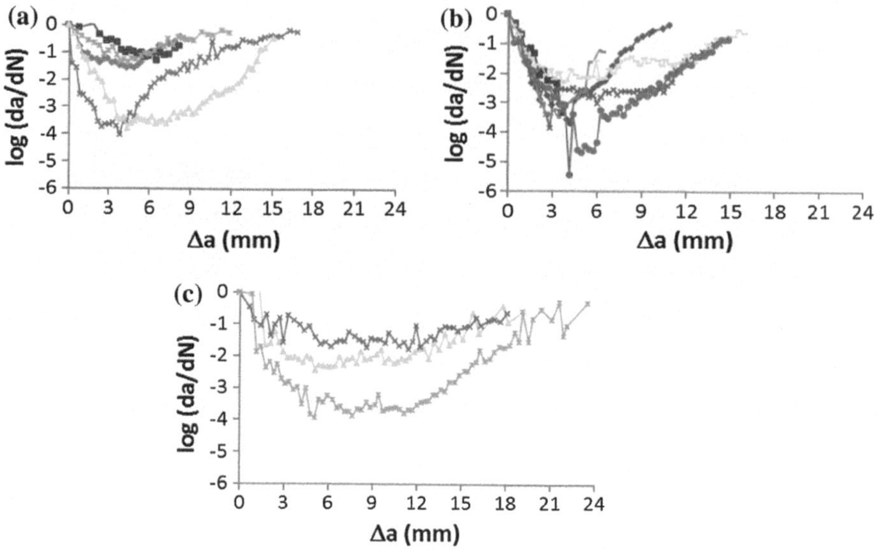

Fig. 2 Crack rate versus crack extension for 100 mm beams under **a** constant, **b** variable, and **c** random loading

Table 2 Summary of test results

Specimen Size	50 mm			100 mm		
	N	log*C* (%)	*n* (%)	N	log*C* (%)	*n* (%)
Constant load	4	−13.2 ± 24	10.5 ± 27	5	−18.4 ± 58	13.4 ± 58
Variable load	10	−23.1 ± 39	18.8 ± 18	9	−24.1 ± 38	18.3 ± 40
Random load	3	−23.4 ± 1	19.4 ± 2	3	−18.6 ± 42	14.1 ± 42
All specimens	17	−20.9 ± 40	16.9 ± 40	17	−21.5 ± 44	16.1 ± 45

using inverse analysis at the last cycle prior to failure. A summary of the Paris parameters determined with and without the use of the fatigue crack resistance curve for both specimen sizes is provided in Table 2. The average tensile strength and crack opening displacement for the 50 mm specimen was: f_t': 4.72 MPa ± 10 %, w_c: 0.34 mm ± 25 %, and 100 mm specimens: f_t': 4.39 MPa ± 14 %, w_c: 0.34 mm ± 28 %. An Analysis of Variance (ANOVA) analysis was then conducted to assess the statistical significance of the specimen size, *D*, and loading sequence on the Paris law parameters that were obtained both with and without the fatigue crack resistance curve. Note, the Paris parameters obtained without the use of the fatigue resistance curve means that $K_R^F = 0$ and the modified Paris law is equal to the original Paris law. Under this condition, only data in the accelerating crack region (positive slope regions in Fig. 2) can be evaluated.

The results of the statistical analysis suggest the Paris parameters do not depend on the size or loading sequence when using the modified law (p-value: log*C*, *D*: 0.887, *Load*: 0.100; *n*, *D*: 0.719, *Load*: 0.071). However, if the fatigue resistance curve is not considered and the Paris law is used, the parameters will depend on specimen size (p-value: log*C*, *D*: 0.029, *n*, *D*: 0.042) using a 95 % confidence level.

5 Conclusion

The focus of this research was to develop a modified Paris law capable of predicting both crack deceleration and acceleration and also account for size effects. The insertion of a crack resistance curve into the Paris Law can be used to predict non-linear crack growth and account for size effect. When the equivalent fatigue resistance curve is used, on average, the Paris constants are statistically equivalent for two different size specimens under constant, variable, and random amplitude loading, with the fracture parameters $\log C = -20.8$, $n = 16.9$ for 50 mm beams, and $\log C = -21.5$, $n = 16.1$ for 100 mm beams. If the resistance curve is not used, the two Paris constants are not statistically significant under constant, variable, and random amplitude loading, and the fracture parameters are $\log C = -23.71$, $n = 15.60$ for 50 mm beams, and $\log C = -34.3$, $n = 22.0$ for 100 mm beams. It should be noted that the values obtained in this research to calibrate the model are valid for one concrete mix and beam sizes tested. Testing of other mixes and specimen sizes would need to be carried out to check the validity of the approach presented.

References

Bazant ZP, Cedolin L (1984) Approximate linear analysis of concrete fracture by R-curve. ASCE J Structural Engineering 110(6):1336-1355

Bazant ZP, Kazemi MT (1990), Determination of fracture energy, process zone length and brittleness number from size effect, with application to rock and concrete. Int J Fracture 44:111-131

Bazant ZP, Planas J (1998) Fracture and size effect in concrete and other quasi-brittle materials. Boca Raton and London, CRC Press

Brake NA, Chatti K (2012) Prediction of transient and steady state flexural fatigue crack propagation in concrete using a cyclic R-curve. J Engineering Mechanics 138(4):371-378

Brake NA, Chatti K (2013) Prediction of size effect and non-linear crack growth in plain concrete under fatigue loading. Engineering Fracture Mechanics, 109:169-185

Cox BN, Marshall DB (1991) Stable and unstable solutions for bridged cracks in various specimens. Acta Metallurgica et Materiallia 39:579-589

Fett T, Munz D, Geraghty RD, White KW (2000). Bridging stress determination by evaluation of the R-curve. J European Ceramic Society 20(12):2143-2148.

Foote RL, Mai YW, Cotterell B (1986) Crack growth resistance curves in strain softening materials. J Mech Phys Solids 34:593-608
Gallops S, Fett T, Ager III JW, Kruzic JJ (2011) Fatigue threshold R-curves predict small crack fatigue behavior. Acta Materialia 59:7654-7661
Jenq YS., Shah, SP (1985) Two parameter fracture model for concrete. ASCE J. Engineering Mechanics 111(10):1227-1241
Kruzic JJ, Cannon RM, Ager III JW, Ritchie RO (2005) Fatigue threshold R-curves for predicting reliability of ceramics under cyclic loading. Acta Materialia 53(9):2595-2605
Li VC, Matsumoto T (1998) Fatigue crack growth analysis of fiber reinforced concrete with effect of interfacial bond degradation. Cement and Concrete Composites 20:339-351
Mai YW (2002) Cohesive zone and crack-resistance (R)-curve of cementitious materials and their fiber-reinforced composites. J Engineering Fracture Mechanics 69(2):219-234
Miner M (1945) Cumulative damage in fatigue. J. Appl Mechanics 67:A159-A164
Morel S (2007) R-curve and size effect in quasi-brittle fractures: Case of notched structures. J Solids and Structures 44(13):4272-4290
Roesler J, Paulino GH, Park K, Gaedicke, C (2007) Concrete fracture prediction using bilinear softening. Cement Concrete Composites 29:300-312
Subramaniam KV, Oneil EF, Popovics JS, Shah SP (2000) Crack propagation in flexural fatigue of concrete. J Engineering Mechanics 126(9):891-898

A New Approach for the Study of Fatigue Cracking Phenomenon in Bituminous Mixtures

Fernando Moreno-Navarro and Mª Carmen Rubio-Gámez

Abstract In spite of the fact that many studies have been carried out to improve the understanding of fatigue cracking in bituminous materials, there are still some aspects that should be clarified and which prevent an accurate analysis. During cyclic loading, fatigue damage co-exist with other phenomena (plastic deformations, thixotropy, heating, etc.) that could distort its measurement and make more difficult its precise quantification. In addition, the damage caused by fatigue loads has global and local effects that cannot be separated. Because of this fact, it is not easy to establish a homogeneous failure criterion. Based on these considerations, this study presents a new approach that allows the identification of the different phenomena appeared during cyclic loading and therefore, to establish a homogenous failure criterion. In addition, using this approach is possible to make an accurate quantification of the cracking damage produced in bituminous mixtures, considering global and local effects.

Keywords New approach · Bituminous mixture · Fatigue damage · UGR-FACT

1 Introduction

Fatigue damage in bituminous mixtures can be considered as a global process which involves three main phenomena (Di Benedetto et al. 2011): (i) accumulation of permanent deformations; (ii) reversible degradation (thixotropy or heating) and initiation of irreversible damage (micro-cracks); (iii) crack propagation (the coalescence of micro-cracks produce the localization and propagation of macro-cracks). During the study of the fatigue behaviour of these materials, the occurrence

F. Moreno-Navarro (✉) · M.C. Rubio-Gámez
LabIC.UGR Laboratory of Construction Engineering, University of Granada, Granada, Spain
e-mail: fmoreno@ugr.es

M.C. Rubio-Gámez
e-mail: mcrubio@ugr.es

A. Chabot et al. (eds.), *8th RILEM International Conference on Mechanisms of Cracking and Debonding in Pavements*, RILEM Bookseries 13,
DOI 10.1007/978-94-024-0867-6_27

of these phenomena can be identified by the changes produced on the phase angle and modulus. Nevertheless, it has been demonstrated that exist some aspects which could limit this analysis (Di Benedetto et al. 2004).

One of these aspects are the permanent deformations appeared in the material previously to the fatigue damage. This phenomenon cannot be considered as real damage, however its appearance change the visco-elastic properties of the material making it more elastic and rigid, due to the strain hardening phenomenon (Darabi et al. 2013). Thus, despite permanent deformations do not cause damage, their effect in the mechanical properties of the bituminous mixture cannot be neglected in the analysis of fatigue processes.

Another aspect that could limit the analysis of fatigue phenomenon in bituminous mixtures is the co-existence of reversible visco-elastic phenomena (heating or thixoptropy) with damage (Pérez-Jimenez et al. 2012). In the stages where these phenomena co-exist, it is very difficult to distinguish which of them is causing the changes in the mechanical properties of the material, and in the same way, how to quantify the changes due only to real damage (Shan et al. 2011).

Finally, to find a stable failure criteria is other of the aspects that makes complicated the analysis of the fatigue response of bituminous mixtures. Fatigue damage is a random and tridimensional phenomena, and thus the amount and distribution of micro-cracks, as well as the way of propagation of the macro-crack will directly depend on many variables: the type of bituminous mixture evaluated, the tests conditions, the type of materials used, the geometry of the specimen and type of test used, etc. Because of this fact, it is very difficult to clearly identify and establish a damage limit which define a homogenous failure criteria that could be generalized in any fatigue study (Mollenhauer and Wistuba 2013).

Based on these considerations, this manuscript presents a new approach to study fatigue phenomenon in bituminous mixtures that could lead to develop a more accurate analysis of this pathology.

2 Previous Considerations

Based on bitumen fatigue cracking kinetics, the mechanical response of bituminous mixtures to repeated loads is highly influenced by the temperature (Netzel 2006). At low temperatures, the non-polar fractions of the bitumen cannot move and they create a rigid network. The mechanical response of this network is controlled by the stretching and bending of intermolecular bonds (Sperling 2006), and thus by chain scission (brittle fracture). Therefore, as plastic deformations are limited under these circumstances, the energy introduced in each load cycle is dissipated through fatigue damage (stress excite a main-chain bond to the state of separation), which propagates by molecular rupture (low molecular weight fragments are created by this scission event, which align with ruptured ends normal to the crack surface, Rosen 1964). At higher temperatures, chains separation due to thermal expansion or mechanical agitation allow the sliding of the molecular network (Cheung and

Cebon 1997; Gaskin 2013), which cause the appearance of plastic deformations in the material (viscoelastic response to load). In this case, the energy of each load cycle is dissipated by the maltenes by friction (thermal energy) during plastic flow, and store in the covalent bonds (strain hardening), until the dislocation of microstructural segments is avoided and molecular scission is produced (ductile fracture).

According to this, asphalt mixtures tested at lower temperatures will offer a reduced molecular mobility, and the fatigue failure under these circumstances would be the appearance of thinner and un-ramified cracks (brittle fracture with reduced plastic deformations before the failure). On the contrary, bituminous materials tested at higher temperatures will have a higher molecular mobility, which is mainly manifested in form of plastic deformations before the appearance of a ductile fracture with wide and ramified cracks.

3 Description of the New Approach

A new approach for the analysis of fatigue damage in bituminous mixtures has been developed using the UGR-FACT method (Moreno-Navarro et al. 2015). This test method reproduces the conditions that lead to the appearance of fatigue cracking in pavements, by using a simple device composed of a platform that is composed of two sloping surfaces with two rails (that allow for the sliding of two supports), and of two vertical spindles that are used to measure vertical deformations in the upper part of the test specimen (Fig. 1). The two supports are composed of a carriage and a support plate (to which the test specimen is attached) where the horizontal deformation gauges are located. Under these support plates, two elastic elements are introduced in order to allow for the flexion of the specimen and a spring that simulates the foundation layers.

Four LVDT's (one vertical and one horizontal in each face of the specimen) are used in order to control the vertical and horizontal displacements produced in the material in each load cycle (Fig. 1). Based on the measures made, two different types of displacements can be observed in each direction and load cycle: a

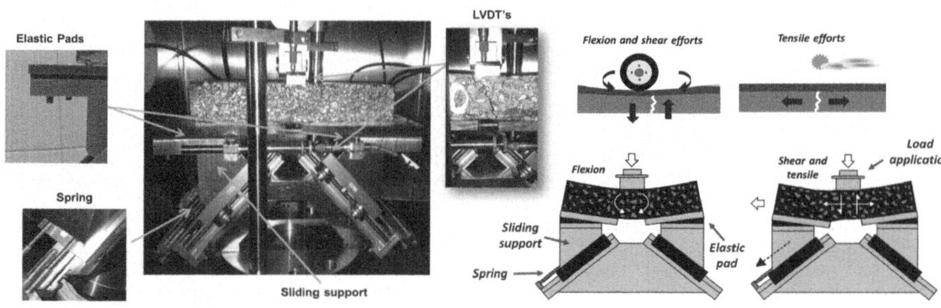

Fig. 1 UGR-FACT device

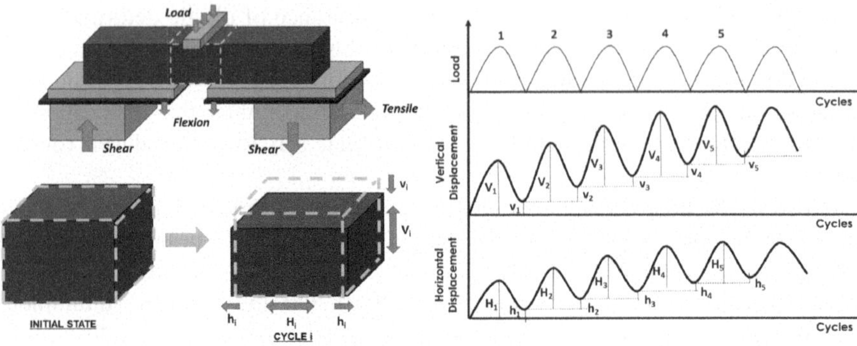

Fig. 2 Sketch of the displacements and efforts produced on bituminous specimen during the UGR-FACT test

"permanent" one (hi, vi) that remains after the load cycle and is related to the non-recoverable deformations or the damage produced in the material; and a "relative" one (Hi, Vi) that is related to the consistency (stiffness) of the material in the given cycle (Fig. 2).

In this respect, if both types of displacements produced in each load cycle are used for a combined analysis, a precise evaluation of the evolution of the different phases appeared during fatigue damage process can be done. On the one hand, the hysteresis loop described due to the "relative" displacements (Hi, Vi) produced in the material are used to define the dissipated energy in each load cycle (Eq. 1).

$$\omega_i = \omega_{hi} + \omega_{vi} \tag{1}$$

where ω_i is the dissipated energy in cycle i (in J/m^3); ω_{hi} is the horizontally dissipated energy in cycle i (in J/m^3); and ω_{vi} is the vertically dissipated energy in cycle i (in J/m^3).

On the other hand, the "permanent" displacements (h_i, v_i) can be used to define de variation of the geometry ($\Delta\varepsilon_i$) of the material in the zone where the fatigue phenomenon takes place. This variation (measured in percentage) is calculated from the changes produced in the dimensions of the material in the vertical and horizontal directions (Eq. 2).

$$\Delta\varepsilon_i = \frac{(\delta_{hi} \cdot \delta_{vi}) - (\delta_{hi-1} \cdot \delta_{vi-1})}{(\delta_{hi-1} \cdot \delta_{vi-1})} \cdot 100 \tag{2}$$

where $\Delta\varepsilon_i$ is the variation of the geometry of the specimen in the cycle i; δ_{hi} and δ_{vi} are the horizontal and vertical dimensions of the specimen in the cycle i; and δ_{hi-1} and δ_{vi-1} are the horizontal and vertical dimensions of the specimen in the previous cycle.

Figure 3a shows a common representation of these two parameters (ω_i in the "x" axis, and $\Delta\varepsilon_i$ in the "y" axis), as well as the load cycles (secondary "y" axis). As can

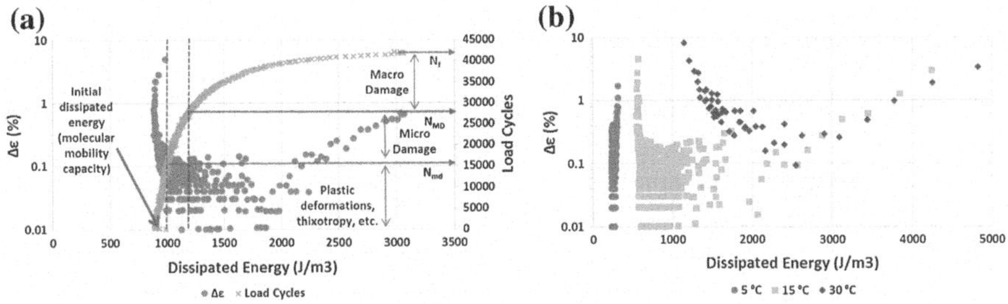

Fig. 3 **a** Example of the values obtained in the representation of Δε and load cycles as a function of the dissipated energy, **b** example of the change produced in molecular mobility of the same material tested at different temperatures

be observed, the different stages of the fatigue process can be clearly identified. During the first part of the test, high variations are produced in the geometry of the material due to their molecular mobility, that are reduced as the number of load cycles applied increases (strain hardening phenomenon). In spite of this variation of the geometry, the dissipated energy measured in each load cycle during this first part does not change. This fact means that the variations produced in the material due to the cyclic loads do not exert damage on it, as its dissipated energy remains similar to the initial one (which represent the visco-elastic response of the undamaged material). The value of the initial dissipated energy obtained represents the molecular mobility capacity of the material under the test conditions used. As this initial dissipated energy decrease, the material shows a lower molecular mobility, reducing the variations produced in the geometry of the specimen due to plastic deformations, before the molecular scission (Fig. 3b).

On the second part of the test, the variations of the geometry of the material becomes very small (due to the strain hardening phenomenon), and the energy introduced in each load cycle is mostly absorbed at molecular bonding level. After that, when the molecular bonds start to fracture (due to the repeated loads), the dissipated energy measured in each additional cycle becomes to increase (N_{md}). In this part, the mixture has consumed its ability to deform, and the energy introduced in each additional load cycle start to produce micro damage due to bonding breakage. Because of this fact, the values measured of $\Delta\varepsilon_i$ still small, and the dissipated energy become to increase. Finally, after a certain number of cycles, the dissipated energy began to increase considerably by maintaining the changes produced in the geometry (N_{MD}, initiation of the macro-crack due to the coalescence of micro-cracks), until a point where the values of $\Delta\varepsilon_i$ began to increase again (propagation of the macro crack), until the total failure of the specimen is reached (N_f).

4 Conclusions

This paper proposes a new approach which could offer a valuable tool to solve some of the current troubles that difficult the study of fatigue phenomenon in bituminous mixtures. This approach allows for a global evaluation of the fatigue phenomenon, from the appearance of permanent deformations to the macro-crack propagation. For this purpose, it combines the study of the changes produced in the geometry of the material and the energy dissipated by it in each load cycle. Thus, it provides a tool to develop a proper analysis of fatigue damage in bituminous materials, without considering the effect of the other phenomena that are not damage. In the same way, using this approach is possible to establish a homogenous failure criteria that can consider a similar level of damage in each material studied, even under different test conditions.

References

Di Benedetto H, De La Roche C, Baaj H, Pronk A, Lundstrom R (2004) Fatigue of bituminous Mixtures. Materials and Structures 37: 202-216.

Di Benedetto H, Nguyen QT, Sauzéat C (2011) Non-linearity, heating, fatigue and thixotropy during cyclic loading of asphalt mixtures. Road Materials and Pavement Design 12(1): 129-158.

Darabi MK, Abu Al-Rub RK, Masad EA, Little DN (2013) Cyclic hardening-relaxation viscoplasticity model for asphalt concrete materials. Journal of Engineering Mechanics 139(7): 832-847.

Cheung, CY, Cebon, D (1997) Deformation mechanisms of pure bitumen. Journal of Materials in Civil Engineering, Vol. 9, pp. 117-129.

Gaskin, J. (2013). On bitumen microstructure and the effects of crack healing. University of Nottingham, Doctoral Thesis.

Mollenhauer K, Wistuba MP (2013) Comparison of fatigue test procedures for asphalt mixes. Proceedings of te 5th Int. Conference of the European Asphalt Technology Association (EATA), Braunschweig, Germany.

Moreno-Navarro F, Sol-Sánchez M, Rubio-Gámez MC (2015) Exploring the recovery of fatigue damage in bituminous mixtures: the role of healing. Road Materials and Pavement Design DOI: 10.1080/14680629.2015.1029706.

Netzel DA (2006) Apparent activation energies for molecular motions in solid asphalt. Energy and Fuels 20: 2181-2188.

Pérez-Jimenez F, Botella R, Miró R (2012) Damage and Thixotropy in Asphalt Mixture and Binder Fatigue Tests. Transportation Research Record: Journal of the Transportation Research Board 1: 8-17.

Rosen, B (1964) Fracture processes in polymeric solids: phenomena and theory. New York: Wiley, 835 pp.

Shan L, Tan Y, Underwood B, Kim YR (2011) Separation of thixotropy from fatigue process of asphalt binder. Transportation Research Record: Journal of the Transportation Research Board 2207: 89–98.

Sperling, LH (2006) Introduction to physical polymer science. New Jersey: John Wiley & Sons, 845 pp.

Bump Formation on a Semi-Rigid Pavement: Interpretation and Modeling Using the Thick Level Set (TLS) Approach for Damage Growth

O. Chupin, J.-M. Piau, H. Odéon, A. Salzman and N. Moës

Abstract This paper is focused on the understanding of the origin of a bump a few centimeters high which suddenly popped up at the surface of a semi-rigid pavement composed of a bituminous layer placed over a cement concrete layer of 20 cm in thickness. Field investigations have shown that this bump developed right above a cement joint which had already undergone maintenance operations consisting in particular in the digging of trenches and resulting in a decrease of the cement layer thickness at the joint location. It was also noticed that the bump appeared during a warm day and in a wet area, which are both source of dilation. In this context, we suspect the mechanism at the origin of the bump to be triggered by compressive forces at the joint location whose effect is the bending of the cement layer and eventually failure by means of crack propagation in mode I (opening mode) due to excessive tensile stress. This paper aims at providing an accurate description of this field case which is modeled using the Thick Level Set (TLS) approach that proposes a straightforward transition from damage growth to fracture. The numerical implementation of the TLS approach is handled within the X-FEM framework. The results from the computations tend to confirm the suspected mechanism.

Keywords Semi-rigid pavement · Bump · Crack propagation · TLS approach

O. Chupin (✉) · J.-M. Piau
IFSTTAR, MAST, LUNAM Université, 44341 Bouguenais, France
e-mail: olivier.chupin@ifsttar.fr

J.-M. Piau
e-mail: jean-michel.piau@ifsttar.fr

H. Odéon
CEREMA, Dter Est, Strasbourg, France
e-mail: hugues.odeon@developpement-durable.gouv.fr

A. Salzman · N. Moës
GeM Institute, Ecole Centrale Nantes/Université de Nantes/CNRS,
1 Rue de La Noë, 44321 Nantes, France
e-mail: alexis.salzman@ec-nantes.fr

N. Moës
e-mail: nicolas.moes@ec-nantes.fr

A. Chabot et al. (eds.), *8th RILEM International Conference on Mechanisms of Cracking and Debonding in Pavements*, RILEM Bookseries 13,
DOI 10.1007/978-94-024-0867-6_28

1 Introduction and Context of the Disorders Observed

The bump under consideration (Fig. 1a) occurred on a pavement that had previously gone through rehabilitation works consisting in milling the original pavement and then refilling with asphalt concrete (Fig. 1b). These works had been carried out because of reflective cracking which had developed right above joints of the cement concrete slabs. As can be noticed in Fig. 1b only a partial milling of the cement concrete layer had been performed. The disorders on the pavement after rehabilitation were observed during a warm period in areas of high humidity and right above the location of the maintenance operations. In this context, we suspect that these disorders may have been triggered by horizontal compressive forces, developing in the cement concrete layer due to blocked dilation, subsequently leading to cracking of the milled part (thinner part) of the concrete course (see Sect. 2). These transverse cracks are believed to play the role of a hinge leading to rotation of the thin section around the crack location, which eventually led to the symmetric pattern of Fig. 1c. A bump would then result from the induced vertical displacement of the thin cracked sections (several centimeters high) whose effect at the surface of the pavement would be smoothed due to visco-elasto-plastic deformation of the asphalt concrete. The cracking mechanism envisaged for the bump formation will be justified in the next section using a simple beam theory. This case study also provides the opportunity of testing the TLS approach and its high potential to model damage growth in the framework of pavement mechanics.

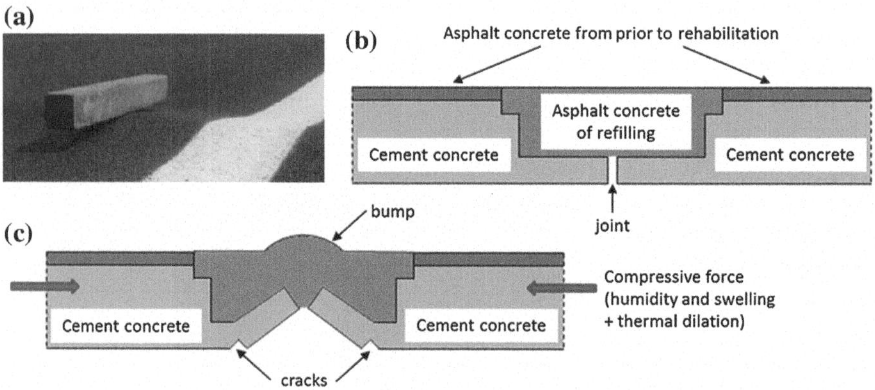

Fig. 1 **a** Picture of the bump observed on site, **b** sketch of the longitudinal section of the pavement resulting from previous milling operation prior to occurrence of the disorders, **c** mechanism envisaged for the bump formation

2 Response of the Pavement to Initial Compressive Stress Using a Simple Beam Theory

In this section, a simple beam theory is used to advocate for the possible occurrence of cracking in the milled part of the concrete slab which is assimilated to two beams of different heights (Fig. 2) and of unit length along the pavement width. The own weight of the slabs is neglected as well as the stiffness of bituminous materials considered small as compared to that of cementitious materials, especially for the slow loading rate imposed by the origin of the loading involving moisture and temperature.

In the calculations, the horizontal displacements at both ends of the beam are assumed to be zero and we consider only the horizontal normal efforts and the bending moments in the two beams resulting from the initial stress loading caused by moisture and temperature (Fig. 2). This combination is assimilated to a unique thermal loading with amplitude θ zeroed once the normal effort has become compressive. These calculations are used to estimate the order of magnitude of θ for which the maximal admissible stress in the cement concrete is reached.

The normal effort by unit length in the beams can be expressed as follows:

$$N = -E\alpha\theta\, h_1 h_2 (l_1 + l_2)/(h_1 l_2 + h_2 l_1) \tag{1}$$

in which α in the coefficient of thermal expansion, E the Young modulus of the cement concrete layer. The minus sign indicates a compression effect. Besides the bending moment in beam 1 arising from the shift between the neutral axes of the two beams is found to be:

$$M = E\alpha\theta(h_2 - h_1)h_1^4 h_2 (l_1 + l_2) l_2/(2(h_1 l_2 + h_2 l_1)(h_1^3 l_2 + h_2^3 l_1)) \tag{2}$$

The positive sign corresponds to the convex curvature of beam 1 whose bottom is in extension. The corresponding stress (σ_{xx}^b) along the bottom fiber is constant and results from the superimposition of effects due to N and M:

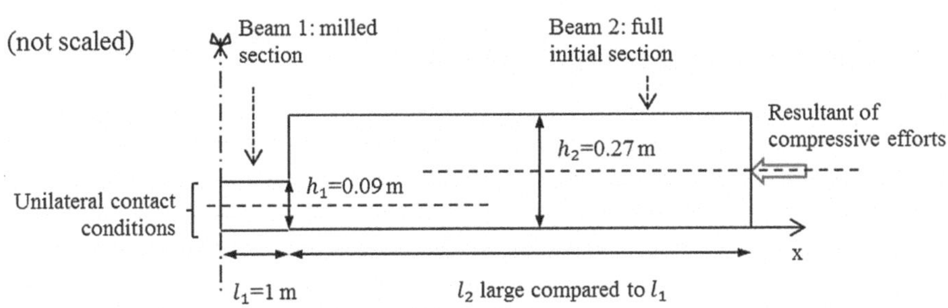

Fig. 2 Sketch of the milled cement concrete layer represented by two beams of length l_1 and l_2: longitudinal section with unilateral contact conditions in the x-direction on the symmetry axis

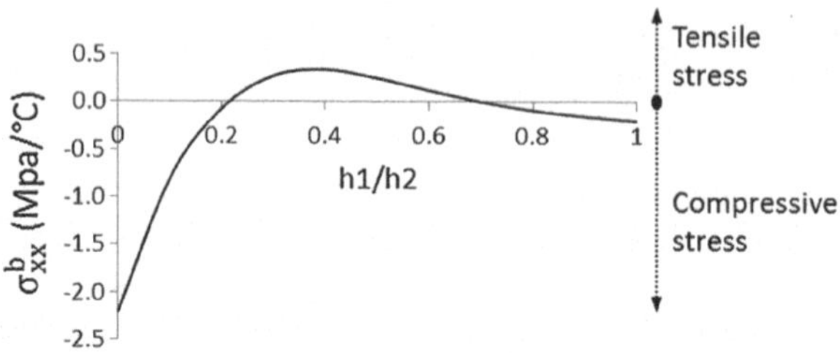

Fig. 3 Stress at the bottom of the milled section for $\theta = 1\,°\text{C}$ as a function of ratio h_1/h_2

$$\sigma_{xx}^{b} = E\alpha\theta h_2(l_1 + l_2)\left(3h_1^2 h_2 l_2 - 4h_1^3 l_2 - h_2^3 l_1\right)/\left[(h_1 l_2 + h_2 l_1)\left(h_1^3 l_2 + h_2^3 l_1\right)\right] \quad (3)$$

σ_{xx}^{b} is plotted as a function of h_1/h_2 in Fig. 3 which highlights the fact that for a range of values of h_1 between approximately $h_2/5$ and $4h_2/5$, the stress at the bottom of beam 1 switch from compressive to tensile stress with a maximum located around $h_1 = 0.35h_2$. This happens to be closed to the pavement case under consideration for which $h_1 = h_2/3$. Besides, substituting this ratio in Eq. (3) yields $\sigma_{xx}^{b} \approx 0.32 \times \theta(MPa)$, indicating that for θ above 5–10 °C the tensile strength of the material is likely to be exceeded, possibly causing the transversal cracking of the milled section. Once the cracks have occurred symmetrically, basic geometrical considerations show that the thermal expansion of beam 2 gives rise to the "hat pattern" of Fig. 1 with an upheaval $f = \sqrt{2l_1 l_2 \alpha\theta}$ which can reach several centimeters.

Note that for $h_1/h_2 < 0.2$ compression is again obtained in beam 1 but in this case one should rather examine the risk of failure based on the compressive strength of the material and/or the possible buckling of the beam.

It is worth noting that the initiation of loading cannot be located accurately using this simple model since the stress profile in beam 1 is independent upon the x-direction. In the next section, the finite element method combined to the TLS approach will make it possible to better account for the change in section of the cement concrete layer and to confirm the cracking mechanism and its location.

3 Analysis Using the TLS Approach

The TLS approach introduced by Moës et al. (2011) is a new way to model damage growth in solids and to regularize the local damage models which are known to suffer from spurious localization. In this approach implemented in a numerical program developed at ECN and validated against benchmarks (Moës et al. 2011; Bernard et al. 2012), the undamaged and damaged zones are distinguished through a level set ϕ (Fig. 4a). In the level set positive domain (damaged zone) the damage

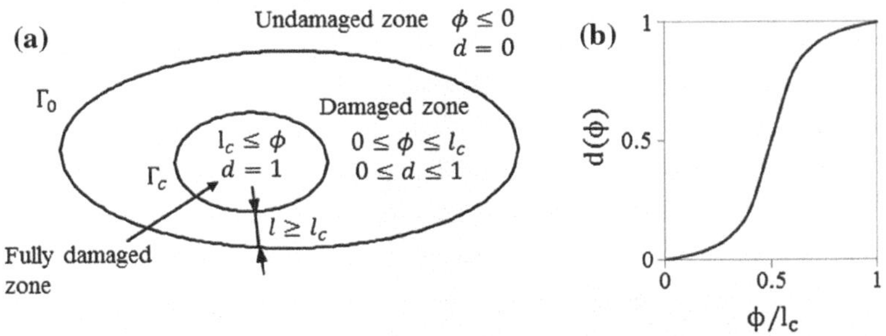

Fig. 4 **a** Example of damage distribution in the TLS approach, **b** function $d(\phi)$

variable depends explicitly on the value of the level set through a function $d(\phi)$ which represents an input for the model (Fig. 4b). Beyond a critical length the material is totally damaged and fracturing develops in a zone of truly zero stiffness delimited by $d = 1$, handled by the X-FEM technique. After onset of cracking in the damaged zone, the minimum geometrical thickness between the two fronts $\phi = 0$ and $\phi = l_c$ is greater or equal to l_c (characteristic length of the model).

The enrichment of the FEM functions considered in this study is the so-called ramped Heaviside enrichment (Bernard et al. 2012). The free energy considered in the computation of the damaged band is asymmetric in traction-compression:

$$\psi(\epsilon, d) = \sum_{i=1}^{i=3} \mu(1 - \beta_i d)\epsilon_i^2 + \lambda(1 - \beta d)tr(\epsilon)^2/2 \quad (4)$$

in which ϵ_i are the eigenvalues of the strain tensor, λ and μ the Lame coefficients and the (β, β_i) parameters are used to convey a different behavior in traction and compression. The damage growth is characterized by the propagation of the level set which is carried out using an explicit algorithm. For a given level set configuration, this algorithm consists in: (i) the calculation of a weighted average of the damage dual variable $\tilde{Y}(s)$ as a function of the curvilinear abscissa s along contour Γ_0 (after solving the elastic problem for a reference load), (ii) the calculation of the load factor μ defined as $\max_{s \in \Gamma_0}(\mu^2 \tilde{Y}(s)) = Y_c$ for a constant threshold Y_c (material parameter), (iii) the setting of the front advance $a(s)$ according to a function $f(\tilde{Y}(s)/Y_c, \mu)$ whose maximum value cannot exceed a_{max} (≈size of an element). More details can be found in the references already cited.

The TLS approach was applied to the structure Ω depicted in Fig. 2. However, at the time of the computations, initial stress loading implying the stiffness modulus as a function of the damage variable was not implemented yet. As a consequence, we replaced the volumetric loading $\int_\Omega (3\lambda(d) + 2\mu(d))\alpha\theta\hat{\varepsilon}_{ii} dV$ by the equivalent contour loading $\int_{\partial\Omega} (3\lambda_{d=0} + 2\mu_{d=0})\alpha\theta\hat{u}_i n_i dS$ neglecting the contribution of the damaged zone. The validity of this assumption can be verified a posteriori from the computation results. $\partial\Omega$ in the contour integral denotes the outer boundary to Ω.

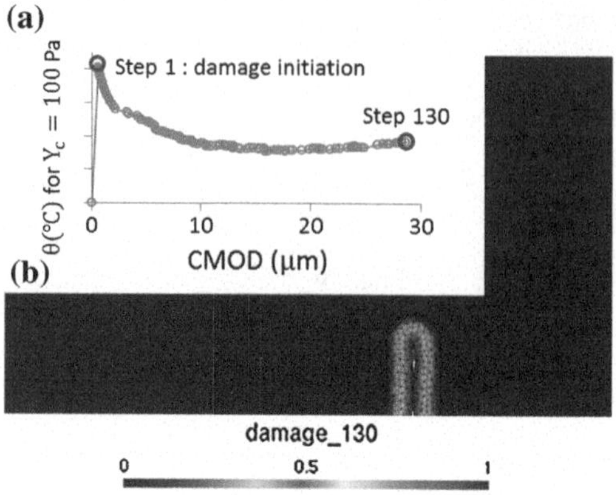

E (Mpa)	20000
ν	0.2
Y_c (Pa)	100
l_c (m)	0.02
α (strain/°C)	10^{-5}

Material parameters
Typical E, ν, α values for cement road subbase
Y_c deduced from $Y_c = E\varepsilon_{crack}^2$ with
$\varepsilon_{crack} = 70\mu def$; $\sigma_{crack} = 1,4MPa$
l_c : approximate dimension of grain size

Fig. 5 **a** Crack mouth opening displacement (CMOD) versus θ curve, **b** close-up over the damage pattern computed by the TLS approach and plotted on the initial geometry

Figure 5a shows CMOD as a function of θ computed with the TLS code and for $Y_c = 100$ Pa supposed to be representative of a low-strength concrete. For the model, damage initiates at step 1 for about $\theta = 4\,°\mathrm{C}$ and the propagation is mathematically observed for decreasing values of θ ending up at step 130 whose corresponding damage pattern is shown in Fig. 5b.

The transposition of these results to the real structure indicates that the collapse of beam 1 is unstable and brutal after the onset of cracking. Note that the order of magnitude of θ above the threshold that brings the structure in compression is related to the estimated value of Y_c which has not been measured for the real material. However, it seems realistic. These results confirm the beam theory approach which predicts the occurrence of cracking for $\theta = \sigma_{crack}/\sigma_{xx}^b \approx 5\,°\mathrm{C}$. They make it possible to specify the crack location and finally seem to validate the mechanism believed to be at the origin of the disorders.

4 Conclusion

Consistent elements between the physical interpretation of the bump formation at the surface of the semi-rigid pavement considered and the two modeling approaches based on the beam theory and the TLS implementation developed at ECN were obtained in this paper. From a practical point of view this study has showed that one should be cautious in the rehabilitation of semi-rigid pavements especially when milling of the concrete layer is involved. Concerning the TLS, the complete potential energy derivation, to account for the volumetric thermal loading, is under implementation in the numerical code.

References

Moës N, Stolz C, Bernard PE, Chevaugeon N (2011) A level set based model for damaged growth: The thick level set approach. Int. J. Numer. Meth. Engng 2011: 86: 358-380

Bernard P-E, Moës N, Chevaugeon N (2012) Damage growth modeling using the Thick Level Set (TLS) approach: Efficient discretization for quasi-static loadings. Comput. Methods Appl. Mech. Engrg. 233–236 (2012) 11–2

Modelling of the Fatigue Damage of Geogrid Reinforced Asphalt Concrete

Cyrille Chazallon, Ioana Maria Arsenie and Jean-Louis Duchez

Abstract This paper presents a modelling approach to predict the fatigue damage of glass fibre grid reinforced asphalt concrete in the fully reverse four point bending test (4PB) configuration. The objective of the presented study is to predict the fatigue damage of a composite material made of asphalt concrete and geogrid, using a damage model proven for asphalt mixes. The non-local damage model of Bodin, implemented in a finite element code, is used. Parameter identification is performed with 4PB test results of non-reinforced asphalt concrete beams. A unique model is developed and three analyses are performed for both non-reinforced asphalt (NRA) and geogrid reinforced asphalt (GRA-100 for a 100 kN/m grid reinforced asphalt and GRA-50 for a 50 kN/m grid reinforced asphalt) in the 4PB fully reverse fatigue test configuration with five-layer beams. A comparison between the predicted and the experimental fatigue damage curves (GRA-100 and GRA-50) indicates that the finite element model is able to predict the fatigue behaviour of geogrid reinforced asphalt mixtures.

Keywords Grid · Fatigue damage · 4PBT · FE modelling

C. Chazallon (✉) · I.M. Arsenie
Laboratory ICUBE—UMR 7357, CNRS, INSA de Strasbourg,
24 Boulevard de la Victoire, 67084 Strasbourg, France
e-mail: cyrille.chazallon@insa-strasbourg.fr

I.M. Arsenie
e-mail: ioanamaria.arsenie@insa-strasbourg.fr

I.M. Arsenie
Eiffage Infrastructures, 12 Rue de Molsheim, 67120 Wolxheim, France

J.-L. Duchez
Epsilon Ingénierie, Parc de Ruissel—Avenue de Lossburg, 69480 Anse, France
e-mail: laboepsilon@wanadoo.fr

A. Chabot et al. (eds.), *8th RILEM International Conference on Mechanisms of Cracking and Debonding in Pavements*, RILEM Bookseries 13,
DOI 10.1007/978-94-024-0867-6_29

1 Introduction

Characterization of the fatigue behaviour of grid reinforced asphalt concrete is required for the understanding of pavement behaviour, in order to find a way to take into consideration the presence and the efficiency of the grid in the pavement design. Modelling of the fatigue damage of grid reinforced asphalt concrete would reduce significantly the number of fatigue tests necessary for determining the fatigue characteristics of the materials.

The Bodin damage model (Bodin et al. 2004) is an elastic isotropic continuum damage model. This model is used in this study to predict the fatigue damage of asphalt concrete and of the composite material made of asphalt concrete and glass fibre grid.

The study is organized in five sections. Section 1 presents the materials and the experimental background. Section 2 presents the Bodin damage model. Section 3 presents the modelling used for the non-reinforced and geogrid reinforced asphalt concrete and the modelling results.

2 Materials and Experimental Background

This study is based on laboratory investigation of fully reverse four point bending (4PB) fatigue tests, performed on non-reinforced (NRA) and geogrid reinforced (GRA) asphalt beams (Arsenie 2013). The beams were made in three layers of 25, 50 and 25 mm. Bitumen emulsion was used between the asphalt concrete layers for both NRA and GRA specimens. The materials and the experimental program are briefly presented in this section.

The studied asphalt mixture is a semi-coarse asphalt concrete (SCAC 0/10 mm), according to the European classification (EN 13108 2007), made with 5.39 % binder content by weight of the asphalt mixture. The binder is a bitumen class 35/50, according to the European Classification (EN 12591 2009).

The studied geogrids are coated glass fibre grids whose strengths are 100 and 50 kN/m. The warp and filling yarns are made of continuous E fibre glass roving and styrene butadiene styrene (SBS) resin. The Young's modulus of the 100 kN/m geogrid is 44 and 35 GPa for the 50 kN/m geogrid. The endurance limit of both types of yarns in cyclic tensile fatigue displacement controlled tests is larger than 2 ‰.

The fully reverse 4PB bending test is performed with a sinusoidal waveform, creating a symmetrical bending around the original position (null loading). A total of 54 fully reverse 4PB fatigue tests were performed with displacement control (constant strain), at T = 10 °C with f = 25 Hz. Among the 54 specimens there were 18 NRA specimens, 18 GRA-100 specimens and 18 GRA-50 specimens. The NRA specimens were loaded at 3 strain levels (150, 135 μm/m and the range (111, 116) μm/m) with six replicates per strain level. The GRA specimens were tested at 3 strain levels (150, 135 μm/m and the range (126, 129) μm/m), with six replicates per strain level.

3 Bodin Model

The Bodin model is a fatigue damage model, which expresses the decrease of dynamic modulus with cyclic loading. The damage evolution is achieved by Bodin's three regimes law (Bodin et al. 2004), which simulates the three phases of the damage evolution of asphalt mixtures. The damage law is implemented in the finite element code cast3M (cast3M 2010). In Eq. (1), the damage variable is introduced as a scalar D whose range is [0, 1]. The damage D = 0 for virgin material and D = 1 for material failure.

$$\sigma_{ij} = (1 - D) \cdot C^{O}_{ijkl} \cdot \varepsilon_{kl} \tag{1}$$

where $\sigma_{ij}, \varepsilon_{kl}, C^{O}_{ijkl}$, are the components of the stress tensor, strain tensor, and elasticity tensor. The rate of damage growth is defined as a function of the equivalent strain by Eq. (2):

$$\dot{D} = f(D) \cdot \bar{\varepsilon}^{\beta} \cdot \langle \dot{\bar{\varepsilon}} \rangle \tag{2}$$

where f(D) is a function of three material parameters ($\alpha_1, \alpha_2, \alpha_3$) defined in Eq. (3), $\bar{\varepsilon}$ and $\dot{\bar{\varepsilon}}$ are the average of the equivalent strain defined in Eq. (4) and its strain rate. The parameter β is calculated directly from the slope p of the fatigue curve in log–log coordinates, as follows $\beta = p - 1$:

$$f(D) = \frac{\alpha_2}{\alpha_1 \cdot \alpha_3} \cdot \left(\frac{D}{\alpha_2}\right)^{1-\alpha_3} \cdot \exp\left(\frac{D}{\alpha_2}\right)^{\alpha_3} \tag{3}$$

$$\bar{\varepsilon}(x) = \frac{1}{V_r(x)} \cdot \int_{\Omega} \psi(x - s) \cdot \tilde{\varepsilon}(s) \cdot ds \tag{4}$$

where $V_r(x)$ is the representative volume at a point of coordinate x calculated in Eq. (5) and is the equivalent strain, calculated with Eq. (6):

$$V_r(x) = \int_{\Omega} \psi(x - s) \cdot ds \tag{5}$$

$$\tilde{\varepsilon} = \sqrt{\sum_{i=1}^{3} \left(\frac{\langle \sigma_i \rangle}{E \cdot (1 - D)}\right)^2} \tag{6}$$

where Ω is the studied volume, $\psi(x - s)$ is the weight function, s is a relative coordinate, $\langle \sigma_i \rangle$ are the principal tension stresses, E is the dynamic modulus (10 °C, 25 Hz). The Macauley brackets are used to account only the positives values.

4 Modelling

Three different plane stress models were made for the NRA, GRA-100 (100 kN/m grid reinforced asphalt), GRA-50 (50 kN/m grid reinforced asphalt) specimens using the software cast3M and five-layer models. The fatigue damage of 4PB fully reverse loading is modelled with the Bodin model.

The five-layer models were named 5NRA, 5GRA-100 and 5GRA-50. The five-layer models have the following structure: the first layer of 24 mm, the second layer of 1 mm, the third layer of 50 mm, the fourth layer of 1 mm, and the fifth layer of 24 mm. The model 5NRA is made of five asphalt concrete layers. The models 5GRA are made of three asphalt concrete layers (the first, the third and the fifth layer) and two geogrid layers (the second and the fourth layer).

The dynamic modulus (10 °C, 25 Hz) $E_i = 13.5$ GPa and $\nu = 0.35$ were taken into consideration for 5NRA model. In the 5GRA-100 model (respectively 5GRA-50 model), the geogrid layers are characterised by $E_i = 44$ GPa and $\nu = 0.35$ (respectively $E_i = 35$ GPa and and $\nu = 0.35$).

The hypothesis of plane stresses is used to model half of the NRA and GRA beams ($315 \times 100 \times 100$ mm^3) and all the layers are bonded. The geometry and the limit conditions of the models are presented in Fig. 1. The loading is strain driven. The elements are triangular with 3 nodes per element (maximum size 10 mm), the internal length parameter is set at 30 mm.

The fatigue damage parameters $(\alpha_1, \alpha_2, \alpha_3)$ of the Bodin model were calibrated with the experimental 4PB test results of NRA specimens on the average fatigue damage curve ($E/E_i - N$) at the strain level $\varepsilon = 135$ μm/m, then all the strain levels were predicted with 5NRA model. The parameter β was determined from the NRA fatigue curve equation.

The same parameter β was considered for both 5NRA and 5GRA models. The same set of parameters (α_2, α_3) was used for 5NRA, 5GRA-100 and 5GRA-50 models. The parameter α_1 was adjusted (Table 1).

Figure 2 presents the strain level used for the calibration for each of the three structures.

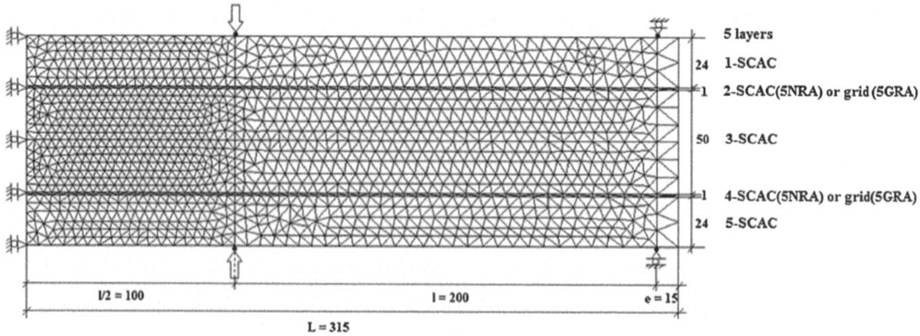

Fig. 1 Geometry and boundary conditions of five-layer models (5NRA and 5GRA)

Table 1 α_1 parameters adjustment for the strain level $\varepsilon = 135$ μm/m

	5NRA	5GRA-50	5GRA-100
α_1	$2.1\ 10^{-16}$	$2.6\ 10^{-16}$	$3.2\ 10^{-16}$

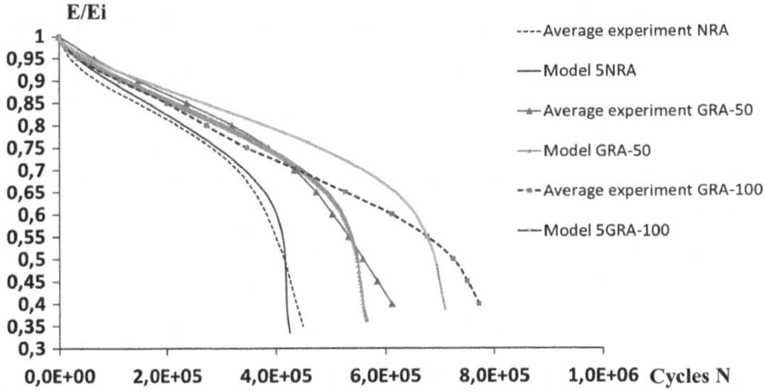

Fig. 2 Fatigue curves of the models 5GRA and 5NRA versus experimental average fatigue damage curves at $\varepsilon = 135$ μm/m for calibration

Then all the strain levels were predicted with 5GRA-100 and 5GRA-50 models (not presented). The fatigue life predictions made by the two GRA models are satisfying compared to experimental average fatigue life obtained in the 4PB fatigue tests (not presented here).

5 Conclusion and Perspectives

In this study, our objectives were to model the fatigue behaviour of a composite material made of asphalt concrete and fibre glass grid subjected to fully reverse four point bending (4PB) fatigue tests. We have used the Bodin model which has been developed for the modelling of fatigue damage of asphalt concrete material. The challenge was to use this model to predict the fatigue damage of asphalt concrete and of the composite material made of asphalt concrete and fibre glass grids.

Three models were tested, the first one for the non-reinforced asphalt concrete (5NRA) and the two others for the geogrid reinforced asphalt concrete (5GRA) for the two strength of the grid (100 and 50 kN/m). The parameters identification was performed with fully reverse four point bending (4PB) fatigue tests results (Arsenie 2013) obtained on non-reinforced and geogrid reinforced asphalt beams. We have to mention that only two parameters differ from one modelling to another: the Young's modulus E_i of the grid and α_1.

The calibrated fatigue damage curves are close to the experimental average fatigue damage curves, which confirm the accuracy of the Bodin model. The predicted fatigue damage curves of the geogrid reinforced asphalt concrete (5GRA-100 and 5GRA-50) are situated between the minimum and the maximum fatigue damage curves (not presented), which show that the model is able to predict damage of reinforced asphalt mixtures. The fatigue life predictions made by the three models are close to experimental fatigue life obtained in the 4PB fatigue tests (not presented).

References

Arsenie, I.M. (2013). Study and modelling of the pavement reinforcement with glass fibre grids under fatigue loading. PhD Thesis. INSA de Strasbourg, France.

Bodin, D., Pijaudier-Cabot, G., De la Roche, C., Piau, J. M., Chabot, A. (2004). Continuum Damage Approach to Asphalt Concrete Fatigue Modeling. Journal of Engineering Mechanics (ASCE), Vol. 130 (6): 700-708.

Cast3M, (2010). "cast3M is a research FEM environment; its development is sponsored by the French Atomic Energy Commission". http://www-cast3m.cea.fr.

EN 13108-1 (2007). European standard. Spécifications des matériaux, Partie 1: Enrobés bitumineux.

EN 12591 (2009). European standard Spécifications des bitumes routiers. Bitumes et liants bitumineux.

Predicting the Performance of SAMI Systems

Nick Thom, Thomas Chong and David Markham

Abstract This paper presents a method of predicting the resistance to reflective cracking of an overlay system incorporating a Stress Absorbing Membrane Interlayer (SAMI). The method is based on simulation of crack growth, both top-down and bottom up, over a joint or crack in an underlying layer, utilizing a strain-based crack propagation law. It uses an incremental system such that the strain in the region of a crack has to be re-calculated after each increment of crack growth. Its simplified nature means that the method is suited to being incorporated into a spreadsheet. In support of this method of crack prediction, a series of laboratory tests are described. These comprise wheel track tests on rubber-supported beams representing a scaled-down pavement construction—a cracked lower layer, a SAMI (where present) and an overlay. The data from these tests is input into an adapted version of the crack prediction system as the first step in the process of validation. Validation is then taken a step further by making reference to a particular case study that compares a concrete pavement overlaid with and without a SAMI, and shows the pavement with a SAMI to outperform that without. The magnitude of the improvement, though as yet not fully quantified since failure of the SAMI system has not yet occurred, is compatible with the predictions made. Finally the prediction method is applied to selected design cases and the types of pavement that lend themselves to SAMI usage are identified and approximate performance benefits estimated.

Keywords SAMI · Performance prediction · Beam tests · Case studies

N. Thom (✉)
University of Nottingham, Nottingham, UK
e-mail: nicholas.thom@nottingham.ac.uk

T. Chong · D. Markham
Tarmac, West Midlands, UK
e-mail: thomas.chong@tarmac.com

D. Markham
e-mail: david.markham@tarmac.com

A. Chabot et al. (eds.), *8th RILEM International Conference on Mechanisms of Cracking and Debonding in Pavements*, RILEM Bookseries 13,
DOI 10.1007/978-94-024-0867-6_30

1 Introduction

Stress Absorbing Membrane Interlayers (SAMIs) comprise an important set of tools available to those maintaining pavements, particularly in countering reflective cracking. It is understood that the purpose of a SAMI is to form a crack barrier against propagation from below and also to lessen the stress and strain concentration that would otherwise occur in the overlying asphalt material, and experience is that SAMIs can be very effective. Nevertheless, pavement designers in general do not have any trustworthy design method that they can use to predict performance in advance for individual cases, and it is the purpose of this paper to point the way towards the development of such a method.

Although this paper is based on experience with one particular SAMI system, a highly polymer modified fine-grained asphalt, it is considered that the approach described would apply equally to other systems.

2 Review of Experience from Site

The particular system upon which this paper is based has now been used since 2008 in several locations in the UK, typically over cracked and jointed concrete. In most cases it is too early to do other than report that the system is performing well so far. One case (Hall Quay, Great Yarmouth: 115 mm asphalt over 25 mm SAMI) showed minor reflective cracking over particularly damaged concrete after 6 years. However the most informative case to date is the A45 at Billing near Northampton, where 50 mm of asphalt was used over a 25 mm SAMI and compared directly to a control section with 75 mm of asphalt; the overlaid pavement was jointed concrete. After 6 years of relatively heavy traffic cracking had started to occur in the control section but not where the SAMI was used.

Nevertheless these experiences do no more than offer clues regarding performance; for design it is necessary to simulate performance in an appropriate way.

3 Computational Approach Taken

Conventional analytical pavement design is based on relatively simplistic assumptions. In design against fatigue cracking it is commonly assumed that the strain calculated at the base of the asphalt layers can be directly related to the life of the pavement by an equation of the form:

$$\text{Strain}(\varepsilon) = k \times \text{Number of load applications to failure } (N)^{-n} \qquad (1)$$

This approach has been found to be reasonably satisfactory for fully flexible pavements, although the constants have to be determined from calibration against actual performance. However, it is not directly applicable to composite pavements, overlays to cracked pavements, or prediction of thermally-induced cracking. It is also unsuitable in cases where different asphalt layers with very different fatigue resistance properties are combined, as in the case of overlays incorporating SAMIs.

The key simplification that limits the use of this approach is that 'failure' is effectively treated as a sudden event. A more accurate representation of reality would be to relate the strain in the material to a crack propagation rate, as follows:

$$\text{Strain}(\varepsilon) = \text{A} \times [1/\text{Crack propagation rate (dc/dN)}]^{-\text{n}} \quad (2)$$

The authors believe strain to be the appropriate parameter to use rather than stress intensity factor, as in conventional fracture mechanics of metals, because bitumen is the cracking medium. Parameters 'A' and 'n' can be found from a fatigue test. The test has to be modelled, tracing the propagation of cracking through several increments and re-calculating the strain value after each increment. Since asphalt is a composite material and the cracking is predominantly occurring through the amorphous bitumen phase, the appropriate strain is the average strain in the region of the crack tip. The strain near the crack tip changes during the test due to the presence of the crack. Figure 1 illustrates this effect for the 2-point bending test.

In fact the 'dead' zones shown in Fig. 1 have been modelled by working in terms of a 'damage' parameter (0 = full modulus; 1 = zero modulus—i.e. a crack) rather than as a sharp division between active and inactive areas of asphalt. This analysis, applied to the test data, allows parameters 'A' and 'n' to be deduced.

This crack propagation equation and computation approach can be applied directly, without any shift factor, to an incremental analysis of pavement cracking, so long as appropriate strains can be calculated. For this study, the OLCRACK and THERMCR systems (Thom 2000) have been adopted and adapted and Fig. 2 illustrates the different modes of pavement damage that have been evaluated.

Mode A is the development of top-down cracking due to tensile strains around the edge of the tyre-surface contact, which are calculated approximately. Mode B

Fig. 1 Stress/strain distribution during a fatigue test

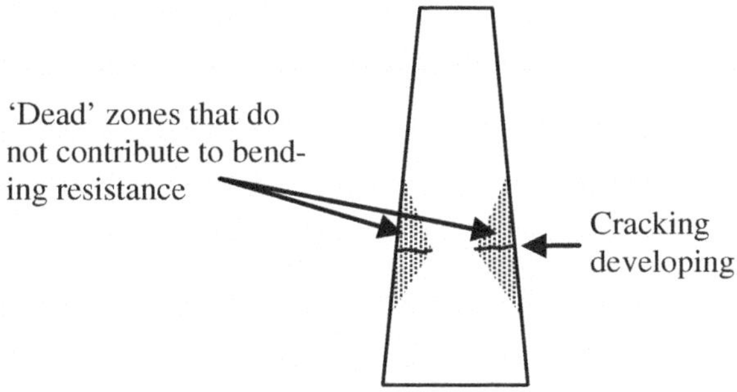

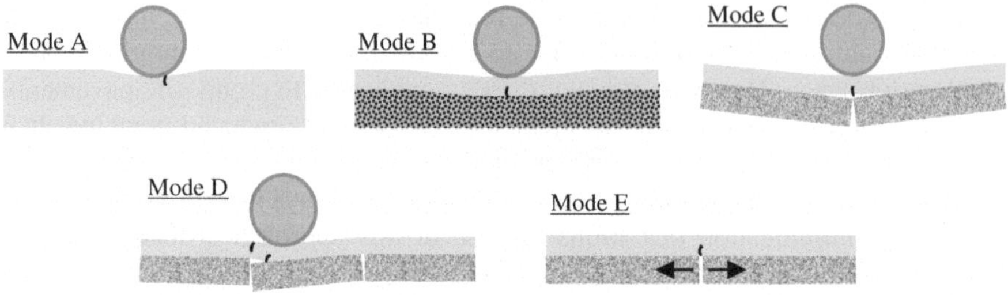

Fig. 2 Basic pavement damage models simulated

represents the simple flexible pavement type conventionally analysed; in this case the strain at the base of the asphalt layer is calculated using the method of equivalent thicknesses (Odemark 1949). Mode C represents an overlay to a cracked pavement. The two-dimensional situation illustrated, with linear elastic foundation and debonding between asphalt and base, can be solved exactly. An approximate conversion to a three-dimensional solution is then applied. Mode D represents the same pavement type but with the load offset from an underlying crack. It too can be solved exactly in two dimensions and is then converted approximately to three dimensions. Finally, Mode E is the thermally induced cracking case, and here it is necessary to calculate a crack opening in the base according to temperature change inputs. A temperature dependent viscosity is assigned to the asphalt and to the asphalt-base interface and the development of tensile strain in the asphalt during a single night is computed. Viscosity is based on the penetration and softening point of the binder using approximate relationships.

Certain further refinements are necessary: re-calculation of the effective bending stiffness of the asphalt layer assuming a zone of 'dead' material adjacent to each crack, as in Fig. 1 (Modes B–D); a correction to the effective spacing between cracks in the base depending on base bending stiffness (Modes C and D); and an adjustment to the resistance to shear movement across the cracks due to continuity of support provided by a bound sub-base (Mode D).

In order to accommodate the effects of a SAMI, a number of further refinements have had to be made. In Modes B–D the bending stiffness of the composite asphalt layer is adjusted according to the relative stiffness moduli of the overlay material and the SAMI. In Modes B and C cracks are allowed to initiate both at the bottom of the SAMI and at the bottom of the overlying asphalt. In Mode E the SAMI replaces the (very thin) binder layer that would otherwise be assumed to separate the asphalt from the base.

In reality the effects of all five modes have to be combined. In an overlay to a cracked pavement, Modes A, C and D combine to give crack propagation rates from the top and bottom of the overlay as well as from the bottom of the SAMI. Mode B only comes into operation if the cracks in the base are so closely spaced that support to the asphalt can be considered as continuous. The combined result of Modes A–D is a prediction for the number of wheel load repetitions until a crack has penetrated through the whole overlay. Mode E will give a prediction for the life of the overlay in terms of a number of thermal cycle repetitions. Since the same crack is being propagated both by both traffic and thermal action, the two predicted lives have to be converted from load repetitions to years and then combined to give an overall life before reflective cracking 'failure'.

4 Calibration/Validation/Utilisation

Predictions based on the traffic dependent damage elements (Modes A–D above) have been compared to standard UK pavement designs from Highways England documents and reasonable agreement has been found.

For thermally-induced damage (Mode E) an asphalt surface temperature range of −3 °C to +26 °C has been selected to represent a 'standard UK night', in a similar way to the use of a 'standard axle' for traffic characterization. In fact damage rate predictions were made numerous measured daily surface temperature ranges throughout a year, which were then summed, and the real year was found to be equivalent to approximately 26 applications of −3 °C to +26 °C, an extreme spring-time case. Predictions were then compared with experience from an overlaid concrete site (without SAMI) and found to be reasonable.

The first approach to validating predictions with a SAMI present involved producing a series of 300 mm long, 50 mm wide beams representing scaled-down pavement elements, which were then trafficked under a moving wheel with a continuous rubber support. The pavement elements comprised a 35 mm thick 15 pen asphalt concrete base and 45 mm of overlay. The overlay either comprised 45 mm of 50 pen asphalt concrete or 35 mm of 50 pen asphalt concrete over a 10 mm SAMI. Light applications of tack coat were applied between layers. The base layer of each slab was sawn throughout its depth across the middle of the slab to represent a crack. Trafficking was carried out at 20 °C under different wheel loads and both top-down and bottom-up cracking recorded.

The prediction approach described above (without the 2D:3D correction) was then applied, using moduli and fatigue characteristics determined from laboratory testing, but adjusted to take account of wheel speed (0.25 ms^{-1}). Values for tyre contact length (40 mm) and rubber modulus (3 MPa) were also determined. The resulting predictions are plotted against measurements in Fig. 3, where it may be noted that agreement is relatively good, particularly at higher load levels.

The second path to validation was by comparing predictions to performance on real roads, and in this case only the A45 site mentioned previously is able to give meaningful data. Assigning reasonable parameters for the asphalt overlay (which included a modified binder) and both concrete and asphalt expansion coefficients, a predicted life for the control case of 3–6 years can readily be obtained (depending on the effectiveness of the modification), matching what has been observed. The principal mechanism causing cracking is thermally driven in this case. When the

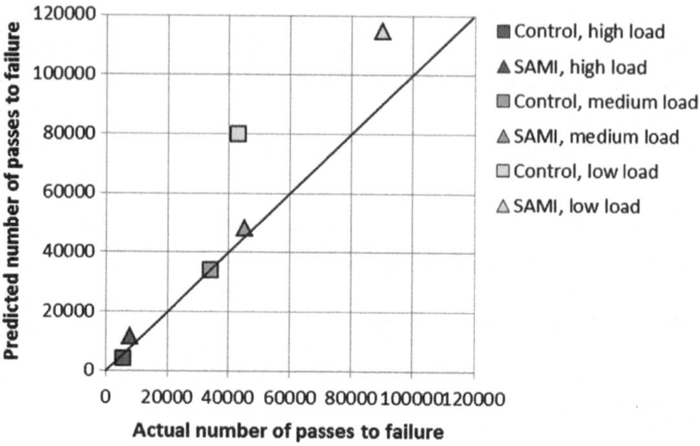

Fig. 3 Comparison of measurements and predictions, beam tests

lower 25 mm is replaced by the SAMI (with known properties), the prediction is for a life of 11–20 years, the dominant mechanism becoming traffic driven.

This pavement illustrates a particular common design case, typically giving a life enhancement factor of 4 or 5. However, when the method is applied to the case of overlays to a badly cracked asphalt road, enhancement factors many times as great (e.g. >50) can be predicted. The same is true for overlaid concrete airfield pavements, where numbers of load applications are much less than on roads.

5 Conclusion

The work described herein has outlined a rational approach to predicting the life of an overlay including a SAMI system. Naturally much more validation data will have to be obtained, preferably from different SAMI systems, before the approach can be claimed to be reliable, but at this stage it appears to comprise a useful design tool for maintenance decision making, potentially allowing a designer to target SAMIs for cases where their benefit is greatest.

References

Odemark N. (1949) Undersökning av elasticitetegenskaperna hos olika jordarter samt teori för beräkning av belägningar eligt elasticitesteorin. Statens Väginstitute, Stockholm.

Thom NH (2000) A simplified computer model for grid reinforced asphalt overlays. Paper presented at the 4th RILEM Conference on Reflective Cracking, Ottawa, p 37-46.

Reflective Cracking in Asphalt Overlays Reinforced with Geotextiles

Paulina Leiva-Padilla, Luis Loria-Salazar, Jose Aguiar-Moya and Fabricio Leiva-Villacorta

Abstract The use of overlays in pavement rehabilitation is one of the most common techniques in pavement management systems. It can be used when a pavement structure is still in adequate structural condition and the observable distresses can be corrected, ensuring that pavement condition can be maintained for an additional service period. The main issue related to the performance of this technique is the reflection of existing cracks. In order to delay the propagation process, interlayer systems can be used (e.g. geosynthetics). Although research on this area has been conducted previously, it is still necessary to better understand the mechanical behavior of the hybrid system to improve the efficiency of its performance. The following paper shows the results obtained from the measurements and modeling based on core samples extracted from an experimental test section built using geotextiles as an interlayer system prior to overlaying. The samples were evaluated in the laboratory by means of the Overlay Tester. The fracture mechanics and viscoelasticity properties were determined from Indirect Traction Tests, in order to define the Cohesive Zone Model (damage), and Dynamic Modulus Tests (viscoelastic Prony Series parameters). The results helped describe the mechanical behavior associated with the reduction in the reflective cracking process when geotextile materials are used as interlayer system: (i) increase in fatigue life (up to 260 %, due to the energy dissipation capabilities of the material), and (ii) crack propagation trend (which generally follows geotextile-asphalt layer interface, and depends on the amount of binder content, testing conditions and existing material).

P. Leiva-Padilla (✉) · L. Loria-Salazar · J. Aguiar-Moya · F. Leiva-Villacorta
National Laboratory of Materials and Structural Models of the University of Costa Rica,
San José, Costa Rica
e-mail: paulina.leivapadilla@ucr.ac.cr

L. Loria-Salazar
e-mail: luis.loriasalazar@ucr.ac.cr

J. Aguiar-Moya
e-mail: jose.aguiar@ucr.ac.cr

F. Leiva-Villacorta
e-mail: fabricio.leiva@ucr.ac.cr

A. Chabot et al. (eds.), *8th RILEM International Conference on Mechanisms of Cracking and Debonding in Pavements*, RILEM Bookseries 13,
DOI 10.1007/978-94-024-0867-6_31

Keywords Reflective cracking · Geotextiles · Cohesive zone · Finite element

1 Introduction and Motivation

Since 1982, during the "Resurfacing Portland Cement Concrete Pavements" symposium organized by the Highway Research Board (Sherman 1982), reflective cracking was recognized one of the most influent factors affecting the effectiveness of asphalt mix overlay systems.

This process is associated with the propagation of cracks from the existent asphalt layer to the constructed overlay, as a result of traffic loading and environmental changes.

According to Marchand and Goacolou (1982) crack propagation in overlay systems is dependent upon the direction of crack growth (vertical or horizontal) and the bound material. Based on the previous two types of cracked structures can be defined: bonded interface and debonded interface (with centered or displaced crack).

Although the use of geosynthetics is one of the most common techniques used to delay crack propagation between an existing and a new layer, there is no unified criteria for the design and the construction. The previous confirms the need for further research in this area.

The following paper shows results obtained from cores evaluated using the Texas Overlay test [OT, (Ma 2014)], proposed by Claros and Lytton in 1970 (Lytton 1989; Zhou 2007). The cores were extracted from a rehabilitated pavement section that incorporated the use of geotextiles. Additionally, asphalt mixture properties were characterized in the laboratory based on the dynamic modulus test, indirect tensile test and in field with a falling weight deflectometer (FWD). The geotextile properties were obtained based on a tensile test.

Based on the measured material properties, a finite element model of the OT sample was developed to study the mechanical performance of the geosynthetic in the overlay systems.

2 Dissipated Energy Theory

The dissipated energy concept started to be used in 1975, after Van Dijk proposed a mathematical relationship between accumulated energy and failure condition. This unique relationship does not depend on loading mode, stresses level, frequency or resting period (Ma 2014).

In dynamic loading tests, dissipated energy (DE) can be computed as the integral of the curve stress-strain (Eq. 1). In linear-elastic range, these curves are constant, because the material do not dissipate energy; however, outside of this range, the curves are hysteretic cycles, and the area within the curve corresponds to the dissipated energy that is transferred as heat or damage (Shen and Carpenter 2007).

$$DE = \pi\sigma_i\varepsilon_i sen\phi_i \quad (1)$$

where, σ_i: stress at cycle i, ε_i: strain at cycle i, and ϕ_i: phase angle from stress and strain during the ith cycle.

In 2000, Ghuzlan and Carpenter proposed the use of dissipated energy ratio (RDEC) to analyze damage and performance in asphalt mixtures. RDEC is the ratio between the change in dissipated energy from cycle i to cycle $i + 1$ (ΔDE) and dissipated energy up to cycle i (DE in i cycle) (Eq. 2) Ghuzlan and Carpenter (2000).

$$RDEC_a = \frac{DE_a - DE_b}{DE_a * (b - a)} \quad (2)$$

where, $RDEC_a$: average energy ratio during cycle a, and; DE_a, DE_b: dissipated energy in cycle a and b.

During testing, RDEC decreases up to a constant Plateau Value (PV, Eq. 3), because of the micro-structural realignment of the material. Once the material fails (with failure defined as a 50 % reduction in stiffness), the PV increases.

$$PV = \left[1 - (1 + 100/N_{f,50})^f\right]/100 \quad (3)$$

where, $N_{f,50}$: number of cycles to 50 % reduction in stiffness; and, f: slope from linear relationship between DE and number of cycles over $N_{f,50}$.

2.1 Texas Overlay Test

The first OT was designed in 1970 by Claros and Lytton. The system consists in gluing a sample (obtained from a 150 mm diameter core with a height of 62 mm) to two steel plates separated by a 2 mm gap as shown in Fig. 1. A constant displacement of 6.35 mm is applied to simulate the opening and closing of cracks.

During testing, three phases of cracking are identified (Zhou and Scullion 2003): (i) Phase 1—Crack initiation and early propagation: During this phase, load and displacement have similar shapes; as maximum displacement is reached, load increases too, (ii) Phase II—Late crack propagation: After the maximum load, crack starts to propagate and load decreases, load and displacement are still in the same

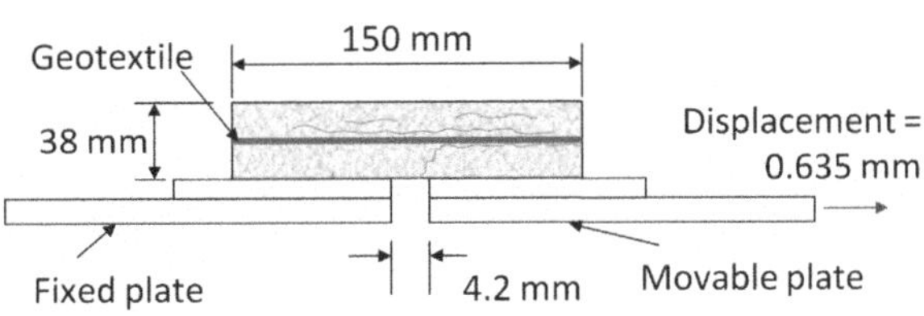

Fig. 1 Overlay test system

phase; and (iii) Phase III—Specimen Failure: In this phase the crack extends to the top of the specimen. The amount of load remaining is associated with the minor adhesion.

Based onTex-248-F testing protocol, failure criteria occurs when a 93 % reduction in the load is achieved (moment where the crack is expected to have traveled thru the full thickness of the specimen).

Ma (2014) proposes the use of ASTM D7460 (Standard Test Method for Determining Fatigue Failure of Compacted Asphalt Concrete Subjected to Repeated Flexural Bending), to analyze fatigue failure. ASTM D7406 defines fatigue failure as the maximum point of the normalized load curve (NLC, Eq. 4) as a function of the number of cycles.

$$NLC_i = P_i N_i P \tag{4}$$

where, NCL_i: normalized load at cycle i, P: maximum load after first cycle, N_i: cycle i, and P_i: maximum load at cycle i.

3 Test Methods

The study consisted of testing and simulating the reflective cracking phenomenon in the OT device under testing protocol Tex-248-F.

Nine cores were extracted from a section of an overlay rehabilitated project, where 50 mm of the existent asphalt layer was milled, a nonwoven geotextile was placed as interlayer system, followed by a 50 mm of asphalt mixture overlay. The existing pavement structure was as follows: 80 mm asphalt layer, 100 mm granular base, 150 mm granular subbase over a subgrade soil.

The mechanical properties of the existing materials were described from FWD backcalculation (Table 1). Additional testing was performed on the overlay asphalt mixture: dynamic modulus testing to describe mixture modulus, indirect tensile tests to define CZM (Cohesive Zone Model) properties of the asphalt mixture, and traction GRAB tests (ASTM D4632) to compute geotextile modulus.

Using the previous results and geometry dimensions of the OT, a tridimensional finite element model was built. The model consists of 128,336 elements and 69,466 nodes. A displacement of 0.635 mm is applied to the movable face, as shown in Fig. 2.

Table 1 Material mechanical characterization properties

Material	Material type	Test	Property	
Existent asphalt mixture	Lineal elastic	FWD	Elastic modulus = 2722 MPa (25 °C)	
New asphalt mixture	Viscoelastic	Dynamic modulus	$\tau_{relaxation}$(s)	Prony parameters, Ei (MPa)
			1.00E − 06	2.01E + 00
			1.00E − 05	8.20E + 01
			1.00E − 04	1.04E + 03
			1.00E − 03	2.72E + 03
			1.00E − 02	3.19E + 03
			1.00E − 01	2.97E + 03
			1.00E + 00	4.18E + 03
			1.00E + 02	1.51E + 03
			1.00E + 03	3.41E + 02
			1.00E + 04	2.33E + 02
			1.00E + 05	9.34E + 01
New asphalt mixture	Cohesive zone	Indirect tensile	Elastic modulus = 2722 MPa (25 °C) Max principal stress = 1.3 MPa Displacement at failure = 0.0209	
Nonwoven geotextile	Linear elastic	Traction GRAB	Tension GRAB = 0.9 kN εlong = 0.7 mm/mm Elastic modulus = 184 MPa	

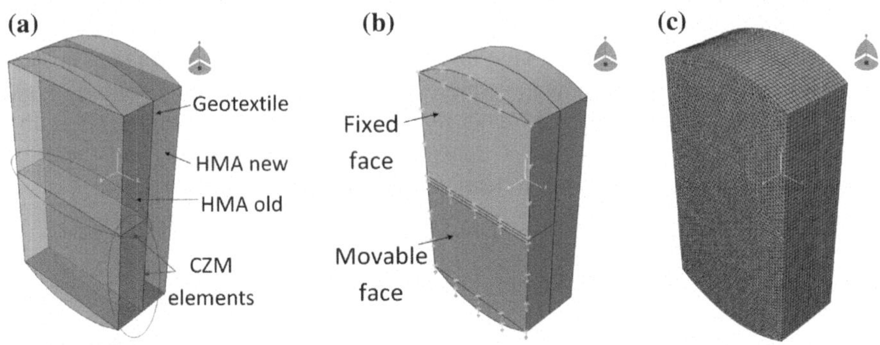

Fig. 2 OT sample finite element model

4 Results

Based on the dissipated energy concept, fatigue performance was analyzed based on the OT. Figure 3a shows how the dissipated energy drops with the number of cycles because of loss of capacity in the material.

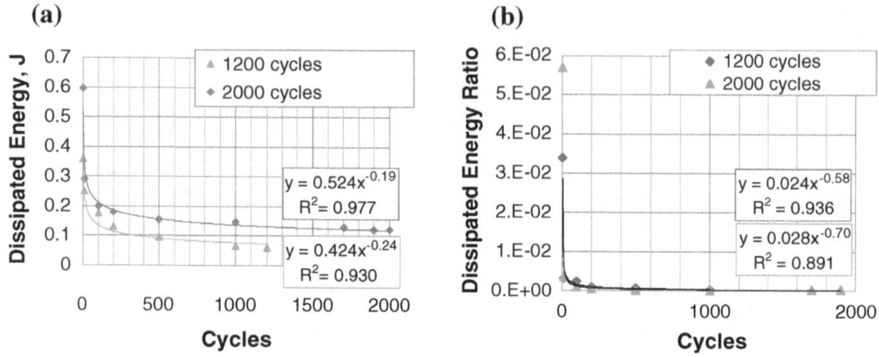

Fig. 3 Overlay tester results **a** dissipated energy and **b** dissipated energy ratio

Additionally, from the RDEC plot shown in Fig. 3b, it is evident that complete failure does not occur during testing. However, the PV is near to zero, which implies a lower damage per cycle and longer fatigue life (Vargas-Nordcbeck et al. 2014; Prowell et al. 2010; Shen and Carpenter 2007).

According to Ma (2014), rich bottom asphalt mixtures require approximately 750 cycles to fail in the OT. Consequently the analyzed specimens have a remaining life of over 260 % prior to reaching failure. This could be associated to the energy absorption by the geotextile.

As a second step, the OT finite element model was developed. Material properties and used geometry are as described in Table 1 and Fig. 2.

Results obtained shows concentration of tensile stresses at the interface between the new and the existent asphalt mix layer (Fig. 4a), causing a gradual loss of the bond capacity between the geotextile and the existent layer. The previous reflects what was observed on laboratory samples (Fig. 4c). The gradual loss of bond capacity generates stretching and higher tensile stresses in the geotextile (Fig. 4b), reducing the tensile stressed that is transmitted to the upper layer, and therefore delaying the development of the reflective cracking mechanism.

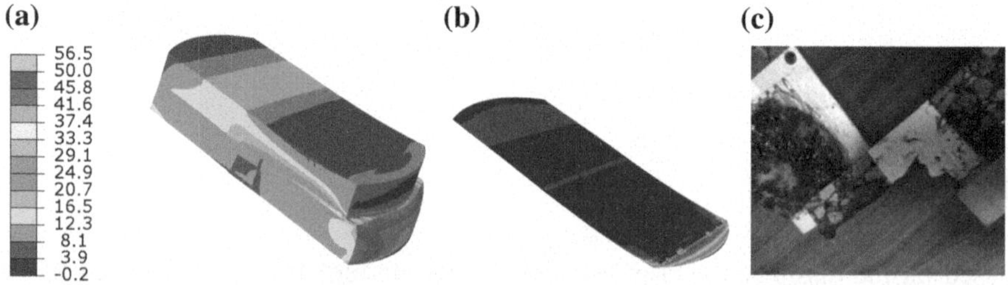

Fig. 4 Finite element modeling results. Maximum principal stresses **a** in the specimen and **b** in the geotextile. **c** Failed OT specimen

5 Conclusions and Recommendations

The main conclusions derived from this study are the following:

- In overlay systems reinforced with geotextile and tested in the OT, crack progression follows the geosynthetic-existent asphalt layer interface. This can be attributed to mechanical work by the geotextile.
- It is important to standardize an adequate correct construction practice to guarantee the adequate functionality of the geotextile: appropriate adhesion between layers and application rate of the bond coats.
- Finite element modeling was able to describe the mechanical performance of the system when geosynthetic are used, showing it could be better to change the settings of the normal test to obtain better results.
- Placing a geosynthetic material in overlay systems improved the overall fatigue resistance in the laboratory. However it is still necessary to transfer the results to field conditions, in order to developed shift or calibration factors to predict structural performance.
- The dissipated energy approach is an useful tool to study fatigue phenomenon in asphalt mixtures. It is recommended to extend the study in order to reach total failure of the system.

References

Sherman G. (1982). Minimizing reflection cracking of pavement overlays. Transportation Research Board, National Research Council.

Marchand J.P. and Goacolou H. (1982). Cracking in wearing courses. Proceedings of the 5th International Conference on the Structural Design of Asphalt Pavements, Delf, 741-777.

Lytton R (1989) Use of geotextiles for reinforcement and strain relief in asphalt concrete. Geotextiles and Geomembranes (Geotextiles and Geomembranes, Vol. 8, 217-237), Vol. 8, 217-237.

Ghuzlan K., & Carpenter S. (2000). Energy-derived, damage-based failure criterion for fatigue testing. Transportation Research Record: Journal of the Transportation Research Board, (1723), 141-149.

Zhou F, Hu S, and Scullion T (2007) Development and verification of the overlay tester based fatigue cracking prediction approach. Austin: Texas Department of Transportation, Research and Technology Implementation Office. Report 9-1502-01-8.

Vargas-Nordcbeck A, Aguiar-Moya JP, Leiva-Villacorta F, and Loria-Salazar LG (2014) Evaluation of Fatigue Life of Asphalt Mixes Through the Dissipated Energy Approach. 94th Annual Meeting of the Transportation Research Board.

Ma W. (2014). Proposed Improvements to Overlay Test for Determining Cracking Resistance of Asphalt Mixtures. (Doctoral dissertation, Auburn University).

Shen S and Carpenter S (2007) Dissipated energy concepts for HMA performance: fatigue and healing. Illinois: Federal Highway Administration.

Zhou F., and Scullion T. (2003). Upgraded overlay tester and its application to characterize reflection cracking resistance of asphalt mixtures. Austin: Texas Department of Transportation and Federal Highway Administration. Report 0-4467-1.

Prowell B, Brian D, Brown R, Anderson M, Daniel J, Swamy A, Von Quintus H, … &Maghsoodlo S (2010) Validating the Fatigue Endurance Limit for Hot Mix Asphalt. NCHRP Report 646. National Cooperative Highway Research Program. Transportation Research Board.

Part IV
Cracking in Asphalt Materials: Thermal Crack Mechanisms

Understanding the Effects of Ageing and Temperature on the Fatigue Cracking Resistance of Bituminous Mixtures

Fernando Moreno-Navarro, Miguel Sol-Sánchez, Gema García-Travé and Mª Carmen Rubio-Gámez

Abstract Bituminous mixtures are visco-elastic materials whose mechanical performance is highly influenced by temperature. In addition, during their service life, these type of materials suffer an ageing process which induce an increase in bitumen viscosity and therefore, a variation in the response of the mixture. The changes produced by these phenomena are mainly manifested in terms of stiffness, which could exert a considerable influence on the fatigue cracking resistance of asphalt materials. In this respect, this paper attempts to provide a better understanding of the effects caused by temperature and ageing on the resistance of bituminous mixtures to the propagation of cracking phenomena. For this purpose, the rheological response of the binder has been correlated with the mechanical performance offered by the bituminous mixture.

Keywords Ageing · Temperature · Rheology · UGR-FACT

1 Introduction

Bituminous mixtures are visco-elastic materials whose mechanical properties highly depend on the temperature of service (Lee and Kim 1998), as well as on the asphaltenes and maltens content (Zhang and Greenfield 2007). At high temperatures, bituminous materials behave in a more viscous way (ductile fracture, offering

F. Moreno-Navarro (✉) · M. Sol-Sánchez · G. García-Travé · M.C. Rubio-Gámez
LabIC, UGR Laboratory of Construction Engineering, University of Granada, Granada, Spain
e-mail: fmoreno@ugr.es

M. Sol-Sánchez
e-mail: msol@ugr.es

G. García-Travé
e-mail: labic@ugr.es

M.C. Rubio-Gámez
e-mail: mcrubio@ugr.es

A. Chabot et al. (eds.), *8th RILEM International Conference on Mechanisms of Cracking and Debonding in Pavements*, RILEM Bookseries 13,
DOI 10.1007/978-94-024-0867-6_32

high values of phase angle δ) and they are more susceptible to flow. On the contrary, when the temperature of service is low, these materials behave in more elastic way (brittle fracture, with low values of phase angle δ) and thus they have more capacity to support the stresses without flow. Therefore, temperature will play an important role in the mechanical resistance of bituminous mixtures to fatigue cracking.

In this sense, ageing also has a considerable influence on how bituminous mixtures are able to resist fatigue cracking (Arega et al. 2013). This phenomenon occurred during the manufacture and the service life of the mixtures (mainly due to their exposition to temperature, oxygen and UV photo-radiation), and it includes loss of volatiles, exudative hardening, and oxidation process (Read and Whiteoak 2003). These processes lead to a modification in the asphaltenes/maltens relationship, which cause an increase in bitumen viscosity, becoming more hard and brittle (Lesueur 2008). Therefore, ageing induce an increase in the stiffness and a decrease in the ductility of the bituminous mixtures, which could reduce its resistance to fatigue cracking (Abu Al-Rub et al. 2103).

The objective of this paper has been to analyse the combined effect of ageing and temperature on the fatigue cracking behavior of bituminous mixtures. For this purpose, the rheological response of the binder has been correlated with the mechanical performance offered by the mixture. In this sense, several tests at binder and mixture level have been carried out using different temperature conditions and ageing processes.

2 Methodology

2.1 Materials

An asphalt concrete mixture (AC, EN 13108-1) manufactured with a hard bitumen B 20/25 (penetration EN 1426: 15–25 0.1 mm; softening point EN 1427: 61–71 °C) and a maximum limestone aggregate size of 16 mm, has been used to carry out this study. In this respect, a low penetration grade binder has been selected in order to assess the grade of influence of ageing in bituminous materials which already have a high asphaltens/maltens relationship.

After the manufacture of the mixture, it was divided into two groups: an un-aged group (AC-16 UA, whose specimens were not conditioned after their manufacture) and an aged group (AC-16 A, whose specimens were introduced in an oven at 90 °C during 8 days). In the same way, to analyse the effect of ageing in the rheological response of the binder, the bitumen was tested without any treatment (B 20/25) and after being aged in the laboratory by means of the standardized combined procedures of the RTFOT (Rolling Thin Film Oven Test) and PAV (Pressure Ageing Vessel) to simulate long-term ageing (B 20/25 RTFOT + PAV).

2.2 Testing Plan

The rheological response of the un-aged (B 20/25) and aged (B 20/25 RTFOT + PAV) binders was analysed using the DSR frequency sweep test (EN 14770) at various temperatures (from 10 to 80 °C). Further, in order to analyse the influence of this parameter on the viscoelastic response of the mixture, the results for a fixed frequency (5 Hz) at different temperatures are displayed.

The mechanical performance of the mixtures was tested using the UGR-FACT method at different temperatures (10, 20 and 30 °C). This test method reproduces the efforts that lead to the failure of the pavement (Moreno-Navarro and Rubio-Gámez 2013). The test device is composed of a base (Fig. 1a), two supports where the specimen is fixed (Fig. 1b), and a load application plate (Fig. 1c), which is also provide of a flat surface for the vertical deformation gauges (Fig. 1f). The base has a platform that is composed of two sloping surfaces with two rails that allow for the sliding of the supports, and of two vertical spindles that are used to measure vertical deformations in the upper part of the test specimen (Fig. 1d). The two supports are composed of a carriage that is adapted to the shape of the rail at the base, and a support plate (to which the test specimen is attached with epoxy resin) where the horizontal deformation gauges are located (Fig. 1e). Furthermore, under these support plates, two elastic elements are introduced in order to allow for the flexion of the specimen (Fig. 1f) and a spring that simulates the foundation layers (Fig. 1g).

The simple geometry of the test device is capable of generating horizontal as well as vertical deformations in the test specimen, which try to simulate the bending and shear stresses due to traffic loading, as well as tensile strains due to the effect of thermal gradients. For this purpose, a combined load function is used (Fig. 1): a main up and down ramp that simulates thermal effects (with lower frequencies and higher amplitude) and a secondary versine that represents traffic loads (with higher frequencies and lower amplitudes).

Thus, the test method is able to generate and propagate a controlled fatigue cracking process, and its evolution is studied through the changes produced in the dissipated energy and in the geometry of a representative volume of the specimen

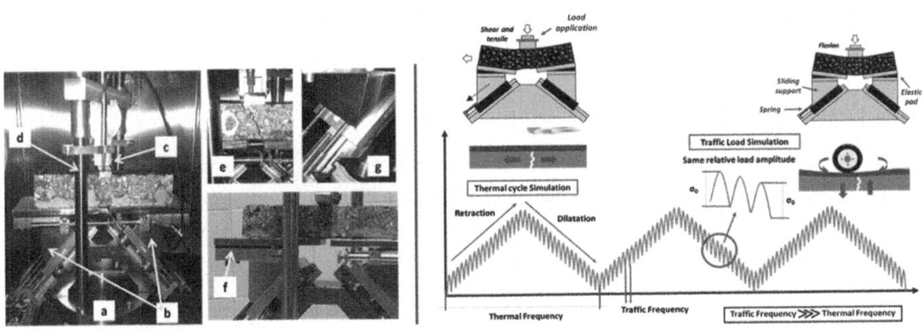

Fig. 1 UGR-FACT device and loading function

where the phenomenon takes place. On the one hand, the hysteresis loops described by the material in each direction are used to define the dissipated energy in each load cycle (Eq. 1).

$$\omega_i = \omega_{hi} + \omega_{vi} \tag{1}$$

Where ω_i is the dissipated energy in cycle i (in J/m^3); ω_{hi} is the horizontally dissipated energy in cycle i (in J/m^3); and ω_{vi} is the vertically dissipated energy in cycle i (in J/m^3). The cumulative RDEC (Ratio of Dissipated Energy Change, Eq. 2) is used to analyze the evolution of the damage produced in the specimen, and the Mean Damage Parameter (γ, Eq. 3) is a value used to establish a reference for the resistance to fatigue cracking of the pavement (Moreno-Navarro and Rubio-Gámez 2014).

$$\mathrm{RDEC}_{n+1} = \frac{\omega_{n+1} - \omega_n}{\omega_n} \tag{2}$$

$$\gamma = \frac{\sum_{\mathrm{i}=1}^{N_f} \mathrm{RDEC}_{\mathrm{i}}}{N_f} \tag{3}$$

Where ω_n is the energy dissipation produced in loading cycle n (in J/m^3); ω_{n+1} is the energy dissipation in loading cycle n + 1 (in J/m^3); and N_f is the failure cycle of the specimen. On the other hand, the permanent displacements measured in the specimen in each load cycle are used to define de variation of the geometry ($\Delta\varepsilon_i$) of the material in the zone where the fatigue phenomenon takes place in the vertical and horizontal directions (Eq. 4).

$$\Delta\varepsilon_i = \frac{(\delta_{\mathrm{hi}} \cdot \delta_{\mathrm{vi}}) - (\delta_{\mathrm{hi}-1} \cdot \delta_{\mathrm{vi}-1})}{(\delta_{\mathrm{hi}-1} \cdot \delta_{\mathrm{vi}-1})} \cdot 100 \tag{4}$$

Where $\Delta\varepsilon_i$ is the variation of the geometry of the specimen in the cycle i (in percentage); δ_{hi} and δ_{vi} are the horizontal and vertical dimensions of the specimen in the cycle i; and δ_{hi-1} and δ_{vi-1} are the horizontal and vertical dimensions of the specimen in the previous cycle. Table 1 summarizes the tests carried out during this research study.

Table 1 Testing plan

	Material tested	Temperature (°C)	Number of specimens	Observations
DSR frequency sweep	B 20/25	From 10 to 80	3	0.1 % strain; from 0.1 to 20 Hz
	B 20/25 RTFOT + PAV	From 10 to 80	3	
UGR-FACT	AC-16 UA	10, 20 and 30	9	Stress amplitude 0.8 MPa; Frequency 5 Hz
	AC-16 A	10, 20 and 30	9	

3 Analysis of Results

Figure 2 shows the results obtained in the DSR frequency sweep tests. As can be observed, ageing clearly increase the complex modulus of the binder and decrease the phase angle for the same test conditions (doing it more hard and brittle). In addition, as the temperature increases the effect of ageing becomes more notable, inducing more differences in the visco-elastic response of the material. Similarly, the aged binder is less susceptible to temperature changes.

Figure 3 shows the mean results obtained in UGR-FACT at different temperatures.

As observed in binder tests, the aged mixture is less susceptible to temperature changes, but it is less resistant to fatigue cracking than the un-aged one (it has a higher mean damage parameter at any temperature). In addition, as the temperature

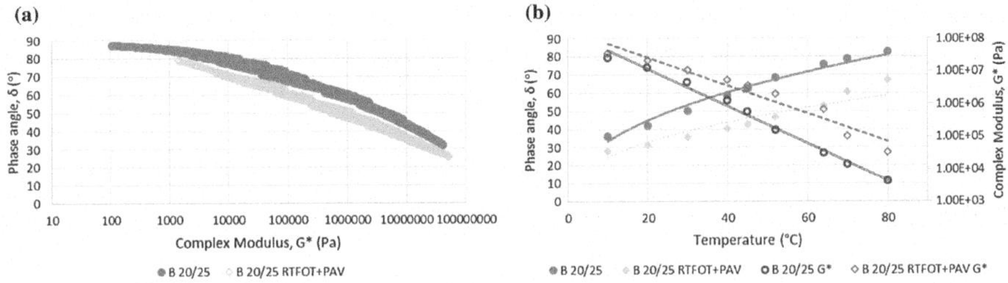

Fig. 2 Mean results obtained in DSR tests. **a** frequency sweep, **b** constant frequency 5 Hz

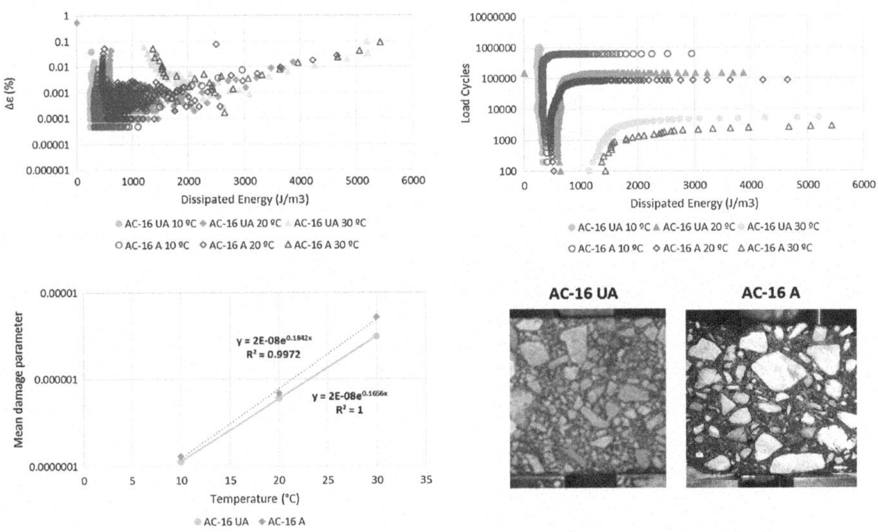

Fig. 3 Mean results obtained in the different specimens during the UGR-FACT test

increases, the effect of ageing becomes more remarkable. Nonetheless, it can be observed that for the binder tested in this study, the influence of the temperature in the mechanical response of the mixture is higher than the influence of ageing.

4 Conclusions

This paper evaluates the influence of ageing and temperature in the fatigue cracking resistance of bituminous mixtures. In order to highlight the real effect of ageing, a low penetration grade binder (which is less susceptible to ageing, as it already has a high asphaltenes content) has been studied. Results obtained showed that binder rheology has a direct relationship with the long-term mechanical performance of the mixture. In this sense, binder ageing has a negative impact in fatigue cracking resistance of asphalt mixtures, which is more remarkable as the temperature of service increases. Nonetheless, when using binders that could have a low rate of ageing (as low penetration grade bitumens), service temperature has a more notable influence on the fatigue cracking resistance of asphalt mixtures.

References

Abu Al-Rub R, Darabi M, Kim S, Little D, Glover C (2103) Mechanistic-based constitutive modeling of oxidative ageing in ageing susceptible materials and its effect on the damege potential of asphalt concrete. Construction and Building Materials 41: 439-454.

Arega Z, Bhasin A, De Kesel T (2013) Influence of extended ageing on the properties of asphalt composites produced using hot and warm mix methods. Construction and Building Materials 44: 168-174.

Lee HJ, Kim YR (1998) Viscoelastic constitutive model for asphalt concrete under cyclic loading. Journal of Engineering Mechanics 124: 32-40.

Lesueur D (2008) The colloidal structure of bitumen: consequences on the rheology and on the mechanisms of bitumen modification. Advances in Colloid and Interfase Science 145(1-2): 42-82.

Moreno-Navarro F, Rubio-Gámez MC (2013) UGR-FACT test for the study of fatigue cracking in bituminous mixes. Construction and Building Materials 43: 184-190.

Moreno-Navarro F, Rubio-Gámez MC (2014) Mean damage parameter for the characterization of fatigue cracking behavior in bituminous mixes. Materials and Design 54: 748-754.

Read J, Whiteoak D (2003) The Shell Bitumen Handbook. 5th ed. London, England: Thomas Telford Publishing, Shell Bitumen UK.

Zhang L, Greenfield ML (2007) Relaxation time, diffusion and viscosity analysis of model asphalt system using molecular simulation. Journal of Chemical Physics 127(9). DOI:10.1063/1.2799189.

Multiplicative Viscoelastic-Viscoplastic Damage-Healing Model for Asphalt-Concrete Materials

Romain Balieu, Nicole Kringos, Feng Chen and Enrique Córdoba

Abstract A viscoelastic-viscoplastic model based on a thermodynamic approach is developed under finite strain in this paper. By introducing a damage evolution, the proposed model is able to reproduce the behavior of Asphalt-Concrete materials until the complete fracture. Moreover, a recoverable part of the degradation is introduced to reproduce the self-healing observed under a sufficiently long rest period. The proposed model is implemented into a Finite Element code and good correlations between the numerical responses and the experiments have been observed.

Keywords Viscoelasticity · Viscoplasticity · Damage-Healing · Finite strains

1 Introduction

In the context of computational modeling, the incorporation of degradation and recovery in the form of self-healing of material properties is still a challenge particularly when the material is submitted to large deformations/rotations. In this work, a finite strain viscoelastic-viscoplastic model coupled to anisotropic damage is developed through a thermodynamics admissible framework. A hyperelastic-based model using the multiplicative decomposition of the deformation gradient tensor into elastic and plastic parts is used and the resulting plastic intermediate

R. Balieu (✉) · N. Kringos · F. Chen · E. Córdoba
Department of Civil and Architectural Engineering, KTH Royal Institute of Technologies, Stockholm, Sweden
e-mail: balieu@kth.se

N. Kringos
e-mail: kringos@kth.se

F. Chen
e-mail: feng.chen@abe.kth.se

E. Córdoba
e-mail: ecl@kth.se

A. Chabot et al. (eds.), *8th RILEM International Conference on Mechanisms of Cracking and Debonding in Pavements*, RILEM Bookseries 13,
DOI 10.1007/978-94-024-0867-6_33

configuration is connected with a fictitious undamaged configuration. A second order degradation tensor is thus introduced to connect nominal and effective configurations. Furthermore, this degradation tensor is also decomposed into permanent and non-permanent parts. As the permanent degradation part represents the classical irreversible damage process, the non-permanent part introduces naturally the self-healing during unloading and rest-period. Through this framework, the evolution equations associated to the dissipative processes (viscoelasticity, viscoplasticity, damage and healing) are derived from the restrictions imposed by the second laws of thermodynamics. The accuracy of the proposed formulation is demonstrated with several applications performed on asphalt-concrete under different kinds of loadings and rest-periods.

2 Kinematics

A material point in reference configuration of the body Ω noted $\mathcal{B}_o$ with position vector $\mathbf{X}$ occupies position $\mathbf{x}$ at time t in a deformed configuration $\mathcal{B}_t$. The position vector of a material point $\mathbf{X}$ is transformed to spatial position vector $\mathbf{x} = \phi(\mathbf{X}, t)$ with the deformation gradient, i.e. $\mathbf{F} = \mathrm{Grad}\,\phi$, with $\det \mathbf{F} > 0$. By using the deformation gradient, the right and left Cauchy-Green tensors (covariant) can be introduced such that $\mathbf{C} = \mathbf{F}^{\mathrm{T}}\mathbf{F}$. In this work, the multiplicative decomposition of the deformation gradient into elastic $\mathbf{F}^{\mathrm{e}}$ and plastic $\mathbf{F}^{\mathrm{p}}$ parts is used, i.e. $\mathbf{F} = \mathbf{F}^{\mathrm{e}}\mathbf{F}^{\mathrm{p}}$, with $\det \mathbf{F}^{\mathrm{e}} > 0$ and $\det \mathbf{F}^{\mathrm{p}} > 0$. The covariant measure defined in the plastic intermediate configuration $\overline{\mathcal{B}}_p$, resulting from the multiplicative decomposition, given by $\bar{\mathbf{C}}^{\mathrm{e}} = \mathbf{F}^{\mathrm{eT}}\mathbf{F}^{\mathrm{e}}$. In order to define the proposed model such as viscoelastic-(visco) plastic, the elastic part of the deformation (at equilibrium) $\mathbf{F}^{\mathrm{e}}$ is also decomposed into non-equilibrated elastic and viscous parts, i.e. $\mathbf{F}^{\mathrm{e}} = \mathbf{F}^{\mathrm{e}}_{\mathrm{i}}\mathbf{F}^{\mathrm{v}}_{\mathrm{i}}$. A new intermediate configuration $\widehat{\mathcal{B}}_i$ is thus resulting from this decomposition. The subscript i corresponds to a particular branch of the multiplicative decomposition. The different strain measures (covariant) associated with this decomposition are given in their respective configurations such as $\bar{\mathbf{C}}^{\mathrm{v}}_{\mathrm{n}} = \mathbf{F}^{\mathrm{vT}}_{\mathrm{n}}\mathbf{F}^{\mathrm{v}}_{\mathrm{n}}$ and $\widehat{\mathbf{C}}^{\mathrm{e}}_{\mathrm{n}} = \mathbf{F}^{\mathrm{eT}}_{\mathrm{n}}\mathbf{F}^{\mathrm{e}}_{\mathrm{n}}$.

In order to achieve the coupling between multiplicative viscoelasticity-viscoplasticity with continuum damage, a fictitious configuration $\widetilde{\mathcal{B}}_p$ relied to the plastic intermediate configuration by means of a "degradation tensor" $\mathbf{F}^{\phi}$. The contravariant damage tensor $\bar{\mathbf{B}}^{\phi}$ defined in the plastic intermediate configuration is also defined such as $\bar{\mathbf{B}}^{\phi} = \mathbf{F}^{\phi}\mathbf{F}^{\phi\mathrm{T}}$. The pull-back operation of the contravariant damage tensor into the fictitious configuration $\widetilde{\mathcal{B}}_p$ leads to the identity tensor, i.e. $\tilde{\mathbf{B}}^{\phi} = \mathbf{F}^{\phi-1}\mathbf{F}^{\phi}\mathbf{F}^{\phi\mathrm{T}}\mathbf{F}^{\phi-\mathrm{T}} = \mathrm{I}$ (the superscript "h-1" and "v-T" represent the inverse and the inverse transpose, respectively). Consequently, the fictitious intermediate configuration $\widetilde{\mathcal{B}}_p$ may be interpreted as an "effective configuration" in the sense of a configuration where the damage is fictitiously removed. Since the damage is represented by a second order tensor the proposed model deals naturally with an anisotropic damage evolution (Menzel and Steinmann 2003).

Finally, in order to introduce a recovery part of the degradation, the damage deformation tensor $\mathbf{F}^{\phi}$ is multiplicatively decomposed into recoverable $\mathbf{F}^{h}$ and permanent degradation $\mathbf{F}^{d}, \mathbf{F} = \mathbf{F}^{h}\mathbf{F}^{d}$. As usual, a contravariant recoverable degradation tensor $\bar{\mathbf{B}}^{h}$ can be defined in the plastic intermediate configuration such as $\bar{\mathbf{B}}^{h} = \mathbf{F}^{h}\mathbf{F}^{hT}$.

3 Evolution Equations of the Dissipative Process

In this work, the free energy stated in the plastic intermediate configuration is given by

$$\psi = \bar{\psi}(\bar{\mathbf{C}}^{e}, \bar{\mathbf{C}}^{v}_{i}, \ldots, \bar{\mathbf{C}}^{v}_{n}, \bar{\mathbf{B}}^{\phi}, \bar{\mathbf{B}}^{h-1}, p), \tag{1}$$

where p is an isotropic hardening variable controlling the change in size of the yield surface. Combining the two first laws of thermodynamics (i.e. energy imbalance and balance of entropy), the free-energy imbalance, also called Clausius-Duhem inequality, of the proposed model can be expressed in the plastic intermediate configuration such as

$$\bar{\mathbf{M}}^{e} : \bar{\mathbf{L}} - \frac{\partial\psi}{\partial\bar{\mathbf{C}}^{e}} : \dot{\bar{\mathbf{C}}}^{e} - \sum_{i}^{n}\frac{\partial\psi}{\partial\bar{\mathbf{C}}^{v}_{i}} : \dot{\bar{\mathbf{C}}}^{v}_{i} - \frac{\partial\psi}{\partial\bar{\mathbf{B}}^{\phi}} : \dot{\bar{\mathbf{B}}}^{\phi} + \frac{\partial\psi}{\partial\bar{\mathbf{B}}^{h-1}} : \dot{\bar{\mathbf{B}}}^{h-1} - \frac{\partial\psi}{\partial p}\dot{p} \geq 0, \tag{2}$$

where $\bar{\mathbf{M}}^{e}$ is the Mandel stress tensor and $\bar{\mathbf{L}}$ is the velocity gradient tensor defined in $\bar{\mathcal{B}}_{p}$. According the defined kinematics of the model, the velocity gradient can be decomposed into an equilibrated elastic and plastic part, i.e. $\bar{\mathbf{L}} = \bar{\mathbf{L}}^{e} + \bar{\mathbf{L}}^{p}$ or $\bar{\mathbf{L}} = \bar{\mathbf{L}}^{e}_{i} + \bar{\mathbf{L}}^{v}_{i} + \bar{\mathbf{L}}^{p}$.The sign + before the dissipation term concerning the recoverable degradation part appears because of the use of the inverse of the covariant degradation $\bar{\mathbf{B}}^{h-1}$. Using the symmetry properties of $\bar{\mathbf{C}}^{e}, \bar{\mathbf{C}}^{v}_{i}, \bar{\mathbf{B}}^{\phi}.\bar{\mathbf{B}}^{h-1}$ and by eliminating the energetic parts of Eq. 2 the dissipation inequality of the proposed formulation can be rewritten such as $\sum_{i}^{n}\bar{\mathbf{M}}^{v}_{i} : \bar{\mathbf{L}}^{v}_{i} + \bar{\mathbf{M}}^{e} : \bar{\mathbf{L}}^{p} - \bar{\mathbf{M}}^{\phi} : \bar{\mathbf{L}}^{\phi} + \bar{\mathbf{M}}^{h} : \bar{\mathbf{L}}^{h-1} - Y^{p}\dot{p} \geq 0$, (3)

where $\bar{\mathbf{M}}^{\phi}, \bar{\mathbf{M}}^{h}$ and $\bar{\mathbf{M}}^{v}_{i}$ are Mandel-like stress measures defined such as

$$\bar{\mathbf{M}}^{\phi} = 2\frac{\partial\psi}{\partial\bar{\mathbf{B}}^{\phi}}\bar{\mathbf{B}}^{\phi}, \bar{\mathbf{M}}^{h} = 2\frac{\partial\psi}{\partial\bar{\mathbf{B}}^{h-1}}\bar{\mathbf{B}}^{h-1}, \bar{\mathbf{M}}^{v}_{i} = 2\bar{\mathbf{C}}^{e}\mathbf{F}^{v-1}_{i}\frac{\partial\psi}{\partial\bar{\mathbf{C}}^{v}_{i}}\mathbf{F}^{v-T}_{i}, \tag{4}$$

Y^{p} is the thermodynamics force work-conjugate with p, i.e., $\bar{\mathbf{L}}^{\phi}$ and $\bar{\mathbf{L}}^{h}$ are the degradation and healing velocity tensors, respectively The maximum dissipation principle which satisfies automatically the positivity of the dissipation (Eq. 4) is used to deduce the evolution laws associated to the dissipative processes. A dissipation potential decomposed into viscoelastic, viscoplastic, damage and healing parts is proposed, i.e. $F = \sum_i^n F_i^v + F^p + F^\phi + F^h$. Furthermore, motivated by the fact that during a deformation of a material where the degradation process take place, the current loading can be applied only on the undamaged portion, the dissipative potential is stated in the fictitious effective configuration $\widetilde{\mathcal{B}}_p$. As shown by Balieu and Kringos (2015), the classic pull-back operation of a Mandel-like stress measure which has a mixed contravariant-covariant nature is not compatible with the concept of effective stress. The covariant pull-back of the Mandel-like stress measures is therefore used as proposed by Balieu and Kringos (2015). The effective Mandel-like stress tensors are thus given by $\tilde{\mathbf{M}} = \mathbf{F}^{\phi-1}\bar{\mathbf{M}}\mathbf{F}^{\phi-T}$. In the proposed model, the viscoelastic, plastic damage and healing dissipation potentials are respectively defined such as

$$F_i^v = \frac{1}{2\eta_i^d}\mathrm{dev}\tilde{\mathbf{M}}_i^v : \mathrm{dev}\tilde{\mathbf{M}}_i^v + \frac{1}{9\eta_i^v}\left(\tilde{\mathbf{M}}_i^v : \mathbf{I}\right)^2, \tag{5}$$

$$F_i^p = (1-\alpha)\sqrt{3/2\mathrm{dev}\tilde{\mathbf{M}}^e : \mathrm{dev}\tilde{\mathbf{M}}^e} + \frac{\alpha}{3}\tilde{\mathbf{M}}^e : \mathbf{I} - Y_0^p - Y^p, \tag{6}$$

$$F^\phi = \left(\frac{\zeta^\phi\sqrt{\mathrm{dev}\tilde{\mathbf{M}}^\phi : \mathrm{dev}\tilde{\mathbf{M}}^\phi} + \tilde{\mathbf{M}}^\phi : \mathbf{I}}{Y_0^\phi}\right)^s, \tag{7}$$

$$F^h = \zeta^h\frac{1}{\varsigma^d}\frac{\mathrm{dev}\tilde{\mathbf{M}}^h\left(\mathbf{I}-\bar{\mathbf{C}}^\phi\right) : \mathrm{dev}\tilde{\mathbf{M}}\left(\mathbf{I}-\bar{\mathbf{C}}^\phi\right)}{Y^h} + \frac{1}{\varsigma^v}\tilde{\mathbf{M}}^h : \left(\mathbf{I}-\bar{\mathbf{C}}^\phi\right), \tag{8}$$

where η_i^d and η_i^v are material parameters controlling the viscoelasticity, α is a parameter controlling the pressure sensitivity on the yield surface, ζ^ϕ and ζ^h are parameters controlling the anisotropic evolution of the degradations, ς^d and ς^v control the rate of healing and Y_0^p, Y_0^ϕ, Y^h are material parameters related to plasticity, damage and healing. The symbol "dev" represent the deviatoric part of a second order tensor. By using the maximum dissipation principle with Eq. DISSSS, the evolution equations associated to the viscoelasticity, viscoplasticity, damage and healing are given in $\overline{\mathcal{B}}_p$ by

$$\bar{\mathbf{L}}_i^v = \frac{\partial F}{\partial \bar{\mathbf{M}}_i^v}, \bar{\mathbf{L}}^p = \dot{\lambda}\frac{\partial g}{\partial \bar{\mathbf{M}}^e}, \bar{\mathbf{L}}^\phi = -\dot{\lambda}\frac{\partial F^\phi}{\partial \bar{\mathbf{M}}^\phi}, \bar{\mathbf{L}}^h = \dot{\lambda}\frac{\partial F^h}{\partial \bar{\mathbf{M}}^h}, \text{and } Y^p = -\dot{\lambda}\frac{\partial F^p}{\partial p}. \tag{9}$$

where $\dot{\lambda}$ is the plastic multiplier. A non-associative evolution where the parameter α is replaced by β (Eq. 6) is used for the plastic flow rule. Moreover, a Norton-type viscoplastic evolution is used in the proposed model, i.e. $\dot{\lambda} = \langle \frac{F^p}{K} \rangle^N$, where K and N are strain rate parameters and the symbol $\langle \cdot \rangle$ is the Macauley brakets.

4 Numerical Results and Future Works

In the following, the proposed model is used to simulate the behavior of a Dense Bitumen Macadam. For the simulations, this particular form of the free is used:

$$\psi = \frac{\lambda}{2}\left(\frac{J^2-1}{2} - \ln(J)\right) + \frac{\mu}{2}J^{-2/3}\left(\mathrm{tr}(\bar{\mathbf{C}}^e\bar{\mathbf{B}}^\phi\bar{\mathbf{B}}^{h^{-1}}) - 3\right) + \sum_{i=1}^{n}\frac{\lambda_i}{2}\left(\frac{J_i^2-1}{2} - \ln(J_i)\right)$$
$$+ \sum_{i=1}^{n}\frac{\mu}{2}J_i^{-2/3}\left(\mathrm{tr}(\bar{\mathbf{C}}^e\bar{\mathbf{C}}_i^{v^{-1}}\bar{\mathbf{B}}^\phi\bar{\mathbf{B}}^{h^{-1}}) - 3\right) + k_1\left(p + \frac{1}{k_2}\exp(-k_2 p)\right) \tag{10}$$

The material parameters used for the numerical simulations are given in Table 1. In sake of simplicity, the anisotropic parameters ζ^ϕ and ζ^h are set to 0, leading to a spherical evolution of the degradation and healing tensors, i.e. an isotropic damage and healing evolution. As shows Fig. 1, the responses of the constitutive model are compared to experiments carried out at different strain rates 5E-4, 5E-3 and 5E-2 1/s under compression loading. Good agreement is observed between the proposed model and the experiments (See Fig. 1).

In order to understand the healing model, a creep recovery test (a pressure of 1 MPa is applied during 800 s followed by an unloading (50 s) and a rest period up to 3000 s) is simulated with the proposed model. Different values of the parameter ς^v are used in order to evaluate its influence on the degradation evolution. As expected, an increase of the degradation $\bar{\mathbf{B}}^\phi$ is observed during the loading phase followed by a decrease which occurs during the unloading and rest period (See Fig. 2).

Table 1 Material parameters used for the numerical simulations

μ	μ_1	μ_2	μ_3	μ_4	μ_5	μ_6	μ_7
8.60972	2.882	408	232	125	45.698	10.672	0.13828
λ	λ_1	λ_2	λ_3	λ_4	λ_5	λ_6	λ_7
85.523	0.01	579	233	0.01	77.447	135	0.01
	η_1^d	η_2^d	η_3^d	η_4^d	η_5^d	η_6^d	η_7^d
	0.002882	8.16	46.4	250	1827.92	10672	1382.8
	η_1^v	η_2^v	η_3^v	η_4^v	η_5^v	η_6^v	η_7^v
	0.00001	11.58	46.6	0.02	3097.88	135000	100
	Y_0^p	k_1	k_2	α	β	K	N
	0.05	0.35	50	0.25	0.2	32.8	2.61
	Y_0^ϕ	s	ς^v				
	0.2	0.8	1.00E-01				

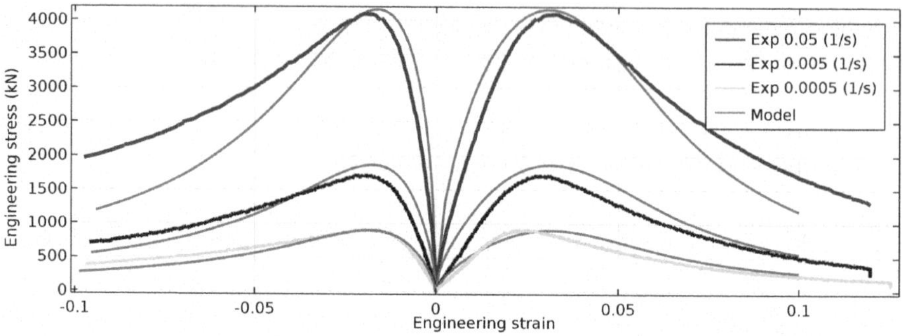

Fig. 1 Comparison between the proposed model and experiments under compression (Axial strain are positives)

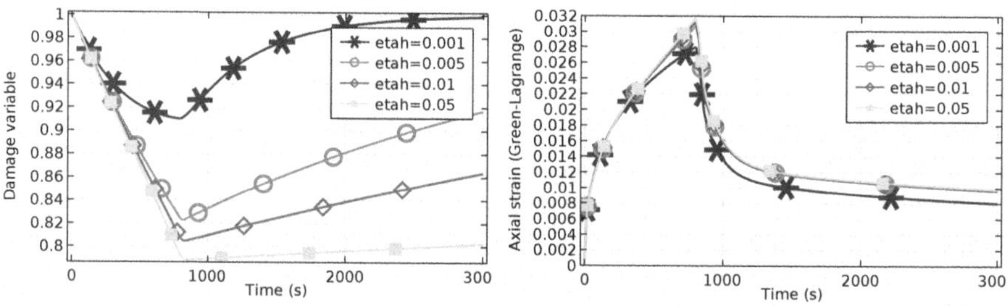

Fig. 2 Evolution of the degradation of the proposed model under creep-recovery load (*left*). Influence the healing parameter on the axial strain resulting of the creep-recovery test

In this paper, a viscoelastic-viscoplastic model coupled to damage and healing model developed under finite strain is presented. The anisotropic damage with different evolution under tension/compression proposed by will be integrated to the model. Regarding the healing evolution, further works are necessary to characterize its material parameters which depend also of the kinds of loads.

References

Balieu R. Kringos N (2015) A new thermodynamical framework for finite strain multiplicative elasto-plasticity coupled to anisotropic damage. Int. J. Plast. 70:126-150.

Menzel A. Steinmann P (2003) A theoretical and computational framework for anisotropic continuum damage at large strains. Int. J. Solids Struct. 38:74-80.

Self-Healing of Dense Asphalt Concrete by Two Different Approaches: Electromagnetic Induction and Infrared Radiation

Harith Ajam, Pedro Lastra-González, Breixo Gómez-Meijide and Álvaro García

Abstract Cracks in asphalt mixture can self-heal if enough resting time (hours or days) is given. This is a viscosity dependent phenomenon that can be accelerated by increasing the temperature of asphalt mixture. In the present paper, the healing performance of asphalt mixture heated using infrared heating to simulate the natural solar radiation, and induction heating, a new method to increase the temperature of asphalt pavements, were compared in terms of time and healing temperature. Healing was defined as the relationship between the 3-point bending strength of an asphalt beam before and after healing. The results show that both methods reach similar and satisfactory healing rates at around 90 %. However, induction heating is more energy-efficient, since the effect is concentrated on the binder, instead of heating the whole mix. This means much shorter healing times to reach the same healing rate than with infrared radiation. Finally, it was found an optimal radiation energy, from which on, infrared radiation reduces the material healing properties.

Keywords Self-healing · Induction heating · Infrared radiation · Road materials

H. Ajam (✉) · B. Gómez-Meijide · Á. García
Nottingham Transportation Engineering Centre (NTEC), Department of Civil Engineering, University of Nottingham, Nottingham NG7 2RD, UK
e-mail: evxha13@nottingham.ac.uk

B. Gómez-Meijide
e-mail: ezzbg@nottingham.ac.uk

Á. García
e-mail: Alvaro.Garcia@nottingham.ac.uk

P. Lastra-González
GITECO Research Group, Universidad de Cantabria, E.T.S. Ingenieros de Caminos, Canales y Puertos de Santander, Av. de los Castros s/n, 39005 Santander, Spain
e-mail: pedro.lastragonzalez@unican.es

A. Chabot et al. (eds.), *8th RILEM International Conference on Mechanisms of Cracking and Debonding in Pavements*, RILEM Bookseries 13,
DOI 10.1007/978-94-024-0867-6_34

1 Introduction

Asphalt mixture is a self-healing material. When a cracked asphalt road is exposed at a temperature above a certain level, between 40 and 70 °C, bitumen can flow through the cracks, filling them (Garcia et al. 2012). Bitumen drains from the asphalt mixture into the cracks until the pressure and surface tension of bitumen filling the cracks equals that in the asphalt mixture (Garcia et al. 2015). Furthermore, cracks can self-heal if asphalt roads are not exposed to traffic loads. However, this process may require several days to complete. In addition, asphalt self-healing is influenced by the viscosity of bitumen, as bitumen with low viscosity accelerates the self-healing process, and the temperature of the road, as higher temperatures reduce the viscosity of bitumen (Liu et al. 2011).

In order to accelerate asphalt self-healing induction heating can be used. This method involves the addition of different electrically conductive particles to the mixture (for example, steel wool fibres) and their heating by means of the action of an alternating magnetic field provided by an electric coil (Liu et al. 2010).

Moreover, although natural solar radiation can produce temperatures similar to those obtained with induction heating, it is unclear why asphalt pavements exposed to the Sun do not heal as efficiently as those exposed to infrared heating. To answer this question, cracks in asphalt beams have been repaired by using induction heating and simulated solar radiation. With the purpose of simulating solar radiation, cracked asphalt beams have been exposed to different times under infrared lamps and compared to identical samples healed by induction heating.

2 Materials and Methods

2.1 Materials

During the present investigation, all the tests were carried out with identical samples. The gradation is corresponding to a continuous and dense mix, with a target voids content of 4.5 %. The natural aggregate was limestone, while the conductive component was a metal grit with uniformed size of 1 mm. The latter was introduced in the mix by replacing part of the natural aggregate in this fraction. The volumetric content of metal grit in the mix was fixed in 4 % (11.2 % in weight). The selected binder was a 40/60 pen and the content was 4.7 %.

2.2 Specimen's Preparation

The gradation was batched by blending samples of limestone with different gradations and with the metal grit. The asphalt concrete was mixed in a laboratory

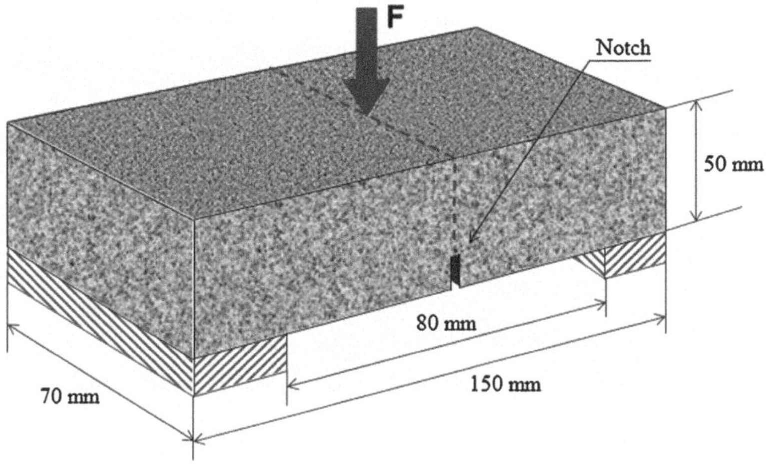

Fig. 1 Design of the 3-point bending test to break the samples

mixer at 160 °C for 2 min, and then compacted as slabs by means of roller compactor until it reached the target void content of 4.5 %. The dimensions of the slabs were 310 × 310 × 50 mm^3. Then, the slabs were cut with a radial saw to obtain 8 150 × 70 × 50 mm^3 prismatic samples from each slab. Finally, a notch was cut at the midpoint from the central axis of the beams, with a thickness of about 2 mm and a depth of about 10 mm (Fig. 1).

2.3 *Mechanical Tests*

The samples were first tested under three-point bending at −20 °C. The tests were carried out under strain control condition, with an increasing load ramp at a deformation rate of 50 mm/min. During each test, a clear crack of approximately 200 μm width was produced, crossing vertically the samples, from the notch to the load application point (Fig. 1).

Once the crack was produced and the sample was split in two different halves, they were put together again and healed by means of one of the following methods:

- Induction heating: The samples were exposed to induction heating for 19 different times between 15 s and 240 s. No longer times were tested as the bitumen reached its burning temperature. The distance from the upper side of the sample to the coil was 1.5 cm, the current 80 A, frequency 348 kHz and the power used 2800 W (Fig. 2, left).
- Infrared radiation: In order to simulate the effect of the Sun, the samples were placed at a distance of 30 cm below four infrared lamps. The samples were embedded in white porous sand with the exception of the upper side to prevent

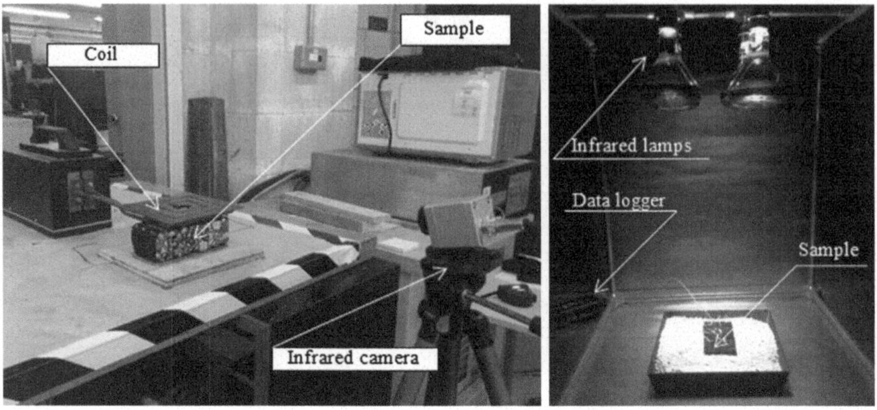

Fig. 2 Healing procedures by induction heating (*left*) and infrared radiation (*right*)

them to receive infrared radiation from any other side and to avoid the deformation caused by high temperature (Fig. 2, right). The samples were exposed to infrared heating for 42 different times between 5 and 5760 min.

The temperature of the samples was constantly monitored by using an infrared camera for asphalt induction heating and thermocouples installed in the top and the bottom of the test specimens in the case of infrared heating.

Once the healing process was finished, the samples were cooled at −20 °C and tested again under three-point bending. The healing rate (*S*) of asphalt samples was defined as the relationship between the ultimate force of the test specimens during a three point bending test, F_i, and the ultimate force measured in the prismatic specimens after the healing process F_b.

$$S = F_b/F_i \qquad (1)$$

3 Results

As it can be seen in Fig. 3, the heating with infrared was much slower than the heating with induction. Moreover, in the case of infrared, the steady state temperature was reached at 101 °C, while in the case of induction heating the steady-state temperature could not be reached because test samples were damaged. Furthermore, infrared radiation needed much more time than induction in order to obtain the same temperature. This can be attributed to the fact that induction heats the metal grit very fast by the Joule principle, which is enough to heal the cracks. However, infrared radiation heats progressively the whole sample by diffusion.

The healing rates obtained can be seen in Fig. 4 (depending on the maximum temperature reached by the samples) and Fig. 5 (depending on the healing time).

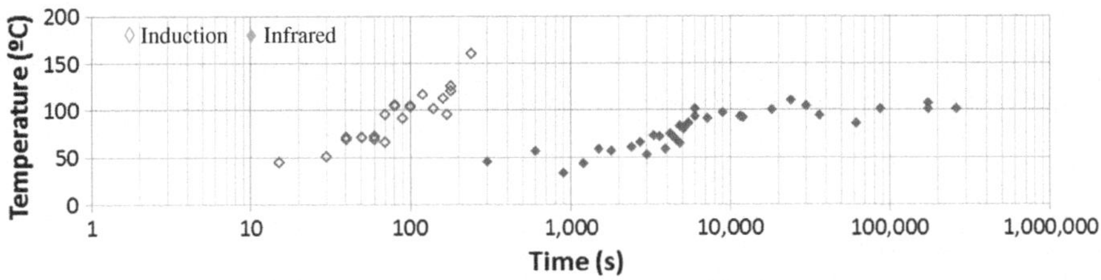

Fig. 3 Maximum temperature (°C) reached by the samples after different healing times

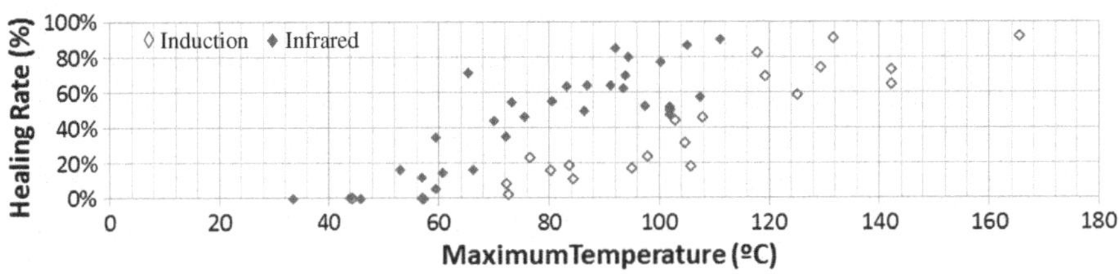

Fig. 4 Relationship between healing rate (%) and maximum temperature reached by samples

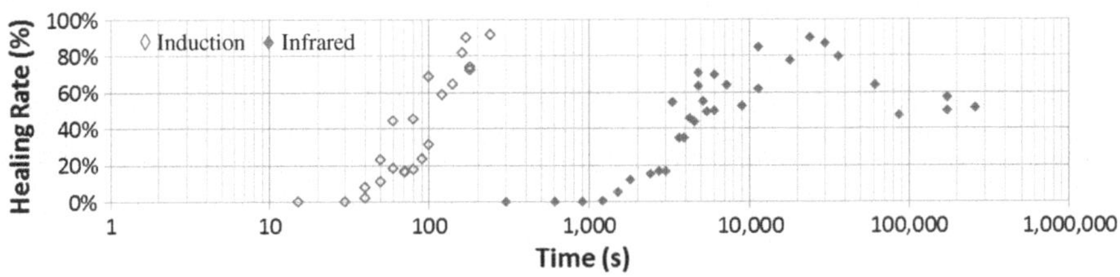

Fig. 5 Relationship between healing rate (%) and healing time with different methods

First, it is noticeable that both methods can reach similar healing rates (around 90 %). Furthermore, for equal maximum temperatures, infrared radiation produced higher healing rates. However, it can also be seen that the amount of time necessary to reach this values is much higher when using infrared radiation.

Moreover, the healing rates of induction method never stopped increasing with time and temperature, while with infrared the healing rates increased only until they reached the steady state temperature, and from this moment they reduced to values close to 50 %. Therefore, although both temperature and time affect the healing, they do it in different ways: Temperature increases can be always translated into healing improvements. However, maintaining the same temperature for longer times produce a detrimental effect on the healing, probably due to the aging of bitumen and internal redistribution of voids, binder and aggregates. This behaviour explains why the roads exposed almost every day to sunlight and high temperatures (for instance, in desert climates) are fully healed during the warm season. Instead of this, it exists an optimal radiation point for asphalt self-healing. Once overcome, further radiation produces nothing but damage in the material.

4 Conclusions

- Induction heating applies the heat directly into the bitumen. Therefore, it is more efficient than infrared radiation and the healing times are much shorter.
- When applying infrared radiation, the temperature increases until reached the steady state temperature of 101 °C. However, with induction heating, the temperature could be increased until the bitumen burned at almost 250 °C.
- Both methods produced similar healing rates up to 90 %. However, the induction heating needed much less energy for self-healing.
- There is an optimal infrared radiation energy for asphalt self-healing. Once this value is overcome, further infrared radiation damages the material. This explains why cracks in roads of very warm and sunny environments can develop cracks that do not heal during the warm seasons.
- Temperature increases can be always translated into healing improvements. However, maintaining the same temperature for longer times produce a detrimental effect on the healing rate.

Acknowledgements The authors acknowledge the funding by the projects EPSRC EP/M014134/1 and INFRAVATION-ERA-NET Plus, HEALROAD (grant agreement no. 31109806.0003).

References

Arpaci VS, Selamet A and Kao S-H (2000) Introduction to Heat Transfer. NJ: Prentice Hall.

Garcia A (2012) Self-healing of open cracks in asphalt mastic, Fuel 93 264-272.

Garcia A, Norambuena-Contreras J, Bueno J, Partl MN (2015) Single and multiple healing of porous and dense asphalt concrete. Journal of Intelligent Material Systems and Structures 26(4):425-433.

Liu Q, Garcia A, Schlangen E, van de Ven M (2011) Induction healing of asphalt mastic and porous asphalt concrete. Construction and Building Materials 25:3746–3752.

Liu Q, Schlangen E, Garcia A, van de Ven M (2010) Induction heating of electrically conductive porous asphalt concrete. Construction and Building Materials 24:1207–1213.

Mechanisms in Healing of Bitumen and the Impact of Normal Force

G.A. Leegwater, A. Scarpas and S.M.J.G. Erkens

Abstract Damage in pavements is known to reduce over time when the material is left to rest, this phenomenon is known as healing. It has been shown that healing is an important influence factor in pavement performance. However, an accepted method to assess the healing capability of a pavement is currently not available. Healing of cracks is assumed to be the sum of two processes, cracked surfaces coming into contact (wetting) and strength gain of surfaces in contact (intrinsic healing). The paper describes influencing parameters of these two processes. The healing potential of bitumen is assessed using a novel test method. In this method two pieces of bitumen are brought together and left to heal under controlled conditions. After healing the amount of healing is assessed by testing the specimens using a direct tensile test. From the results it can be seen that normal force has a significant impact on the observed healing, indicating that the process of two surfaces coming into contact (wetting) has a significant impact on healing behavior of the bitumen.

Keywords Healing test · Bitumen · Intrinsic healing · Wetting

1 Introduction

Asphalt concrete has the advantageous ability to heal. Measurements on asphalt concrete show a regain of strength and stiffness after rest periods. The phenomenon is called healing and was first described by Bazin and Saunier (1967) and it is studied ever since. Almost all research shows that the level of healing increases

G.A. Leegwater (✉) · A. Scarpas
Delft University of Technology, Delft, The Netherlands
e-mail: g.a.leegwater@tudelft.nl

A. Scarpas
e-mail: a.scarpas@tudelft.nl

S.M.J.G.Erkens
Delft University of Technology and Rijkswaterstaat, Delft, The Netherlands
e-mail: s.m.j.g.erkens@tudelft.nl

A. Chabot et al. (eds.), *8th RILEM International Conference on Mechanisms of Cracking and Debonding in Pavements*, RILEM Bookseries 13,
DOI 10.1007/978-94-024-0867-6_35

with longer resting periods, higher healing temperatures (Bonnaure et al. 1982; Qui 2012). However, the mechanisms behind healing of asphalt are not fully understood, consequently many observed trends in healing behavior remain unexplained (Bhasin et al. 2008; Qui 2012). Fundamental understanding of the healing mechanism can be used to optimize the healing properties of asphalt to increase the performance of pavements.

This paper will continue by presenting a popular model for healing of asphalt, which separates observed healing in different processes. The model describes healing as the combination of surfaces coming into contact and the ability of surfaces in contact to transfer loads. Also influencing parameters for both processes will be described. Next, a novel test method is presented that aims to quantify the different healing processes. The paper presents some results obtained with this novel test method. After this implications of the test results are discussed in light of the presented model.

2 Healing Model

2.1 Conceptual Healing Model

In order to understand the mechanisms driving the healing process, researchers in asphalt have adopted the healing model used in polymers (Wool and O' Conner 1981; Bhasin et al. 2008). This model describes healing of cracks as the combined effect of two processes, which are schematically shown in Fig. 1. Firstly, surfaces

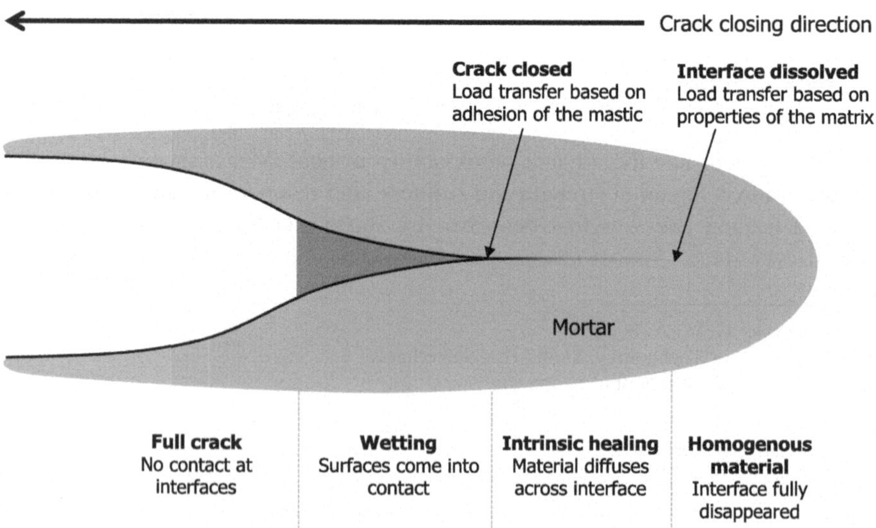

Fig. 1 Schematic representation of the closing of a crack

can only transfer loads if they are in contact. The process of two surfaces coming into close contact is referred to as *wetting*. Secondly, the ability of surfaces in contact to transfer loads changes over time. There is an initial ability to transfer loads from the moment of first contact based on adhesion (or cohesion as the two sides of the crack are of the same material). Subsequently, the load bearing capacity of the interface in contact, increases over time as molecules diffuse across the interface. This diffusion leads to full homogenization of the interfaces if time is infinite. This second process describing how load bearing capacity of the interfaces increases over time is named *intrinsic healing*.

Following the definition that intrinsic healing originates from molecular motion at the crack interface, is can be concluded that the level of intrinsic healing should not be affected by normal force (Wool and O' Conner 1981).

Figure 1 represents the healing of a single crack, however on a macroscopic level a multitude of surfaces are in different stages of wetting and each interface is in its own stage of intrinsic healing. Consequently the observed healing on a macroscopic level is the convolution of the wetted area and the load transfer ability over time.

Influencing factors of wetting are; the external load that brings the surfaces together; the geometry (roughness of the surface); the viscoelastic properties of the material and the adhesion of the material (Schapery 1989).

2.2 Mathematical Healing Model

The conceptual model described in the previous paragraph has a mathematical representation (1), also formulated by (Wool and O' Conner 1981). In order to address the regain of strength it uses a convolution integral to sum up the macroscopic healing based on all the separate areas that are healing.

$$R = \int_{\tau=-\infty}^{\tau=t} R_h(t-\tau)\frac{d\phi(\tau, X)}{d\tau}d\tau \quad (1)$$

In Eq. 1 R is the ratio of the healed performance compared to the original performance, it ranges from 0 to 1. The formula consists of a wetting function (τ, X) and an intrinsic healing function R_h, τ is the running variable on the time axis.

The wetting function is expected to have a sigmoidal shape, based on the physical process which stands at its basis (Wool and O' Conner 1981). The intrinsic healing function is expected to have an initial level R_0 that originates from intermolecular forces, and a part related to diffusion over time which is expected to be a power function of time. A graphic overview of the mathematical model and the shape of the expected healing processes is presented in Fig. 2.

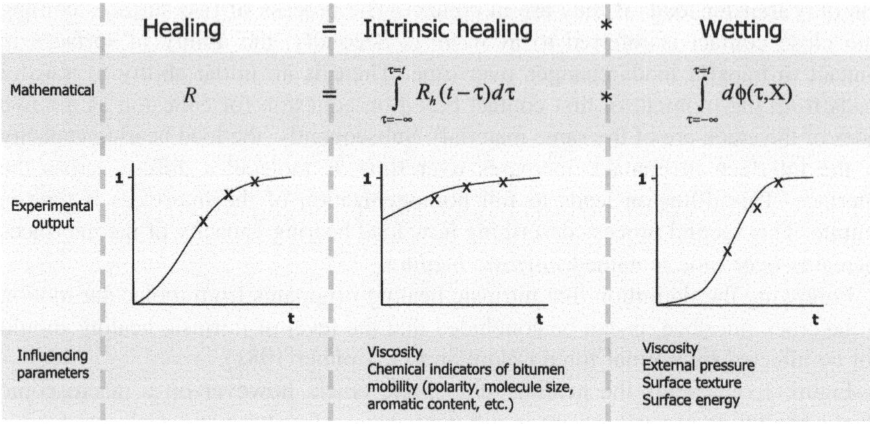

Fig. 2 Graphical overview of healing model

3 Direct Tensile Test to Asses Healing

To quantify the relative importance of the processes that play a role in healing, as described in the previous chapter, a novel test method has been developed to asses healing. The damage in asphalt concrete is assumed to be a (micro) discontinuity in the material. The design of the test method is aimed to investigate the most extreme version of a discontinuity; two separate pieces of bitumen. These two pieces of bitumen are brought together and after a period of healing, they are pulled apart again, testing the amount of tensile strength that has built up during the healing period. The global set-up of the test method is shown in Fig. 3.

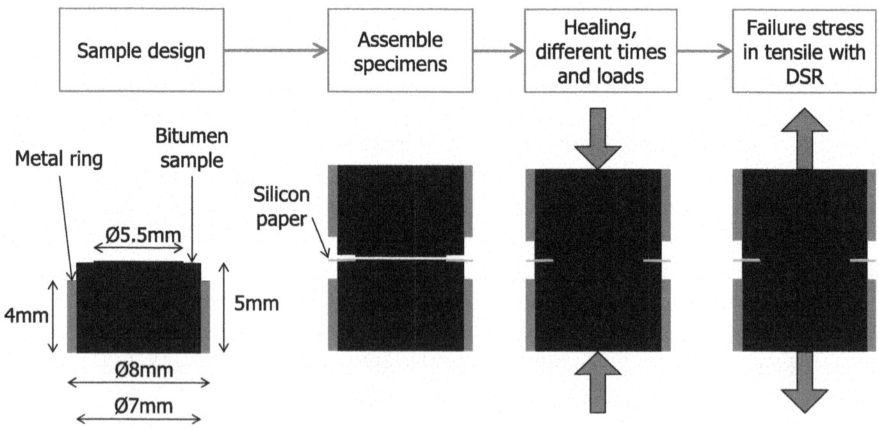

Fig. 3 Schematic impression of sample design, preparation and testing

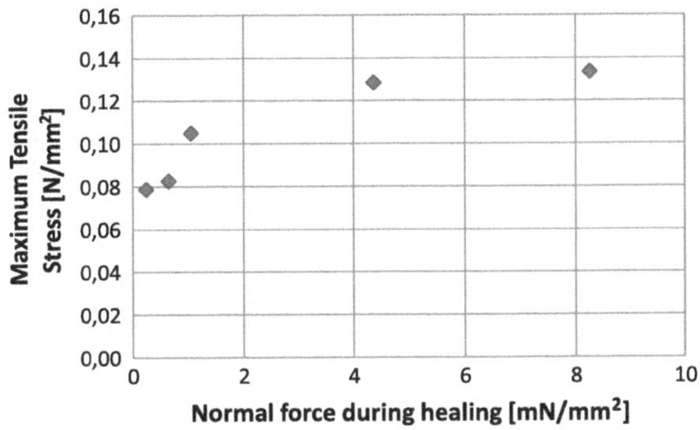

Fig. 4 Impact of normal force on the maximum tensile stress

Special bitumen test samples are designed. A small piece of bitumen is cast inside a stainless steel ring, this ring enables handling and fixation during preparation and testing. The size and design of the ring and bitumen samples are shown in Fig. 3 on the left. Two pieces of bitumen are brought together and left to heal for a specific amount of time while controlling temperature and the normal force perpendicular to the contact area. The contact area is controlled by a piece of silicon paper with a small hole of Ø5.5 mm, which is placed in between the two bitumen samples. After healing, the specimens are tested in *direct tension* using a DSR equipped with a temperature chamber and a normal force load cell (Anton Paar, EC Twist 502).

4 Test Program and Results

Past research (Bazin and Saunier 1967) pointed towards normal force as an important factor for healing, this study aims to expand on this insight, by varying the normal force during healing to evaluate the impact on healing.

The samples are made of a relatively soft bitumen, pengrade 70/100, which is known to have good healing properties. Samples were left to heal for 1 h, at a temperature of 13 ± 1 °C. During healing five load levels were applied; 0.24, 0.65, 1.06, 4.36, 8.26 mN/mm^2. The smallest load level is the self-weight of the sample paced on top. The results are presented in Fig. 4.

5 Discussion, Conclusion and Outlook

Figure 4 presents the maximum tensile stress after the healing period as an indicator for the amount of healing. From the graph it can be seen that up until a load level of 4.36 mN/mm^2 the amount of normal force has a significant impact on the amount of healing that is observed, increasing the observed healing up to 70 %.

Following the definitions of intrinsic healing and wetting presented in the Chap. 2 of this paper, wetting is influenced by the load level, while intrinsic healing is not, as the latter is governed by diffusion. The clear impact of the load level on the observed healing demonstrates the large influence of the wetting process in these short time frames.

If the presented result is extrapolated to asphalt, it is likely that the stress state during healing of asphalt is of significant importance for the observed healing.

The test method and program presented in this paper are the start of a larger investigation into healing of bitumen and mortaric, which aims to improve fundamental insight into healing behavior of asphalt. A large tests program is foreseen to explore the relative impact of wetting and healing. In this test program healing conditions such as temperature, time and load level will be varied, also different types of bitumen and mortar will be tested.

References

Bazin, P., & Saunier, J. (1967, January). Deformability, fatigue and healing properties of asphalt mixes. In Intl Conf Struct Design Asphalt Pvmts.

Bhasin, A., Little, D. N., Bommavaram, R., & Vasconcelos, K. (2008). A framework to quantify the effect of healing in bituminous materials using material properties. Road Materials and Pavement Design, 9(sup1), 219-242.

Bonnaure, F. P., Huibers, A. H. J. J., & Boonders, A. (1982). A laboratory investigation of the influence of rest periods on the fatigue characteristics of bituminous mixes. Journal of the Association of Asphalt Paving Technologists, 51, 104-128.

Schapery, R. (1989) On the mechanics of crack closing and bonding in linear viscoelastic media. International Journal of Fracture 39(1-3): 163-189.

Qiu, J., (2012) Self-Healing of Asphalt Mixtures, PhD Thesis, Technical University Delft: Delft.

Wool, R. P., & O'Connor, K. M. (1981). A theory crack healing in polymers. Journal of Applied Physics 52, 5953. http://dx.doi.org/10.1063/1.328526.

Assessment of Healing Properties of Asphalt Mixtures

Ivan Isailović, Michael P. Wistuba and Augusto Cannone Falchetto

Abstract Fatigue failure due to repeated traffic loading is a significant distress for asphalt pavements. Nevertheless, the damage associated to this phenomenon can be partially or entirely recovered during rest periods between loading repetitions, potentially resulting in a longer pavement service life. In this paper, the results from a comprehensive investigation on healing potential of asphalt materials are presented based on cyclic uniaxial tension compression fatigue tests under stress-control mode with and without rest periods. Then a new healing index parameter is proposed and evaluated using a testing procedure with different rest period durations. The analysis of the experimental results reveals that the newly developed healing index effectively asses the mixtures healing properties with an increasing trend for longer rest periods. A limit plateau value is then achieved when the duration of the rest period does not any longer affect the asphalt healing properties. The proposed healing index provides a simple and practical tool for optimal pre-selection of pavement materials, which can be used on a routine basis.

Keywords Material healing · Healing index · Rest period · Fatigue

1 Introduction

As an asphalt pavement is not continuously exposed to traffic loading, there is a possibility for regeneration of the developed damage, which is known as material healing.

I. Isailović (✉) · M.P. Wistuba · A.C. Falchetto
Pavement Engineering Centre, Technische Universität Braunschweig,
Beethovenstraße 51 b, 38106 Brunswick, Germany
e-mail: i.isailovic@tu-bs.de

M.P. Wistuba
e-mail: m.wistuba@tu-bs.de

A.C. Falchetto
e-mail: a.cannone-falchetto@tu-bs.de

A. Chabot et al. (eds.), *8th RILEM International Conference on Mechanisms of Cracking and Debonding in Pavements*, RILEM Bookseries 13,
DOI 10.1007/978-94-024-0867-6_36

Healing properties of binder and hot mix asphalt (HMA) have been investigated since many years and several healing indexes were introduced with the purpose of healing evaluation. In order to assess the influence of the rest period on fatigue life extension, discontinuous and continuous fatigue tests were mostly compared. The main issue in this comparison is associated to high results scatter in continuous fatigue tests, since it can lead to erroneous healing interpretation.

The objective of this work is to propose a new and unique healing index which is not dependent on the experimental scatter in continuous tests, and which allows more accurate healing evaluation. For that purpose uniaxial tension-compression fatigue tests in stress control including a single rest period were used.

2 Material Composition

Healing tests were performed on an asphalt mixture for surface course, i.e. asphalt concrete of the type AC 11 D S used in Germany for highest road category. According to the German Technical Specifications these asphalt mixtures were prepared with aggregate having maximum grain size of 11 mm and with a bitumen with penetration grade 50/70 (acc. to EN 1426) in percentage of 5.9 % by total mixture weight. The tested specimens had an air voids content of 2.5 %. Test specimens were cut from slabs made with a large Rolling Sector Compactor.

3 Healing Index

Healing analysis of the selected HMA was performed using uniaxial tension-compression test, which creates almost homogenous stress and strain fields in the middle portion of specimen. A sinusoidal cyclic stress with amplitude of 1.05 MPa was imposed under load control mode at a frequency of 10 Hz and at a test temperature of 20 °C.

In order to obtain a reference value for the healing evaluation and compare results from continuous and discontinuous (i.e. containing a rest period) fatigue tests, tests without rest periods were firstly performed (Fig. 1). For the evaluation of the number of loading repetitions at failure the energy ratio (*ER*) approach proposed by Hopman et al. (1989) was used. *ER* represents the ratio between the initial dissipated energy (W_0), to the dissipated energy at cycle n (W_n), multiplied by the load cycle value n (Eq. 1). By plotting *ER* versus the number of loading cycles, fatigue life is defined as the number of loading cycles (N_{Makro}) for which *ER* achieves the maximum (*ER100*).

$$ER_n = \frac{n \cdot W_0}{W_n}, \tag{1}$$

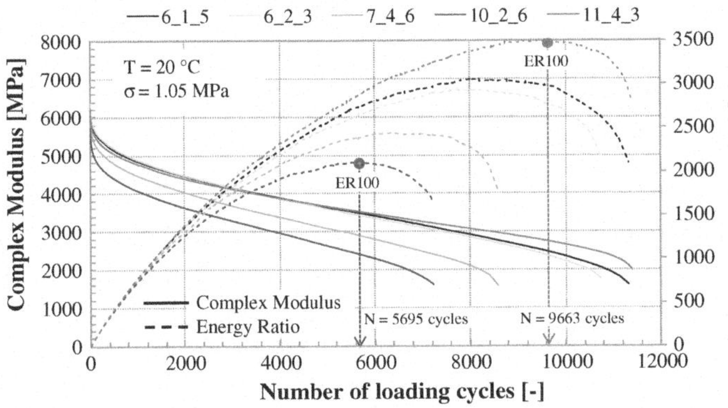

Fig. 1 Evolution of absolute value. of complex modulus and energy ratio over the number of loading cycles in continuous uniaxial tension-compression test at 20 °C and 1.05 MPa stress amp

Although the continuous tests were performed at the same loading conditions, a high scatter in the number of loading cycles at failure is observed (from 5695 to 9663 cycles at failure, Fig. 1). Therefore, the evaluation of the healing properties based on this reference results is quite inconvenient and can lead to erroneous healing results and interpretation Wistuba et al. (2013). By plotting the relative values of energy ratio (ER100 = 100 %, Fig. 1) over relative values of complex modulus (E_0 = 100 %, where E_0 is absolute value of complex modulus at 100th cycle) it can be seen that the values of all continuous tests follow almost the same curve (Fig. 2).

It implies that no matter how many loading cycles HMA experienced at failure (at the same loading conditions), a unique curve representing the material response can be drawn. Based on this finding, evaluation of the materials healing properties can be performed based on the relative energy ratio approach (see Fig. 2) and avoiding the high scattering observed when using the number of loading cycles.

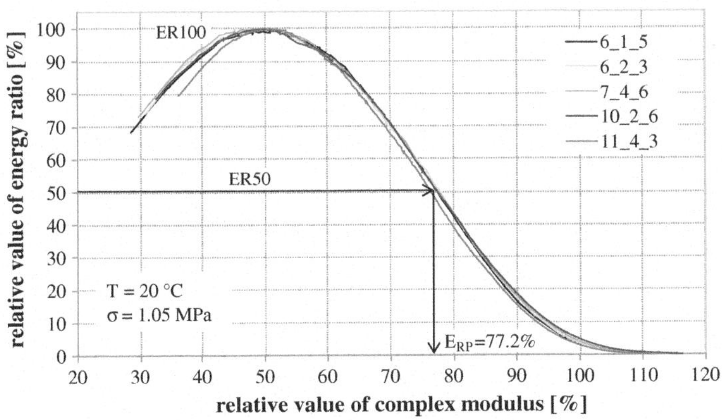

Fig. 2 Evolution of the relative values of the energy ratio over the relative values of complex modulus for five continuous uniaxial tension-compression fatigue tests presented in Fig. 1

Fig. 3 Evolution of the complex modulus in one healing test; test protocol used to characterize the healing capacity of HMA: loading phase 1 + rest period (RP) + loading phase 2

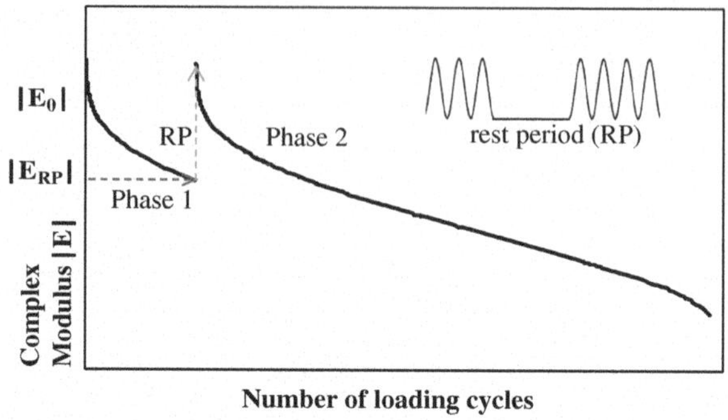

The healing tests were performed with a single rest period, which was introduced after the material reached 50 % of its maximum energy ratio, or a corresponding decay of the complex modulus down to the $E_{RP} = 77.23$ % (see Figs. 2 and 3, Phase 1). During the rest period specimen is stress free. After the rest phase specimen is loaded with a new loading sequence till failure (Fig. 3, Phase 2). It is noticeable, that as a consequence of the material recovery the value of the complex modulus after rest period significantly increased, inducing prolonged material′s fatigue life.

The used testing procedure with one rest period does not simulate real loading conditions in the field, however, it provides the opportunity to better estimate the influence of the rest period on material healing potential compared to procedures with rest periods after each cyclic repetition.

Figure 4 shows an example of the energy ratio evolution in the first and in the second loading phase (full line). Taking into consideration that specimen is continuously loaded in the first phase until the HMA experienced 50 % of max. energy ratio, using approach shown in Fig. 2 which can be assumed as a unique "material law" it is possible to calculate maximal value of the energy ratio (ER100, Fig. 4, red point) for an extended first phase without rest period (represented as the continuous fatigue test, Fig. 4, dashed line). In such a way there is a possibility to

Fig. 4 Evolution of energy ratio in the first and second phase in one healing test (*full line*) and calculated part for extended first phase without rest period (*dashed line*)

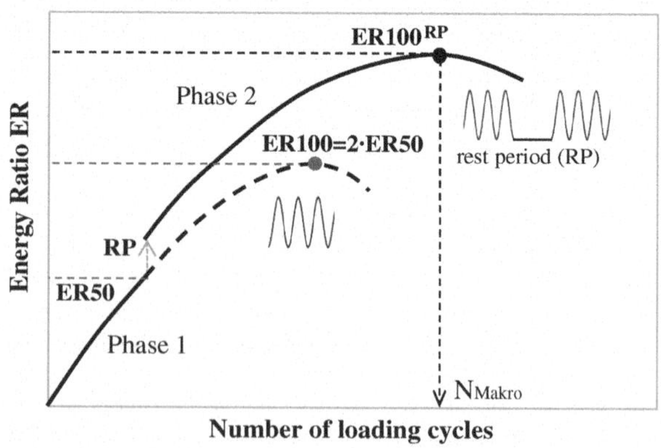

compare the max. energy ratio from tests with rest period (Fig. 4, $ER100^{RP}$, black point) with the calculated energy ratio from continuous tests (Fig. 4, ER100, red point), conducted on the same HMA specimen. The relative difference between these two values represents the new healing index (see Eq. 2). The higher the increase in energy ratio, the higher is the material healing capacity.

$$\text{Hi} = \frac{\text{ER100}^{\text{RP}} - \text{ER100}}{\text{ER100}} \cdot 100\,[\%], \tag{2}$$

where Hi is the new healing index, $ER100^{RP}$ is max. energy ratio from a test with rest period, ER100 is max. energy ratio from the extended first loading phase (in a continuous test).

4 Healing Index Validation

Past research efforts on healing have shown that the healing property is highly dependent on the duration of the rest period, and it increases as the time for material recovery increases (Little et al. 1999; Carpenter and Shen 2006). In order to validate the newly developed healing index (Eq. 2), different rest period durations were introduced in the testing procedure (500, 1000, 7500, 11000, and 30000 s.).

The healing results are shown in Fig. 5. It can be seen that the developed healing index provides a very good estimation of the HMA healing properties, showing an increase for longer rest periods. Figure 5 also shows that beyond a certain rest period duration there is no influence on the recovery capacity of HMA and a plateau stage is reached. Most likely, the recovery mechanism during this time is completed and any additional time for recovery does not affect the healing capacity of the material. For the HMA used in this research the plateau value was reached after approx. 14000 s.

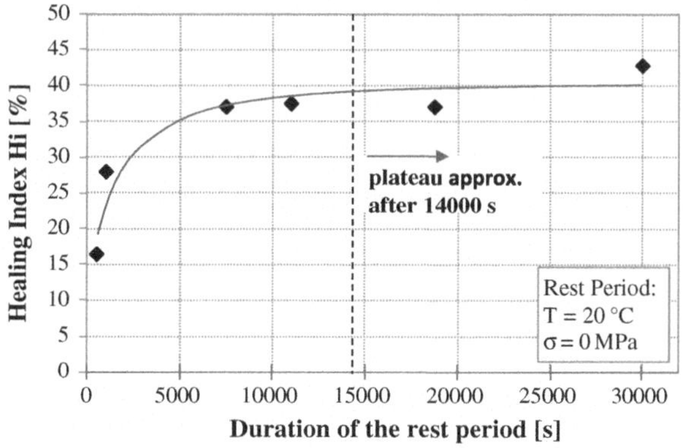

Fig. 5 Healing index as a function of rest period duration

5 Conclusions

In this study the healing properties of one HMA for surface course were evaluated using uniaxial stress controlled tension-compression fatigue test with single rest period. Based on energy ratio calculation, a new healing index is proposed, which relays on the results from continuous and discontinuous fatigue tests. A unique material curve is obtained, which excludes high result scatter from tests without rest period and makes healing evaluation more accurate. The validation of the healing index has shown that it provides effective and reasonable assessment of the healing and recovery properties, with increasing values as the duration of the rest period increases up to a plateau limit. The newly proposed healing index represents a simple and practical tool for optimal pre-selection of pavement materials, which can be used on a routine basis. It should be noted that, the proposed index also include a number of other phenomena such as thixotropy, heating due to energy dissipation and nonlinearity.

References

Carpenter, SH, Shen, S (2006) Dissipated energy approach to study hot-mix asphalt healing in fatigue. Transportation Research Record No. 1790, pp. 178-185. Washington D.C.

Hopman, P, Kunst, P, and Pronk, A (1989) A Renewed Interpretation Model for Fatigue Measurement. Verification of Miner's Rule. 4th Eurobitumen Symposium, October, 1989, Madrid.

Little, DN, Lytton, R L, Williams, D, Kim, Y R (1999) An analysis of the mechanism of microdamage healing based on the application of micromechanics first principles of fracture and healing. Asphalt Paving Technology, pp. 501-537. Chicago.

Wistuba, M, Alisov, A, Isailović, I (2013). Ansprache und Steuerung von Healing-Effekten bei Asphalt. Schlussbericht, FE 07.0251/2011/ERB, i. A. des Bundesministeriums für Verkehr, Bau und Stadtentwicklung, Institut für Straßenwesen, Technische Universität Braunschweig.

Part V
Cracking in Asphalt Materials: Material Crack Performance

Cracking Performance of Lower Asphalt Binder Coarse Hot Mix Asphalt Mixes

Eshan V. Dave, Chelsea Hoplin and Benjamin Helmer

Abstract Historically, asphalt mixtures in Minnesota have been produced to be fine-graded in nature. Recently, there are increasingly more coarse-graded mixtures being produced with relatively low asphalt binder content as compared to fine-graded mixtures. The performance of the coarser low asphalt content mixes is not known; therefore, the objective of this study is to determine if the coarse asphalt mixes with lower binder content are prone to a reduced service life and increased performance problems. The present study focused on 13 low asphalt binder content, coarse-gradation mixes from actual field projects constructed mostly in the last 1–5 years. The performance of the pavement sections was quantified using Minnesota Department of Transportation's (MnDOT) pavement management systems (PMS). Field cracking performance was compared to laboratory tests on core samples procured from the field sections. The field procured core samples were tested using the disk-shaped compact tension (DCT) fracture energy test and permeability measurements. The lab results were also analyzed with each pavement section's field performance. The results from the study will not only provide answers to this question but also provide estimates of the pavement performances for low binder content mixes. Final outcomes will help revise future asphalt mix specifications. The modifications will ensure that pavements provide the service for as designed time duration and have minimal performance problems that reduce the future rehabilitation and maintenance costs.

Keywords Coarse mix · Binder content · Pavement performance · Cracking

E.V. Dave (✉)
Department of Civil and Environmental Engineering, University of New Hampshire,
33 Academic Way, Durham, NH 03820, USA
e-mail: eshan.dave@unh.edu

C. Hoplin · B. Helmer
Department of Civil Engineering, University of Minnesota Duluth,
1405 University Drive, Duluth, MN 55812, USA
e-mail: hopli007@d.umn.edu

B. Helmer
e-mail: helme065@d.umn.edu

A. Chabot et al. (eds.), *8th RILEM International Conference on Mechanisms of Cracking and Debonding in Pavements*, RILEM Bookseries 13,
DOI 10.1007/978-94-024-0867-6_37

1 Introduction and Background

In recent years, there has been a large number of relatively coarse graded mixes being produced and used in the northern state of Minnesota in United States. These coarse graded mixes typically have low total asphalt binder content as compared to the fine graded ones. The cracking performance of the coarser low asphalt content mixes is not well known. Some preliminary testing has shown that these mixes might be prone to premature cracking (Dave et al. 2015b). Furthermore, the use of coarser mixes with lower asphalt content increases the permeability of the mix making them more prone to moisture induced damage (Cooley et al. 2002). The increased permeability of mix is counter to pavement design assumption of dense graded surface layer that drains water over surface away from underlying granular layers. Thus pavements with permeable asphalt mix will be more susceptible to moisture damage as well as other distresses due to the reduction of unbound layer modulus values.

This study quantifies the performance effects and pavement service life of these lower asphalt binder coarse mixes. The main objective of this study is to quantitatively and qualitatively determine if the low asphalt binder coarse mixes are prone to performance issues and make recommendations regarding potential solutions to alleviate any identified problems. This project evaluated 13 low asphalt binder content mixes from actual field projects. The field samples were subjected to a battery of tests to predict the pavement performance using, fracture energy and permeability measurements. The lab results were analyzed to predict the distress severity and life expectancies of the pavements.

2 Pavement Section Cracking Performance

The study sections were selected to encompass a breadth of variables in asphalt pavements in terms of their location, pavement types and years in service. Each section had field procured samples that were used in various laboratory tests as well as pavement management system (PMS) data available to be used as a performance indicator. A number of previous researchers have developed and used various cracking performance measures in order to quantify pavement performance with respect to its lifetime. While only three are discussed here, Dave et al. (2015b) presented several measures and the pros and cons of each, and for brevity, this discussion was not presented in this paper. A brief description of three measures used in this paper are listed in Table 1. The field sections that are being studied here are listed in Table 2 with their corresponding cracking performance illustrated in Fig. 1. In several instances, a field section had more than one pavement profile, for example, part of the section is construction as overlay on milled pavement and the other part is over reclaimed base. This table also presents the year of construction and the type of pavement construction.

Table 1 Cracking performance measures

Measure	Description	Unit
Average transverse cracking rate (ATCR)	Sum of the transverse cracking normalized by life	% cracking/year
Maximum transverse cracking rate (MTCR)	Maximum increase in transverse cracking amounts between any two consecutive years of service	% cracking/year
Total transverse cracking (TCTotal)	Sum of the transverse cracking work (area under cracking versus time) normalized over the service life	% cracking/year

Table 2 Summary of pavement sections

Section	Specimen	Construction year	Construction type
TH 220	K	2012	8 cm M/O
CH 10	L	2012	4 cm O/L
TH 27	M	2010	8 cm M/O
TH 27	N	2010	8 cm M/O
TH 9	O	2011	8 cm O/L on FDR
TH 9	P	2011	8 cm O/L on FDR
TH 28	Q	2012	11 cm M/O
TH 28	R	2012	11 cm M/O
TH 6	S	2010	4 cm M/O
TH 10	T	2013	9 cm M/O
CH 30	U	2012	15 cm M/O
TH 10	V	2005	10 cm M/O
TH 10	W	2005	10 cm M/O

M/O Mill and overlay; *AC* Asphalt Concrete; *FDR* Full depth reclamation; *O/L* Overlay

3 Material Properties

The properties of the wear course mixtures of the various pavement sections under study are presented in Table 3. The table shows the pertinent parameters that are typically used for purposes of characterizing asphalt mixtures. The asphalt binder grade used for the virgin binder component of the mixture is shown along with the amount of binder contribution to the mixes from recycled sources. The three most commonly used volumetric parameters for characterization and specification of asphalt mixtures in practice at present are adjusted asphalt film thickness (Adj. AFT), voids in mineral aggregates (VMA) and voids filled with asphalt (VFA). These are also reported in Table 3. Since a major focus of this study is to evaluate the impact of low asphalt binder content of the mixtures on its cracking performance, comparisons plots are generated between cracking performance measures and design asphalt contents. The results showing comparisons between the design binder content and various cracking performance measures is shown in Fig. 2 (left).

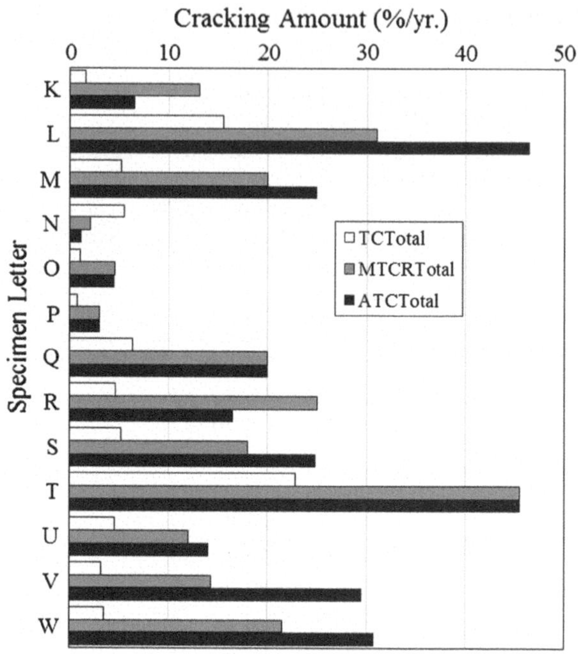

Fig. 1 Cracking performance of pavement sections

Table 3 Section mix design properties

Section	Specimen letter	PG grade	PG spread	Asphalt content (%)	Recycled asphalt content (%)	Adj. AFT	Voids in mineral aggregate (VMA) (%)	Voids filled with asphalt (%)
TH 220	K	58-28	86	4.20	23.80	9.5	13.50	70.30
CH 10	L	58-28	86	4.30	23.30	9.1	13.50	70.40
TH 27	M	58-28	86	4.30	37.20	8.8	13.60	70.60
TH 27	N	58-28	86	4.30	37.20	8.8	13.60	70.60
TH 9	O	58-34	92	4.20	26.20	8.9	13.10	69.60
TH 9	P	58-34	92	4.20	26.20	8.9	13.10	69.60
TH 28	Q	58-34	92	4.20	23.80	9.4	12.50	68.10
TH 28	R	58-34	92	4.20	23.80	9.4	12.50	68.10
TH 6	S	58-28	86	4.40	36.40	9.2	13.90	71.20
TH 10	T	58-28	86	4.30	23.30	8.9	13.70	70.80
CH 30	U	64-34	98	4.40	11.40	9.0	13.40	70.20
TH 10	V	64-28	92	5.30	45.30	7.8	14.40	72.30
TH 10	W	64-28	92	5.30	45.30	7.8	14.40	72.30

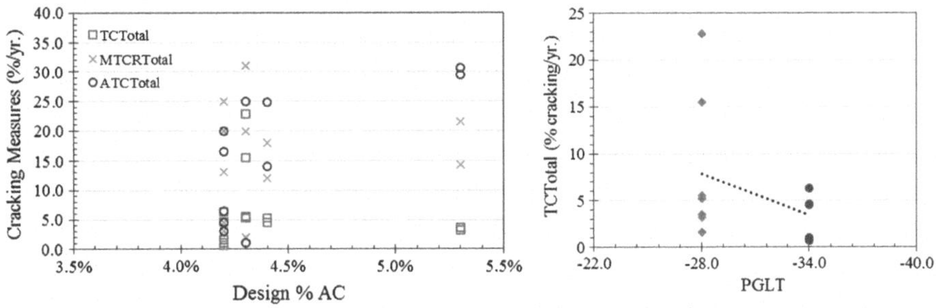

Fig. 2 Cracking measures versus design AC (*left*) and PG low-temperature grade (*right*)

It can be seen from the plot that the design asphalt content by itself may not be a good indicator of the pavement's cracking performance as no clear trends are evident. A similar analysis was done looking at asphalt content versus AFT and VMA which also showed a weak trend. This further confirms that use of volumetric measures as a predictor of asphalt mixture's field cracking performance may not be adequate by itself.

The low temperature performance grade (referred to as PGLT) of the binders are compared with TCTotal; this is presented in Fig. 2 (right). It can be seen from the fitted trend-lines that the PGLT has an effect on the cracking performance, with −34 graded binders showing significantly better cracking performance.

The permeability of asphalt mixtures has been hypothesized to have significant effect on the durability and performance. The cause of high permeability is primarily presence of interconnected voids. The use of permeability over air void level has been recommended by researchers in the past as a better measure of asphalt mixture durability, for example the work by Cooley et al. (2002). The lab measured permeability (using Karol-Warner permeameter) for the cored specimens is presented in Table 4. As it can be seen from the results the permeability varied quite significantly for the asphalt mixtures. The typical permeability range for these mixtures (all of them are 12.5 mm sized) would be between 1E-05 to 1E-06 cm/s range. In Fig. 3 (left), this is denoted by the shaded region.

The mixtures that have permeability greater than this typical range are indicated. It can be seen that more than half of the mixtures have permeability that is significantly higher than the typical range. These mixtures are thus prone to inferior durability. In general, it can be seen that as the permeability increases the cracking performance deteriorates. The comparison between TCTotal and permeability in Fig. 3 (left) show that of the eight mixtures with permeability greater than typical range for dense graded asphalt mixtures, six have very high average cracking rates [greater than 5 % cracking per year, denoted by the dashed line in Fig. 3 (left)]. The DCT test is standardized as ASTM D7313-13 for measurement of fracture

Table 4 Permeability and DCT fracture energy of various asphalt mixtures

Section	Specimen letter	Permeability (cm/s)	Comparison to typical range (1E-05–1E-06 cm/s)	Fracture energy (J/m^2)
TH 220	K	6.28E-04	Very high	182.86
CH 10	L	4.81E-05	High	379.81
TH 27	M	1.48E-05	Borderline high	334.65
TH 27	N	2.33E-07		272.34
TH 9	O	8.18E-06		351.99
TH 9	P	9.69E-05	High	270.99
TH 28	Q	7.86E-06		291.47
TH 28	R	–		227.12
TH 6	S	1.26E-05	Borderline high	260.18
TH 10	T	5.39E-05	High	211.89
CH 30	U	5.65E-07		566.72
TH 10	V	5.23E-05	High	270.21
TH 10	W	5.53E-05	High	238.35

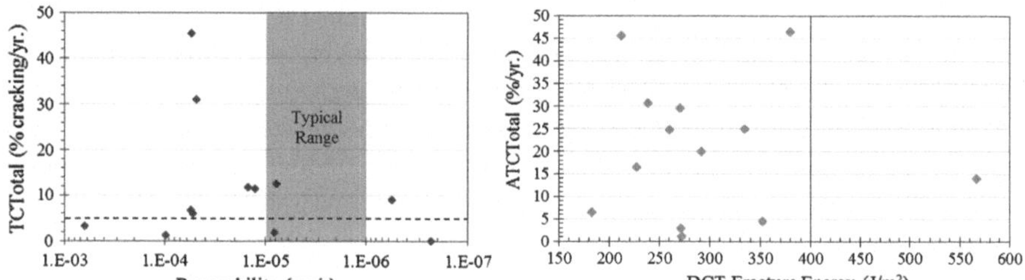

Fig. 3 TCTotal versus permeability (*left*) and ATCTotal versus DCT fracture energy (*right*)

energy of asphalt mixtures. The test has been shown to reliably predict thermal cracking resistance an asphalt mixture (Marasteanu et al. 2012; Dave et al. 2015b). All sections in this study, along with the majority of the State of Minnesota, undergo extensive low temperature climatic conditions.

The DCT fracture energy for the cored specimens is presented in Table 4. The comparison between DCT fracture energy and average transverse cracking rate (ATCTotal) is plotted in Fig. 3 (right). Once again it can be seen eight mixes have very high average cracking rates. It should be noted that only one mixture meets the recommended threshold of 400 J/m^2 for this parameter (Marasteanu et al. 2012; Dave et al. 2015a). Thus, indicating substantially low fracture energies for these coarse, low asphalt binder content mixes.

4 Summary of Findings and Conclusion

This paper reports field cracking performance for 13 pavements study sections on using data from MnDOT pavement management system as well as site visits by the researchers. The cracking performance results are presented using three performance indicators developed through a previous research study.

The cracking performance is compared with various asphalt mix design parameters (such as, asphalt content and binder type). Comparisons are also drawn between cracking performance and the DCT fracture energy as well as the permeability.

On basis of the results and the discussion following observations can be made:

- From perspective of asphalt mix designs, the only parameter showing a strong trend to cracking performance is the low temperature grade of the asphalt binder. The binders with −34 Superpave performance grade show approximately 12 % average transverse cracking rate per year as opposed to approximately 26 % for mixtures with −28 low temperature grade.
- The typically used volumetric measures for ensuring the performance of asphalt mixtures, i.e. asphalt film thickness and voids in mineral aggregates did not show a consistent trend with cracking performance. This is in agreement with previous research results of MnDOT studies.
- Both DCT fracture energy and the permeability show a relatively poor performance (low fracture energy and high permeability) indication for the coarse mixtures with low asphalt binder contents.

References

Cooley, Jr. L.A., Prowell B.D., Brown E.R. (2002) "Issues pertaining to the permeability characteristics of coarse-graded mixes." Nat. Center for Asphalt Tech. NCAT Report 02-06

Dave E.V., Hanson C.E., Helmer B., Dailey J., Hoplin C.M. (2015a) Laboratory Performance Test for Asphalt Content. Minnesota Dept. of Trans. Contract No. 99008, Work Order No. 40

Dave E.V., Helmer B., Hanson, C., Munch, J. (2015b) "Implementation of Laboratory Testing to Predict Low Temperature Cracking Performance of Asphalt Pavements," 8th RILEM International Symposium, Volume 11 RILEM Bookseries, pp 993-1003, Springer.

Marasteanu et al. (2012) "Investigation of Low Temperature Cracking in Asphalt Pavements," Report No. MN/RC 2012-23, Minnesota Department of Transportation.

Combined Effect of SBS and Devulcanized Rubber (DVR) Modification on Performance Grade and Fatigue Cracking Resistance of Asphalt Binders

Salih Kocak and M. Emin Kutay

Abstract Polymer modification of asphalt binders has gained quite popularity in many transportation agencies, primarily due to the superior crack- and rut-resistant performance. However, added cost of polymer modification results in an appreciable increase in the initial cost of an asphalt pavement. There are more economical and sustainable alternatives to polymers, such as the so-called "De-Vulcanized Rubber (DVR)". Primary advantage of DVR technology is that, it is made from scrap tires and when mixed with asphalt binder, the particles completely dissolve within the binder. Therefore, the final product is a complete fluid, not a suspension. The main objective of this study was to investigate the relative performances of the SBS polymer and DVR in an asphalt binder. The impact of the modifications on performance grade (PG) and fatigue cracking resistance was investigated. The results have shown that more sustainable modification of asphalt binders can be achieved by replacing the entire or some amount of SBS with DVR.

Keywords Tire rubber · Devulcanized rubber · Polymer · Fatigue cracking

1 Introduction

Asphalt pavements have been experiencing more heavy traffic volume as the population of the world keeps growing in recent years (Nahas et al. 1990). Conversely, the funding shortages for maintenance and higher costs of pavements have been leading to thinner pavements, hence decrease in the service life

---

S. Kocak · M. Emin Kutay (✉)
Civil and Environmental Engineering Department, Michigan State University, East Lansing, MI, USA
e-mail: kutay@msu.edu

S. Kocak
e-mail: kocaksal@msu.edu

A. Chabot et al. (eds.), *8th RILEM International Conference on Mechanisms of Cracking and Debonding in Pavements*, RILEM Bookseries 13,
DOI 10.1007/978-94-024-0867-6_38

(Lewandowski 1994). The use of polymer-modified binders have been seen as one of the alternatives to address these problems Bulatovic et al. (2014) . Polymer modified binders typically provide more crack- and rut-resistant asphalt mixtures Johnson CM (2010). While they increase the initial cost of the asphalt mixture by 15–20 % (per ton of mixture), the superior performance of polymer modified binders may lead to decrease in pavement thickness, in turn, possible decrease in overall cost.

Similar to polymer-modified binders, crumb rubber-modified asphalt binders have been used by many roadway agencies due their crack-resistant performance. However, the current Performance Grading (PG) system is currently not approved for use in traditional crumb rubber modified binders (e.g., CRTB—Crumb Rubber Terminally Blend and CRWet—Crumb Rubber Wet processes). This is because the PG system was developed primarily for conventional asphalt binders. Typical crumb rubber modified binders are suspensions, i.e., the rubber particles are visible in the asphalt binder. Therefore, the PG system is still not approved by the AASHTO and alternative and more empirical testing methods are used to specify them. One of the recently invented crumb rubber modified asphalt technologies is the so-called "De-Vulcanized Rubber (DVR)". Primary advantage of DVR is that, when mixed with asphalt binder, the rubber particles completely dissolve within the binder. The final product, i.e., the DVR modified binder is a complete fluid, not a suspension. As a result the PG system can easily be applied to specify DVR binder. The unit cost of DVR binder is less than the typical polymers, therefore, if they can be used to partially replace the polymer, more economical and sustainable pavements can be constructed.

2 Objective and Scope

The main objective of this study was to investigate the relative performances of the SBS polymer and DVR in an asphalt binder. The scope of this study included testing one type of binder (PG58-28), one type of polymer (SBS) and the DVR obtained from a manufacturer located in Ohio, U.S. Polymer modified binders were prepared at three percentages, 1, 2 and 3 % by weight of binder. The SBS polymer was a linear block copolymer based on styrene and butadiene with bound styrene 31 % by mass. DVR modified binders were prepared at three percentages 3, 6 and 9 %. In addition, DVR+SBS modified binders were tested at the following combinations: 1 % SBS+3%DVR, 1 % SBS+6%DVR, 2 % SBS+3%DVR, and 2 % SBS+6%DVR. The tests included determination of continuous PGs (at high, intermediate and low temperatures) and fatigue resistance using the Linear Amplitude Sweep (LAS) test.

3 Sample Preparation and Testing

In order to prepare SBS and/or DVR modified binders, the neat binder was first heated to 163 °C. Fine dispersion and size reduction of SBS and/or DVR pellets was accomplished by using a slotted stator head in the high shear mixer. The SBS polymer and/or DVR were shear milled into the binder by using the rotor-stator at 5000 revolutions per minute (rpm). The temperature of the system was stable although no external heat was provided to the system. This was because of the internal heat produced during milling of the polymer and/or DVR pellets. Following the 30 min high shear milling process, the binder was transferred to a low shear mixer and mixed for another 120 min at 180 °C. A stable 180 °C temperature was achieved using either circulatory heating oil bath or an adjustable heating mantle. Rotation of the mixer was 1000 rpm. In addition, cross-linking agent (liquid Sulphur) was introduced into the mixture during the last 30 min of low shear mixing. Once the mixing was finalized, the binder was transferred to a preheated oven at 163 °C and kept for another 16 h for static aging. AASHTO M320 was followed to determine the PG of the binders (AASHTO M 320-10, 2010). Linear amplitude sweep (LAS) tests were performed on PAV aged samples, at equi-stiffness condition (AASHTO TP 101-12, 2015). At least two modified binder batches were prepared and two replicates from each batch were sampled (i.e., total 4 replicates) were tested.

4 Results of Performance Grade (PG) Linear Amplitude Sweep (LAS) Tests

Table 1 shows a summary of the PG and LAS test results. It is noted that the continuous low PG values were obtained by linear interpolation based on creep stiffness "S" and "m-value" at 60 s measured at different temperatures on PAV residue Jamrah A et al. (2015). The low continuous PG was determined based on the smaller temperature (in absolute value) corresponding to S = 300 MPa or m-value = 0.300. It is also noted that the sample denoted with Aged* underwent to high and low shear mixing cycle without the SBS (or DVR), and static aging in the oven to simulate the aging that the binders go through while preparing the SBS or DVR modifications. Based on the Table 1, the following observations can be made:

- Aged* binder slightly increased the high PG and decreased the low PG, enough to reduce the binder grade from PG58-28 to PG58-22
- 1 %SBS modification increased the high PG by about 5 degrees and binder PG changed from PG58-28 to PG64-28
- 2 %SBS modification increased the high PG by about 9° and binder PG changed from PG58-28 to PG64-28 (but very close to PG70-28).
- There was no significant difference in PG when SBS amount increased from 2 to 3 %.

Table 1 PG test results

Modification	Continuous PG				
	PG	High	Inter.	Low	High PG − Low PG
		°C	°C	°C	°C
Unmodified	58-28	60.6	17.0	−28.7	89.3
Aged*	58-22	61.7	18.0	−27.4	89.1
1 % SBS	64-28	65.5	17.3	−28.5	93.9
2 % SBS	64-28	69.9	17.6	−28.7	98.6
3 % SBS	70-28	70.3	16.2	−29.7	99.9
3 % DVR	64-28	66.8	16.7	−28.6	95.4
6 % DVR	64-28	69.2	16.5	−27.8	97.0
9 % DVR	70-28	72.7	16.6	−28.5	101.2
1 % SBS + 3 % DVR	64-28	67.1	16.0	−29.0	96.1
1 % SBS + 6 % DVR	64-28	68.3	15.3	−29.7	98.0
2 % SBS + 3 % DVR	64-28	69.2	14.9	−30.0	99.2
2 % SBS + 6 % DVR	70-28	71.6	14.9	−29.5	101.1

Notes Aged* = aged under the condition of mixing polymer (or DVR) and static aging in the oven, *SBS* styrene-butadiene-styrene copolymer, *DVR* devulcanized rubber

- It appears that, in general, every 3 % DVR created the effect of 1 % SBS. It is noted that the current price of DVR is about 1/3rd of the SBS, therefore the price of 3 % DVR is about the same as 1 % SBS.
- Looking at the continuous PGs:
  - Continuous high PG of 3 % DVR (66.8 °C) was better than that of 1 % SBS (65.5 °C). Continuous low PGs were about the same.
  - Continuous high PG of 6 % DVR (69.2 °C) was slightly worse than that of 2 % SBS (69.9 °C). Continuous low PG of 6 % DVR (27.8 °C) was also slightly worse than that of 2 % SBS (28.7 °C).
  - Continuous high PG of 9 % DVR (72.7 °C) was better than that of 3 % SBS (70.3 °C). However, continuous low PG of 6 % DVR (28.5 °C) was slightly worse than that of 2 % SBS (29.7 °C).
- Regarding the combinations of the SBS and DVR, it appeared that 3 % DVR did not make the same effect of 1 % SBS (or vice versa) for high PGs. In other words:
  - High PG of 2 % SBS (69.9 °C) > High PG of 1 % SBS + 3 % DVR (67.1 °C)
  - High PG of 3 % SBS (70.3 °C) > High PG of 1 % SBS + 6 % DVR (68.3 °C)
  - High PG of 3 % SBS (70.3 °C) > High PG of 2 % SBS + 3 % DVR (69.2 °C)
- For low PGs, the combinations of SBS and DVR were as good or better than the SBS alone. In other words:
  - Low PG of 2 % SBS (−28.7 °C) > Low PG of 1 % SBS + 3 % DVR (−29 °C)

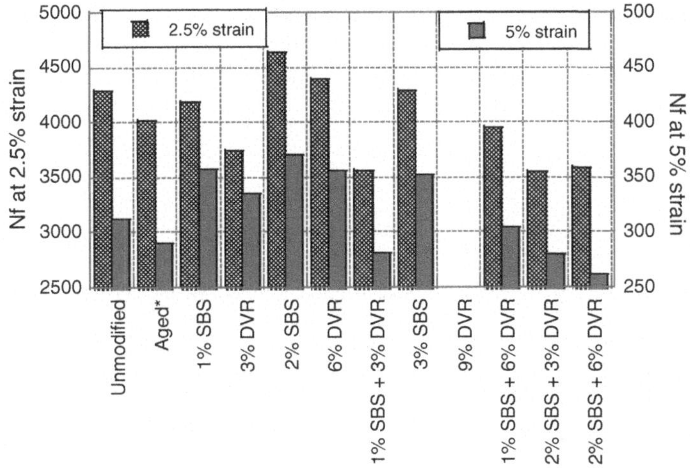

Fig. 1 Results of Linear Amplitude Sweep (LAS) tests. N_f = number of cycles to failure

- Low PG of 3 % SBS (−29.7 °C) = Low PG of 1 % SBS + 6 % DVR (29.7 °C)
- Low PG of 3 % SBS (−29.7 °C) > Low PG of 2 % SBS + 3 % DVR (−30 °C)

Figure 1 shows the results of the LAS tests. Based on the Fig. 1, the following observations can be made:

- Except 2 % SBS, none of the SBS contents and DVR contents improved the fatigue life at 2.5 % strain level
- At 5 % strain level, all SBS and DVR contents improved the fatigue life, however, combinations of SBS and DVR reduced the fatigue life as compared to the original binder.
- 1 % SBS performed better than the 3 % DVR
- 2 % SBS performed better than 6 % DVR, and 6 % DVR performed better than the 1 % SBS+3%DVR combination
- 3 % SBS performed better than the 1 % SBS+6%DVR and 2 % SBS

5 Conclusions

In this study, the effects of binder modification with SBS and/or DVR at different combinations and concentration levels were investigated. First, the effect of modification on PG of binder was investigated. Second, fatigue resistance of behavior of the same modified binders were studied by using the Linear Amplitude Sweep (LAS) procedure at equi-stiffness temperatures. It was observed that, in general, every 3 % DVR created the effect of 1 % SBS in PG testing. It is noted that the current price of DVR is about 1/3rd of the SBS, therefore the price of 3 % DVR is about the same as 1 % SBS. On the other hand the LAS (fatigue) test results indicated that the SBS performed slightly better than the DVR, at the equivalent percentages (i.e., at 1 % SBS = 3 % DVR).

References

AASHTO M 320-10, (2010), Standard Specification for Performance-Graded Asphalt Binder, American Association of State and Highway Transportation Officials.

AASHTO TP 101-12, (2015), Standard Method of Test for Estimating Fatigue Resistance of Asphalt Binder Using the Linear Amplitude Sweep, American Association of State and Highway Transportation Officials.

Bulatovic VO, Rek V, Markovic KJ (2014) Effect of Polymer Modifiers on the Properties of Bitumen. Journal of Elastomers and Plastics 46(5) 448-469. doi: 10.1177/0095244312469964

Jamrah A, Kutay ME, Varma S (2015) Backcalculation of Swollen Crumb Rubber Modulus in Asphalt Rubber Binder and its Relation to Performance. Transportation Research Record

Johnson CM (2010) Estimating Asphalt Binder Fatigue Resistance Using and Accelerated Test Method. Dissertation, University of Wisconsin

Lewandowski LH (1994) Polymer Modification of Paving Asphalt Binders. Rubber Chemistry and Technology 67(3)447-480. doi: 10.5254/1.3538685

Nahas NC, Bardet J, Eckman B et al (1990) Polymer Modified Asphalt for High Performance Hot Mix Asphalt Pavement Binders. Association of Asphalt Paving Technologists

Determining the Allowable Content of RAP in HMA with Regard to Fatigue Resistance

Chiara Riccardi, Pietro Leandri and Massimo Losa

Abstract The determination of the allowable percentage of Reclaimed Asphalt Pavement (RAP) that could be used in Hot Mix Asphalt without compromising fatigue cracking is a crucial task. The principal issue is related to the evaluation of the fatigue resistance of the aged binder contained in the RAP and how does it influence the fatigue resistance of the fresh and aged binder blend. In order to solve this issue a new approach is proposed to estimate the allowable RAP binder content in the asphalt mixture, with regard to fatigue resistance, by considering the Superpave parameter $G^{*}\sin\delta$. This parameter is determined by using a recently proposed methodology, which avoids the extraction and recovery of the RAP binder to calculate the complex modulus and phase angle master curves of the blend of fresh and RAP binders. This procedure allows evaluating the rheological properties of the bituminous blend by the results of Dynamic Shear Rheometer tests carried out on mortars composed of a selected fine fraction of the RAP, totally passing the 150 μm sieve, mixed with fresh binder at different volumetric fractions of the selected RAP. These properties are estimated by using the Nielsen model specifically modified to take into account the influence of lower frequencies on the stiffening effects of fine particles in bitumen mortars. The allowable content of RAP, determined by this way, is then validated performing time sweep tests on mortars composed of RTFOT+PAV aged and fresh binders, confirming the reliability of the proposed procedure.

Keywords RAP binder · Mortars · Fatigue resistance · Time sweep

C. Riccardi · P. Leandri · M. Losa (✉)
Department of Civil and Industrial Engineering (DICI), University of Pisa, Pisa, Italy
e-mail: losa@ing.unipi.it

C. Riccardi
e-mail: chiara.riccardi@unifi.it

P. Leandri
e-mail: p.leandri@ing.unipi.it

A. Chabot et al. (eds.), *8th RILEM International Conference on Mechanisms of Cracking and Debonding in Pavements*, RILEM Bookseries 13,
DOI 10.1007/978-94-024-0867-6_39

1 Introduction

The use of high content of Reclaimed Asphalt Pavement (RAP) in binder and base layers is quite a standard practice; conversely, the use of RAP in surface layers isn't so common, especially at higher contents (more than 10 %) and in high performance surface layers. This is the case of Stone Matrix Asphalt (SMA) that is a gap-graded Hot Mix Asphalt (HMA) that maximizes rutting resistance and durability with a stable stone-to-stone skeleton held together by a relevant Asphalt Content (AC), filler, and stabilizing agents such as fibers and/or asphalt modifiers. Due to their high performance levels required, their construction needs the use of high quality materials whose costs are only partially compensated by money. In fact, being bitumen and aggregates un-renewable resources, their costs have to be valued in terms of sustainability too. Therefore, there could be a significant economic advantage and a well-balanced solution between eco sustainability and technology in using RAP in SMA, if it could be used without sacrificing the excellent performance of SMA mixtures. The main purpose of this study is based on the above concepts; in fact, it aims to define the maximum percentage of RAP that could be added in a high performance surface layer, specifically a SMA mixture, without compromising the fatigue resistance. This procedure avoids the extraction and recovery of the RAP binder, and it is based on rheological tests on binders and mortars instead of tests on asphalt mixtures samples that are more costly and time consuming.

2 Background

The complex modulus of the bituminous blend composed of fresh and RAP binders in mortars can be estimated by using a procedure proposed recently (Leandri et al. 2015), which is based on a modification of the Nielsen model (Landel and Nielsen 1993) to calculate the stiffening effect of mineral particles in bituminous mortars.

By using data of rheological tests carried out on mortars composed of aggregate particles passing the 150 μm and bitumen, the parameters of the Modified Nielsen model, expressed in Eq. 1, can be determined:

$$\left(\frac{G_m^*}{G_b^*}\right) = \frac{1 + A \cdot B \cdot V_p}{1 + B \cdot \Psi \cdot V_p} - (a \cdot \ln(f) - b)V_p^c \tag{1}$$

where:

G_m^* is the complex modulus of the mortar (aggregate particles and bitumen);
G_b^* is the complex modulus of the bitumen;
V_p is the volume fraction of aggregate particles.

A is a constant that is equal to: $A = K_E - 1$ where K_E is the generalized Einstein coefficient that is an indicator of the physical chemical contribution to stiffening;

this parameter is determined, from the Eq. (1), by non-linear curve fitting of the measured stiffening ratio versus the volume fraction V_p at frequency equal to 20 Hz and it was found to vary exponentially with temperature.

B accounts for the relative moduli of particles and bitumen phases and it is equal to: $B = (G_p^*/G_b^* - 1)/(G_p^*/G_b^* + A)$, where G_p^* is the modulus of particles.

The coefficient ψ is given by $\psi = 1 + ((1 - \varphi)/\varphi^2)V_p$ where φ is the maximum volumetric packing fraction that results equal to 0.63.

The coefficients a, b and c can be determined by a regression fitting method, considering all the data at different V_p and different temperatures; the details of the Modified Nielsen model and determination of the parameters are reported in the above mentioned paper.

3 Experimental Program and Materials Used

One RAP material source was used in this study along with a two component fresh binder: this is a blend of a Hard bitumen (H) 35/50 Penetration Grade, 58-22 Performance Grade, and a Soft bitumen (S) characterized by a viscosity of 8000 mm^2/s at 60 °C. The S bitumen, mixed with the RAP binder, acts as a rejuvenator of the aged binder whilst the H bitumen acts as virgin binder, added to obtain the design Asphalt Content (AC) of the mixture. The fresh binder was mixed with RAP material passing the #100 sieve, called Selected RAP (SRAP), to produce mortar samples. The #100 sieve was selected to assure the reasonableness of the test procedure considering that the DSR gap should be at least ten times bigger than the maximum aggregate particle size (Liao et al. 2013). The Asphalt Content (AC) of the SRAP is determined by using the Soxhlet extractor and it results 9.89 % respect to the aggregate. The SRAP is burned in the ignition oven and the resulting aggregates, called Burned Selected Reclaimed Asphalt Pavement (BSRAP), are tested to determine the aggregate size gradation and the maximum volumetric packing fraction φ, that can be calculated as the ratio between the true volume of aggregate particles, determined using the Rigden Voids apparatus (EN 1097-4 2008), and the apparent volume of the aggregate particles, determined in accordance to EN 1097-7: 2008. In total 18 mixes are composed considering three different V_p for both the BSRAP and SRAP (20, 40, 60 %), and three different percentage combinations of the H and S bitumen (90 %H + 10 %S, 80 % H + 20 %S, 70 %H + 30 %S); the specimens are prepared by mixing the preheated particles and binder for two hours at 140 °C in order to allow the diffusion process, as demonstrated by Rad et al. (2014). In order to have the same ageing of the binders, the BSRAP mortars are prepared in the same way. All the mortars and binders are tested with the DSR in the Linear Viscoelastic Range (LVE) in the condition summarized in Table 1.

Table 1 DSR test condition

Plate/gap (mm)	Strain (%)	Frequencies (Hz)	Temperature (°C)	Conditioning time (min)
8/2	0.05	0.2–20	0, 10, 20, 30, 40	50, 40, 30, 20, 10

Table 2 Parameters of the modified Nielsen model

Asphalt binder	a	b	c	B	V_p	Ψ	K_E	R^2
70 %H + 30 %S	1.30	4.06	1.20	1	0.2	1.18	4.57 $e^{0.0177\ T}$	0.992
					0.4	1.37		
					0.6	1.56		
80 %H + 20 %S	1.30	4.06	1.20	1	0.2	1.18	4.12 $e^{0.020\ T}$	0.992
					0.4	1.37		
					0.6	1.56		
90 %H + 10 %S	1.30	4.06	1.20	1	0.2	1.18	2.75 $e^{0.026\ T}$	0.980
					0.4	1.37		
					0.6	1.56		

All the parameters of the Modified Nielsen model are summarized in Table 2.

Considering that the aggregate particles are the same in both the BSRAP and SRAP mortars, we can assume that the stiffening ratio (G_m^*/G_b^*) is the same in both the mortars; therefore, using the same coefficients of the Modified Nielsen model for the SRAP mortar too, the complex modulus of the bituminous matrix G_b^* could be calculated. G_b^* represents the complex modulus of the bituminous blend composed by the virgin and RAP binder. In Fig. 1 the back-calculated master curve of the complex modulus, at T_0 = 20 °C, of the bituminous blend composed by the 80H + 20S binder and different RAP binder contents are reported.

As shown, the stiffness increases substantially as the RAP binder content increases, especially in the range of low frequencies. This is due to the higher complex modulus of the RAP binder compared to the fresh binder.

4 Determination of the Allowable Percentage of RAP Binder Without Compromising the Fatigue Resistance

Since the aim of this study is to determine the allowable RAP binder in SMA mixture, thin layers HMA are considered. Therefore the Superpave parameter $G^*\sin\delta$ could be used to measure the binder resistance to fatigue cracking. In fact, as suggested in the development of the Superpave binder specifications, the $G^*\sin\delta$ index is directly proportional to dissipated energy assuming that the analysis is conducted at a constant strain; this type of loading is considered applicable only to thin asphalt concrete layers (Papagiannakis and Masad 2007). In this study, the

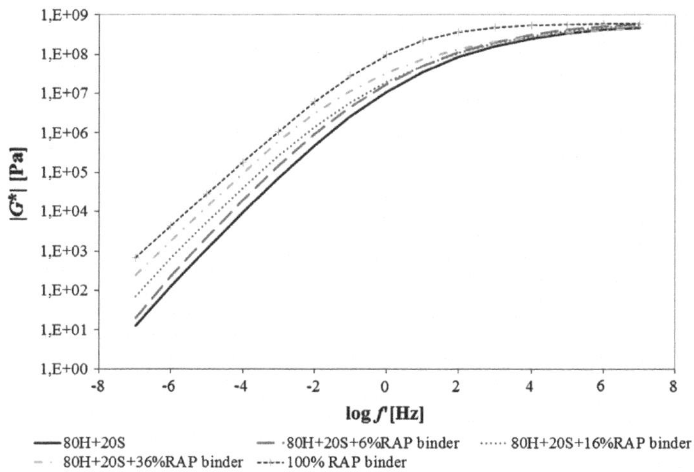

Fig. 1 Complex modulus master curve of the bituminous blend composed by 80H + 20S and different percentage of RAP binder

value of $G^*\sin\delta = 2200$ kPa, measured at 25 °C and at 10 rad/s for a bitumen 50/70, characterized by a PG 64-22, is kept as reference value, since SMA mixtures composed with bitumen 50/70 don't show problems related to fatigue resistance (Leandri et al. 2012). Therefore, the $G^*\sin\delta$ values, at 25 °C and at 10 rad/s, are calculated from the complex modulus and phase angle master curves of the bitumen blends composed of fresh and RAP binder by using the simple method described in Riccardi et al. (2015). Plotting the calculated $G^*\sin\delta$ versus the percentage of RAP binder (Fig. 2a), the maximum percentage of RAP binder that could be added without compromising the fatigue resistance could be found. For this specific case, it results equal to 23.3 % as reported in Fig. 2a. This percentage corresponds to 33.5 content of RAP that could be added in the SMA mixture with an optimum AC equal to 7.5 %; this is obtained by considering the RAP binder content is 5 %.

5 Validation of the Test Procedure

In order to check that the reference value of $G^*\sin\delta = 2200$ kPa is a good predictor of the fatigue resistance, time sweep tests at five different strain levels and at 25 °C were performed on RTFO and PAV aged mortars composed by 70H + 30S binder and $V_p = 48.7$ % from SRAP; this proportion was selected in order to achieve a mortar with the same RAP binder content equal to 23.3 %. Time sweep tests were also performed on RTFO and PAV aged mortars composed of 50/70 bitumen and $V_p = 48.7$ % of BSRAP. The fatigue laws corresponding to the two mortars were determined using the failure criteria based on the 20 % deviation from the initial

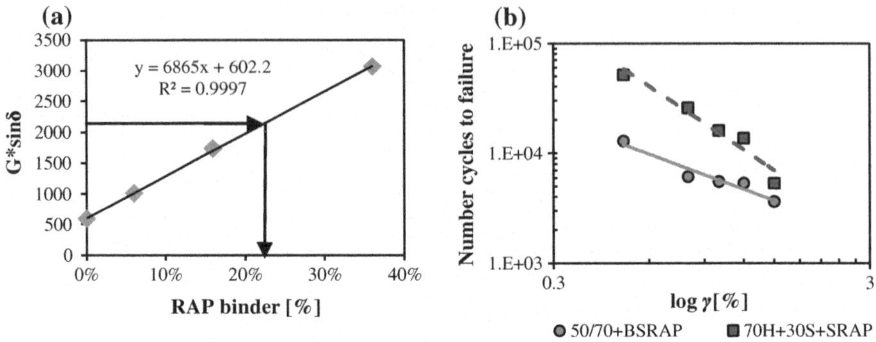

Fig. 2 **a** $G^*\sin\delta$ versus RAP binder percentage; **b** Fatigue laws of the mortars composed by 70H + 30S + SRAP materials and the mortar composed by 50/70 + BSRAP

trend of the Dissipated Energy Ratio (DER) and it was verified, as reported in Fig. 2b, that the fatigue law of the SRAP mortar is above that of the BSRAP mortar, that represents the fatigue law benchmark.

6 Conclusions

In the present study, a simple method to determine the allowable percentage of RAP that could be added in a SMA mixture without compromising the fatigue resistance is presented. This is achieved from DSR frequency sweep tests on the fresh binder and on BSRAP and SRAP mortars, avoiding tests on asphalt mixture samples. The proposed procedure has been validated by time sweep tests on RTFO and PAV aged mortars. More work is needed to check the procedure with other RAP sources and fresh binders; other tests after RTFO and PAV aging are under development in order to find the $G^*\sin\delta$ index after aging, thus the limit of 5000 kPa of the PG specification could be used and a comparison between the RAP binder percentage calculated by using the two different approaches could be analyzed.

References

Landel, R. and Nielsen, L., "Mechanical Properties of Polymers and Composites," CRC Press, Boca Raton, 1993.

Leandri P., Cuciniello G., Losa M.(2012) Study of Sustainable High Performance Bituminous Mixtures. Procedia: social&behavioral sciences, Vol. 53, pp. 495-503. doi:10.1016/j.sbspro.2012.09.900.

Leandri P, Riccardi C, Losa M (2015) A new approach to estimating rheological properties of the rap binder at intermediate temperatures. Road Material Pavement Design. 16 (1): 280-299. doi:10.1080/14680629.2015.1029695.

Liao M C, Airey G, Chen J S (2013) Mechanical properties of filler-asphalt mastics. Int J Pav Res Technol 6(5): 576-581. doi:10.6135/ijprt.org.tw/2013.6(5).576.

Papagiannakis, A., T., Masad, E., A.(2007). Pavement Design and Materials. John Wiley & Sons, Inc., Hoboken, New Jersey.

Riccardi, C., Leandri, P., Losa, M., (2015) Determining the allowable content of RAP in HMA using the blending charts and RAP mortar properties.Conference 6th International conference on bituminous mixtures and pavements, Tessaloniki, Greece. CRC Press/Balkema.

Yousefi Rad, F., Roohi Sefidmazgi, N., Bahia, H. U. (2014).Application of Diffusion Mechanism to Study Degree of Blending Between Fresh and RAP Binder in Dynamic Shear Rheometer. TRB. Vol. 2444, pp.71-77.

Effect of Bitumen Type and Content on the Cracking Resistance of Asphalt Mixtures at Different Temperatures

Rodrigo Miró, Adriana Martínez, Félix Pérez-Jiménez and Ramón Botella

Abstract Bitumen type and content are two of the variables that most influence on the cracking resistance of an asphalt mixture, since both determine a more or less ductile fracture. Furthermore, this cracking resistance will be different depending on the temperature at which the mixture is exposed and is especially critical at low temperatures. The effect of bitumen type and content on the cracking resistance and fracture energy of an asphalt concrete mixture is analysed in this paper, by means of a new direct tensile test, Fénix test. With the aim of covering a wide range of performances, three different bitumens were used: a conventional 50/70 penetration bitumen, considered as the reference binder; a crumb rubber modified bitumen, which gives flexibility of the mixture; and a low penetration bitumen, which provides more stiffness to the mixture. The test was carried out at different temperatures to evaluate the effect of low temperatures at which the mixture may be critical against cracking phenomenon.

Keywords Cracking resistance · Fénix test · Fracture energy · Toughness index

R. Miró (✉) · A. Martínez · F. Pérez-Jiménez · R. Botella
Universitat Politècnica de Catalunya BarcelonaTech, Jordi Girona, 1-3, Edificio B1, 08034 Barcelona, Spain
e-mail: r.miro@upc.edu

A. Martínez
e-mail: adriana.martinez@upc.edu

F. Pérez-Jiménez
e-mail: edmundo.perez@upc.edu

R. Botella
e-mail: ramon.botella@upc.edu

A. Chabot et al. (eds.), *8th RILEM International Conference on Mechanisms of Cracking and Debonding in Pavements*, RILEM Bookseries 13,
DOI 10.1007/978-94-024-0867-6_40

1 Introduction

One of the most frequent and characteristic failures of bituminous mixtures in pavements is cracking caused by repeated traffic loads and thermal stresses occurring at low temperatures (Li et al. 2008). However, despite its importance, the cracking resistance property is not taken into account when designing a bituminous mixture. For example, the properties considered in the design stage according to the Spanish specifications are air void content, resistance to permanent deformation and water sensitivity.

A cracking process can also lead to other types of pavement damage, allowing access of water, leading to debonding between layers, weakening the support of the lower layers and thus compromising the structural integrity of the pavement. In order to evaluate this property it must be considered that the response of each mixture depends on its characteristics, being bitumen type and content among the most important ones (Read and Whiteoak 2003).

The objective of this study is to analyse the effect of binder type and content on the cracking resistance of a dense mixture from Fénix test, carried out at different temperatures, with the aim of including the evaluation of cracking resistance in the design stage of the mixtures from this test.

2 Methodology

Fénix test is a monotonic tensile test performed at constant displacement rate of 1 mm/min (Pérez-Jiménez et al. 2013; Miró et al. 2014). A tensile stress is applied on a semicylindrical specimen by two plates fixed to the loading platen of the press, which are attached to the specimen in its diametral plane. The specimen has a small notch between the two plates to induce cracking. The force applied during the test is recorded as a function of the imposed displacement. From the resulting load-displacement curve, a number of parameters related to mechanical characteristics of the mixture can be defined: tensile stiffness index, fracture energy and toughness index.

The results obtained with the Fénix test on a semi-dense mixture are presented below. The mixture has a maximum size of 16 mm and is cantered within the grading envelope (AC16S type, according to Spanish specifications), Table 1. Regarding the components of the mixture, the aggregates nature is limestone and three different bitumens were considered: a 50/70 penetration bitumen (B 50/70), a crumb rubber modified bitumen (BC 50/70) and a low penetration bitumen or high

Table 1 Mixture AC16S gradation

Sieve (mm)	22	16	8	4	2	0.5	0.25	0.063
% Passing	100	95	67.5	42.5	31	16	11	5

Table 2 Characteristics of the bitumens used

Bitumen	Unit	Standard	B 50/70	BC 50/70	B 15/25
Penetration at 25 °C	0.1 mm	UNE EN 1426	65	61	20
Softening point R&B	°C	UNE EN 1427	50.2	53.6	66.5
PI	–	NLT-181	−0.49	0.2	0.26
Elastic recovery	%	NLT-329	–	28	–

modulus bitumen (B 15/25). The basic characteristics of these bitumens are shown in Table 2.

For each type of bitumen, series of specimens were manufactured with different bitumen contents (3, 4, 5 and 6 % by weight of mixture) in the gyratory compactor using 100 gyrations, a vertical pressure of 0.6 MPa and an internal angle of gyrations of 0.82°.

To evaluate the effect of temperature, the specimens were tested at different temperatures: 20, 5 and −5 °C. The variation of the different parameters obtained from Fénix test with bitumen type and content, as well as with temperature, is shown in Figs. 1, 2, 3 and 4.

3 Results and Discussion

3.1 Tensile Stiffness Index (TSI)

At 20 °C a clear difference in stiffness between the high modulus bitumen and the two other bitumens is observed, Fig. 2; high modulus bitumen, B 15/25, is the one with the highest values of TSI. The behaviour of the other two bitumens is similar to each other, conventional bitumen, B 50/70, being stiffer than crumb rubber modified bitumen, BC 50/70. TSI tends to decrease with increasing bitumen content, except for the high modulus bitumen, for which, given its low penetration, TSI tends to increase until the bitumen content is high enough to decrease the mixture

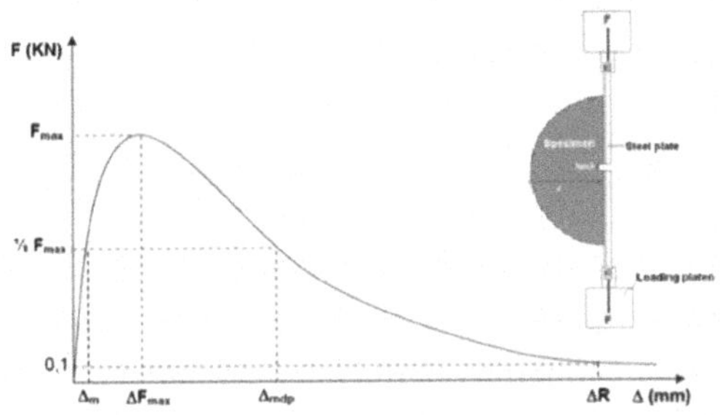

$$TSI = \frac{F_{50} - F_{25}}{(d_{50} - d_{25})}$$

$$G_F = \frac{\int_0^{d_f} F(x)dx}{S}$$

$$TI = \frac{\int_0^{d_f} F(x)dx}{S} \cdot (d_{0,5Fmax} - d_{Fmax})$$

Fig. 1 Fénix test set-up, load-displacement *curve* and parameters obtained

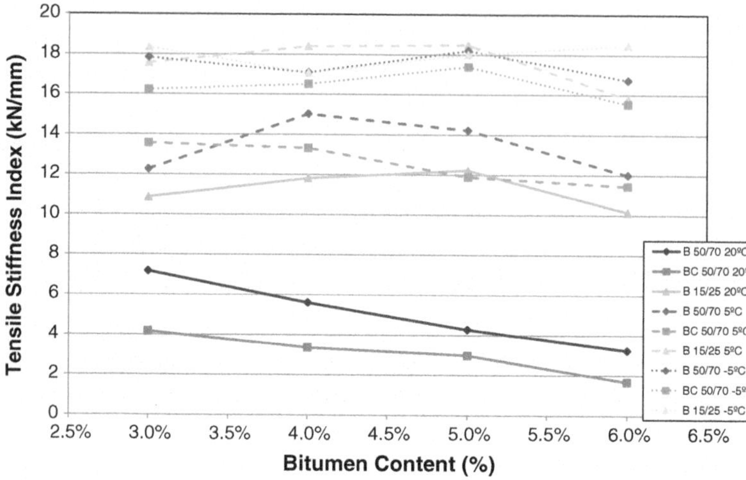

Fig. 2 Variation of tensile stiffness index with bitumen type and content

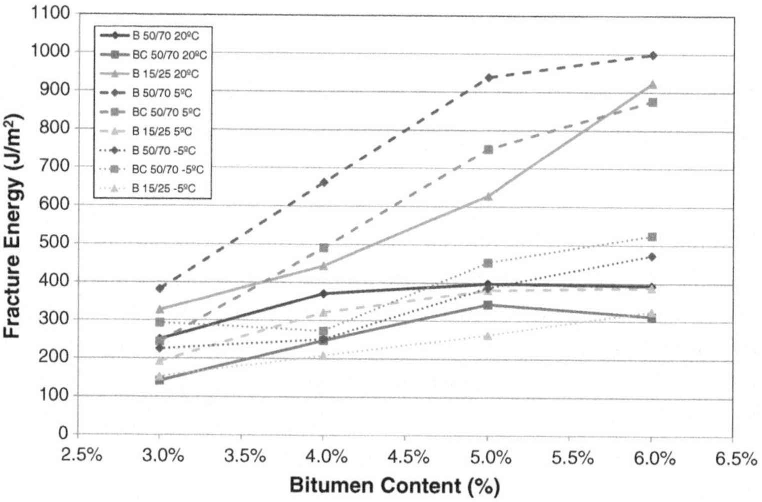

Fig. 3 Variation of fracture energy with bitumen type and content

stiffness, thus showing a maximum value at intermediate an bitumen content between 4 and 6 %.

As temperature decreases, 5 °C, the stiffness curves of the three bitumens are closer to each other, until once reached −5 °C, the values of TSI are more alike for both bitumen type and content. The effect of bitumen content is only noticeable at 20 or 5 °C, depending on the bitumen type.

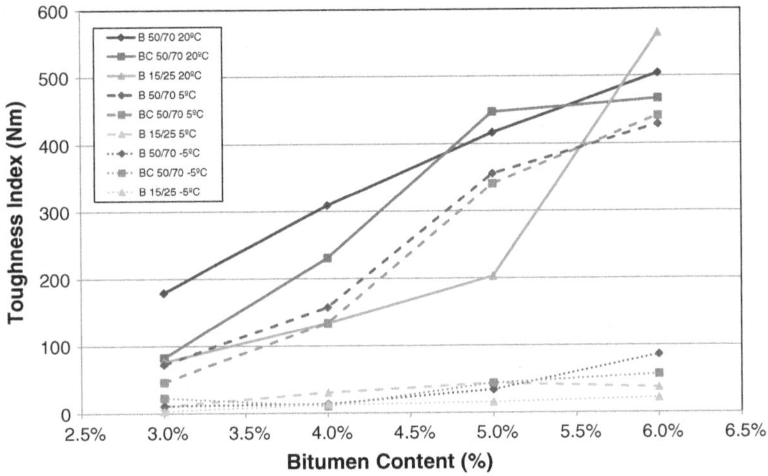

Fig. 4 Variation of toughness index with bitumen type and content

3.2 *Fracture Energy (G_D)*

Figure 3 shows that fracture energy at 20 °C of 50/70 penetration bitumens (both conventional and crumb rubber) tends to present a maximum value by varying the bitumen content, while that of high modulus bitumen (with lower penetration and consequently higher stiffness) is higher and increases with bitumen content, potentially reaching a maximum value above 6 % of bitumen.

When the temperature drops to 5 °C, the fracture energy of the 50/70 penetration bitumens increases considerably (the greater the bitumen content the higher the energy) and decreases in the case of the high modulus bitumen, reversing the trend shown at 20 °C.

At −5 °C the differences among the bitumens are very small and the results are similar, since stiffness and strength, as well as fracture type, are also very alike.

In any case, the fracture energy is shown as a sensitive parameter to bitumen content throughout the temperature range studied (even at −5 °C).

3.3 *Toughness Index (TI)*

Regarding the toughness, Fig. 4 shows how the high modulus bitumen at 20 °C presents generally lower TI than the other two bitumens, although the variation of this parameter with the bitumen content is similar for all bitumens, increasing as bitumen content increases.

At 5 °C the differences between the high modulus bitumen and the other two bitumens are even more pronounced. The TI of high modulus bitumen is very low and is scarcely sensitive to the variation of bitumen content.

At −5 °C all bitumens behave almost equally, so that the toughness index is a sensitive parameter to the bitumen content at temperatures between 20 and 5 °C.

3.4 *Displacement at 50 % of Post-Peak Load*

The trends observed in the displacement at 50 % of post-peak load show even better than IT trends, the differences between bitumens tested, Fig. 5. At 20 °C the crumb rubber modified bitumen presents the greatest ability to deform after maximum load, especially when the bitumen content is high; followed by the conventional bitumen and finally, by the high modulus bitumen, which is the one with the lowest displacement at peak load (due to its higher stiffness and fragility). Values of displacement at peak load increase significantly when bitumen content increases.

As temperature decreases, the differences between bitumens also diminish. At 5 °C differences between the high modulus bitumen and the other two bitumens are still observed, but at −5 °C the differences among them are practically nil and the value of displacement at peak load barely changes with bitumen content.

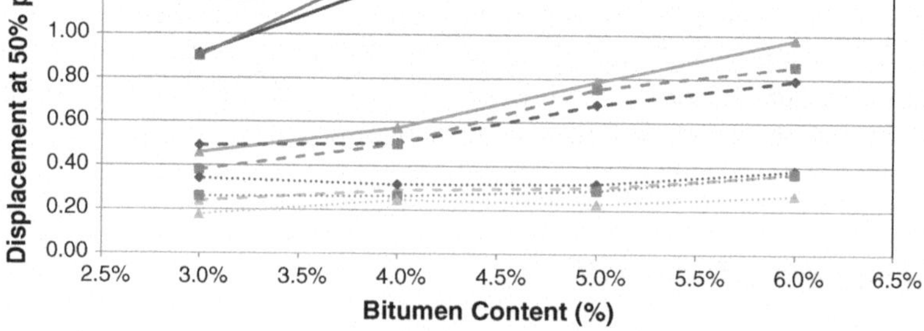

Fig. 5 Variation of displacement at 50 % of post-peak load with bitumen type and content

4 Conclusions

The sensitivity shown by the Fénix test allows differentiating the behaviour of mixtures with different bitumen type and content. At 20 °C tensile stiffness index (TSI), fracture energy (G_D) and toughness index (TI) are significantly different by varying the bitumen type and content, although the observed differences are reduced at low temperatures.

The combination of TSI, G_D and TI obtained from the Fénix test gives an idea of the performance of the mixture against cracking. In principle, high tensile stiffness index and high fracture energy will be beneficial to the mixture, provided that a minimum toughness index is maintained. The requirement of minimum values for these parameters, along with the usual volumetric specifications (air voids content), allow to design tough and cracking resistant mixtures. So, the assessment of cracking resistance could be included in the mixture design stage, so far not considered.

References

Li X, Braham A, Marasteanu M, Buttlar W, Williams R (2008). Effect of factors affecting fracture energy of asphalt concrete at low temperature. Road Mater. Des., Vol. 9, pp. 397-416 (special issue).

Miró R, Martínez A, Pérez-Jiménez F, Botella R (2014) Analysis of cracking resistance of bituminous mixtures using Fenix test. Const. Build. Mater., Vol. 59, pp. 32-38.

Pérez-Jiménez F, Botella R, Martínez A, Miró R (2013) Analysis of the mechanical behaviour of bituminous mixtures at low temperatures. Const. Build. Mater., Vol. 46, pp. 193-202.

Read J, Whiteoak D (2003) Shell Bitumen Handbook (5th Edition). ICE Publishing.

Effects of Recycled Shingle and Virgin Asphalt Binder Mixing on Mixture Performance

Brian Hill, He Wang and William G. Buttlar

Abstract The use of recycled asphalt shingles (RAS) in asphalt concrete leads to questions regarding the mixing of recycled and virgin asphalt binders. These questions concentrate on: the degree of mixing, whether mixing or blending of the asphalt binders occurs, and how the degree of mixing affects performance. Researchers to date have focused efforts on vetting out the degree of blending through various asphalt binder and mixture tests. Research at the University of Illinois used mixture performance tests to evaluate blending. Mixture designs were developed using traditional Superpave methods with 2.5 and 5.0 % RAS. Then, samples with the same virgin aggregate structure and RAS pulp (material remaining post-extraction) and completely mixed asphalt binder were created for high and low temperature evaluation using Hamburg wheel and Disk-Shaped Compact Tension (DCT) tests. In addition, the recently developed Hamburg-DCT Performance Space Diagram was employed to consider mixture property shifts. Research findings demonstrated that up to 20 % asphalt binder replacement met volumetric and performance specifications for both high and low temperature tests. Furthermore, the completely mixed asphalt binder and RAS pulp mixtures performed approximately similar to the traditionally mixed samples at both high and low temperatures.

Keywords DCT · Hamburg · RAS

B. Hill (✉) · H. Wang · W.G. Buttlar
The University of Illinois at Urbana-Champaign, Champaign, IL, USA
e-mail: bchill2@illinois.edu

H. Wang
e-mail: hewang4@illinois.edu

W.G. Buttlar
e-mail: buttlar@illinois.edu

A. Chabot et al. (eds.), *8th RILEM International Conference on Mechanisms of Cracking and Debonding in Pavements*, RILEM Bookseries 13,
DOI 10.1007/978-94-024-0867-6_41

1 Introduction

The sustainability revolution in the paving industry over the past 20 years has led to a proliferation of approaches towards meeting asphalt mixture design, performance, and sustainability standards and goals. RAS is particularly attractive due to the high content of asphalt cement present within the material. Research in RAS has led to industry questions revolving around the amount of available RAS binder and the effect of RAS modification on laboratory measured performance properties.

Early RAS studies showed that the additional RAS did not have an evident effect on volumetric properties and improved compactability (Watson et al. 1998; Foo et al. 1999; Mallick et al. 2000). Watson et al. (1998) and Mallick et al. (2000) showed that the gradation and volumetric properties of RAS-modified mixtures did not change appreciably in reference to the control virgin aggregate mixtures. However, RAS was found to have appreciable effects on high temperature performance with respect to rutting as shown by Foo et al. (1999) and Maupin (2010).

This increase in rutting resistance and thus modulus of the bulk material has created concern with respect to low temperature cracking. Research conducted by Arnold (2014) completed an initial examination of the low temperature fracture behavior of RAS-modified asphalt mixtures using the DCT test. However, research to date has not examined mixed/blended asphalt binder systems at the mixture level in terms of high and low temperature performance. This worst-case scenario of complete blending tested in the laboratory could provide insight into the true low temperature cracking potential of RAS asphalt mixtures.

This RAS study focused on low temperature cracking evaluation and the application of the Hamburg-DCT Performance Space Diagram to RAS mixtures with and without blended asphalt binders. This performance plot space, developed by Buttlar et al. (2015), provides an avenue to examine how modifiers shift asphalt mixture performance at both and high low temperatures. RAS mixtures with 2.5 and 5.0 % RAS were designed to meet Superpave requirements for volumetric properties and maintain equivalent VMA and effective asphalt content such that valid mixture comparisons could be made. Next, DCT and Hamburg tests were completed on the RAS mixtures with and without blended asphalt binder in addition to virgin aggregate PG 64-22 and PG 58-28 mixtures to consider the low and high temperature mixture performance properties of the mixtures. Finally, application of the Hamburg-DCT plot was completed to determine the effects of RAS use on performance plot shifts.

2 Materials and Testing Methods

The components of the asphalt mixtures included: PG 64-22 and PG 58-28 asphalt cement, RAS (with a PG 112 + 2 grade), CM16 (crushed dolomitic limestone coarse aggregate), FM20 (crushed dolomitic limestone sand), and FM02 (natural

Table 1 Mixture volumetrics (Superpave N90)

Volumetric property	Virgin	2.5 % RAS	5.0 % RAS
Total asphalt content (%)	6.6	6.6	6.6
VMA (%)	15.2	15.3	15.2
VFA (%)	74.0	73.8	73.7
Effective asphalt content (%)	4.9	4.9	4.9

sand). The aggregates were sampled from a paving plant in Champaign, IL. The mixture design portion of the study did not include blended extracted RAS and virgin binder mixtures. It was assumed that the blended binder mixtures did not significantly affect volumetric properties.

Mixing and compacting of asphalt mixture specimens occurred at a temperature of 150 °C in accordance with IDOT specifications. All aggregate and asphalt cement samples were heated for approximately 4 h prior to mixing to ensure temperature consistency. The mixtures contained 0.0, 2.5, and 5.0 % RAS to evaluate those with approximately 0, 10.6, and 21.2 % asphalt binder replacement (ABR). The volumetric properties of the asphalt mixtures (assuming 100 % blending of the RAS asphalt binder) are shown in Table 1. These results demonstrate that the total asphalt content at 4.0 % air voids remains equal while holding the VMA, VFA, and P_{be} approximately constant.

To characterize the low temperature cracking behavior of the asphalt mixtures, DCT fracture tests were performed. Generally, temperature-induced transverse (or thermal cracking) in asphalt pavements is thought to predominantly occur in a Mode I opening manner. The DCT test evaluates the fracture energy associated with propagating a crack perpendicular to the applied load through the asphalt mixture using ASTM D7313 at −12 °C.

The Hamburg Wheel Tracking test was used to evaluate the permanent deformation characteristics of the asphalt mixtures investigated. The Hamburg test, specified in AASHTO T-324, is conducted in a water immersed state at 50 °C to induce both permanent deformation and moisture damage. PG 58-28 mixtures are considered satisfactory in terms of permanent deformation resistance if they can withstand 5000 wheel passes prior to reaching a 12.5 mm rut depth in order to conform to IDOT standards in central Illinois.

3 Hamburg Wheel Tracking Results

The Hamburg test results are provided in Fig. 1. As shown, the virgin asphalt mixtures did not meet the Hamburg testing requirement developed by IDOT. On the other hand, the RAS mixtures both exceeded the minimum requirement of 5000 wheel passes prior to reaching 12.5 mm rut depth. This result demonstrated that aggregates with lacking strength coupled with a softer virgin asphalt binder

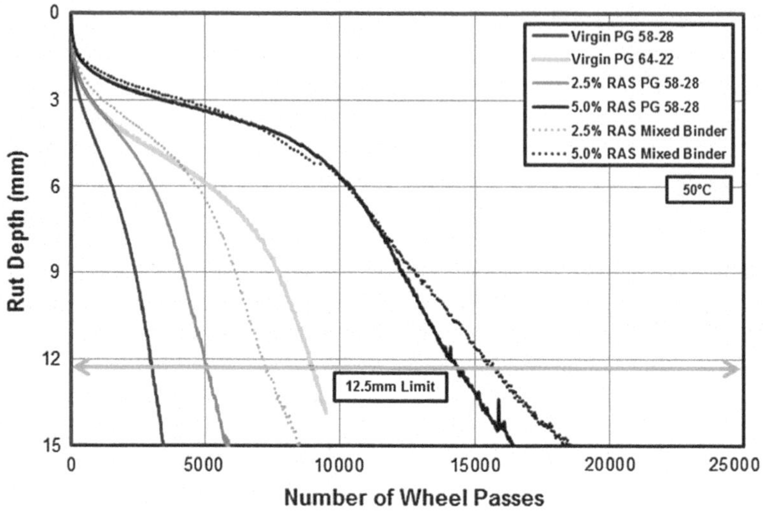

Fig. 1 Hamburg wheel tracking results

benefited from the inclusion of RAS. The improvement in Hamburg wheel tracking performance from 2.5 to 5.0 % RAS showed that an increase in ABR from 10.6 to 21.2 % significantly improved the bulk stiffness properties of the mixture at high testing temperatures.

The 2.5 and 5.0 % RAS mixed binder Hamburg results demonstrate the increased stiffness created by mixing virgin and extracted RAS binder. Research by Mogawer et al. (2013) found that complex modulus of traditional RAP-RAS mixtures yielded a lesser modulus as compared to a fully mixed RAP-RAS binder mixture. Although blended (mixed) asphalt binders yielded improved rutting performance, low temperature fracture properties may be negatively affected.

4 DCT Test Results

The DCT results are shown in Fig. 2. The PG 58-28 virgin asphalt binder was used in this study to offset the potential stiffening provided by the RAS. In this study, it was found that the use of PG 58-28 and two levels of RAS did not lead to poor performance in either low temperature test. Furthermore, the RAS mixtures outperformed the PG 64-22 virgin asphalt mixture. Therefore, the presence of RAS did not lead to low temperature cracking susceptible mixtures. In fact, the only mixtures that met the high and low temperature performance requirements were the 2.5 and 5.0 % RAS mixtures.

Additional DCT tests (shown with red bars) were completed on mixtures containing RAS pulp and mixed virgin and extracted RAS binders. In both RAS

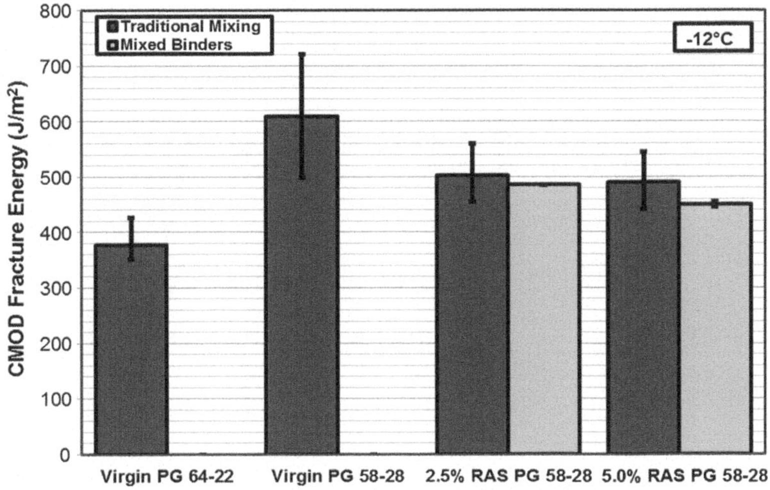

Fig. 2 DCT test results at −12 °C

mixtures, the binder mixing led to decreases in fracture energy, but said fracture energies did not fall outside the statistical range of the DCT results for the traditionally mixed samples. Furthermore, the mixed binder mixtures exhibited improved DCT fracture energy as compared to the reference PG 64-22 mixture. Thus, in the worst case scenario in which the asphalt binders complete mix, grade bumping in this case allowed the mixtures to meet the high and low temperature performance requirements.

5 Hamburg-DCT Plot Results

Data generated during the current study was plotted in the Hamburg-DCT plot space in Fig. 3. Hamburg data plotted in this figure were the values at 5000 wheel passes corresponding to the IDOT minimum value. The use of RAS led to a trade-off of low temperature performance for rutting resistance and movement toward the upper left corner of the Hamburg-DCT Plot. The movement was advantageous as the rutting resistance was improved with the use of RAS without a significant loss of fracture energy. As such, the use of 5.0 % RAS led to sufficient performance in both Hamburg and DCT tests. In fact, the addition of RAS to a PG 58-28 mixture led to improved performance as compared to the reference virgin PG 64-22 mixture. This is demonstrated by the rotation toward the upper-right section of the plot which may have occurred due to the presence of fibers in the RAS material.

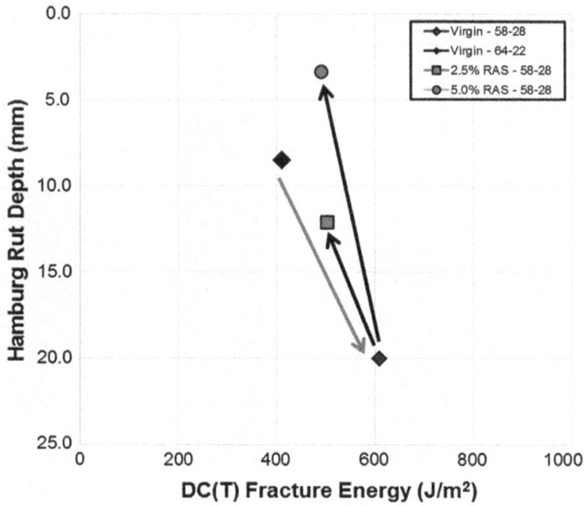

Fig. 3 Hamburg-DCT plot for RAS mixtures

6 Concluding Remarks

The current study evaluated the effects of two RAS contents (2.5 and 5.0 %) on the high and low temperature performance of Illinois asphalt mixtures. The following findings were drawn from this study:

- The inclusion of RAS led to improvements in Hamburg wheel tracking performance. In particular, weaker virgin aggregates such as those used in this study may need the inclusion of RAS to meet performance criteria.
- The increased stiffness of the RAS mixtures did not lead to poor DCT fracture performance. RAS mixtures met and exceeded the performance of a virgin PG 64-22 asphalt mixture.
- The RAS mixtures in the current study were the only mixtures which passed the bracketed performance criteria set for Hamburg and DCT tests. Consequently, mixtures using local Illinois aggregates may benefit from the use of RAS.

References

Arnold, J.W. (2014). Quantitative Evaluation of Low-Temperature Performance of Sustainable Asphalt Mixtures and Binders Containing Recycled Asphalt Shingles (RAS) and Rejuvenators. University of Illinois at Urbana-Champaign.

Buttlar, W.G., Hill, B.C., Wang, H., and Mogawer, W.S. (2015) Performance-Space Diagram for the Evaluation of High and Low Temperature Asphalt Mixture Performance, Accepted for Publication in the Association of Asphalt Paving Technologists.

Foo, K.Y., Hanson, D.I. and Lynn, T.A. (1999). Evaluation of Roofing Shingles in Hot Mix Asphalt, Journal of Materials in Civil Engineering, vol. 11, no. 1, pp. 15-20.

Mallick, R.B., Teto, M.R. and Mogawer W.S. (2000). Evaluation of Use of Manufactured Waste Asphalt Shingles in Hot Mix Asphalt, Chelsea Center For Recycling and Economic Development, Chelsea, MA.

Maupin, G.W. (2010). Investigation of the Use of Tear-Off Shingles in Asphalt Concrete, Virginia Department of Transportation, Richmond, VA.

Mogawer, W.S., Booshehrian, A., Vahidi, S., and Austerman, A.J. (2013) Evaluating the Effect of Rejuvenators on the Degree of Blending and Performance of High RAP, RAS, and RAP/RAS Mixtures, Road Materials and Pavement Design, Vol. 14, Sup. 2, pp. 193-213.

Watson, D.E., Johnson, A. and Sharma, H.R. (1998). Georgia's Experience with Recycled Roofing Shingles in Asphaltic Concrete, Transportation Research Record: Journal of the Transportation Research Board, vol. 1638, no. 0361-1981, pp. 129-133.

Impact of Additives on the Cracking Resistance of Asphalt Mixtures

Peter E. Sebaaly and Elie Y. Hajj

Abstract This paper presents the results of long term investigations that evaluated the impacts of various additives on the resistance of asphalt mixtures to the two types of cracking distresses; fatigue and thermal. Even-though these investigations did not evaluate all additives on the same asphalt mixture and to the same cracking distress, the cracking resistance of asphalt mixtures with and without each type of additive are compared within the same asphalt mixture. Resistance to fatigue and thermal cracking were evaluated using flexural beam fatigue and uniaxial thermal stress and strain test (UTSST), respectively. Since cracking is a problem occurring in the latter part of pavement life, all asphalt mixtures were subjected to long-term oven aging prior to performance testing. The analysis of the data generated from this extensive study indicates that the cracking resistance of AC mixtures is impacted by several factors, including: aggregate absorption, type of WMA additive, asphalt binder modification process, and anti-strip additive.

Keywords Warm mix · Polymer · Tire rubber · Cracking resistance

1 Introduction

Due to the significant impact of cracking distresses on the long-term performance of rehabilitated pavements, only expensive treatments such as recycling or reconstruction are effective methods to rehabilitate cracked asphalt concrete (AC) pavements. Structural overlay can be used only if the appropriate actions have been taken to prevent reflective cracking which traditionally have not been very effective. The pavement engineering community has recognized the seriousness of cracking failures of AC pavements which has led to the incorporation of three out of

P.E. Sebaaly (✉) · E.Y. Hajj
University of Nevada, Reno, USA
e-mail: psebaaly@unr.edu

E.Y. Hajj
e-mail: elieh@unr.edu

A. Chabot et al. (eds.), *8th RILEM International Conference on Mechanisms of Cracking and Debonding in Pavements*, RILEM Bookseries 13,
DOI 10.1007/978-94-024-0867-6_42

four cracking failures of AC pavements (except block cracking) in the American Association of State Highway and Transportation Officials (AASHTO) Mechanistic-Empirical pavement design method (M-E Design). The National Cooperative Highway Research Program (NCHRP) funded two research projects on top-down fatigue cracking and reflective cracking of AC pavements. The results of these studies are currently incorporated or in the process of being incorporated into the AASHTOWare Pavement M-E Design software (AASHTO 2015). The Asphalt Research Consortium (ARC) funded by the Federal Highway Administration (FHWA) conducted fundamental investigations to develop constitutive models of fatigue damage of AC pavements (Darabi et al. 2013).

2 Impact of WMA Additive on Aggregate Type

This study was funded by the South Dakota Department of Transportation (SDDOT) in the United States to evaluate the properties of warm-mix asphalt (WMA) mixtures under SD conditions (Sebaaly et al. 2015a, b). Table 1 summarizes the experimental program for this research. All mixtures were prepared with a PG64-28 polymer-modified asphalt binder. The resistance to fatigue cracking of the natural gravel source was not evaluated due to lack of materials.

2.1 Resistance to Fatigue Cracking

The resistance of the various mixtures to fatigue cracking was evaluated using the flexural beam fatigue test (AASHTO T321-03). The 64 × 50 × 380 mm flexural beam specimen is subjected to a four-point bending with free rotation and horizontal translation at all load and reaction points. This produces a constant bending moment over the center portion of the specimen. Constant strain tests were conducted at different strain levels; using a repeated sinusoidal load at a frequency of 10 Hz, and a test temperature of 22 °C. A total of 8 beams were tested to develop the fatigue relationship for each mixture. It should recognized that the repeatability of the AASHTO T321 test has been estimated around 25 %. Initial flexural stiffness was measured at the 50th load cycle. Fatigue life or failure was defined as the number of cycles corresponding to a 50 % reduction in the initial stiffness.

Table 1 Laboratory experimental program for South Dakota study

Aggregate Type	Warm-mix technology			
	None	Advera	Evotherm	Foaming
Quartzite	X	X	X	X
Limestone	X	X	X	X
Natural gravel	X	X	X	X

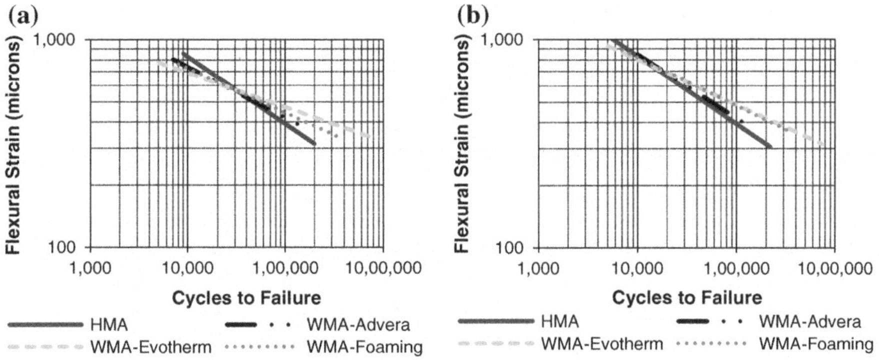

Fig. 1 Fatigue characteristics of asphalt mixtures: **a** Quartzite aggregates, **b** Limestone aggregates

Figure 1a and b present the fatigue characteristics of the quartzite and limestone mixtures, respectively.

Examination of the fatigue cracking model presented in Fig. 1a and b leads to the following conclusions:

- Quartzite Mixtures: WMA additives improved the fatigue resistance of the AC mixture at strain levels lower than 600 μm where Evotherm showed the highest improvement followed by foaming while Advera showed the least improvement.
- Limestone Mixtures: WMA additives Evotherm and Foaming improved the fatigue resistance of the AC mixture while Advera did not show any improvement.

The impact of aggregate type on the fatigue resistance of the AC mixtures can be assessed by evaluating the data presented in Table 2. The fatigue resistance of the limestone WMA-Foaming mix is significantly higher than the quartzite mix. The fatigue resistance of the quartzite WMA-Evotherm mix is significantly higher than the limestone mix at low strain level.

Table 2 Impact of aggregate type on fatigue characteristics of mixtures at 22 °C

Aggregate type	Fatigue life, Nf	HMA	WMA-Advera	WMA-Evotherm	WMA-foaming
Quartzite	Nf at 300 μm	232,783	398,460	1605,475	622,739
	Nf at 500 μm	47,924	48,869	70,704	56,936
	Nf at 700 μm	16,921	12,267	9040	11,778
Limestone	Nf at 300 μm	230,880	300,441	915,417	778,466
	Nf at 500 μm	47,727	56,275	87,633	84,588
	Nf at 700 μm	16,897	18,671	18,685	19,606

2.2 *Resistance to Thermal Cracking*

The uniaxial thermal stress and strain test (UTSST) was used to determine the low-temperature cracking resistance of the various asphalt mixtures (Hajj et al. 2013). The test consists of cooling down a beam 50 × 50 × 250 mm specimen at a rate of 10 °C/h while restraining it from contracting. While the beam is being cooled down, tensile stresses are generated due to the ends being restrained. The AC mixture would fracture as the internally generated stress exceeds its tensile strength. The fracture temperatures are summarized in Table 3 representing the field temperatures under which the pavements will experience thermal cracking.

Examination of the thermal cracking data presented in Table 3 leads to the following observations:

- Overall, WMA mixtures exhibited fracture temperatures that are 2 °C cooler than the HMA fracture temperatures for all aggregate type.
- There is a strong correlation between the binder absorption and fracture temperature for both HMA and WMA mixtures; the higher the binder absorption the warmer the fracture temperature. Binder absorption was calculated based on the difference between the effective and bulk specific gravities of the aggregate. The natural gravel had the highest binder absorption with the warmest fracture temperature while the limestone aggregate had the lowest binder absorption with the coldest fracture temperature.

3 Impact of WMA Additives on Modified Asphalt Mixtures

This study was funded by the Alon Asphalt Company to evaluate the properties of WMA mixtures with polymer and tire rubber modified PG64-28 asphalt binders (Sebaaly et al. 2015a, b). The asphalt mixtures were treated with two WMA additives; Evotherm and water foaming and with two anti-strip additives; hydrated lime and liquid. Figures 2, 3 and 4 present the fatigue characteristics of the un-treated, lime-treated, and liquid-treated asphalt mixtures, respectively. Examination of the data presented in Figs. 2, 3 and 4 leads to the following observations.

- Impact of WMA additives: The WMA-Evotherm mixtures showed similar fatigue resistance to the HMA mixtures for all cases of un-treated, lime-treated, and liquid-treated. On the other hand, the WMA-Foaming mixtures showed lower fatigue resistance than the HMA mixtures for all cases of un-treated, lime-treated, and liquid-treated.

Table 3 Thermal cracking characteristics of SD mixtures (UTSST fracture temperatures)

Aggregate type	Fracture temperature, °C					
	Optimum binder content, %	Binder absorption, %	HMA	WMA-Advera	WMA-Evotherm	WMA-foaming
Quartzite	6.0	0.45	−29	−31	−32	−32
Limestone	4.9	0.21	−32	−34	−34	−32
Natural gravel	5.5	0.91	−26	−28	−28	−27

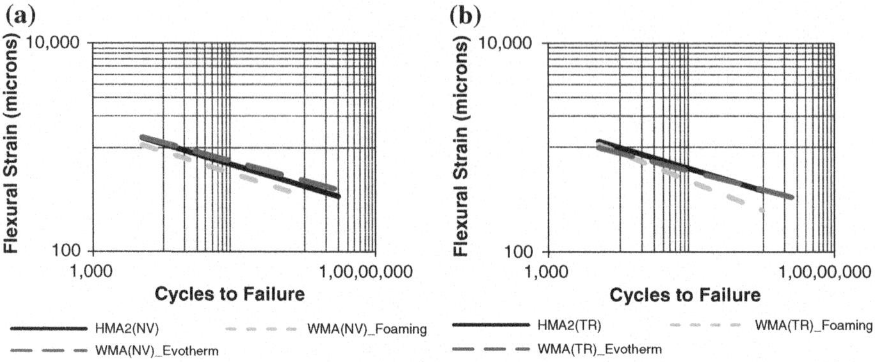

Fig. 2 Fatigue characteristics of un-treated mixtures at 22 °C: **a** polymer-modified, **b** tire rubber modified

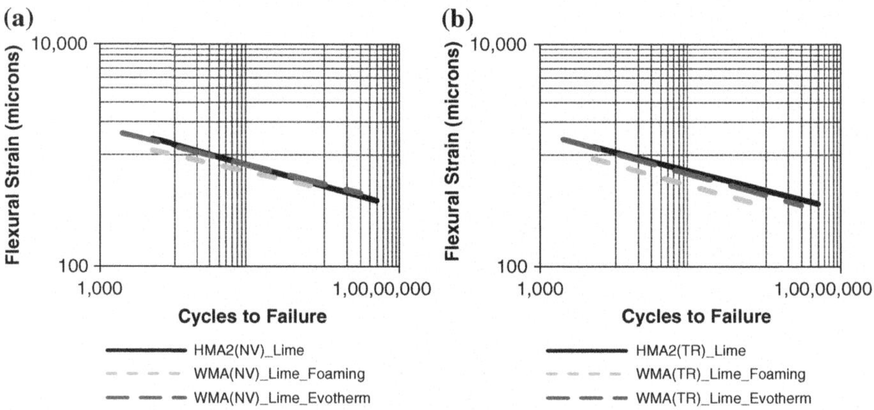

Fig. 3 Fatigue characteristics of lime-treated mixtures at 22 °C: **a** polymer-modified, **b** tire rubber modified

- Impact of anti-strip additives: the lime-treated mixtures showed better fatigue resistance than the un-treated and liquid-treated for both the polymer and rubber modified mixtures.
- Impact of modification: the polymer-modified and tire rubber-modified mixtures showed similar resistance to fatigue, except in the case of the liquid-treated mixtures where the polymer-modified mix showed a steeper fatigue curve than the tire rubber modified mix.

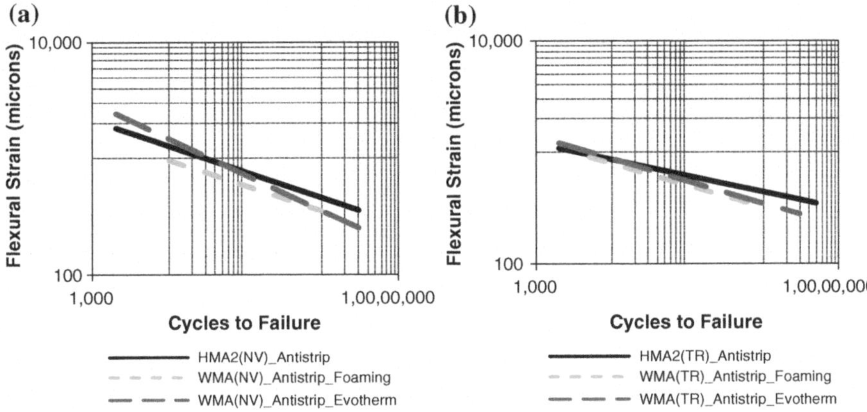

Fig. 4 Fatigue characteristics of liquid-treated mixtures at 22 °C: **a** polymer-modified, **b** tire rubber modified

4 Summary and Recommendations

The analysis of the data generated from this extensive study indicates that the cracking resistance of AC mixtures is impacted by several factors, including: aggregate absorption, type of WMA additive, asphalt binder modification process, and anti-strip additive. This leads to the conclusion that every AC mix can be designed to resist cracking as long as the optimum combination of these critical factors is achieved through the proper combination of mix design and evaluation.

The future direction of this research activity is to continue the assessment of impact of additional factors on the cracking resistance of AC mixtures by incorporating recycled asphalt pavements (RAP) and recycled asphalt shingles (RAS) into the asphalt mix. It is greatly anticipated that the incorporation of RAP and RAS into asphalt mixtures will lead to reduced resistance to cracking. Hence, it is the responsibility of the asphalt pavement engineering community to identify technologies that reduces the negative impacts of RAP and RAS on the cracking resistance of asphalt mixtures.

References

American Association of State Highway Transportation Officials (2015) AASHTOWare Pavement ME Design, www.aashtoware.org, Washington DC, USA.

Darabi MK, Abu Al-Rub RK, Masad EA, Little DN (2013) Constitutive modeling of fatigue damage response of asphalt concrete materials. J of TRR 2373: 11–21. doi 10.3141/2373-02.

Hajj EY, Alavi MZ, Morian NE, Sebaaly PE (2013) Effect of Select Warm-Mix Additives on Thermo-Viscoelastic Properties of Asphalt Mixtures. J of RMPD 14, Supp 1: 175–186. doi:10.1080/14680629.2013.774754

Sebaaly PE, Hajj EY, Souliman MI (2015) Evaluation of warm mix asphalt concrete pavement in South Dakota conditions. Final Research Report, Department of Transportation, Pierre, South Dakota, USA.

Sebaaly PE, Hajj EY, Maharjan R, Hitti E (2015) Performance of tire rubber-modified asphalt binder in warm mix asphalt mixtures. Proc. of RAR2015: 245–266, Oct 4–7, 2015, Las Vegas, Nevada, USA.

Mixture and Production Parameters Affecting Cracking Performance of Mixtures with RAP and RAS

Reyhaneh Rahbar-Rastegar and Jo Sias Daniel

Abstract Many methods and approaches have been developed to evaluate the asphalt performance in the laboratory. The laboratory testing can be conducted on mix design (lab produced) and QA/QC (plant produced) stages. To predict the asphalt field performance, an understanding of differences between the behavior of lab and plant produced mixtures is required. There are many factors in the design and production of recycled mixtures that can potentially impact the performance of the in-place material. The project involves testing of both laboratory and plant produced mixtures containing 20–30 % recycled materials (RAP and RAS) by total weight. The mixtures have different gradations (12.5 and 19 mm NMSA), and contain different virgin binder PG grade (PG 58-28, and PG 52-34). The characteristic damage curves and failure criterion diagrams of mixtures are developed based on fatigue testing following the simplified viscoelastic continuum damage (SVECD) protocols. The results show that the most of lab produced mixtures have better fatigue properties in damage characteristics, but they do not follow a consistent trend in fatigue life.

Keywords HMA · Cracking · SVECD · Recycled materials

1 Introduction

The asphalt industry is moving towards the performance based designs. Asphalt performance is intended to be evaluated in the laboratory, based on developed methods and approaches. Different handling, mixing, and compaction methods in the lab can make a difference between the measured properties of lab produced mixtures and those properties in the mixtures produced at plants. Therefore, a good

R. Rahbar-Rastegar (✉) · J.S. Daniel
University of New Hampshire, Durham, USA
e-mail: rr10@unh.edu

J.S. Daniel
e-mail: jo.daniel@unh.edu

A. Chabot et al. (eds.), *8th RILEM International Conference on Mechanisms of Cracking and Debonding in Pavements*, RILEM Bookseries 13,
DOI 10.1007/978-94-024-0867-6_43

understanding about the correlation between the behavior of lab produced and plant produced mixtures is important. To understand this distinction, the behavior of plant and lab produced mixtures should be evaluated in a variety of mixtures with different properties.

One of the most important issues in flexible pavements is cracking, which causes bad riding quality and water penetration from pavement surface to underneath layers. Mixture type and materials (binder grade and source, aggregate size, and recycled materials type and content), volumetric properties (air void, asphalt content), climate conditions, and pavement structure are the most important factors affecting the cracking performance of asphalt pavement.

Considering the increasing use of recycled materials in asphalt pavements, the effect of these materials on cracking has gained more attention in the recent years. Reclaimed asphalt pavement (RAP) and recycled asphalt shingles (RAS) are the most common types of materials used for recycling. While the contractors and departments of transportation (DOTs) are interested in incorporating higher percentages to realize savings of costs and natural resources, one of the concerns with using higher amounts of RAP or RAS is the increased potential for cracking, as the materials tend to be stiffer and less workable.

The objective of this study is the comparison of cracking potential in the plant and lab produced mixtures, varied in binder PG grade, nominal maximum aggregate size (NMAS), and recycled materials type and content, and the evaluation of impacts of these factors on cracking resistance using experimental testing.

2 Materials and Testing

This study includes testing on 11 plant produced and 8 lab produced mixtures, as shown in Table 1. Specimens were fabricated from plant produced mix without reheating or fabricated from raw materials in the laboratory. The RAS material was primarily from tear-off shingles that could not be graded in the laboratory. The RAP binder had a continuous grade of 81.3–19.3 °C. The percentages of binder replacement presented in Table 1 are rounded to 20 and 30 % in results and discussion section.

The asphalt binders from both plant and laboratory mixed materials were extracted and recovered and then fully characterized for PG grading and critical cracking temperature. Complex modulus testing (AASHTO T 342) was conducted on the mixtures. The results of the binder testing and linear viscoelastic characterization of the mixtures are presented elsewhere (Rahbar-Rastegar 2015).

Uniaxial tensile fatigue testing and analysis using the simplified viscoelastic continuum damage (S-VECD) approach was conducted on four specimens of each mixture following AASHTO TP 107 (2014).

Damage analysis for each mixture was performed and damage characteristic curves (C-S) were obtained using models available within ALPHA-Fatigue software. Also, the fatigue cracking resistance was assessed by fatigue failure criterion

Table 1 Mixtures types and properties

Binder PG grade	Binder source	NMSA (mm)	%Total binder replacement (% RAP/ % RAS)	Specimen type
58-28	McAsphalt	12.5	18.9 (18.9/0)	Plant & Lab
	McAsphalt	12.5	18.5 (7.4/11.1)	Plant
	McAsphalt	12.5	28.3 (28.3/0)	Pant & Lab
	McAsphalt	19	20.8 (20.8/0)	Plant
	Avery Lane	19	20.4 (8.2/12.2)	Plant & Lab
	Avery Lane	19	31.3 (31.3/0)	Plant & Lab
52-34	McAsphalt	12.5	18.9 (18.9/0)	Plant& Lab
	McAsphalt	12.5	18.5 (7.4/11.1)	Plant
	McAsphalt	12.5	28.3 (28.3/0)	Plant & Lab
	Suncor	19	20.4 (8.2/12.2)	Plant & Lab
	Suncor	19	31.3 (31.3/0)	Plant & Lab

of asphalt mixtures versus number of cycles ($G^R - N_f$). G^R is the rate of change of the averaged released pseudo strain energy (per cycle) throughout the test, and is calculated from the Eq. 1.

$$G^R = \frac{\int_0^{N_f} W_C^R}{N_f^2} \tag{1}$$

where W_C^R is released pseudo strain energy, and N_f is the number of cycles before failure (Sabouri and Kim 2014).

3 Results and Discussion

Figures 1 and 2 compare the damage characteristic curves of the different plant and lab produced mixtures for 19 and 12.5 mm NMSA, respectively. Generally, this curve shows how the material integrity of the specimen decreases as damage is growing. The mixtures that have C-S curves further up and to the right would be expected to perform better, since they are able to maintain their integrity better during the test. However, the cracking performance of a mixture in the field depends on pavement structure as well.

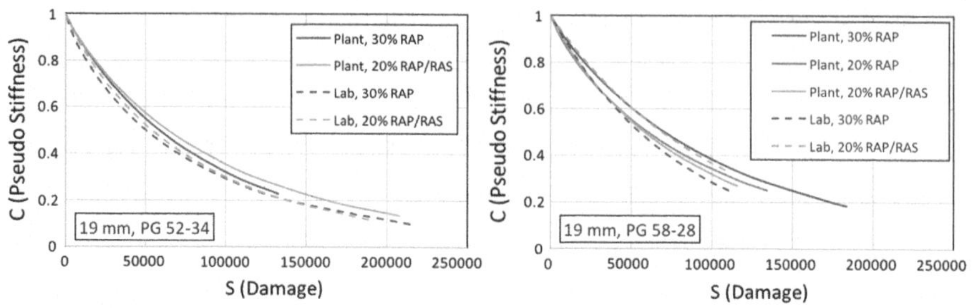

Fig. 1 Damage characteristic curves of 19 mm mixtures

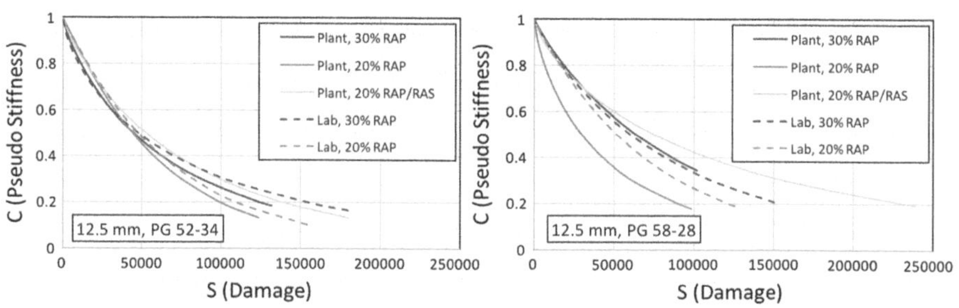

Fig. 2 Damage characteristic curves of 12.5 mm mixtures

The C-S curves of lab produced mixes are very close to, or higher than plant produced ones for all 12.5 mm mixtures, indicating lab produced specimens might show better performance during the fatigue cyclic test, while most of the 19 mm plant produced mixtures have slightly better fatigue resistance than lab produced mixtures with showing higher pseudo stiffness in a same amount of damage.

One interesting point is that there is not a significant difference between the damage characteristic curves of 19 mm mixtures, while 12.5 mm mixtures show a larger distinction between mixtures. 20 % RAP mixtures show mainly a rapid decrease in material integrity compared to 30 % RAP and 20 % RAP/RAS mixtures, while the material integrity of 20 % RAP/RAS mixtures decreases with a lower slope in most cases. The order of 20 % RAP/RAS, 30 % RAP, and 20 % RAP is the same for most 12.5 mm mixtures, but it is not always true for 19 mm mixtures. 30 % RAP and 20 % RAP/RAS curves of 19 mm mixtures are too close to each other to be distinguished.

Another observation is that in most cases, the last point of C-S curves indicating pseudo stiffness in failure (C_F) increases by using higher percentage of RAP. This point has been suggested by Hou et al. (2010) and observed in another study conducted at North Carolina State University (Norouzi et al. 2014) that in the mixtures with more RAP, the C_F value goes higher. In this study, the C_F values of 30 % RAP mixes are higher than those of 20 % RAP, that is in accordance with the previous studies, and showing the more brittle mixtures, which might fail at a

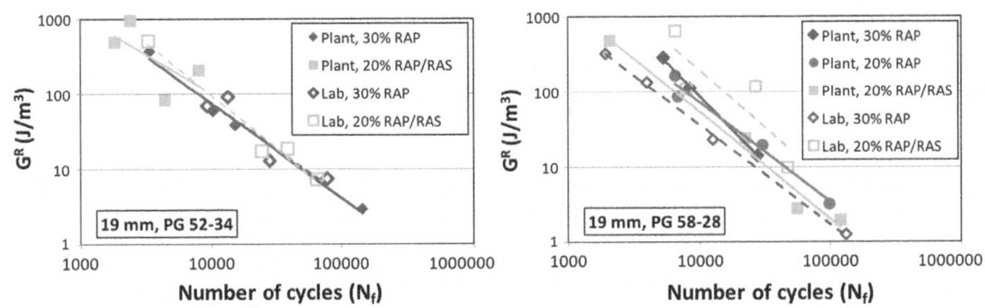

Fig. 3 Fatigue failure criterion versus number of cycles for 19 mm mixtures

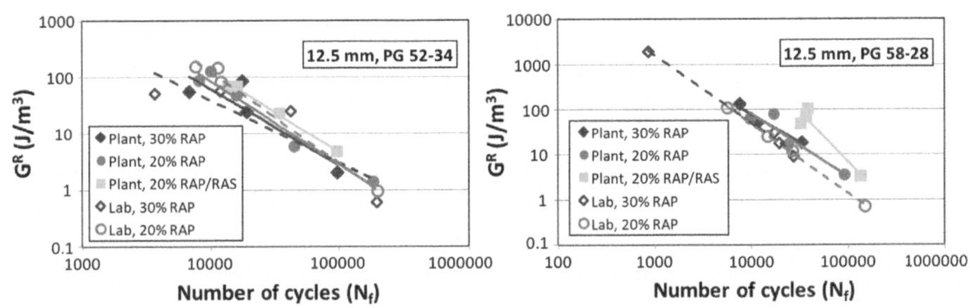

Fig. 4 Fatigue failure criterion versus number of cycles for 12.5 mm mixtures

higher integrity. C_F values of 20 % RAP/RAS mixtures are very close to 30 % RAP mixtures.

Figures 3 and 4 compare the fatigue failure criterion (G^R) versus the number of cycles (N_f) of plant produced and lab produced mixtures for 19 and 12.5 NMSA, respectively. Generally, the higher G^R values at the same number of cycles (N_f) indicates better fatigue performance.

In most cases, there is a strong relationship (high coefficient of determination, R^2) between fatigue failure criterion (G^R) and the number of cycles to failure (N_f). Although, a consistent trend could barely be found between the fatigue life of plant and lab produced mixtures, 30 % RAP lab produced mixtures seem to perform worse than 30 % plant produced ones, and 20 % RAP/RAS mixtures show better fatigue life in lab produced mixtures.

20 and 30 % RAP mixtures with both 12.5 and 19 mm NMSA show a similar behavior, while incorporating RAS improves the asphalt fatigue life. However, adding RAS might result in a more brittle mixture and sudden failure at high G^R values, as occurred in 19 mm, PG 52-34 mixture.

4 Conclusion

The main objective of this study was to compare the laboratory measured cracking properties of plant and laboratory produced mixtures and evaluate the impacts of different parameters on cracking potential. The following conclusions were drawn from the results of testing and analysis:

- Generally, the difference between C-S curves for lab produced mixtures is less than those for plant produced ones. Also, most 12.5 mm lab produced mixtures show more integrity than their corresponding plant produced mixes during the fatigue testing, while the integrity of 19 mm plant produced mixtures are higher.
- The variation of damage characteristic curves for 12.5 mm mixtures is greater than 19 mm mixtures. Generally, 30 % RAP mixes show the lower pseudo stiffness as damage grows than 20 % RAP/RAS mixtures, followed by 20 % RAP mixtures, indicating best damage characteristics for 20 % RAP/RAS mixtures and worst for 20 % RAP.
- The fatigue properties of most PG 58-28 mixtures seems to be better than PG 52-34 mixtures. Although, the sources and rheological parameters of binders are recommended to be considered as well, it shows using a softer binder does not necessarily improve the fatigue properties.
- There is no evident trend between $G^R - N_f$ of plant and lab produced mixtures.
- 20 % RAP and 30 % RAP mixtures show similar behavior in $G^R - N_f$ diagrams, while incorporating RAS can increase G^R values. It should be noted that higher amounts of RAP or RAS might improve the fatigue performance but make it more brittle that increases the probability of cracking at higher material integrity.

References

Rahbar-Rastegar, R., (2015), "Correlation between Laboratory and Plant Produced High RAP/RAS Mixtures", Presented at North East Asphalt User/Produced Group (NEAUPG) Annual Meeting, 2015, Burlington, VT

Sabouri, M. and Kim Y. R. (2014), "Development of Failure Criterion for Asphalt Mixtures under Different Modes of Fatigue Loading", Transportation Research Record: Journal of Transportation Research Board.

Hou, T., Underwood, B. S., and Kim, Y. R., (2010), "Fatigue Performance Prediction of North Carolina Mixtures Using Simplified Viscoelastic Continuum Damage Model. Journal of the Association of Asphalt Paving Technologists", Vol. 79, 2010, pp. 35–80.

Norouzi, A., Sabouri, M., and Kim, Y. R. (2014), "Evaluation of the Fatigue Performance of Asphalt Mixtures with High RAP Content", Journal of Tylor and Francis Group, pp. 1069–1077.

American Association of State Highway Transportation Officials (AASHTO), AASHTO TP 107, "Determining the Damage Characteristic Curve of Asphalt Concrete from Direct Tension Cyclic Fatigue Tests" (2014)

Shrinkage Characteristics of Alkali-Activated Slag Mortar with SAP

Young Cheol Choi, Gyu Don Moon, Sungwoo Oh, Sang Hwa Jung and Kwang Myong Lee

Abstract The advantages of alkali activated slags (AAS) with adequate types and dosages of activators are high strength development, rapid setting, lower permeability, excellent durability, low hydration heat, high early-age strength and high resistance to chemical attack. The hydration products found in AAS are C-S-H with a low Ca/Si ratio related to GGBFS and the components of activators used. However, high autogenous and drying shrinkages of alkali-activated slag concrete are frequently reported. Therefore, researchers have continued their efforts to mitigate shrinkage strains and stresses. Two types of alkali activators and superabsorbent polymers were used to investigate autogenous shrinkage, and porosity characteristics of AAS. To mitigate shrinkage of AAS mortars, superabsorbent polymers (SAP) were applied to create an internal curing effect on the mixtures. As a result, autogenous shrinkage strains and AAS mortars that SAP were applied was significantly decreased, therefore, it is concluded that SAP played an important role on shrinkage mitigation of all mixtures.

Keywords AAS · Internal curing · Autogeneous shrinkage · SAP

Y.C. Choi (✉) · G.D. Moon · S. Oh · S.H. Jung
High-Tech Construction Materials Center, Korea Conformity Laboratories, Seoul, South Korea
e-mail: zerofe@kcl.re.kr

G.D. Moon
e-mail: mgd0123@kcl.re.kr

S. Oh
e-mail: ohsungwoo@kcl.re.kr

S.H. Jung
e-mail: jsh2593@kcl.re.kr

K.M. Lee
Sungkyunkwan University, Seoul, South Korea
e-mail: leekm79@skku.edu

A. Chabot et al. (eds.), *8th RILEM International Conference on Mechanisms of Cracking and Debonding in Pavements*, RILEM Bookseries 13,
DOI 10.1007/978-94-024-0867-6_44

1 Introduction

Over the past decades, many researches on finding alternative materials that can be substituted for cement have been conducted: particularly, geopolymer and alkali activated slag (AAS), which are polymerized by using alkaline activators with fly ash and metakaolin, and slag activations, have been investigated to substitute to Portland cement. AAS with adequate types and dosages of alkali activators are able to perform high strength development, lower permeability, excellent durability, low hydration heat, high early strength and high resistance to chemical attack (Bakharev et al. 2000; Haha et al. 2011). However, high shrinkage one of the drawbacks of alkali activated slag and still frequently reported, impeding the application of alkali activated slag (AAS) concrete in the construction practice. In general, AAS activated by sodium silicate generated more shrinkages than sodium hydroxide activated slag, and the shrinkage of AAS increases with increasing dosage of activators as well as slag fineness.

Internal curing is one of the methods to mitigate autogenous shrinkage by additional water supply in the mixtures. During internal curing process, a water reservoir is used, commonly either superabsorbent polymers or lightweight aggregates (LWA) such as pumice and expanded clay. Sakulich and Bentz (2012) showed that the autogenous shrinkage of alkali-activated slag mortar could be reduced by using pre-wetted lightweight aggregate due to water contents in the LWA that play important role in further hydration, which is called "internal curing effect".

The purpose of the research is to investigate autogenous shrinkage characteristics of alkali-activated slag and the factors affecting it. Two AAS mixtures and a plain mixture were used, and SAP was used to evaluate the effects on shrinkage reduction. A mercury intrusion porosimetry (MIP) test was conducted to find out the relationship between the mesopore volume and shrinkage of AAS mortars.

2 Material Results and Discussion

Ground granulated blast furnace slag (GGBFS) and OPC were used in this research. The chemical compositions and physical characteristics of binder materials are listed in Table 1. The bulk polymerized SAP (provided by LG chemistry) of irregular particle shape was provided in Fig. 1.

Table 1 Chemical and physical properties of ground granulated blast slag and OPC

	Chemical compositions (wt%)										
	CaO	SiO_2	Al_2O_3	Fe_2O_3	MgO	K_2O	Na_2O	SO_3	LOI	Density (g/cm^3)	Blaine fineness (cm^2/g)
OPC	37.2	34.0	16.4	0.50	6.29	0.45	1.33	2.71	0.17	3.15	3400
GGBS	61.9	21.2	4.64	2.91	1.87	1.22	0.29	2.31	2.48	2.94	4100

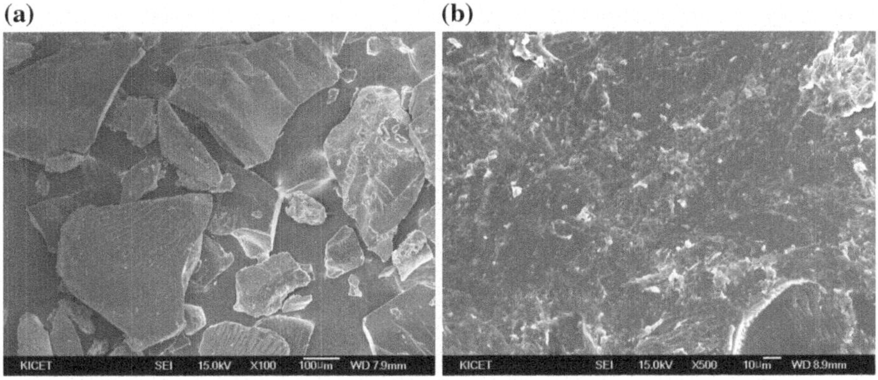

Fig. 1 SEM images of SAP. **a** SAP (×100). **b** SAP (×500)

Table 2 Mixture proportions

	Materials used (g)						
	OPC	GGBFS	Sand	Na_2SiO_2	NaOH	Na_2CO_3	SAP
Plain	100		237				
ANN		100	237	16	4.6		
ANC		100	237			6.8	
Plain-S	100		214				0.3
ANN-S		100	214	16	4.6		0.3
ANC-S		100	214			6.8	0.3

Alkali activators were prepared with sodium hydroxide (99.0 % purity) and sodium silicate solution (Na_2O: 9.3 wt%, SiO_2: 28.2 wt%) for ANN, and sodium carbonate (99.5 % purity) was utilized to produce ANC. All alkali activators were produced to contain 4 wt% of Na_2O by binder, and left 1 day to cool down to room temperature. Table 2 shows mixture proportions of AAS and OPC mortars. Water to binder ratio is constant at 0.4 for all specimens, and additional water, which is approximately 0.04 of water to binder ratio, applied to the mixture with SAP to equalize the mini slump with the specimens without SAP. SAP can take up to 30 times of its dry weight, so only 0.3 g of SAP was applied to the specimens.

OPC and AAS mortar specimens, 40 by 40 by 160 mm, were prepared for compressive strength. After 24 h, the prism specimens were demolded and placed in a chamber at 90 ± 5 % relative humidity and 20 ± 3 °C temperature. The compressive strength was measured at the ages of 1, 3, 7, and 28 day by Universal Testing Machine (UTM) in accordance with ASTM C109.

The linear autogenous deformation of samples was measured by using the corrugated tube method in compliance with ASTM C 1698. As soon as specimens were casted, they were immediately sealed with the end plugs and placed in the steady temperature and humidity room at 20 ± 3 °C and 50 ± 5 % for real time measurement.

Mercury Intrusion Porosimetry(MIP) was conducted to investigate porosity characteristics with the effects on SAP in AAS. All paste specimens were produced at 7 days, and 1 g of specimens were made and dried for 24 h at 60 °C. Autopore IV 9500 by Micromeritics was utilized to analyze pore distribution of all mixtures, which applied the pressure up to 33,000 psia. Total porosities and pore size distributions were drawn to utilize for pore characteristics of AAS with and without SAP.

3 Result and Discussion

At 28 days, compressive strengths for Plain and ANN mixtures were developed almost the same level, which were over 50 MPa, but ANC mixture at 28 days was only developed up to 40 MPa. The specimens with SAP developed low compressive strength at the early age, but experimental result indicated that compressive strengths at 28 days for AAS with SAP and without SAP were almost identical (Hasholt et al. 2010).

Figure 2a and b show the results from SEM-EDS analysis for hydration products of ANN and ANC. From Fig. 2c, Ca/Si of hydration products of ANN and ANC are 1.87 and 2.47, respectively. Generally, AAS containing hydration products with low Ca/Si ratio is likely to cause high shrinkage generation.

Autogenous shrinkages of AAS and OPC mortars were evaluated, and shrinkage data were collected in real time after the final setting time for 360 h, which is 15 days. According to the Fig. 3, 120 με were obtained in Plain, while deformations of ANN and ANC were 1100 and 250 με, respectively. This is because of the different hydration products activated by different activator types (Gebregziabiher et al. 2015).

High reduction of autogenous shrinkage was observed in all specimens with SAP, and even Plain-S specimens slightly expanded due to the internal curing effect, which were 140 με reductions. Also, 80 % of shrinkages were reduced in ANN-S, and half reduction in shrinkage was obtained in ANC-S. Sakulich and Bentz also noticed that more than half strain reductions were observed in the mixtures with SAP (Sakulich and Bentz 2012).

As described in Fig. 4a, the pores of ANN-S increased in Zone A because of the volume of SAP in hardened specimen, average pore sizes are 600 μm in dry

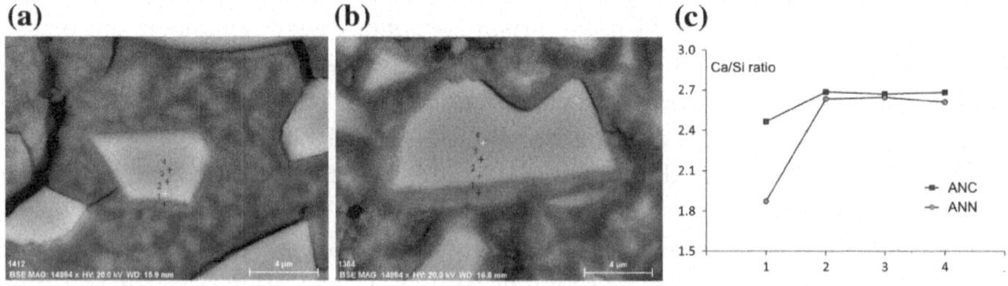

Fig. 2 SEM images at 3 days and EDS analysis of hydrated products. **a** ANN image. **b** ANC image. **c** Ca/Si ratio

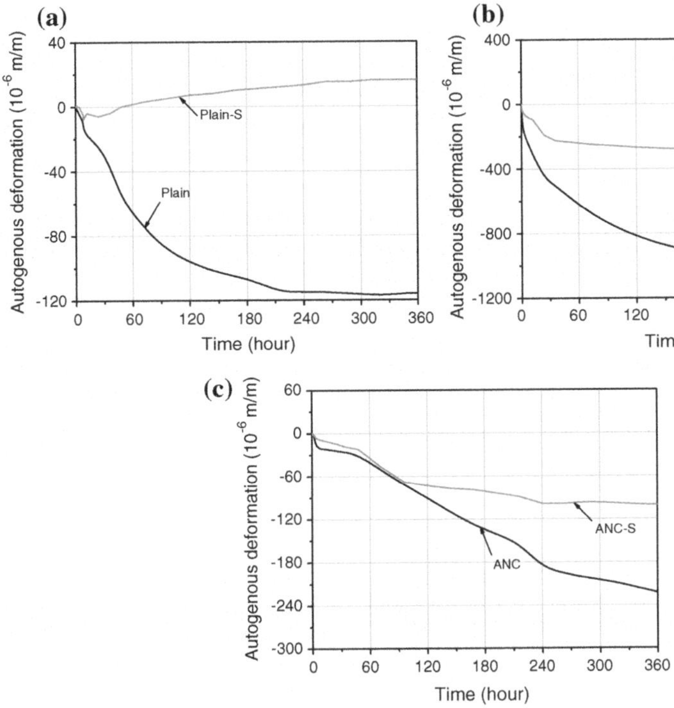

Fig. 3 Test results of autogenous shrinkage

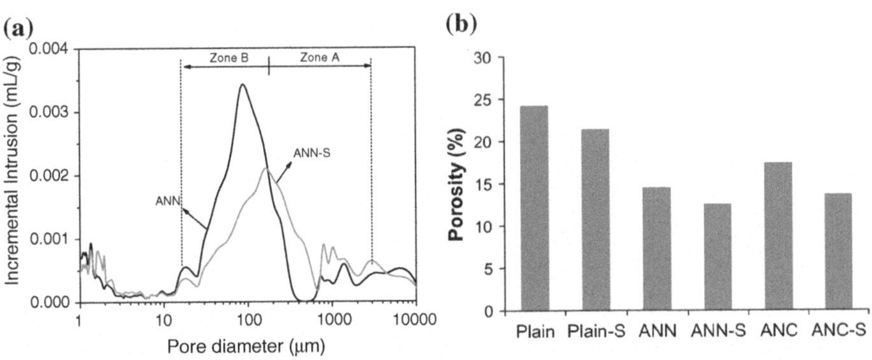

Fig. 4 Pore size distribution and porosity with/without SAP

condition. However, the pores of ANN-S significantly decreased in Zone B because of internal curing effect by SAP. Figure 4b provides the porosities of six mixtures, and the porosities of ANN series were the lowest. Internal curing by SAP played an important role in decrease in porosities in all specimens.

4 Conclusion

The compressive strengths of AAS and OPC mortars with SAP were equivalent level of those without SAP at the age of 28 days. AAS mortars generated large amount of shrinkages, and AAS with hydration products with low Ca/Si ratio produced more deformations as described in the experimental results. Moreover, high shrinkage reduction of AAS with SAP was observed due to the internal curing. SAP played an important role in decrease in pore volume of AAS, but larger pores were generated due to the water-released SAP.

Acknowledgements This research was supported by Grant (Code 11-Technology Innovation-F04) from Construction Technology Research Program (CTIP) funded by Ministry of Land, Infrastructure and Transport.

References

Sakulich, A. R. and Bentz, D. P. (2012) Mitigation of autogenous shrinkage in alkali activated slag mortars by internal curing, Rilem, DOI 10.1617/s11527-012-9978-z

Bakharev, T., Sanjayan, J.G., Cheng, T.-B. (2000) Effect of admixtures on properties of alkali-activated slag concrete, Cement and Concrete Research, 30, 1367-1374

Haha, M.B., Saout, G.L., Winnefeld, F., Lothenbach, B. (2011) Influence of activator type on hydration kinetics, hydrate assemblage and microstructural development of alkali activated blast-furnace slags, Cement and Concrete Research, 41, 301-310

Gebregziabiher, B.S., Thomas, R., Peethamparan S. (2015) Very early-age reaction kinetics and microstructural development in alkali-activated slag, Cement and Concrete Composites, 55, 91-102

Hasholt, M.T., Jespersen, M.H.S., Jensen, O.M. (2010) Mechnical properties of concrete with SAP - Part 1: Development of compressive strength, International RILEM conference on Use of Superabsorbent Polymers and Other New Additives in Concrete, 15-18 August

The Impact of Homogeneity of High Modulus Asphalt Concrete Layer on Low-Temperature Cracking

Jozef Judycki, Piotr Jaskula, Bohdan Dołzycki, Marek Pszczola, Mariusz Jaczewski, Dawid Rys and Marcin Stienss

Abstract During winter season of 2012 numerous transverse cracks developed in high-modulus asphalt concrete (HMAC) base of newly constructed motorway. Pavement cracked both in transverse joint locations and in the area between them. Research which was conducted during investigation of the causes and mechanisms of cracking consisted of: field examination, laboratory testing of specimens cored out of the existing pavement, computational analyses and effect of pavement homogeneity on transverse crack frequency. This paper focuses mainly on impact of homogeneity of asphalt layer on number of transverse cracks. The field investigation of analyzed motorway section includes visual assessment of homogeneity and number of cracks. Laboratory test conducted on specimen cored out of the pavement allowed to assess volumetric properties: binder content, voids content and compaction degree, and mechanical properties: indirect tensile stiffness modulus and strength. Analyses of HMAC layer properties revealed their impact on the number of transverse low-temperature cracks observed in field. It was found that a less effective compaction contributes to increase in the number of cracks. Quality

J. Judycki · P. Jaskula · B. Dołzycki · M. Pszczola · M. Jaczewski · D. Rys (✉) · M. Stienss
Department of Highway Engineering, Gdansk University of Technology,
Narutowicza Str. 11, 80-233 Gdansk, Poland
e-mail: dawid.rys@wilis.pg.gda.pl

J. Judycki
e-mail: jozef.judycki@wilis.pg.gda.pl

P. Jaskula
e-mail: piotr.jaskula@wilis.pg.gda.pl

B. Dołzycki
e-mail: dolzycki@pg.gda.pl

M. Pszczola
e-mail: marek.pszczola@wilis.pg.gda.pl

M. Jaczewski
e-mail: mariusz.jaczewski@wilis.pg.gda.pl

M. Stienss
e-mail: marcin.stienss@wilis.pg.gda.pl

A. Chabot et al. (eds.), *8th RILEM International Conference on Mechanisms of Cracking and Debonding in Pavements*, RILEM Bookseries 13,
DOI 10.1007/978-94-024-0867-6_45

and homogeneity of pavement courses have a considerable effect on mechanical properties of HMAC. As analysis showed, intensity of cracks is well correlated with mechanical properties and homogeneity of asphalt layer.

Keywords High modulus · Low temperature · Transverse cracking · Hard bitumen

1 Introduction

The paper presents investigation of pavement in newly constructed, 61 km long section of motorway A1 in Kujawsko-Pomorskie region, in the middle part of Poland. During the winter season the transverse cracks developed in the base courses made of high modulus asphalt concrete (HMAC) which were constructed in the summer and the autumn before. The HMAC courses were not covered by wearing nor binder courses before the onset of winter. The air temperature in February 2012 dropped to the range from −6 to −10 °C during the day and from −14 to −20 °C during the night and such temperatures persisted for almost two weeks. The field investigation of HMAC base revealed the large number of low temperature transverse cracks, which developed in that time. Variations in the quality and homogeneity of asphalt layer were also found. The problem was investigated and was described by Judycki et al. (2015).

The main objective of this paper is to assess the influence of asphalt layer homogeneity on the number of low-temperature cracks.

2 Field Investigation

The analyzed section of motorway A1 was divided into 49 short, continuous, construction sections with length from 130 to 2760 m in accordance with construction time-schedule. All of construction sections were built from the same HMAC composition and complied with the polish requirements. However the construction conditions were the same in all of sections, in some places variations in HMAC base surface appearance were observed. In these places layer structure was more open, both at the edges, coinciding with the edge zone of a paver, and in its middle zone, what is presented in Fig. 1. These places were classified as non-homogeneous. The main reasons of the non-homogenous surface appearance were construction shortcomings like: inadequate compaction, too low mix temperature during compaction, paver stoppage and to a lesser extent mix segregation.

Summary of the visual investigation of the pavement homogeneity is presented in Table 1.

Fig. 1 Examples of the HMAC base with non-homogeneous surface texture

Table 1 Summary of visual investigation of A1 motorway HMAC base layer homogeneity

Carriageway	Length L (km)	No. of sections	No. of homogeneous sections	No. of non-homogeneous sections	No. of LTC cracks	No. of JLTC cracks
Left	19.88	22	14	8	84	37
Right	20.56	27	16	11	73	22

During investigation of 49 sections of HMAC layer, following type of cracks were observed:

- Transverse cracks of asphalt base in areas between transverse joints, hereinafter low-temperature cracks (abbreviated as LTC).
- Cracks coinciding with the construction transverse joints of the asphalt base, hereinafter joint low-temperature cracks (abbreviated as JLTC).

It should be explained that transverse construction joints, especially if not properly constructed, are the weakest points in the base layer and they crack first if high tensile longitudinal forces are induced by thermal shrinkage during cooling. The cracking at joints (JLTC) is not related to the homogeneity of asphalt layer built in between these joints but is solely related to the quality of joint formation. On the contrary, cracks developed in between joints (LTC) are greatly dependent on the homogeneity of mechanical and thermal properties of the material built in the layer. The non-homogeneous material cracks first during cooling.

3 Laboratory Tests

Properties of all constructed layers and quality of the earthworks were assessed in laboratory tests. Tests results excluded the probability of development of reflective cracks, caused by joints in subgrade or subbase layers thus transverse cracks had typically low-temperature character.

Table 2 Volumetric and mechanical properties for specimens of homogeneous and non-homogeneous sections

Section homogeneity class	Average air voids (%)	No. of specimens	Test temp. (°C)	ITSM			ITS		
				Average (MPa)	Max (MPa)	COV (%)	Average (MPa)	Min (MPa)	COV (%)
Homogeneous	5.15	21	−10	19,054	19,867	2.48	4.64	4.15	6.10
		21	−20	24,102	25,798	4.09	4.97	4.08	10.38
		21	−30	22,996	23,942	2.39	4.29	3.65	8.66
Non-homogeneous	6.16	42	−10	16,254	18,796	9.30	3.67	2.78	14.35
		42	−20	20,788	25,228	12.70	4.19	2.59	22.71
		42	−30	20,866	24,586	10.60	3.61	2.41	19.82

Tests of asphalt mixtures were conducted on specimens cored out from 12 different locations. Each of locations were visually assessed and classified as either homogeneous or non-homogeneous. For each specimen following properties were identified: content of voids/asphalt, indirect tensile stiffness modulus (ITSM) and indirect tensile strength (ITS). Tests were conducted according to EN Standards with some minor modifications. Summary of tests results are presented in Table 2 separately for groups of homogeneous and non-homogeneous sections.

Specimens cut out from sections classified as non-homogeneous have higher average void content indicating more open surface texture observed during visual investigation. Higher coefficient of variation of ITSM and ITS indicates at worse homogeneity of asphalt layer. Maximum and minimum values of ITSM and ITS were calculated from t-distribution with probability of 90 %.

It is worth to note that both stiffness moduli and strength have a significant influence on low-temperature cracking (Marasteanu et al. 2012). Thermal stresses in asphalt layer increase proportionally to the modulus increase (i.e. Hiltunen and Roque 1994; Shahin and McCullough 1973; Vervaecke and Vanelstraete 2008). However maximum values of ITSM are comparable both for homogenous and non-homogenous sections, minimum ITS values are much lower for non-homogenous sections what contribute to the increase of LTC. Also greater variation of strength increases the numbers of weak points in asphalt layers, where thermal cracks can develop.

4 Relationships Between Mix Properties and Cracking Index

The results of field investigation are presented as a classification of the homogeneity of the HMAC base according to visual assessment and the number of low-temperature cracks on particular road sections. To express crack intensity the cracking index was introduced. Cracking index expresses the number of cracks per one kilometer of ready-constructed HMAC base and it is presented in Table 3 separately for sections classified as homogeneous and non-homogeneous.

Homogeneity of the base course had a considerable effect on the cracking index in the case of LTC and had no effect at all in the case of JLTC. In non-homogeneous

Table 3 Relationship between cracking index and homogeneity of HMAC base

Type of crack	Homogeneity class	Total number of cracks	Total length (km)	Cracking index (no./km)
JLTC	Non-homogeneous	16	11.45	1.4
	Homogeneous	43	28.98	1.5
LTC	Non-homogeneous	79	11.45	6.9
	Homogeneous	78	28.98	2.7

sections the number of transverse cracks between the construction joints (LTC type) was almost three times higher as compared to the sections where the surface texture was rated as homogeneous.

Mean values and COV of test results were correlated with cracking index for given road sections and these relationships are presented in Figs. 2, 3 and 4. Each point in Figs. 2, 3 and 4 represents one continuous section with paved asphalt base. Sections assessed visually as homogeneous and non-homogeneous during field investigation are marked. Laboratory tests of binder content and voids content were carried out for 40 of 43 analysed sections. Mechanical properties of asphalt layers were assessed for a limited number of 10 randomly chosen sections.

Figure 2a indicates that less effective compaction contributes to increase of the number of LTC. Figure 2b shows that binder content did not have influence on the number of LTC. Figures 3a and 4a shows that index of LTC relates evidently to ITSM and ITS. Mean values of ITS and ITSM were lower for sections classified as non-homogeneous. It has been found out that index of LTC increases with the

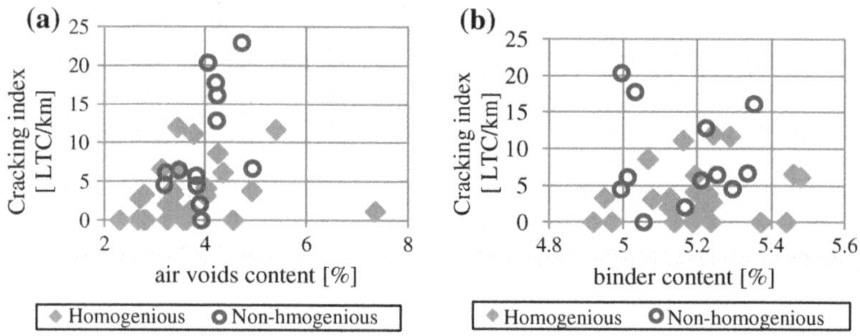

Fig. 2 Relationships between cracking index and mix properties of HMAC (**a**) mean values of voids content (**b**) mean values of binder content

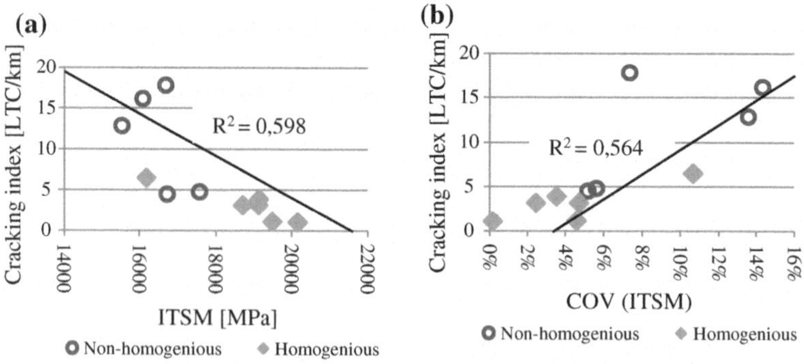

Fig. 3 Relationships between cracking index and (**a**) mean values of indirect tensile stiffness modulus (ITSM) (**b**) coefficient of variation of ITSM (test temperature −10 °C)

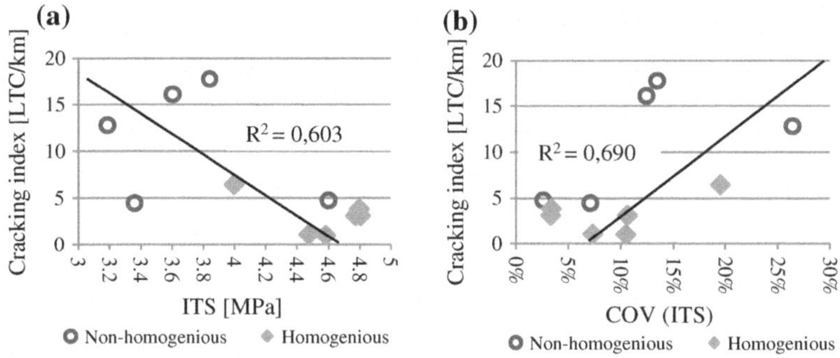

Fig. 4 Relationships between cracking index and (**a**) mean values of indirect tensile strength ITS (**b**) coefficient of variation of ITS (test temperature −10 °C)

increase of COV of ITSM (Fig. 3b) and ITS (Fig. 4b). Higher COV values were obtained for sections assessed as non-homogeneous what confirms accordance of visual field investigation with mechanical properties of asphalt layers.

5 Conclusions

The field investigations as well as the properties of the HMAC assessed in the laboratory tests allow to conclude that: (1) The transverse cracks in the HMAC base had the characteristics of low-temperature cracking; (2) Intensity of cracks varied on particular sections despite the same pavement structure and climatic conditions; (3) More homogeneous internal structure of asphalt as well as more homogeneous and closed surface texture were the factors limiting the number of cracks; (4) Quality and homogeneity of pavement courses have a considerable effect on mechanical properties of asphalt concrete and consequently on the number of low-temperature cracks that develop on a given road section.

References

Hiltunen D.R., Roque R., (1994). A mechanical-based prediction model for thermal cracking of asphaltic concrete pavements", Journal of the Association of Asphalt Paving Technologists, 63, 81-108

Judycki J., Jaskula P., Dolzycki B., Pszczola M., Jaczewski M., Rys D., Stienss M., (2015). Investigation of low-temperature cracking in newly constructed high-modulus asphalt concrete base course of a motorway pavement, Road Materials and Pavement Design, 16(1)

Marasteanu M.O. et al., (2012). Investigation of Low Temperature Cracking in Asphalt Pavements, National Pooled Fund Study - Phase II, Final Report Minnesota Department of Transportation, Report No. MN/RC - 2012-23

Shahin M.Y., McCullough B.F., (1973). A damage model for predicting temperature cracking in flexible pavements, Centre for Highway Research, The University of Texas at Austin

Vervaecke F., Vanelstraete A., (2008). Resistance to Low Temperature Cracking of High Modulus Bituminous Binders and Mixtures, Road Materials and Pavement Design, 9(1), 163-176

The Influence Mineral Additions on the Failure Properties of Bitumen Emulsion Mortars

Carlotta Godenzoni, Andrea Graziani and Valeria Corinaldesi

Abstract The use of both cold asphalt and cold recycled asphalt mixtures produced with bitumen emulsion is constantly increasing. The failure properties of these low-energy materials are normally improved by using mineral additions or active fillers. These additions are a key component of the bituminous mortar, which binds the coarse aggregate fraction of the mixture. The objective of the present study is to investigate the evolution of the failure properties bitumen-emulsion mortars obtained by using three different types of additions: Portland cement (active filler), hydrated lime and calcium carbonate (inactive filler). The mortars were prepared with standard sand and a fixed bitumen/addition rate. Cylindrical samples were compacted with a shear gyratory compactor, the selected procedure also allowed evaluating the effect of the different additions and water content on workability. The indirect tensile strengths were measured after 7, 14, 28 and 100 days of curing in two conditions: sealed (i.e. no evaporation) and unsealed, at 50 % relative humidity. Each bituminous mortars was also characterized in terms of low temperature fracture resistance. Results showed the influence of the mineral additions on the emulsion breaking and the different curing conditions allowed highlighting the impact of both evaporation and cement hydration on strength and stiffness evolution.

Keywords Composite mortar · Mineral addition · Curing · Failure

C. Godenzoni (✉) · A. Graziani
Dipartimento di Ingegneria Civile Edile e Architettura,
Università Politecnica delle Marche, Via Brecce Bianche, 60131 Ancona, Italy
e-mail: c.godenzoni@pm.univpm.it

A. Graziani
e-mail: a.graziani@univpm.it

V. Corinaldesi
Dipartimento di Scienze e Ingegneria della Materia, dell'Ambiente ed Urbanistica,
Università Politecnica delle Marche, Via Brecce Bianche, 60131 Ancona, Italy
e-mail: v.corinaldesi@univpm.it

A. Chabot et al. (eds.), *8th RILEM International Conference on Mechanisms of Cracking and Debonding in Pavements*, RILEM Bookseries 13,
DOI 10.1007/978-94-024-0867-6_46

1 Introduction

The use of both cold asphalt and cold recycled asphalt mixtures (CR) produced with bitumen emulsion is constantly increasing. The mechanical performance (stiffness, resistance to permanent deformation and failure properties) of these low-energy materials are normally improved by using active or inactive fillers (Hodgkinson et al. 2004). Such mineral additions affect emulsion breaking and setting, thus controlling the curing process of the bitumen component. Moreover, cement, even though it is used in small dosages (Giuliani 2001), behaves as active filler creating additional hydraulic bonds that improve mixture strength.

Cold-mixtures similar to hot-mix asphalt (HMA) may be schematized as a composite where the bitumen disperses mainly amongst the finest particles, resulting in a bitumen-rich mortar binding the coarse particles. Based on these considerations, the present work focused on the mechanical characterization of cold bituminous mortars (CBM), aiming to upscale these properties to the mixture scale (Underwood et al. 2013). In particular, the objective of the present study was to investigate the evolution of the failure properties of CBM composed by fine aggregate, bituminous emulsion, water and three types of additions: Portland cement, hydrated lime and calcium carbonate. Each CBM was characterized in terms of ITS and low temperature fracture resistance.

2 Experimental Program

2.1 Materials

A fluvial sand with maximum dimension 0.5 mm, particle density 2.71 Mg/m^3 and absorption 1 % [UNI EN 1097-6], was adopted to prepare the CBM. A cationic slow-setting bituminous emulsion designated as C 60 B 10 [EN 13808] was selected in this research and its formulation guarantee high workability during the mixing and compaction phase. Three mineral additions were selected:

- Portland limestone cement II A/LL, strength class 42.5 R [UNI EN 197-1];
- Calcium carbonate, finely ground, $CaCO_3$;
- Hydrated lime obtained by the turn off of quicklime, $Ca(OH)_2$.

Portland cement, calcium carbonate and hydrated lime were characterized determining the Rigden voids and the fineness (specific surface) using the Blaine method (Table 1).

Table 1 Characterization of mineral additions

Parameter		Cement	Calcium carbonate	Hydrated lime
Rigden voids (%)	UNI EN 1097-4	40.18	31.14	53.87
Specific surface (m^2/g)	UNI EN 196-6	0.41	0.51	0.345

The adopted dosage of bituminous emulsion was 16.67 % by dry aggregate weight, corresponding to 10 % of residual bitumen. This amount was selected considering the specific surface of the mortars and that of a mixture with a maximum dimension of 20 mm (Attia et al. 2009). The same dosage was adopted for the mineral additions (ratio between bitumen and mineral addition, $B/M_A = 1$), as usually adopted for CR mixtures (Godenzoni et al. 2016).

The mortars were coded as CEM, CC and HL according to the mineral addition used (Portland cement, calcium carbonate and hydrated lime).

2.2 Testing Methods

The specimens employed for mechanical testing were compacted by means of shear gyratory compactor adopting the following protocol: constant pressure of 600 kPa, gyration rate of 30 rpm and angle of inclination of 1.25° [UNI EN 12697-31].

The ITS was measured at 20 °C, with constant rate of deformation of 50 ± 2 mm/min, according to UNI EN 12697-23. The test configuration was modified adding two inductive displacement transducers in order to measure the horizontal deformation.

Semi Circular Bending (SCB) tests were performed to evaluate the low temperature fracture resistance according to UNI EN 12697-44, usually adopted for HMA. Tests were carried out at 10 °C using a constant vertical deformation rate of 5 mm/min.

2.3 Testing Program

The first part of the experimental program was focused on the development of the compaction and mixing protocol and on the evaluation of the effect of the mineral addition on the workability of the cold mortars. The aim was to determine the mixing water content, W_{des} and the compaction energy N_{des}, that guarantee similar volumetric properties, avoiding any material loss during compaction. The adopted W_{des} was 8 % for mortars CEM and CC and 9 % for mortar HL, whereas the adopted N_{des} was 100 gyrations for mortars CEM and HL and 50 gyrations for mortar CC.

In the second part of the experimental program, the influence of curing time (7, 14, 28 and 100 days) and humidity (DRY or WET curing condition) was evaluated in terms of ITS. The low temperature fracture resistance was also investigated on samples cured for 100 days at 50 % of humidity (DRY condition). A unique curing temperature of 20 °C was employed. The weight loss of all specimens was continuously monitored during the entire curing period (before mechanical testing).

3 Results and Analysis

3.1 Indirect Tensile Strength

ITS values, measured at different curing times and conditions, are depicted in Fig. 1. Each value is the average of three replicates. In WET curing condition, both mortars HL and CC exhibited very low strength as compared with DRY curing conditions. In DRY conditions, for mortar CC, ITS increased from 0.23 MPa (7 days) to 0.4 MPa (14 days) stabilizing at longer curing periods. This behavior can be related to the evaporation process and water evaporation. In fact, the weight loss exhibited a similar trend and stopped after 14 days. For mortar HL, ITS showed a constant trend over the observed curing time and the ITS average value was 0.3 MPa. As can be observed, ITS values measured for HL-DRY are lower with respect to CC-DRY. This result could be related to the higher Rigden voids obtained analyzing hydrated lime (Table 1). In addition, the mortar HL was characterized by less compactability and this led to a reduction of ITS.

As expected, the adoption of cement resulted in higher ITS at any curing time and both DRY and WET curing conditions. This increase was the consequence of the additional contribution given by cement hydration. Until 28 days, mortar CEM-DRY showed higher ITS compared with that obtained for CEM-WET. This suggested that the evaporation process controls the early-stage of curing (curing of bituminous fraction).

At 100 days of curing, ITS was equal to 0.77 and 0.92 MPa for CEM-DRY and CEM-WET respectively. For CEM-DRY, the evaporation and the ITS increasing stopped as previously observed for HL-DRY and CC-DRY. The ITS development

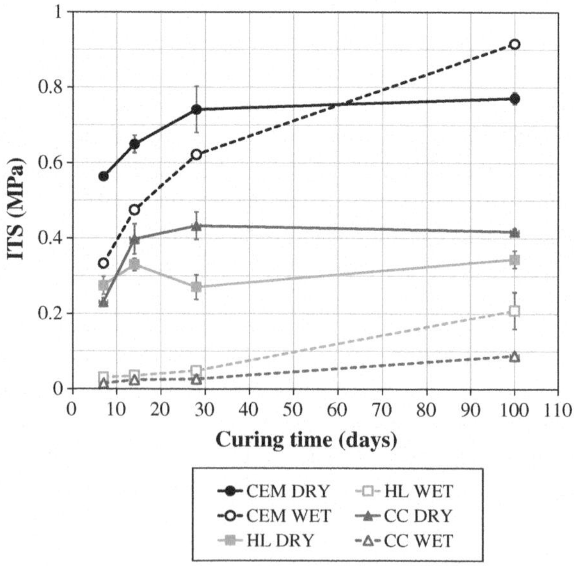

Fig. 1 Evolution of ITS at 20 °C with curing time for mortars CEM, HL and CC, DRY and WET curing conditions (*error bars* represent the maximum and minimum values)

continued for CEM-WET thanks for the water that was available in the environment. This highlights the contribution given by the complete development of hydration process (Godenzoni et al. 2016).

3.2 *Resistance to Crack Propagation*

SCB tests were carried out in order to evaluate the crack propagation aptitude in relation to the mineral addition adopted in the mortar. During the test, the crack starts to propagate from the tip of the artificial notch where the stress concentration is the highest; then the crack propagates (Frigio et al. 2013).

Figure 2a reports the typical load-deformation curves. As can be observed, the relationship was non-linear and this response was more pronounced if hydrated lime or calcium carbonate were used in the mortars. As expected, the mortar CEM-DRY had higher maximum load values with respect to HL-DRY and CC-DRY. Mortar CEM-DRY was characterized by a brittle behavior: low deformation at maximum load and a rapid decline afterward. Otherwise for HL-DRY and CC-DRY, a high deformation at maximum load and a slow decline afterwards was observed (typical plastic behavior).

Fracture toughness *k* and the fracture energy *e* were calculated according to EN 12697-44. The mean values are shown in Fig. 2b along with the minimum and the maximum values (error bars). CEM-DRY showed high fracture toughness, $k = 13.44\ \text{N/mm}^{3/2}$, but small fracture energy, $e = 0.41\ \text{kJ/m}^2$ (cement-like behavior). Differently, mortar CC-DRY exhibited lower fracture toughness, $k = 7.23\ \text{N/mm}^{3/2}$, and higher fracture energy, $e = 1.24\ \text{kJ/m}^2$, with respect to CEM-DRY (asphalt-like behavior). Mortar HL-DRY was characterized by low

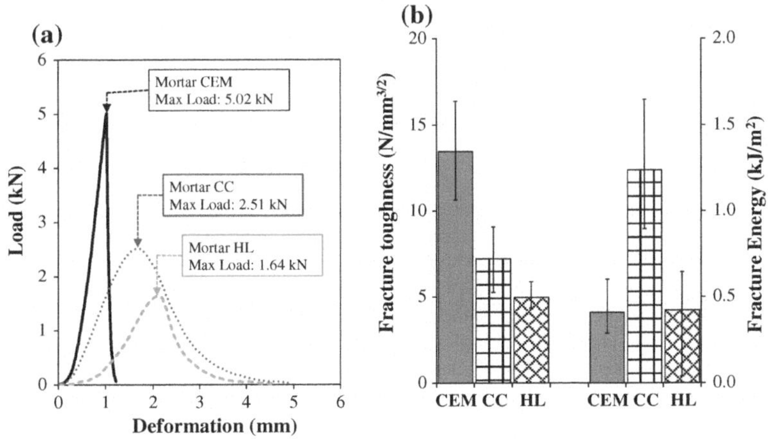

Fig. 2 Results of SCB tests: (**a**) Typical load-deformation curves and (**b**) average value of fracture toughness *k* and fracture energy *e* for mortars CEM, HL and CC, in dry curing condition

mechanical performance in terms of both fracture toughness (k = 4.96 $N/mm^{3/2}$) and fracture energy (e = 0.42 kJ/m^2).

4 Conclusion

In the present experimental study, the evolution of the failure properties of cold-bituminous mortars, obtained by using three different types of additions (Portland cement, hydrated lime and calcium carbonate) and cured at different curing conditions (time and humidity), was investigated. ITS values measured for mortar CEM increased over the considered curing period, with different rates depending on the curing conditions (DRY or WET). At the early stage, CEM-DRY had higher ITS with respect to CEM-WET. This indicates that the curing of bituminous component (emulsion breaking and evaporation) controls the curing process. Otherwise, at long term curing (100 days), CEM-WET had higher ITS with respect to CEM-DRY. This suggests that the curing of cementitious phase (cement hydration) prevails over the curing of bituminous component because, in dry condition, the water was almost completely evaporated. Both mortars HL-WET and CC-WET exhibited very low strength as compared with DRY curing conditions. This result was due to the absence of the resistance contribution related to the bituminous emulsion breaking and water evaporation. Otherwise, for mortars CC-DRY and HL-DRY, the ITS development was exclusively connected to these phenomena.

Regarding to the resistance to crack propagation, results confirmed the observations carried out with ITS test. Mortar CEM-DRY showed high fracture toughness but small fracture energy (cement-like behavior). Differently mortars CC-DRY exhibited low fracture toughness and high fracture energy (asphalt-like behavior). Mortar HL-DRY was characterized by low mechanical performance in terms of fracture toughness and fracture energy.

References

Attia, M. I., Abdelrahman, M. A., Molakatalla, U., & Salem, H. M. (2009). Field evaluation of asphalt film thickness as a design parameter in Superpave mix design. International Journal of Pavement Research and Technology, 2(5), 205.

Frigio, F., Pasquini, E., Ferrotti, G., & Canestrari, F. (2013). Improved durability of recycled porous asphalt. Construction and Building Materials, 48, 755-763.

Giuliani, F. (2001, October). X-ray diffraction method for studying cement-modified bitumen-emulsion mixtures in asphalt pavement cold recycling. In International Symposium on Subgrade Stabilisation and In Situ Pavement Recycling Using Cement, 1st, 2001, Salamanca, Spain.

Godenzoni, C., Cardone, F., Graziani, A., & Bocci, M. (2016). The Effect of Curing on the Mechanical Behavior of Cement-Bitumen Treated Materials. In 8th RILEM International

Symposium on Testing and Characterization of Sustainable and Innovative Bituminous Materials (pp. 879-890). Springer Netherlands.

Hodgkinson, A., & Visser, A. T. (2004). The role of fillers and cementitious binders when recycling with foamed bitumen or bitumen emulsion. In 8th Conference on Asphalt Pavements for Southern Africa (CAPSA'04).

Underwood, B. S., & Kim, Y. R. (2013). Effect of volumetric factors on the mechanical behavior of asphalt fine aggregate matrix and the relationship to asphalt mixture properties. Construction and Building Materials, 49, 672-681.

Part VI
Cracking in Asphalt Materials: Structural Crack Characterisation

Influence of SAMI on the Performance of Reinforcement Grids

Martin Arraigada, Christiane Raab, Manfred N. Partl, Federico Perrotta and Gabriele Tebaldi

Abstract Over the last decades, reinforcement grids have been used to prolong the service life of pavements. Although there is a broad consensus about their overall efficiency, there are still many open questions about the best alternatives regarding the materials to use, the shapes of the grid, their installation method or their position in the pavement among others. One controversial aspect is the use of Stress Absorbing Membranes Interlayers (SAMI) together with the application of a grid. Many manufacturers claim that the SAMI helps sealing any preexisting cracks in the underlying layer, retarding their propagation to the surface. This paper reports about the performance of a reinforcement grid installed with SAMI in a pavement with artificial cracks. One of the parameters analyzed to evaluate the performance, is the propagation of artificial cracks from the base course to the pavement surface. Other important consideration is the rutting development. To that end, the same cracked pavement, reinforced with a fiber grid and SAMI and without any reinforcement is loaded with identical accelerated trafficking, carried out with the traffic simulator MLS10. Results show that, although the grid with SAMI is able to slow down the

M. Arraigada · C. Raab (✉) · M.N. Partl
Empa, Duebendorf, Switzerland
e-mail: christiane.raab@empa.ch

M. Arraigada
e-mail: martin.arraigada@empa.ch

M.N. Partl
e-mail: manfred.partl@empa.ch

F. Perrotta
University of Parma, Parma, Italy
e-mail: perrottafederico2@gmail.com

G. Tebaldi
Department of Civil and Environmental Engineering and Architecture,
University of Parma, Parco Area delle Scienze, 181/A, I-43124 Parma, Italy
e-mail: gtebaldi@unipr.it

G. Tebaldi
E.S.S.I.E., Department of Civil and Coastal Engineering,
University of Florida, Gainesville, USA

A. Chabot et al. (eds.), *8th RILEM International Conference on Mechanisms of Cracking and Debonding in Pavements*, RILEM Bookseries 13,
DOI 10.1007/978-94-024-0867-6_47

formation of cracks in the surface, the density of cracks at the end of the tests is as high as if no grid was used. Moreover, resulting rutting in the reinforced pavement is higher than in the pavement without grid, indicating that the use of SAMI has a counterproductive effect on the rutting performance of the reinforcement.

Keywords Reinforcement grid · Interlayer (SAMI) · Crack propagation · Accelerated testing

1 Introduction

Stress Absorbing Membranes Interlayers (SAMI) have been widely used to delay reflective cracking, which is a consequence of the propagation of cracks from the bottom layers to the surface of the pavement. They are mostly used to act as a barrier against the infiltration of water to the lower layers once the cracks reached the surface of the pavement. Further, due to their flexibility, they are able to absorb the large movements in the vicinity of cracks (Vanelstraete and de Bondt 1997). Reinforcement grids are also used to stop reflective cracking by taking tension forces and reducing the concentration of stresses around the already existing cracks (Jaecklin 1993). It is argued that the use of reinforcement grids and SAMI would improve performance of the pavement as a result of the synergetic combination of both interlayer systems (Montestruque et al. 2012; Pasquini et al. 2015). This work presents a selection of results obtained with in an accelerated pavement test (APT) to evaluate the performance of a pavement with artificial cracks in the binder course, using different reinforcement systems. Specifically in this paper, focus is given to the joint application of a glass fiber grid and a SAMI (grid+SAMI) and its comparison to a pavement with no reinforcement grid (reference). The performance indicators used in this work are the surface cracking and rutting performance under the accumulation of repetitive traffic-like loading.

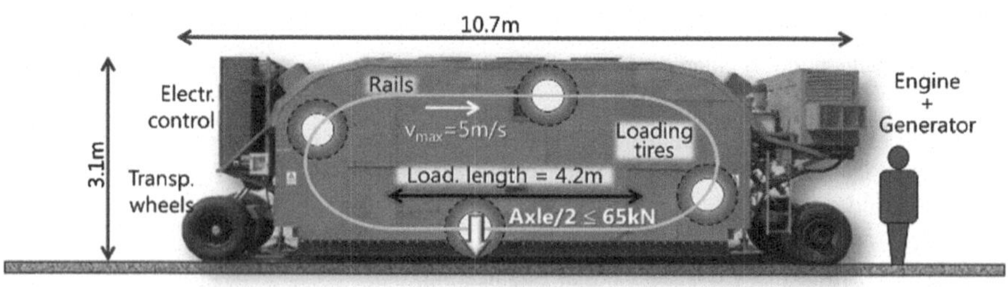

Fig. 1 Traffic load simulator MLS10

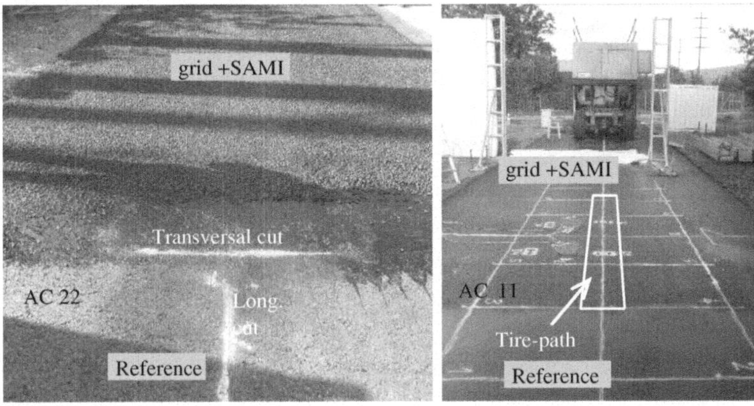

Fig. 2 On the *left*, a view of the binder course AC 22 before laying down the wearing course AC 11. The outlook of the testing site ready for loading is presented on the *right* side of the figure

2 Experimental Procedure

The field evaluation was carried out by loading an ad hoc pavement with the traffic load simulator MLS10 (Fig. 1). The pavement consisted of a wearing course of 40 mm AC 11 and a binder course of 65 mm AC 22 on top of a 600 mm unbound gravel layer. The materials were selected according to the Swiss Standard SN 640 431-1 NA (VSS 2013). For the experiment, the MLS10 was set to apply 5000 unidirectional load cycles per hour over a length of 4.2 m, using truck super single tires. The load used was 65 kN, which correspond to a 130 kN axle. The speed of the loading wheels was established at 5 m/s and channelized loading was used. More information about the device can be found elsewhere (Arraigada et al. 2014).

The binder course was cut in different locations across its entire thickness to simulate a crack, as shown in Fig. 2. Cuts were performed perpendicular and in the same direction of MLS10 trafficking. Then, half part of the pavement was covered with the glass fiber grid and the SAMI. Finally, the wearing course was lay down. During the construction, several sensors including thermocouples, strain gauges and accelerometers were installed to account for the climate conditions and the pavement response under loading.

As the tests were carried out in a special testing area with no temperature control capabilities, and due to the fact that the length of the MLS10 tire-path is only 4.2 m, both pavements (reference and grid+SAMI) were trafficked at the same time, i.e. the machine was set perpendicular to the transition of both pavement sections. This was considered to have the same temperature conditions, which is important to compare the distress performances of the studied sections.

A total of 300,000 load cycles were applied to the pavement. The tests were carried out in autumn to account for a mild temperature profile. After finishing with the loading phase, cores were taken to carry out laboratory tests. Apart from monitoring the temperature and the response of the pavements with the sensors mentioned before, several measurements of the permanent deformation were carried out using a surface profilometer. In this paper, only results obtained from visual inspection and rutting profiles are presented.

3 Results

3.1 Surface Cracking

Trafficking with MLS10 was interrupted every 50,000 load cycles and the surface of the pavement was scanned looking for emerging cracks. New cracks were marked with white paint and pictures were taken. Figure 3 shows the photos of the surface through the course of the loading phase, with emphasis on fist observation of cracks on each pavement. The reference section is shown in the upper side of the pictures, whereas the reinforced pavement is in the bottom part.

The first transversal crack was observed in the reference pavement after 130,000 load cycles (the figure shows the pictures taken after 150,000 load cycles). The position was coincident with one of the artificial cracks cut underneath. The first longitudinal crack appeared in the same pavement section following the 200,000 load cycles interruption. These distresses progressed into a dense pattern of transversal cracks at the end of the tests. A similar pattern of transverse cracks was observed on the pavement with grid reinforcement and SAMI, but after 30,000 load cycles. On one hand, the reinforcement system succeeded in delaying the progression of surface cracking in up to more than two times. On the other hand, the amount and length of cracks were almost similar than in the reference section.

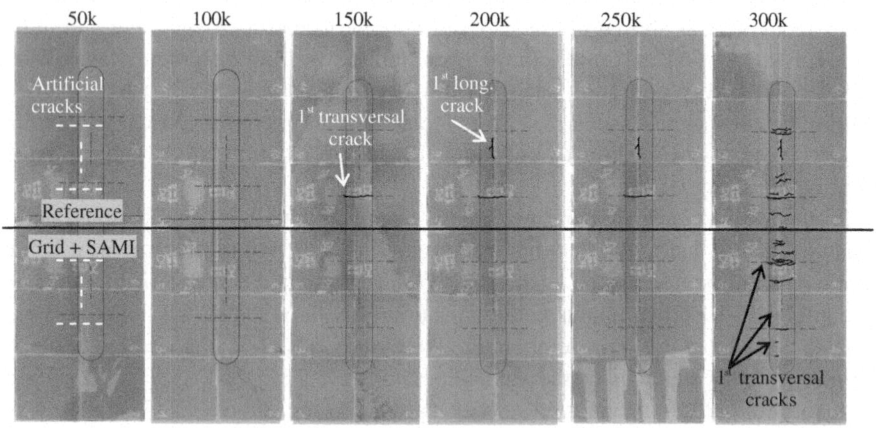

Fig. 3 Surface crack patterns

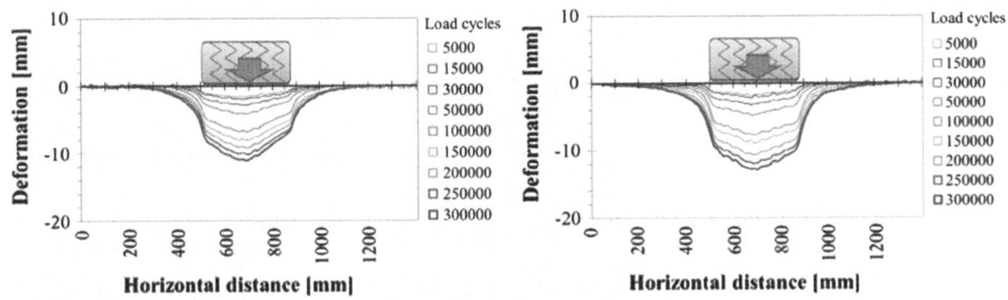

Fig. 4 Permanent deformation profiles taken in the reference section (*left*) and in the grid+SAMI section (*right*)

3.2 Rutting Performance

Figure 4 presents the transversal profiles taken after a number of accumulated number of MLS10 load cycles, to evaluate the rutting performance of both pavement sections under study. Each transversal profile was taken in a length of 1.4 m. The curves clearly show the effect of the channelized loading. Further, the fact that the curves do not present any elevation at the edges of the rut, indicate that in both cases there was no lateral shoving of asphalt material from underneath the tire-path to the sides and rutting is produced mostly by the post-compaction of the 600 mm thick unbound gravel layer directly under the wheel path.

Rutting depth, calculated as the difference between the highest upward peak in the profile and the averaged rut depth under the tire, is depicted in Fig. 5. In the same figure, the average temperatures obtained every 50,000 load cycles is also presented. The figure puts into evidence that the rutting in the grid+SAMI section is greater than in the reference pavement.

This poor rutting behavior can be a consequence of the plastic deformation of the SAMI, which can be partially explained by the amount of binder used for its construction and the poor compaction obtained at the bottom of wearing course. This was observed in the cores taken from the pavement, as shown in Fig. 6.

Fig. 5 Permanent deformation profiles taken in the reference section (*left*) and in the grid+SAMI section (*right*)

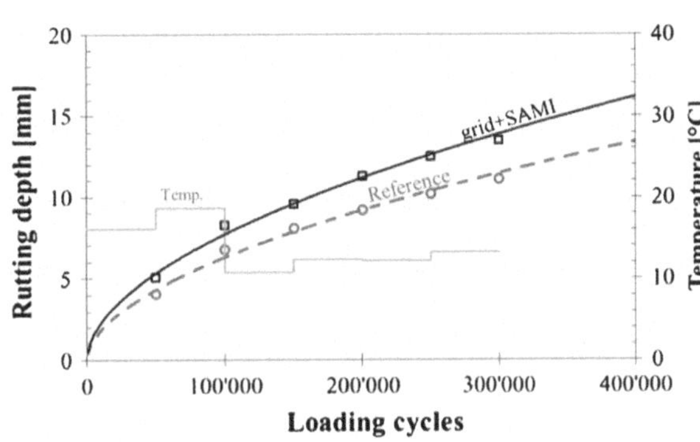

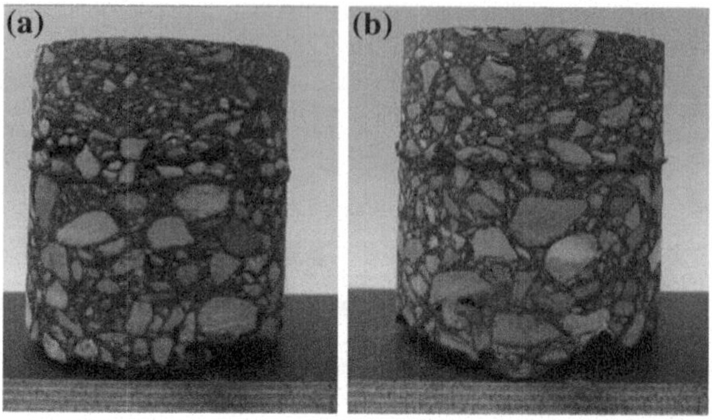

Fig. 6 Cores taken from the grid+SAMI section outside (**a**) and in (**b**) the tire-path area

4 Summary and Conclusions

This paper presents a selection of results from an accelerated pavement test to evaluate the performance of a pavement with fiber grid and SAMI reinforcement. The study shows that this type reinforcement is able to retard cracking on the pavement surface, but after a number of loading cycles the amount and length of cracks are similar to those observed in the same pavement without any reinforcement. On the other hand, it is evident from profiles measurements that the rutting of these types of reinforcements performs worse than a pavement without reinforcement due to the presence of the SAMI.

References

Arraigada M, Pugliessi A, Partl MN, Martínez F (2014) Effect of full-size and down scale accelerated traffic loading on pavement behavior, Mater. Struct. 47, 1409-1424

Jaecklin FP (1993) Geotextile use in asphalt overlays-design and installation techniques for successful applications, in Reflective Cracking in Pavements: State of the Art and Design Recommendation ed. by L. Francken, E. Beuving, A.A.A. Molenaar s, 83-101

Montestruque G, Bernucci L, Fritzen M, Goretti da Motta L (2012) Stress Relief Asphalt Layer and Reinforcing Polyester Grid as Anti-reflective Cracking Composite Interlayer System in Pavement Rehabilitation A. Scarpas et al. (Eds.), 7th RILEM International Conference on Cracking in Pavements, Volume 4, 1189–1197

Pasquini E, Pasetto M, Canestrari F (2015) Geocomposites against reflective cracking in asphalt pavements: laboratory simulation of a field application: Road Materials and Pavement Design, DOI: 10.1080/14680629.2015.1044558

Vanelstraete A and de Bondt H (1997) Crack prevention and use of overlay systems, in Prevention of Reflective Cracking in Pavements, RILEM Report 1 8

Schweizerischer Verband der Strassen- und Verkehrsfachleute VSS (2013) Asphaltmischgut – Mischgutanforderungen– Teil 1: Asphalt-beton (Asphalt mixes – requirements – Part 1 Asphalt concrete) Swiss Standard N 640431-1 NA (in German)

Investigating SAMI System Performance Through Pilot-Scale Testing

Nick Thom, John Richardson and Olumide Ogundipe

Abstract The effectiveness of SAMI systems in combatting cracking in pavements is frequently questioned and is often found to vary greatly from case to case. This paper will present the results of two different sets of simulative tests: the first set comprises pilot-scale wheel track tests of an overlaid cracked pavement; the other set is of slow strain tests that mimic the effects of a thermal cycle in an overlay over an underlying crack. In both cases the apparent mechanism provided by the SAMI has been investigated, based on the evidence of crack pattern and speed of crack growth, allowing deductions to be made regarding the way the presence of the SAMI changes the stresses and strains present in the system as a whole. These observations also include a limited appraisal of the differences in effectiveness given by different types of SAMI, allowing comment on the role of both the thickness and the viscosity of the SAMI to be made. In general terms it was found that all the SAMI types investigated gave enhanced performance, both against traffic-induced damage and against thermal action. In the case of effectiveness against thermal action, the critical parameter was the viscosity of the material; in the case of resistance to traffic damage, the main influencing factor was fatigue resistance.

Keywords SAMI · Pilot-scale · Trafficking · Thermal simulation

N. Thom (✉)
University of Nottingham, Nottingham, UK
e-mail: nicholas.thom@nottingham.ac.uk

J. Richardson · O. Ogundipe
Colas, Mid Sussex, UK
e-mail: john.richardson@colas.co.uk

O. Ogundipe
e-mail: momide2002@yoahoo.com

J. Richardson · O. Ogundipe
Ekiti-State University, Ado Ekiti, Nigeria

A. Chabot et al. (eds.), *8th RILEM International Conference on Mechanisms of Cracking and Debonding in Pavements*, RILEM Bookseries 13,
DOI 10.1007/978-94-024-0867-6_48

1 Introduction

Many different laboratory test configurations have been used to study the effectiveness of Stress Absorbing Membrane Interlayers (SAMIs) and in most cases it has been found that SAMI systems increase the number of cycles to failure.

Field trials have produced more mixed results, perhaps because laboratory tests sometimes lack true realism. For example Storsteen and Rumpca (2000) and Engle (2001) evaluated geosynthetic SAMIs over jointed concrete and alligator cracked asphalt pavements respectively; neither showed clear benefit. However, Kretov and Gorelysheva (2000) reported measurably slower development of reflective cracks with an organomineral mix as a SAMI beneath an asphalt overlay to a rigid base. Similarly Ellis et al. (2002) found that an open-textured asphalt SAMI used over a cracked flexible composite pavement reduced reflective cracking by about 80 % compared to an equivalent control area. Palacios et al. (2008) evaluated the use of a fibre-reinforced asphalt SAMI and found no reflected cracks in the SAMI section whereas cracking had started in the control section. And Vervaecke et al. (2008), who studied SAMIs on a jointed concrete road with a 50 mm overlay, found that a mastic asphalt and a non-woven geotextile both gave a considerable delay in crack initiation.

Clearly many variant SAMI systems exist. However, crack-resistant mixes such as mastics and fibre-reinforced asphalts are consistently reported to work well. Examples of these SAMI types have therefore been selected in this project and an attempt has been made to apply genuinely realistic test regimes.

2 Materials Used

In this project, the following SAMI systems were studied:

- SAMI A: a sand asphalt with 0/4 mm aggregate and an SBS modified binder.
- A conventional sand asphalt with a 160/220 penetration grade binder.
- SAMI B: chopped glass fibres in bitumen emulsion with granite chippings.
- SAMI C: as SAMI B except with a polymer modified bitumen in the emulsion.

It was intended that this selection of SAMIs would allow comparative performance of different systems to be sensibly evaluated.

3 Behaviour Under Trafficking

The pavement test facility (PTF, Fig. 1) at the University of Nottingham can apply a load magnitude of up to 12 kN and maximum speed of 14.5 km/h, although the load in this study was limited to 9.6 kN. Tyre pressure is approximately 550 kPa. The PTF has length, width and depth of 5.0, 2.4 and 1.5 m, respectively.

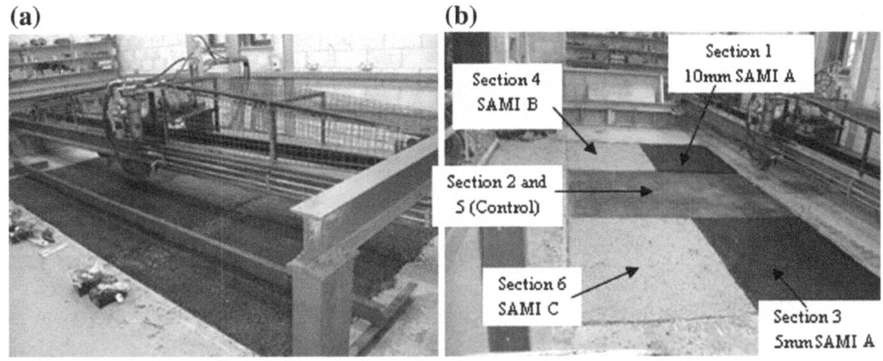

Fig. 1 Pavement test facility: **a** during trafficking; **b** under construction

For purposes of this project, a very soft clay (CBR 1–2 %) was overlaid by a granular sub-base (CBR 14–21 %), and a 60 mm thick 10 mm asphalt concrete layer with 40/60 penetration grade bitumen, together representing an existing pavement. The area was then divided into six sections (3 sections in each of two wheel paths) saw cuts made (see Fig. 2) to simulate existing cracks. Overlays including selected SAMI systems and control cases were then applied, all with the same overall thickness of 40 mm above the surface of the existing pavement. The main overlay material was the same 10 mm asphalt concrete as used for the lower layer.

During trafficking (at 20 °C) two stages of cracking failure were identified, namely the first appearance of a reflective crack and full cracking, i.e. all cracks fully reflected. Furthermore, each section had two underlying crack situations, namely a single transverse centre-line crack and a region of closely-spaced cracks at the section edge. Thus, four different lives were recorded for each section. For the control sections, with a region of close-spaced cracks at each end, the average of the two ends was taken. Control sections sustained 14,000 (Sect. 2) and 20,000

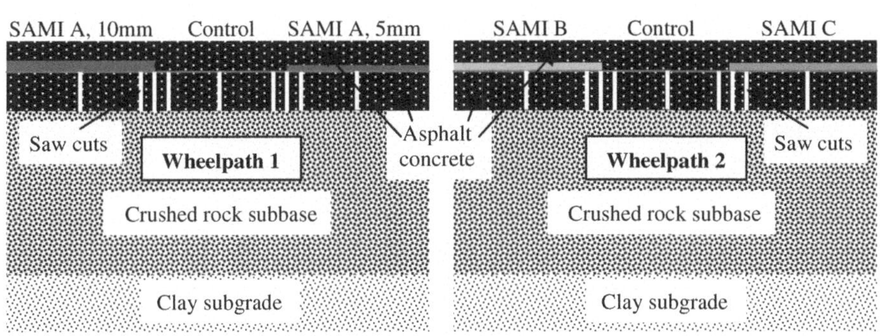

Fig. 2 Schematic longitudinal sections through test pavements

Table 1 Number of load cycles to crack formation as a ratio of the control sections (20 °C)

Section	Close-spaced cracks		Central crack	
	First appearance	Full cracking	First appearance	Full cracking
1. SAMI A, 10 mm	1.33	1.27	2.26	2.26
3. SAMI A, 5 mm	1.12	1.77	3.26	4.53
4. SAMI B	1.24	1.36	4.51	2.92
6. SAMI C	1.31	1.29	2.70	1.93

(Sect. 5) wheel passes before full cracking. Results for SAMI sections are given in Table 1, expressed as ratios of the life of the relevant control section.

From Table 1 it is clear that the SAMIs were more effective in the case of an individual crack than in the multiply-cracked regions. Furthermore the 5 mm SAMI A performed better than the 10 mm (although note that the test is small scale). The overall degree of benefit in the case of a discrete crack was typically a factor of 2–4. In the case of the multiply-cracked regions, cracks appeared more or less directly over underlying cracks. In the case of individual transverse cracks it was observed that on all SAMI sections the first appearance at the surface was of cracks about 100 mm offset from the underlying saw cut. These were presumed to be top-down in nature. Final failure, however, always occurred with a crack directly above the saw cut, presumed to be bottom-up. Thus two different mechanisms appeared to be in operation. In neither case was layer de-bonding observed.

4 Including the Effect of Thermal Action

A schematic of the thermal cracking testing device is shown in Fig. 3 (Ogundipe et al. 2013). Test specimens comprised beams of length 1000 mm and width 125 mm. The lower layer (simulating an existing pavement) was a 40 mm thick 10 mm asphaltic concrete with 10/20 penetration grade bitumen, above which was laid a further 40 mm comprising a SAMI (where present) and an asphalt overlay of

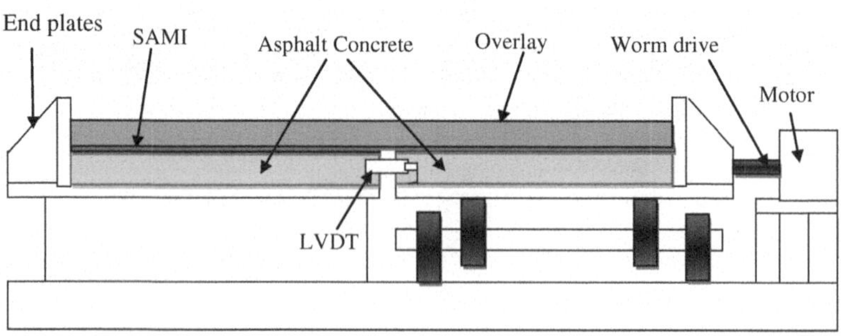

Fig. 3 Thermal cycling device

10 mm asphaltic concrete with 40/60 penetration grade bitumen. Pips were glued 50 mm apart along the centre-line of the surface layer to allow surface strain measurement and the surface was painted white to enable cracks to be monitored easily.

The test temperature chosen for this project was −3 °C, simulating a cold UK night. The tests were conducted by opening the movable part of the rig for a period of 6 h and closing it for another 6 h. The crack openings used for the test were 0.5, 1, 1.5, 2, 3, and 5 mm, applied to each specimen in sequence, or until failure occurred.

Since the PTF tests indicated that SAMIs are most effective against trafficking when thin, and considering the 5 mm thick SAMI A in the PTF to equate to a 10 mm thick layer in a real pavement, it was decided to test SAMI A as a 10 mm layer. SAMIs B and C remained unchanged. The overlaying asphalt concrete was 30 mm thick. Figure 4 presents the results and typical crack patterns, including the maximum measured strain at the surface of the specimen.

These results are most revealing. Based on experience from fatigue testing one would expect cracking to propagate rapidly once the strain level reached greater than about 2000–3000 microstrain—and this does indeed appear to have been the case. SAMIs B and C have clearly been successful in distributing the strain along the length of the specimen, thus preventing any damaging strain concentration even at several millimetres of crack opening. SAMI A has not been quite so successful—it failed at 3 mm opening—although it has clearly been more successful that the sand asphalt and it allowed a crack opening approximately three times greater than the control case. It would appear that viscosity is a key factor affecting performance since the fibre-reinforced SAMIs were of very low viscosity.

With respect to resistance to traffic loading, an attempt was made to model the crack development using the OLCRACK system (Thom 2000), assigning crack propagation parameters for both the SAMI and the asphalt overlay based on test data, and this gave close agreement for the two control sections (13,300 and 22,300

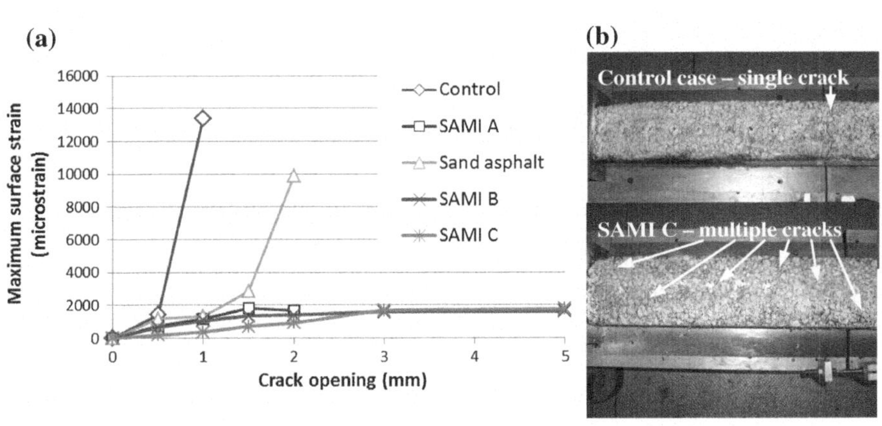

Fig. 4 Thermal cycling tests: **a** data; **b** typical failure patterns

cycles predicted compared to 14,105 and 20,185 recorded). Yet the same procedure applied to the SAMI reinforced cases, while it correctly predicted the better performance of 5 mm SAMI A compared to 10 mm, gave significant under-prediction (by 42–57 %) of absolute lives. The real reason for this is not known, but it suggests that the assumption inherent in OLCRACK that crack propagation commences from the first cycle with no initiation phase may need to be re-assessed for very highly crack-resistant materials like SAMIs.

5 Conclusion

This paper has presented results that demonstrate the effectiveness of three different SAMI systems and a conventional sand asphalt under laboratory conditions reasonably simulative of the pavement. The fibre-reinforced products performed particularly well against thermally-induced cracking; the polymer-modified SAMI excelled against traffic-induced cracking. Furthermore, modelling of crack propagation under traffic indicates that SAMI effectiveness may be a function of an initial and fairly large resistance to crack initiation rather than simply propagation.

References

Ellis SJ, Langdale PC, Cook J (2002) Performance of Techniques to minimize Reflection Cracking and Associated Developments in Pavement Investigation for Maintenance of UK Military Airfields. Federal Aviation Administration Airport Technology Transfer Conference.

Engle E (2001) Field Evaluation of Engineering Fabrics for Asphalt Concrete Resurfacing - Audubon County. Final Report 8-93 to 6-99, Iowa Department of Transportation, USA.

Kretov VA, Gorelysheva LA (2000) Influence of Thin Interrupting layer to Development of Reflected Cracks. Paper presented at the 4th RILEM Conference on Reflective Cracking, Ottawa, p 291–295.

Ogundipe OM, Thom NH, Collop AC (2013) Evaluation of the performance of stress absorbing membrane interlayers (SAMI'S) against reflective cracking under thermal loading. Paper presented at the 12th Int. Conference on Sustainable Construction Materials, Pavement Engineering and Infrastructure, Liverpool.

Palacios C, Chehab GR, Chaignon F, Thompson M (2008) Evaluation of Fibre Reinforced Bituminous Interlayers for Pavement Preservation. Paper presented at the 6th RILEM Conference, Chicago, p 721–729.

Storsteen M, Rumpca H (2000) Evaluation of Geosynthetics in Asphalt Overlays of Jointed Concrete Pavements. South Dakoda Department of Transportation, SD95–23-X.

Thom NH (2000) A simplified computer model for grid reinforced asphalt overlays. Paper presented at the 4th RILEM Conference on Reflective Cracking, Ottawa, p 37–46.

Vervaecke F, Maeck J, Vanelstraete A (2008) On Site Validation and Long Performance of Anti-Cracking Interfaces. Paper presented at the 6th RILEM Conference, Chicago, p 761–768.

Fast and Easy Road Renovation Through a Steel Based Anti-reflective Cracking Interlayer for Asphalt Overlays

F. Vervaecke, H. Cornelus and P. Straubinger

Abstract Anti-reflective cracking interlayers for the renovation of roads have been extensively used for the last decades. Today different anti-cracking interface systems exist: SAMI's, non-woven geotextiles, geogrids (plastic, glass or carbon), combigrids and steel reinforcing nettings. The different interlayer systems all have their specific properties with advantages and disadvantages. Moreover the performance of a specific product is not only depending on the product properties and the origin of the cracks but also on the proper installation of the product. Past research revealed that steel is an ideal reinforcement material for an anti-reflective cracking interlayer. Steel nettings as they are known today are very rigid and require a good fixing by nails or a slurry seal. Due to their dimensions, it is also advisable to use an overlayer thickness of at least 5 cm in order to prevent the reflection of the crack/joint through the asphalt surface. In this paper we present a new steel based anti-reflective cracking interlayer with improved installation properties compared to traditional steel nettings, enabling a fast and easy installation.

Keywords Reflective cracking · Asphalt concrete · Interlayer · Steel

1 Introduction

Cracking of asphalt concrete roads is a widely spread phenomenon which deteriorates the road surface and which reduces the driving comfort and the life time of the total structure. Reflective cracking originates when cracks from a base layer (rigid, semi-rigid or flexible) propagate through the surface via the asphalt overlay

F. Vervaecke (✉) · H. Cornelus · P. Straubinger
NV Bekaert SA, Bekaertstraat 2, 8550 Zwevegem, Belgium
e-mail: Frederik.Vervaecke@bekaert.com

H. Cornelus
e-mail: Henk.Cornelus@bekaert.com

P. Straubinger
e-mail: Peter.straubinger@bekaert.com

A. Chabot et al. (eds.), *8th RILEM International Conference on Mechanisms of Cracking and Debonding in Pavements*, RILEM Bookseries 13,
DOI 10.1007/978-94-024-0867-6_49

(Sanders 2001). This is mostly due to horizontal and/or vertical movements caused by traffic combined with environmental conditions, such as seasonal and daily temperature variations (Perfetti and Sangster 1988). In order to minimize this problem, anti-reflective cracking interlayers are widely used for the rehabilitation of cracked asphalt pavements Francken and Vanelstraete (1993).

A wide range of anti-cracking interlayers exist, i.e. SAMI's, non-woven geotextiles, geogrids (plastic, glass or carbon), combigrids and steel reinforcing nettings. The properties of the raw materials, their possible coating as well as their appearance in the geotextile (i.e. mesh dimensions, mesh shape, etc.) are diverse which makes it very difficult to compare these products. The performance evaluation of these products has been subject to both laboratory and field testing and showed a delay of the crack propagation through the surface (Norambuena-Contreras and Gonzalez-Torre 2015; Pasquini et al. 2013; Vervaecke and Maeck 2008). The installation of these products is really dominant in achieving good anti-reflective cracking behaviour. Some products might have a high strength but due to their brittle nature or temperature sensitive properties, the high strength might not be available anymore after the installation process (Gonzalez-Torre et al. 2014). Furthermore it is also important to follow the installation instructions of the producers and install these products flat and with the correct amount of tack coat in order to achieve a good performance.

Steel is widely used in construction as reinforcement in concrete, masonry and even asphalt. The current generation steel products for road reinforcement are hexagonal meshes with reinforcement bars in the transversal direction. These meshes are very rigid and in order to install them properly they need to be fixed by nails or preferentially with a slurry seal. They are known for their excellent anti-cracking properties thanks to the high Youngs modulus and good anchorage. Some disadvantages of the current generation products is their multiple step installation process and their time intensive removal process at the end of their life.

In this paper a new steel-based anti-cracking interlayer is presented. A first section describes the properties of the steel-based geotextile product. The second section describes some particularities towards installation. In a last section performance data are elaborated.

2 Product Definition

The presented steel-based anti-cracking interlayer differs from existing products by the half-product as well as on the shape of the mesh. The new mesh has a rectangular shape and is made from flexible steel cord. The steel cord mesh is kept in position on a plastic carrier (i.e. a low weight plastic grid or non-woven). The steel is galvanized in order to ensure a lifetime of the product in the application of at least 15 years. The properties of the product compared to traditional hexagonal steel mesh and a classical glass-based geotextile are presented in Table 1. The tensile strength is determined according to ASTM D6637 Method 1 (single rib test).

Table 1 Properties of metal and non-metal geotextiles used as anti-cracking interlayer

		Mesh properties	Mechanical properties
Existing steel mesh	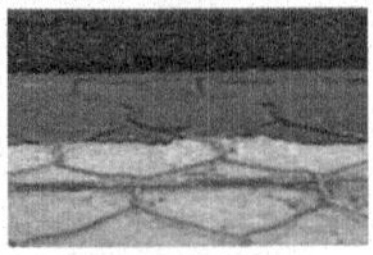	Steel wire 2.45 mm Hexagonal 118 × 80 mm Trans. wire 3 × 7 mm ±1.7 kg/m^2	Strength: 40 × 58 kN/m EA: 22,600 × 32,500 kN/m
Existing glass mesh	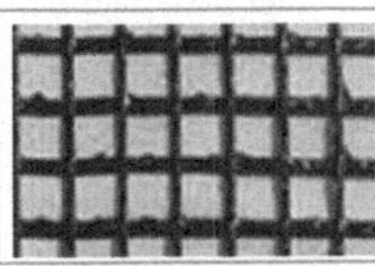	Glass roving Square 35 × 35 mm Plastic coating ±330 g/m^2	Strength: (70 × 100)kN/m EA: (2800 × 4000)kN/m
New steel mesh	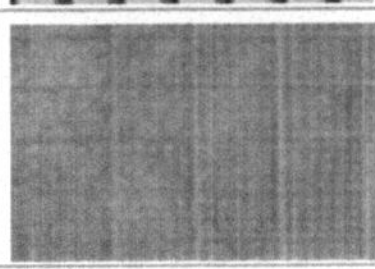	Steel cord (1.05 mm) Rectangular 40 × 30 mm ±335 g/m^2	Strength: 38 × 50 kN/m[a] EA: 3100 × 4400 kN/m

[A]The embrittled zones are considered as 0 kN

Although tensile strength is a dominant characteristic of geosynthetics, tensile strength is less important in the final application. The dominant parameters for obtaining good anti-cracking behaviour are "stiffness" EA (material cross section A * material modulus E), the secant modulus and the adhesion with the pavement layers (Gonzales-Torre et al. 2014 and references). Since steel has a high Youngs modulus, the cross section of the reinforcement material can be reduced. The new mesh has an EA in the range of existing glass-based geosynthetics.

Note that the cord half-product is foreseen of embrittled zones at predetermined positions. These embrittlements are applied in order to ensure that the product can be milled together with the asphalt at the end of the pavements lifetime. Magnets enable afterwards to recover the steel from the asphalt ensuring pure waste streams and the complete re-use of both the steel and the asphalt. This is a unique property of the steel and this is not achievable with any other material.

3 Installation

The new steel-based anti-cracking interlayer can be installed in a very easy and simple way. The existing road surface should be prepared by milling the top layer, the treatment of the cracks, the cleaning by high pressure rinsing and brushing. Subsequently, the appropriate type of tack coat is applied in a minimum amount of 300 g/m^2 (residual weight). The new steel mesh is then applied in the fresh tack coat in order to assure an immediate interaction between tack coat and carrier. The system is then covered with at least 3 cm of asphalt concrete. The flexible and low weight steel cord makes that this product can easily be unrolled manually as well as automatically from a truck. Moreover, the weight of the steel compared to the

Fig. 1 Installation process of the new flexible steel-based anti-cracking interlayer

weight of the carrier, ensures a flat installation on the surface and the carrier on the other hand ensures a good contact with the tack coat. Some impressions of the installation process are shown in Fig. 1.

Traditional geosynthetics made from plastic, glass or carbon are sensitive to damage during the installation process. Gonzalez-Torre et al. 2014 investigated several mechanisms to assess the damage during the installation process, i.e. mechanical damage (ISO 10722: 2007), real installation and installation in laboratory. Depending on the material, damages of 10–90 % are reported in tensile strength and secant modulus. Steel was however not investigated in that research.

Steel half product, i.e. rectangular and round wire from the traditional steel product was tested on damage during installation. The material was placed on an asphalt layer, subsequently fresh asphalt was installed by a paver and the fresh layer was compacted. Before the complete cooling of the fresh asphalt the steel wires were removed from the asphalt (Fig. 2).

Virgin as well as material which underwent the installation process were tensile tested according to EN ISO 6892-1: 2010. The normalized values of tensile strength, Youngs modulus and secant stiffness at 0.2 % are presented in Fig. 2.

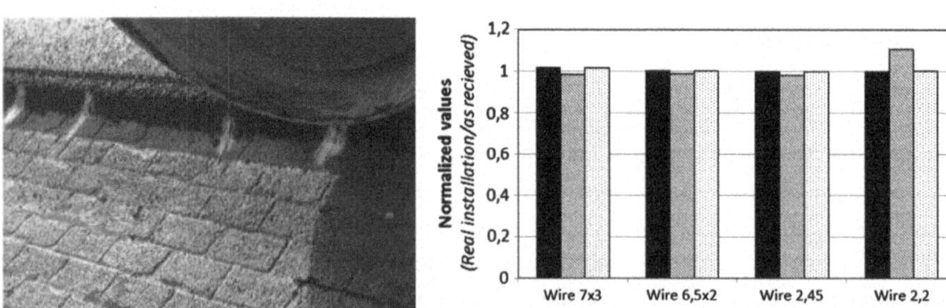

Fig. 2 Left: *White paint* indicates steel pieces installed under 4 cm of asphalt; right: normalized values for Fm, Youngs modulus and secant modulus at 0.2 % from different steel half product

There is no decrease visible in the mechanical properties of the steel during the asphalt installation. Steel is not sensitive to mechanical damage by stone aggregates and is thermally stable at temperatures up to 700 °C. As such steel can be assessed as ideal material for asphalt reinforcement. Moreover due to its stability during asphalting, it can be applied on both a smooth or rough/milled surface.

4 Performance

There is no standard test available for the evaluation of the performance of anti-cracking interlayers in the asphalt application. Several research groups have developed their own system to simulate a certain fracture mechanism. Note that since reflective cracking is a fatigue mechanism, cyclic testing is very important. Apart of this the dimensions of several products can be very different and they have to be taken into consideration during the interpretation of the test results. In the early 90s, BRRC developed a horizontal plate test to simulate the crack formation by thermal movement (De Visschere and Vanelstraete 2010). This test was found to be relevant and shows a good differentiation between different products and how they are installed.

The horizontal plate test is performed on samples of 600 × 175 × 140 mm^3. The samples consist of a 70 mm thick concrete base layer with a rough surface finishing (brushed) and an artificial notch (7 mm) in the middle. The anti-cracking interlayer is installed according to the producers recommendations and covered by 40 mm of asphalt. The samples are conditioned for 12 h at −10 °C, the bolts are tensioned and the test is started. During the test the notch is opened and closed by 1 mm at a very slow rate by the thermal expansion of steel bars holding the sample. The maximum force is recorded and the crack propagation is followed as a function of the number of cycles and the total testing time.

In this test series 2 different anti-cracking interlayers (glass and new steel product from Table 1) are tested together with a reference sample without any reinforcement. The samples were all prepared the same way with the same amount of tack coat 300 g/m^2 residual. The results of the horizontal plate test are presented in Table 2. One can conclude that an anti-cracking interlayer increases the maximum

Table 2 Properties of metal and non-metal geotextiles used as anti-cracking interlayer

	F_{max} (kN)	Crack initiation		End of test		Remarks
		Cycles	Time (h)	Cycles	Time (h)	
Reference	8	1	2	5	10	Crack
New steel grid	9	Na	Na	Na	130	No cracks
	8.7	35	126	35	126	Delamination
	9.5	15	42	24	90	Crack
Glass mesh	9.6	8	32	11	48	Delamination + crack
	9.3	28	108	28	108	Delamination

force take-up by the asphalt samples and delays the occurence of the first crack in the asphalt. Both the crack initiation and the crack propagation are better for the steel product compared to the glass product. This is a striking result since both products have a similar EA and the steel product has only half the strength of the glass grid. The reason for the performance difference can be found in the better anchorage of the steel in the asphalt. The round cords are nicely anchored in the asphalt and work as such as a real reinforcement while the flat glass is only adhering to the asphalt via the coating. This hypothesis is also demonstrated by the failure mechanism of the glass product which is preferential delamination.

5 Conclusions

A new steel grid made from steel cords has been presented in this article with some specific advantages compared to traditional geosynthetics and existing hexagonal steel mesh. First of all, the flexible steel cords enable a fast and easy installation of the anti-cracking interlayer. The steel is stable during the installation process and its properties are not reduced to mechanical or thermal interaction with the asphalt. And finally, the anti-cracking properties clearly outperform other glass based anti-cracking interlayers with a similar EA and a higher tensile strength.

References

De Visscher J, Vanelstraete A (2010) Essai de fissuration thermique. Asphalt roads and other bituminous applications of Belgian Road Research Center. Available via http://www.brrc.be/fr/article/f670_01. Accessed 11 Sept 2015

Francken L, Vanelstraete A. On the thermorheological properties of interface systems, Proceedings of the 2nd international RILEM conference on reflective cracking in pavements; (1993) p. 206–19.

Gonzalez-Torre I, Calzada-Perez MA, Vega-Zamanillo A, Castro-Fresno D, Damage evaluation during installation of gesynthetic used in asphalt pavements, Geosynthetics International, 3, 21, No. 6 (2014) 377–386

Norambuena-Contreras J, Gonzalez-Torre I, Influence of geosynthetic type on retarting cracking in asphalt pavements, Construction and building Materials 78 (2015) 421–429

Pasquini E, Bocci M, Ferrotti G, Canestrari F, Laboratory characterization and field validation of geogrid-reinforced asphalt pavements, Road materials and Pavement Design 14 No.1 (2013) 17–35

Perfetti J, Sangster T, The use of a textile based system to control pavement cracking, Geotextiles and Geomembranes 7 (1988) 165–178

Sanders P (2001) Reinforced asphalt overlays for pavements – PhD Thesis. University of Nottingham, department of civil engineering, Nottingham

Vervaecke F, Maeck J, On site validation and long term performance of anti-cracking interfaces, Proceedings of 6th Rilem International Conference on Cracking in Pavements, Chicago, USA, 16–18 June 2008

Reflective Cracking Included into Routine Design of New Asphaltic Pavements

J.G.F. Schrader and A.H. de Bondt

Abstract During the past decades, it has been experienced that reflective cracking is a very complex phenomenon. Not only which one of the possible mechanisms behind the reappearance of cracks in new pavement surfaces (traffic, temperature variations in time or uneven settlements) is dominant, depends on the typical circumstances of a specific project, but also a variety of maintenance solutions often seems to be applicable. Examples are: (combinations of) thick overlays, use of modified asphaltic mixtures, application of stress-relieving systems or the incorporation of reinforcement. At motorway and airfield (maintenance and rehabilitation) projects there usually is time, budget and information. At those large projects it pays tribute to include reflective cracking into the routine design, because a cement treated base or concrete slabs are quite often present. From these specific pavement layers, cracks or joints propagating into and through the asphaltic overlay is the dominant mechanism. Ooms Civiel has developed an analysis method to include the crack driving mechanisms temperature variation and traffic into the routine design of new asphaltic pavements. The temperature influence is analysed by means of ARCDESO®; the traffic loading requires finite elements simulations. This paper explains the design method, which meets the challenge mentioned above. The method has been calibrated with long term field experience (crack mapping data) and is being validated continually.

Keywords Pavement design · FEM-analysis · Reflective cracking · Polymer modified asphalt

J.G.F. Schrader (✉) · A.H. de Bondt
Ooms Civiel Bv, Scharwoude, The Netherlands
e-mail: JSchrader@ooms.nl

A.H. de Bondt
e-mail: AdeBondt@ooms.nl

A. Chabot et al. (eds.), *8th RILEM International Conference on Mechanisms of Cracking and Debonding in Pavements*, RILEM Bookseries 13,
DOI 10.1007/978-94-024-0867-6_50

1 Introduction

During the maintenance or rehabilitation planning of motorways and airfields a cement treated base (CTB) or concrete slabs (PCC) often imply challenges. In the CTB cracks will be present, as well as joints in between the PCC-slabs. The PCC can also contain cracks. Lastly, also new pavements are designed with a (pre-) cracked CTB. In general, in (current) design software a CTB or an old PCC is simulated as a uniform base layer with a specific Young's modulus (E). This material modulus can be reduced to an equivalent layer stiffness to incorporate the effect of cracks and joints. However, reflective cracking from the joints into the asphalt layers has not been implemented into standard design software yet. For many clients and contractors this is the reason to exclude the use of a CTB or an old PCC into their pavement design.

According to Ooms Civiel it is very interesting to use these pavement materials as they are cheap, widely available and in general they possess a higher Young's modulus than unbound granular base materials. This results into thinner asphalt layers on top of the base layer and/or a thinner base layer itself. Ooms Civiel has recognized the risk of reflective cracking by adding two additional pavement design analyses to the traditional multi-layer thickness design approach.

First of all the pavement structure has to be designed by using the traditional design software. Afterwards the structure has to be analysed by means of the software programme ARCDESO for the risk of reflective cracking caused by temperature induced (horizontal) slab movements. These analyses are described in Sect. 2. The vertical slab movements caused by traffic loadings are analysed by means of the finite elements programme CAPA-3D and will be described in Sect. 3. In Sect. 4 the results are superimposed.

2 Temperature Analyses

During the hardening of a CTB, (restraint) shrinkage can initiate cracking. These cracks can propagate into the asphalt layers on top of the CTB because of horizontal movements of the CTB. The horizontal movements are caused by the (daily) temperature drops and these movements can initiate cracking in the CTB as well. In the end a random crack pattern shall develop with an average crack distance of more than 6 m. By pre-cracking over a shorter distance, the asphalt overlay is relaxed and a regular crack pattern shall arise. The width of the pre-cracks will be less than the width of the natural cracks and the seasonal/daily widening of the pre-cracks will also be less as the temperature induced movements of shorter CTB-slabs will be smaller.

Ooms Civiel has developed the (web-based) software programme ARCDESO to design pavement structures with respect to the failure mechanism thermal induced reflective cracking (www.ARCDESO.nl). ARCDESO stands for Anti-Reflective

Cracking DEsign SOftware and the project was with the (partial) financial support of ADFORS Saint-Gobain. ARCDESO is a mechanistic empirical program that is based on a) the PhD. research by Arian de Bondt at Delft University of Technology from 1989 to 1996/1999, entitled "Anti-Reflective Cracking Design of (Reinforced) Asphaltic Overlays" (de Bondt 1999) later work by the R&D-department of Ooms Civiel (de Bondt 2012).

The software development took a decade (from 2001 to 2011) and afterwards Ooms has updated the programme, by validating and calibrating it on numerous long-term field and laboratory data. ARCDESO is (one of) the first type(s) of design software in the world that deals with reflective cracking in asphalt layers. The viscoelastic behaviour of the asphalt layers is indirectly taken into account, and the amount of fatigue in the structure is controlled by Miner's law and presented as the Cumulative Damage Factor ($CDF_{TEMPERATURE}$). Variability in material parameters and thicknesses are taken into account automatically in the analysis. Please remind that according to EU COST action 348 (de Bondt 2006) crack initiation is dominant versus crack propagation in case of temperature cycles.

In ARCDESO the pavement details have to be inputted, including the distance between (a) the (pre-)cracks of the CTB-base layer, (b) the cracks in an old asphalt layer or (c) the joints in the PCC-slabs. Also the monthly temperature profile has to be inputted. ARCDESO transforms this input into mean daily temperature variation cycles, combined with one extreme temperature cycle per month, and one seasonal temperature cycle (summer–winter drop). ARCDESO calculates the reflective cracking behaviour from the cracks for the given (hundreds of) maintenance options. The maintenance exists of an asphalt concrete overlay of a certain thickness and an optional addition of (glass fibre) asphalt reinforcement that could be combined with a tack coat, SAMI or a PMB bond coat, via a spring model. There is also the possibility to change the conventional asphalt into polymer modified asphalt. In ARCDESO the optimum slab size can be determined given one maintenance option, but also the maintenance option can be optimized for a certain slab length. In common, the maintenance option to start with corresponds with the result of the general (thickness) design software. In Fig. 1 an example of these optimizations is shown.

From Fig. 1 it becomes clear that a shorter notch distance results into less reflective cracking. Also the positive effects of applying PMB-asphalt and glass fibre reinforcement are obvious.

An important parameter in this part of the pavement design is the shear stiffness of the interface between the asphalt concrete and the underlying CTB. As there are relatively few temperature loadings (only one cycle per day) and the temperature loading (shear rate) is very slow, it is not likely that fatigue of the shear stiffness at the interface will occur. Of course, the value of the actual shear stiffness depends on the temperature, the loading time and the pavement structure characteristics (which materials are e.g. used as tack coat or bond coat?). ARCDESO contains a wide data base with (tested) materials and interfaces.

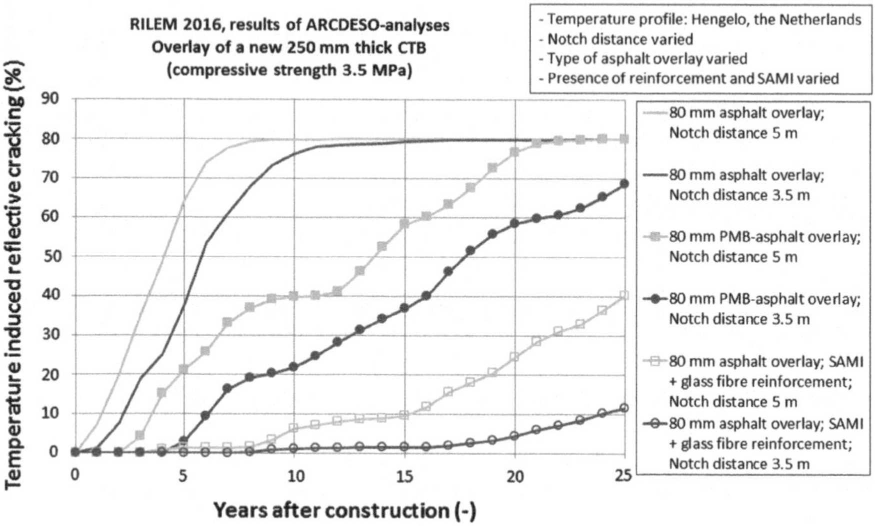

Fig. 1 ARCDESO optimization overlaying a new 250 mm thick CTB-base layer

3 Traffic Load Analyses

The traffic loads can also initiate reflective cracking from the notches into the asphalt layers. In general the effect of the traffic loads reduces as the thickness of the asphalt concrete layers increases (better: the flexural deformation decreases). However, there is no rule of thumb at which thickness the effect of traffic loading cannot be ignored. This means that case dependent, the effect of traffic loading has to be analysed. For this Finite Element Modelling (FEM) software is necessary. Ooms Civiel performs this FEM-analyses by using the CAPA-3D software, developed at the section Structural Mechanics of Delft University of Technology (Scarpas and Karsbergen 1999). Figure 2 shows a section of a created element model that is loaded by a wheel load located exactly on top of the notch. The deformations in Fig. 2 have been enlarged 800 times.

In Fig. 2 the upper part of the CTB is shown, including the opened notch in the middle. On top of the CTB the asphalt concrete is visible.

During the analyses the wheel load has been moved over the asphalt concrete pavement and the corresponding stresses and strains in all individual elements have been analysed. The first step in the analysis is to determine the critical wheel load position(s) and therefore the maximum strain at the bottom of the asphalt concrete has to be determined. Figure 3 shows that the maximum principal strain is caused by wheel load position 21. This position equals to the wheel load exactly on top of the notch, which means that bending of the asphalt concrete is the normative failure mechanism for this specific case (project).

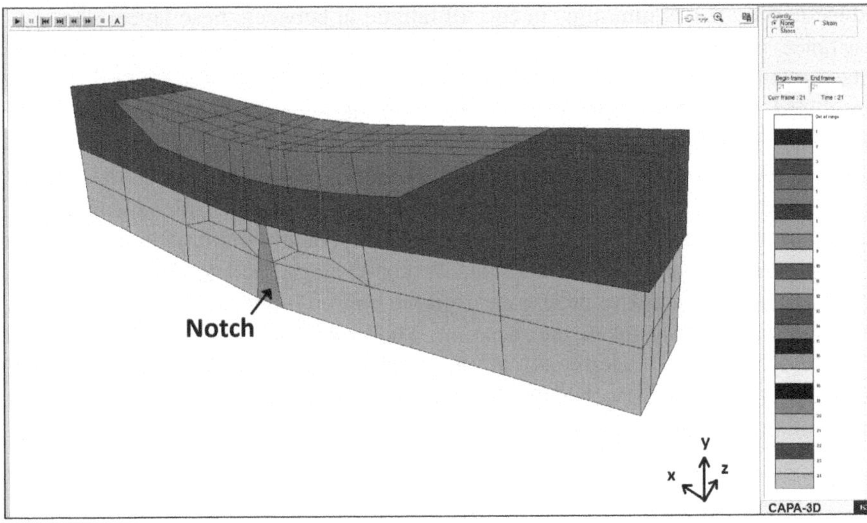

Fig. 2 Vertical deformations in FEM-model caused by wheel load on top of the notch

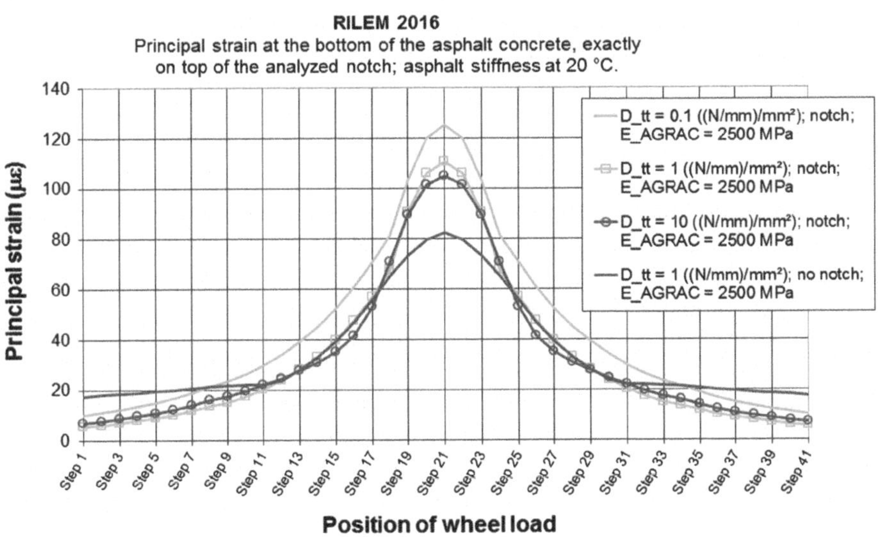

Fig. 3 Principal strain at the bottom of the asphalt concrete above the notch caused by wheel load

Figure 3 shows that the shear stiffness (D_tt) of the interface between the asphalt concrete and the CTB (modelled as a 1 mm thick, separate horizontal layer in between the cubes) is indeed an important input parameter for the traffic load analysis. In general this shear stiffness varies from 1 to 10 N/mm/mm^2, but it might reduce to only 0.1 N/mm/mm^2 in case of fatigue in between these layers (the shear resistance is generated by a combination of dry friction which is constant and adhesion which shall reduce in time). The calculations without a notch have been introduced to calibrate this design method to the traditional pavement design methods (when layers are homogenous).

When the normative failure mechanism (bending or shear) has been determined, the corresponding $CDF_{TRAFFIC}$ can be calculated for several interface shear stiffness values. In this calculation amongst others the temperature, the fatigue performance of the asphalt concrete (determined via CE type-testing), the lateral wander and the time required for a crack to propagate from the bottom of the asphalt concrete to the pavement surface, are taken into account. The crack propagation is based on the time delaying factor determined in the PhD. research by M. Jacobs at Delft University of Technology (Jacobs 1995). According to EU COST action 348 crack propagation is as dominant as crack initiation for traffic loadings.

4 Design Life Determination

The determined CDF-values in Sects. 2 and 3 should be superimposed to determine the design life of the pavement structure. In principle, this combined CDF represents only the expected design life of the asphalt concrete on top of the notches in the CTB. So a 100 % value of the combined CDF means that cracks from all notches have propagated to the pavement surface. Given for example a notch distance of 3.5 m and an average cracked area of 0.1 m width, this results in an overall damage of 3 % of the pavement area.

However, in CTB layers with a low fatigue strength, the combination of traffic and temperature loads could result into additional cracking of the CTB-layer itself and additional crack propagation. There is then a chance that cracking at the asphalt bottom might originate also at other locations than above the notch.

For these two reasons Ooms Civiel has decided to limit the combined CDF to a maximum of 0.5. Given this criterion it implies that structures with a small interface shear stiffness (near 0.1 N/mm/mm^2) often do not “survive”. In other words, an adequate interface shear resistance (but not too high) is required.

5 Conclusions

- Ooms Civiel has performed thermal and traffic load analyses to determine the expected lifetime of a pavement structure including a (CTB-)layer, by amongst others verifying the resistance of the structure against reflective cracking from a notch or (pre-)crack.
- The new or existing pavement design review on reflective cracking is possible to complete within a period of three weeks, for a given project.

- The results of the FEM-analysis without a notch correspond to the results of the traditional pavement design programmes and therefore this analysis method is adequate.
- It is possible to eliminate the chance of reflective cracking by pre-cracking the CTB and applying enough asphalt concrete (with the right quality) on top.
- Reflective cracking can be included into the routine design of (new) asphaltic pavements by adding a thermal ARCDESO analysis and an FEM-analysis.

References

de Bondt A (1999) Anti-Reflective Cracking Design of (Reinforced) Asphaltic Overlays. PhD-Thesis, Delft University of Technology, the Netherlands

de Bondt A (2006) EU COST ACTION 348 - REINFORCEMENT OF PAVEMENTS WITH STEEL MESHES AND GEOSYNTHETICS Work Package 4: Selection of Design Models and Design Procedures, 3 January 2006

de Bondt A (2012) 20 years of research on asphalt reinforcement – Achievements and future needs. Paper presented at the 7th RILEM Conference on Cracking in Pavements, Delft, the Netherlands, 20–22 June 2012

Jacobs M (1995) Cracking in Asphaltic Mixes. PhD-Thesis, Delft University of Technology, the Netherlands

Scarpas A, Karsbergen C (1999) CAPA-3D User's Manual, Delft University of Technology, the Netherlands

M4-5n Numerical Solution Using the Mixed FEM, Validation Against the Finite Difference Method

Hanan Nasser, Jean-Michel Piau, Olivier Chupin and Armelle Chabot

Abstract The final aim of this work is to build a tool dedicated to the calculation of the mechanical fields in pavements incorporating possible vertical cracks in some layers or partial debonding at the interface between layers. The development of this tool is based on a specific layer-wise modeling of the structure so-called M4-5n. In this model the stress fields are approached through polynomial approximations in the vertical direction for each layer. Its construction is based on the Hellinger-Reissner (H-R) variational principle of continuum mechanics. One advantage of the M4-5n is to reduce by one the dimension of the problem. Moreover this model leads to finite values of the generalized interface stresses at the crack lips of the structures studied. This approach is thus particularly adapted to parametric studies and might be considered for analyzing crack growth in layered structures such as pavements. The contribution of the present paper to this model is focused on the computation of its numerical solution by means of the mixed Finite Element Method (FEM). The developed method is based on the maximum of the complementary energy theorem using Lagrangian multipliers to ensure the equilibrium equations. The resulting formulation is equivalent to the H-R variational principle applied to the generalized displacement and stress fields. This approach is applied to a beam structure composed of four elastic homogenous layers resting on Winkler's springs. Vertical cracks across some layers are introduced. The results obtained are compared with those from an earlier approach using the Finite Difference Method (FDM).

Keywords M4-5n · Mixed FEM · Cracking · Debonding

H. Nasser (✉) · J.-M. Piau · O. Chupin · A. Chabot
LUNAM Université, IFSTTAR, CS4, 44344 Bouguenais cedex, France
e-mail: hanan.nasser@ifsttar.fr

J.-M. Piau
e-mail: jean-michel.piau@ifsttar.fr

O. Chupin
e-mail: olivier.chupin@ifsttar.fr

A. Chabot
e-mail: armelle.chabot@ifsttar.fr

A. Chabot et al. (eds.), *8th RILEM International Conference on Mechanisms of Cracking and Debonding in Pavements*, RILEM Bookseries 13,
DOI 10.1007/978-94-024-0867-6_51

1 Introduction

A pavement structure is a multilayer structure resting on a soil. It can be damaged due to many phenomena resulting in different types of degradation among which vertical cracking across layers or partial/total debonding at the interface between two layers. To study these mechanisms, the existing advanced models are generally based on 3D continuum mechanics using the fracture or the damage theories (Pommier et al. 2009). Recently Chabot et al. (2005) proposed to apply one of the Multi-Particle Models of Multilayer Materials (M4) to the pavement field. Among the different formulations proposed for this family of models, the M4-5n is especially adapted to the analysis of delamination in composite materials. It is particularly adapted to parametric studies (Chabot et al. 2013). This specific formulation developed for linear elasticity considers five generalized displacement fields per layer (n: total number of layers). In particular the use of generalized stress fields with finite values avoids the problem of singularity near cracks. Hence this formulation seems also appropriate for modeling cracking and debonding in pavements. The use of M4-5n makes it possible to reduce the problem by one dimension avoiding the explicit presence of the z-coordinate in the equations. This is an interesting feature with regards to numerical modeling (mesh generation and computation time). As a consequence, a 3D pavement structure is represented by plate geometry. In each layer of the M4-5n, the usual stress tensor is replaced by the generalized stresses $M^i_{\alpha\beta}(x,y)$, $Q^i_\alpha(x,y)$, $N^i_{\alpha\beta}(x,y)$ (bending, shear and normal components) per layer i ($\alpha, \beta \in \{1,2\}$, $i \in \{1,n\}$). The related generalized displacements are $\Phi^i_\alpha(x,y), V^i(x,y), U^i_\alpha(x,y)$ (rotation, vertical and horizontal displacements). The link between two consecutive layers, i and i + 1, is ensured by the interface shear $\tau^{i,i+1}_\alpha(x,y)$ and the normal $\nu^{i,i+1}(x,y)$ stresses. Each M4-5n layer has its own set of equations: compatibility, equilibrium and elastic constitutive law. The introduction of a vertical crack across a layer is performed by zeroing some of the generalized stresses along the crack contour projected on the (x, y) plane, in accordance with the assumptions made in terms of load transfer. On the other hand the debonding of interface (i, i + 1) can be taken into account by zeroing, on the relevant area, $\tau^{i,i+1}_\alpha(x,y)$ and possibly $\nu^{i,i+1}(x,y)$ depending on the contact conditions considered in the vertical direction. Discontinuities in the model are then induced from one side to the other of the contour of cracks or delaminated areas. The direct derivation of the M4-5n mathematical equations leads to a linear system of partial differential equations of second order with boundary conditions. This system involves a large number of unknown variables. The whole set of equations related to the pavement modeling using the M4-5n and the Winkler soil can be found in (Nasser and Chabot 2015). In this paper we present the main guidelines for the development of a mixed FEM method to solve more generally this set of equations. This method is then applied to a beam geometry example including a vertical crack across a layer and its implementation is validated by comparison with earlier results obtained with the FDM.

2 The Mixed FEM to Solve the M4-5n Equations

The developed approach is inspired from the derivation of the H-R principle (Reissner 1950) for 3D elastic problems that we transpose to the generalized fields of stresses and displacements of M4-5n. To accomplish this, we start from the theorem of maximum complementary energy, using the elastic density energy related to the generalized stresses of the M4-5n layers. The equilibrium equations and the stress boundary conditions are taken into account through fields of Lagrangian variables which, for an appropriate choice, appear to be the generalized displacements fields of M4-5n. The expression of a Lagrangian (L) is thus obtained. This can be integrated by parts to lower the derivation order of the generalized stresses. Then the mixed formulation is obtained by expressing the stationarity of L with respect to all fields ($\delta L = 0$). The resulting variational formulation is similar to that of H-R principle expressed in terms of the generalized displacement and stress fields. Then this formulation can be solved using a mixed FEM in which the generalized stress and displacement fields are discretized. A special attention must be paid to the dimensions of the interpolation spaces of the different fields to avoid undetermined system of algebraic equations (Lagrangian spaces to be chosen with lower dimension than those of the dual generalized stresses).

3 Example of 2D Cracked Pavement Structure

To test the method described in the previous section, a specific FE numerical program along the x direction was developed for the computation of the M4-5n structures located in the real 2D (x, z) plane. All the M4-5n fields depend on x only (13n-4 for this "beam" geometry). Hereafter the results obtained with this numerical program for some given data set are compared to those from simulations based on FDM (Nasser and Chabot 2015). The structure under consideration is taken similar to a 2D composite pavement composed of three layers: a surface layer of 0.08 m of semi-coarse asphalt concrete (BBSG, E^1 = 5400 MPa, υ^1 = 0.35); a base layer of 0.15 m of gravel-bitumen (GB3, E^2 = 9300 MPa, υ^2 = 0.35); a foundation layer composed of 0.23 m of cement-treated base (GC3, E^3 = 23000 MPa, υ^3 = 0.25). A fourth layer resting on Winkler springs (k) represents the soil (E^4 = 120 MPa, υ^4 = 0.35, k = 45.6 MPa/m) (Fig. 1).

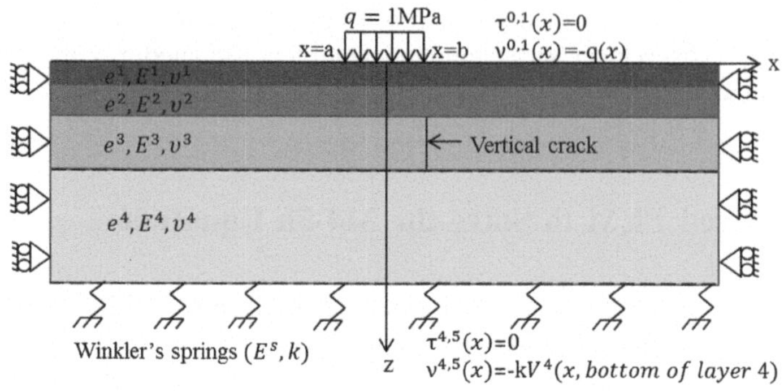

Fig. 1 Sketch of the structure considered for the comparison between the mixed FEM and FDM

The third layer is assumed to be crossed by a vertical crack simulating the shrinkage of the cement material. A unit pressure load (1 MPa), q(x) = H (x − a) − H(x − b), is applied at the surface of the structure over interval [a, b] with (b − a) = 0.15 m. The right edge of the load is taken coincident with the x location of the crack. The length of the structure is taken equal to 6 m. The horizontal displacements $U^i(x)$ and rotations $\Phi^i(x)$ are blocked on the lateral boundaries, whereas the shear efforts per layer i, $Q^i(x)$ are set to zero. For this example, after integration by parts and here omitting the contribution of the crack for the sake of simplification, the Lagrangian writes as (Eq. 1):

$$
\begin{aligned}
& L(\widehat{N}^1, \widehat{M}^1, \widehat{Q}^1, \underbrace{\tau^{0,1}}_{=0}, \underbrace{\nu^{0,1}}_{-q(x)}, \widehat{U}^1, \widehat{\Phi}^1, \widehat{V}^1, \widehat{N}^2, \ldots, \hat{\tau}^{1,2}, \hat{\nu}^{1,2}, \ldots, \underbrace{\tau^{4,5}}_{=0}, \hat{\nu}^{4,5}, \ldots, \widehat{V}^4) = \\
& - \int_{-l}^{l} (w_{5n}^* + w_{Winkler}^*) dx + \sum_{i=1}^{4} \int_{-l}^{l} \left[\widehat{N}^i \widehat{U}'^i + \widehat{M}^i \hat{\Phi}'^i + \widehat{Q}^i \widehat{V}'^i + \widehat{Q}^i \hat{\Phi}^i \right] dx \\
& - \sum_{i=2}^{3} \int_{-l}^{l} [(\hat{\tau}^{i,i+1} - \hat{\tau}^{i-1,i}) \widehat{U}^i + (\hat{\nu}^{i,i+1} - \hat{\nu}^{i-1,i}) \widehat{V}^i + \frac{e^i}{2} (\hat{\tau}^{i-1,i} + \hat{\tau}^{i,i+1}) \hat{\Phi}^i] dx \\
& - \int_{-l}^{+l} [\hat{\tau}^{1,2} (\widehat{U}^1 - \frac{e^1}{2} \hat{\Phi}^1) - \widehat{V}^1 (q(x) + \hat{\nu}^{1,2}) + \hat{\tau}^{3,4} (\widehat{U}^4 + \frac{e^4}{2} \hat{\Phi}^4) + \widehat{V}^4 (\hat{\nu}^{4,5} \\
& - \hat{\nu}^{3,4})] dx - \sum_{i=1}^{4} [\widehat{N}^i \widehat{U}^i + \widehat{M}^i \hat{\Phi}^i]_{-l}^{+l}
\end{aligned}
\tag{1}
$$

The symbol $\widehat{()}$ is used to distinguish the virtual generalized fields from the solution of the problem. w_{5n}^* and $w_{Winkler}^*$ are the density of the elastic stress energy of the M4-5n and the energy of Winkler's springs. These are expressed as follows (Eqs. 2–3):

$$w^*_{5n}\left(\widehat{N}^1, \widehat{M}^1, \widehat{Q}^1, \underbrace{\tau^{0,1}}_{=0}, \underbrace{\nu^{0,1}}_{-q(x)}, \widehat{N}^2, \ldots, \hat{\tau}^{1,2}, \hat{\nu}^{1,2}, \ldots, \underbrace{\tau^{4,5}}_{=0}, \hat{\nu}^{4,5}\right) = \frac{1}{2}\sum_{i=1}^{4}\left[\frac{1-\upsilon^{i^2}}{e^i E^i}\widehat{N}^{i^2}\right.$$
$$+\frac{12\left(1-\upsilon^{i^2}\right)}{e^{i^3}E^i}\widehat{M}^{i^2}+\frac{12(1+\upsilon^i)}{5e^iE^i}\widehat{Q}^{i^2}]+\sum_{i=2}^{3}[\frac{13e^i}{35E^i}\left(\hat{\nu}^{i-1,i^2}+\hat{\nu}^{i,i+1^2}\right)+\frac{9e^i}{35E^i}\hat{\nu}^{i-1,i}\hat{\nu}^{i,i+1}$$
$$\left.+\frac{4e^i(1+\upsilon^i)}{15E^i}\left(\hat{\tau}^{i-1,i^2}+\hat{\tau}^{i,i+1^2}\right)-\frac{2e^i(1+\upsilon^i)}{15E^i}\hat{\tau}^{i-1,i}\hat{\tau}^{i,i+1}-\frac{2(1+\upsilon^i)}{5E^i}\widehat{Q}^i(\hat{\tau}^{i-1,i}+\hat{\tau}^{i,i+1})\right]$$
$$+\frac{13e^1}{35E^1}\left(q(x)^2+(\hat{\nu}^{1,2})^2\right)+\frac{9e^1}{35E^1}q(x)\hat{\nu}^{1,2}+\frac{4e^1(1+\upsilon^1)}{15E^1}(\hat{\tau}^{1,2})^2-\frac{2(1+\upsilon^1)}{5E^1}\widehat{Q}^1\hat{\tau}^{1,2}$$
$$+\frac{13e^4}{35E^4}\left((\hat{\nu}^{3,4})^2+(\hat{\nu}^{4,5})^2\right)+\frac{9e^4}{35E^4}\hat{\nu}^{3,4}\hat{\nu}^{4,5}+\frac{4e^4(1+\upsilon^4)}{15E^4}(\hat{\tau}^{3,4})^2-\frac{2(1+\upsilon^4)}{5E^4}\widehat{Q}^4\hat{\tau}^{3,4} \tag{2}$$

$$w^*_{Winkler} = \frac{1}{2}\frac{(\hat{\nu}^{4,5})^2}{k} \tag{3}$$

The solution fields are those which make the Lagrangian be stationary and check the condition $\delta \mathrm{L} = 0$ for all variations of the virtual fields. The discretization of this condition by the FEM leads to a mixed algebraic system having the typical shape given in (Eq. 4). Σ and U denote the vectors of nodal values of the generalized stresses and displacements, respectively. F_Σ, F_U are the vectors of nodal forces related to q(x).

$$\begin{bmatrix} K_{\Sigma\Sigma} & K_{\Sigma\mathrm{U}} \\ {}^tK_{\Sigma\mathrm{U}} & 0 \end{bmatrix}\begin{pmatrix}\Sigma \\ U\end{pmatrix} = \begin{pmatrix}F_\Sigma \\ F_\mathrm{U}\end{pmatrix} \tag{4}$$

As expected this linear system is symmetric since it derives from a quadratic form for the Lagrangian [excepted the terms including the pressure load q(x)]. All the fields are interpolated by Lagrange polynomials ($\wp$(x)) with the following orders (Eq. 5) ensuring the determination of the system:

$$\begin{aligned}&\dim(\wp_N) = 5,\ \dim(\wp_M) = 3,\ \dim(\wp_Q) = 4,\ \dim(\wp_\tau) = 4,\ \dim(\wp_\nu) = 4,\\ &\dim(\wp_U) = 4,\ \dim(\wp_\lrcorner) = 2,\ \dim(\wp_V) = 3\end{aligned} \tag{5}$$

Figure 2 shows the typical 1D finite element used to calculate the elementary matrix between two consecutive nodes j and j + 1. A local condensation method is used to eliminate the internal nodes and keep only the end nodes.

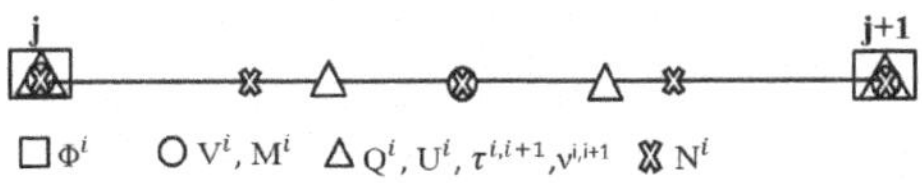

Fig. 2 1D M4-5n mixed finite element for 2D problems with multiple nodes related to the order of interpolation of each field and multiple degree of freedom related to the number of layers

The global matrix is then obtained as usual by assembly of the elementary condensed matrices for the whole 1D mesh. To introduce the vertical crack in a systematic way, we first double the nodes at the crack location and then all the degrees of freedom (dof) for these nodes. The continuity of stresses and displacements in the non-cracked layers is ensured by first penalty terms. The nil values of $\widehat{N}^3, \widehat{M}^3, \widehat{Q}^3$ in layer 3 at the crack edge (Fig. 1) are accounted for by other penalty terms. The nodal values of the interfacial stresses $\hat{\tau}^{2,3}$, $\hat{\tau}^{3,4}$, $\hat{\nu}^{2,3}$, $\hat{\nu}^{3,4}$ as well as the displacements of the cracked layer $\widehat{\mathrm{U}}^3$, $\widehat{\Phi}^3$, $\widehat{\mathrm{V}}^3$ are left free without any special treatment and thus can reflect discontinuities. By doing so the Lagrangian of Eq. 1 becomes an augmented Lagrangian leading to some additional coefficients with large values in the final algebraic system.

4 Results and Validation

We compared the mixed FEM described in this paper with the FDM defined in (Nasser and Chabot 2015) on the 2D composite pavement example presented in Sect. 3 with a vertical crack across layer 3 (Fig. 1). The structure is discretized into 160 finite elements of the same length (0.0375 m). Figure 3 shows the good agreement obtained between the two methods. We can also note on these graphs the discontinuities induced by the crack (Fig. 3b). Moreover, as illustrated in Fig. 3c–d, the M4-5nW leads to finite values of the shear and the normal stress at interfaces and near cracks. These values could be used directly to evaluate the risk of a debonding phenomenon in a mixed mode fracture condition.

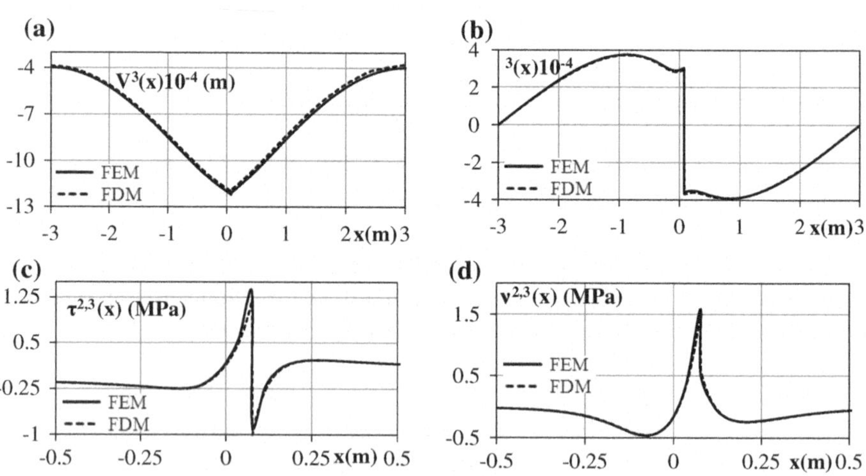

Fig. 3 Comparison between the mixed FEM and FDM: **a** deflection $V^3(x)$; **b** rotation $\Phi^3(x)$; **c** interfacial shear stress $\tau^{2,3}(x)$; **d** interfacial normal stress $\nu^{2,3}(x)$

5 Conclusion

This paper shows the advantages of the mixed variational H-R principle applied to M4-5n in terms of generalized stresses and displacements. This leads to a compact writing of the problem, making it possible to derive systematic numerical solving methods which can be extended in particular to plate geometry domains with cracks and partial delamination areas between two layers. The implementation of this extension will be our next step in order to apply the M4-5n to the modeling of 3D pavement structures (27n-6 unknowns for the plate geometry). The numerical developments will be based on the use of specialized open source software.

References

Chabot A, Tran QD, Ehrlacher A (2005) A simplified modeling for cracked pavements, in BLPC, ISSN 1269-1496, (258-259): 105-120, 2005

Chabot A, Hun M, Hammoum F (2013) Mechanical analysis of a mixed mode debonding test for composite pavements. Construction and Building Materials 40:1076–1087

Nasser H, Chabot A (2015) Two-Dimensional Software for Analysing Mechanical Fields in Elastic Cracked Pavements, in J. Kruis, Y. Tsompanakis, B.H.V. Topping. CC2015 Stirlingshire, UK, Paper 214, 1-4 september 2015. doi:10.4203/ccp.108.214

Pommier S, Gravouil A, Moës N, Combescure A (2009) La simulation numérique de la propagation des fissures: milieux tridimensionnels, fonctions de niveau, éléments finis étendus et critères énergétiques. Edition Hermes – Lavoisier

Reissner E (1950) On a Variational Theorem in Elasticity. J Math. Phys., 29: 90-95

Part VII
Cracking in Asphalt Materials: Pavement Fatigue Performance

A Comprehensive Study About Stresses in Upside-Down Pavements

Glauco Tulio Pessa Fabbri and Ana Paula Furlan

Abstract This paper presents a study about the effect of thicknesses and resilient moduli of upside-down pavement layers on the stresses that occur on surface course. For this, a full factorial experiment was developed considering three layers upside-down pavement structures with thicknesses and resilient moduli variations, generating 20,160 experimental conditions, whose elastic behaviors were simulated using Elsym5 software. Factorial statistical analysis was used to evaluate the results and another simulation set was made evaluating variable effect separately. The results indicated that there is increase on tensile stresses on surface course when (i) there is decrease of the surface course thickness or increase of granular layer thickness; (ii) an increase of resilient modulus of surface course or a decrease of resilient modulus of granular base course. Cemented base course thicknesses, resilient modulus and the subgrade modulus have no or slight influence on surface coarse induced stresses.

Keywords Upside-down pavement · Cracking · Simulations · Tensile stress

1 Background

Cemented materials have been used as pavement layers to deal with the problems associated to the increasing of traffic volume. A higher strength of cemented materials leads to smaller layer thicknesses in the pavement and a better stress distribution on the top of subgrade. However, cracking is an inherent characteristic

---

G.T.P. Fabbri · A.P. Furlan (✉)
Department of Transportation Engineering, Engineering School of Sao Carlos, University of Sao Paulo (STT-EESC-USP), Av. São-Carlense, 400, São Carlos, São Carlos, SP CEP 13560-970, Brazil
e-mail: afurlan@sc.usp.br

G.T.P. Fabbri
e-mail: glauco@sc.usp.br

A. Chabot et al. (eds.), *8th RILEM International Conference on Mechanisms of Cracking and Debonding in Pavements*, RILEM Bookseries 13,
DOI 10.1007/978-94-024-0867-6_52

of the cemented base courses due to shrinkage (Yoder e Witczak, 1975; Ceratti 1991; Adaska and Luhr 2004).

Materials stabilized with Portland cement are susceptible to cracking in function of the cement curing and hydration processes and when cracks connect themselves result in block cracking which can reflect in the asphalt surface into approximately rectangular shape.

To avoid the premature failure of the asphalt mixes due to shrinkage cracking reflection, one alternative is to add a granular layer between the surface course and cement-treated base (Yoder e Witczak 1975; Huang 1993; Adaska and Luhr 2004). Upside-down pavement is one of the design solutions into stress relief category whose purpose is to prevent the cracking propagation from cemented base to asphalt surface. However, unbound granular material increases the vertical deflection of granular base course and increase the tensile stress on the asphalt surface (Adaska and Luhr 2004).

Rasoulian et al. (2000) have evaluated distress and surface conditions of Louisianna low-traffic pavements. It was observed that a certain pavement composed by 220 mm soil-cement base-course and 90 mm hot mix asphalt (HMA) surface course presented higher distress incidence and more severe than another one built up 170 mm soil-cement subbase 100 mm granular base, 90 mm HMA. These authors concluded that a well-designed granular interlayer course increased pavement life cycle in approximately five times.

Granular layers exhibit low stiffness and, under repeated loading, induce tensile stresses at the bottom of the surface layer. Therefore, in upside-down pavement, fatigue can be a deterioration mechanism due to the alternate flexion of the HMA surface course. This is a problem since the design solution used to prevent the reflection cracking can induce fatigue cracking.

Ren et al. (2009) have studied typical upside-down pavements and found that the cemented base course resilient modulus effect on the induced tensile stress to surface course is not significant. In the other hand, the granular base course resilient modulus effect on the induced stress was quite important, i.e.: as granular base course modulus increases, the induced tensile stress on the bottom of surface course decreases. These researchers depicted that granular base and subbase moduli can guide design and construction solutions of the upside-down pavements.

It is important to understand what thicknesses and/or stiffness granular layers can be used without promoting fatigue. Therefore, to know the tensile stress in base-surface interface can help to reach thickness and stiffness combinations less detrimental to this type of pavements.

2 Experimental Program

A full factorial experiment was developed considering three-layer inverted pavements and load characteristics as is presented in Fig. 1a and b. The input load data were dual wheel assembly of 20 kN and tire inflation pressure equals to 0.56 MPa.

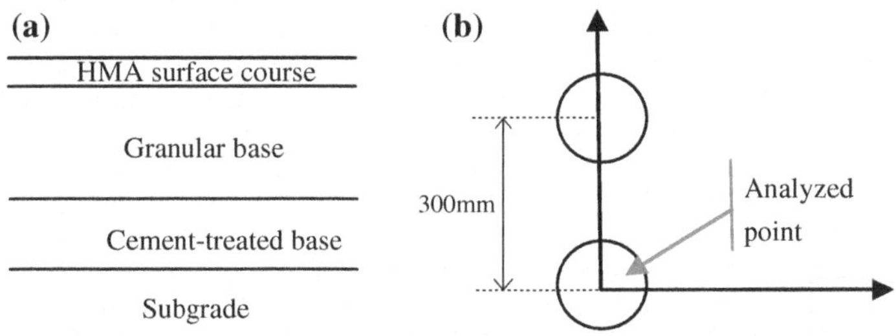

Fig. 1 Typical upside-down pavement structure (**a**) and load applied scheme (**b**)

Elsym5 software was used to compute structural responses. Since stresses and strains depend on the pavement structure configuration, different pavements were tested considering different thicknesses and resilient moduli. The material characteristics were adopted in order to include typical values, resulting in an experimental program with seven factors and different levels:

- factor 1: surface course thicknesses (T_{SC} = 30, 50, 70, 90, 110, 130 and 150 mm)
- factor 2: surface course RM (RM_{SC} = 2000, 5000 and 10,000 MPa);
- factor 3: granular base course thickness (T_{GBC} = 50, 100, 150 and 200 mm);
- factor 4: granular base course RM (RM_{GBC} = 150, 300 and 450 MPa);
- factor 5: cemented-treated course thickness (T_{CTBC} = 100, 150, 200 and 250 mm);
- factor 6: cemented-treated RM (RM_{CTBC} = 1000, 5000, 9000 and 13,000 MPa);
- factor 7: subgrade RM ($RM_{Subgrade}$ = 50, 150, 250 and 350 MPa).

Poisson coefficient adopted values were 0.30, 0.35, 0.20 and 0.40, for the surface course, granular base course, cement-treated base layer and subgrade, respectively.

3 Results and Findings

3.1 Effect of Variables in Mean Terms

Figure 2 presents the main effect graphs, in mean terms, of all experiment factors on the stress in X-direction. It can be noted that all factors influenced on the stress (σ_{XX}) in different intensity, except the variable related to subgrade, where the effect is practically null. It is important to remember that in this type of analysis, the values represent the mean of all parameter's values (stress or strain in X-direction) in a particular condition; for instance, a surface course thickness of 50 mm has

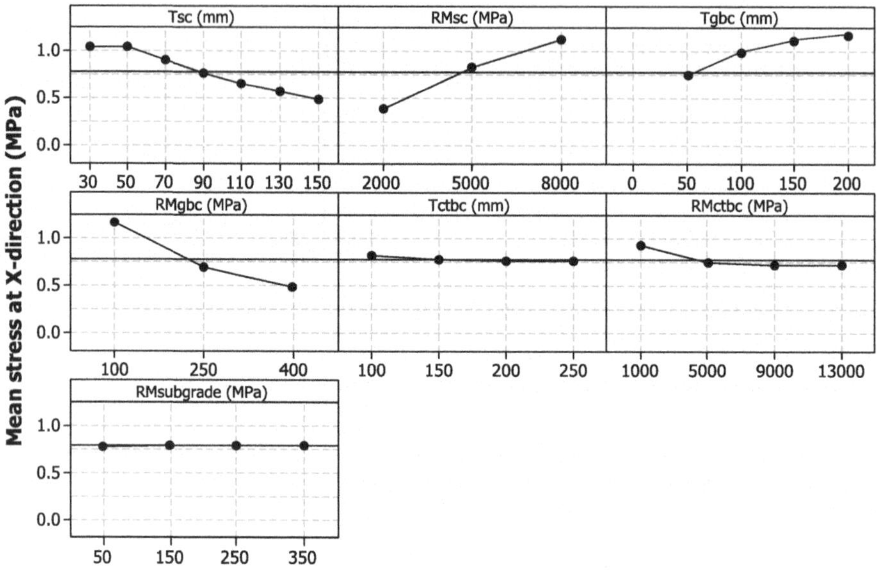

Fig. 2 Effects of thickness and resilient modulus on mean stress at X-direction

2880 experimental conditions (3 RM_{SC} × 5 E_{GBC} × 3 RM_{GBC} × 4 E_{CTBC} × 4 RM_{CTBC} × 4 $RM_{Subgrade}$ = 2880 conditions).

Graphs in Fig. 2 indicate that the factors which produced higher effect intensity in the stress were those related to surface course and to granular base course. Considering all the pavement structure combinations, the mean stress value in surface course was 0.80 MPa.

For the surface course, both thickness and resilient modulus were important in the stress values and the stress variation interval was 0.7 MPa. For thinner surface course (40 mm), the maximum mean stress value was 1.00 MPa. For more rigid asphalt mixes (RM_{SC} = 8000 MPa) the stress reaches 1.10 MPa.

Observing the factors related to granular course it can be depicted that both thickness and resilient modulus produced important changes in the stress values but in opposite sense, i.e., the increase of granular base course thickness leads to stress increase in surface course, while the increase of granular base course stiffness reduces it. The effect intensity of these factors is significant and produced a stress variation range of 1.3 MPa. The granular base course thickness effect is more pronounced than the stiffness effect on the surface course stress.

In relation to the effects of the cement-treated base course and the subgrade, it was noted that they did not influence significantly the stress values on surface course.

3.2 *Effects of Isolated Variable*

Figures 3 and 4 are presented to clarify the isolated effects of the variables. Figure 3 presents the stress in X-direction versus surface course thickness of pavement structures with characteristics as follow: RM_{SC} = 2000 MPa; RM_{GBC} = 100, 150 e 200 MPa (b); and T_{GBC} = 150 and 300 mm. Observing Fig. 3 it can be noted that the curves shift within a particular range, depending on RM values. For T_{GBC} = 150 mm, stresses varies from 0.50 to 1.10 MPa and T_{GBC} = 300 mm the range is from 0 to 0.50 MPa.

The effect of the resilient modulus variation can be noted by the curves proximity, indicating that the effect of the granular base course thickness variation is higher when compared to the effect of the granular base resilient modulus variation. In general, higher induced stresses are observed in combination with lower resilient modulus and lower thicknesses.

Figure 4 presents curves of stress in the X-direction of pavements composed by hypothetical stiffer asphalt mixes with RM_{SC} of 5000 and 10,000 MPa and granular base coarse with MR_{GBC} of 150, 300 and 450 MPa and constant thickness.

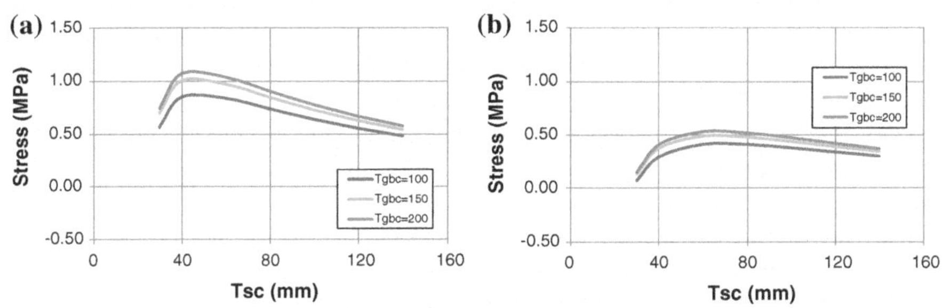

Fig. 3 T_{GBC} e RM_{GBC} influence in X-direction stress (RM_{SC} = 2000 MPa). **a** T_{GBC} = 150 mm. **b** T_{GBC} = 300 mm

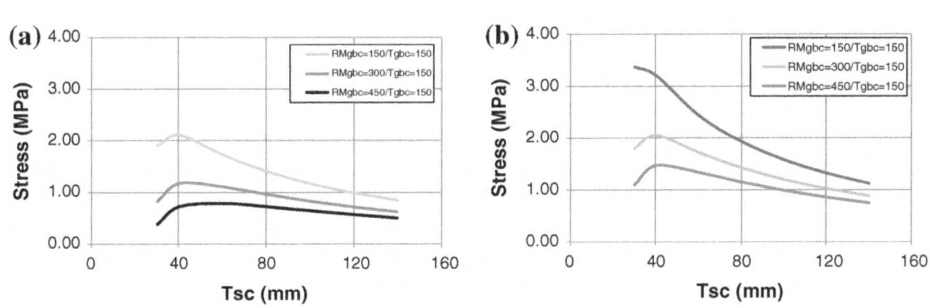

Fig. 4 Influence of RM_{GBC} (variable) and T_{GBC} (constant) in X direction stress. **a** RM_{SC} = 5000 MPa. **b** RM_{SC} = 10000 MPa

In general, Fig. 4a and b shows similar findings observed in structures with lower RM_{SC}, thus, once again, it can be noticed that the tendency of stress increasing due to the RM_{GBC} reduction, and the presence of the stress peak associated to surface course thickness equals to 40 mm, excepting for the case RM_{GBC} = 150 MPa and MR_{SC} = 10,000 MPa, where the peak is not evident.

For RM_{SC} = 5000 MPa, the stress peaks were 0.75, 1.70 and 2.20 MPa for RM_{GBC} of 150, 300 e 450 MPa, respectively; while to RM_{SC} = 10,000 MPa, the stress peaks were 1.50, 2.00 and 3.50 MPa for RM_{GBC} of 150, 300 e 450 MPa.

Both graphs demonstrate that induced stress depends on stiffness of granular base course, and that the curves convergence shows the reduction of the induced stress as the surface course increases.

Figure 4a and b exhibits absolute stress values from 1.50 to 3.30 MPa; and this is a concerning question because, actually, TS values of HMA do not reach such high values and hardly ever it could also be met due to the temperature effect.

4 Conclusions

Results indicated that in upside-down pavement, the factors which promote the induction as well as the intensity of the tensile stress in HMA are related to the thickness and to the surface and granular course stiffnesses, but the stiffness effect is more pronounced than that caused by thickness.

Thus, thicker and stiffer surface courses in upside-down pavements with granular layer (even in small thickness) may lead to highly unfavorable combination that can contribute to premature fatigue of the surface course.

The variation of cemented course and subgrade factors did not produce significant effects in the surface coarse stress values.

It is important to observe that all the research's findings based on simulation and in service pavements can present significantly different elastic characteristics than the adopted ones. Nevertheless, the upside-down pavement option will imply always to the occurrence of induced stresses in the surface course independently of the adopted configuration, and, this fact have to be considered in the design phase.

References

Adaska, W.S., Luhr, D.R. (2004). Control of reflective cracking in cement stabilized pavements. 5th International Rilem Conference, Limoges, France.

Ceratti, J. A. P. (1991). Estudo do comportamento a fadiga de solos estabilizados com cimento para utilização em pavimentos. PhD. Universidade Federal do Rio de Janeiro, Rio de Janeiro.

Huang, Y. H (1993). Pavement analysis and design. Englewood Cliffs. Prentice Hall.

Rasoulian M.; Becnel B. and Keel, G. (2000). Stone interlayer pavement design. In Transportation Research Record: Journal of the Transportation Research Board, No. 1709. Transportation Research Board of the National Academies, Washington, D.C., pp. 60-68.

Ren, R.; Hiwen, L.; Wang, Z. (2009). Theoretical analysis of design parameters on semi-rigid asphalt pavement with flexible base. Critical Issues in Transportation Systems Planning, Development and Management. ASCE, p.2381-2386.

Yoder, E. and Witczak, M. *Principles of pavement design* (1975) 2a. Edição: John Willey & Sons, New York.

Developing an Indicator for Fatigue Cracking in Hot Mix Asphalt Pavements Using Viscoelastic Continuum Damage Principles

David J. Mensching, Jo Sias Daniel and B. Shane Underwood

Abstract Fatigue cracking in asphalt pavements results in decreased ride quality, decreased fuel economy, and provides an avenue for intrusion of water. Design of the mixture is predominantly focused on volumetrically balancing the amount of air, asphalt, and aggregate in the system. Such an approach only implicitly considers performance and so the natural progression is to design through direct material property assessment and correlations to field performance. The objectives of this paper are to relate mixture stiffness, fatigue, and pavement system characteristics together for use in performance-based mixture design; identify a Simplified-Viscoelastic Continuum Damage model output parameter which produces the most separation between poorly and satisfactorily performing structures when combined with dynamic modulus and phase angle information; and evaluate the impact of reclaimed asphalt pavement on the performance of the indicator. Results show a relationship between fatigue life of the pavement system and an energy-based index. This approach holds promise because of its reliance on material attributes that can be derived on one testing machine. The constitutive model parameters can be found from the direct tension cyclic fatigue test and can be incorporated into prediction software, further enhancing the appeal of a performance specification.

Keywords Asphalt mixture · Fatigue · Viscoelastic · Continuum damage

D.J. Mensching (✉)
National Research Council Research Associate, Federal Highway Administration, Office of Infrastructure Research and Development, Mclean, VA, USA
e-mail: dave.mensching.ctr@dot.gov

J.S. Daniel
Civil and Environmental Engineering, University of New Hampshire, Durham, NH, USA
e-mail: jo.daniel@unh.edu

B. Shane Underwood
Civil, Environmental, and Sustainable Engineering, Arizona State University, Tempe, AZ, USA
e-mail: shane.underwood@asu.edu

A. Chabot et al. (eds.), *8th RILEM International Conference on Mechanisms of Cracking and Debonding in Pavements*, RILEM Bookseries 13,
DOI 10.1007/978-94-024-0867-6_53

1 Introduction

In asphalt pavements, design of the mixture is predominantly focused on volumetrically balancing the amount of air, asphalt, and aggregate in the system. However, such an approach only implicitly considers performance and so the natural progression is to design through direct material property assessment and correlations to field performance. This material property-driven performance-based methodology can then be applied to production and acceptance, which enhances its ability to act as a pay factor specification for agencies. This approach holds potential for increased flexibility from a contractor perspective, as producers can focus on meeting the thresholds set by material property-performance relationships as opposed to many volumetric boundary conditions. This is promising when considering the opportunity to add more recycled and resource-conscious materials to mixtures.

One scheme that shows potential for a performance-based application is the Viscoelastic Continuum Damage (VECD) model. This constitutive model relates material integrity to an internal state variable of damage and is well-featured in the literature (Kim and Little 1990; Lee and Kim 1998; Daniel and Kim 2002; Chehab et al. 2002). Recently, the calculation and testing procedure has been adjusted for use in the Asphalt Mixture Performance Tester (AMPT) and is known as the Simplified-Viscoelastic Continuum Damage model (S-VECD) (Underwood et al. 2010). The AMPT has been identified as a potential Standard Performance Tester for hot mix asphalt and there should be value in coupling this piece of equipment with performance-based methodologies for other major distresses.

Due to the mechanisms governing fatigue cracking, it is critical that the pavement system as a whole be accounted for in any performance rankings or assessments. Linear elastic (LEA) structural response models have been developed to determine system characteristics, such as the WinJULEA platform used in the Mechanistic-Empirical Pavement Design Guide (ARA, Inc. 2004). A combination of asphalt mixture and pavement system characteristics should allow for a more accurate performance-based approach to design and construction and increase communication between pavement designers and mixture designers. Additionally, if the methodology is simplistic and inexpensive in the long-term, it brings with it a higher probability of being implemented by owner agencies.

2 Objectives

The primary objectives are to: (1) relate mixture stiffness, fatigue, and pavement system characteristics together for use in performance-based mixture design; (2) identify an S-VECD output parameter which adequately differentiates among poorly performing structures and satisfactory performing structures when combined with dynamic modulus (|E*|) information; and (3) evaluate the impact of reclaimed asphalt pavement (RAP) on the indicators.

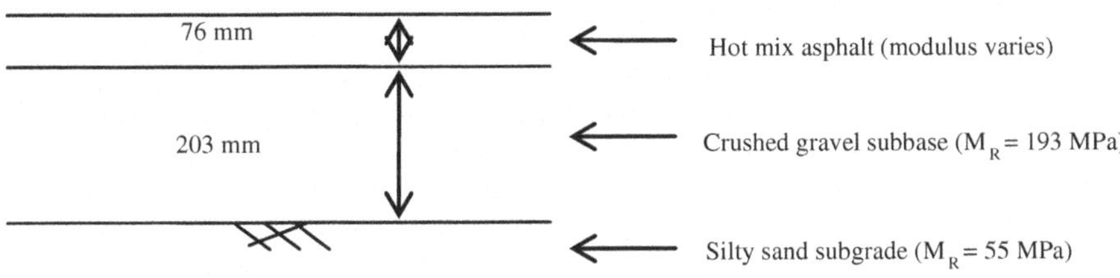

Fig. 1 Simulated pavement cross-section

3 Materials and Methods

The materials used in this paper are part of a recent study to evaluate high percentage RAP mixes. The mixes come from three different states: New York (NY), New Hampshire (NH), and Vermont (VT). The naming convention for the materials is simply "state-binder grade-RAP content by weight of mixture". Details on the materials can be found in Mensching et al. (2014). In this study, |E*| and direct tension cyclic testing (at 25 °C) was conducted (American Association of State Highway and Transportation Officials (AASHTO) 2012).

To represent a typical pavement structure in the northeast United States, a Long Term Pavement Performance project cross-section from New York was selected to model base and subgrade layers. This project was selected due in part to the availability of resilient modulus data for input into WinJULEA and its relatively simple structure. Figure 1 details the simulated pavement structure.

The sub-layering and asphalt modulus calculation techniques follow the method used in the National Cooperative Highway Research Program Project 1-37A final report. To calculate effective temperature for the study mixtures, climate sites from the Enhanced Integrated Climatic Model near to the production plant were used (ARA, Inc. 2004). The effective temperature method stems from Odemark's transformation method and is related to the critical depth and weather details (i.e., percent annual sunshine). For S-VECD analysis, the relaxation modulus (E(t)), damage relationship, and failure criterion were found as detailed in AASHTO TP 107 and Sabouri and Kim (2014) (AASHTO 2014). The loading mode-independent failure criterion, G^R, is the average released pseudostrain energy per cycle and is calculated from the experimental data and the work potential theory in the VECD formulation.

4 Results

The mixture damage characteristic curves (DCCs) explain the relationship between a quantity representing material integrity (C) with an internal state variable, damage (S). From the experiments, the NY mixture curves are above the others. A curve with a higher C value at a given S represents a better degree of material integrity. However, it is not optimal to assign objective mixture resistance ratings based only on the DCC because of the lack of a failure criterion at this stage of the analysis, the undefined physical meaning of the damage internal state variable, and absence of the controlling mechanism (stress or strain) (Mensching 2015).

To determine parameters for a performance-based mixture design framework, three elements must be considered for the fatigue distress mode: (1) a value from the S-VECD model since this type of testing would be prescribed; (2) a value that captures mixture stiffness; and (3) a value that relates to components of the pavement system. Given the approach described above, the mixture stiffness and pavement system can be described by the critical microstrain (measured at bottom of asphalt layer under the center of the single-tire loading point) from LEA analysis. For thin pavements, it is expected that the material's flexibility or relaxation capabilities will influence the response under loading rather than strictly stiffness. The predicted strains, along with the predicted cycles to failure based on the traditional fatigue laws, are shown in Table 1. The strain level does not directly correlate to the rank or the magnitude of the cycles to failure in controlled strain mode ($N_{f,\ strain}$). This reinforces the need for a fatigue-related parameter.

As mentioned, G^R classifies fatigue resistance through a combination of stiffness and energy dissipation principles. Readers should note that a higher G^R comes with the supposition of better performance. Generally, the VT-5234 materials exhibit the highest G^R for a particular N_f, with the NH-6428-00 mixture dissipating the least

Table 1 Tensile strain, N_f at G^R = 100 J/m^3 values, and cycles to failure at bottom of asphalt layer

Mixture	Critical microstrain	N_f at G^R = 100 J/m^3	Cycles to failure ($N_{f,\ strain}$)	Rank by cycles to failure
NH-6428-00	426	11,838	59,767	10
NY-5828-30	389	16,491	62,201	9
NY-6422-00	364	19,014	101,512	6
NY-6422-20	365	14,203	21,376	14
NY-6422-30	367	18,061	51,320	11
NY-6422-40	358	25,248	222,100	3
VT-5234-00	486	18,224	433,867	2
VT-5234-20	453	20,302	121,393	5
VT-5234-30	442	25,101	818,948	1
VT-5234-40	446	19,171	191,919	4
VT-6428-00	411	17,776	48,608	12
VT-6428-20	414	19,366	94,178	7
VT-6428-30	404	18,818	68,340	8
VT-6428-40	408	14,085	26,298	13

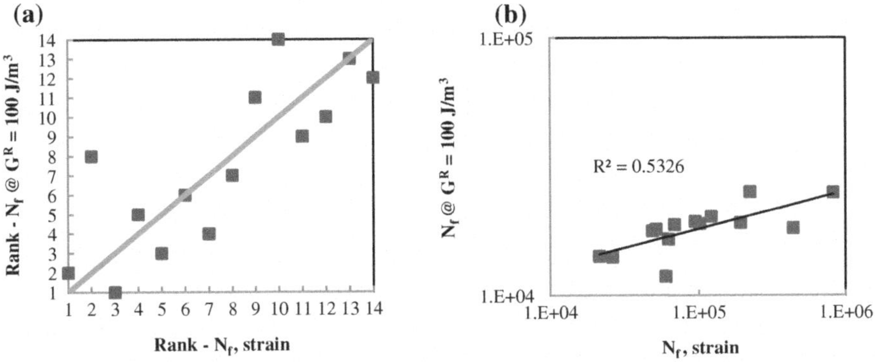

Fig. 2 Energy-based index versus fatigue life by (**a**) rank and (**b**) numerical values

amount of energy in the N_f range shown. The N_f at G^R = 100 J/m^3 index $\left(N_{G^R_{100}}\right)$ is used to separate the mixture performance from the predicted performance of the pavement structure (which is based on $N_{f,\ strain}$).

The next step is to determine if there is a correlation between the proposed parameter to capture fatigue effects and the parameter to capture material property, stiffness, and structural effects ($N_{f,\ strain}$). Figure 2 shows the expected trend that as $N_{G^R_{100}}$ increases (rank goes to 1), more energy is dissipated per loading cycle and the pavement's overall fatigue life should increase (rank goes to 1). Despite a modest correlation among the numerical data values, the general trend is confirmed that the higher $N_{G^R_{100}}$ leads to a higher $N_{f,\ strain}$. The correlation among the rank of the parameter with respect to the fatigue life rank is also captured through Kendall's tau coefficient (τ_k), which varies from −1 (complete disagreement) to 1 (complete agreement), with a value of zero corresponding to no relationship. The τ_k is 0.626, a significant agreement (at α = 0.05) with fatigue life.

Based on the results above, the $N_{G^R_{100}}$ term shows promise as it relates to pavement fatigue life in controlled strain mode over a spread of nearly 800,000 cycles to failure from the worst performing to best performing mixture. Other parameters were compared to pavement system fatigue life, including a C at failure criterion, a model term relating to the E(t) master curve, and |E*|-based terms to quantify energy dissipation through combination with phase angle, with the $\left(N_{G^R_{100}}\right)$ term exhibiting the best relationship with the overall pavement system performance.

5 Conclusions

The research conducted in this paper is intended to be a first step in the process of defining an index for the identification of poorly performing fatigue mixtures. The authors provide the following conclusions:

- When evaluating the DCC and G^R failure criterion, there was not a clear trend between RAP increases and a change in fatigue performance of the mixture. Generally, the stiffer mixtures produced less tensile strain at the bottom of the asphalt layer, but this was not necessarily a function of RAP percentage.
- The G^R term was used to rank performance. The G^R was set to 100 J/m^3 and the predicted cycles to failure was selected as a measure of the mixture's fatigue performance. When compared to the N_f prediction for the pavement system there was a degree of separation and a general trend stating that an increase in dissipated energy results in an increase in fatigue life.
- This approach in general holds promise because of its reliance on material attributes that can be derived from one testing machine. While the S-VECD test is time-consuming, the constitutive model parameters that can be found are powerful and applicable to fatigue and low temperature performance.

In closing, the implications of this work for the industry are evident when considering general fatigue analyses in asphalt materials research. Upcoming research may help to reinforce the power behind the S-VECD model for cracking characterization, which will allow for more accurate performance prediction.

Acknowledgements The authors thank the project sponsors: New Hampshire, Maryland, New Jersey, New York, Pennsylvania, Rhode Island, and Virginia Departments of Transportation, and the Federal Highway Administration. Thanks are given to Callanan and Pike Industries for supplying materials and Thomas Bennert of Rutgers University and Michael Elwardany of North Carolina State University for conducting the |E*| and S-VECD fatigue testing, respectively.

References

AASHTO (2012) Standard method of test for determining the dynamic modulus and flow number for hot mix asphalt using the Asphalt Mixture Performance Tester. AASHTO TP 79, Washington, D.C.

AASHTO (2014) Determining the damage characteristic curve of asphalt concrete from direct tension cyclic fatigue tests. AASHTO TP 107, Washington, D.C.

ARA, Inc. (2004) Guide for mechanistic-empirical design of new and rehabilitated pavement structures. Proj 1-37A, National Cooperative Highway Research Program, Washington, D.C.

Chehab GR, Kim YR, Schapery RA, Witczak MW, Bonaquist R (2002) Time-temperature superposition principle for asphalt concrete with growing damage in tension state. J Assoc of Asphalt Paving Technologists, 71 559-593

Daniel JS, Kim YR (2002) Development of a simplified fatigue test and analysis procedure using a viscoelastic damage model. J Assoc of Asphalt Paving Technologists, 71 619-650

Lee HJ, Kim YR (1998) Viscoelastic constitutive model for asphalt concrete under cyclic loading. J Eng Mech, 124:1 32-40

Kim YR, Little DN (1990) One-dimensional constitutive modeling of asphalt concrete. J of Eng Mech, 116:4 751-772

Mensching DJ, Daniel JS, Bennert T, Medeiros, Jr. MS, Elwardany MD, Mogawer W, Hajj EY, Alavi MZ (2014) Low temperature properties of plant-produced RAP mixtures in the Northeast. J Road Materials and Pavement Design, 15 Supp 1 1-27

Sabouri M, Kim YR (2014) Development of failure criterion for asphalt mixtures under different modes of fatigue loading. In press, Transportation Research Record

Underwood BS, Kim YR, Guddati MN (2010) Improved calculation method of damage parameter in Viscoelastic Continuum Damage model. Intl J of Pavement Eng, 10:6 459-476

Effect of Heavy Traffic Loading on Predicted Pavement Fatigue Life

Yared H. Dinegdae and Björn Birgisson

Abstract Fatigue cracking is one of the dominant failure modes of asphalt concrete pavements. There are a number of analysis and design methods that can be used to optimize pavement sections for this kind of distress. Most of these methods incorporate advanced material property predictive models. However, traffic loading, which has been identified as a primary contributing factor in causing fatigue cracking, is characterized relatively simplistically. There is a concern in light of recent advancement in traffic characterization, and tire inflation pressure surveys that existing methods might not be adequate. The objective of this paper is to evaluate and quantify the effects of truck traffic characterization in axle load spectra and high tire inflation pressure levels on predicted fatigue cracking performance. This was achieved by evaluating a number of pavement sections using the mechanics-based fatigue cracking analysis framework. The studied traffic characterization approaches are ESALs, axle load spectra with and without traffic seasonal variations and three levels of tire inflation pressures. It is evident from the result that higher tire inflation pressure and traffic characterization using axle load spectra induce more damage and subsequently early crack initiation time.

Keywords Mechanics-based · Fatigue · Traffic · Tire-pressure

Y.H. Dinegdae (✉)
Civil and Architectural Engineering, KTH, SE-100 44 Stockholm, Sweden
e-mail: dinegdae@kth.se

B. Birgisson
School of Engineering and Applied Science, Aston University, B4-7ET, Birmingham, UK
e-mail: bjorn.birgisson@aston.ac.uk

A. Chabot et al. (eds.), *8th RILEM International Conference on Mechanisms of Cracking and Debonding in Pavements*, RILEM Bookseries 13,
DOI 10.1007/978-94-024-0867-6_54

1 Introduction

Top-down fatigue cracking (i.e., cracking that initiates at the surface of the asphalt concrete layer (AC) and propagates downwards) is one of the dominant failure modes in flexible pavements. Wang et al. (2006) consider factors such as traffic induced lateral or horizontal stress, near surface age hardening, temperature induced stiffness-gradient, low fracture resistance and soft unbound layers as primary factors that contribute for the onset and propagation of top-down fatigue cracking. There are a number of mechanistic-empirical (M-E) analysis and design procedures that can be used to evaluate pavement sections for this kind of distress (Wang et al. 2006; NCHRP 2010; Dinegdae et al. 2015; Dinegdae and Birgisson 2015). These procedures incorporate advanced material property predictive models which take into account factors such as mixture aging and healing. However, they utilize rather a simplistic traffic characterization in terms of equivalent single axle load (ESAL) for damage evaluation. Characterizing the complex traffic phenomena with a single factor might provide erroneous designs, thus limiting the applicability of these analysis and design procedures.

Instead of aggregating the whole traffic into a single factor, the recently developed Mechanistic-Empirical pavement design guide (MEPDG) recommends expected traffic to be characterized with inputs that require the magnitude, configuration and frequency of axle loads (ARA 2004). By characterizing the traffic in a more precise approach, the MEPDG can analyze systematically the effect of traffic on pavement performance and deliver optimum designs. Moreover, it allows the design engineer to evaluate and assess the effects of seasonal traffic variations and vehicle class composition.

Tire inflation pressure plays a critical role in pavement performance analysis as it controls the interface conditions between the tire and the pavement. Studies have shown that in addition to tire inflation pressure, the type of tire used by trucks also contributes to the corresponding applied pressure distribution (e.g., Chen et al. 1986; Roque et al. 2006). A steady increase in tire inflation pressure has been observed in the last two decades, which is attributed to advancements in tire construction technology. De Beer (1998) reported that tire inflation pressures have increased from 556 kPa in 1960s to approximately 750 kPa in 1995. Another tire inflation pressure survey has shown that over 86 % of axle tire inflation pressure measurements exceed 700 kPa (Morton et al. 2004). Moreover, there has been a shift from bias-ply tires to conventional width radial tires and super single wide-base radial tires. This change in tire technology has further increased tire inflation pressure to 900–965 kPa (Al-Qadi et al. 2004).

In this study, the effects of various tire inflation pressure levels were investigated in conjunction with truck traffic characterization approaches. The traffic characterization approaches which were selected for this study are ESALs and axle load spectra with and without traffic seasonal variations. Field pavement sections that have a wide range in design inputs and quality laboratory and field data were analyzed using a recently developed mechanics-based fatigue cracking analysis framework. The significance of each traffic characterization parameter was established by comparing the fatigue cracking initiation time of the studied pavements.

2 Truck Traffic Characterization

The two most commonly used traffic load characterization approaches are ESALs and axle load spectra. ESAL for a given axle represents the relative damage induced in comparison with a standard single axle with dual wheel. It has its origin in the 1950s American Association of State Highway Officials (AASHO) road test and has been implemented since in many M-E design procedures (AASHTO 1993). Axle load spectra, on the contrary, represent for a specified axle type (single, tandem, tridem and quad) and vehicle class, the percentage of total axle load applications within a given load interval. According to Swan et al. (2008), the axle load spectra and the number and types of trucks are the most significant factors affecting pavement performance. The MEPDG recommends three levels of traffic inputs which are based on the accuracy of traffic data. Level 1 and 2 inputs are developed based on site and regional Weigh in Motion (WIM) measurements, whereas Level 3 is developed based on national WIM data.

Most pavement design procedures evaluate pavement performance by assuming a uniform vertical pressure distribution over a circular contact area. The magnitude of this contact pressure is assumed to be equivalent to the tire inflation pressure level.

For this study, the mixed traffic stream was converted into an 80 kN ESAL by using the equivalent axle load factors (EALF) provided by Huang (2004). The principle of superposition was used to convert tandem and tridem axles into single axle. For the load spectra characterization, the MEPDG default level 3 inputs were used. These inputs include axle load spectra, the number of axle per truck (NAPT) and traffic seasonal variation factors. For vehicle class distribution, TTC group 3 was used to determine the composition of expected traffic. Tire inflation pressure levels included in the study were 689, 800 and 900 kPa.

3 Mechanics-Based Analysis and Design Framework

The mechanics-based analysis framework, which is a mechanistic-empirical approach and predicts fatigue cracking initiation (CI) time by employing hot mix asphalt fracture mechanics (HMA-FM) and morphology-based material property predictive models, was used in this study. The calibrated framework has been verified through predicting crack initiation times that are in agreement with observed performances in the field. Pavement performance (PE) according to the mechanics-based analysis and design framework is evaluated by comparing two dissipated creep strain energy-based parameters (Dinegdae et al. 2015).

4 Results and Discussion

Five pavement sections from the State of Florid, USA were used in this study. Table 1 presents a summary of information about these sections. More information regarding input parameters can be obtained from Dinegdae et al. (2015).

The normalized error (NE), which normalizes the difference between two predictions, was used to compare and establish the significance of each characterization:-

$$\mathbf{NE} = \frac{|\mathbf{CI_x} - \mathbf{CI_{BM}}|}{CI_{BM}} \cdot \mathbf{100\ \%} \quad (1)$$

where CI_{BM} predicted CI time from the bench mark cases and CI_X predicted CI time from other traffic characterizations. For the traffic load characterization, the prediction by the ESALs for every pressure condition was taken as a benchmark and the prediction by load spectra with and without seasonal traffic variations were evaluated accordingly. For the tire inflation pressure evaluation, the prediction by 689 kPa was used as a bench mark.

The comparison between load spectra (LS) and load spectra with seasonal traffic variations (LSSV) with respect to ESALs for the three tire inflation pressure levels are presented in Fig. 1a–c. For the same tire inflation pressure level, load spectra with and without seasonal variations induce more damage than ESALs. Furthermore, the results have shown that load spectra with seasonal traffic variation induce slightly more damage than load spectra.

Figure 2a, b compares tire inflation pressure level effect on predicted fatigue performance for the ESALs and load spectra with seasonal traffic variations characterizations. As can be seen, the effect of tire inflation pressures is the same for both traffic characterizations and the 900 kPa pressure has caused more damage in comparison with the 689 kPa pressure.

Table 1 Summary of information of used pavement sections

	County	HAC (cm)	EB (MPa)	Coring year traffic ($ESALs \cdot 10^3$)
TPK2	St. Lucie	15.5	234	166
NW39-2	Alachua	8.4	434	182
SR483	Volusia	8.9	407	145
SR823	Broward	10.2	655	162
US1	Monroe	10.2	276	421

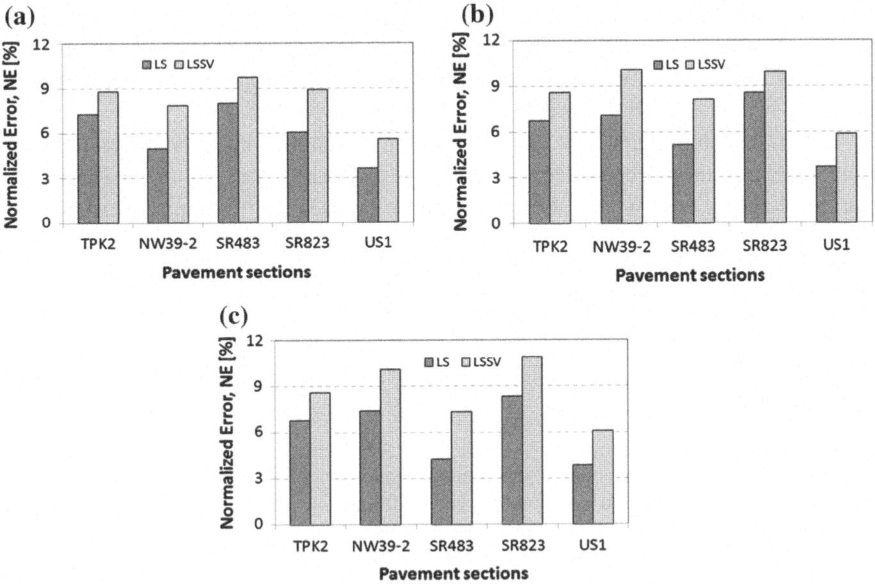

Fig. 1 Normalized error for load spectra with and without traffic seasonal variation for tire inflation pressure levels of **a** 689 kPa, **b** 800 kPa, **c** 900 kPa

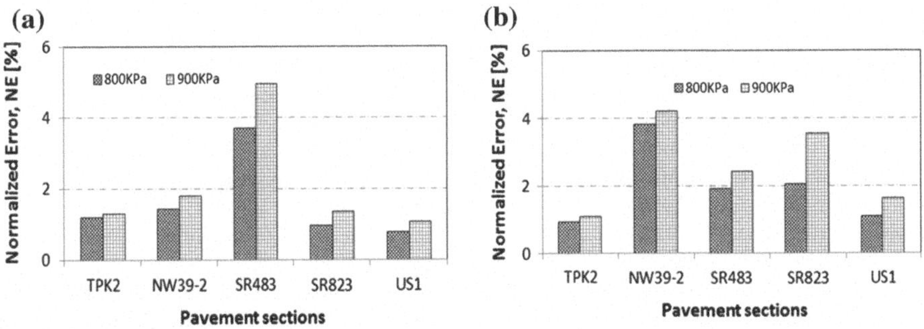

Fig. 2 Normalized error for tire inflation pressure evaluation **a** ESALs, **b** load spectra with seasonal traffic variations

5 Conclusions and Recommendations

This paper presented the effects of traffic load characterization in axle load spectra and tire inflation pressure levels on predicted fatigue cracking performance. It is apparent from the presented results that irrespective of tire inflation pressure levels, load spectra with and without seasonal traffic variations induce more damage and subsequently early crack initiation time. Moreover, the predicted crack initiation time difference between the load spectra with and without traffic seasonal variations was found to be insignificant. The tire inflation pressure analyses have shown that the higher the pressure the more severe the resulting damage would be. This was shown to be true for all the traffic characterizations studied.

Further research that takes into account the non-uniform vertical and lateral contact stresses should be undertaken to fully understand and quantify the influence of traffic loading and tire inflation pressure on predicted fatigue performance.

Acknowledgements The authors would like to thank Trafikverket and SBUF for providing the financial support for the project under which this work was performed.

References

Wang J., Birgisson B., & Roque R. (2006) Implementation of the Florida cracking model into the AASHTO 2002 Mechanistic-empirical pavement design. Gainesville, FL: Department of Civil and Coastal Engineering, University of Florida.

NCHRP Web-Only Document 162 (2010) Top-down cracking of hot-mix asphalt layers: Models for initiation and propagation. Washington, DC: National Cooperative Highway Research Program, Transportation Research Board.

Dinegdae, Y. H., Onifade, I., Jelagin, D., & Birgisson, B. (2015) Mechanics based top-down fatigue cracking initiation prediction framework for asphalt pavements. Journal of Road Materials and Pavement Design doi:10.1080/14680629.2015.1055335

Dinegdae, Y. H., & Birgisson, B. (2015) Reliability-based calibration for a mechanics-based fatigue cracking design procedure. Journal of Road Materials and Pavement Design.doi:10.1080/14680629.2015.1055335

ARA Inc., ERES Consultants Division (2004) Guide for Mechanistic–Empirical Design of New and Rehabilitated Pavement Structures. NCHRP Project 1-37A Final Report, Transportation Research Board of the National Academies, Washington, D.C

Chen et al., (1986) Effects of Truck Tire Contact Pressure Distribution on the Design of Flexible Pavements: A Three-Dimensional Finite Element Approach, Transportation Research Record No. 1095, TRB, pp. 72-86.

Roque, R., Birgisson, B., Drakos, C., Myers, L. & Zhang, Z. (2006)Evaluation of surface-initiated longitudinal wheel path cracking, University of Florida, Gainesville

De Beer M., (1998) Development of the Vehicle - Road Surface Pressure Transducer Array Mark III Stress - In - Motion (SIM) system (VRSPTA Mark III SIM): Contract Report CR-98/047 CSIR, Transportek, Pretoria, South Africa.

Morton, B.S., et al., (2004) The effect of axle load spectra and tyre inflation pressures on standard pavement design method [online] In: Proceedings of the 8th conference on asphalt pavements for Southern Africa (CAPSA'04), 12–16 September, Sun City, South Africa, 15 pp., ISBN: 1-920- 01718-6. Available from: http://asac.csir.co.za/capsa/documents/047.pdf [Accessed on 4 May 2015].

Al-Qadi, I.L., Elseifi, M., & Yoo, P.J. (2004) "Pavement Damage due to Different Tire Tires and Vehicle Configurations," Final Report Submitted to: Michelin Americas Research and Development Corporation

AASHTO (1993) Guide for Design of Pavement Structures, American Association of State Highway Transportation Officials, Washington, D.C

Swan, D.J., Tardif, R., Hajek, J., & Hein, D.K. (2008) Development of regional traffic data for the mechanistic-empirical pavement design guide (Transportation Research Record, No. 2049) (pp. 54–62) Washington, DC: Transportation Research Board

Huang, Y.H., (2004) Pavement analysis and design Upper Saddle River, New Jersey: Pearson Prentice Hall

Effect of Temperature and Traffic Speed on the Asphalt Moduli for Fatigue Cracking and Pavement Structural Design Considerations

Didier Bodin, Olivier Chupin and Erik Denneman

Abstract Asphalt pavements exhibit strongly temperature-dependent viscoelastic behaviour resulting in response to load varying with both temperature and traffic speed. To design against fatigue cracking for structural pavement design applications the linear-elastic material response is typically assumed to simplify the calculation of the critical strains in the pavement structure layers. To be representative the simplified pavement model needs to be defined in such a way that it accurately reflects the effect of both temperature and traffic speed on the critical strains used to compare with the material's tolerable strains. Against this background, this paper presents a method to determine an equivalent asphalt modulus (EAM) for the asphalt layer which represents the effect of temperature and loading speed on the critical tensile strains. The EAM is determined from viscoelastic modelling. Two thick asphalt pavement configurations representative of typical French pavement designs are considered. Results show expected trends of the equivalent asphalt modulus increasing with increasing traffic speed and decreasing with increasing temperature. The application of the asphalt material shift factors allowed building pseudo-master curves for the EAM dataset. Finally, the effect of temperature and pavement structure on the relationship between traffic speed and complex modulus frequency is examined. Results support the use of 10 Hz for 70 km/h at intermediate temperatures currently used for pavement designs in France.

Keywords Asphalt pavement · Temperature · Traffic speed · Viscoelastic modelling

D. Bodin (✉)
ARRB Group Ltd., 500 Burwood Hwy, Vermont South, VIC 3133, Australia
e-mail: didier.bodin@arrb.com.au

E. Denneman
ARRB Group Ltd., 123 Sandgate Rd, Albion, QLD 4010, Australia
e-mail: erik.denneman@arrb.com.au

O. Chupin
MAST, LUNAM Université, IFSTTAR, 44341 Bouguenais, France
e-mail: olivier.chupin@ifsttar.fr

A. Chabot et al. (eds.), *8th RILEM International Conference on Mechanisms of Cracking and Debonding in Pavements*, RILEM Bookseries 13,
DOI 10.1007/978-94-024-0867-6_55

1 Introduction

Both resilient response and fatigue cracking performance of asphalt are affected by in-service pavement temperature. Temperature susceptibility and the viscoelastic nature of bituminous binders affect asphalt pavement behaviour in the spectrum of in-service pavement temperature which can range from 40 °C and above in warm climates to temperatures below zero for winter seasons in cold climates. In this extent of temperature variation the asphalt modulus can change from one to more than 10–30 GPa. To a lesser extent, the loading speed affects pavement response and should be considered in slow speed areas (intersections and inclines) as opposed to highway conditions where traffic speed can range between 70 and 100 km/h.

Pavement design needs to consider relevant asphalt moduli values which reflect the effect of environment and loading conditions in the calculation of the critical strains used for pavement design purposes.

Previous research that has investigated the relationship between the asphalt modulus measured in the laboratory and the modulus values assigned to the asphalt layer in equivalent multilayered elastic analysis includes work by Chupin et al. (2014), Jameson and Hopman (2000). Recently Denneman et al. (2015), Bodin et al. (2015) re-examined the effects of temperature and traffic speed using a comparable framework to evaluate the relationship used in the Australian asphalt pavement thickness design (Austroads 2012). The relationships used to correct the asphalt design modulus for temperature and traffic speed effect appeared appropriate to the mostly common design situations. At the same time other research has used Fourier transformation of viscoelastic strain pulses to evaluate the equivalent asphalt layer modulus from the main frequency (Ulloa et al. 2013). This technique could offer means of linking the asphalt layer modulus with laboratory testing frequency under complex modulus testing.

This paper briefly recalls the methodology developed (Denneman et al. 2015) and presents its application to two typical French thick asphalt pavement configurations. Effect of both temperature and loading speed are examined through a parametric study.

2 Methodology

2.1 Equivalent Asphalt Modulus and Calculation Principle

The equivalent asphalt modulus approach is based on defining an equivalent pavement system where the viscoelastic asphalt layer is replaced by an elastic layer of the same thickness with a modulus which leads to the same calculated peak strain under the same static load as the viscoelastic layer. This modulus has been termed the equivalent asphalt modulus (EAM). The EAM was calculated for each asphalt temperature and traffic speed condition. For a given layer thickness and underlying structure, the equivalent asphalt moduli back-calculation was based on the following steps:

- The peak responses at the bottom of the asphalt layer using the viscoelastic model were calculated at a particular asphalt temperature and traffic speed.
- Elastic pavements with different asphalt moduli were calculated to determine the relationship between the asphalt moduli and the peak strain. The inverse of this relationship was used to calculate the elastic moduli (EAM) which provides the same peak strain as calculated in step 1.

The EAM is used throughout this paper to assist in comparing the behaviour of the viscoelastic model to the results of simplified elastic pavement models. Viscoroute 2.0© (Chabot et al. 2010) software was used to perform the analysis.

2.2 Pavement Configurations Examined

Two thick asphalt pavements were constructed in Viscoroute 2.0© for this paper (Table 1). The predicted pavement responses were calculated following a parametric study defined by asphalt temperatures from 5 to 40 °C assumed constant with depth and traffic speeds from 30 to 110 km/h.

In this analysis the moving load used was equivalent to the French reference load for pavement design defined by one dual-wheel load under 65 kN with a uniform contact stress applied to two circular areas (radius = 125 mm) at 375 mm centre-to-centre spacing.

For the model the wearing course characteristics are not differentiated from the asphalt structural layer. The full asphalt thickness is 260 and 160 mm respectively for structure 1 and 2. The asphalt material under consideration is a typical structural dense graded asphalt—French Grave Bitume (GB) specifications. The material was prepared with pure 35/50 pen grade bitumen and compacted to an air void content around 5 %.

The asphalt complex modulus was measured in the laboratory under flexural conditions at different frequencies (0.1–40 Hz) and temperatures (0–40 °C). To characterise the viscoelastic behaviour of the material, the rheological model based on the complex modulus expression "Eq. 1" was used (Huet 1963; Sayegh 1967).

Table 1 Pavement thickness for the two asphalt pavement configurations

Structure	Material	Structure 1	Structure 2
Wearing course	BB	60 mm	
Asphalt structural layer	GB3	200	100
Granular subbase (Eg = 300 MPa)	GNT	200	300
Subgrade (Es = 120 MPa)	Sand subgrade	Semi-infinite (E = 120 MPa)	

Table 2 Huet-Sayegh's model parameters obtained for the asphalt

Material	E_0 (MPa)	E_∞ (MPa)	k	h	δ	A_0	A_1	A_2
GB3	20	22,980	0.219	0.686	2.121	3.896	−0.378	0.001582

$$E^*(\omega,\theta) = E_0 + \frac{E_\infty - E_0}{1 + (i\omega\tau(\theta))^{-h} + \delta(i\omega\tau(\theta))^{-k}} \quad \text{with} \quad \tau(\theta) = \exp(A_0 + A_1\theta + A_2\theta^2) \tag{1}$$

The model parameters E_0, E_∞, k, h, δ and A_i were fitted to the measured complex modulus data using Viscoanalyse software developed and provided by IFSTTAR and are given in Table 2.

In the modelling, the asphalt, granular and subgrade materials were modelled assuming isotropic behaviour and their Poisson ratios have been set to 0.35. All the layers are assumed fully bonded according to typical pavement design assumptions between asphalt/granular and granular/subgrade layers.

3 Results

Calculated equivalent asphalt modulus are shown in Fig. 1a. Results plotted for the two pavement thicknesses show a decrease of the EAM with an increasing asphalt layer thickness. This result is consistent with the findings of Jameson and Hopman (2000). Similarly to the material master curve construction, Fig. 1b shows the EAM modulus data translated using the complex modulus shift factors $a(T)$. The translated curves of the EAM can be seen to form a single curve similar to a master curve.

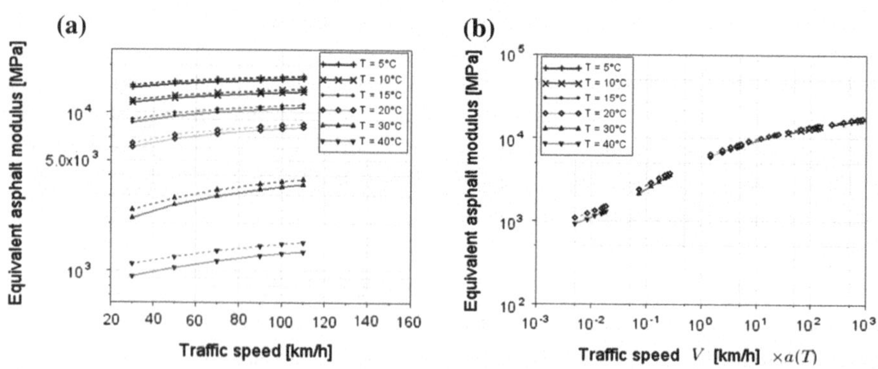

Fig. 1 Calculated equivalent asphalt modulus (EAM) *solid* and *dashed lines* plots for structures 1 and 2 respectively: **a** effect of loading speed and temperature, **b** EAM pseudo-master curve

However the EAM is not a fundamental asphalt characteristic and depending on the pavement composition, the pseudo-master curve property might not translate to more complex pavement configurations.

4 Traffic Speed Versus Laboratory Testing Frequency Relationship

The equivalent asphalt modulus master curve from the viscoelastic rheological model can be used to determine which frequency would give the same modulus value at a given asphalt temperature under complex modulus testing in the laboratory. More details are provided in (Bodin et al. 2015). This relationship was found to be dependent on the asphalt thickness however, some methods assume this phenomenon to compensate for the change in asphalt layer temperature with depth (Jameson and Hopman 2000; Bodin et al. 2015).

Using the Huet-Sayegh rheological model, the frequencies f_{lab} which would give moduli values equating to the EAM are plotted against the traffic speeds examined in the parametric study. The plots in Fig. 1a are valid for both pavement structures under study.

Another finding of the study is that, at intermediate temperatures, the traffic speed versus frequency relationship can be modelled by a linear relationship. This result is consistent with the findings from Ulloa et al. (2013). In practice, when the frequency is known for a given traffic speed it can be easily extrapolated to other speed values. For the two structures examined here at average pavement temperature in France (T = 15 °C) the frequency which would lead to the EAM value in the laboratory is between 8 and 12 Hz for the traffic speed of 70 km/h. Similarly, Chupin et al. (2014) obtained frequencies around 10 Hz for pavement structures with softer subgrade support (E = 50 MPa) and no granular subbase. These simulations confirm the validity of the '10 Hz' rule used in the French pavement design method (Corté and Goux 1996) for a traffic speed of 70 km/h which gives acceptable estimated asphalt moduli for intermediate temperatures (10–30 °C) as shown in Fig. 2b.

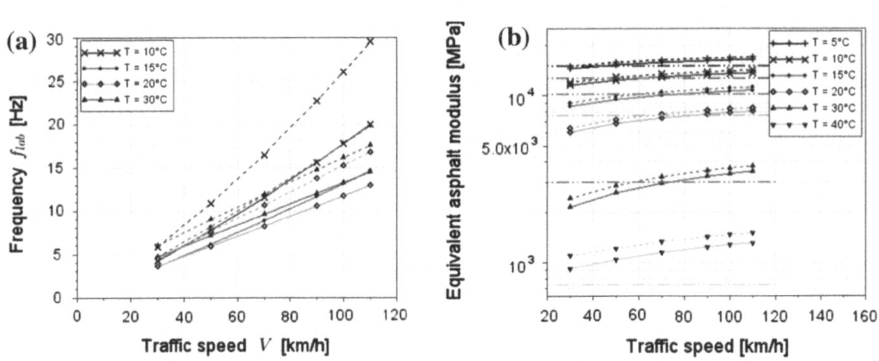

Fig. 2 **a** Traffic speed laboratory frequency relationships: *solid* and *dashed lines* for structures 1 and 2 respectively, **b** EAM compared with material moduli at 10 Hz (*double dot lines*)

5 Conclusion

The paper presents a method to derive asphalt layer moduli and critical tensile strains at the bottom of the asphalt layer as function of traffic speed and pavement temperature from a simplified multilayer elastic pavement response to load model.

The defined equivalent asphalt moduli (EAM) was computed for a large range of traffic speed ($V = 30–110$ km/h) and environmental conditions ($T = 5–40$ °C). Using the asphalt shift factors to traffic speed, the EAM values exhibited a single curve similar to complex modulus master curves.

Finally, the relationship between the traffic speed and the laboratory frequency leading to the EAM values is linear. Calculated frequencies support the '10 Hz rule' used for 70 km/h in the French design approach. These findings can be examined further to assist pavement practitioners with the determination of asphalt design moduli for design against fatigue cracking. It is recommended that the results of this study be validated against pavement loading time response measurements in the field.

References

Austroads (2012) Guide to Pavement Technology - Part 2: Pavement Structural Design, 3rd ed., Sydney: Austroads

Bodin D, Aguiar L, Chupin O, Denneman E (2015) Temperature and Traffic Speed Effects on Asphalt Pavement Response and the Elastic Asphalt Modulus, Paper presented at the 16th AAPA International Flexible Pavements Conference, Gold Coast, Australia, 2015

Chabot A, Chupin O, L. Deloffre L, Duhamel D (2010) Viscoroute 2.0: a tool for the simulation of moving load effects on asphalt pavement. Road Mat. Pav. Design, 11(2): 227-250 doi:10.3166/rmpd.11.227-250

Chupin O, Chabot A, Bodin D, Piau JM (2014) Determination of an equivalent elastic system to a multilayer viscoelastic structure: application to the case of thick flexible pavement. Paper presented at the 12th conference on asphalt pavements, Raleigh, North Carolina, 2014

Corté JF, Goux MT (1996) Design of pavement structures: the French technical guide. Transp. Res. Rec., 1539:116-124

Denneman E, Beecroft A, Bodin D, Griffin J (2015) Improved design procedures for asphalt pavements: outcomes for Year 2 of 3. Report AP-T296-15, Austroads, Sydney

Huet C (1963) Etude par une méthode d'impédance du comportement viscoélastique des matériaux hydrocarbonés, PhD thesis (in French), Paris University

Jameson G, Hopman P (2000) Austroads Pavement Design Guide Chapter 6: Development of Relationships between Laboratory Loading Rates and Traffic Speed, ARRB, Vermont South

Sayegh G, Viscoelastic properties of bituminous mixes. Paper presented at the 2nd International Conference on the Structural Design of Asphalt Pavements, Ann Arbor, Michigan, 1967

Ulloa A, Hajj EY, Siddhartan RV, Sebaaly PE (2013) Equivalent loading frequencies for dynamic analysis of asphalt pavements. J. Mat. Civ. Eng. 25: 1162-1170, doi 10.1061/(ASCE)MT.1943-5533.0000662

Flexural Properties of Cemented Granular Materials for Pavement Design

Arooran Sounthararajah, Leslie Wong, Nhu Nguyen, Ha Hong Bui, Jayantha Kodikara and Peerapong Jitsangiam

Abstract Cementation stabilization of unbound granular materials often offers a feasible solution for the strengthening of existing degraded unbound pavements. The primary determination mode of cemented pavement materials is fatigue cracking. Flexural properties including flexural modulus and tensile strain at break are incorporated into the fatigue criteria of cemented materials. However, there are no universally accepted testing protocols available for cemented pavement materials to determine the aforementioned properties in the laboratory. The four-point bending test is chosen in this paper to study the flexural properties of two different cemented pavement materials as it more closely simulates the stress/strain gradients generated in service. The results from this study revealed that, the strain at break could not be determined with sufficient precision as the tensile strain at the bottom of the specimen just prior to the point of fracture increases uncontrollably without further increase in tensile stress. Based on the response of cemented materials to different loading conditions, an equation is tentatively proposed for determination of the modulus for pavement design.

Keywords Pavement materials · Cement stabilization · Flexural properties

A. Sounthararajah (✉) · L. Wong · N. Nguyen · H.H. Bui · J. Kodikara
Department of Civil Engineering, Monash University, Melbourne, Australia
e-mail: arooran.sounthararajh@monash.edu

L. Wong
e-mail: leslie.wong@monash.edu

N. Nguyen
e-mail: nguyen.nhu@monash.edu

H.H. Bui
e-mail: Ha.Bui@monash.edu

J. Kodikara
e-mail: jayantha.kodikara@monash.edu

P. Jitsangiam
Department of Civil Engineering, Curtin University, Perth, Australia
e-mail: P.Jitsangiam@curtin.edu.au

A. Chabot et al. (eds.), *8th RILEM International Conference on Mechanisms of Cracking and Debonding in Pavements*, RILEM Bookseries 13,
DOI 10.1007/978-94-024-0867-6_56

1 Background

It is widely known that cement stabilization of pavement bases is a cost-effective and environmentally friendly technique for the highway construction, which provides extra capacity for increased traffic loads and frequency. The commonly considered distress mechanisms of stabilized pavement materials are tensile fatigue and shrinkage cracking (Chakrabarti et al. 2002). Fatigue cracking is considered to be the primary deterioration mode, and the fatigue life (N) of a CTB layer is usually considered to be a function of either the traffic induced tensile stress (σ), the traffic induced tensile strain (ε) or a ratio of these responses to the flexural strength (σ_b) or breaking strain (ε_b) of the material (Austroads 2014; Corte and Goux 1996; NCHRP 2004; Theyse et al. 1996). The general form of the fatigue relationships can typically be expressed as shown in Eq. 1.

$$\mathrm{Log(N)} = f_n\left[\frac{\sigma}{\sigma_b} \text{ or } \frac{\varepsilon}{\varepsilon_b}\right] \tag{1}$$

In the South African mechanistic pavement design approach, the breaking strain of the cemented materials is incorporated into the fatigue relationship (Theyse et al. 1996). However, difficulties were encountered in determining breaking strain of cemented materials from the flexural strength test data as the mid span deflection of the specimens just prior to the point of fracture increased uncontrollably. Litwinowicz and Brandon (1994) proposed that the breaking strain to be defined as the maximum tensile strain at 95 % of the ultimate failure load, but they did not justify this definition rationally. Flexural modulus is commonly considered as one of the main inputs for the mechanistic pavement design of CTB layers. The standard flexural modulus test method adopted in Australia does not specify a single load level at which testing should be conducted. Austroads (2014) examined the flexural modulus dependency on the peak tensile strain applied over a wider tensile strain range and concluded that for an increase in peak tensile strain of one micro-strain the flexural modulus of cemented materials decreases about 40 MPa. Incorrect choice of flexural properties in pavement design could have a major impact on the performance of CTB layers in service under heavy axle loading. This paper investigates the response of cemented materials to different loading conditions, and also provides recommendations for pavement design.

2 Experimental Program

2.1 Materials and Specimens

Two different pavement materials (Holcim Road Base and quartzite) were selected for this research study. Holcim Road Base (HRB) is commonly used as a base course material for Western Australian roads, and it was sourced from Holcim Gosnells quarry, Western Australia. The source rock of this pavement material was Granite (acid intrusive igneous rock). Quartzite has been used extensively in both unbound and cement stabilized pavement base construction around Adelaide, and it was sourced from Boral Para Hills quarry, South Australia. The source rock of this pavement material was fine quartzite (very low-grade regional metamorphic rock). These pavement materials were treated with 3 % General Purpose (GP) cement (percentage by dry mass of parent material). Material grading consisted of two categories of maximum particle size: material passing sieve sizes 19.00 and 12.70 mm. The Steel molds of dimensions 50 by 50 by 200 mm were used to manufacture the beam specimens for flexural modulus and flexural strength tests. The specimens were compacted in one layer with modified compaction effort; after a few compaction trials, a procedure was established to achieve 98–100 % relative compaction effort. The beam specimens were covered to prevent moisture evaporation and cured in the molds for two days at ambient room temperature (23 °C) and were then cured in the fog room (95 ± 5 % relative humidity, 23 °C) until tested.

2.2 *Flexural Strength and Flexural Modulus*

The standard modulus of rupture test was adopted for the flexural strength testing based on Standards Australia (2000). This test involved increasing the load monotonically at a rate of 1 MPa/min extreme fiber stress (0.833 kN/min for the 50 mm square beam sections) until the specimen fails. The mid-span deflection was measured using two linear variable differential transducers (LVDTs) fitted with a support frame resting on the beam specimen over the support rollers. The experimental setup is illustrated in Fig. 1.

The flexural modulus test was performed using stress control with a cyclic haversine load pulse of 500 ms (250 ms loading period and 250 ms rest period),

Fig. 1 Flexural beam test setup (four-point bending test)

and the loading was continued until several hundred load pulses. A seating load of 0.05 kN was applied to the specimens in order to eliminate the effect of noise in data transferring. The cyclic load applied to the specimen and the cyclic mid span deflection of the specimen associated with each load pulse were recorded. A range of loading magnitudes was used in this study (20–95 % of the ultimate breaking load of the specimen). The maximum plastic strain of the specimen at each load pulse were measured for only few beam specimens using optical fiber sensors mounted on the bottom surface of the beam specimen.

3 Test Results and Discussion

The average tensile strain at break and the average tensile strain at 95 % of the ultimate failure stress calculated by using Eq. (2) from the experimental data are summarized in Table 1. The average tensile strain at 95 % of the ultimate failure stress is about 50–59 % of the average tensile strain at break (Table 1). Thus, the tensile strain at 95 % of the ultimate failure stress is not an appropriate definition of breaking strain for cemented materials, as it appears to be.

$$\varepsilon = \frac{108\delta\,\mathrm{H}}{23\,\mathrm{L}^2} \tag{2}$$

Table 1 Flexural strength test results of cemented pavement materials

Material	Curing age (days)	Average tensile strain at break (S1) (με)	Average tensile strain at 95 % of the ultimate failure stress (S2) (με)	S2/S1 (%)
Cemented quartzite (12.7 mm)	7	326	166	51
Cemented quartzite (12.7 mm)	40	355	176	50
Cemented HRB (12.7 mm)	7	332	175	53
Cemented HRB (12.7 mm)	40	386	207	54
Cemented HRB (19 mm)	7	305	176	58
Cemented HRB (19 mm)	40	246	145	59

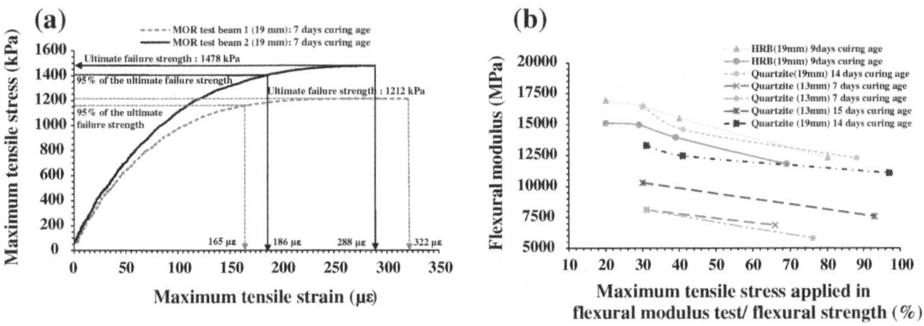

Fig. 2 **a** Example flexural strength results of cemented HRB material, **b** Example of flexural modulus dependency on stress over a wider stress range

where ε is the maximum tensile strain, δ is the mid-span deflection, H is the mean beam height and L is the span length.

Equation (2) was derived assuming that the material is linear elastic; however, it may not be valid at the point of fracture as plastic deformation increases with the increasing load magnitude after a certain load magnitude.

The tensile strain at the bottom of the specimen just prior to the point of fracture increases uncontrollably without further increase in tensile stress as it can clearly be seen in Fig. 2a. The information on macro-crack initiation and propagation during monotonic loading forms the basis for the breaking strain estimation; however, it is very difficult to monitor the macro-crack initiation and propagation during testing, as it requires more sophisticated testing facilities. The data given in Fig. 2b show the average flexural modulus values of the first hundred load cycles for individual beams. As it can be seen, the flexural modulus decrease with increasing tensile stress magnitude applied, and it seems that values are more or less constant at lower tensile stress levels (20–30 % of the flexural strength of the specimen). The maximum plastic strain variation of beam specimens in the flexural modulus test is depicted in Fig. 3a and indicates plastic strain accumulation continues until the first 30–50 load cycles even at 20 % stress level, this attributes to the shakedown behavior of the cemented materials.

The beam specimens tested at higher stress levels (70–90 % of the flexural strength of the specimen) showed a rapid reduction in flexural modulus within a few hundred load cycles, and the amount of reduction increased with increasing stress level (Fig. 3b).

4 Conclusions and Recommendations

The four-point bending test was chosen in this paper to study the flexural properties of two different cemented pavement materials under different test conditions. The results from this study revealed that, the strain at break could not be determined

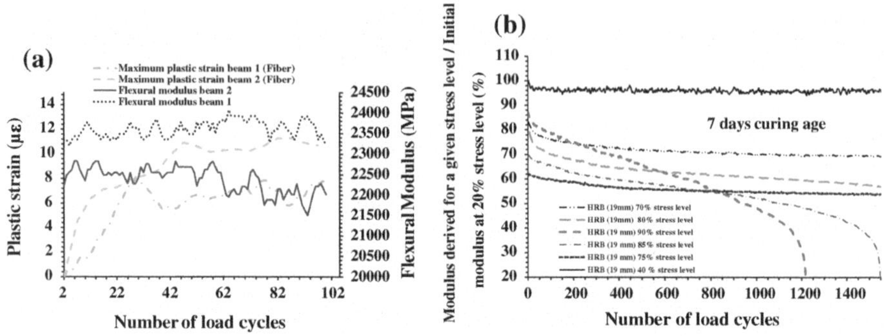

Fig. 3 **a** Plastic strain variation at 20 % stress level for cemented HRB (40 days curing age), **b** Modulus degradation curve for individual beams tested at different stress levels

with sufficient precision as the tensile strain at the bottom of the specimen just prior to the point of fracture increases uncontrollably without further increase in tensile stress. The accuracy of the measured breaking strain of cemented materials used in a fatigue relationship is extremely essential as inaccuracies in the breaking strain may have significant impact on the derived fatigue life. Consequently, it is argued that breaking strain may not be suitable for use in fatigue relationships of cemented materials. On the basis of limited flexural modulus test results, Eq. 3 is tentatively proposed for the estimation of modulus of cemented materials taking into its initial rapid reduction for mechanistic pavement design.

$$E_{design} = k \times E_{20\%} \tag{3}$$

where E_{design} is the design modulus, $E_{20\%}$ is the average flexural modulus of the second 50 consecutive load cycles at 20 % stress level and k is a coefficient, which mainly depends on the stress level in service, material, age of curing and initial stage of the compacted material. However, this concept needs to be further examined with additional testing and research. In addition, it is tentatively suggested that the first 50 load cycles should not be used to calculate the flexural modulus ($E_{20\%}$) of the beam specimen as the plastic strain accumulates at a higher rate within the first 30–50 load cycles affecting the calculated modulus.

Acknowledgements This work is a part of a research project sponsored by the Australian Research Council (ARC), IPC Global Pty Ltd, Queensland Department of Transport and Main Roads (QDMR), Golder Associates Pty Ltd and Hong Kong Road Research Laboratory (HKRRL).

References

Austroads (2014) Cemented materials characterisation. Austroads research report, AP-R462-14. Austroads, Sydney, p 40-41

Chakrabarti S, Kodikara JK, Pardo L (2002) Survey Results on Stabilisation methods and performance of local government roads in Australia. Road and Transport Research 11:3-16

Corte JF, Goux MT (1996) Design of Pavement Structures: The French Technical Guide. Transportation Research Record 1539:116–124

Litwinowicz A, Brandon AN (1994) Dynamic flexure testing for prediction of cement: treated pavement life. Proceedings of the 11th ARRB conference, Vol 17. Australian Road Research Board (ARRB), p 229-247

NCHRP (2004) Guide for Mechanistic-Empirical Design of New and Rehabilitated Pavement Structures- Part 3 Design Analysis. Final report, Transportation Research Board, Washington

Standards Australia (2000) Methods of testing concrete: method 11: determination of the modulus of rupture. AS 1012.11-2000(R2014), Standards Australia, Sydney

Theyse HL, Beer De M, Rust FC (1996) Overview of South African Mechanistic Pavement Design Method. Transportation Research Record 1539:6–17

Mechanistic Evaluation of the Long Term Performance Characteristics of Warm Mix Additives in Modified Asphalt Mixtures

Zahi Chamoun and Mena Souliman

Abstract The immediate benefit of producing Warm Mix Asphalt (WMA) mixtures is the reduction in the consumption of energy required to produce the traditional hot mix asphalt (HMA). WMA additives also improve workability and compactability of mixtures to a point where they can be produced at lower temperatures. With the decreased production temperatures comes the benefit of reduced emissions, fumes, dust production and odors, as well as an extended mix haul distance. This study evaluated the long term performance of two different WMA technologies with neat, polymer modified and terminal blend tire rubber asphalt mixtures, designed according to Nevada Department of Transportation (NDOT) and California Department of Transportation (Caltrans) specifications using the Hveem mix design, using one source of aggregates sampled in two different batches from a local quarry. An experimental program was designed to cover the impact of WMA additives on the performance characteristics of the different mixtures using the flexural beam fatigue test and a mechanistic analysis of mixtures in simulated pavements using a 3D-Move analysis on two pavement structure; thin (100 mm) and thick (200 mm). The findings will be presented in the paper below.

Keywords Fatigue · Rutting · Mechanistic · Warm mix

1 Introduction

Warm Mix Asphalt (WMA) additives allow reduction in the temperatures at which asphalt mixtures are produced and placed. The immediate benefit of producing WMA mixtures is the reduction in the consumption of energy required to produce

Z. Chamoun
Terracon Inc, Nashville, TN 37211, USA
e-mail: Zahi.Chamoun@terracon.com

M. Souliman (✉)
Departments of Civil Engineering, The University of Texas, Tyler, USA
e-mail: msouliman@uttyler.edu

A. Chabot et al. (eds.), *8th RILEM International Conference on Mechanisms of Cracking and Debonding in Pavements*, RILEM Bookseries 13,
DOI 10.1007/978-94-024-0867-6_57

the traditional hot mix asphalt (HMA). WMA additives also improve workability and compactability of mixtures. A total of four WMA technologies are available. The organic additives or waxes and chemical additives, the water bearing Zeolites, and finally the water based technology or Foaming. It was noticed that past research have focused on evaluating the use of WMA technologies with unmodified binders, therefore, it was found necessary to assess the performance of modified binder mixtures with WMA technologies. One of the major concerns this research paper is trying to address is: would the reduced production temperature lead to premature permanent deformation? This paper will attempt to answer this question by performing a mechanistic analysis that explores the long term performance of Sasobit and Advera WMA additives along with polymer modified and tire rubber modified asphalt binders.

Sebaaly et al. (2012) conducted study using Advera and Sasobit additives along with polymer modified and tire rubber modified asphalt binders. The study concentrated on the resistance to moisture damage of different mixtures based on the indirect tensile strength test (ITS), Tensile Strength Ratio (TSR), dynamic modulus |E*| as well as the dynamic modulus |E*| ratio, along with a preliminary mechanistic analysis. It was observed that the use of terminal blend tire rubber-modified binder WMA mixtures along with lime treatment will significantly improve the resistance to moisture damage.

From other available literature (Mogawer et al. 2012; Shivaprasad et al. 2012), it can be realized that minimal mechanistic analysis research work was conducted on WMA mixtures with polymer or tire rubber modified asphalt binder.

The objective of this study was to evaluate the use of select WMA additives with polymer modified and terminal blend tire rubber asphalt mixtures, designed according to Nevada Department of Transportation (NDOT) and California Department of Transportation (Caltrans) specifications using the Hveem mix design. The concern is whether or not the addition of WMA additives to the modified mixtures will affect the observed good performance. An experimental program was designed to assess the resistance to fatigue cracking of WMA mixtures with and without asphalt binder modification and compared to their corresponding HMA control mixtures. The paper presented below assess the impact of WMA additives on the performance characteristics of the mixtures via addressing and evaluating the resistance to fatigue cracking by performing a mechanistic fatigue analysis using both developed fatigue models as well as 3D-Move analysis software (Asphalt Research Consortium 2013) in order to evaluate the effect of asphalt binder modifications on the long term performance of WMA mixtures.

2 Mix Designs

All mixtures were designed to meet the NDOT and Caltrans specifications following Hveem mix design method for heavy traffic with a nominal maximum aggregate size of 19 mm (NDOT 2001). As per the Hveem mix design method, HMA samples were mixed at mixing temperatures determined based on temperature-viscosity charts as described in AASHTO T 312 (AASHTO 2012), cured for 16 ± 1 h at 60 °C before compaction at 110 °C. The optimum binder contents were determined for the HMA mixtures and verified for the WMA. The WMA mix designs were verified following the mix design procedure proposed by NCHRP 9-43 (Bonaquist 2004) for WMA accompanied with the implementation of the Hveem mix design. The mixing and compaction temperatures of the WMA mixtures were determined per NCHRP 9-43 (Bonaquist 2004). The mixture conditioning period was amended to two hours at compaction temperature as suggested by the draft WMA mix design procedure (Bonaquist 2004). Mixing temperatures for HMA mixtures ranged from 153.3 to 160 °C while mixing temperatures for WMA mixtures ranged from 123.9 to 143.3 °C. The compaction temperature for the WMA mix design was kept at 110 °C.

3 Mechanistic Analysis

To combine the impact of the dynamic modulus |E*| of the mixtures with their performance characteristics in terms of resistance to fatigue cracking, a 3D-Move software (Asphalt Research Consortium 2013) was utilized to perform a detailed mechanistic analysis on two pavement structures; thin (100 mm) and thick (200 mm). To estimate the number of load repetitions to fatigue failure N_f, fatigue models were previously developed using the beam fatigue test method along with the tensile strain calculated by 3D-Move. A dynamic and a static analysis were performed for both structures. For the dynamic analysis, the vehicle speed was assumed to be 40 mph. The analyzed traffic configuration consisted of dual tires. The response points at the bottom of the asphalt layer were taken at three different positions; under the center of the tire, at the edge of the tire and in between the dual tires.

The pavement responses based on the dynamic analysis were estimated using 3D-Move software and on the master curve developed for each mixture. Thus, in order to compare the response calculated through a static analysis to the ones obtained through the dynamic analysis, proper dynamic modulus was assigned to each pavement structure.

For the thin pavement, the assigned dynamic modulus value of each mixture was at 30 Hz and 21.1 °C, while for the thick pavement static analysis, the dynamic modulus value of each mixture at 10 Hz and 21.1 °C was used. Those frequencies were used based on the MEPDG recommendation (Guide for Mechanistic-Empirical Design 2004). The resistance to fatigue cracking of the different mixtures was then evaluated using the fatigue model developed through lab testing. It was observed that the K_2 factor had a significant effect on the fatigue resistance of the mixtures as well as the strains obtained from the 3D-Move analysis

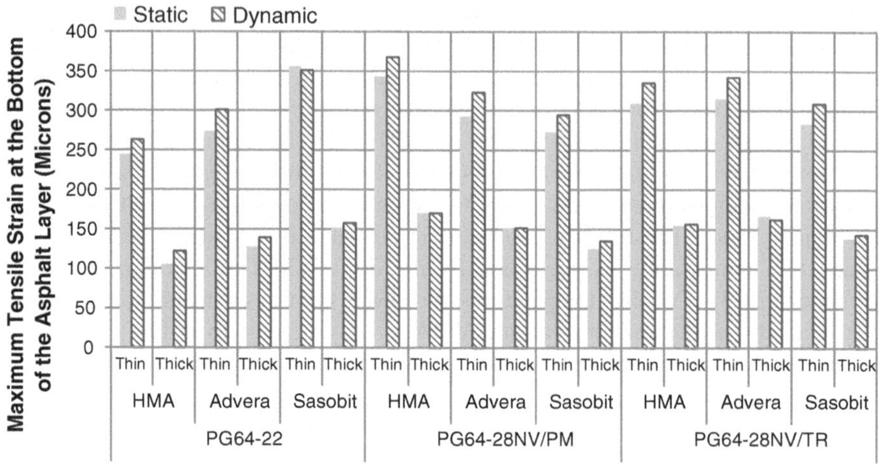

Fig. 1 3D-Move analysis of the maximum tensile at bottom of asphalt layer

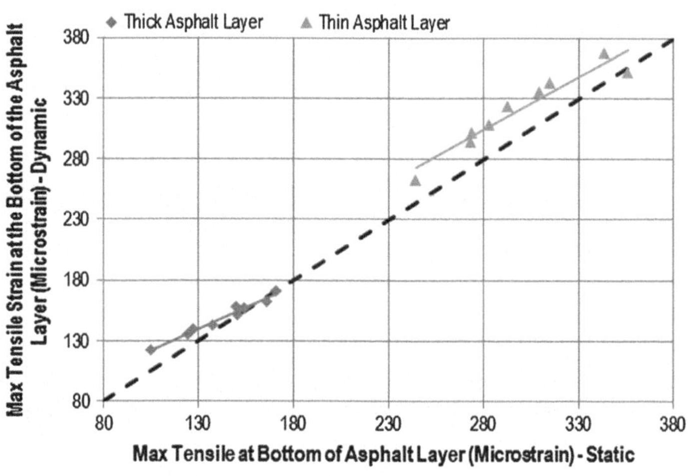

Fig. 2 Maximum tensile strain at bottom of asphalt layer: static vs. dynamic analysis

(Figs. 1 and 2). The following model was used to characterize the fatigue behavior of the various mixtures:

$$N_f = k_1 \left(\frac{1}{\varepsilon}\right)^{k_2}$$

where N_f is the fatigue life (number of load repetitions to fatigue damage), ε is the applied tensile strain and k_1 and k_2 are experimentally determined coefficients. The following observations were made:

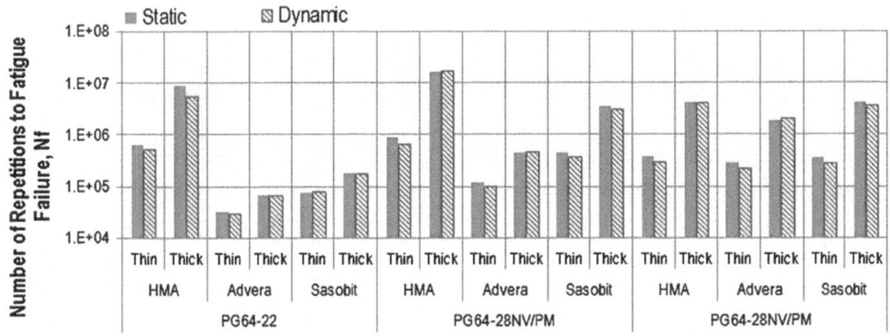

Fig. 3 N_f comparison of all mixtures using 3D-Move static and dynamic analysis

- Based on 3D-Move analysis, polymer modified WMA mixtures, with both thin and thick structures, developed lower tensile strains at the bottom of the asphalt layer compared to its corresponding HMA mixture while tire rubber modified mixtures exhibited similar tensile strains (Fig. 1).
- Overall, the thick pavement structure had a lower tensile strains compared to the thin pavement structure (Fig. 1).
- The 3D-Move analysis of the maximum tensile strain at the bottom of the asphalt layer was only affected by the dynamic modulus values, a higher $|E^*|$ will lead to a lower tensile strain value and vice versa (Fig. 1).
- Calculated tensile strain values based on the Dynamic analysis were higher than their corresponding tensile strain values using the static analysis. It is believed that the dynamic analysis uses the dynamic modulus master curve of the mixture to perform the tensile strain calculation whereas for the static analysis, the user has the flexibility to assign any given value of the dynamic modulus at a certain frequency (Fig. 2).
- Upon the completion of the 3D-move analysis, the calculated tensile strains at the bottom of the asphalt layer of each mixture along with their corresponding fatigue model, analysis of the fatigue life of different mixtures presented in Fig. 3. The following observations can be made.
- Both thin and thick structures showed a similar behavior where the HMA mixtures had the highest resistance to fatigue cracking followed by the WMA Sasobit mixture then WMA Advera mixture.
- Polymer modified HMA mixture exhibited the highest resistance to fatigue cracking among all evaluated mixtures.
- Rubber tire modification improved the fatigue life of Sasobit mixture allowing it to reach similar fatigue life to its corresponding HMA mixture.

Overall, the fatigue life behavior of the various evaluated mixtures has to be explained by the fact that the fatigue resistance is not merely dependent on the calculated maximum tensile strain at the bottom of the pavement layer, it is also highly affected by the K_2 factor obtained from the fatigue prediction model.

4 Summary and Conclusions

This laboratory research documented the resistance of various untreated mixtures to fatigue cracking using the flexural beam fatigue test and by running a 3D-Move analysis. The HMA mixtures with PG64-28NV showed the better resistance to fatigue cracking than other mixtures. To better evaluate the fatigue life of the different mixtures, the dynamic modulus effect was incorporated in the 3D-Move fatigue analysis. The addition of a WMA additive will decrease the fatigue life. Also it was observed that WMA mixtures with Sasobit always exhibited a higher resistance to fatigue cracking than the Advera mixtures. HMA and WMA mixtures with PG64-28NV/TR have exhibited a close behavior in relation to fatigue resistance. In addition, the measured dynamic modulus was incorporated to evaluate the strain at the bottom of the flexible pavement layer. Polymer modified HMA mixtures exhibited the best resistance to fatigue cracking among all evaluated mixtures followed by the unmodified and rubber modified HMA mixtures, respectively. In summary, by identifying an optimum binder type, binder content, anti-strip additive type, along with the WMA additive, a sustainable asphalt pavement can be designed to efficiently resist the moisture damage, permanent deformation, and fatigue cracking.

To strengthen the concluded results, more WMA technologies are recommended to be evaluated as well as including different aggregate sources since multiple studies showed that different chemical compositions of the aggregates may have an impact on the behavior of the mixtures.

References

Asphalt Research Consortium. 2013. www.arc.unr.edu/Software.html#3DMove.

Bonaquist, R. (2004). Mix Design Practices for Warm-Mix Asphalt. NCHRP Report 691, Champaign, Illinois: Transportation Research Board.

Guide for Mechanistic-Empirical Design. (2004). Champaign, Illinois: Transportation Research Board. www.trb.org/mepdg/guide.htm

Mogawer, W., Austerman, A., Kluttz, R., and Roussel, M. (2012). High-Performance Thin-Lift Overlays with High Reclaimed Asphalt Pavemetn Content and Warm-Mix Asphalt Technology Performance and Workability Characteristics. In Transportation Research Record No. 2293, Washington, D.C., pp. 16-28.

Sebaaly, P., Hajj, E., Hitti, E. (2012). Performance of Recycled Rubber Modified Binders in Warm Mix Asphalt Mixtures. In The Asphalt Rubber Conference, Munich.

Shivaprasad, P., Xiao, F., Amirkhanian, S. (2012). Evaluation of moisture sensitivity of stone matrix asphalt mixtures using polymerised warm mix asphalt technologies. In International Journal of Pavement Engineering, Vol. 13 No.2, pp. 152-165.

Performance Assessment of JPCP and CRCP Rigid Pavements Implementing M-E Analysis

Michele Agostinacchio, Donato Ciampa and Saverio Olita

Abstract The Mechanistic-Empirical approach in rigid pavements design, allows to achieving the superstructure damage by calculating incremental degradation. This study has examined two different pavements typologies: JPCP (Jointed Plain Concrete Pavements) and CRC (Continuously Reinforced Concrete Pavements). Referring to the AASHTO Design Guide the superstructure performances are evaluated in terms of Joint Faulting and Transverse Cracking for JPCP, Punchouts for CRCP, and IRI (International Roughness Index) for both pavements typologies. The performances have been determined using ME-PDG software that is able to evaluate, for JPCP design, the structural fatigue distresses related to Transverse Cracking of PCC slabs and differential deflection related Transverse Joint Faulting. For CRCP, the principal structural distress considered is edge Punchouts. The authors propose the comparison between the different typologies of pavements with the objective of identifying the design solutions more effective with equal materials performance and traffic conditions. These comparisons were carried out by varying the soil class, the climatic conditions and the type of cement concrete. The numerous analyzes performed have enabled to evaluate the influence of different design parameters and then to define useful suggestions that can be used by rigid pavement designers to reduce the occurrence of premature cracking, so as to increase the service life of pavement system.

Keywords Rigid pavements · Transverse cracking · Punchouts · M-E design

---

M. Agostinacchio · D. Ciampa · S. Olita (✉)
School of Engineering (SI-UNIBAS), University of Basilicata, 85100 Potenza, Italy
e-mail: saverio.olita@unibas.it

M. Agostinacchio
e-mail: michele.agostinacchio@unibas.it

D. Ciampa
e-mail: donato.ciampa@unibas.it

A. Chabot et al. (eds.), *8th RILEM International Conference on Mechanisms of Cracking and Debonding in Pavements*, RILEM Bookseries 13,
DOI 10.1007/978-94-024-0867-6_58

1 Introduction

In this paper the Mechanistic-Empirical Pavement Design method introduced by the "AASHTO Guide 2002" was applied. The M-E PDG software, developed within the NCHRP 1-37A project (AASTHO 2004) was used. The Mechanistic-Empirical methodology implemented by the software implies three different levels of input, according to the criticality of the project, the availability of data and the required level of reliability. Since the numerical comparison analyses among the different structural typologies refer to the same materials in standard conditions, the third level was adopted. The research was focused on two aspects:

- analysis and optimization of different pavement typologies (JPCP, CRCP) of primary and secondary extra-urban roads with high traffic at parity of materials and in standard conditions;
- evaluation of the effects produced varying the performances of the used materials at parity of design solution and operative conditions.

The reference initial JPCP and CRCP sections adopted in this study were selected from the Italian Catalogue of Pavements (CNR 1995), precisely from the 3RG card, referring to JPCP and 3RC to CRCP. These sections are defined referring to a number of commercial vehicle transits equal to 10,000,000 on the most loaded lane hypothesizing a service life of thirty years, estimated on the basis of the pavement typology and the traffic spectrum corresponding to the road category. The AADTT obtained dividing the number of commercial vehicle transits on the most loaded lane (10,000,000) by the days of service life (10,950) and multiplying by the directions number (two) was 1830. The vehicle class distribution was defined both comparing the FHWA Classification used by the M-E PDG software with the Italian Classification, and comparing the Truck Traffic Classification (TTC) groups and the corresponding distribution used by the Design Guide with the CNR Catalogue Traffic Spectra, assessing the TTC 14 group (FHWA 2001) approached better the Italian distribution corresponding to the extra-urban roads.

2 Performance Parameters, Limits and Reliability

The JPCP design is based on Transverse Slab Cracking, Transverse Joint Faulting, and pavement smoothness (IRI). For CRCP design, Crack Width and Crack LTE, Punchouts and smoothness (IRI) are the key performance indicators. In order to investigate the effectiveness of the different construction approaches in relation to the major distress modes, the following performance indexes were assumed:

- IRI (International Roughness Index)—The IRI model is able to foresee the roughness variation on the basis of the estimated distresses, calculated by specific forecast models, of the initial IRI index and of other parameters referring to the situ (Agostinacchio 2014). The distress target of 271 cm/km (172in/mi), with a reliability target of 90 %, was adopted.

- Transverse Slab Cracking (JPCP)—The performance criterion for transverse cracking defines the maximum allowable percentage of cracked slabs at the end of design life and determines the level of slab cracking that may occur over the design period. The distress target of 15 %, with a reliability of 90 %, was assumed.
- Transverse Joint Faulting (JPCP)—It is desirable to limit mean faulting to ensure that a pavement will remain smooth over the design period. The performance criterion for joint faulting defines the allowable amount of mean joint faulting at the end of the design life and determines the level of joint faulting over the design period. The distress target of 3 mm (0.12in) was adopted.
- Crack Width and LTE (CRCP)—This should be limited to 0.51 mm (0.02in) or less. The crack LTE should be limited to greater than 95 percent throughout the design life. However, it is acceptable to use lower values but still higher than 70 % (Gucunski 2009). The distress target of 75 % was defined.
- Punchouts (CRCP)—Typical values of allowable CRCP punchouts range from 6.2 to 12.4 per km (10–20 per mile). The distress target was assumed equal to 10 per mi, with a reliability target of 90 %.

3 Pavement Sections Optimization

Since the environmental conditions influence the pavement performance all analyses were referred to three different places: "Washington DC" (Elev: 94 m, Lat: 38.52°, Long: −77.02°), "Elkins, WV" (Elev: 604 m, Lat: 38.53°, Long: −79.51°), "Colorado Springs, CO" (Elev: 1884 m, Lat: 38.49°, Long: −104.43°). The localities were chosen so as to have the same latitude but which were placed at a different altitude so as to have a different seasonal climatic trend, and in particular the first was chosen on sea level, the second at hilly elevation and the third in the mountains. All places parameters are reported in the database of the EICM (Enhanced Integrated Climatic Model) routine. The selected latitude is characterized by a climatic seasonal trend near enough to many areas of the Italian inland.

The pavement sections adopted in this study consisted of the following layers succession, placed on a natural soil AASHTO Class A-4 (silty soil):

- compacted soil AASHTO Class A-3 (fine sand) characterized by a thickness of 40 cm;
- unbounded crushed gravel foundation of 15 cm;
- stabilized cement base layer characterized by a thickness of 15 cm;
- non-reinforced concrete slab thickness of 24 cm for the JPCP section;
- reinforced concrete layer thickness of 24 cm for the CRCP section (aggregate type: limestone, steel 0.7 %, bar diameter 16 mm, steel depth 10 cm);

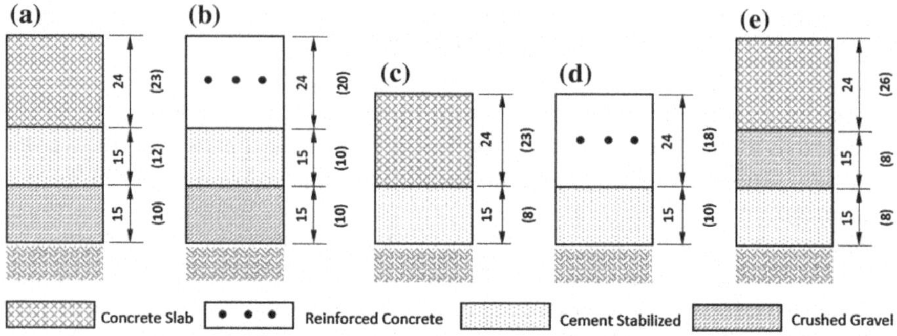

Fig. 1 Pavements typologies (Note: the optimized thicknesses are shown in parentheses). **a** JPCP **b** CRCP **c** FDS-1 **d** FDS-2 **e** IS

Starting from these sections called JPCP (Fig. 1a) and CRCP (Fig. 1b), the following pavements typologies were defined:

- Full Depth Sections, FDS-1 and FDS-2 (Fig. 1c, d), obtained eliminating the foundation from the previous sections;
- Inverted Section IS (Fig. 1e), obtained inverting the position of the unbounded layer with the base.

Once defined typologies, sizes and materials characteristics of the trial sections, the optimization was made using the M-E PDG code: the layers' thicknesses were set right on the basis of the performance parameters (see Sect. 2) referring to the different pavement typologies (Fig. 1).

4 Evaluation of Effects Induced by the Design Parameters

The optimized sections (Fig. 1) are approximately equivalent in terms of reliability referring the considered performance parameters (see Sect. 2). The second step of the study consisted of evaluation of the effect produced varying the performance and/or the typologies of materials, at parity of design solution and operative conditions. In particular, the clime, the performance characteristics of compacted soil layer and of the natural soil subgrade and the type of cement for the slabs were taken into account. Figure 2 summarize all the parameters analyzed in the study. With reference to the climate it was decided to evaluate the effects that it has on superstructure by choosing three additional locations with the same altitude ($\sim$ 600 m) but different latitude so that to have a different seasonal climatic trend, and in particular the first has been chosen with a latitude of 47.41°, the second at a latitude of 38.53° and the third at a latitude of 31.21°. These locations are: "Spokane, WA" (Elev: 603 m, Lat: 47.41°, Long: −117.19°), "Elkins, WV" (Elev: 604 m, Lat: 38.53°, Long: −79.51°), "San Angelo, TX" (Elev: 580 m, Lat: 31.21°,

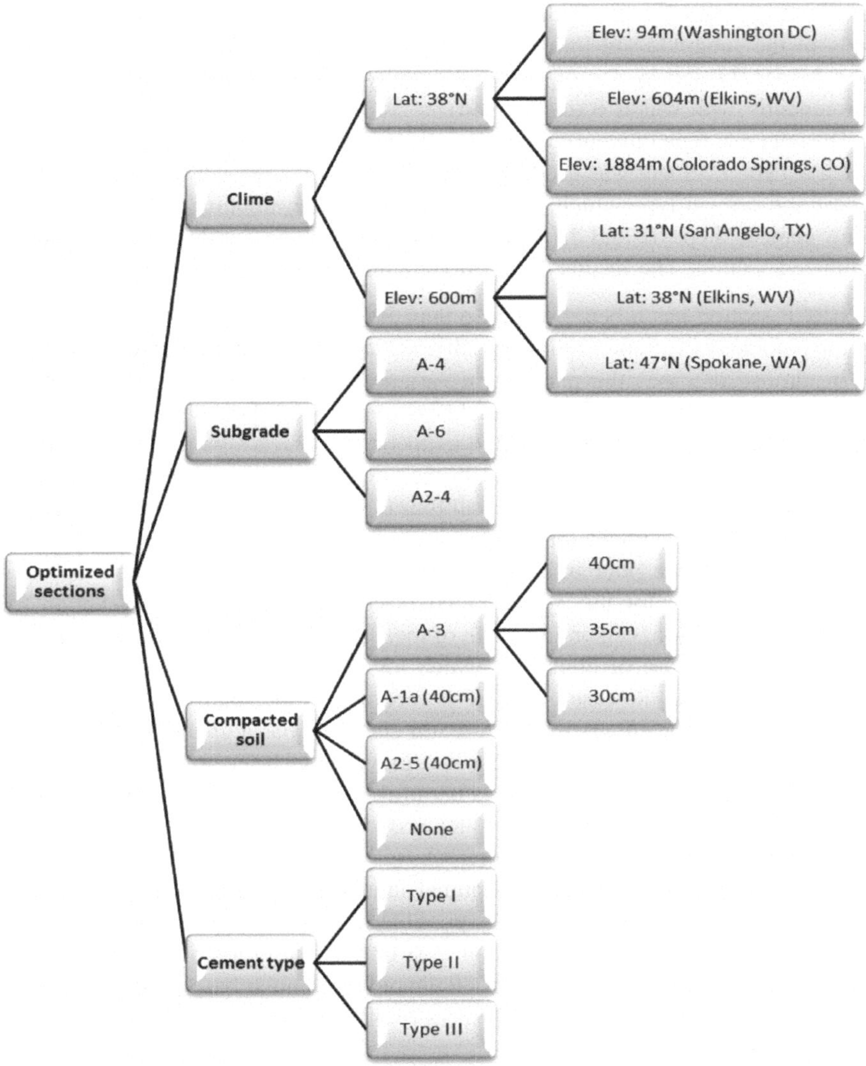

Fig. 2 Parameters analyzed

Long: −100.29°). The influence analysis of the compacted soil layer, preliminarily, was carried out by replacing the compacted layer (40 cm) of AASHTO soil class A-3 with soil class A-1a (stone fragments, gravel and sand) and A2-5 (silty or clayey gravel and sand) of same thickness.

It was then considered the possibility that this layer is not present. Finally, it has evaluated the thickness influence of the A-3 compacted soil layer by analyzing the pavement section behavior with compacted soil thicknesses of 35 and 30 cm. The study of the subgrade soil influence was carried out by replacing the subgrade of type AASHTO soil A-4 with soil A2-4 (silty or clayey gravel and sand) and subsequently with soil A-6 (clayey soil).

The concrete type influence analysis was conducted on the optimized sections, in which a Type I cement was used. Within the software is present the opportunity to choose the cement type which will be used for the construction of the slab. Then, using cement Type II and III, the analysis was redone starting from the JPCP optimized section. The results obtained are summarized in the following:

- CRCP section has a better response than JPCP, reflecting a minor influence of the parameters variation (climate, subgrade, compacted soil layer, cement type). In almost all cases the parameters variations produce a minimum reliability deviations and then the layers' thickness remains substantially constant;
- the climate parameters, altitude and temperature, affects significantly the layers' thickness, which become larger with the elevation increase (i.e. as the temperature decreases). This trend unites all four types of analyzed sections (JPCP and CRCP with and without foundation);
- the latitude, has a poor influence on the layers' thickness which are approximately the same for the 3 considered places, having the same elevation but placed in the north, center and south of the USA. The faults that occur consist in a minimal increase of thickness, which for JPCP registers at the intermediate locality, while in the case of CRCP is recorded in the south locality;
- with regard to the subgrade, the A-4 soil class one has a greater reliability, and with it is possible to avoid the crushed gravel base layer both in JPCP that in CRCP. In fact, the granular base is harmful in the presence of a good subgrade (A-4) causing an increase in the thickness of the cross-section. Its use is necessary where the subgrade is poorer (A-6);
- for the compacted soil layer, the optimal situation is with the A-3 soil class with a thickness of 40 cm. For JPCP is possible to reduce the layer thickness up to 30 cm, while the use of a poorer soil class or failed insertion of this layer, causes a section failure. In the case of CRCP devoid of granular base (FDS) is essential to use a soil class A-3 with a thickness of 40 cm because in all other cases the section is not verified;
- with regard to the different types of cement, the influence of the one or the other type involves a minimum variation of the slab thickness (about 1 cm). The best type of cement results the "Type II" for all analyzed sections;
- finally, the IS analysis has shown an increase of section thickness, therefore the inversion, very advantageous for flexible pavements (La Agostinacchio 2008), does not have the same advantages but is harmful for rigid pavements.

5 Conclusions

In this paper the Authors, implementing M-E PDG software, point out the results obtained analyzing different rigid pavements typologies in the same operative conditions. Starting from sections defined by the Italian Catalogue, design solutions were developed in order to inhibit the main distress modes. Then, the influence of several parameters (presence of the compacted soil layer, use of a different soil class, variation of layer thickness of the compacted soil layer, clime variation and cement type) on the optimized sections performance was evaluated, achieving useful suggests for the design (see Sect. 4). It is evident that in the paper we could not return the full results obtained from all the numerical processing performed due to the limited paper space available.

References

AASTHO (2004) Guide for Mechanistic-Empirical Design. NCHRP Project 1-37A. USA.

Agostinacchio M, Ciampa D, Olita S (2014) The vibrations induced by surface irregularities in road pavements - a Matlab® approach. European Transport Research Review, 6 (3), pp. 267-275. doi:10.1007/s12544-013-0127-8.

CNR (1995) Catalogo delle pavimentazioni stradali, CNR-BU 178/95 Italy.

FHWA (2001) Guide to LTPP Traffic Data Collection and Processing. Washington DC. USA.

Gucunski N, Vitillo N, Zaghloul S (2009) Development of FWD Procedures Manual – Final Report, FHWA-NJ-2009-005.

La Agostinacchio M, Ciampa D, Olita S (2008) Cracking response and service life prediction of flexible and semi-rigid road pavements implementing M-E PDG 2002 code, Pavement Cracking: Mechanisms, Modeling, Detection, Testing and Case Histories, pp. 201-210. ISBN 978-0-415-47575-4.

Weak Interlayers Found in Flexible and Semi-flexible Road Pavements

Morris De Beer

Abstract Weak interlayers within the upper structural layers of road pavements are specifically prohibited in most road-building specifications. However, such layers are extremely common and often lead to premature pavement distress. Heavy Vehicle Simulator (HVS) evaluation indicates that the presence of such layers within the structural layers of a flexible/semi-flexible pavement is far more deleterious than is commonly appreciated. These effects are modelled using examples based an HVS testing and simplistic mechanistic pavement analyses. In particular, weak upper base courses of lightly cemented pavement under thin bituminous surfacing may lead to severe surfacing and upper base failure within a matter of months after opening to traffic, not excluding failure during construction. The causes of these adverse conditions, together with simple detection methods during construction and analyses of their effects on the structural capacity of flexible and semi-rigid (cemented) road pavements, are briefly discussed. Methodologies are available to detect and investigate the existence of these weak layers in cemented pavement layers. Analysis were done on a typical recycled hot mix asphalt (HMA) pavement, cemented base pavement and a granular base pavement, with and without these weak layers and interface conditions. The analyses focused on the strain energy of distortion (SED) as a pavement response parameter indicating the potential for structural damage. SED shows some potential for quantifying relative distortional effects of these weak layers and/or weak interfaces within flexible and semi-flexible pavements. In this paper a very brief overview of above is given.

Keywords Weak layers · Modelling · Interlayers · Semi-flexible

M. De Beer (✉)
CSIR Built Environment, P.O. Box 395, Pretoria, South Africa
e-mail: mbeer@csir.co.za

A. Chabot et al. (eds.), *8th RILEM International Conference on Mechanisms of Cracking and Debonding in Pavements*, RILEM Bookseries 13,
DOI 10.1007/978-94-024-0867-6_59

1 Introduction

In general, this paper discusses mechanisms and models used to illustrate adverse effects of weak layers, interlayers, laminations and/or weak interfaces in flexible and semi-flexible pavements. These rural class pavements incorporating lightly cemented layers [3–5 % cementitious binder, with Unconfined Compressive Stress (UCS) ~1.5–3 MPa, sees COLTO (1996)]. Owing to shrinkage cracking (block cracking) of highly cementitious road bases (5–7 % cementitious binders) its use was discouraged, and replaced with relatively "lightly" cemented layers. However, the risk of low durability, relatively low-strength, weak layers, interlayers, laminations and/or weak interfaces increases, especially owing to poor construction. Nevertheless, such effects on bases and/or subbases of flexible and semi-flexible road pavements are specifically prohibited in most specifications in southern Africa. Premature pavement distress in the form of rippling, acute slippage, pumping, cracking or shoving of the bituminous surfacing and shallow base failures of cemented base pavements are not rare in southern Africa. (see Netterberg and De Beer 2012; De Beer et al. 2012). It is therefore important to discuss these observations and associated conditions in more detail in order to improve the quality of road construction in southern Africa. Examples of full-scale testing with the Heavy Vehicle Simulator (HVS), including the so-called traffic-associated "crushing failure" found in the lightly cemented base layers as well as typical fatigue cracking are discussed.

2 Full-Scale HVS Tests on Pavements

Netterberg and De Beer (2012) showed that using instruments such as the full scale Heavy Vehicle Simulator (HVS) together with the Dynamic Cone Penetrometer (DCP) could be successfully used to illustrate the rather critical importance of avoiding a weak interlayer between the road base and the surfacing layer. In Fig. 1a

Fig. 1 Full scale HVS tests on cemented base pavement (**a**) and recycled HMA (**b**) during wet conditions. Note the fatigue cracks (**b**) and weak upper cementitious subbase layer underneath the recycled HMA shown in **c**—typical "sandwich" structure

an illustration is given of a typical full-scale HVS test (wet) on a cracked cemented base layer with weak interlayers within and underneath the base. Note: thin (~12 mm) asphalt seal surfacing, and ~25 mm weak layer at bottom of ~80 mm cemented base. Figure 1b, c illustrates an HVS test example, where a relatively stiff hot mix asphalt base layer (~70 % recycled HMA) on a relatively weak upper cemented subbase "sandwiched" (Fig. 1c) between HMA and lower cementitious subbase were tested. Typical "fatigue cracking" failure is clearly shown (Fig. 1b) after an HVS test in wet conditions by spraying water on the test surface, as well as gravity in-depth water feed (top 500 mm of pavement) next to the HVS test sections.

3 Crushing Failure Mechanism: Cemented Base Pavements

Lightly cemented base/subbase layers may also suffer from traffic-associated "crushing failure" (Litwinowicz and De Beer 2013). This failure mode could result from a relatively weak upper portion in cemented bases, but could even occur in relatively dry conditions. This "weakness" in the layer is to be seen relative to the applied tyre contact stress it needs to carry from especially truck tyres. See rather self-explanatory images Fig. 2a, b, as well as Fig. 3a, b. In this paper, it is assumed that the tyre inflation pressure is and equal to the vertical tyre contact stress. However, a multitude of studies have indicated that actual three-dimensional (3-D) tyre-pavement contact stresses are not simple, and special provision should, ideally, be made to incorporate these complex contact stresses into pavement design. In these days of higher tyre inflation pressures (which result in actual tyre contact

Fig. 2 Crushing failure resulting in a pothole of a cemented base pavement in (**a**), and failure in top of a lightly cemented base layer—after HVS testing in (**b**)

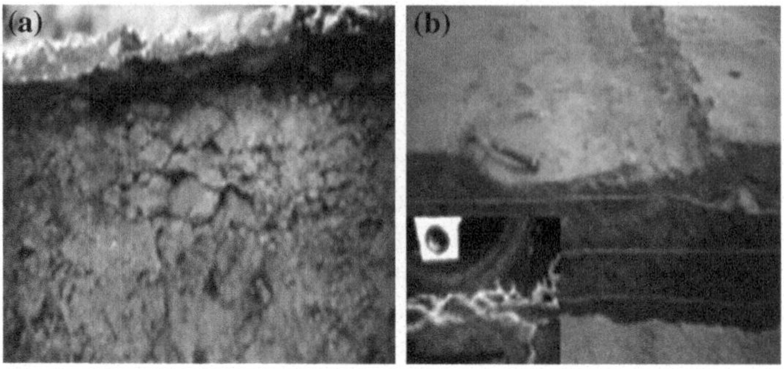

Fig. 3 Crushing of a lightly cemented dry base layer: result of HVS trafficking in (**a**). Final state is shown in (**b**), after HVS testing in the wet state. *Insert* in (**b**): signs of wet pumping under HVS test tyre after water sprayed on surface

stresses of up to 1000 kPa or more), the importance of avoiding a relatively weak upper base is well illustrated (Netterberg and De Beer 2012; De Beer et al. 2012). Even in the absence of relatively weak layers, or interlayers, cementitious base layers may suffer "crushing failure", of which empirical damage laws were developed. See recent work by Litwinowicz and De Beer (2013).

In this section simplistic mechanistic analyses of various types of road pavement structures with and without interlayer slip conditions are briefly illustrated. Simple Multiple Layer Linear Elastic (MLLE) with variable interlayer slip (a compliance model), as well as horizontal tyre loading was used, as proposed by De Beer et al. (2012) and Maina et al. (2007). The Strain Energy of Distortion (SED), a combination of normal and shear stresses in an elastic system (Timoshenko and Goodier 1951) was used as the primary pavement response parameter, quantitatively indicating "hot spots" for initiation of failure or distortion within the pavement structure. "Hot spots" are defined as peak SED response values.

A four layered "linear elastic" pavement with 15 mm surfacing was modelled by using dual tyres reporting SED responses from the pavement. The load configuration is the standard loading per tyre, which is 20 kN and a value of 520 kPa for the contact stress were used. See insert in Fig. 4. This figure also shows the different layer thicknesses and the depth of the lamination in the cemented base layer, which was modelled at a depth of L = 65 mm (i.e. 15 mm + 50 mm). See Fig. 4, note the relatively high value ($\sim$130 N-m/m^3) of SED at depth (L) = 65 mm with full slip ($\sim$single lamination at L = 65 mm) directly under the tyre. In Fig. 5 note the relatively high value of SED at L = 65 mm with full slip ($\sim$single lamination at L = 65 mm) and relatively weak layer (t = 50 mm) on top of the base layer, between the two tyres (Position B).

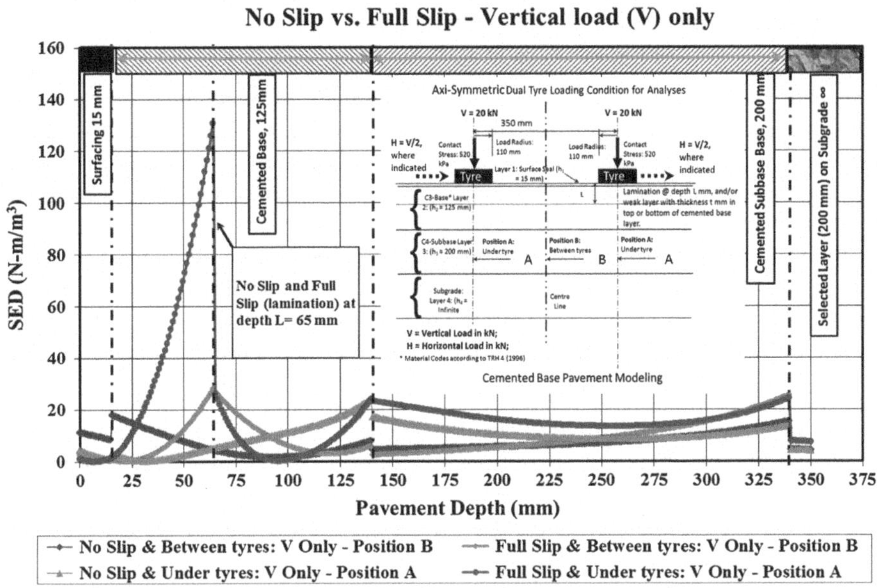

Fig. 4 Strain energy of distortion (SED) response: cemented base pavement, with and without slip at L = 65 mm. *Note insert* Position A = between, Position B underneath tyres. *V* Vertical loading. *V* Vertical load, *H* horizontal load

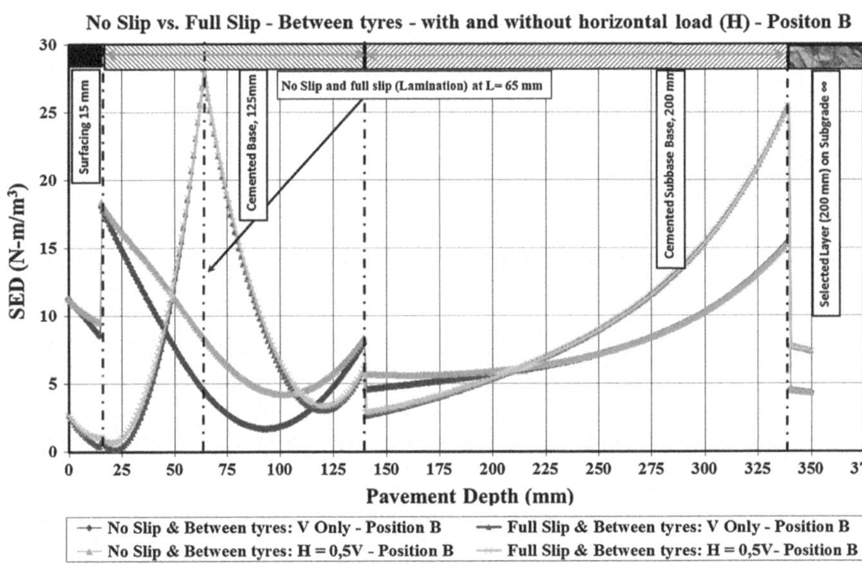

Fig. 5 Strain energy of distortion (SED) response of cemented base pavement between the two tyres (see *insert* Fig. 4), with and without horizontal (H) load

4 Conclusions and Recommendations

It is concluded that the effects of weak interlayers, laminations and/or weak interfaces can potentially be modelled mechanistically to illustrate and quantify potential failure conditions (or distortion) and their locations. The effect of potential crushing failure of cemented pavement base/subbase layers can also be quantified, albeit empirically.

These conditions in the structural pavement layers should (and could) be avoided, especially during construction, in order to ensure quality structural performance of pavements in the long term. For the design and modelling of durable flexible and semi-rigid pavements in the context of this summary paper, the following two aspects are recommended:

- The mechanistic methodology should include an analysis based on the Strain Energy of Distortion (SED) response parameter, which may greatly improve engineering understanding of the failure mechanisms, structural behaviour and associated potential for *premature* failure (or distortion) of flexible and semi-rigid pavement system and its layers.
- Full-scale pavement studies should be performed and evaluated to calibrate the SED as a potential failure or "distortion" parameter, especially for roads in the conditions described above.

Acknowledgements This paper is published with the approval of the Executive Director of the CSIR Built Environment unit, Dr C Ruiters.

References

De Beer M, Maina J W & Netterberg, F (2012). Mechanistic modelling of weak interlayers in flexible and semi-flexible road pavements – Part 2. Journal of the South African Institution of Civil Engineering, Volume 54 Number 1 April 2012, pp43-54.

COLTO (Committee of Land Transport Officials) (1996). Structural design of flexible pavements for interurban and rural roads. Draft TRH 4: 1996, Pretoria: Department of Transport.

Litwinowicz, A., and De Beer., M., (2013). Long-term crushing performance of lightly cementitious pavement materials – update to the South African procedure, Road Materials and Pavement Design, DOI:10.1080/14680629.2012.755934.

Maina, J W, De Beer, M & Matsui, K (2007). Effects of layer interface slip on the response and performance of elastic multi-layered flexible airport pavement systems. Proceedings, 5th International Conference on Maintenance and Rehabilitation of Pavements and Technological Control (MAIREPAV5), Park City, Utah, US, 8–10 August.

Netterberg, F & De Beer, M (2012). Weak interlayers in flexible and semi-flexible road pavements – Part I. Journal of the South African Institution of Civil Engineering, Vol 54 No 1, April 2012, Pages 32–42, Paper 762-1.

Timoshenko, S., Goodier, J. N. (1951). Theory of Elasticity, 2nd Ed., New York, McGraw- Hill, Inc., 1951.

EME2 Fatigue Properties and Pavement Design in a Sub-tropical Climate

Laszlo Petho and Peter Bryant

Abstract Enrobé à module élevé Class 2 (EME2) is currently being introduced into Queensland which has a warm sub-tropical climate. While the French design system for flexible pavements directly applies inputs from performance-based mix design, immediate implementation of such a system in Queensland was inhibited by the existing Australian pavement design methodology which does not make such links. In addition, there has been limited application of EME2 in similar climates as in Queensland. The execution of a manufacturing and paving trial, including collection of pavement temperature at various depths over a one-year period, provided vital input into the technology transfer. Flexural modulus temperature-frequency sweep tests, and a series of asphalt fatigue tests at 10, 20 and 30 °C were carried out. The results provide input into the LCPC-developed methodology for calculating asphalt fatigue properties at a given equivalent temperature, and now is utilized and validated for a warm sub-tropical climate.

Keywords EME2 asphalt · Temperature distribution · Performance-based · Fatigue properties

1 Background

Enrobé à module élevé (EME) was developed in France over 30 years ago. EME2, which is the highest class of EME in the French standards, is currently being introduced into Queensland, which has a warm sub-tropical climate. EME2 is primarily intended to reduce the thickness of full depth asphalt pavements whilst

---

L. Petho (✉)
ARRB Group, Brisbane, Australia
e-mail: laszlo.petho@arrb.com.au

P. Bryant
Department of Transport and Main Roads, Brisbane, Australia
e-mail: peter.n.bryant@tmr.qld.gov.au

A. Chabot et al. (eds.), *8th RILEM International Conference on Mechanisms of Cracking and Debonding in Pavements*, RILEM Bookseries 13,
DOI 10.1007/978-94-024-0867-6_60

still providing sound pavement performance through a combination of high modulus, and superior fatigue, deformation and moisture resistance.

Mix design requirements have been developed at a national level (Austroads 2014a). Readily available performance-based test methods in the existing Australian Level 2 mix design system (Austroads 2014b) provided the basis for the Australian EME2 mix design framework. These also provide the platform for establishing links to pavement design, which is the main focus of this paper.

A research project was carried out through the Department of Transport and Main Roads (TMR) and ARRB National Asset Centre of Excellence (NACoE) to develop structural design procedures for EME2 pavements. Based on this work, interim pavement design guidelines were developed by Petho and Bryant (2015) and published in Technical Note 142 (Department of Transport and Main Roads 2015).

To enable immediate implementation, the interim method adopted the current Austroads pavement design procedure which, unlike the French methodology, does not make direct links to performance-based mix design. This paper details some of the more recent developments aimed at removing this disconnect, including detailed analysis of both pavement temperature and fatigue properties of EME2. Analysis of the results also enabled the development of transfer functions for pavement structures containing EME2.

2 Pavement Temperatures Under Queensland Climatic Conditions

Detailed pavement temperature measurement was conducted in Australia in the 1970s (Dickinson 1981). Unfortunately the complete data from those experiments is no longer available for detailed analysis. To gain a better understanding of the relationship between pavement temperature and performance, temperature sensors were installed in the pavement at the first Australian EME2 trial in Brisbane, as detailed in Austroads (2014a). The sensors collected pavement temperatures at 50, 70, 110, 190, 290 and 390 mm depth at a frequency of 10 min.

More than 50,000 temperature readings per sensor were collected and analyzed for the period between 8 May 2014 and 9 May 2015. Table 1 shows that the average pavement temperature throughout a year is unchanged with depth, and the standard deviation slightly decreases with increasing depth.

Table 1 Pavement temperature analysis at different depths in Brisbane, Queensland

Depth from the road surface (mm)	50	70	110	190	290	390
Average pavement temperature (°C)	30.4	30.4	30.4	30.4	30.4	30.4
Standard deviation pavement temperature (°C)	9.5	8.7	7.8	7.1	6.8	6.7

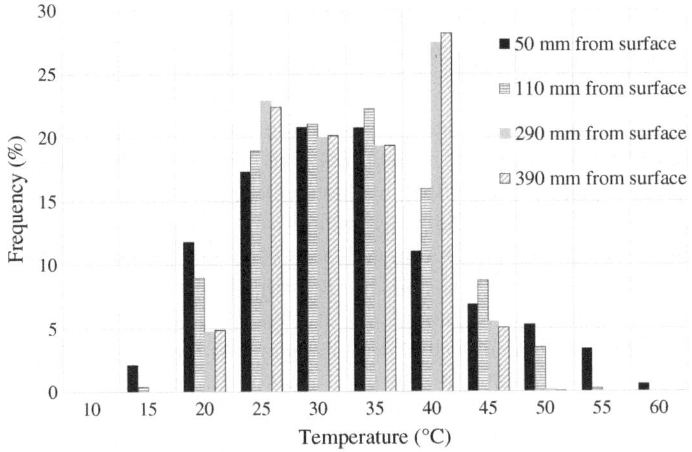

Fig. 1 Temperature distribution in Brisbane, Queensland between May 2014 and May 2015

Figure 1 shows the temperature frequency distribution at selected depths. Both Table 1 and Fig. 1 show that with increasing depth the temperature range decreases. The distribution shape also changes from a lognormal distribution at 50 mm to a normal distribution at 110 mm. At greater depths of 290 and 390 mm the distribution is bi-modal. Hence, it is reasonable to assume a normal distribution for upper layers, but histograms should be used for lower layers.

3 Calculation of the Allowable Strain at Different Temperatures

EME2 pavement design in metropolitan France uses an equivalent temperature of 15 °C. Equation 1 shows the Laboratoire Central des Ponts et Chausees (LCPC) method for accounting for the dependency of the modulus E and the strain at given temperature, and can be used over a broad range of temperatures (LCPC 1997). By using this equation the fatigue properties at any given equivalent temperature can be calculated using results from the standardized fatigue test, and flexural stiffness results from a temperature-frequency sweep.

$$\varepsilon_6(\theta) \times E(\theta)^n = constant \quad (1)$$

where $\varepsilon_6(\theta)$ is the fatigue resistance of the asphalt mix, determined at 10^6 loading cycles at the equivalent temperature θ, $E(\theta)$ is the stiffness of the asphalt material at the equivalent temperature θ, and n is a material constant.

3.1 Laboratory Validation of the Fatigue Properties at Different Temperatures

Bodin et al. (2010) presented a series of two-point bending fatigue tests at different temperatures, using two different asphalt materials, which provides validation of the model in Eq. 1; they estimated that by using the exponent of 0.5 the laboratory fatigue behavior of EME2 at higher temperatures is overestimated.

For local validation, flexural fatigue tests were carried out on an Australian EME2 mix at 10, 20 and 30 °C and 10 Hz using four-point bending (Austroads 2015). The fatigue tests were carried out at three strain levels, with results summarized in Table 2. The results show that the slope of the fatigue curves (classical power function) at different temperatures is different and the residual standard deviation decreases with increasing temperature, providing a more reliable data set. Four-point bending flexural stiffness testing (Austroads 2015) was also undertaken, at a range of frequencies and temperatures of 5, 15, 25 and 30 °C.

As was expected, the calculated strain at 1 million cycles (ε_6) increased with increasing temperature; another observation is that the residual standard deviation, which is an indication of the reliability of the correlation, decreased with increasing temperature. Figure 2a shows the ε_6 value as a function of the temperature and the calculated trendline for these three values. By using the trendline equation the (ε_6) was extrapolated beyond the test temperatures, which is shown in Fig. 2b; the

Table 2 Flexural 4-point-bending fatigue test results at 10, 20 and 30 °C

Property	10 °C, 10 Hz	20 °C, 10 Hz	30 °C, 10 Hz
Air voids, average (%)	3.9	3.4	3.4
Strain at 1 million cycles (ε_6)	138	170	243
S_N (residual standard deviation)	0.30	0.26	0.14

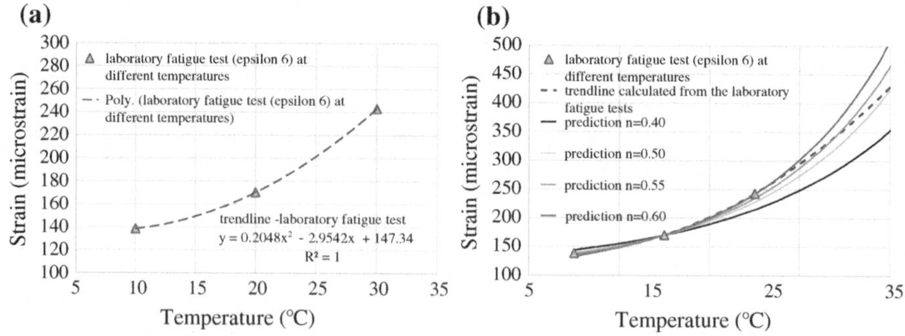

Fig. 2 Fatigue test results at different temperatures: **a** laboratory fatigue test results, **b** estimated fatigue properties using different exponents

figure also shows ε_6 calculated according to Eq. 1, with the exponent n selected at 0.40, 0.50, 0.55 and 0.60.

The error was calculated for each exponent of Fig. 2b over the temperature range of 10 to 45 °C to determine which exponent would correspond best with real fatigue test results. Between 30 and 45 °C the fatigue data was extrapolated using the trendline equation in Fig. 2a. For n of 0.40, 0.50, 0.55 and 0.60 the error was calculated as 113, 34, 38 and 86 respectively. This means that using Eq. 1 with the exponent of n = 0.50, the difference between the calculated and the measured (ε_6) value is minimal. Therefore the fatigue properties at different temperatures can be reliably estimated for Australian test conditions by using Eq. 1 with n = 0.5. By using a value of n = 0.5, Eq. 1 can be reorganized as shown in Eq. 2.

$$\varepsilon_6(\theta_i) = \varepsilon_6(20\,^\circ\mathrm{C}; 10\,\mathrm{Hz}) \times \sqrt{\frac{E(20\,^\circ\mathrm{C}; 10\,\mathrm{Hz})}{E(\theta_i; 10\,\mathrm{Hz})}} \tag{2}$$

3.2 Calculation of the Fatigue Damage

To demonstrate the calculation of the elementary damage, two pavement structures were considered comprising 110 and 290 mm thick (full depth) EME2 asphalt supported by a granular foundation layer and subgrade with CBR 7 %.

The pavement response (asphalt strain) was calculated using the CIRCLY linear elastic software package. The results, assuming a fatigue curve slope of 0.2 (the most commonly used fatigue characteristic for asphalt), are shown in Table 3 and Fig. 3.

The elementary damage, the weighted elementary damage—calculated using the temperature distribution in Fig. 1—and the total damage were calculated in accordance with LCPC (1997), as shown in Fig. 3. The equivalent temperature is determined where the elementary damage equals the total weighted damage; this is 34 °C for 110 mm EME2 and 35 °C for 290 mm EME2. Further refinement is needed to accurately calculate the equivalent temperature for Eq. 2.

Table 3 Data input for calculating the fatigue damage

Temperature (°C)	10	15	20	25	30	35	40	45	50
$E_{asphalt}$ (MPa)	16,127	13,358	10,637	8112	5905	4096	2708	1709	1034
$\varepsilon_{asphalt}$—110 mm EME2	115	132	155	188	233	297	384	499	643
$\varepsilon_{asphalt}$—290 mm EME2	30	35	42	51	65	85	115	158	217

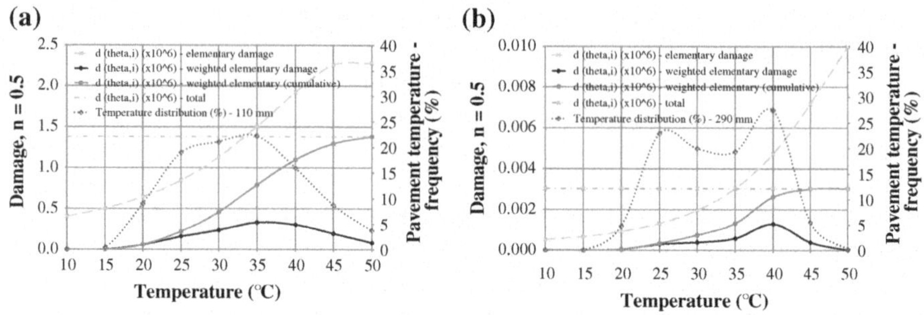

Fig. 3 Calculated fatigue damage: **a** 110 mm EME2 asphalt, **b** 290 mm EME2 asphalt

4 Conclusions

This paper explores recent developments in the transfer of EME2 technology to Queensland, specifically relating to the linking of pavement design with performance-based mix design, and its applicability in a sub-tropical climate.

Key conclusions from this work include: (a) it is reasonable to assume that pavement temperature over one year is normally distributed for upper pavement layers, but histograms should be used for lower layers; (b) an exponent of $n = 0.50$ in Eq. 1 is suitable for use in Queensland for the mix considered; and (c) the equivalent temperature in Brisbane was determined for two example pavement designs.

References

Austroads (2014a), High modulus high fatigue resistance asphalt (EME) technology transfer, AP-T283/14, Austroads, Sydney, NSW.

Austroads (2014b), Guide to pavement technology: part 4b: asphalt, AGPT04B-14, Austroads, Sydney, NSW.

Austroads (2015), Characterisation of flexural stiffness and fatigue performance of bituminous mixes, test method, AGPT-T274, Austroads, Sydney, NSW.

Bodin, D, Terrier, J, Perroteau, C, & Hornych, P (2010), 'Effect of temperature on fatigue performances of asphalt mixes', 11th International conference on asphalt pavements, Nagoya, Japan, ISAP, USA.

Department of Transport and Main Roads (2015), Technical Note 142 High Modulus Asphalt (EME2) Pavement Design, TMR, Brisbane, QLD.

Dickinson, EJ (1981), Pavement temperature regimes in Australia, special report SR 23, Australian Road Research Board, Vermont South, Vic.

Laboratoire Central des Ponts et Chaussées (LCPC) (1997), French design manual for pavement structures, LCPC, Paris, France.

Petho, L and Bryant, P (2015), High modulus asphalt (EME2) pavement design in Queensland, AAPA International flexible pavements conference, 16th, 2015, Gold Coast, Queensland, Australia.

Comparative Assessment of Pavement Fatigue Life Prediction Approaches for Use in Performance Based Specification

Mirkat Oshone and Jo Sias Daniel

Abstract There is a need to incorporate long term pavement performance evaluation into the current specification through development of performance-related specification (PRS). PRS calls for the prediction of pavement life and development of pay adjustment factors based on expected life differences between as-design and as-built conditions. Different distress models have been developed and are in use by different researchers. Testing and analysis features, simplicity, test equipment and knowledge availability play a significant role in selection of the model for use. Nevertheless, to date, a comparative assessment of how the different analysis approaches affect pavement performance evaluation has not been undertaken. In this study fatigue performance evaluation is done on pavements with different surface mixture condition, varying asphalt and air void content to replicate a range of conditions in the field. Material characterization is done by performing complex modulus and fatigue testing. Pavement response is computed using three layer elastic analysis (LEA) programs. Fatigue life evaluation is done using three different fatigue life prediction models. The results from dynamic modulus testing show that there is an increase in stiffness with a decrease in asphalt and air void content which infers a reduction in tensile strain at the bottom of the asphalt layer. The tensile strain outputs from the different LEA programs were comparable and there is about a 20 % increase in strain for a change in bond condition from full bond to full slip. There is a distinction between predicted fatigue life due to the use different models. The highest variation is observed at low air void content level whereas the impact was lower at optimum and high air void content levels.

Keywords Performance specification · Fatigue performance · Elastic analysis · Cracking models

M. Oshone (✉) · J.S. Daniel
University of New Hampshire, Durham, USA
e-mail: mto1@unh.edu

J.S. Daniel
e-mail: jo.daniel@unh.edu

A. Chabot et al. (eds.), *8th RILEM International Conference on Mechanisms of Cracking and Debonding in Pavements*, RILEM Bookseries 13,
DOI 10.1007/978-94-024-0867-6_61

1 Introduction

State agencies use pay adjustment in their contracts typically based on percent-within limits or percent defective specifications of field test results. Three common parameters that are used to determine the acceptance of the constructed work and the assigned level of payment are: in-situ air void, asphalt content, and roughness of the final pavement surface. These can be easily determined after construction, however even when these values are within the specification, the longevity of the pavement isn't necessarily ensured. There is, therefore, an exigent need to develop better long term pavement performance evaluation models and PRS. The implementation of PRS helps state agencies to avoid failure for a given period of time based on predicted lives. Contractors also get motivated by the incentive payment to adjust their mix, and produce substantially high performing pavements (Jeong 2010). Nevertheless, a great deal of research has to be carried out before the transition is realized.

PRS calls for the prediction of pavement life and development of pay adjustment factors based on expected life differences between as-design and as-built conditions. Long term pavement performance in terms of predicted pavement life using simple mathematical predictive models is originated from Weed (1996). Chamberlin (1995) also emphasized on the development of efficient pavement performance prediction models to be used as a basis for developing pay adjustment factors. Models, with particular strengths and weaknesses, have been developed to predict the response and performance of pavements. Nonetheless, there is very limited information on how the predictions made by the different models compare with each other and which ones are appropriate for a particular set of circumstances. This paper therefore examines how the analysis approach would impact the prediction of pavement performance with respect to fatigue cracking. Material characterization is done by performing complex modulus and fatigue testing. Three LEA programs (WinJULEA, WESLEA and KENPAVE) are used to compute pavement response considering both full bond and full slip conditions. Pavement fatigue performance evaluation is done using three fatigue life prediction models that include Asphalt institute (MS-1) and MnDOT transfer functions and analysis using the simplified viscoelastic continuum damage (S-VECD) fatigue testing and analysis approach.

2 Materials and Testing

For this study, a typical 12.5 mm nominal maximum aggregate size (NMAS) Rhode Island mixture was selected for evaluation. Specimens were fabricated using a gyratory compactor at three asphalt content levels (5.2, 5.6, and 6.1 %) and three air void content levels (4.0, 7.0, and 9.0 %) to replicate the range of conditions in the field.

Complex modulus testing was performed on three replicate specimens following AASHTO TP 62 and master curves were constructed at a reference temperature of

21.1 °C. Fatigue testing in uniaxial tension mode was carried out following AASHTO TP 107. The S-VECD approach developed by Underwood and Kim (2010) was used to analyze the fatigue test results.

3 Results

Figure 1 shows the impact of changing asphalt and air void content on the dynamic modulus (|E*|) master curves. As expected, the stiffness decreases with an increase in asphalt and air void content. For low air void content mixture, the low and

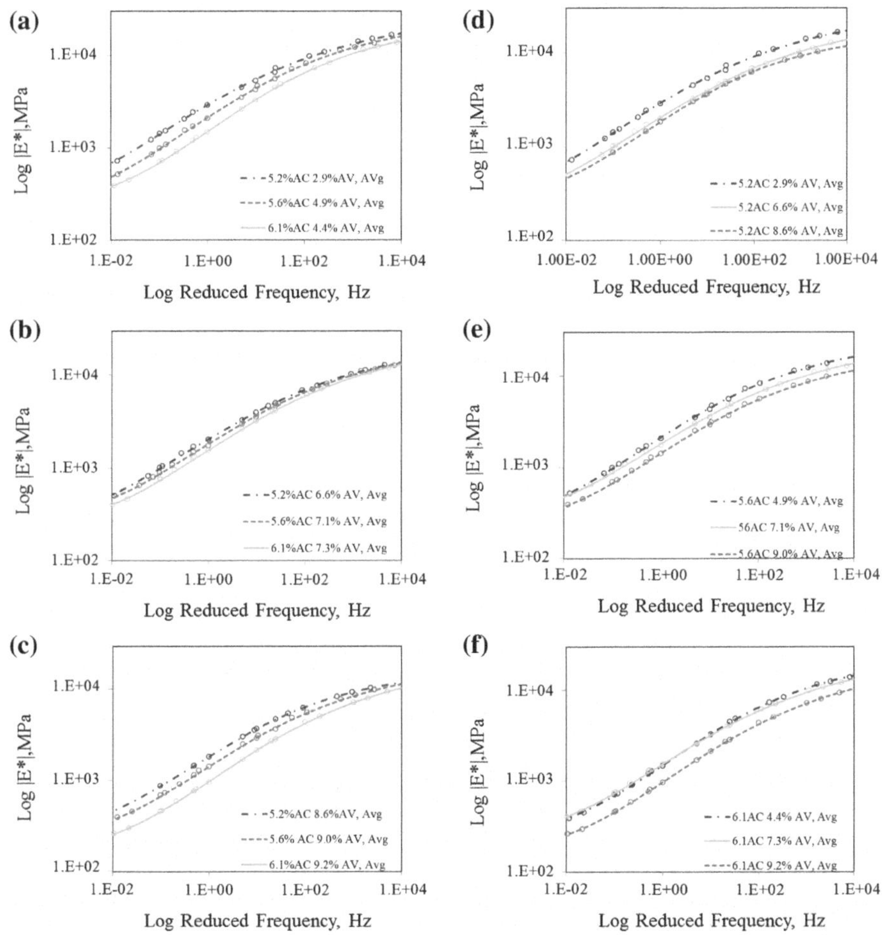

Fig. 1 Effect of variation of asphalt and air void content on mixture stiffness. **a** Low air void content, **b** optimum air void content, **c** high air void content, **d** Low asphalt content, **e** optimum asphalt content, **f** high asphalt content

optimum asphalt content master curves are significantly different at low temperature. For high air void content mixture, the optimum and high asphalt content master curves are significantly different at intermediate and high temperatures and the low and optimum asphalt content master curves are significantly different at low temperature. In addition, a significant difference is observed among the low and high asphalt content master curves at all temperatures. For low asphalt content mixture, there is a statistically significant difference between |E*| master curves of low and optimum air void content mixture and between the low and high air void content |E*| master curves at all test temperatures. For optimum asphalt content mixtures, the optimum and high air void content |E*| master curves and the low and high air void content |E*| master curves are significantly different at low and intermediate temperatures. For high asphalt content mixture, the low and high air void content |E*| master curves are significantly different from each other at all test temperatures whereas the optimum and high air void content |E*| master curves are significantly different at intermediate and high temperatures.

4 Pavement Evaluation

The pavement cross-section used for this study comprises asphalt concrete (AC) surface layer (2 in), base course (10 in), sub base (12 in) and sub grade (semi-infinite). Material properties and structure of the base course, sub base and sub grade are kept constant for all conditions. To account the effect of the different pavement conditions due to variation in % asphalt and % air void content, the |E*| values at the average effective temperature and frequency corresponding to each condition was used for the surface layer.

The tensile strain at the bottom of the asphalt layer was computed using three different LEA programs considering both full bond and full slip conditions between layers. The strains calculated from all three programs are nearly equal, Fig. 2. There is approximately a 20 % increase in strain as the bond condition changes from full bond to full slip. The result also shows that there is an inverse relationship between tensile strain and stiffness of AC surface layer, as expected.

In order to demonstrate the response of mixtures to fatigue damage, the consideration of the all layers materials and the pavement structure is crucial. Three different approaches to predicting the fatigue life were compared in this study using the tensile strain calculated from LEA programs.

The first (Eq. 1) is a fatigue transfer function developed by Minnesota Department of Transportation (MnDOT) (Timm et al 1999). The equation was originally developed by Illinois Department of Transportation (Thompson 1987) and then calibrated with data from the Minnesota Road research (Mn/Road). The second is the Asphalt Institute (MS-1) model (Asphalt Institute 1982) which is based on a constant stress criterion. The model is calibrated with national data and shown in Eq. (2). The third (Eq. 3) is from the S-VECD model and the traditional fatigue coefficients k_1 and k_2 in the equation are obtained from S-VECD analysis.

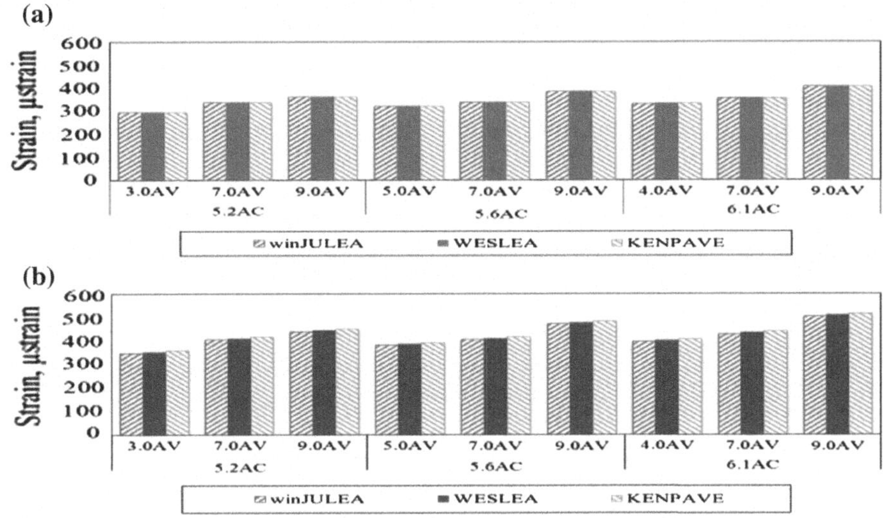

Fig. 2 Tensile strain at the bottom of the asphalt layer. **a** Tensile strain, full bond condition, **b** tensile strain, full slip condition

The failure criterion in S-VECD model is based on the amount of the rate of change of the averaged released pseudo strain energy and is represented by G^R.

$$N_f = 2.83 \times 10^{-6}\left(\frac{10^6}{\varepsilon_t}\right)^{3.148} \tag{1}$$

$$N_f = 0.07566 * C * \beta_{f1} * k_1'\left(\frac{1}{\varepsilon_t}\right)^{3.291*\beta_{f2}}\left(\frac{1}{E}\right)^{0.854*\beta_{f3}} \tag{2}$$

$$N_f = K_1\left(\frac{1}{\varepsilon_t}\right)^{K_2} *(|E*|)^{K_3} \tag{3}$$

The N_f computed from the three models is displayed in Fig. 3a. The MnDOT transfer function seems to be insensitive and doesn't show the variation that is expected due to the different asphalt and air void content level in the surface mix. The N_f from the MS-1 model indicates that at all asphalt content levels the low air void content resulted in a high fatigue resistant pavement where as the high air void content resulted in a lower fatigue resistant pavement. The N_f from S-VECD shows the same trend as MS-1 model at low and optimum asphalt content levels. However at high aspahlt content level, a significant variation between the two predictions is seen.

When it comes to implementing the predicted fatigue life from the different approaches in to PRS, the N_f due to variation in volumetric properties are normalized with respect to the as-design N_f. Figure 3b shows the expected life

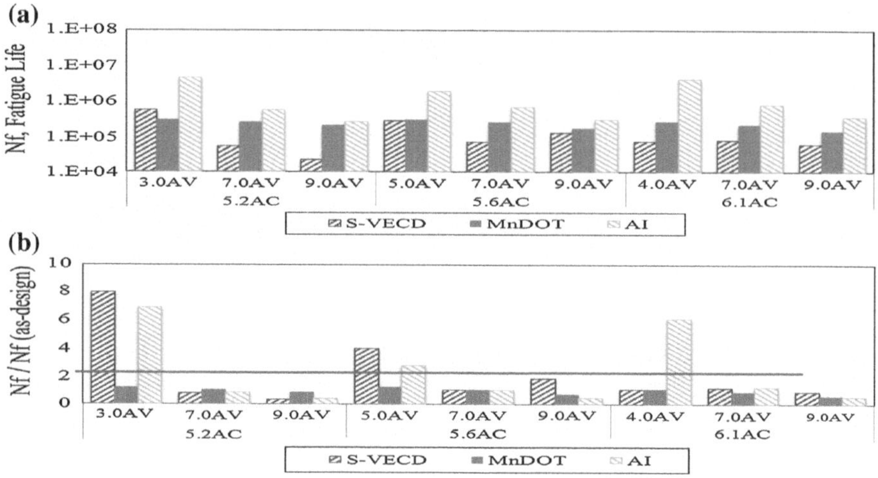

Fig. 3 **a** Pavement fatigue life, N_f, **b** expected fatigue life differences, $N_f/N_{f(as\text{-}design)}$

differences interms of $N_f/N_{f(as\text{-}design)}$ ratio. The red line in the graph represents a ratio of 1, indicating equal N_f values of field target condition (5.6 % asphalt content and 7.0 % air void content). It should be noted that expected life differences computed from the models serve as a basis for pay adjustment factor as part PRS. The observation of the expected life differences from the different models indicate that there is a considerable variation. The MnDOT values are highly distinct than S-VECD and MS-1 values at low and optimum asphalt content and low air void content level. There is also a considerable variation among S-VECD and MS-1 predictions at high asphalt content level and low air void content level. Generally the difference observed between the models is lower at optimum and high air void content levels.

5 Conclusions

In this study the effect of analysis approach on fatigue performance evaluation of mixtures with a range of asphalt and air void content was evaluated. Mixtures were evaluated with complex modulus and S-VECD fatigue testing. Pavement response and fatigue life computations are done using three LEA and fatigue life prediction models. The expected life differences, a basis for pay adjustment factor development, due to the different mixture condition computed from the three models were compared. The complex modulus testing showed a decrease in stiffness with an increase in air void and asphalt content which infers an increase in tensile strain. The LEA programs displayed similar tensile strain output with a 20 % increase for full slip condition than the full bond condition between layers. The use of the different fatigue life prediction models caused a distinction among the expected life differences. The effect is pronounced at low air void content level and was minimal at optimum asphalt content level. A transition to PRS requires carrying out different performance tests, use of different LEA and pavement performance prediction models. This study showed that the selection of different methods will have an impact on the prediction of the expected pavement life differences which in turn impacts the pay adjustment factor, a key parameter in PRS. Tremendous effort has to be made in refining the methods before implementation.

Acknowledgment The authors would like to acknowledge the Rhode Island Department of Transportation for sponsoring the project and supplying the mixtures.

References

Asphalt Institute (1982) Research and development of the Asphalt Institute's thickness design manual (MS-1). Asphalt Institute, College Park

Chamberlin, W.P. (1995) NCHRP Synthesis 212: Performance-Related Specifications for Highway Construction and Rehabilitation, Washington, D,C

Jeong, M.G. (2010) Implementation of a Simple Performance Test Procedure in a Hot Mix Asphalt Quality Assurance Program. Ph.D. thesis, Arizona State University

Weed, R.M (1996) Quality Assurance Software for the personal computer. Transportation Research Record 1544, pp.116-124

Underwood, B.S., and Y.R. Kim (2010) Improved calculation method of damage parameter in viscoelastic continuum damage model. International Journal of Pavement Engineering, Vol. 11, p 459-476

Thompson, M.R. (1987) ILLI-PAVE based full-depth asphalt concrete pavement design procedure. Sixth International Conference on Structural Design of Asphalt Pavements. Ann Arbor, MI

Timm, D.H., Newcomb, D.N., Birgisson, B., and Galambos, T.V. (1999) Incorporation of Reliabilty into the Minnesota Mechanistic-Empirical Pavement Design, Final report MN/RC-1999-35. Minnesota Department of Transportation, Minnesota

Coupling a Multi-linear Fatigue Model and a Pavement Model to Estimate Truck Aggressiveness

Farah Homsi, Jean-Maurice Balay, Didier Bodin, Denys Breysse and Sylvie Yotte

Abstract The fatigue performance of asphalt materials is a key parameter for asphalt pavement structural design. It is determined in the laboratory as a function of the loading strain magnitude under standard continuous sinusoidal signals. However, real loading signals measured in asphalt pavements differ in shape from sinusoidal signals and many studies showed that the shape of the loading signal has an important effect on the fatigue life of asphalt mixtures. Moreover, to improve road transport efficiency and reduce its environmental impact, new regulations allow longer trucks to circulate on highways and allow current trucks to transport heavier loads which implies a need of evaluating the impact of these changes on pavement fatigue performance. In this paper, a multi-linear fatigue model taking into account the shape of the loading signal is coupled with a viscoelastic model to evaluate the fatigue life of asphalt pavements and determine the effect of loading conditions and trucks configurations.

Keywords Asphalt pavements · Fatigue · Multiple axles · Pavement design

F. Homsi (✉)
Faculty of Engineering, Beirut Arab University,
P.O. Box 11-5020 Riad El Solh, Beirut 11072809, Lebanon
e-mail: f.homsi@bau.edu.lb

J.-M. Balay
Université Nantes Angers Le Mans, IFSTTAR, Route de Bouaye,
CS4, 44341 Bouguenais cedex, France
e-mail: jean-maurice.balay@ifstaar.fr

D. Bodin
ARRB Group Ltd, 500 Burwood Highway, Vermont South,
VIC 3133, Australia
e-mail: didier.bodin@arrb.com.au

D. Breysse
Département Génie Civil et Environnement (GCE), Université Bordeaux 1,
I2M UMR CNRS 5295, Bât B18, Av des Facultés, 33405 Talence cedex, France
e-mail: denis.breysse@u-bordeaux1.fr

S. Yotte
Université Limoges, GEMH, Boulevard Jacques Derche, 19300 Egletons, France
e-mail: sylvie.yotte@unilim.fr

A. Chabot et al. (eds.), *8th RILEM International Conference on Mechanisms of Cracking and Debonding in Pavements*, RILEM Bookseries 13,
DOI 10.1007/978-94-024-0867-6_62

1 Introduction

The repeated passes of loading signals on pavements cause their fatigue damage which is a key issue for their design. The French asphalt pavements design method is based on the European standard laboratory fatigue test that consists of the application of continuous sinusoidal signals at a given temperature, a fixed frequency (10 °C–25 Hz in the French context) and different strain levels on the top of a trapezoidal specimen clamped at its base. The number of loading signals before failure, characterized by the loss of half of the initial stiffness, is determined and the fatigue life model giving the fatigue life as function of the strain level is written:

$$\log(Nf) = a\log(\varepsilon) + b \tag{1}$$

where Nf is the number of loading cycles to failure, ε the strain magnitude and a and b the relationship parameters.

Real truck loadings differ in shape from sinusoidal signals (Chatti and El-Mohtar 2004) which may have an impact on the pavement fatigue damage. In the current French design method, this effect is taken into account by using load equivalency factors calculated using Miner's law (Miner 1945). Several authors have estimated the effect of the loading signals shape on the fatigue life of bituminous mixtures either by means of laboratory tests (Francken 1979; Raithby and Sterling 1972; Gillespie et al. 1992; Breysse et al. 2002; Chatti and El-Mohtar 2004; Kogo and Himeno 2008; Bodin et al. 2009), or by means of in situ measurements (Salama et al. 2006). Research results show that signal parameters: rest time between signals, frequency and number of peaks have an influence on the fatigue damage of asphalt mixtures. In addition, the standard trucks for international freight in France (T2S3, 40 tons on 5 axles) are now allowed to carry a load of 44 tons without adding extra axles and European Modular Systems (EMS), which are heavier trucks (48–60 tons) including more axle-group configurations are now allowed to circulate in Europe Bernadet (1995). On the one hand, decreasing the number of trucks running on a pavement by increasing the number of axles and increasing their allowable load could result in environmental benefits if it reduces gas emissions. On the other hand, the effect on the fatigue life of asphalt pavements under multiple axle configurations must be evaluated.

2 The Multi-linear Fatigue Life Model and Its Coupling with a Pavement Model

The classical fatigue model (Eq. 1) being unable to take into account the shape of the loading signal, a multi-linear fatigue model was developed from laboratory fatigue tests under a range of loading pulses (Homsi et al. 2012a, b). It gives the fatigue life of bituminous mixtures as a function of four independent parameters:

$$\log(\mathrm{Nf}) = -4.58\log(\varepsilon) - 0.84\log(\mathrm{N_p}) + 1.31\hat{\mathrm{A}}_\mathrm{n} + 1.76\bar{\mathrm{D}} + 15.22 \tag{2}$$

where ε is the strain magnitude, N_p the number of peaks in the loading signal, $\hat{A}_n$ the area under the loading signal normalized by both the strain level and the loading duration and $\bar{D}$ is the loading duration normalized by the number of peaks.

The importance of energy parameters having been pointed out by different authors (Chatti and El-Mohtar 2004; Kogo and Himeno 2008), a variant of the multi-linear fatigue model was also used, replacing the strain level by the dissipated energy by cycle W_d calculated using the Huet-Sayegh model, as in Eq. 3:

$$\log(\mathrm{Nf}) = -2.29\log(\mathrm{W_d}) + 1.22\log(\mathrm{N_p}) - 1.78\hat{\mathrm{A}}_\mathrm{n} + \bar{\mathrm{D}} + 24.42 \tag{3}$$

One can notice that the input parameters for Eqs. 2 and 3 are not straightforward and modifying one loading condition can affect different shape parameters at the same time. For this purpose, these two fatigue models are combined with a viscoelastic pavement model (Viscoroute 2.0 ©) that evaluates the pavement response under a moving load at a constant speed (Duhamel et al. 2005; Chabot et al. 2010) using the Huet-Sayegh viscoelastic rheological model. Once the loading signal is determined, its shape parameters can be computed and the corresponding fatigue life calculated.

3 Effect of Loading Variables and Truck Configurations on the Fatigue Life

A parametric study consisting of 252 different loading cases was performed. For each calculation, longitudinal and transverse stress and strain signals were extracted at the bottom of the asphalt layer. The shape input parameters of the two fatigue models (Eqs. 2 and 3) are processed and the four possible fatigue lives (using the models (2) and (3) in the longitudinal and transverse directions) are computed and the lowest value, which was in the majority of the cases (237 of 252) the fatigue life in the longitudinal direction using the fatigue life model (3), is kept to represent the critical pavement damage under the considered loading configuration. The results show that all loading parameters have an effect on the fatigue life, even if the magnitude of their relative impact varies.

In this paper, the effect of the number of axles and the vehicle velocity are presented. The analysis of the axle grouping effect is performed using the total carried load (TCL, in tons) before failure: $TCL = LPW \times NA \times N_f$, where LPW is the load per wheel (in tons), NA the number of axles (1, 2 or 3) which is also the number of peaks of the signal (N_p in Eq. 2), and N_f the number of cycles before failure. The results Table 1 and Figs. 2, 3 represent the fatigue lives N_f and the TCL of a pavement structure (8BB/31GB3/PF3) under single, tandem and tridem

Table 1 Effect of the loading variables on the fatigue life

Nb axles	Velocity (km/h)	Fatigue life (10^6 cycles)	*TCL* (10^6 tons)
1	30	14.6	47.6
1	70	23.3	75.6
1	110	32.5	106.0
2	30	7.7	50.0
2	70	13.4	87.0
2	110	18.2	119.0
3	30	5.50	53.6
3	70	9.3	90.3
3	110	11.9	116.0

axle configurations having a load per wheel of 3.25 t, at velocities of 30, 70 and 110 km/h and an inter-axle distance of 1.4 m.

The effect of the axle configuration can be evaluated by comparing the TCL under different configurations at a velocity of 70 km/h. The results (Table 1; Figs. 2, 3) show that the pavement can carry respectively 75.6, 87 and 90.3 million tons with a single, tandem and tridem axle configurations before conventional pavement failure. Therefore, for the identical pavement configuration, the calculated life for tandem and tridem axle groups allows transporting respectively 1.15 and 1.19 times the loads carried using isolated single axles. These results agree with the results found by (Chatti and El-Mohtar 2004) who showed that multiple axle configurations have higher load equivalency factors than single axle.

The effect of traffic speed was also evaluated. The results (Table 1; Figs. 1, 2) show that the traffic speed increase results in an increase of the fatigue life. This effect is confirmed for the different single, tandem and tridem axle groups. For all configurations the *TCL* increases with the increasing loading velocity.

Fig. 1 Effect of speed on the TCL for different axle groups

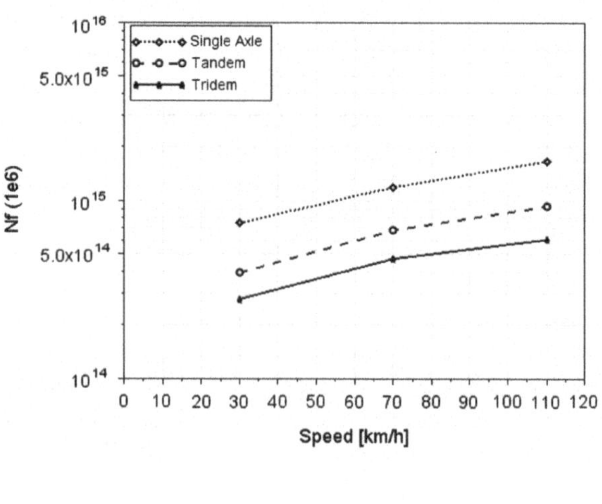

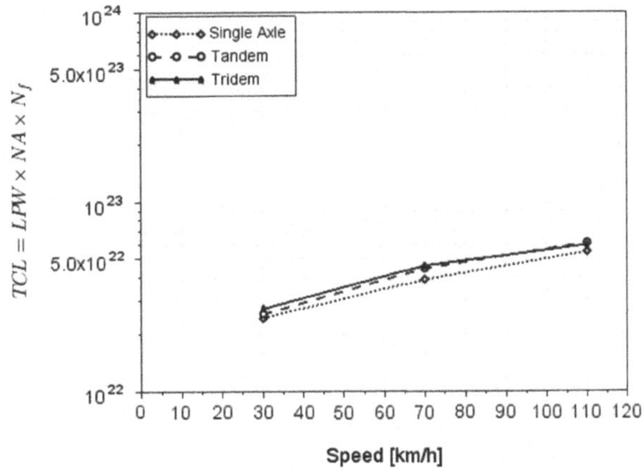

Fig. 2 Effect of speed on the number of cycle to failure for different axle groups

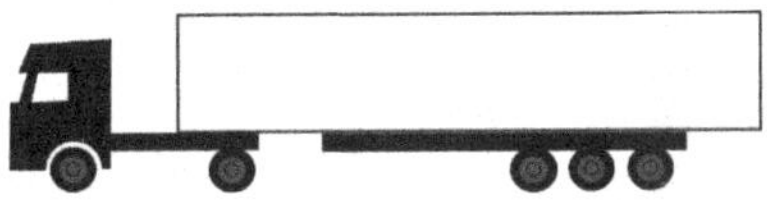

Fig. 3 Standard heavy truck, T2S3

Table 2 AC, N_f and *TCL*, T2S3 at 40 and 44 tons

Truck	AC	N_f (10^6 cycles)	*TCL* (10^6 tons)
T2S3, 40 t	4.19	2.04	81.7
T2S3, 44 t	5.75	1.49	65.5

The same approach was used to evaluate the effect of different truck configurations on the fatigue life of asphalt pavements. The standard truck T2S3 (Fig. 3) carrying 40 t and 44 t was simulated at a velocity of 90 km/h and with an inter-axle distance of 1.3 m. The results are given in terms of the fatigue life of the pavement (N_f), *TCL* before failure and aggressiveness coefficients (AC). Results show that increasing the load carried by the T2S3 by 10 % increases the aggressiveness coefficient by 37 %. In addition, the results show that if the load per truck increases from 40 to 44 tons, the TCL to failure decreases by 20 % approximately (Table 2).

The fatigue life of pavements under different European Modular Systems (EMS) was calculated using the same approach. Two EMS were simulated: one carrying 46 tons on 7 axles (Fig. 4) and another carrying 50 tons on 8 axles (Fig. 5). The objective was to determine how the addition of an axle and the increase of the carried load by 4 tons affect the aggressiveness of the EMS. Comparison of the results of the two truck configurations (Table 3) shows that the

Fig. 4 EMS 1, 46 tons on 7 axles

Fig. 5 EMS 2, 50 tons on 8 axles

Table 3 Fatigue life and total transported load before failure, EMS 1 and EMS 2

	N_f (10^6 loading cycles)	*TCL* (10^6 tons)
EMS 1	2.46	113
EMS 2	4.21	211

EMS 2 can transport 86 % more load before the failure of the pavement. This significant difference is mainly due to the third axle, for which the total load is reduced from 18 tons (for EMS 1) to 16.5 tons (for EMS 2) which has a very beneficial impact in reducing the critical strain magnitudes. No results on the aggressiveness of these 2 EMS were found in the literature.

4 Conclusions

This paper presents an approach for determining asphalt pavement damage and fatigue life under different grouped axles or truck configurations. The framework provides a means of predicting the fatigue life of asphalt pavements taking into account the effect of the loading signal in a more detailed way compared to current peak-strain practices. Among the range of potential applications, two were presented in this paper. The first was a parametric study where the effects of the number of axles and the vehicle velocity on the fatigue damage of asphalt pavements were evaluated. It was shown that, with the same load per wheel, the multiple axle configurations enabled the transport of more loading before the failure of the pavement and that the fatigue life of asphalt pavements increased when the velocity of the vehicle increased. The second application was a comparison of the aggressiveness of different truck configurations. The results showed that the standard truck T2S3 carrying 44 tons was 37 % more aggressive than the same truck carrying 40 tons and that the effective total carried load was lower. In addition, the fatigue lives under two European Modular Systems (EMS) were compared. The results show that an EMS carrying 50 tons on 8 axles is less aggressive than an EMS carrying 46 tons on 7 axles.

References

Francken,L, "Fatigue performance of a bituminous road mix under realistic test conditions". *La technique routiere, Bruxelles*, 24:1-26, 1979.

Miner, M.A., "'Cumulative damage in fatigue", *J. Appl. Mech.* **12**, 1945, A159-A164

Kogo K., Himeno K., 2008, "The effect of different waveforms and rest period in cyclic loading on the fatigue behavior of the asphalt mixtures", *Pavement cracking - Al Qadi, Scarpas & Loizos (eds)*, pg 509-517.

Raithby K. D., Sterling A. B. 1972, "Some Effects of Loading History on the Performance of Rolled asphalt", *TRRL-LR* 496, Crowthorne, England.

Bodin D., Merbouh M., Balay J.-M., Breysse D. and L. Moriceau, 2009. "Experimental study of the waveform shape effect on asphalt mixes fatigue", *7th Int. RILEM Symp. on Advanced Testing and Characterization of Bituminous Materials,* Rhodes.

Bernadet, 1995; "Les enjeux du 44 tonnes à 5 essieux.", Transports, 372:237- 245, 1995.

Breysse D., C. De La Roche, J. Chauvin and V. Domec, 2002, "Influence of rest time on recovery and damage during fatigue tests on bituminous composites: balance between damage and recovering. ", *15th ASCE Engineering Mechanics Conference*, Columbia University, New York, 2002.

Chabot, A., Chupin, O., Deloffre, L. and Duhamel, D., 2010. "ViscoRoute 2.0, A tool for the simulation of moving load effects on asphalt pavement", *Road Materials and Pavement Design*. Volume 11- No. 2/2010, pages 227 to 250.

Chatti, K., and El-Mohtar, "Effect of different axle configuration on fatigue life of asphalt concrete mixtures", *Transport Research Record 1891*, pages 121–130, 2004.

Duhamel, D., Chabot, A., Tamagny, P. and Harfouche, L., "Viscoroute, logiciel de modélisation viscoélastique des chaussées bitumineuses.", *Bulletin de liaison des laboratoires des ponts et chaussées*, 258-259, p. 89-103, 2005.

Gillespie, T., Karamihas, S., Cebon, D., Ans, M. S., Nasim, M. A., Hansen, W., and Ehsan, N., "Effects of heavy vehicle characteristics on pavement response and performance." *Technical report, The University of Michigan, Transportation Research Institute*, 1992.

Homsi, F., Bodin, D., Breysse, D., Yotte, S. and Balay, J.M. 2012,a, "A Multi-linear Fatigue Life Model of Flexible Pavements under Multiple Axle Loadings", *7th Rilem International Conference on Cracking in Pavements*, June 2012, Delft, Netherlands, pp 697-706.

Homsi, F., Bodin, D., Yotte, S., Breysse, D. and Balay, J.M., 2012,b "Fatigue life modelling of asphalt pavements under multiple axle loadings", *Road Materials and Pavement Design*, Volume 13, Issue 4, December 2012, pages 749-768.

Salama, H., K. Chatti, & R. Lyles 2006, "Effect of heavy multiple axle trucks on flexible pavement damage using in service pavement performance data", *Journal of transportation engineering*, pages 763–770.

Prediction of Thermally-Induced Reflective Cracking Using Full-Scale Test Data

Hao Yin

Abstract Reflective cracking is not addressed in the current Federal Aviation Administration (FAA) Advisory Circular for asphalt concrete (AC) overlaid rigid pavements. This paper presents the development of a reflective cracking model using full-scale test data. The prediction model reflected all three-stage process involving crack initiation, propagation, and final failure. Regression models were established to relate varies parameters including overlay temperature, cracking strain, crack propagation rate, number of loading cycles, and crack length. Based on the validation study results, it was shown that a reasonable prediction of the overlay fatigue life can be forecast. It was also demonstrated that lack of considering fracture healing effect during rest periods could result in substantial overestimate of the crack propagation rate and therefore premature overlay failure.

Keywords Reflective cracking · HMA overlay · Crack propagation · Full-scale test

1 Introduction

The LEDFAA overlay design model incorporated a crude estimate of reflection crack growth (1 year per inch), but this was removed in FAARFIELD because it was found to produce unsatisfactory or illogical design thicknesses (FAA 2009). The current FAARFIELD model does not explicitly consider reflection cracking at all, which is likewise unacceptable. An ongoing project sponsored by the FAA led to a series of successfully full-scale experiments (Phase I, II, III, and IV) at the FAA National Airport Pavement Test Facility (NAPTF). Extensive test data, particularly, crack propagation rates, under controlled loading conditions to mimic the temperature cycles occurring in nature were obtained (Yin and Barbagallo 2013; Yin 2014; Yin and Ishee 2015).

H. Yin (✉)
Gemini Technologies, Inc, Warrington, USA
e-mail: hao.yin@gemitek.com

A. Chabot et al. (eds.), *8th RILEM International Conference on Mechanisms of Cracking and Debonding in Pavements*, RILEM Bookseries 13,
DOI 10.1007/978-94-024-0867-6_63

2 Objective

The research objective was to phenomenologically predict thermally-induced reflective cracking of a given HMA overlay structure and materials using full-scale test data. The failure criterion was defined as the appearance of the first thorough reflection crack on the HMA overlay surface.

3 Test Overlays, Instrumentation, and Data

Prior to overlay construction, a thin tack coat of straight PG 64-22 asphalt was applied on the milled concrete surface to prevent interface slippage and secondary cracks. A wood form was then set to divide the overlay into two 1.5 m-wide strips with a 0.6 m gap. Finally, a 50 mm overlay was paved with standard FAA P-401 materials (PG 64-22). To simulate temperature cycles mechanically, the overlay bottom temperature was maintained at 0 °C. The temperature variations were approximated by a haversine load waveform defining the relationship between the concrete joint opening and cycle time. Both embedded strain gages (EG) and surface strain gages (SG) were installed. To better monitor the reflection crack development between two strain gages, a crack detector (CD) was inserted. Pavement temperatures were recorded by thermocouples (T) installed at the same depth as strain gages.

4 Fatigue Life Prediction

Although previous test data revealed that reflection cracks propagated in both vertical (thickness) and horizontal (along joint) directions, the crack channeling phenomenon was not considered in this study. Test data collected the North and South sections of Phase II overlay, and the Control section of Phase III overlay were used to predict the overlay fatigue life.

The overlay temperature data were collected from three depths: 0.0, 50, and 114 in below the overlay surface. Full-scale tests utilized a 40-ton chiller and refrigeration grids to maintain the overlay bottom temperature between −2.8 and −2.2 °C. This bottom-up cooling scheme constantly resulted in higher temperatures in the upper portion of the overlay. The following linear regression model was selected to predict the overlay temperature profile for its simplicity applicable to a given overlay structure and materials.

$$T = a_1 * D^2 + a_2 * D + a_3 \tag{1}$$

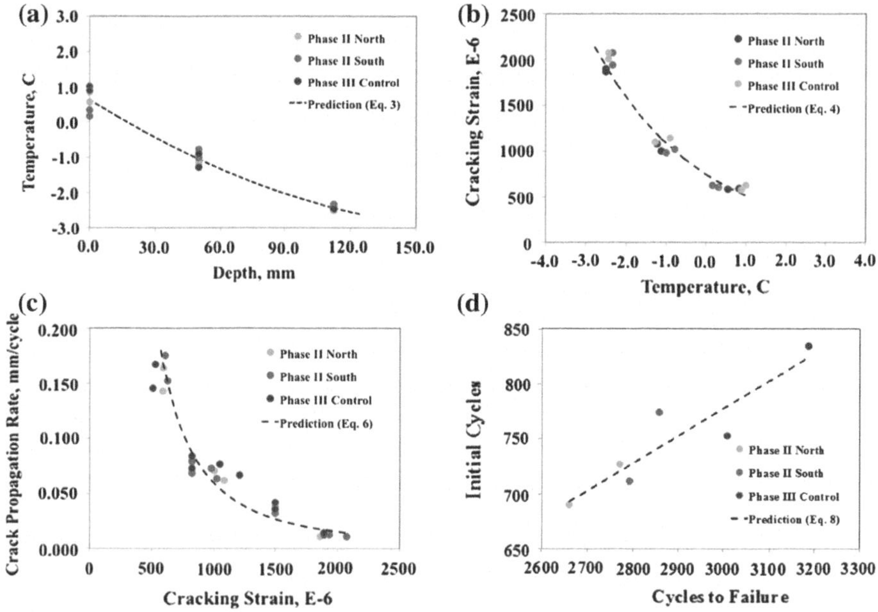

Fig. 1 Prediction of fatigue life. **a** Overlay temperature prediction, **b** cracking strain prediction, **c** crack propagation rate prediction, **d** initial loading cycle prediction

where T is temperature, D is depth of interest, and a_1, a_2, and a_3 are model coefficients. A *p-value* of 0.102 indicated an accurate fit of a second-order polynomial to the temperature data, as revealed in Fig. 1a.

Figure 1b presents the cracking strain and overlay temperature at multiple depths. The cracking strain steadily declines with the temperature rise but the magnitude of reduction rate becomes smaller and smaller, the cracking strain eventually approaches to a constant value (asymptote). Thus, an exponential regression function was selected to describe the dependency of the response variable, cracking strain, on the predictor variable, temperature:

$$\varepsilon_c = b_1 * EXP(b_2 * T) \tag{2}$$

where ε_c is cracking strain, T is temperature, and b_1 and b_2 are model coefficients. A *p-value* of 0.184 suggested an excellent agreement between measured and predicted cracking strains, as presented in Fig. 1b.

In the case of repeated loading, crack evolution consists of a sequence of incremental crack propagations. Given a macrocrack develops slowly within the stage of stable crack growth, individual crack propagation was assumed as a fixed crack length (*CLi*) with a constant crack propagation rate (*CPRi*). The number of loading cycles for each crack propagation was determined from the timestamp on the strain gage and crack detector responses. Figure 1c plots the measured cracking strain and crack propagation rate. As exhibited in Fig. 1c, the crack propagation became more aggressive once the reflection crack penetrated into the upper portion of the overlay. A Power (convex) function similar to Paris' law (Paris and Erdogan 1963; Schapery 1984) was employed to relate the crack propagation rate (*CPR*) to the cracking strain under a cyclic loading condition:

$$\frac{dL}{dN} = c_1 * (\varepsilon_c)c_2 \tag{3}$$

where L is crack length, N is number of loading cycles, c_1 and c_2 are model coefficients. A *p-value* of 0.219 suggested the appropriateness of chosen model form even though only 15 data points were used (see Fig. 1c).

The Crack Propagation model implies a series of successive analysis, in each one the crack propagation rate is calculated. In this study, the reflection crack was simplified as a tension driven large crack. At this scale, the crack propagation would be fairly insensitive to the microstructure of HMA materials. Therefore, the cycles to failure (N_f) can be calculated by integrating Eq. 3. To complete the prediction of fatigue life, crack initiation must be considered. Since the relationship between the cycles to failure and initial loading cycles (N_{ini}) appeared to be a straight line on the log-log scale plot, the following regression function was proposed to correlate these two variables:

$$N_f = d_1 * (N_{ini})^{d_2} \tag{4}$$

Figure 1d suggests that the larger number of cycles required to transmit a reflection crack to the surface the more initial loading cycles to initiate it at the overlay bottom. Consequently, the overlay fatigue life can be expressed as the summation of crack initiation and propagation life.

5 Validation Study

To examine the predictive capacity and accuracy, full-scale test data from the North section of Phase IV overly were chosen due to the same overlay thickness, materials, and instrumentations.

Figure 2a reveals that the temperature profile predicted from Eq. 1 compared marginally with the measured temperatures. Temperatures were significantly underpredicted at all three overlay depths and could have resulted in erroneous subsequent predictions of crack propagation and fatigue life. A calibration factor was therefore introduced in Eq. 1 as a multiplier to the coefficient a_3. As shown by the grey circles in Fig. 1a, after applying a calibration factor of 0.051, Eq. 1 yielded a much improved temperature prediction.

A comparison of measured and predicted crack propagation rates is provided in Fig. 2b. At lower crack propagation rates, the predictions matched the

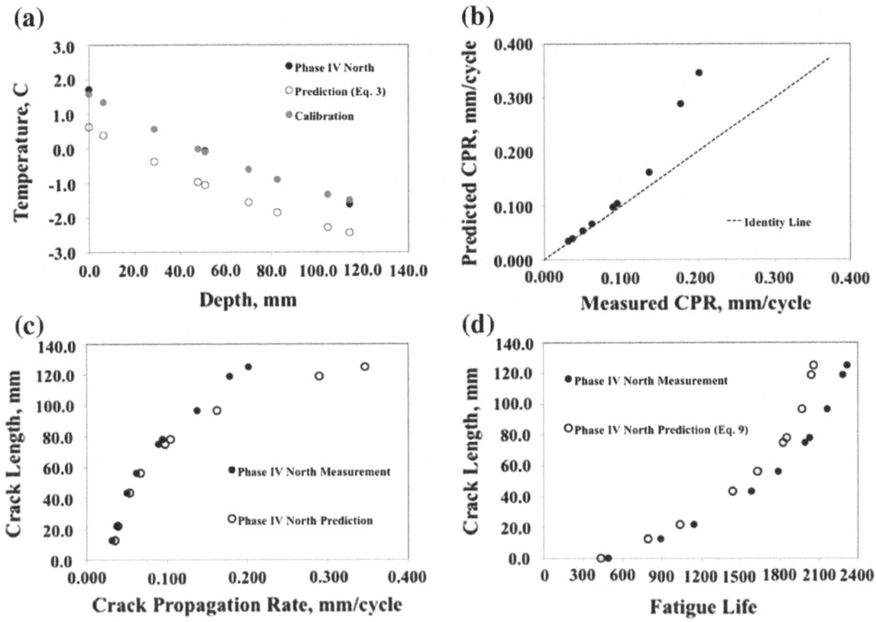

Fig. 2 Validation study. **a** Overlay temperature calibration, **b** measured versus predicted crack propagation rate, **c** measured versus predicted crack length, **d** measured versus predicted overlay fatigue life

measurements reasonably well. The predictions began to deviate from measurements once the *CPR* exceeded 0.127 mm/cycle. A further examination of the impact of *CPR* on the crack evolution is given in Fig. 2c. Although the reflection crack advanced as both predicted and measured *CPR* increased, the crack progressed substantially divergent once it penetrated through the bottom lift of the overlay. To fracture the top 28.6-mm overlay, the predicted *CPR* boosted from 0.165 to 0.35 mm/cycle. On the other hand, a growth of only 35 % in the measured *CPR* projected the appearance of a reflection crack on the surface. One explanation for the above discrepancy would be the fracture healing (recovery of crack damage) occurred during the rest periods between each loading cycles.

As illustrated in Fig. 2d, the Fatigue Life model always predicted less number of cycles than the full-scale measurements. Shorter lives in the prediction were most likely due to higher crack propagation rates as it was a function of the cracking strain that could have experienced a gradual reduction as the material weakened in the field. The predicted crack initiation occurred 57 loading cycles earlier than the measurement. For the 125 mm thick Phase IV overlay, the measured fatigue life was about 11 % longer due to the combined effect of fracture healing and underestimate of the initial loading cycles. In addition, no noticeable fatigue life extension was observed when the crack length exceeded the 75 mm benchmark. This is to say, increasing overlay thickness may not be an effective way to prolong the fatigue life (delay crack propagation) for the overlay structure and materials considered in the study.

6 Conclusions and Recommendations

This paper presents the development of a phenomenological prediction model for thermally-induced reflective cracking using full-scale test data. Regression analyses were employed to relate varies variables, including overlay temperature, cracking strain, crack propagation rate, number of loading cycles, and crack length. Crack initiation, propagation, and final failure were all taken into consideration. This study has shown that there are areas that need to be investigated to provide a better understanding of the effect of fracture healing on the reflection crack propagation.

References

Federal Aviation Administration (2009) AC 150/5320-6E, Airport Pavement Design and Evaluation.

Yin, H. and Barbagallo, D. (2013) Development of Full-Scale Reflective Cracking Test at FAA NAPTF, Transportation Research Record No. 2368.

Yin, H. (2014) Full-Scale Test of Thermally-Induced Reflective Cracking in Airport Pavements, International Journal of Road Materials and Pavements Design, Vol 16.

Yin, H. and Ishee, C. (2015) Evaluation of Strain-Relieving Interlayer to Retard Thermally Induced Reflective Cracking, Journal of the Association of Asphalt Paving Technologists, Vol 84.

Paris, P. C. and Erdogan, E. (1963) A Critical Analysis of Crack Propagation Laws, Journal of Basic Engineering, Transaction of the American Society of Mechanical Engineering, vol. 85.

Schapery. R. A. (1984) Correspondence principles and a generalized J-integral for large deformation and fracture analysis of viscoelastic media, International Journal of Fracture, Vol 25.

Part VIII
Interface Debonding Behavior: Bond Characterization

Influence of Specimen Dimension and Test Speed on the Shear Strength of Bituminous Interfaces

Musab Abuaddous, Francesco Canestrari, Andrea Graziani and Gilda Ferrotti

Abstract In view of the fact that pavements are multilayer systems, achieving high bonding between layers is a key element to increase service life. Interface debonding is mainly responsible for the slipping failure of pavements that leads to high rehabilitation and maintenance costs. The bonding between asphalt layers is usually evaluated by testing the interlayer shear strength and is affected by several parameters such as test speed, test temperature, normal stress applied and specimen diameter. This paper focuses on the effect of test speed and specimen diameter on the shear strength evaluated through the Leutner equipment, for a typical dense graded asphalt mixture. Leutner tests were carried out on double-layered specimens with a diameter of 100 and 150 mm and with interlayer deformation rates corresponding to nominal test speeds of 1, 2.5, 5, 10, 25 mm/min. The effective interlayer deformation rate was calculated by measuring the deformation through an external transducer in order to perform a reliable data analysis. Results showed a steady increase in the shear strength with the increase in the interlayer deformation rate. Moreover, a clear scale effect was observed at any test speed resulting in higher values for shear strength measured on specimens with diameter of 100 mm.

Keywords Shear strength · Leutner test · Scale effect · Time dependent behavior

M. Abuaddous (✉) · F. Canestrari · A. Graziani · G. Ferrotti
Polytechnic University of Marche, via Brecce Bianche, 60131 Ancona, Italy
e-mail: m.abuaddous@pm.univpm.it

F. Canestrari
e-mail: f.canestrari@univpm.it

A. Graziani
e-mail: a.graziani@univpm.it

G. Ferrotti
e-mail: g.ferrotti@univpm.it

A. Chabot et al. (eds.), *8th RILEM International Conference on Mechanisms of Cracking and Debonding in Pavements*, RILEM Bookseries 13,
DOI 10.1007/978-94-024-0867-6_64

1 Introduction

Interlayer bonding has attracted attention in recent years due to the need to increase pavement performance properties (Raab and Partl 2009). It is well known that interlayer bonding is influenced by many parameters such as temperature (Uzan et al. 1978), tack coat type and dosage (Canestrari et al. 2005), interface characteristics (Santagata et al. 2008), mixture type (West et al. 2005). Moreover, interlayer shear strength (ISS) measurements are strongly affected by the employed testing device and procedure (Santagata et al. 2009; Canestrari et al. 2013) and by characteristic parameters such as test speed and specimen diameter (Canestrari et al. 2013).

This study focuses on the evaluation of the influence of test speed, test temperature and specimen diameter on the ISS, with the same material and without application of normal stress.

2 Experimental Program

2.1 Materials

A dense graded asphalt mixture with a bitumen content of 5.5 % (by aggregate weight) and a maximum aggregate size of 10 mm was used to prepare double-layered square slabs (305 × 305 mm^2), compacted in the laboratory through a steel roller compactor compliant with EN 12697-33. The lower layer was compacted with a thickness of 60 mm and a target air void content of 6 %. It was left for 6 h at room temperature for cooling and no tack coat was spread at the interface. Then, the upper layer was compacted with a thickness of 35 mm and the same target air void content was considered. Four cylindrical samples (two with 150 mm-diameter and two with 100 mm-diameter) were cored from each slab and tested after 21 days at room temperature.

2.2 Testing Method

Shear tests were performed with the Leutner equipment (Fig. 1a), which is a shear device able to test cylindrical specimen with 100 and 150 mm nominal diameter. It was installed on a servo-mechanic press frame able to apply displacement rates up to 50 mm/min.

Figure 1b shows the scheme of the Leutner equipment, which is divided into two parts. The lower part is connected to the loading plate of the press that moves upward, whereas the upper part is positioned in contrast with the loading frame. An interface gap of about 5 mm is left between the two parts.

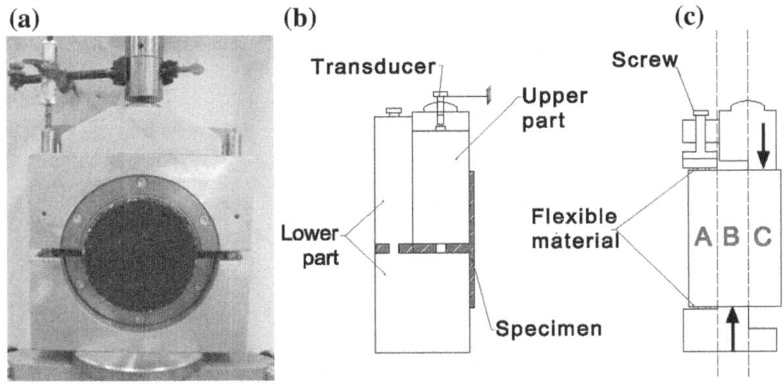

Fig. 1 Leutner equipment; **a** front view; **b** side view (scheme); **c** cross section normal to the shear plane

In order to highlight how the shear load is applied to the specimen, a section of the Leutner equipment, normal to the shear plane, is shown in Fig. 1c. The specimen can be ideally divided into three sections. In correspondence of section A, the specimen is fixed to the lower part of the Leutner equipment through a screw that is manually locked in order to ensure an adequate contact. In correspondence of section B, the lower part of the specimen is in contact only with the lower part of the equipment, whereas its upper part is unconfined. In correspondence of section C, the lower part of the specimen is unconfined whereas on the top it is in contact with the upper part of the Leutner equipment.

An external transducer was added to measure the relative displacement between the lower and the upper part of the Leutner equipment. Such displacement was used to calculate the interlayer shear deformation rate.

For each test, the device returns a data-set where the displacement of the loading plate, the load measured by the load cell and the deformation measured by the external transducer, are reported as functions of time.

As shown in Fig. 2, the measured displacement increases linearly during the test because of the imposed constant displacement rate (nominal test speed). Load and deformation show an initial flat phase which is due to an adjustment of the specimen inside the Leutner equipment. Afterwards, both load and deformation start to increase contemporaneously, indicating that the measured load is due to the relative displacement imposed to the specimen on opposite sides of the interface (upper and lower layers).

The applied load could also cause a small rotation of the specimen that, together with the compliance of the testing apparatus (Leutner equipment and press), explains the difference between displacement (applied by the press) and deformation (read by the transducer), as shown in Fig. 2.

Since the deformation was variable and lower than the displacement, the actual deformation rate was lower with respect to the nominal test speed. In particular, the average speed reduction increases with test speed and with the decrease of the

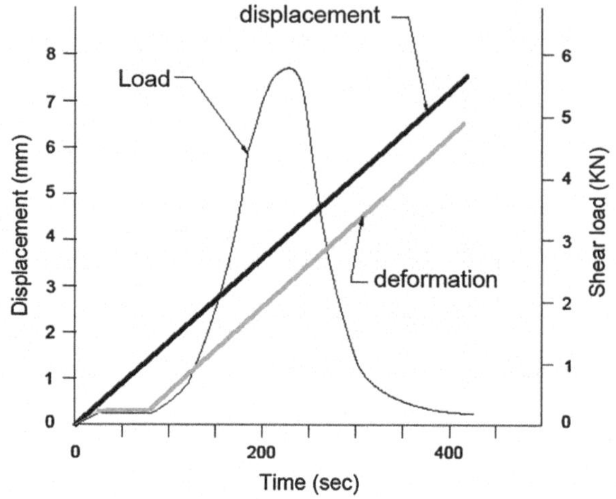

Fig. 2 Leutner test result for the single specimen

specimen diameter. In fact, for 100 and 150 mm specimen diameter the average speed reduction was 24 and 21 %, respectively.

In this paper, all results will be shown considering the ISS (τ_{peak}) calculated as ratio between maximum load and specimen cross section area and the average deformation rate at failure (v), calculated as ratio between deformation and time at failure (i.e. in correspondence of the maximum load).

2.3 *Testing Program*

Double-layered samples were subjected to shear tests using a full factorial experimental design, considering one interface conditions (without tack coat), two specimen diameters (D = 150 and D = 100 mm), five nominal test speeds (1, 2.5, 5, 10, 25 mm/min) and three test temperatures (5, 20 and 40 °C). Four replicate specimens were tested for each case.

3 Result and Analysis

In Fig. 3a, ISS results at 20 °C are reported as function of deformation rate. For both specimen diameters, results showed an overall increase of shear strength with deformation rate. Such trend was fitted using a power-law model:

$$\tau_{\text{peak}} = a \cdot v^b$$

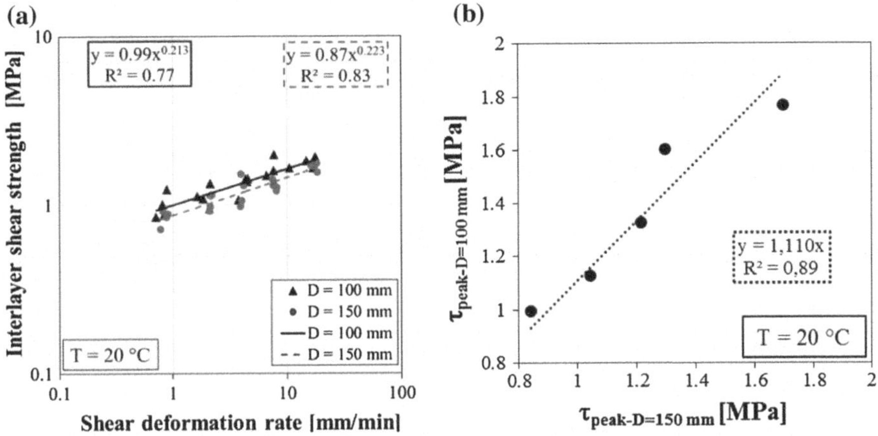

Fig. 3 Leutner results at 20 °C; **a** ISS versus shear deformation rate for D = 100 mm and D = 150 mm; **b** average values of ISS for D = 150 mm versus average values of ISS for D = 100 mm

where *a* and *b* are regression parameters. In particular, *b* represents the slope of the linear regression which is obtained in a log-log plane. It can be seen that the slopes, 0.213 and 0.223, obtained for D = 100 mm and D = 150 mm, respectively, are very close. In addition these values are in agreement with the value *b* = 0.22 reported by Piber et al. (2009).

Moreover, it should be noted that samples with D = 100 mm provide, in general, higher shear strength with respect to samples with D = 150 mm. In Fig. 3b, the average values of τ_{peak} for both diameters are reported. A linear relationship with a regression coefficient equal to 1.11 can be found. This parameter can be directly associated with the scale effect and it is similar to the value 1.07 reported by Piber et al. (2009).

However, the relationship between τ_{peak} for D = 100 mm and τ_{peak} for D = 150 mm strongly depends on the test temperature, as shown in Fig. 4, where results for 5 and 40 °C are reported. At each temperature, the slope of the regression lines for D = 100 mm and D = 150 mm is different, suggesting that the mixed effects of speed and temperature are related to the specimen dimension. It is worth noting that test results at 5 and 40 °C are more scattered with respect to 20 °C (Fig. 3a) and this also affects the value of R^2. A more in depth investigation is necessary in order to study how specimen dimensions influence the effect of temperature and deformation rate on ISS.

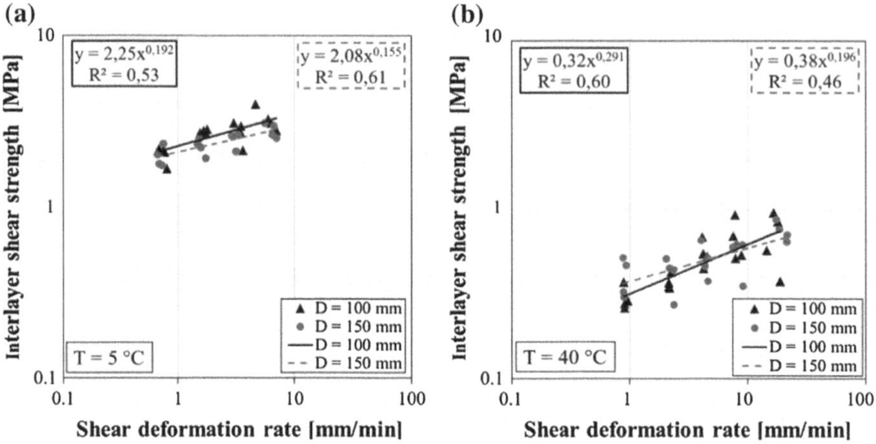

Fig. 4 ISS versus deformation rate for D = 100 mm and D = 150 mm: **a** 5 °C; **b** 40 °C

4 Conclusions

This study focuses on the evaluation of the influence of test speed, test temperature and specimen diameter on the ISS measured with the Leutner equipment. In particular, an external transducer was added to measure the relative displacement between the lower and the upper layer of the specimen. Such displacement was used to calculate the interlayer shear deformation rate in place of the nominal test speed.

Results showed that, for specimen with diameter of 100 mm and diameter of 150 mm, ISS increases with deformation rate and decreases with temperature. At each temperature, a power-law model was fitted to measured data. The slope of the regression line, which was obtained in a log-log plane, was a function of temperature.

Moreover, it was observed that specimens with diameter of 100 mm achieved higher shear strength compared to specimen with diameter of 150 mm. This can be explained with a scale effect and highlights that the mixed effects of speed and temperature are related to the specimen dimension.

References

Canestrari F, Ferrotti G, Partl MN, Santagata F (2005). Advanced testing and characterization of interlayer shear resistance. Transport Res Record 1929:68–78.

Canestrari F, Ferrotti G, Lu X, Millien A, Partl MN, Petit C, Phelipot-Mardelé A, Piber H, Raab C (2013) Mechanical testing of interlayer bonding in asphalt pavements. In: Advances in Interlaboratory Testing and Evaluation of Bituminous Materials, Springer, pp 303–360

EN 12697-33 (2003) Bituminous mixtures – test methods for hot mix asphalt – part 33: Specimen prepared by roller compactore. CEN-Committee for Standardization, Brussels, Belgium

Piber H, Canestrari F, Ferrotti G, Lu X, Millien A, Partl MN, Petit C, Phelipot-Mardelé A, Raab C (2009) Interlaboratory test on interlayer bonding of asphalt pavements. RILEM 2009 : advanced testing and characterisation of bituminous materials

Raab C, Partl MN (2009) Interlayer bonding of binder, base and subbase layers of asphalt pavements: Long-term performance. Constr Build Mater 23(8):2926-2931. doi:10.1016/j.conbuildmat.2009.02.025

Santagata F.A., Partl M.N., Ferrotti G., Canestrari F., Flisch A. (2008) Layer Characteristics Affecting Interlayer Shear Resistance in Flexible Pavements Proceedings of Association of Asphalt Paving Technologists 77:221–256

Santagata FA, Ferrotti G, Partl MN, Canestrari F (2009) Statistical investigation of two different interlayer shear test methods. Materials and structures 42(6):705–714

West R., Zhang J., Moore J., (2005) Evaluation of Bond Strength between Pavement Layers. NCAT (National Center for Asphalt Technology) Report 05-08.

Uzan J, Livneh M, Eshed Y (1978) Investigation of adhesion properties between asphaltic–concrete layers. Proceedings of Association of Asphalt Paving Technologists 47:495–521

Investigation of the Impact of Conditions of Application of Tack Coats on the Interlayer Bond Strength

Alexandra Destrée and Joëlle De Visscher

Abstract An effective and durable bond between the various constituent layers is an absolute prerequisite for the durability of a road pavement. To ensure the adhesion between successive pavement layers, cationic bitumen emulsions are the most frequently used type of tack coats. The intrinsic characteristics of tack coats play an important role in the adhesion between layers, but the conditions of application of these coats are equally crucial. In this context, the Belgian Road Research Centre actively participates in a Belgian working group on tack coats initiated by the Walloon federation of road contractors and the Walloon public service. The objective of this joint working group is to carry out a "field" study about adhesion between layers while evaluating the influence of different parameters—such as type and rate of spread of emulsion, spraying equipment, nature and preparation of the binder course, breaking and curing times of the emulsion before overlaying, etc. With a view to this objective, a test site was constructed in the summer of 2014, consisting of four test sections differing in type of tack coat, milling speed and cleaning operation of the binder course. The bond strengths were investigated by direct shear test. This article describes the conditions of application, the measurements made on site (tack coat application rate, binder course texture and cleanliness, etc.) and the results of the interlayer adhesion test performed in the laboratory on specimens taken from the four test sections.

Keywords Tack coat · Milling speed · Cleaning operation · Shear test

A. Destrée (✉) · J. De Visscher
Belgian Road Research Centre (BRRC), Woluwedal 42, 1200 Brussels, Belgium
e-mail: a.destree@brrc.be

J. De Visscher
e-mail: j.devisccher@brrc.be

A. Chabot et al. (eds.), *8th RILEM International Conference on Mechanisms of Cracking and Debonding in Pavements*, RILEM Bookseries 13,
DOI 10.1007/978-94-024-0867-6_65

1 Introduction and Objective

Pavements are multi-layered structures and their overall bearing capacity and lifetime depend not only on the thickness and stiffness of each individual layer, but also on the bond between them. The purpose of a tack coat is to provide the necessary bond between layers. The intrinsic characteristics of these tack coats play an important role in the adhesion between layers, but the conditions of application of these coats are equally crucial Destrée et al. (2015).

In this context, the objective of this field study was to investigate some of the factors that influence the adhesive bond provided by the tack coat at the interface between pavement layers. These factors included the type of tack coat, the macrotexture and the cleanliness of the milled binder courses.

The bond strengths were investigated on field cores by shear bond test according to pre-standard prEN 12697-48 "Bituminous mixtures—Test methods for hot mix asphalt—Part 48: Interlayer Bonding". This article describes the conditions of application, the measurements made on site and the results of the shear bond tests.

2 Shear Bond Test

Shear bond tests were carried out on field cores with a diameter of 150 ± 2 mm. They were conditioned and tested at 20 °C, using a Leutner shear test device with a 5-mm gap between the shearing rings. The test is conducted displacement-controlled at a rate of 50 ± 2 mm/min. The shear strength is calculated as the average of three specimens. Germany (FGSV 2003) and Switzerland (SN 640430 2012) specify minimum average shear strength specifications for tack coat in function of the tested interface on 150 mm diameter cores and a temperature of 20 °C (Table 1). These values are currently criticized and some authors (Stöckert 2002; Raab and Partl 1999) made new recommendations for higher limits which are based on their experience and their research (Table 1).

Table 1 Shear bond test specifications and recommendations for Germany and Switzerland (for an interface between a surface course and a binder course)

Source	Minimum average shear strength in MPa (shear force in kN)
Specifications	
FGSV (2003)	0.85 (15)
SN640430 (2012)	0.85 (15)
Recommendations	
Stöckert (2002)	1.41 (25)
Raab and Partl (1999)	1.3 (23)[a]

[a]For asphalt concrete (AC) and stone mastic asphalt (SMA)

3 Test Site and Measurements on Site

The test site was constructed in the summer season of 2014 on the regional road N975 at Hymiée in Belgium, consisting of four test sections (Table 2). The pavement of these four test sections was bi-layered:

- A 50 mm thick surface course in stone mastic asphalt (SMA-14)
- A milled asphalt concrete binder course. Two milling speeds of the binder course were used in the field testing to investigate the effect of milling on

Table 2 Experimental conditions and results of shear bond tests on specimens cored from the four test sections

[Section] Tack coat and rate[a]	Sub-sections	Pressure (bars)	MTD[b] (mm)	Cleanliness in g/m^2	Shear strength (MPa)		CV (%)
					Mean	St dev.	
[1] C60B3 50/70 213 g/m^2 (fine texture)	**1.1.a**	40	2.10	12 (E)	1.25	0.02	2
	1.1.b		2.03	21 (E)	1.18	0.13	11
	1.2.a	100	1.97	16 (E)	1.19	0.09	8
	1.2.b		2.01	20 (E)	1.52	0.16	11
	1.3.a	150	1.99	20 (E)	1.43	0.07	5
	1.3.b		1.96	25 (E)	1.17	0.04	3
[2] C60B3 50/70 262 g/m^2 (coarse texture)	**2.1.a**	40	3.47	18 (E)	1.00	0.07	7
	2.1.b		3.34	37 (G)	0.97	0.20	21
	2.2.a	100	3.55	27 (G)	1.80	0.15	8
	2.2.b		4.36	29 (G)	2.15	0.32	15
	2.3.a	150	3.87	22 (E)	1.35	0.01	1
	2.3.b		3.42	41 (G)	1.91	0.20	10
[3] C60B3 257 g/m^2 (fine texture)	**3.1.a**	40	1.61	8 (E)	0.86	0.07	8
	3.1.b		1.67	10 (E)	0.70	0.07	10
	3.2.a	100	1.53	12 (E)	1.01	0.20	20
	3.2.b		1.64	16 (E)	1.25	0.17	14
	3.3.a	150	1.51	17 (E)	1.46	0.06	4
[4] C60B3 276 g/m^2 (coarse texture)	**4.1.a**	40	3.08	23 (E)	1.02	0.17	17
	4.1.b		3.38	25 (E)	1.23	0.06	5
	4.2.a	100	3.15	15 (E)	0.90	0.09	10
	4.2.b		3.30	18 (E)	1.13	0.10	9
	4.3.a	150	3.15	15 (E)	1.27	0.14	11
	4.3.b		3.30	20 (E)	1.19	0.20	17

[a]Expressed in measured residual binder
[b]Mean texture depth
G Good, *E* Excellent

interface bonding: 10 m/min (creating a fine texture) and 20 m/min (creating a coarser texture). The obtained milled surfaces will be hence referred to as fine textured (sections 1 and 3) and coarse textured (sections 2 and 4), respectively. These milled binder courses were then cleaned with a high-pressure suction sweeper combining brushing with high-pressure water cleaning: 40, 100 and 150 bars.

To obtain results that are representative of the laying conditions on the four test sections, the cores required for the shear bond tests were taken in two subsections (a and b) for each test section and for each water pressure level. For each subsection, the three cores required for the shear bond tests were taken 10 cm apart.

The milled binder courses in the four test sections were characterized in terms of:

- Cleanliness, using a protocol developed at Ministère des Transports du Québec-MTQ and based on the gravimetric evaluation of particles recovered by aspiration. According to Table 2, the cleanliness of the milled binder courses could be rated in most cases as excellent (between 0 and 25 g/m^2) and sometimes good (between 25 and 50 g/m^2) on the scale currently proposed by MTQ (Lavoie 2015). The differences between subsections a and b (named in the order of cleanliness) are not systematically significant in view of the variability of the cleanliness test.
- Texture by standard EN 13036-1. For each section, the mean texture depth is in general the same in both subsections. This value is between 1.5 and 2.1 mm for the fine-textured sections and between 3.1 and 4.4 mm for the coarse-textures sections. In most cases, according to Table 2, the differences between subsections a and b are not systematically significant in view of the precision of the sand-patch method.

Two types of cationic bituminous emulsions were applied to the cleaned milled binder courses of the four test sections at a same target residual binder rate of 250 g/m^2 prior to the SMA-14 overlay construction: C60B3 50/70 with harder bitumen (sections 1 and 2) and C60B3 with soft bitumen (sections 3 and 4). High attention was paid to the tack coat application so that the measured tack coat application rate (according to EN 12272-1) was within a reasonable range of the target rate. In general, they were relatively close to 250 g/m^2 (Table 2). The curing time (one night) was the same for the two types of tack coat.

4 Results of Shear Bond Tests and Discussion

The shear bond tests results are given in Table 2 listing the mean value of the shear bond strength (Mean), the standard deviation (St dev.) and the coefficient of variation (CV) for each test section. The analysis of the results leads to the following findings:

- Whatever the type of tack coat used, the measured application rate, the water cleaning pressure, the cleanliness and the macrotexture of the milled binder courses analysed:
  - The minimal required average shear strength specified in Table 1 (0.85 MPa) is met for all sections, except subsection 3.1b, although for subsections 3.1a and 4.2a. The low strengths on these three subsections could be explained by the state of deterioration of the milled binder course itself. Indeed, the visual inspection of the tested cores after failure indicates a both crumbly and fragile appearance of the milled binder course.
  - Only six subsections satisfy the most severe requirement of 1.41 MPa specified in Table 1; these subsections were all cleaned with a high pressure water of 100 bars or 150 bars. In addition, ten subsections reach the lower requirement specified in Table 1 (1.3 MPa); the tack coat used on these subsections was in most cases the emulsion prepared with harder bitumen (C60B3 50/70).
  - The mean shear bond strength values for the tack coat prepared with harder bitumen (C60B3 50/70) are in most cases higher in comparison with the tack coat with soft bitumen (C60B3).
  - In comparison with the fine textured binder course, it could not be demonstrated that a coarser underlayer has a positive effect on the shear strength. We note nevertheless that the three highest shear strength values were obtained in section 2 with the coarse-textured milled binder course.
- It is not possible to find a direct causal relationship between the cleaning conditions (40, 100 and 150 bars) and the quality of interlayer bonding. This is not surprising in this case, as the different cleaning conditions did not lead to significant differences in cleanliness. When comparing the two subsections a and b for each test section and each high pressure water, the differences in shear strength are not in most cases statistically different. In fact, the difference in cleanliness was also not significant in view of the reproducibility of the cleanliness measurement.
- The variability in the shear bond test results is probably due to the variability of other critical factors on this site, e.g. level of compaction of the surface course, state and macrotexture of the milled binder course, tack coat application rate, state of the tack coat before overlaying…

5 Conclusions

On the basis of the test results obtained from testing of field specimens from the regional road N975 at Hymiée, the following conclusions are drawn:

- **The Shear Bond Test (SBT)** is a good laboratory method to investigate the interlayer bond strength between pavement layers. With the exception of one subsection, all the field conditions lead to mean shear bond strengths that reached and even exceed the requirement from Swiss and German standards of 0.85 MPa.

- **Tack coat type**. The mean shear bond strength values for the tack coat prepared with harder bitumen (C60B3 50/70) are in most cases higher in comparison with the tack coat with soft bitumen (C60B3).
- **Binder course texture**. On this test site, the milling speed (leading to fine or coarser texture) does not have a significant influence on the shear strength.
- **The cleanliness measurement protocol** is considered as a simple method to quantify cleanliness. When a high-pressure suction sweeper is used, the cleanliness of the milled surface is rated as good to excellent.
- **High pressure level**. Even if it could not be demonstrated in this case that a specific high pressure level in water cleaning is more favorable for a more efficient bonding between the pavement layers, it is highly recommended to use a high-pressure suction sweeper on a milled existing pavement to avoid possible negative effects of dust and other particles.
- **Adhesion performance of multilayer pavements**. Many parameters may influence adhesion in a significant way and these parameters also interfere with each other (e.g. state and surface characteristics, tack coat type and rate…). To ensure a strong and homogeneous adhesion, it is essential that all of these parameters are equally well controlled.
- **Recommendation limits**. At this stage of the study, it is too early to give recommendations for mean shear strength limits for new surface/milled binder course. More field studies are needed to evaluate correctly these recommendation limits.

References

(FGSV 2003) Forschungsgesellschaft für straβen und verkehrswesen, 2003.

(SN 640430 2012) Swiss Standard SN 640430, Walzasphalt, VSS, Zürich, 2012.

(Stöckert 2002) Ein Beitrag zur Festlegung von Grenzwerten für den Schichtenverbund im Asphaltstraßenbau, Stöckert U., TUD, 2002.

(Raab and Partl 1999) Methoden zur Beurteilung des Schichtenverbunds von Asphaltbelägen, Raab,C., Partl, M.N, Eidgenössisches Departement für Umwelt, Verkehr, Report No 442, November, 1999

(Lavoie 2015) Formation sur la mise en œuvre des enrobés. Formation interne. Ministère des Transports du Québec (MTQ)

(EN 12272-1 2012) European standard EN 12272-1: Surface dressing—Test methods — Part 1: Rate of spread and accuracy of spread of binder and chippings.

(Destrée et al. 2015) Field study to investigate the impact of conditions of application of tack coats on the interlayer bond strength, A. Destrée, J. De Visscher, N. Piérard, A. Vanelstraete, 8th International RILEM Symposium, Ancona, Italy, 7-9 October 2015

Damage Modeling of Asphaltic Pavement Rough Interfaces Under Tensile and Shear Loading

Rahma Ktari, Fazia Fouchal, Anne Millien and Christophe Petit

Abstract The interface mechanical behavior in asphaltic pavement structures represents a key parameter for the computational design. Several degradation mechanisms are observed in surface layers due to the de-bonding despite the implementation of tack coats or reinforcing systems. The interface modeling becomes more important in the pavement field during the last ten years, even if this concept is widely studied in composites, masonry structures… Furthermore, the experimental research highlights the sensitivity of the imperfect interface behavior in respect to the underlayer roughness. Due to these concerns, parametrical studies are proposed in this paper, by using a macro scale cohesive zone model. This model allows taking into account the damage behavior of the interface, in pure or mixed mode. Damage mechanisms are governed by internal variables. The numerical results based on a finite element method are compared with experimental tensile and shear data. The geometrical periodic roughness (triangular, crenel) and the mixed mode failure are investigated under monotonic loading. A parametric numerical analysis is presented and shows clearly the roughness influence of these structures.

Keywords Interface · Pavement · Damage · Cohesive zone model (CZM)

R. Ktari (✉) · F. Fouchal · A. Millien · C. Petit
GEMH-GCD, University Limoges, EA 3178, 19300 Egletons, France
e-mail: rahma.ktari@unilim.fr

F. Fouchal
e-mail: fazia.fouchal@unilim.fr

A. Millien
e-mail: anne.millien@unilim.fr

C. Petit
e-mail: christophe.petit@unilim.fr

A. Chabot et al. (eds.), *8th RILEM International Conference on Mechanisms of Cracking and Debonding in Pavements*, RILEM Bookseries 13,
DOI 10.1007/978-94-024-0867-6_66

1 Introduction

Interfaces between bituminous pavement layers represent a very important parameter for the computational design (Roffe and Chaignon 2002). Some experimental researches (Santagata et al. 2008; Raab et al. 2012; D'Andrea et al. 2013) show the roughness influence between the different layers on the interface constitutive law in these kind of structures. New pathologies of road surfaces require today rational methods for design by taking into account the interfaces behavior. In the literature interface modeling is considered in the most cases as perfectly bonded with a local behavior based on empirical fatigue laws (Petit et al. 2009), on fracture mechanics such as elasto-plastic laws based on the Mohr-Coulomb criterion (Romanoschi and Metcalf 2001) or on elasticity laws (Hun et al. 2012; De Bondt 1999; Ozer et al. 2008) have given a particular attention to interface roughness modeling with saw-tooth type of model by zero thickness interface element within the concept of Finite Element (FE) theory. In other fields the interface modeling is more studied with no linear behavior such as in composite materials (Gornet and Ijaz 2011), masonry (Fouchal et al. 2014). In addition, cohesive zone models (Song et al. 2006) seems to be a promising approach to treat this problem. The current study is based on damage cohesive zone models. This paper is organized as follows, in Sect. 2, the classical damage model proposed by (Allix et al. 1995) for the prediction of delamination in laminated composites is recalled. The Sect. 3 consists to study the influence of the interface parameters through a numerical analysis on the mechanical behavior and adhesion between different layers. Simulations results are presented. Experimental roughness measurements are input as parameters for geometrical modeling at the macroscale and simulation results are successfully compared with available experimental data. Finally some concluding remarks are given in Sect. 4.

2 Interface Damage Model

The model selected in this study is based on the classical static damage evolution law proposed by (Allix et al. 1995). The displacement jump at the interface between two layers (cf. Figure 1a) can be written as:

$$[\mathrm{U}] = [\mathrm{U_n}]\mathrm{n} + [\mathrm{U_t}]\mathrm{t} \quad (1)$$

where $[\mathrm{U_n}]$ and $[\mathrm{U_t}]$ are the normal and tangential displacement jumps at the interface.

The interface damage is taken into account with internal variable $\mathrm{d} \in [0, 1]$. The constitutive law of the interface can be writhen as:

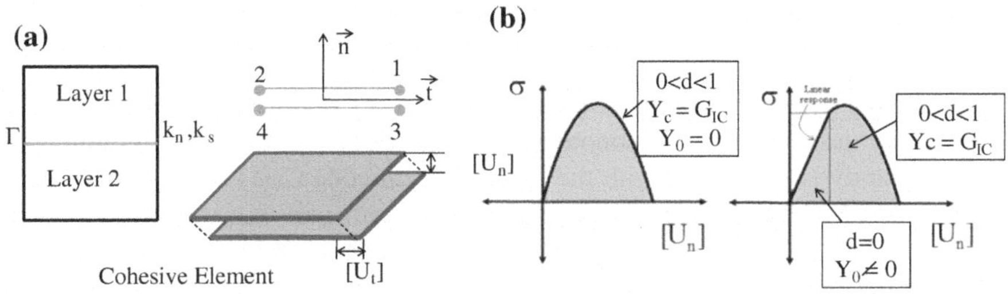

Fig. 1 **a** Interface element between two layers, **b** stress versus displacement curve for mode I

$$\begin{pmatrix} \sigma \\ \tau \end{pmatrix} = \begin{pmatrix} k_n(1-d) & 0 \\ 0 & k_s(1-d) \end{pmatrix} \begin{pmatrix} [U_n] \\ [U_t] \end{pmatrix} \tag{2}$$

where σ and τ are the normal and tangential stress respectively, k_n and k_s are the normal and tangential interface stiffness associated to damage variables.

For 2D interface model, (Allix et al. 1995) suppose two different damage variables corresponding to two failure modes which are governed by equivalent single energy release rate function as follows:

$$\underline{Y}(t) = Sup|_{\tau \leq t}\left[(Y_{d_n})^{\alpha} + (\gamma Y_{d_s})^{\alpha})^{1/\alpha}\right] \tag{3}$$

where $Y_{d_n} = \frac{1}{2}\frac{\langle\sigma\rangle_+^2}{k_n^0(1-d)^2}$ and $Y_{d_s} = \frac{1}{2}\frac{\tau^2}{k_s^0(1-d)^2}$ are the thermodynamic forces in mode I and II respectively, γ and α are material parameters which governs the damage evolution in mixed mode.

The interface quasi-static damage function is defined as follows:

$$d = \omega(\underline{Y}) = \left[\frac{n}{n+1}\frac{\langle \underline{Y} - Y_0\rangle_+}{Y_c - Y_0}\right]^n \quad \text{if } d < 1, d = 1 \text{ otherwise} \tag{4}$$

where Y_0 is a threshold damage energy, Y_c is the critical damage energy (cf. Figure 1b), n is characteristic function of material (higher values of n corresponds to brittle interface).

According to (Allix et al. 2003), the main assumptions of the delay model are the evolution of damage due to the force variations is not instantaneous. The introduction of a delay effect leads to a new damage evolution law which can be written as:

$$\dot{d} = k\langle\omega(\underline{Y}) - d_n\rangle_+^m \quad \text{if } d < 1d = 1 \text{ otherwise} \tag{5}$$

where k is the maximum damage rate and m is a second brittleness parameter.

3 Simulations and Results of Interface Damage Model

Direct Tensile (DTT) and Double Shear (DST) tests are performed in this work. The samples are tested under a monotonic displacement rate of 0.5 mm/min. The modeling study is performed with the finite element code Cast3 M assuming plane strain (Cast3 M). The calculation is nonlinear incremental with a time step optimized at 0.1 s.

Geometry, boundary conditions and mesh of specimens are presented in the Fig. 2. The tested material is provided from the experimental pavement structures of the RILEM-SIB project (Canestrari et al. 2013). Specimens are bi-layered asphalt concrete AC12 with Carbon Fiber grid at the interface. The AC12 material has 12.5 mm maximum aggregate dimension and 50/70 penetration bitumen dosed at above 5 % by aggregate weight. A previous study analyzed with Digital Correlation Images (DIC) allowed us to characterize materials and the interphase/interface properties, and to identify the uncoupled interface stiffness, damage evolutions and critical energy release rate. The model parameters are regrouped in Table 1.

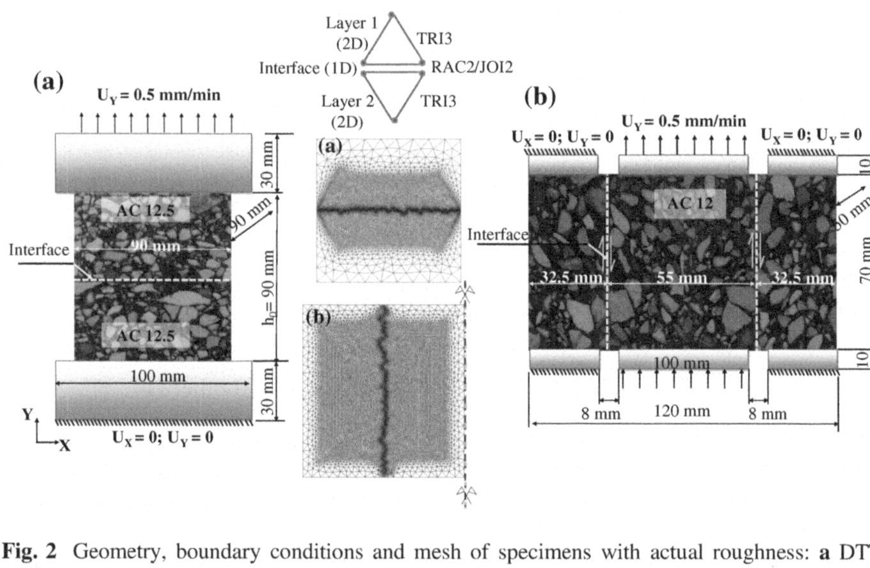

Fig. 2 Geometry, boundary conditions and mesh of specimens with actual roughness: **a** DTT, **b** DST

Table 1 Materials properties and model parameters for smooth interface

Material properties		Interface parameters						
E	υ	k_n	k_s	Y_0	Y_c	α	γ	n
(MPa)		(MPa/mm)	(MPa/mm)	(J/m^2)	(J/m^2)			
2000	0.4	500	100	0	30	1	0.1	0.01

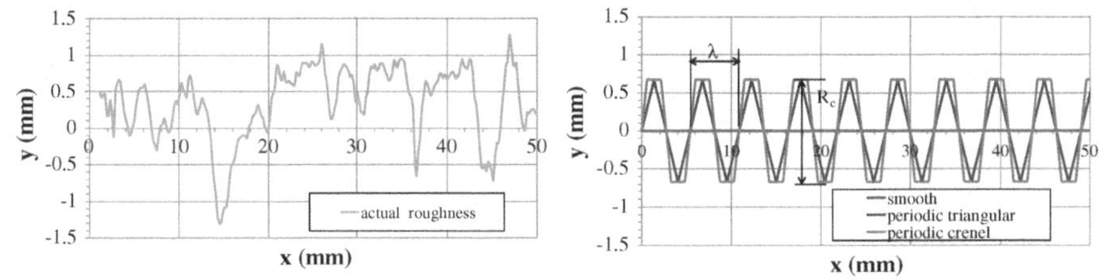

Fig. 3 Profiles roughness

In the first step, the simulations are done for a planar (smooth) interface. Results in mode I and II are compared to experimental data, for different damage rate values. The obtained results indicate the sensitivity of the interface constitutive law in respect to the non intrinsic k parameter. For tensile test, with $k = 1\ s^{-1}$, the general shape of the curve is very close to the experimental one. However, the value $k = 10\ s^{-1}$ is only optimal at the early damage for shear test.

In the second step, the roughness influence is simulated for different profiles: actual roughness, periodic triangular, periodic crenel and smooth (Fig. 3). The actual and periodic roughness parameters values such as $\lambda = 5.45$ mm and $R_c = 1.35$ mm, are obtained experimentally with the phase shifting projected fringe profilometry method. For the mode I (Fig. 4a), roughness has no significant influence on the behavior unlike results for mode II (Fig. 4b).

In this case, in order to take into account the mesoscale properties, the parameters of the interface model are modified with $k_{S\mu} = 50$ MPa/mm, $k_{n\mu} = 140$ MPa/mm and a coupling parameter $\gamma = 0.2$. The stiffness of the interface as well as the shear strength are dependent on roughness for the shear mode. The actual profile is situated between the periodic crenel and periodic triangular shapes. The results with this roughness are in very good agreement with the experimental results. This confirms

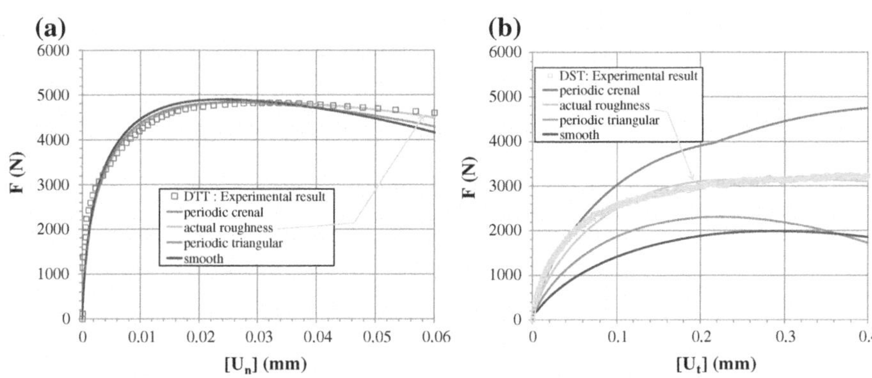

Fig. 4 Roughness effect for: **a** DDT ($k = 1\ s^{-1}$), **b** DST ($k = 10\ s^{-1}$)

that the geometric roughness model coupled with the CZM is well-suited to describing the mechanical behavior of the interfaces in asphaltic pavement structures.

4 Conclusion

In conclusion, this article presents a comprehensive interface model including delay effect, based on damage energy release rate. The effectiveness of the proposed model is tested by performing finite element simulations for two different tests (DTT and DST) under monotonic loading conditions on bi-layered asphaltic specimen AC12. The damage behaviors predicted by the proposed model for pure mode I, pure mode II and for mixed-mode with taking into account of roughness are found in good agreement with experimental results. Finite element simulations were conducted for two main results: (a) the roughness has no effect in mode I, (b) the interfacial strength increases significantly in the shear mode II. In the future work, an internal variable can be add to the present cohesive zone model in order to take into account the effect of the geometric roughness in a smooth interface.

References

Allix, O., Ladevéze, P., and Corigliano, A. (1995). Damage analysis of interlaminar fracture specimens. Compos. Struct. 31, 61–74.

Allix, O., Feissel, P., and Thévenet, P. (2003). A delay damage mesomodel of laminates under dynamic loading: basic aspects and identification issues. Comput. Struct. 81, 1177–1191.

Canestrari, F., Belogi, L., Ferrotti, G., and Graziani, A. (2013). Shear and flexural characteriza-tion of grid-reinforced asphalt pavements and relation with field distress evolution. Mater. Struct.

Cast3 M., code d'éléments finis CEA: http://www-cast3m.cea.fr/ Accessed 29 Jan 2016

D'Andrea, A., Tozzo, C., Boschetto, A., and Bottini, L. (2013). Interface Roughness Parameters and Shear Strength. Mod. Appl. Sci. 7, 1–10.

De Bondt, A.H. (1999). Anti-Reflective Cracking Design of (Reinforced) Asphaltic Overlays (Nieuwegein, the Netherlands: Delft University of Technology).

F.A. Santagata, M.N. Partl, G. Ferrotti, F. Canestrari, and A. Flisch (2008). Layer Characteristics Affecting Interlayer Shear Resistance in Flexible Pavements. J. Assoc. Asph. Paving Technol. 222–254.

Fouchal, F., Lebon, F., Raffa, M.L., and Vairo, G. (2014). An Interface Model Including Cracks and Roughness Applied to Masonry. 263–271.

Gornet, L., and Ijaz, H. (2011). A high-cyclic elastic fatigue damage model for carbon fibre epoxy matrix laminates with different mode mixtures. Compos. Part B Eng. 42, 1173–1180.

Hun, M., Chabot, A., and Hammoum, F. (2012). A four-point bending test for the bonding evaluation of composite pavement. RILEM Bookseries 4, 51–60.

Ozer, H., Al-Qadi, I.L., and Leng, Z. (2008). A fracture-based constitutive model for pavement interface characterization. In Pavement Cracking: Mechanisms, Modeling, Detection, Testing and Case Histories, pp. 661–670.

Petit, C., Diakhaté, M., Millien, A., Phelipot-Mardele, A., and Pouteau, B. (2009). Pavement Design for Curved Road Sections Fatigue Performance of Interfaces and Longitudinal Top-down Cracking in Multilayered Pavements. Road Mater. Pavement Des. 10, 609–624.

Raab, C., Abd El Halim, A.O., and Partl, M.N. (2012). Interlayer bond testing using a model material. Constr. Build. Mater. 26, 190–199.

Roffe, J.-C., and Chaignon, F. (2002). Characterisation tests on bond coats: worldwide study, impact, tests, recommendations. In Proceedings of the 3rd International Conference on Bituminous Mixtures and Pavements, Held Thessaloniki, Greece, T.R. Laboratory, N.M.R. Crowthorne House, and B.R. 3GA U.K. Wokingham, eds. pp. p. 603–614.

Romanoschi, S., and Metcalf, J. (2001). Characterization of Asphalt Concrete Layer Interfaces. Transp. Res. Rec. 1778, 132–139.

Song, S.H., Paulino, G.H., and Buttlar, W.G. (2006). A bilinear cohesive zone model tailored for fracture of asphalt concrete considering viscoelastic bulk material. Eng. Fract. Mech. 73, 2829–2848.

Effect of Reinforced Asphalt Pavements on Reflective Crack Propagation and Interlayer Bonding Performance

Christiane Raab, Martin Arraigada and Manfred N. Partl

Abstract Reinforcement of asphalt pavements has become a valuable constructive method for preventing reflective cracking and prolonging the service life of asphalt pavements. Although the advantage of reinforcement overall seems beyond doubt, there is still insufficient information about the actual effect on the prolongation of the pavement's life-span. Further, it is not clear how the different grids affect the bonding performance between layers. This paper investigates the performance of different reinforcement grids on the performance of two layered asphalt pavements, mainly considering: (a) Reflexive crack propagation, (b) Shear bonding strength. The paper compares the performance of different non-reinforced and reinforced pavements, constructed in the laboratory and in the field and trafficked with a down-scaled (Model Mobile Load Simulator MMLS3) and a full-scale traffic load simulator (Mobile Load Simulator MLS10) respectively. Loading the pavements with rolling tires (cyclic loading) could show differences in the durability between the different reinforcement types as well as the improvement in the flexural bonding strength when using reinforcements compared to non-reinforced structures. Grid styles with different mesh opening sizes of 12.5 and 25 mm plus different tack coats or tack films were also considered. The investigation of the interlayer bonding strength was done by means of the layer parallel direct shear tester (LPDS) and 150 mm cores. It was found that the bonding properties do depend on the reinforcement type, but that in most cases the interlayer bond is according to the requirements.

Keywords Asphalt reinforcements · Reflective cracking · Interlayer bonding · Traffic simulators

C. Raab (✉) · M. Arraigada · M.N. Partl
Empa, Duebendorf, Switzerland
e-mail: christiane.raab@empa.ch

M. Arraigada
e-mail: martin.arraigada@empa.ch

M.N. Partl
e-mail: manfred.partl@empa.ch

A. Chabot et al. (eds.), *8th RILEM International Conference on Mechanisms of Cracking and Debonding in Pavements*, RILEM Bookseries 13,
DOI 10.1007/978-94-024-0867-6_67

1 Introduction

Reinforcement of asphalt pavements has become a valuable constructive method for preventing reflective cracking and prolonging the service life of asphalt pavements. Although the advantage of reinforcement overall seems beyond doubt, there is still insufficient information about the actual effect on the prolongation of the pavement's life-span. Further, it is not clear how different reinforcement grids affect the bonding performance between layers (de Bondt 2012; Raab and Partl 2006). This paper investigates the performance of 3 different reinforcement grids on the performance of two layered asphalt pavements, mainly considering their potential regarding the prevention of cracking and debonding comparing the behaviour with the unreinforced situation. Tests are conducted in the laboratory on roller compacted specimens using a 1:3 traffic load simulator as well as in situ by means of a full scale traffic simulator.

2 Material and Sample Preparation

In the investigation beside an unreinforced sample (reference) or test section three different kind of grids were used in the slab or test sections:

(a) Grid A: Glass fiber grid. This grid was installed by placing a Stress Absorbing Membrane (SAMI) between the grid and the asphalt layer.
(b) Grid B: Polyester grid.
(c) Grid C: glass and carbon fiber grid. The carbon fiber was installed in the traffic direction.

All grids were installed on top of the underlying layer according to the manufacturer's specifications using bond coats or heat (Partl et al. 2014). The SAMI used for the glass fiber grid was placed using a spray unit.

For the laboratory tests slabs were prepared using a laboratory steel roller compactor. The slabs were constructed in two layers of AC 8 and AC 11 with a thickness 30 mm of each. The slabs had a dimension of 1800 mm × 435 mm, for each grid two slabs were prepared. After the construction the slabs were turned upside down and two artificial cracks of about 3 mm over the total width of the slab, with a depth of 25 mm and in a distance of 190 mm were cut symmetrical to the middle from the bottom of the slabs.

For the in situ testing 4 identical pavement sections with a length of 9 m and a width of 5 m each were constructed according to Swiss specifications. They consisted of 40 mm AC 11 on 65 mm AC 22 on top of a 600 mm unbound gravel layer.

Before constructing the surface layer cuts were done through the entire thickness of the lower layer (AC 22) in order to simulate existing cracks. They were then filled with sand and sensors were installed.

3 Reflective Crack Testing

3.1 Laboratory Testing

For the laboratory testing the model mobile load simulator MMLS3 (Fig. 1) was used. The MMLS3 is a 1:3 scaled Accelerated Pavement Testing (APT) device that applies a scaled load with four single tires and a speed of 9 km/h (Hugo and Epps 2004). The slabs were installed at 20 °C in a container which allowed for a controlled temperature situation. The slabs were put on two supports in a distance of 1750 mm from each end on top of a concrete plate. In the middle a thin rubber mat was placed. The failure of the slabs was expected to be caused by cracking and/or delamination during the traffic simulation. It was expected that due to the progression of cracking the stiffness of the slabs was going to decrease, leading to an increase in their bending deformation. Therefore, 6 deformation sensors were installed as visible in Fig. 1 in order to monitor the development of cracking. The right part of Fig. 1 shows the whole test set-up.

3.2 In Situ Testing

For the in situ testing testing the mobile load simulator MLS10 was used (Arraigada et al. 2014). In order to optimize time two sections were tested at a time. The first section comprised the reference pavement and grid A. The other section grids B and C. In order to obtain similar temperatures in both fields, the MLS10 was moved between sections every 100,000 cycles. For testing, the MLS10 was equipped with super-single tires, which loaded the pavement with unidirectional tires passing along about 4.2 m. The tires pressure was set to 0.90 MPa. The load amplitude was set to 65 kN. The rolling speed of the tires was set to 18 km/h, reproducing the passing of ca. 5000 half-axles per hour. Overall, 300,000 load cycles were applied on each section of the pavement.

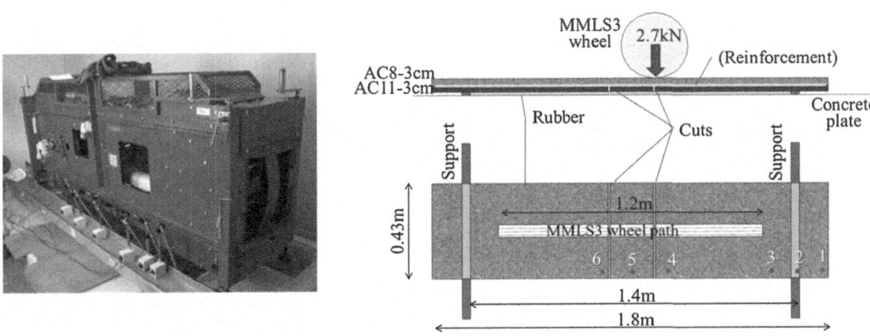

Fig. 1 *Left* MMLS3 with sensors, *right* schematic test set-up

4 Interlayer Bond Testing

Interlayer bond testing was done using the Layer-Parallel Direct Shear (LPDS) test device, a modified version of equipment developed in Germany by Leutner (1979), Raab et al. (2009). The interlayer bond was determined at a temperature of 20 °C. For the laboratory tests special slabs were compacted for the different systems using the compaction device of the LCPC rutting tester with a specially designed steel roller. Cores of 150 mm diameter were then taken and tested. For in situ testing in each section cores (Ø 150 mm) were taken directly from the pavement in the trafficked (under MLS10) and in the untrafficked areas.

5 Results

5.1 Reflective Cracking MMLS3

Figure 2 presents selected results of the deflection amplitudes for each tested system. The bearing capacity limit was established from these curves combined with visual inspections. It was observed that, in general, the deflection amplitude curves show an exponential increase when the cracks reach the top surface of the slab.

According to the results, gird A showed a poor performance compared to the reference slabs. On the other hand, the slab with the grid B slab was able to take almost twice as much load as the reference. The best performance was obtained by the slabs with the grid C which had almost three times more loading endurance than the reference slabs.

5.2 Reflective Cracking MLS10

The most important factor for the analysis of the pavement performance was the capability of the different grids to delay the propagation of the cracks from the

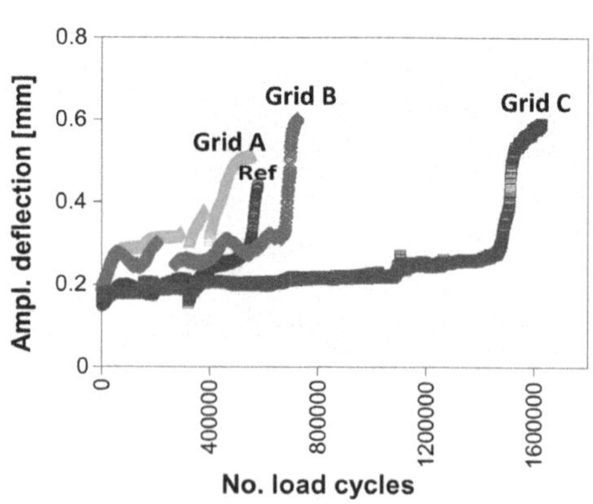

Fig. 2 MMLS3 testing results

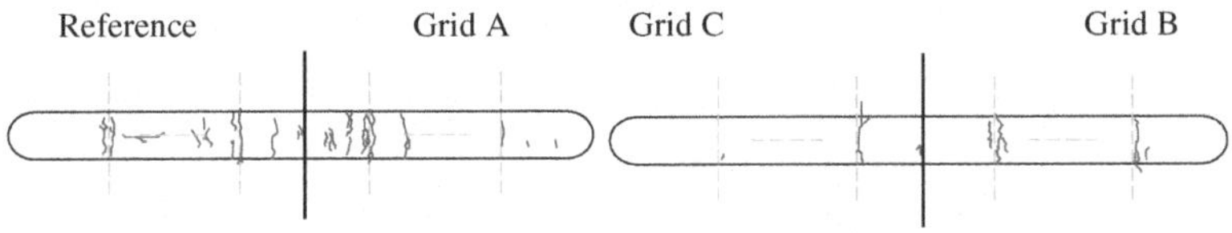

Fig. 3 Surface crack patterns

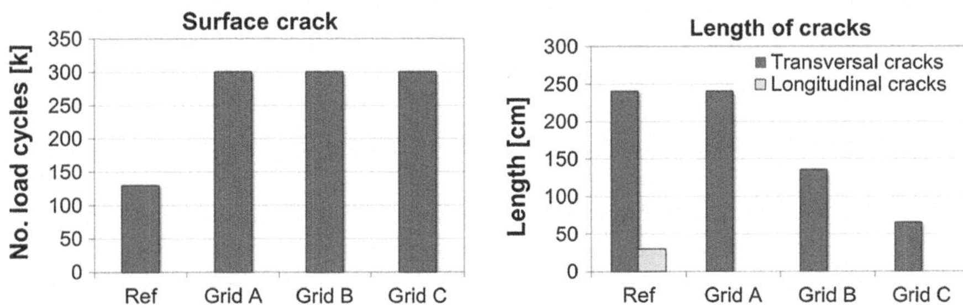

Fig. 4 Summary of crack characteristics

lower layers to the surface. This was carried out through visual inspections. The first transversal crack in the reference pavement was observed at 130,000 load cycles interruption, whereas in the other pavements the first crack appeared at the end of the tests, after 300,000 load cycles. Figure 3 shows the crack distribution schematically at the end of the loading phase. It can be perceived, that the crack density is much higher for the reference pavement and grid A, than for the other two pavement sections. Further, the majority of the reflected cracks are transversal to the loading direction and coincident with the cut in the lower layers, at least at an early cracking stage. All detected cracks had a thickness of less than 1 mm. A comparative summary of the characteristics of the cracking performance of each reinforcement system is presented in Fig. 4. From this figure it can be observed that all systems succeeded in delaying the reflection of cracks more than two times, compared to the pavement without reinforcement.

5.3 Interlayer Bond Testing

Figure 5 shows the interlayer bond test results. In the laboratory the maximum shear forces for all investigated reinforcement systems were very good, for grid B even better than for the non-reinforced specimens. All systems had values well above the Swiss standard requirements. For the in situ cores the situation is different, although the ranking between the systems remains similar: while the shear force values of the reference reflect the laboratory results, grids A and C (especially after trafficking) could be critical regarding the requirement value.

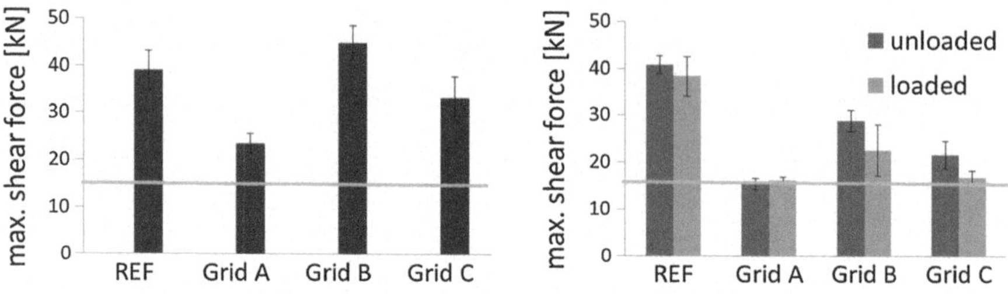

Fig. 5 Interlayer bond test results, left: laboratory, right: in situ

6 Conclusions

As for reflective cracking, from both, laboratory and in situ testing grids B and C showed the best performance resulting in a delay of crack propagation. Although grid A succeeded in delaying the crack propagation in the field, the laboratory results were less favorable. Overall, reinforcement grids are capable of prolonging the pavement life. Regarding interlayer bonding the results show that the bond values of the reference laboratory specimens are similar to the field. However, for the reinforcement grids the in situ bond values are significantly lower, which shows the importance of in situ testing. Further, the MLS10 loading appears to have a negative influence on the bonding properties.

References

Arraigada M, Pugliessi A, Partl MN, Martínez F (2014) Effect of full-size and down scale accelerated traffic loading on pavement behavior. Mater. Struct. 47, 1409-1424 (2014)

de Bondt AH (2012). 20 Years of Research 2 on Asphalt Reinforcement,– Achievements and Future Needs. 7th RILEM Conference on Cracking in Pavements: Mechanisms, Modeling, Testing, Detection and Prevention Case Histories, Volume 4, 327-335

Hugo F, Epps A (2004) Significant Findings From Accelerated Pavement Testing, *NCHRP Synthesis 325*, Washington, TRB, 2004

Leutner, R (1979) Untersuchungen des Schichtenverbunds beim bituminösen Oberbau. Fachartikel. Bitumen 3, 1979

Part MN, Raab C, Arraigada M (2014) Innovative Asphalt Research using Accelerated Pavement Testing. Journal of Marine Science and Technology JMST, Vol. 23, No. 3, pp 269-280, DOI: 10.6119/JMST-014-0326-1

Raab C, Partl M N (2006) Rehabilitation of Concrete pavements with Geotextiles and Steel Meshes", Proc. Int. Conference on Concrete Repair, Rehabilitation and Retrofitting ICCRRR, Cape Town, South Africa, November 21-23, 2005, pp. 889-893, ISBN 0 415 39654 9

Raab, C., Partl, M. N., & Abd El Halim, A. O. (2009). Evaluation of Interlayer Shear Bond Devices for Asphalt Pavements. Baltic Journal of Road and Bridge Engineering, 4 (4), 186-195.

Role of Concrete-Asphalt Interface in Bonded Concrete Overlays of Asphalt Pavements

Angel Mateos, John Harvey, Julio Paniagua, Fabian Paniagua and Angela Fan

Abstract Bonded Concrete Overlay of an Asphalt pavement (BCOA) is a rehabilitation technique consisting of 50–175 mm thickness portland cement concrete overlay on an existing flexible, semi-rigid or composite pavement. This technique, that has also been known as thin (minimum 100 mm) or ultrathin whitetopping (thinner than 100 mm) in the past, relies on the composite action of the concrete and asphalt layers acting together with a third phase of the system being the interface between the two materials. For this study, the stiffness and strength/fatigue resistance of this interface have been characterized by means of a series of laboratory tests conducted on asphalt and composite cylindrical specimens under different loading and environmental conditions. Loading conditions were intended to reproduce the rapid traffic pulses and the slow temperature and shrinkage related actions, in both shear and vertical tensile modes. Environment related testing conditions included wet and dry, and temperatures between 4 and 40 °C, which is a range applicable to asphalt bases located under 100–175 mm thick bonded concrete overlays in California. Several conclusions were extracted that provide insight into the mechanical nature of the interface concrete-asphalt as well as the basis for designing a laboratory protocol for interface mechanical characterization.

Keywords Bonded overlay · Asphalt pavement · Rehabilitation · Concrete-Asphalt bonding

A. Mateos (✉) · A. Fan
University of California Pavement Research Center, Berkeley, CA, USA
e-mail: angel-mateos@berkeley.edu

A. Fan
e-mail: angefan@berkeley.edu

J. Harvey · J. Paniagua · F. Paniagua
University of California Pavement Research Center, Davis, CA, USA
e-mail: jtharvey@ucdavis.edu

J. Paniagua
e-mail: jpaniaguaf@ucdavis.edu

F. Paniagua
e-mail: fpaniaguaf@ucdavis.edu

A. Chabot et al. (eds.), *8th RILEM International Conference on Mechanisms of Cracking and Debonding in Pavements*, RILEM Bookseries 13,
DOI 10.1007/978-94-024-0867-6_68

1 Introduction

A Bonded Concrete Overlay of an Asphalt pavement (BCOA) is a rehabilitation technique consisting of 50–175 mm thickness portland cement concrete overlay on an existing flexible, semi-rigid or composite pavement. This technique, that has also been known as thin (minimum 100 mm) or ultrathin whitetopping (thinner than 100 mm) in the past, relies on the composite action of the concrete and asphalt layers acting together with a third phase of the system being the interface between the two materials (Mateos et al. 2015). This new conception results in a much stronger pavement structure than if the layers were to act alone, in the same way that two independent beams cannot stand as much load as a single beam of double thickness. When well bonded, the underlying asphalt lowers the neutral axis of the concrete overlay, thus reducing tensile stresses at the bottom of the concrete and, consequently, significantly increasing the fatigue life. This composite action requires complete bonding between the two materials, which constitutes the main factor leading the conception, design, and construction of this type of pavement. Numerous studies agree on the critical importance of bonding for BCOA performance and present numerous experimental results that support this conclusion (Rasmussen and Rozycki 2004; Li et al. 2013). However, very little research has been conducted to understand the bonding mechanics and performance, which are mostly unknown at this time.

The objective of the experimental study presented in this paper is to provide insight into the mechanical nature of the concrete-asphalt interface and to provide the basis for a laboratory protocol for mechanical characterization of the interface. Preliminary laboratory results are presented of a three-year research project supported by the California Department of Transportation, one of whose specific goals is understanding the performance of the concrete-asphalt interface in BCOA.

2 Research Approach

Different laboratory mechanical tests of the interface have been conducted in order to achieve the goal of this experimental study. Tests are focused on either horizontal (shear) or vertical bonding performance, and they consider two different time scales: the rapid loads of the traffic vehicles and the slow environment-related actions. Loading times in the order of the tenths of a second are important in the first case, so frequency sweep tests have been used for the characterization. Creep tests have been used to simulate the longer loading times of environment-related actions. Shear performance has been evaluated with the Superpave Shear Tester (SST), where different shear loading patterns are applied to a disc-shaped specimen (Fig. 1—left). A universal testing machine (UTM) has been used to evaluate vertical tensile performance of the interface (Fig. 1—right, Table 1).

Left picture

Shear specimen, ϕ 150mm & h75mm (Concrete 25mm + Asphalt 50mm)
Shear deformation is measured with an LVDT attached to the testing jigs.

Right picture

Tensile test specimen, ϕ 100mm & h190mm (Concrete 100mm + Asphalt 90mm).Left and right LVDTs in the picture measure asphalt strain, and LVDTs in the front and back of the specimen (the last one is not visible in the picture) measure interface opening plus 9.5 mm of asphalt.

Fig. 1 SST (*left*) and UTM (*right*) composite specimens

Table 1 Summary of mechanical characterization tests

	Stiffness		Strength/fatigue resistance	
	Traffic loading	Env. loading	Traffic loading	Env. loading
Shear (SST)	Frequency sweep 4/20/40 °C	Creep test 4/20/40 °C	Fatigue test, 1200 με 10 Hz, 40 °C	
Tensile (UTM)		Creep test 20 °C		Monotonic test 1/10 kPa/s, 20 °C

Specimens were produced in the laboratory by casting the concrete overlay on an asphalt slab. Asphalt slabs were prepared using rolling-wheel compaction. Concrete overlay was produced with Type I/II cement (US standard, mild sulfate resistance), and it was covered with a wet burlap after casting to prevent it from moisture loss. The asphalt was a Superpave gap-graded mix with 12.5 mm nominal maximum aggregate size, 7.4 % asphalt rubber binder content (PG64-16 base, minimum 18 % recycled tire rubber), compacted to 6 % air voids. No surface conditioning was used for the asphalt before casting the concrete overlay. Composite specimens were cored from the concrete-asphalt slabs, and then trimmed to the final height. Figure 1 shows both SST and UTM specimens.

3 Analysis of Laboratory Results

Shear stiffness of the concrete-asphalt interface was determined by subtracting asphalt specimens compliance from the compliance of the composite specimens. Testing was conducted on dry and wet conditions. For the wet conditions, specimens were saturated in water (AASHTO T283) and then wrapped in parafilm for

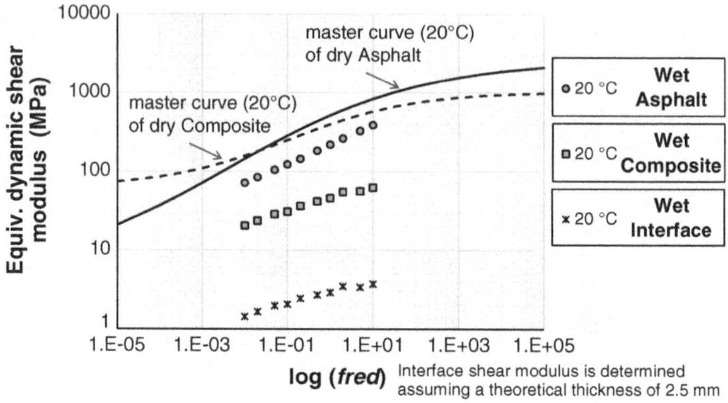

Fig. 2 WET shear stiffness of asphalt and composite specimens and concrete-asphalt interface

testing. The interface stiffness was reduced by more than one order of magnitude after the specimen was subjected to water saturation (Fig. 2). Dry testing results showed little temperature and frequency susceptibility of the interface stiffness. On the contrary, frequency dependency is clear from wet testing (Fig. 2). Both temperature and time dependency are clearly seen after creep shear testing of the dry interface. This testing, whose results are not shown in this paper due to space limitations, indicated that the interface has a tremendous capacity for shear relaxation. Even for a low temperature of 4 °C, interface creep deformation was over 0.25 mm after 100 s.

Repeated shear sinusoidal loading (1200 με peak to peak) was applied on asphalt and composite specimens in order to reproduce damage accumulation under traffic loads at high temperatures (40 °C). Tests were conducted on dry and wet conditions (parafilm wrapped, saturated specimens). This testing revealed that the evolution of dynamic modulus and phase angle during fatigue testing were almost identical in both the asphalt and composite specimens. Furthermore, stiffness recovery after one month at rest was also almost identical in both the asphalt and composite specimens. These results seem to indicate that fatigue damage was mainly related to the asphalt, and not to the interface. This result is in line with experimental results obtained with the Fabac accelerated pavement testing facility at Nantes (Chabot et al. 2008), that concluded asphalt damage took place before debonding in one section tested during summer. Another important observation after repeated shear testing is that specimens that were fatigued in dry conditions recovered most stiffness after the rest period, while recovery of the specimens fatigued under wet conditions was negligible. This result is in line with the importance that several authors confer to water for BCOA performance (Vandenbossche and Barman 2010).

The Iowa shear test was conducted to further compare shear strength of asphalt versus interface. In this test, a composite specimen is sheared at a constant stress rate of 2.8–3.5 MPa/min until failure. Failure surface affected both the asphalt and the interface, but the more to the former than to the latter.

The tensile stiffness of the interface was evaluated by means of creep testing. Creep compliance of the interface (assuming a theoretical interface thickness of 2.5 mm) was around eight times greater than the creep compliance of the asphalt, i.e., its stiffness is around eight times smaller. This ratio is in line with results presented in Fig. 2 for 20 °C. The shapes of the creep functions of asphalt and interface were very similar to each other, which indicates that the mechanical nature of the concrete-asphalt interface is strongly related to that of the asphalt.

The tensile strength of the interface was evaluated at 20 °C by conducting constant load rate tensile strength tests. One of the two tested specimens failed at the interface, while the other one failed in the asphalt as reflected in Fig. 1 (right). These results contradict common belief that concrete does not bond well to new asphalt (Yu and Tayabji 2007).

4 Conclusions

A series of laboratory tests have been conducted on asphalt and composite (concrete on asphalt) cylindrical specimens in order to characterize interface stiffness and strength/fatigue resistance under different loading and environmental conditions. Loading conditions were intended to reproduce the rapid pulses of shear from traffic and the slow tension and shear caused by temperature and shrinkage changes in the concrete. Environment related conditions included wet and dry, and temperatures between 4 and 40 °C, a range applicable to asphalt bases located under 100–175 mm thick concrete overlays in California. Several conclusions can be extracted as a result of this testing, as presented below.

- The mechanical nature of the concrete-asphalt interface was strongly related to that of the asphalt. The interface stiffness showed a clear time and temperature dependency, and it significantly softened—by at least one order of magnitude–under wet conditions.
- The strength and fatigue resistance of the concrete-asphalt bond was comparable to the strength and fatigue resistance of the gap-graded mix with a high (7.4 %) asphalt rubber binder content used in the composite material.
- Fatigue of the composite specimens tested in shear at 40 °C under repeated loading happened mainly in the asphalt, and not in the interface.
- Experimental results from this study do not support the common belief that concrete does not bond well to new asphalt.
- Experimental results from this study are in line with common belief that water is one of the critical factors leading to failure of BCOA sections. Nonetheless, results from this study indicate that the negative action of water happens even before debonding, as the interface and asphalt soften to a point where the composite action of the two layers is lost.

- The stiffness of the interface-asphalt system may be at least one order of magnitude lower under slow environment related actions compared to the stiffness under rapid traffic loads. This relationship has to be acknowledged in BCOA design procedures.

Acknowledgements Results here presented constitute the preliminary laboratory testing of a three-year research project focused on BCOA performance. The research presented in this paper was requested and sponsored by the California Department of Transportation (Caltrans). Caltrans sponsorship of that work is gratefully acknowledged. The contents of this paper reflect the views of the authors and do not necessarily reflect the official views or policies of the State of California, or the Federal Highway Administration. Special thanks are for Nick Burmas, Joe Holland, Deepak Maskey and Robert Hogan for their support for this research project.

References

Chabot A, JM Balay B, Pouteau, F. de Larrard (2008) FABAC accelerated loading test of bond between cement overlay and asphalt layers. 6th Rilem Int. Conf. on Cracking in Pavements, June 16-18 2008, Chicago, US. Taylor & Francis Group Proceedings, 13-23

Li Z, Dufalla N, Mu F, Vandenbossche JM (2013) Bonded Concrete Overlay of Asphalt Pavements Mechanistic-Empirical Design Guide (BCOA-ME). User's Guide, FHWA TFP Study 5 (165)

Mateos AJ, Harvey JC, Paniagua, (2015) Development of Improved Guidelines and Designs for Thin Whitetopping: Literature Review, University of California Pavement Research Center Draft Technical Memorandum: UCPRC-TM-2015-01

Rasmussen RO, Rozycki DK (2004) Thin and Ultra-Thin Whitetopping, NCHRP Synthesis of Highway Practice 338, National Cooperative Highway Research Program, National Research Council, Washington, DC

Vandenbossche JM, Barman M. (2010) Bonded Whitetopping Overlay Design Considerations for Prevention of Reflection Cracking, Joint Sealing, and the Use of Dowel Bars, Transportation Research Record: Journal of the Transportation Research Board, 2155(1)

Yu HT, Tayabji S (2007) Thin Whitetopping—The Colorado Experience, No. FHWA-HIF-07-025

Adhesion Between Asphalt Layers Through the Leutner Shear Test

Matheus de Souza Gaspa, Kamilla L. Vasconcelos and Liedi Légi Bariani Bernucci

Abstract Reflective cracking is a common issue present in recently rehabilitated pavements, especially when hot mix asphalt overlay is applied over the cracked pavement. It can cause the cracks to propagate to the new asphalt layer in a short period of time, wasting most of the resources spent in the rehabilitation project. Among many existing techniques to control reflective cracking, there is the application of an asphalt interlayer capable of relieving the stresses that reach the overlay, also known as Stress Absorbing Membrane Interlayer (SAMI). This solution can be highly efficient but it is essential to ensure adequate adhesion at the interface between layers, otherwise movement at the interface will eventually cause cracking at the pavement surface. The described rehabilitation solution was applied in a test site at BR-116 highway. As for the SAMI, an asphalt mixture produced with fine granite aggregates and Highly Modified Asphalt (HiMA) binder was used. The hot mix asphalt (HMA) overlay was produced using SBS-modified asphalt binder and the same aggregates source. Tack coat was applied between the asphaltic layers to ensure proper bonding. In this study, field samples were taken and the Leutner Shear Test was used to assess the bond strength at the interlayer-overlay interface. Results have shown that the bond strength was adequate in the test section. Therefore, the efficiency of the solution can be studied without influence of any debonding effects. The values obtained for the shear strength were higher than recommended, so lower rates of tack coat could still be effective.

Keywords HiMA · SAMI · Bond strength · Leutner

M. de Souza Gaspa (✉) · K.L. Vasconcelos · L.L.B. Bernucci
Escola Politécnica da Universidade, de São Paulo, Brazil
e-mail: msouzagaspar@gmail.com

K.L. Vasconcelos
e-mail: kamilla.vasconcelos@gmail.com

L.L.B. Bernucci
e-mail: liedi@usp.br

A. Chabot et al. (eds.), *8th RILEM International Conference on Mechanisms of Cracking and Debonding in Pavements*, RILEM Bookseries 13,
DOI 10.1007/978-94-024-0867-6_69

1 Introduction

Cracked asphalt pavements are often rehabilitated by milling the damaged surface and applying a new hot mix asphalt overlay, which should restore the original characteristics of the wearing course. This solution usually has its service life reduced because existing cracks on the underlying pavement propagate to the surface in a short period of time (Ogundipe 2011). One of the existing techniques to control this phenomenon, which is called reflective cracking, is the construction of an asphalt interlayer between the overlay and the existing pavement. Its role is to relieve the stresses that concentrate on the existing cracks, reducing the stress level that reach the overlay. This interlayer is known as Stress Absorbing Membrane Interlayer (SAMI) (Ogundipe 2011). The SAMI can be a highly efficient solution, but it is important to ensure that the interlayer is completely bonded to the overlay, since poor control of the bond between layers is one of the most influential factors in the appearance of cracks.

As a consequence of the importance of proper adhesion, many countries developed different methods to test the interlayer bond of pavement layers. However, there is not an internationally recognized procedure so far. The Leutner shear test, created in 1979, is the most common of the pure direct shear tests (Canestrari et al. 2012), in which no normal stress is applied. In this test, a shear displacement is applied at a constant rate (usually 50 mm/min) and the resulting shear force and displacement are monitored (Collop et al. 2009).

In this study, a 540 m test site at BR-116 highway (São Paulo, Brazil) was rehabilitated using 25 mm thick SAMI underneath 50 mm of HMA overlay. In another 50 m test site section, 55 mm of SAMI was applied as interlayer. Since the effectiveness of the interlayer system could be affected by the bond condition on the interface, core samples were extracted from the test site and the bond strength was evaluated through the Leutner shear test. The results were then compared to others in the literature for a better comprehension.

2 Materials

The materials used in the interlayer and overlay system were two asphalt mixtures, produced with different binders and different gradations of the same granite aggregate source. The SAMI mixture was prepared with fine aggregate (nominal maximum size of 5 mm) and Highly Modified Asphalt (HiMA) binder. The other mixture, a hot mix asphalt (HMA) applied as the overlay, was produced using regular SBS modified asphalt and 19 mm aggregate gradation. The asphalt content in mixture was 7.0 % for the SAMI and 4.2 % for the HMA. Hydrated lime filler was also added to the mixtures, which were produced in the hot mix plant and then taken to the test site. The tack coat material was a polymer-modified asphalt emulsion type CRS-1P, a cationic rapid setting emulsion commonly used in Brazil.

3 Experimental Test Site

The section at BR-116 highway that connects São Paulo to Curitiba, known as Régis Bittencourt highway, is an important route with intense traffic of heavy vehicles. The pavement surface was severely damaged in some sections and reflective cracking would be a problem if regular rehabilitation was done. In this way, a section was chosen to apply the interlayer system as the rehabilitation solution. In 540 m of the road, 25 mm of SAMI was applied underneath 50 mm of HMA overlay, on both lanes. Another section 50 m long received 55 mm thick SAMI under 50 mm of overlay, only on the left lane.

At the interface between the interlayer and the overlay, tack coat was applied in the rate of 0.3 L/m^2 of residual bitumen, following recommendation of the Brazilian standard ES-145 (DNIT 2010). The compaction method was the same for both layers. Since the SAMI is a fine and asphalt-rich mixture, the surface of the interlayer after compaction was very smooth. After the overlay conclusion, 15 core samples were extracted and taken to the laboratory to be tested.

4 Testing Procedure

The adhesion strength between layers was assessed through the Leutner Shear Test. The equipment used to perform this test was a MTS hydraulic testing machine with two semicircular moulds that form a frame of 100 mm diameter (Fig. 1a), as the samples extracted from the field had this diameter.

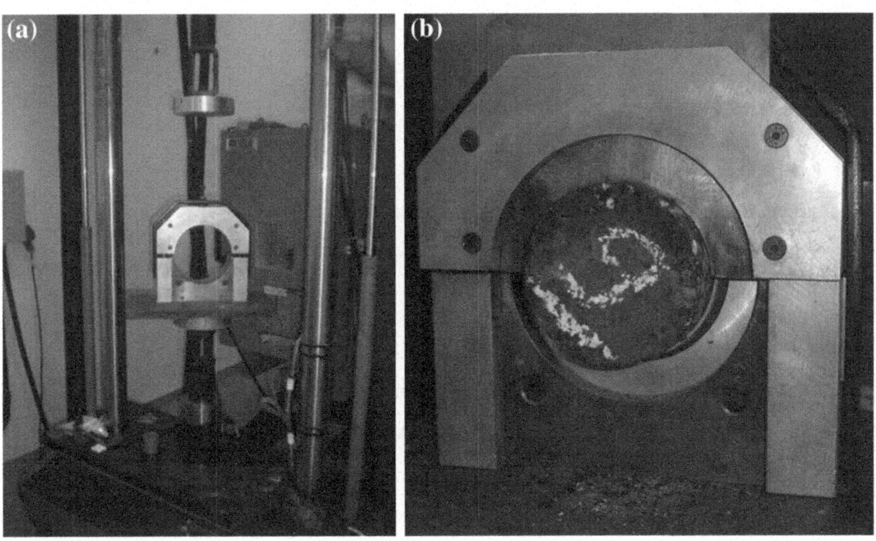

Fig. 1 **a** Equipment used for the Leutner shear test, **b** placement of the core sample in the frame

That is a modification from the standard Leutner test, which uses specimens of 150 mm diameter. Previous studies found that there is an influence of the sample size on the Leutner test results (Guimarães 2013; Canestrari et al. 2012). This happens because the concentration of stresses at the beginning of the shear plane is higher on smaller samples. Guimarães (2013) found peak shear stresses are higher when using 100 mm samples, especially when less tack coat was applied. Canestrari et al. (2012) tested more than 600 samples in different temperatures to study the influence of the specimen diameter, and found out that shear stress on 100 mm specimens is about 14 % higher than the stress on 150 mm samples.

The cores were properly placed into the frame with special care, to ensure that the interface of the sample was aligned with the shear axis of the equipment, as shown in Fig. 1b. The applying rate of the load was 50 mm/min and the samples were conditioned at 20 °C for 12 h before the test. The shear force and displacement were measured and registered during each test, until the layers detached.

5 Results and Discussion

The result of the Leutner Shear test is a typical plot of applied load against shear displacement. From these results, it was possible to calculate the peak shear stresses, dividing the peak load by the area of the interface. The average values of the peak loads and shear strengths are presented on Table 1. The coefficient of variation (COV) is defined as the ratio of the standard deviation to the mean.

It is interesting to compare these values with results from other studies. Guimarães (2013) did the same test on 100 mm field samples, but using SBS-modified binder on both layers. In that case, the thickness of the SAMI and the overlay were 20 and 50 mm, respectively, and the tack coat rate was not registered. It is possible to notice that the bond strength found in the present study, with higher SAMI thickness and the use of HiMA binder, was higher than it was found by

Table 1 Leutner shear test results

SAMI thickness (mm)	Peak load (kN)	Shear strength (MPa)	COV (%)	Number of samples
25	16.7	2.12	13.5	12
55	20.7	2.63	7.1	3

Table 2 Shear strength values for comparison (Guimarães 2013)

Interface	Average shear strength (MPa)
Interlayer-overlay, tack coat only	1.74
Interlayer-overlay with geogrid	1.43
Overlay-base layer	1.82

Table 3 Shear strength values for comparison (Ogundipe et al. 2013)

Interface	Average shear strength (MPa)
30 mm overlay—10 mm SAMI	0.98
40 mm overlay—base layer	0.74
35 mm overlay—5 mm SAMI	0.58

Guimarães (2013) in her study (Table 2). She has also tested samples with geogrids and geotextiles in the interlayer-overlay interface and concluded that the use of these materials reduce the shear bond strength.

In another study by Ogundipe et al. (2013) an accelerated pavement test facility was used to examine the efficiency of different SAMI thicknesses. The standard Leutner shear test was used to evaluate adhesion on the interface. In their study, polymer-modified binder was used only in the interlayer, and the use of tack coat was not mentioned. They found that the thickness of each layer could influence the bond strength at the interface. Indeed, it was found here that the shear strength was higher on the section with 55 mm thick SAMI. Their results are presented on Table 3. The strength values are lower than it was found in the present study, even when the difference caused by the diameter of the samples is considered. Different layer thicknesses, materials and tack coat application contributed for this difference as well.

Collop et al. (2009) gathered from the literature different proposals of limits for shear strength that should be respected. The limits range from 0.85 to 1.41 MPa for the interface between the surface and the binder course, for 150 mm cores. Applying the increase of 14 % that is expected on 100 mm samples, the maximum limit to be respected would be around 1.6 MPa, which is still lower than the shear stresses obtained in this study. The bonding in the studied pavement was, therefore, adequate, respecting the limits found in the literature.

6 Conclusions

It is possible to conclude from the present study that:

- The bond conditions of the interlayer-overlay interface of the studied pavement was adequate, allowing the system to behave as a monolithic layer. It is crucial to have this conviction before studying the efficiency of the interlayer system on controlling reflective cracking, since bad adhesion between the layers may increase stresses and strains at the interface, reducing the pavement life.
- Comparison between the results and other data from the literature showed that the high amount of polymer on the interlayer binder contributed to a better adhesion. In addition, the high shear strengths found are also a consequence of the higher thickness of the layers.

- It was also shown that the use of tack coat is important to ensure better bonding. However, it is important to determine the right application rate to be used, which is related to the surface texture of the layers. Using more tack coat than necessary reduces pavement life and increases costs.

Acknowledgements The authors thank the Arteris Group (*Autopista Régis Bittencourt* and *Centro de Desenvolvimento Tecnológico*) and ANTT (*Agência Nacional de Transportes Terrestres*) for sponsoring this research.

References

Canestrari F, Ferrotti G, Lu X, Millien A, Partl MN, Petit C, Phelipot-Mardelé A, Piber H, Raab C (2012) Mechanical testing of interlayer bonding in asphalt pavements. RILEM State-of-the-art Reports, Volume 9, pp 303-360.

Collop AC, Sutanto MH, Airey GD & Elliot RC (2009) Shear bond strength between asphalt layers for laboratory prepared samples and field cores. Construction and Building Materials 23, issue 6, pp 2251-2258.

DNIT (2010) ES-145/2010 – Pavimentação: pintura de ligação com ligante asfáltico convencional – Especificação de serviço. Departamento Nacional de Infraestrutura de Transportes. Instituto de Pesquisas Rodoviárias, Rio de Janeiro, Brazil.

Guimarães PA (2013) Estudo da aderência entre camadas asfálticas de pavimentos. MSc thesis. Universidade de São Paulo, Brazil.

Ogundipe OM (2011) Mechanical behaviour of stress absorbing membrane interlayers. PhD thesis. University of Nottingham, UK. October 2011.

Ogundipe OM, Thom NH & Collop AC (2013) Evaluation of performance of stress-absorbing membrane interlayer (SAMI) using accelerated pavement testing. International Journal of Pavement Engineering, 14:6, 569-578.

Advanced Characterisation Methods for Interface Shear Resistance for Airport Overlays

Greg White and Tom Gabrawy

Abstract It is recognised that the adequacy of bond between an asphalt surface layer and the underlying pavement material is fundamental to good pavement performance. This is even more important in airport pavements where shear forces imparted by braking and turning aircraft are high. Various measures of interface shear resistance are available to characterise the bond of asphalt surface layers. Advanced test methods were developed to measure the shear resistance of the interface between asphalt layers. These methods include monotonic testing in direct shear, as well as repeated load testing in inclined shear mode. Both methods have been utilised to compare the interface shear resistance of runways of different field performance, as well as assessing the impact of rain on interface shear resistance achieved during runway asphalt overlay work. Further, the repeated load test has been used to identify a shear susceptible asphalt mixture, with poor field performance, using the monotonic test results to compare cores from two runways.

Keywords Shear stress · Shear resistance · Interface fatigue · Airport asphalt

1 Introduction

Interface shear resistance is a measure of the bond between two layers under shear loading. In the context of asphalt surface layers, it is a measure of the durability and adequacy of the interface between the surface layer and the underlying pavement. Bond is a non-specific term that broadly considers adhesion between the layers as well as interlayer friction and mechanical interlocking of the two layers due to aggregate embedment. Interface shear resistance provides a holistic measure of the risk of loss of bond leading to delamination under certain shear loading conditions.

G. White (✉) · T. Gabrawy
Fulton Hogan, Parkville, Australia
e-mail: greg.white@fultonhogan.com.au

T. Gabrawy
e-mail: gabrawy@fultonhogan.com.au

A. Chabot et al. (eds.), *8th RILEM International Conference on Mechanisms of Cracking and Debonding in Pavements*, RILEM Bookseries 13,
DOI 10.1007/978-94-024-0867-6_70

Where the interface shear resistance is exceeded or fatigued by the imposed shear stresses induced, debonding can occur (Mohammad et al. 2009).

The aim of this research was to develop advanced methods of laboratory characterisation of asphalt surface layer interfaces, specifically for airport pavement surfaces. The development of methods for monotonic and repeated load assessment of interfaces is described as well as protocols for the analysis of the results.

2 Background

Achieving adequate and durable interfaces during asphalt surface construction is critical to achieving good pavement performance. Various researchers have demonstrated the impact of poor interface condition on pavement distress and service life (White and Gabrawy 2016). This is particularly important for aircraft pavements, where a typical landing aircraft has been shown to induce maximum shear stresses double those induced by a heavy braking truck (White 2015a).

Laboratory testing of asphalt layer interfaces can be performed on cores recovered from the field or on samples manufactured in the laboratory. Field testing of interface shear strength is not common. Due to its increased reliability and popularity, laboratory testing of cores recovered from the field is preferred. There are a range of test methods and modes available for laboratory testing of interfaces. These can be either monotonic or repeated load in nature.

Interface resistance to shear can be measured by its strength, modulus/stiffness or work/energy. The three concepts are demonstrated using a typical shear load-displacement plot from a direct shear test in Fig. 1. Monotonic strength is the easiest of these to measure and is the most intuitively interpretable. As a result,

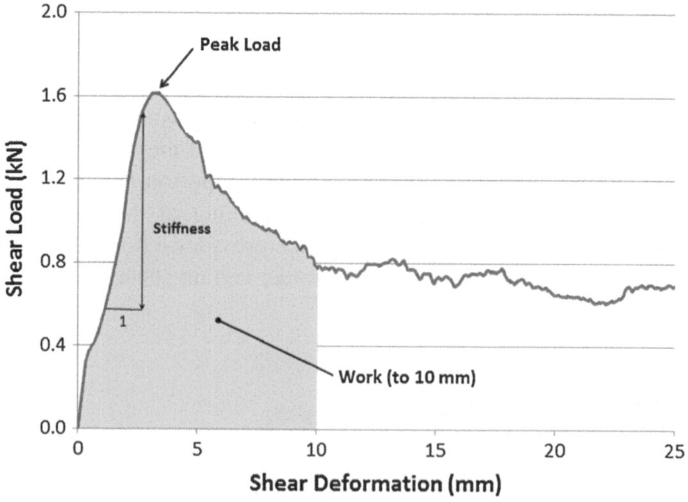

Fig. 1 Example load versus displacement plot and shear resistance indicators

many researchers have compared interface shear resistance based on Interface Shear Strength (ISS) calculated from Eq. 1.

Interface Shear Modulus (ISM) represents the stiffness of the interface, as represented by Goodman's Constitutive Law, as expressed in Eq. 2 (White 2015a). Equation 2 is essentially the gradient of the stress/strain plot. Interface Shear Work (ISW) is the area under the load-displacement plot up to a specific amount of shear deformation, as expressed in Eq. 3 and shown in Fig. 1 as the shaded portion under the graph, in this case for 10 mm of deformation.

$$ISS = L_P/A \tag{1}$$

$$ISM = \Delta L/\Delta d \tag{2}$$

$$ISW = \sum (L \times \Delta d) \tag{3}$$

The test methods and protocols are generally grouped into three main loading mechanisms; axial tension, torsional shear and direct shear. Axial tension tests measure the degree of adhesion between the two layers. Direct tension testing of interfaces is more suited to studies that are interested in comparison of the adhesion between layers. Torsional testing is less frequently reported. Although not reflective of the actual loading scenario in the field, torsion tests are capable of inducing pure radial and tangential shear forces across the interface (Goodman et al. 2002).

Direct shear tests offer a more comprehensive assessment of the full interface strength likely to be achieved in the field with adhesion, friction and interlock all contributing to the resulting interface shear strength. The load is also applied in a more representative manner and direction to that expected in the field. Field cores and laboratory prepared samples can both be readily tested and a number of test methods can accommodate both round and square samples (Santagata et al. 2009). Square samples offer more reliable and uniform contact with the load platen. The most common arrangements for the direct shear test are the shear box test the shear tube test.

It is well established that interface characteristics are influenced by surface condition and preparation, temperature, tack coat material, tack coat curing time and tack coat application rate (Tashman et al. 2008). To this list Mejia et al. (2008) added the rate of loading while Kruntcheva et al. (2006) included traffic loading to the list of important factors. These findings are not inconsistent with those reported by Uzan et al. (1978). Numerous investigations have considered the influence of one or more parameters on the various measures of interface shear resistance using various test modes. Some parameters have been found to be more important than others and not all findings appear consistent if considered in isolation, but generally make more sense when considered in context, emphasizing the need for advanced characterisation methods.

3 Developed Test Methods

Two test methods were developed to characterise the interface between asphalt overlays, specifically for airport runways. One for the measurement of the monotonic characteristics, and the second, a repeated load test to measure the fatigue properties. The intention was to characterise the performance of interfaces constructed in the field during resurfacing using current and potential future construction methods based on testing cores recovered from the field.

3.1 Monotonic Test Method

The Direct Shear (DS) test method was designed to measure the monotonic characteristics of the interface between the surface and the underlying layer. A shear box style test was selected over a shear tube test. Up to six prismatic samples are cut from each single 240 mm diameter core and tested at various normal stresses. Square samples are not commonly used for direct shear strength testing but were selected to avoid any point-loading associated with imperfectly matching circular sections. The samples are nominal 50 mm by 50 mm interface dimensions and nominally 100 mm thick, with the interface at the mid-point.

The applied normal stresses can range from approximately 20–700 kPa, in order to generate Mohr-Coulomb type envelopes. These normal stresses were selected as being indicative of the range of stresses experienced by an interface as an aircraft passes and can be adjusted for the specific investigation requirements.

Direct shear testing is performed on samples conditioned to a temperature representing the mean summer pavement temperature approximately 50 mm below the asphalt surface. For Australia, 55 °C is recommended. The samples are sheared at a constant rate of 50 mm per minute to be consistent with the rate used by other researchers. Both the test temperature and strain rate can be adjusted. During the DS test, temperature, deformation, normal force and shear force are all recorded every 0.1 s throughout the deformation.

For each sample the ISS, ISM and ISW can be calculated. The ISS is calculated using the remaining interface contact area at the time of the peak shear load. The ISM is calculated between 25 and 75 % of the peak shear force. The ISW is calculated as the area under the load-displacement graph over the first 10 mm.

Linear regressions are subsequently performed on the ISS, ISM and ISW results for each core. For the ISS, the y-intercept of the linear relationship represents the interface cohesion, which is provided by the adhesion between the layers resulting from the tack coat. The slope of the regression provides the interface friction angle. Equivalent parameters (intercept and gradient angle) are also calculated for ISM and ISW across the samples tested at different confining stress for each core sample.

3.2 Repeated Load Test Method

The Inclined Repeated Interface Shear (IRIS) test was designed to imitate the cyclic shear stresses expected to occur in an asphalt surface layer in the field during aircraft braking. To induce a shear stress across the interface, the interface is orientated at 45° to the vertical at the mid-height of the sample. Two 75 mm diameter cores are cut from a 240 mm diameter core on a 45° angle to the vertical.

In many cases, the samples are not sufficiently tall to allow the tops to be trimmed at the required sample height. An epoxy of approximately matching stiffness is provided between coring and final trimming to create cylindrical samples. Samples are measured and pre-conditioned to a temperature consistent with the DS testing. The samples are subject to cyclic compression loading, which can be adjusted but is recommended as:

- Load rate/frequency. 0.1 s haversine loading.
- Rest time. 0.9 s.
- Confining stress. 138 kPa.
- Cyclic axial (deviatory) stress. 828 kPa.

Each sample is tested under a sub-maximal cyclic load until 20,000 cycles or deformation exceeds 100,000 με. The sample deformation is logged against load cycles. Following completion of the test, the cores are inspected.

Test temperature, confining pressure, vertical load and axial strain are logged every 0.1 s during the cyclic loading. Axial strain and strain rate are calculated. The strain after 450 and 2000 cycles is reported as well as the number of cycles at which tertiary asphalt flow commenced. The deformation after 20,000 cycles or the number of cycles to 100,000 με is also recorded, depending which triggers termination of the test.

3.3 Application

The developed test protocols have been validated elsewhere (White and Gabrawy 2016) and utilised to compare interface shear strength achieved by different runway surfacing projects (White 2015a) as well as assessing the impact of rain on interface shear resistance during runway asphalt overlay works. Further, the repeated load test has been utilised to identify shear susceptible airport asphalt, exhibiting poor field performance. This utilised the monotonic test results to compare cores from the two runways of significantly different field performance, which the repeated load test identified the asphalt surface layer as the weakest link in the two layered (surface-interface-underlying asphalt) system (White 2015b). Future application includes the impact of different tack coats materials, application rates and interface construction methods of the interface resistance.

4 Conclusions

Interface shear resistance is important to airport pavement performance. Advanced methods of characterising interface shear resistance under monotonic and repeated loading is essential to enabling researchers to better understand the factors that affect interface performance. The methods developed have been verified via testing on a number of typical asphalt surfaces, as reported elsewhere. The test methods can now be utilised to measure the impact of different tack coats materials, application rates and interface construction methods of the interface resistance.

References

Goodman, S, Bekheet, W, Hassan, Y, & Abd El Halim, AO 2002, 'Rapid in-situ shear testing of asphalt pavements for runway construction quality control and assurance, *Proceedings 2002 FAA Airport Technology Transfer Conference*, Atlantic City, USA, May.

Kruntcheva, MR, Collop, AC & Thom, NH 2006, 'Effect of bond condition on flexible pavement performance', *Journal of Materials in Civil Engineering*, vol. 18, no. 3, pp. 467-471.

Mejia, D, Celaya, M, Iyer, S, Rao, C, Shokouhi, P & Nazarian, S 2008, *A Work Plan toward Evaluation of Technologies to Assess Presence and Extent of Delamination of HMA Airfield Pavements*, Research Project 06-04, Centre of Transportation Infrastructure Systems, El Paso, USA, June.

Mohammad, LN, Bae, A, Elseifi, MA, Button, J & Scherocman, JA 2009, 'Interface shear strength characterisation of emulsified tack coats', *Proceedings Asphalt Pavement Technology*, Minneapolis, USA, 15-18 March, Association of Asphalt Pavement Technologists, vol. 78, pp. 250-277.

Santagata, FA, Ferrotti, G, Partl, MN & Canestrari, F 2009, 'Statistical investigation of two different interlayer shear test methods', *Materials and Structures*, vol. 42, pp. 705-714.

Tashman, L, Nam, K, Papagiannakis, T, Willoughby, K, Pierce, L & Baker, T 2008, 'Evaluation of construction practices that influence the bond strength at the interface between pavement layers', *Journal of Performance of Constructed Facilities*, vol. 22, no. 3, pp. 154-161.

Uzan, J, Livneh, M, & Eshed Y 1978, 'Investigation of adhesion properties between asphaltic-concrete layers', *Journal of the Association of Asphalt Paving Technologists*, vol. 47, pp. 495-521.

White, G 2015a, 'Asphalt overlay bond strength', *Proceedings Airport Pavement and Lighting Forum*, Sydney, NSW, Australia, 30 April – 1 May 2015, Australian Airports Association.

White, G 2015b, 'Asphalt tenderness in an Australian airport runway', *Transportation Geotechnics*, article in press doi:10.1016/j.trgeo.2015.08.001.

White, G & Gabrawy, T 2016, 'Airport asphalt surface interface shear resistance', *International Journal of Pavement Engineering and Asphalt Technology*, vol. 17, no. 1.

Analysis of the Draft European Standard on Interlayer Bonding and Understanding of the Influencing Factors

Anne Dony, Imade Koutiri, Bernard Yvinec and Eric Godard

Abstract The pavement lifetime depends in particular on the presence and quality of the tack coat at the interfaces. Various methods are proposed in the draft European project 12697-48:2014. The USIRF, Professional Union of the French Road Industries, by means of two internships (2012–2014) supervised by the ESTP and ENSAM, two French engineering schools, suggested analyzing the relevance of the tests of the draft standard, by realizing a literature review but also a laboratory work. The use of experimental design relieved the tests number. The tests were conducted within the Scientific Campus of Colas on bi-layer cores. The pavement structure was made up of two asphalt layers bonded with a tack coat (French BBSG 0/10 and GB 0/14). Five main factors were taken into account in experimental design: temperature, loading rate, density of the two layers, and type of emulsion, with 2 interactions. The three principal tests methods proposed by the draft standard were applied: Torque Bond Test (TBT), Shear Bond Test (SBT), Tensile Adhesion Test (TAT). The results turn out complementary between them, even sometimes contradictory, with not always the same evolution. Consequently, we have numerous questioning on the relevance of the tests proposed by the draft standard: can we select one and which? Besides, we have noticed feasibility and security

A. Dony (✉)
Université Paris-Est, ESTP-IRC, 28 avenue du Président Wilson,
94230 Cachan, France
e-mail: adony@estp-paris.eu

I. Koutiri
Arts et Métiers ParisTech, PIMM UMR CNRS 8006,
151 Bd de L'Hôpital, 75013 Paris, France
e-mail: imade.koutiri@ensam.eu

B. Yvinec
Campus Scientifique et Technique COLAS, 4 Rue Jean Mermoz,
78114 Magny les Hameaux, France
e-mail: yvinec@campus.colas.fr

E. Godard
USIRF, 9, Rue de Berri, 75008 Paris, France
e-mail: eric.godard@campus.colas.fr

A. Chabot et al. (eds.), *8th RILEM International Conference on Mechanisms of Cracking and Debonding in Pavements*, RILEM Bookseries 13,
DOI 10.1007/978-94-024-0867-6_71

problems on the manual Torque bond test described by the standard. In addition, mechanical Torque bond tests were led, with more reliable results.

Keywords Tack coat · Torque test · Shear test · Tensile test

1 Context of Study

The quality of the layer bonding affects the durability of the road. According to the type of layer (surface, base), shearing, torsion or traction stresses occur with variable intensity but always at the expense of the structure. The draft standard "pr EN 12697-48" offers several tests methods. They are static loading; except one which is a cyclic loading test. The future draft standard provides a particular set of associated parameters and values for each test. Some test parameters and tested material parameters may influence the results. In this work, we focused only on the three static loadings of the normative part (traction, shearing and torsion) proposed in the new draft standard. This aims at better understanding these problematics, with different partners: ESTP and ENSAM (two French engineering school of civil engineering and mechanics), COLAS (French road builder) and USIRF (union of the French industry road). To resolve the problematic of the number of factors, we have decided to use the experimental design. On a first time, we will focus on the literature in the field.

2 Literature Review

The seven test methods presented in the future draft standard come from previous works of research labs which realized tests according to several loading solicitations. It should be noted that lots of works on evaluation of tack coat were listed. Some of them aim at comparing various equipments. Other devices tested the same equipment with variable parameters, either in laboratory or in field, with specimens from fields or lab (Diakhaté 2007). So West et al. (2005) presented in 2005 an interesting work listing at first several equipment used by various labs and analyzing conclusions from studies between 1978 and 2005: direct shear, tensile strength, torsion strength tests at several temperatures, speeds, in different countries. In their experimental study, only a shear type test was tested (the NCAT bond strength); experimental various factors were applied. Results on laboratory specimens indicated the importance of the main factors and some interactions on strength. The observations after field tests conducted on recommendations. Destrée et al. (2012) published a comprehensive study in 2012 comparing the 3 main methods of the standard with a BRRC method. Comparisons between laboratory tests were made. A good correspondence between laboratory and field cores is found but there aren't sufficient field studies to conclude and propose relevant specifications. Finally Canestrari et al. (2012) from the RILEM TG 4 organized an inter-laboratory test in order to compare the different test procedures to assess the interlayer bonding properties of asphalt pavement. Fourteen European laboratories participated in this study and carried out shear or torque tests on 1400 cores. In an overview on interlayer bonding tests, destructive and non-destructive tests were described showing the variation of conditions and so the variation of results. The interpretation of results according to type of test permits to measure influence of parameters. The main conclusion is the good agreement between different unidirectional monotonic static shear tests, the influence of parameters as temperature, speed, size of specimens…No recommendation is given to choose a type of test. This non exhaustive literature review confirms the difficulty to choose one test. The aim of the experimental design presented in this paper is to better understand the effect of several parameters on the three main types of loading proposed in the standard project.

3 Experimental Approach

One of the benefits of the experimental design is the ability to test a lot of parameters with only a few tests. At first repeatability tests are performed with the conditions proposed by the draft standard. Our aim is to check the variation and to have first values of the maximum stress. On a second time, we apply the design of experiments method with the following steps: Define the interesting parameters, choose the type of table to test, perform the different tests and analyze the results.

3.1 Parameters Setting and Choice of Taguchi Table

We have limited the study to five parameters (temperature, loading rate, density of the surface layer, density of the base layer and type of emulsion). The other parameters (size grading or surface appearance) are fixed in order not to introduce variability between the different tests. Each parameter will be tested with two levels for each type of loading.

In order to set up our experimental design and reduce the number of test significantly, we selected a L8 type Taguchi table. This table has five parameters with two interactions: type of emulsion (pure or modified), load speed (2.5 or 90 mm/min), temperature (0 or 40 °C), density of surface layer (85 or 95 %), density of base layer (75 or 95 %), the coupling between the load speed vs the type of emulsion (Interaction 1) and the coupling between temperature vs the type of emulsion (Interaction 2). The aim is to test the link between the type of emulsion and two parameters defined as influent in the literature. For each parameter set, three tests are performed. The conditions are defined via the L8 Taguchi table.

3.2 Specimens Preparation and Experimental Set-up

All of the specimens have been prepared in the COLAS Technical and Scientific Center (production of BBSG, "béton bitumineux semi grenu" and GB, "grave bitume" with a controlled density, application of tack coat at a same rate and coring of plates in 100 mm diameter). All the specimens are manufactured in the same mixer with the same methodology. The different plates have been compacted with a Cooper compactor. The layer coat has been spread with a paintbrush. The plates have been controlled by the Gamma bench to check the density. The mechanical tests have been performed with an Instron type, servo-hydraulic machine, with load sensors. The force was controlled precisely during the tests. The displacement is provided by the cross-head of the machine. For each type of loading, the specific experimental set-up proposed by the Standard is used. It was purchased for the shear loading, manufactured for the tensile loading and provided by EUROVIA, member of USIRF for the torsion loading.

4 Analysis of Results

4.1 Precision

Before applying our experimental design, six tests have been performed to check the repeatability and the dispersion on the different types of loading, with the conditions defined in the standard. The other parameters are fixed: the density of the surface layer (91.5 %), the density of the base layer (85 %), the rate of application (300 g/m^2 with 100 % covering), the type of emulsion (cationic rapid setting emulsion), the temperature of application of the tack coat (45 °C). The results presented on the Table 1 shows coefficients of variation between 5 % for TBT to 15 % for SBT, and a high sensitivity to temperature (TAT). Our conclusion, detailed in RGRA, is the necessity to realize at least three tests for the same set of parameters.

Table 1 Results of the dispersion versus type of loading

	TAT	TAT[a]	SBT	TBT
T° test (°C)	10	20	20	20
Loading speed	200 N/s	200 N/s	50 mm/min	180°/min
σ_{mean} (MPa)	1.64	0.82	1.73	3.39
CV (%)	12	8	15	5

[a]10 °C is one of the standardized temperature for TAT. 20 °C was chosen to compare with the other tests. The standard 0 °C has not been in this study

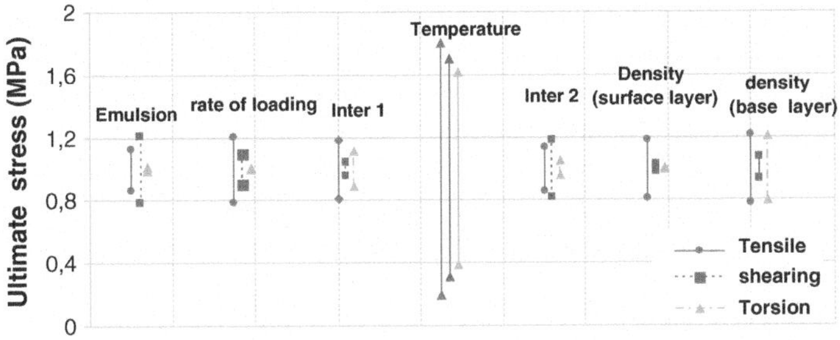

Fig. 1 Graph of effects depending on the type of loading

4.2 Comparison of the Type of Loading

Considering the experimental design and the results on precision, three tests for each parameter are realized and the mean of the maximum stresses was calculated. It was called the response for each level of parameters. The graph of effect is drawn and we can visualize and prioritize the influence of each parameter. Then, an analysis of variance is used to measure the significance of the main effect and interactions (level of significance = 5 %). In order to compare the three types of loading, a normalized graph (Fig. 1) with all the parameters used for each type of loading is presented. It appears that the temperature is the most influential parameter, irrespective of the type of loading. Its effect is four times greater than the other parameters. Looking at other parameters, their influence seems very random in terms of type of loading. For instance, there is a low effect of the density on the ultimate stress for shear and torsion loadings contrary to the tensile loading. The type of emulsion, conventional or modified has no effect on torsion test result, contrary to the other type of loading, etc. Also, for the shear loading, emulsion appears as the second most influent factor after the temperature. Finally the analysis of the graph of effects permit to conclude that, except temperature, the others effects have a random influence according to the type of loading.

4.3 Influence of Parameters and Analysis of Variance

The analysis of variance is used to check if the effect of a factor is significant. A factor is considered influent if its action on the studied response is statistically greater than a level set by the experimenter. The results are the following:

- The temperature is confirmed as an influential parameter for the three loading, logical result explained by the thermos-susceptibility of bitumen. Strength varies between 0.06 MPa/°C until 0.09 MPa/°C according to the type of test.
- ***For the direct tensile mode***, no other parameter is evaluated as significant.

- ***For the shear mode***, the type of emulsion and the interaction between emulsion and temperature stand out as influential. In particular, the results show a better (but not expected) strength of the bonding layer with a conventional emulsion, in comparison with modified emulsion, in our lab condition.
- ***For the torsion mode***, more significant parameters were identified than for the other loading: temperature, interaction between emulsion and temperature, but also interaction between emulsion and loading rate and density of the base layer.

5 Conclusion and Prospects

The draft standard offers three modes including conventional static load. Initial tests allowed to estimate the dispersions according to the modes, although the use of specific devices with the same material and the same technician in the same lab. The second interesting point is the dependence of certain parameters related to certain modes of stress. As expected, the temperature appears to be the most influential parameter and the common denominator of all the modes of loading. A test on site would require a very accurate interface temperature measurement, what is tricky to obtain. A slight variation will have some impact on the result of the failure stress. Regarding the effects of the other parameters, they appear very random according to the type of loading. The choice of a type of loading is complex knowing that none of them reproduce the real cyclic type and level of stresses in the road structure depending on its use (acceleration, braking, turn, heavy loads, etc.). This whole study is presented in detail by Godard et al. (2015) in the RGRA. Given the importance and difficulties of the control of all the parameters on site, a new material will soon be discussed on a prospective national project.

References

Diakhaté M (2007), Fatigue et comportement des couches d'accrochage dans les structures de chaussée, Phd Thesis.

West R, Zhang J, Moore J (2005), Evaluation of bond strength between pavements layers, NCAT Report 05-08

Destrée A, De Visscher J, Vanelstraete A (2012), Evaluation of tack coat performance for thin and ultra-thin asphalt pavements, 5th Eurasphalt & Eurobitume Congress

Godard E, Yvinec B, Dony A, Koutiri I, (2015), Collage des couches d'enrobés bitumineux-Evaluation des méthodes d'essais du projet de norme européenne EN 12697-48, RGRA n° 928

Canestrari,F, Ferrotti G, Lu X, Millien A, Parti M, Petit C, Phelipot ·Mardelé A, Piber H, Raab C (2012) Chapter 6, Mechanical Testing of Interlayer Bonding in Asphalt Pavements Volume 9, series RILEM State-of-the-Art Reports pp 303-360

Comparative Analysis of Interlayer Bonding Behaviour of Different Types of Pavement Interfaces

Ana Cristina Freire, Luís Quaresma and Carla Gil

Abstract The rehabilitation of flexible pavements aims the recovery of its characteristics and is usually accomplished either by the application of additional bituminous mixtures (reinforcement), or by removing the layers whose function is compromised and subsequent placement of new bituminous layers, seeking to increase load capacity, serviceability and extend its life cycle. The bonding between the applied layers is one of the key aspects in pavements performance, reflecting the capability of pavements to work as a whole. A comparative study of the behaviour of the interface between the bituminous top layer and an underlying layer by applying different reinforcement elements has been developed. In order to evaluate the interface bonding behaviour for national conditions, namely in what concerns the type of actions, applied materials and construction techniques, a laboratory study was made with destructive tests such us pull-of test and shear type test allowing the proposition of reference values.

Keywords Interlayer bonding · Mechanical tests · Pre-bituminized grids

1 Introduction

A growing need for an effective and sustainable resource management, whether economic or environmental, is an important facilitator for the development of new reinforcement and rehabilitation techniques of road and airfield pavements. The

---

A.C. Freire (✉)
LNEC—National Laboratory of Civil Engineering, Lisbon, Portugal
e-mail: acfreire@lnec.pt

L. Quaresma · C. Gil
FCT-UNL—Faculty of Science and Technology, Universidade Nova de Lisboa, Lisbon, Portugal
e-mail: lmtq@fct.unl.pt

C. Gil
e-mail: carlaandradegil@gmail.com

A. Chabot et al. (eds.), *8th RILEM International Conference on Mechanisms of Cracking and Debonding in Pavements*, RILEM Bookseries 13,
DOI 10.1007/978-94-024-0867-6_72

rehabilitation of flexible pavements aimed at the recovery of its structural and functional characteristics is usually accomplished either by the application of additional bituminous mixtures (reinforcement), or by removing the layers whose function is compromised and subsequent placement of new bituminous layers, seeking to increase load capacity, serviceability and extend its life cycle.

The bonding between the layers is one of the key aspects in pavements performance, reflecting the capability of pavements to work as a whole. The interlayer bonding conditions can compromise pavement performance and consequently the life cycle. A poor interlayer bonding between bituminous pavement layers leads to major pavement overlay distresses namely premature fatigue, top down cracking, potholes, and surface layer delamination. One of the most common distresses due to poor interlayer bonding is a slippage failure, which usually occurs following the passage of heavy vehicles.

The bonding of bituminous pavement layers is influenced by several causes namely the size of aggregates used in the asphalt mixes, the type of asphalt mix and binder, a poor or excessive tack-coat application, and by the type of construction technology used, the climate conditions during pavement construction and water flow between the layers, among others (Canestari et al (2013; Sutanto 2009). The techniques used for the application of two-dimensional surface elements between the existing layer and the reinforced layer, are usually referred to as anti-crack interface. A more efficient management of existing pavement infrastructures requires new materials and construction techniques. Thus the incorporation of reinforcing grid for the rehabilitation of pavements can promote the respective structural reinforcement and limit the propagation of cracks to the surface.

A comparative study of the behaviour of the interface between bituminous top layers and underlying layers when different reinforcement elements are applied has been developed. In order to evaluate the interface bonding behaviour for national conditions, namely in what concerns the type of application and solicitations, a laboratory study was made with destructive tests such us pull-of test and shear type tests on filed samples, allowing to obtain reference values.

2 Materials and Methods

A laboratory study was performed on bituminous samples (slabs and cores), comprising two layers, collected from pavements, for a comparative analysis of the behaviour of the interface between the top layer and the underlying layer. Different reinforcement elements were applied at the interface, in order to evaluate its influence on the pavement performance and to obtain reference values. Two different types of pre-bituminised grids were selected to be applied: a pre-bituminised carbon grid and a pre-bituminised glass grid, named as Type 1 and Type 2. The selected grids were applied at two different pavement rehabilitations, a road pavement and an airfield runway pavement, identified as Case 1 and Case 2, respectively. The pre-bituminised grids are usually used for increasing the durability of asphalt surfaces by reducing fatigue and thermal cracks. The carbon grid additionally contributes to the structural reinforcement of the pavement. The application methodology is the same for both grids.

A two-layer bituminous sample without any grid was also tested, as reference sample (Case 3). Table 1 presents the studied pavements materials characteristics and Fig. 1 presents several phases of the grid application and taking off of samples for testing.

A laboratory evaluation of the interface bonding was made performing destructive tests namely shear type tests (modified Leutner test) and pull-of tests, according to Appendix A.1 of the Manual of Contract Documents for Highway Works (MCHW) and taking into consideration the prEN 12697-48 "Bituminous mixtures—Test methods—Part 48: Interlayer bonding" and ASTM D4541-09 —"Standard test method for pull-off strength of coatings using portable adhesion testers". The samples were conditioned at 20 °C and the tests were performed at the same temperature.

Figure 2 shows the equipment used in the modified Leutner test placed in the testing machine, and a tested sample and its interlayer appearance after testing. Figure 3 shows the portable equipment used in the pull-off test performed in this study and a sample after testing.

Table 1 Studied materials

	Case 1		Case 2	Case 3	Reference
	Test sectoin 1 (pavement construction)	Test section 2 (experimental section)	(pavement construction)	(pavement construction)	study Marcelino et al (2013)
Pre-bituminized grid	Type 2	Type 2	Type 1	No grid	Type 2 or no grid
Top layer	Base course AC 20 base 35/50[a] envelope A with 10 % RAP	Wearing course AC 14 surf PMB 45/80-65[a]	Base course AC 20 base 35/50[a]	Base course AC 20 base 35/50[a] envelope A with 10 % RAP	Wearing course AC 14 surf 35/50[a]
Underlying layer	Milled existing cracked bituminous layer	Base course AC 20 base 35/50[a] envelope A with 10 % RAP	Base course AC20 base 10/20[a]	Milled existing cracked bituminous layer	Milled existing cracked bituminous layer
Emulsion type	Termoadhesive bituminous emulsion from bitumen asphalt, cationic quick breaking, containing polymers				
	(C60BP4[b])	(C60BP4[b])	(C60BP4[b])	(C60BP4[b])	–
Application dosage (g/m^2)	450	350	400	450	500

[a]According to EN 13108-1
[b]According to EN13808; RAP—Reclaimed Asphalt

Fig. 1 Application of a pre-bituminised grid and sampling procedures for laboratory tests

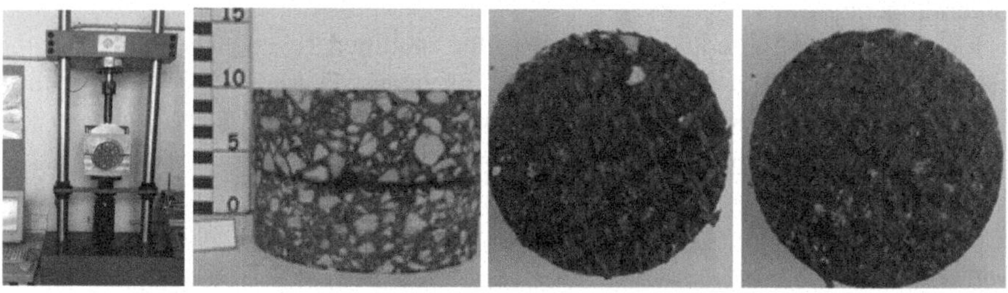

Fig. 2 Phases of the modified Leutner test

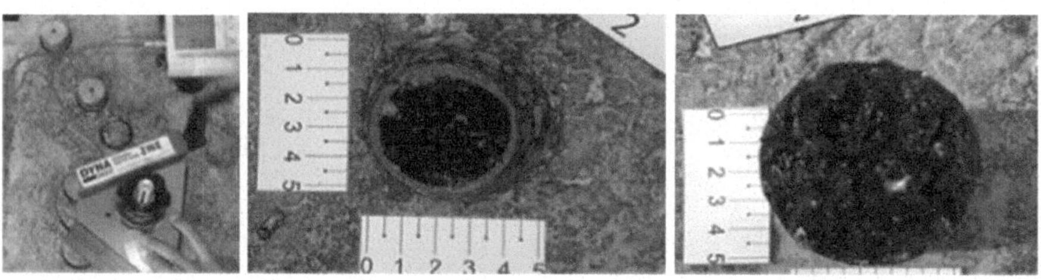

Fig. 3 Aspect of the portable equipment used in the pull-off test and tested sample

3 Results and Discussion

In Fig. 4 the Leutner test results are presented (maximum shear force, Fmax), for the 3 studied Cases and the two pre-bituminised grids type. It is also presented, as reference case, the maximum shear force for interfaces without grids.

The analysis of the values presented in Fig. 4 shows that Fmax varies from 12 kN (interface between course bases) to 50 kN (interface between wearing course and base course), obtained for the different types of bituminous layers, applied grids

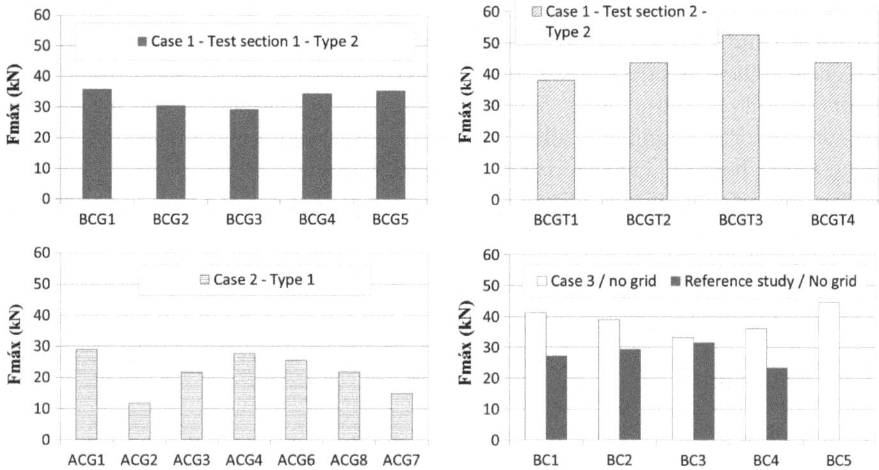

Fig. 4 Leutner test results for case studies and reference study

and emulsion rate. These values meet the limits adopted as national standards in Germany and Switzerland (ZTV Asphalt StB 07 and SN 640 430, respectively).

The results of tests carried out on samples from the experimental section Case 1-Test section 2-Type 2 are the highest, which could result from the fact that this section has been specially built for this purpose. It can also be observed from Fig. 4 that for Cases without the introduction of reinforcing elements, Fmax values are generally higher than those observed in the other studied cases. However it is possible to achieve equivalent values for Fmax with the introduction of grid, as expressed for Case 1–Test section 1–Type 2.

Figure 5 presents a comparative analysis of the performed two tests—pull off and modified Leutner test—through the tensile bond strength and the maximum shear stress, respectively. Figure 5 shows that the pull-off tensile bond strength is always lower than the modified Leutner maximum shear stress.

Fig. 5 Comparative analysis of modified Leutner test and pull-off tests results

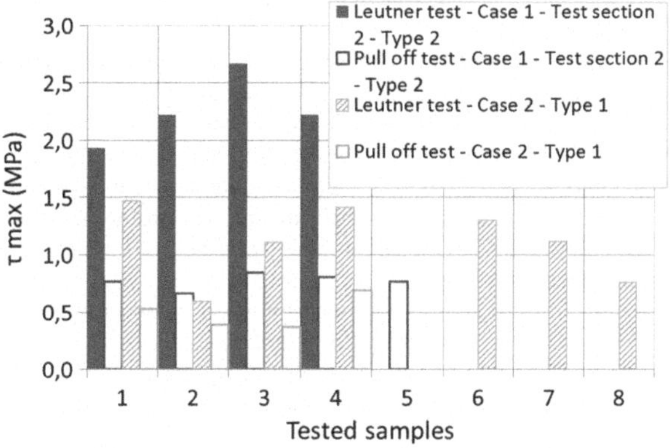

Considering that the diameter of the tested samples was different (50 mm for the pull off test and 150 mm for the modified Leutner test) the results may present some influence of the size effect. Nevertheless, more tests must be performed in order to clearly identify its meaning. It is considered however that performing the pull off test, with portable equipment, allowing in situ tests, could provide indicative figures of the layers bonding characteristics, by an expeditious manner.

The present study is still going on, by performing 4PBT and wheel-tracking tests on the different bituminous samples in order to evaluate the influence of different interlayer solutions on interlayer bonding.

4 Conclusions

The values obtained by the modified Leutner tests performed for different types of bituminous mixtures, reinforcement grids and emulsion rate, satisfy the limits adopted by German and Switzerland normalization that present as a minimum 12 kN, for the modified Leutner test shear force, for the interface between base courses and a minimum of 15 kN for the interface between a wearing course and a base course. So it recommends the adoption and verification of reference values for the shear force to ensure the quality of the interlayer connection.

As presented the pull-off tensile bond strength is always lower than the modified Leutner maximum shear, being necessary to carry out further tests to establish a possible correlation between the two test methods.

Acknowlegments The authors wish to thanks to LNEC, and to BRISA, Auto-Estradas de Portugal, ANA- Aeroportos de Portugal and S&P Clever Reinforcement Iberica companies for the support in this research with all needed equipment, materials and real applications of the studied materials.

References

ASTM D4541-09 (2009) – Standard test method for pull-off strength of coatings using portable adhesion testers.

Canestari, F., et al (2013) Mechanical testing of interlayer bonding in asphalt pavements. In. M. N. Partl et al (ed) Advances in Interlaboratory Testing and Evaluation of Bituminous Materials, State-of-the-Art Reports 9, Springer, New York, p 303 – 360.

MCHW (2008) – Specifications for Highway Works, Volume 1, Series 900.

Marcelino, P., *et al* (2013). Avaliação da interface entre camadas ligadas com aplicação de grelhas de reforço de pavimentos pelo método de Leutner modificado (in Portuguese).

prEN 12697-48 (2014). Bituminous mixtures – Test methods for hot mix asphalt – part 48: Interlayer bonding

Sutanto, M. (2009), Assessment of bond between asphalt layers, Thesis submitted to the University of Nottingham for the degree of Doctor of Philosophy, Nottingham.

SN 640 430 (2013) Enrobés bitumineux compactés. Conception, exécution et exigences relatives aux couches en place. Norme Suisse.

Investigation of Bond Between Asphalt Layers in Flexible Pavement

Ngoc Lan Nguyen, Van Dong Dao, Mai Lan Nguyen
and Duy Huu Pham

Abstract Poor bond between two asphalt layers is one of the main causes of premature failure of pavement surface layers, such as slippage cracking, fatigue cracking, shoving or surface layer delamination. These distresses become more serious in pavement areas often adversely affected by horizontal loads due to traffic wheels. There are many factors affecting the quality of bond between asphalt layers including temperature and application rate of tack coat. In an elemental study, the modified Leutner shear test, which allows determining shear strength and shear stiffness modulus at the interface of double-layered cylindrical specimen, was used. Tests were carried out at a range of temperatures between 20 and 60 °C and with different application rates between 0.0 and 0.9 l/m^2 (non-residual) using a Cationic Rapid Setting 1 (CRS-1) emulsion. Results of the elemental study are then used to establish regression equations, relating bonding strength, temperature and application rate which are used as part of larger studies to investigate the influence of interface bond on premature failures in pavement structures, such as premature fatigue cracking or rutting. Some interesting results of these studies cast additional light on the behavior of bonding between asphalt layers.

Keywords Bonding · Tack coat · Shear test · Premature failure

N.L. Nguyen (✉) · D.H. Pham
University of Transportation and Communications, Hanoi, Vietnam
e-mail: nguyenngoclan@utc.edu.vn

V.D. Dao
The University of Transport Technology, Hanoi, Vietnam
e-mail: dongdv@utt.edu.vn

M.L. Nguyen
LUNAM Université, IFSTTAR, Bouguenais, France
e-mail: mai-lan.nguyen@ifsttar.fr

A. Chabot et al. (eds.), *8th RILEM International Conference on Mechanisms of Cracking and Debonding in Pavements*, RILEM Bookseries 13,
DOI 10.1007/978-94-024-0867-6_73

1 Introduction

Flexible pavement structures are generally designed with several layers and their construction is carried out using a layered paving process. This may lead to weak interfaces and debonding between asphalt layers, contrary to pavement design assumptions. According to Khweir and Fordyce (2003) and Lan and Phuc (2013), failures of asphalt concrete road surface due to poor bond between asphalt layers may reduce the road surface service life by 40 % to over 80 %. Therefore, bituminous emulsions are usually used as adhesive materials during pavement construction to ensure good bond between asphalt layers (Mohammad et al. 2012).

During the service life of the asphalt concrete road surface, it has been shown that the bonding between two asphalt concrete layers depends on many factors, including temperature, type and application rate of tack coat (Destree et al. 2015; Raab et al. 2015). The bond between two asphalt layers can be assessed by interface shear strength τ (ISS) and shear stiffness modulus k (SSM) through the modified Leutner shear test (prEN 12697-48), which was originally developed in Germany (Leutne 1979). Many studies showed that bonding strength between asphalt layers decreased when temperature increased and that temperature was the most important factor affecting the bond strength (Godard et al. 2015). While the trend of variation of bond strength with temperature is very clear, mixed results can be found in the literature concerning the impact of tack coat application rate.

The work presented in this paper is part of larger studies where the bond between asphalt layers is investigated. The objective is to identify the main factors affecting bonding, and to take them into account in the evaluation of rutting of double-layered specimens as well as in the design of pavement structures. Up to now, several studies have been made to characterize the effect of bond condition on pavement design, but only using a theoretical approach (Kruntcheva et al. 2005) or an experimental approach, based on laboratory tests (Diakhate et al. 2011). A complete study, including experimental testing of bond strength, pavement performance analysis and in situ verification needs to be carried out.

This paper presents firstly an experimental work, investigating the behavior of the interface between two asphalt layers at a range of temperatures between 20 and 60 °C and with different application rates of tack coat between 0.0 and 0.9 l/m^2 (non-residual) using a CRS-1 emulsion. These experimental results are then used to establish regression equations that relates bond strength, temperature and application rate. These equations can be used in the ongoing parts of the study which evaluate the effect of bond strength on rutting and pavement design. First interesting results of the ongoing parts are presented in this paper.

2 Experimental Program

2.1 Test Matrix

The testing program included five temperatures (20, 30, 40, 50 and 60 °C) and four application rates (0.0, 0.2, 0.4 and 0.9 l/m^2 non-residual), using the same tack coat emulsion CRS-1. Six test repetitions were performed for each application rate and temperature, leading to a total of 120 tests.

2.2 Materials and Samples Preparation

The two asphalt concretes AC 12.5 and AC 19 were designed according to TCVN 8820:2011. Table 1 presents the mix design results for AC 12.5 and AC 19. The tack coat CRS-1 was characterized according to TCVN 8817.

Cylindrical specimens of 150 mm diameter were prepared using the Troxler 4140 gyratory compactor. The AC 12.5 layer (50 ± 2 mm thick) was compacted on the AC 19 layer (70 ± 2 mm thick) after curing of the tack coat.

2.3 Test Procedure

A modified Leutner shear test with 5 mm gap between the shearing rings, which was designed at the University of Transport and Communication (UTC) and was the first development of this test in Vietnam, was used for the study. The experiment was conducted according to the guidelines of prEN 12697-48 with the displacement rate of 50.8 mm/min. Figure 1 shows the testing equipment and a typical result obtained at 20 °C.

Table 1 Specifications of asphalt mixes

Composition and indicators	Type of asphalt mix		TCVN 8820:2011
	AC 12.5	AC 19	
Marshall stability (kN)	14.85	10.37	Min 8
Flow (mm)	3.10	2.98	2–4
Air voids (%)	4.15	4.47	3–5
Voids in the mineral aggregate (%)	14.13	13.55	Min 13
Voids filled with asphalt (%)	70.63	67.01	65–75

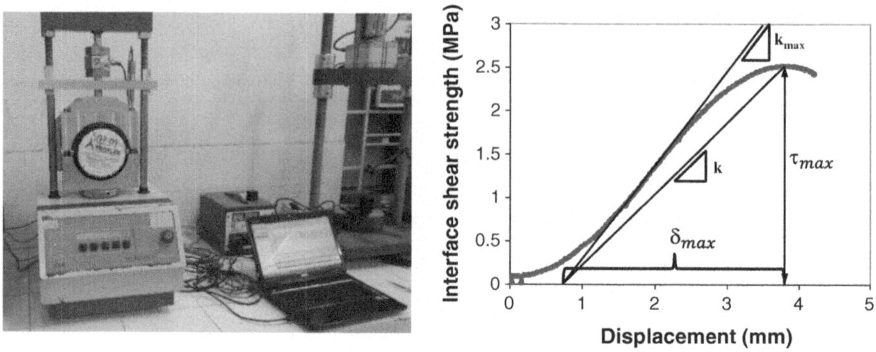

Fig. 1 Shear testing equipment and example of a test at 20 °C

3 Results and Analysis

The results of ISS and SSM are plotted in Fig. 2. These results confirm that when temperature increased from 20 to 60 °C, at all application rates, both ISS and SSM decreased considerably. This effect corresponds to the decrease of the shear modulus of the tack material when temperature increases.

Figure 2 (left) shows that, at the tested temperatures, the influence of the application rate on ISS is much less important. However, for temperatures higher than 20 °C, one can observe a slight reduction of ISS at higher application rates. Figure 2 (right) shows a slightly different trend. At 20 and 30 °C, the highest values of SSM are measured at the application rate of 0.2 l/m^2. At higher temperatures, SSM does not change much with application rate. An interesting result was found here: for tests at 20 °C, an optimum application rate could exist, which is close to 0.2 l/m^2 where both ISS (Fig. 2 left) and SSM (Fig. 2 right, more remarkably) present their maximum values. Moreover, at temperatures higher than 30 °C, the highest values of ISS and SSM were measured on specimens without tack coat.

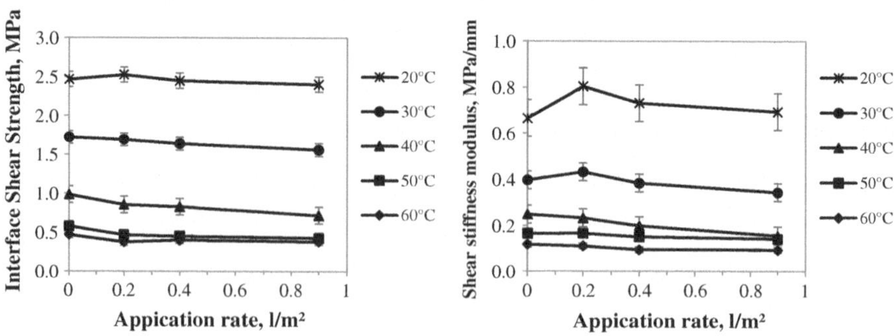

Fig. 2 Correlations between shear strength, shear stiffness modulus and application rate

That means when considering ideal interface conditions (clean and dry surface of the lower layer, as in laboratory preparation), it seems possible to apply directly the upper layer on the lower layer without using a tack coat.

In order to establish regression equations displaying the relationship between the bond strength and test temperature and application rate, an analysis of variance (ANOVA) was applied. The following forms were obtained:

$$\tau = e^{1.865-0.047T-0.016R-0.0042T\times R} \quad (1)$$

$$k = e^{0.539-0.046T-0.063R-0.0044T\times R} \quad (2)$$

where τ is shear strength (MPa), k is shear stiffness modulus (MPa/mm), T is test temperature (°C), R is application rate (l/m^2).

The above equations allow the determination of the most appropriate application rate which needs to be used at the interface between asphalt layers using the same materials tested in this study. It should be noted that ISS and SSM are the bonding parameters determined at the failure state, shear stress levels in a normal pavement condition may remain well below the maximum failure stress. For the practice, these two parameters can be used as control values to ensure a good bonding between asphalt layers or to study influence of different affecting factors.

4 Influences of Interface Bond on Premature Failures

In order to improve the pavement design, the bond condition between layers needs to be taken into account. Some pavement analyses have been conducted for a flexible pavement to investigate if the experimental values of k determined from the Leutner shear tests could be used in a pavement design procedure (Fig. 3). The ALIZE-LCPC software was used for these calculations using an option that takes

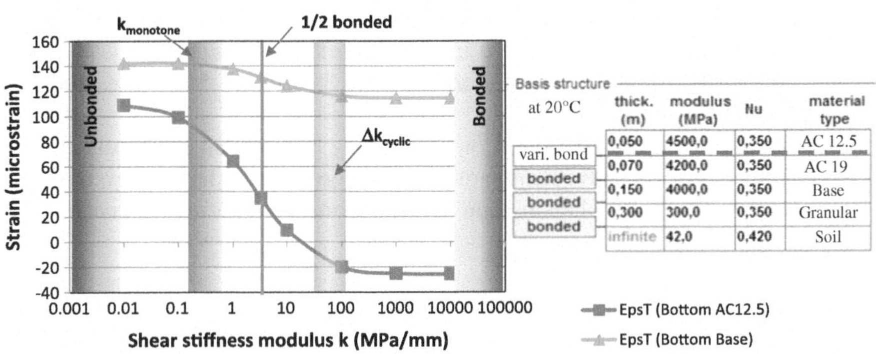

Fig. 3 Calculations using ALIZE-LCPC of horizontal strains at the bottom of the surface layer and at the bottom of the base layer due to a French standard dual-wheel load of 65 kN

into account the bonding condition at the interface by modelling an "interface with an elastic stiffness". First results show that when using the k values determined from the monotonic Leutner shear tests, it corresponds to weak bonding levels of the interface between the asphalt layers which results in a higher tensile strain at the bottom of the base layer, leading to a decrease of the fatigue life of the pavement structure. Particularly the tensile strain at the bottom of the surface layer can increase further, causing fatigue cracking of the surface layer. Moreover, further investigations show that for a normal pavement condition with correct bonding state between the asphalt layers, it is $\Delta k = \Delta\tau/\Delta u$ determined in a cyclic test [such as the shear fatigue test done in the work of Diakhate et al. (2011)] which should be the correct bonding parameter to be taken into account in the pavement design calculations.

In addition to this work, an experimental study using the wheel tracking test to evaluate rutting of double-layered specimens, made with the same asphalt mixes and tack coat used in the bond strength study, is also being undertaken. First result shows that rut depth increases by about 74 % when the bond strength between the asphalt layers decreases from 0.64 MPa to 0.40 MPa (Dong et al. 2015).

5 Conclusions

The results of these studies lead to the following conclusions and perspectives:

- Temperature has a more profound effect on the interface bond strength between asphalt layers; application rate of tack coat has a smaller impact. These experimental results on the studied materials confirm the ones found in the literature on other materials. Moreover these results are being used in broader studies to investigate the influence of the interface bond on premature failures of pavement structures made of the same studied materials, such as premature fatigue cracking or accelerated rutting which show interesting first results.
- A rare result was found in the elemental study: an optimum tack coat application rate could exist, which is close to 0.2 l/m^2 (non-residual), for the tests only at 20 °C. It should be noted that for in situ pavement construction, under actual field conditions, optimum application rates may be greater than those predicted by laboratory experiments.
- Studies on other tack coat types, other affecting factors and their influence on premature failures in pavements are also undergone in order to correctly take into account this bond phenomenon in the pavement performance analysis.

References

Destree A., De Visscher J., Pierard N. and Vanelstraete A. (2015) Field study to investigate the impact of conditions of application of tack coat on the interlayer bond strength. 8th International RILEM SIB Symposium, Ancona, 7-9 October

Diakhate M., Millien A., Petit C., Phelipot-Mardelé A., Pouteau B. (2011) Experimental investigation of tack coat fatigue performance: Towards an improved lifetime assessment of pavement structure interfaces. Construction and Building Materials 25 (2011) 1123–1133

D.V.Dong, N.N.Lan, V.T.Trong, B.T.Q.Anh (2015). Evaluation of Rutting resistance of double-layered asphalt mixes. Communication and Transport Magazine, No. 7/2015

Godard E., Yvinec B., Dony A. and Koutiri I. (2015) Bonding of asphalt layers: Evaluation of the test methods of draft European Standard EN 12697-48. RGRA No. 928

Khweir K. and Fordyce D. (2003). Influence of Layer Bonding on the Prediction of Pavement Life. Proceedings of the Institution of Civil Engineering Transport 156, 73-83

Kruntcheva M.R., Collop A.C. and Thom N.H. (2005) Effect of Bond Condition on Flexible Pavement Performance. Journal of transportation engineering, Vol. 131, No. 11, 880-888

Leutner, R. (1979). Untersuchung des Schichtenverbundes beim bituminösen Oberbau. Bitumen, 41-3, 84–91

Mohammad LN, Elseifi MA, Bae A, Patel N, Button J and Scherocman JA (2012) Optimization of tack coat for HMA Placement, NCHRP Report 712

N.N.Lan and N.Q.Phuc (2013) Investigation of impacts of bonding condition between asphalt layers on performance of asphalt pavement, Communication and Transport Science Magazine, No. 10/2013

prEN 12697-48 (2013) Bituminous mixtures - Test methods for hot mix asphalt. Part 48 Interlayer Bonding.

Raab C., Grenfell J., Abd El Halim A.O. and Partl M.N. (2015) Comparison of Interlayer Bond Behavior Due to Ageing. 8th International RILEM SIB Symposium, Ancona, 7-9 October

Performance-Based Test Regime for High Friction Surfacing Systems on Asphalt Pavements

Jeremy P. Wu, Shaun R. Cook and Philip R. Herrington

Abstract Specialist surfacings such as high-friction and coloured traffic calming surfaces have gained huge popularity since their introduction. However, the reputation of these specialist systems in New Zealand is also plagued by premature failures due to cracking and other related modes. Many of the failure modes are in fact originated from or at least associated with the performance of the underlying pavement substrate. The purpose of this work is to assess test methods where the emphasis is on the performance of the underlying substrate and its interaction with the specialist surfacing systems to ensure best outcome. A number of resin systems, namely epoxy, polyurethane, and methyl methacrylate were tested using ASTM standard test methods such as a pull-off bond test and % elongation measurements. Thermal effects were also investigated by measuring the coefficient of thermal expansion and conducting thermal cycling experiments. Part of the experimental work is to develop tests which can be implemented in the field. Using these results, it is intended that a practical specification can be developed to include a test or tests to better understand the performance of the surfacing and the underlying pavement and thus ensure future surfacing systems meet their life expectancy.

Keywords High friction surface · Pull off test · Cohesive failure

1 Introduction

The concept of using epoxy-resin as the binder in surface treatment was first investigated in the USA in the mid-1950s. A development of the system, in which calcined bauxite aggregate (1.2–2.8 mm) is held in a bitumen-extended epoxy-resin binder, was introduced into the UK in the 1960s (James 1963) to provide improved skid resistance on accident-prone sites. Since then specialist surfacings such as

J.P. Wu (✉) · S.R. Cook · P.R. Herrington
Opus Research, Opus International Consultants Limited, Wellington, New Zealand
e-mail: Jeremy.Wu@opus.co.nz

A. Chabot et al. (eds.), *8th RILEM International Conference on Mechanisms of Cracking and Debonding in Pavements*, RILEM Bookseries 13,
DOI 10.1007/978-94-024-0867-6_74

High Friction Surfacing (HFS) and coloured surfaces have been proven to be very effective (Denning 1978).

As a result of a growing market for specialist surfacing systems, a number of alternative binders are being offered and the number of suppliers that offer resin-based systems is increasing (Nicholls 1998). In New Zealand, HFS has been used to reduce accidents on sites with high crash rates since the 1990s. Similarly, coloured surfacing have become very common for designated traffic lanes such as bus and cycle lanes. Hudson and Mumm (2003) reported the benefit of HFS system on a section of Wainuiomata Hill Road in New Zealand where the number of crashes significantly reduced after the original porous asphalt surface was resurfaced with calcined bauxite HFS system.

Izeppi et al. (2010) investigated a number of HFS systems available in the US market and showed that it is reasonable to expect some of the systems to maintain high friction values for up to 10 years of service based on historical data. Unfortunately, there are some sites which do not last the targeted life for a variety of reasons. Nicholls (1998) investigated the performance of 4 types of resin systems via road trials in the UK. It was found that debonding (delamination) from the asphalt substrate is a potential problem with all HFS systems. Similar modes of failure were observed in New Zealand (Waters 2011) where HFS systems failed prematurely.

The most recognized product certification for these systems is the HAPAS (Highway Authorities Product Approval Scheme) certification from British Board of Agrément (BBA). While the testing procedure for the approval scheme is sound, it can be argued that it focuses mainly (and rightly so) on the surfacing product itself.

In a recent research work conducted by Arnold et al. (2014), two test methods were investigated: A Leutner shear test was conducted to measure bond strength and a flexural beam test was conducted to measure flexibility of specialist surfacing resins used in New Zealand. The shear test showed all resins achieved full bond strength (the same as the asphalt mix shear strength) when the asphalt surface was water cut, suggesting that substrate surface preparation is important for good adhesion bonding. The flexural beam test demonstrated high flexibility of the specialist surfacing resins, indicating that the thin surfacing should cope with high pavement deflections and the underlying asphalt would crack before the specialist surfacing did.

While the two tests showed a promising future, further investigation was recommended to review the failure modes and develop other candidate test methods. The New Zealand Transport Agency (NZTA) seek the development of a practical specification that includes a test to predict potential failure, namely cracking, of the specialist surfacing and its underlying substrate. The intention of a revised specification is to ensure future surfacing systems meeting their life expectancy. The purpose of this work is to assess test methods where the emphasis is on the performance of the underlying substrate and its interaction with the specialist surfacing systems to ensure best outcome.

2 Materials and Methodology

2.1 Materials

2.1.1 Pavement Substrates

Asphalt surfaces are the most commonly used substrates for specialist surfacing systems in New Zealand. While chip seal and concrete surfaces may be used as the underlying substrate, chip seal surfaces require specific resin systems to be able to cope with the viscoelastic nature of the seal. Concrete surfacing is not very common in New Zealand. For the purpose of assessing the sensitivity of the pull-off test method, four different gradations of asphalt mix (Fig. 1) were sampled from an asphalt plant in Wellington, New Zealand. Mix 10, which is now designated as DG7 in the new M/10 specification, Mix 15, Mix 20, and AC20. All of which used an 80–100 penetration graded asphalt binder. Several large cores of the same Mix 20 asphalt were sampled from State Highway 1 near Wellington.

2.1.2 High-Friction Surfacing Systems

Four different resins, namely one epoxy-based, two polyurethane-based, and a methyl-methacrylate resin were tested for their adhesive bond strengths to various asphalt mixes using a pull-off test as well as for their behaviours under tension at various temperatures at a constant loading rate of 50 mm per minute using a tensile test machine. Standard 2–3 mm graded calcined bauxite aggregates were used.

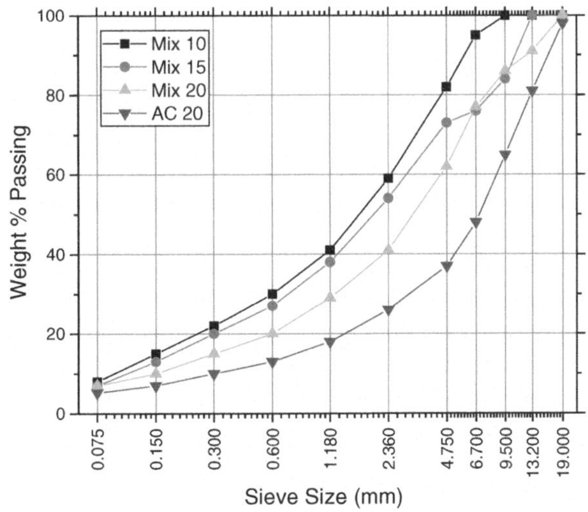

Fig. 1 Asphalt mix gradation chart of various mixes

2.2 *Test Methods*

2.2.1 Pull-off Test

The Pull-off test apparatus at Opus Research consists of a hand-cranked unit with a 12 kN load cell. The apparatus is designed to work with 50 mm diameter aluminium stubs which are glued onto the test substrate. As part of the substrate preparation, a small core cut (using a 50 mm ID core drill) is made to at least 10 mm in depth for each asphalt core. Since the core cut is made with water cooling, each sample is allowed to dry overnight before application of the aluminium stub is glued on using an epoxy-based adhesive (with an adhesive strength of at least 4 MPa). At the point of failure, a digital readout of the force at break is displayed and recorded after converting against a calibration chart.

2.2.2 Tensile Test

Temperature has a huge bearing on the performance behaviour of most materials. Therefore, there is value to investigate how the resins behave under tension at various temperatures. The tensile tester (Shimadzu EZ-TEST series) is equipped with a 5 kN load cell and an environmental chamber controlled by a fluid. The testing temperatures (−10, 0, 15, 30, and 45 °C) were controlled to the nearest 0.1 °C.

3 Results and Discussions

Several parameters were investigated: asphalt mix gradation, adhesive strength of resin systems, bituminous binder age/viscosity, and effect of thermal cycling.

The pull-off test showed that while the failure bond strength varied slightly between the different resin systems, the adhesive bond strengths of all four resin systems tested exceeded that of the cohesive strength of the asphalt substrates. Hence, all of the test failures were within the asphalt substrate regardless of the mix gradation.

The effect of ageing was the most significant as shown in Fig. 2 where the unaged block failed cohesively within the substrate as expected, and the asphalt blocks aged for 40 days at 85 °C oxidized heavily, leading to a cohesively strong asphalt substrate and it failed adhesively at the resin/substrate interface.

The viscosity measurement appears to draw excellent correlation to the cohesive strength of unaged/aged binders (Fig. 3). After the binder viscosity passes a certain threshold, the failure mechanism began to transition from cohesive substrate failure to adhesive failure at the interface between the substrate and the resin system. This reinforces the need to measure substrate properties prior to the application of

(a) (b)

Fig. 2 **a** Cohesive failure within the substrate, and **b** adhesive failure at the resin/substrate interface

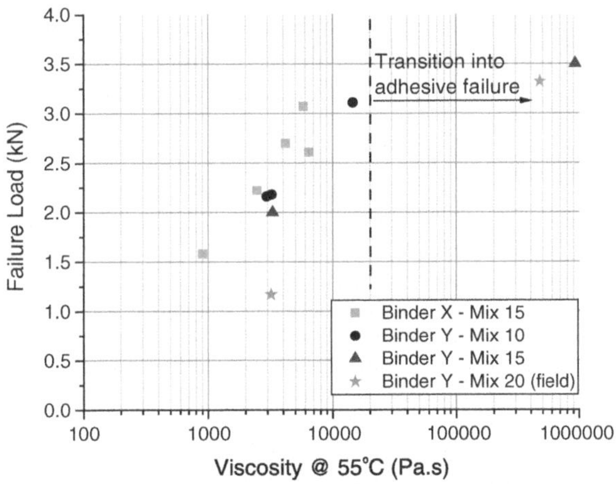

Fig. 3 Pull-off test failure loads versus binder viscosity

specialist surfacing systems in order to prevent premature failures. Every test method has its own advantages and disadvantages. The shear bond test, such as Leutner, provides excellent bond strength information on the substrate.

However, due to the non-uniform stress distribution, the test method tends to develop cracks in the substrate instead of along the interface. On the other hand, tensile bond strength test, such as the pull-off test, is more appropriate for evaluating the adhesive capabilities of a resin on a ceramic substrate.

The pull-off test is designed to detect the weakest bond in a multi-layered structure. One of its strengths is that it can be conducted in a controlled environment as well as in a field. This enables direct measurement of the adhesive and cohesive strength of the pavement layers, and provides good indication of the 'weakest link'. The test method has shown good sensitivity across various materials, and it is simple to conduct. There is a flaw to the test method used in this study, which is the loading rate. The hand-cranked unit is subject to operator consistency. As it was identified in the tensile testing, the failure strength as well as the mechanism are dependent of the loading rate. However, the new generation of pull-off testing apparatus that are currently on the market are mostly automated.

4 Conclusions

The pull-off test offers a simple but sensitive test, which can be appropriately used in the field as well as in more controlled laboratory environment to assess the bond strengths of a multi-layered structure. In order to prevent premature failure of specialist surfacing systems, it is recommended that this test method is conducted prior to the application to make sure that the underlying substrate is of sufficient cohesive strength, as well as subsequently to ensure adhesive bond is achieved.

Acknowledgements The authors would like to thank the New Zealand Transport Agency for funding the research project ART14/28 Cracking in Specialist Surfacings.

References

Arnold, G., Stubbs, A., and Vercoe, J. (2014) Pavement design for specialist surfacings, NZ Transport Agency research report 543

James, J.G. (1963) Epoxy resins as binders for road and bridge surfacings: a review of the present position. Roads & Road Construction 41 (488): 236-43

Denning, J.H. (1978) Epoxy-resin/calcined bauxite surface dressing on A1, Sandy, Befordshire: skid-resistance measurements 1968 to 1977. Laboratory Report LR867, Transport Research Laboratory, Crowthrone

Hudson, K.C. and Mumm, P. (2003) Six years of calcined bauxite, Proceedings of the 21st ARRB and 11th REAAA Conference. Transport. Our Highway to a Sustainable Future. 2003. Cairns, Australia.

Izeppi, E., Flintsch, G.W., McGhee, K.K. (2010) Field Performance of High Friction Surfaces, Final Contract Report, VTRC 10-CR6, Virginia Transportation Research Council, Virginia, US

Nicholls, J.C. (1998) Trials of high-friction surfaces for highways. TRL Report 125, Transport Research Laboratory, Highways Agency, UK

Waters, J. (2011) High Friction Surfacing Failure Mechanism, 3rd International Surface Friction Conference, Safer Road Surfaces – Savings lives, Gold Coast, Australia

Round Robin Tests on Reflective Cracking Laboratory Tools

Martin Antoine, Pouteau Bertrand, Delfosse Frédéric, Chevalier Emmanuel, Perraton Daniel and Bilodeau Francis

Abstract In the context of road network rehabilitation, private companies develop multilayered systems to prevent reflective cracking problems in pavements. These systems are submitted to a mixed mode of cracking propagation due to thermal (mode I) and traffic (modes I and II) effects. In addition, a fatigue mechanism has to be considered under the repetition of these solicitations. In order to evaluate the performances of these anti-reflective cracking systems, several laboratory devices have been developed. In this paper, we focus on two of these tools in order to perform round robin tests on various types of anti-reflective cracking systems. The first test was developed at CEREMA laboratory in Autun (former LRPC). The second was developed at the Ecole de Technologie Supérieure (ETS) in Montréal and is used at Eurovia's Research Centre in Bordeaux for the purpose of the study. This paper first proposes an analysis of the two test procedures which particularly highlights a main difference in the horizontal and vertical ways of solicitations. Horizontally, a monotonic stretching displacement (mode I) is applied for the first test whereas it is cyclic for the other.

M. Antoine · P. Bertrand (✉) · D. Frédéric · C. Emmanuel
Eurovia Research Center, Mérignac, France
e-mail: bertrand.pouteau@eurovia.com

M. Antoine
e-mail: antoine.martin@eurovia.com

D. Frédéric
e-mail: frederic.delfosse@eurovia.com

C. Emmanuel
e-mail: emmanuel.chevalier@eurovia.com

P. Daniel · B. Francis
Ecole de Technologie Supérieure, Montréal, Canada
e-mail: daniel.perraton@etsmtl.ca

B. Francis
e-mail: francis.bilodeau@etsmtl.ca

A. Chabot et al. (eds.), *8th RILEM International Conference on Mechanisms of Cracking and Debonding in Pavements*, RILEM Bookseries 13,
DOI 10.1007/978-94-024-0867-6_75

Vertically, shear loading (mode II) is applied with the second test device whereas it is flexural (mode I) with the first one. The author's analysis will focus on seeking correlations between results of the two tests.

Keywords Reflective cracking · Interlaboratory tests

1 Introduction

Although the overlay method leads to reflective cracking problems, this technique is still widely used in the context of road network rehabilitation. Various solutions, combined to overlays, are then developed to prevent this problem including multilayer systems (Vanelstraete and Francken 2004). Those are dedicated to be used over old cement concrete pavements. Their bottom layers are designed to damp strains induced by existing cracks/joints whereas the upper one is to endure traffic.

These systems are submitted to cracking due to thermal (mode I) and to load effects generated by truck traffic (modes I and II). In addition to this combination of cracking modes, a fatigue mechanism for thermal (up to several thousands of cycles on a 10 years period) and traffic (several millions of cycles on the same period) has to be considered. To assess the performance of these systems, several research teams have developed laboratory tools to simulate these conditions of different manners (Florence et al. 2004; Gallego and Prieto 2006).

In this paper, we aim to perform round robin tests between two laboratory tools on various types of products. The first test was developed at CEREMA laboratory (former LRPC) in Autun (Dumas and Vecoven 1993; Vecoven 1989) and the second at the Ecole de Technologie Supérieure (ETS) in Montréal (Perraton et al. 2008). This paper first provides details on the two test procedures and tested products. The results of the tests (performed to date) are then presented.

2 Comparison of Two Reflective Cracking Laboratory Tests

In this section, we present the two reflective cracking test setups used in this study focusing on their similarities and differences (see Fig. 1). For the sake of simplicity, the test developed at the CEREMA is noted Test-A and the one at ETS Test-B.

2.1 Fabrication of Specimens

The first comparison point is in connection with the fabrication method of specimens. As shown in Fig. 1, the samples are multilayered systems supposed to be

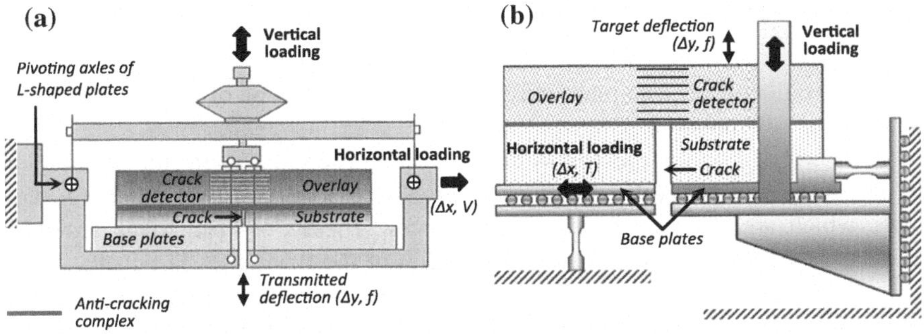

Fig. 1 Schematic diagrams of the reflective cracking devices developed at the: **a** CEREMA (Test-A) and **b** ETS (Test-B)

representative of those used in situ. From bottom to top, they are composed of: (i) a substrate layer made of a reference asphalt concrete material (with sulfur for the Test-A specimens), voluntarily discontinuous to simulate a pre-existing crack or joint in the lower layer of road pavements, (ii) the complex, aiming to reduce or avoid the reflective cracking problem initiated by this pre-crack under the applied solicitations, (iii) a reference asphalt concrete overlay.

Concerning the Test-A method, the specimens are bars cut from test plates of dimensions 600 × 400 × 85 mm^3 (approximate thickness, depends on that of the used complex). A plate is made with the aid of a wheel track compactor by placing and compacting each layer successively to obtain the dimensions given in Table 1. During the substrate fabrication, a rigid cardboard piece is introduced splitting it in two parts to simulate the pre-crack. Three bars can be extracted from one plate after removing edges to avoid undesirable side effects.

Among the different fabrication methods presented in (Perraton et al. 2008), we focus here on that one using asphalt concrete material compacted in laboratory. This method is similar to the previous except the fact that the bars are made individually and that the crack width is set to 5 mm (see Table 1 for dimensions).

Table 1 Specimens size for the two reflective cracking devices

Device	Layer	Length (mm)	Width (mm)	Thickness (mm)
Test-A (CEREMA)	Overlay	560	110	60
	Complex	560	110	Variable
	Substrate	559 (1 mm crack)	110	15
Test-B (ETS)	Overlay	445	120	55
	Complex	445	120	Variable
	Substrate	220 × 2 (5 mm crack)	120	55

2.2 Test Principles

As explained hereafter, the main difference relies on the type of applied horizontal and vertical solicitations.

As shown on Fig. 1a, the Test-A device is composed of a vertical jack acting on the top surface of the tested specimen. This one is glued on two aluminum base plates which are fixed on two L-shaped plates. These ones can rotate around a pivoting axle making possible the transmission of the flexural cyclic displacement (Δy) applied at a frequency (f) on the top surface to the bottom one; simulating a part of the traffic effects (mode I). In addition, the right L-shaped plate is connected to an horizontal jack applying a longitudinal constant stretching (Δx) at a given speed (V) simulating the thermal shrinkage of the structure (mode I). These two simultaneous solicitations are applied under constant temperature conditions by placing the whole device into a thermal chamber.

Concerning the Test-B device, it is composed of two parts as shown in Fig. 1b. Each part is composed of one base plate mounted on bearing and on which the specimen is glued. The first part of the device is dedicated to apply a slowly horizontal stretching cyclic displacement (Δx, sinusoidal cycle period T) in a mechanical way; simulating the thermal shrinkage (mode I). The other part allows applying vertical cyclic displacement at a frequency (f) using a hydraulic jack acting on the base plate to simulate traffic solicitations (mode II). This displacement is controlled in such a way to obtain a vertical target displacement (Δy) measured on the top surface of the overlay. This device is also positioned into a thermal chamber.

For the two tests, different sensors are positioned on the specimen to measure the forces and displacement evolution. The crack propagation is measured with the aid of an electrical wires network, glued on one vertical surface of the overlay right above the pre-crack (see crack detector in Fig. 1), composed of 24 wires (on 80 mm thickness) for Test-A specimens and only 8 (on 30 mm thickness) for Test-B ones. During the crack propagation, the wires alternately break and give the relation between the test time and the cracking height, especially for the anti-crack complex.

3 Round Robin Tests

Round robin tests using these two reflective cracking devices are planned to be performed in this research project. To date, only tests of the Test-A device were realized. The tests with the Test-B device are still ongoing at the Eurovia's Research Centre. So, we present hereafter the two experimental programs and the results of the Tests-A.

Table 2 Tests conditions for each experimental device

	Temperature (°C)	Horizontal solicitations	Vertical solicitations
Test-A	5	$\Delta x = 7.2$ mm, $V = 0.6$ mm/h	$\Delta y = 0.20$ mm, $f = 1$ Hz
Test-B	−7.5	$\Delta x = 2.0$ mm, $T = 3120$ s	$\Delta y = 0.02$ mm, $f = 1$ Hz

3.1 Experimental Programs

Five anti-crack complex systems were considered in this research program (noted from A to E). The technical details of such system are not given explicitly in this paper for reasons of confidentiality. Just note that the system A is set as reference since it is composed of a widely used anti-crack complex system. In accordance with CEREMA recommendations, 6 specimens were made and tested with the Test-A device for each formula (a total of 30 specimens have been tested). For the series on Test-B device, 3 specimens by system were manufactured with the ETS fabrication method to take into account tests dispersion (e.g. 15 specimens). According to the specificities of each experimental device, the test conditions are relatively different as reported in Table 2 (see Sect. 2.2 for notations).

3.2 Tests Results

Figure 2a presents an illustration of the obtained results of Test-A (for the 6 specimens made of the complex system A). Those are given under the form of cracking height time histories. On this Figure, two additional curves are drawn and correspond to some reference specimens made of an asphalt concrete (AC ref.) and an asphalt concrete/coated sand (AC/CS ref.). The first one is devoid of anti-crack

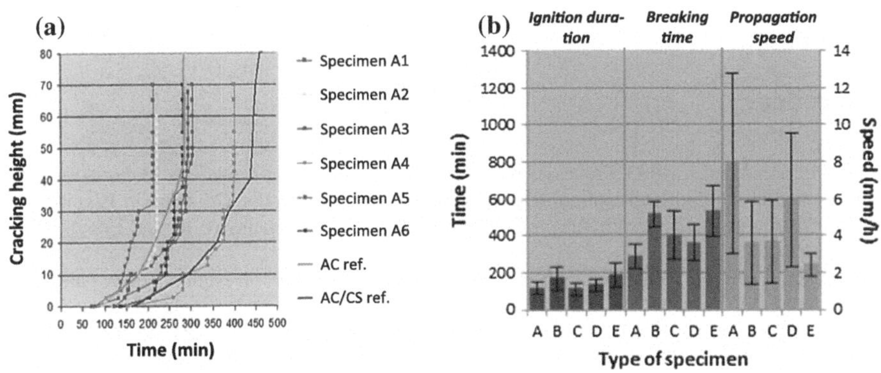

Fig. 2 Overlay cracking propagation results of complex systems tested with the Test-A device: **a** relationship between the cracking height in the overlay and the testing time as measured for the 6 specimens of the complex system A and, **b** synthesis of the whole campaign results (T = 5 °C)

complex and the second is supposed to give relatively good performances. This type of representation provides different information as: the crack initiation duration (delay from the time origin to the first point), the time of whole specimen breaking (materialized by the vertical asymptote) and the average crack propagation speed into the complex (calculated by dividing the complex thickness and the necessary duration so that the crack height reaches this thickness). For each of the complex systems tested in this testing program (A to E), the mean values and standard deviations of these three characteristics were calculated based on the result of each 6 tested specimens per system as reported in Fig. 2b.

In this figure, we can see that system B and E provide a longer duration of crack initiation and breaking time (in reference to the left axis) whereas the others complex systems have smaller ones. In reference to the right axis, the smaller propagation speed is attributed to the system E and the bigger one to the system A.

4 Conclusion

In this study, round robin tests are planned to be performed on two selected reflective cracking devices, e.g. those developed at CEREMA and ETS. The comparison of their experimental procedures in particular reveals differences in the type of applied solicitations. Results show that sensors system used to follow the overlay crack propagation is efficient. It can be used to better understand the general trend of an anti-crack system. As illustrated, depending on the anti-crack system used the crack propagation can be delayed in the overlay to improve their sustainability.

Since the road networks is aging, reflective cracking tests will become increasingly used to develop new products. In relation with this road rehabilitation context, the perspective of the research works related to the study presented in this paper is to provide a reflective cracking device to the Eurovia's Research Centre.

References

Dumas, P., Vecoven, J., 1993. Processes reducing reflective cracking; synthesis of laboratory tests. Presented at the 2nd International RILEM Conference on Reflective Cracking in Pavements - State of the Art and Design Recommendations, Rigo J.M., Degeimbre R. and Francken L., pp. 246–253.

Florence, C., Foret, G., Tamagny, P., Sener, J.Y., Ehrlacher, A., 2004. Design of a new laboratory test simulating the reflective cracking in pavements with cement treated bases. Presented at the 5th International RILEM Conference on Reflective Cracking in Pavements, pp. 147–154.

Gallego, J., Prieto, J., 2006. New Laboratory Equipment for Study of Reflective Cracking in Asphalt Overlays. Transportation Research Record: Journal of the Transportation Research Board 1970, 215–222. doi:10.3141/1970-25

Perraton, D., Guissi, G., Doré, G., Pierre, P., 2008. A new laboratory test for reflective cracking in mode I and/or mode II, in: Pavement Cracking. Presented at the 6th RILEM Conference on Cracking in Pavements, CRC Press, Chicago, pp. 447–456.

Vanelstraete, A., Francken, L., 2004. Prevention of Reflective Cracking in Pavements, RILEM Report 18. CRC Press.

Vecoven, J.H., 1989. Méthode d'étude de systèmes limitant la remontée de fissures dans les chaussées. Presented at the 1st International RILEM Conference on Reflective Cracking in Pavements - Assesment and Control, Rigo J.M. and Degeimbre R., pp. 57–62.

The Effect of Interface Bonding Criterion on Pavement Design Reinforcement

Loizos Andreas, Plati Christina and Tsaimou Christina

Abstract The present paper deals with the effect of interlayer bonding conditions between Hot Mix Asphalt (HMA) and Cement Bound Material (CBM) layers on pavement design reinforcement in a case of a semi-flexible pavement. For this purpose a field experiment was conducted in a motorway by the Laboratory of Pavement Engineering of the National Technical University of Athens (NTUA). Non-Destructive Testing (NDT) was applied using a Falling Weight Deflectometer (FWD) and a Ground Penetrating Radar (GPR) system in order to assess the existing pavement condition and take into account reasonable considerations towards the repair and upgrade of the existing motorway. The pavement analysis was conducted considering two approaches: (a) full-bonding and (b) no-bonding conditions between HMA and CBM layers. The overall investigation indicated that the differences between the two approaches range for the different cases of the new road elevation. Moreover elevation increase results in less significant differences between the two approaches. The related findings and results are displayed and discussed thoroughly.

Keywords Interface bonding · HMA · CBM · Pavement reinforcement

L. Andreas · P. Christina (✉) · T. Christina
Laboratory of Pavement Engineering, School of Civil Engineering,
National Technical University of Athens, Athens, Greece
e-mail: cplati@central.ntua.gr

L. Andreas
e-mail: aloizos@central.ntua.gr

T. Christina
e-mail: ctsaimou@central.ntua.gr

A. Chabot et al. (eds.), *8th RILEM International Conference on Mechanisms of Cracking and Debonding in Pavements*, RILEM Bookseries 13,
DOI 10.1007/978-94-024-0867-6_76

1 Introduction

Interface bonding conditions between pavement layers can be one of the most significant factors affecting pavement service life. Various factors may have impact on the bonding condition at the interface, including tack coat material, tack coat application rate, material aging, temperature, and moisture conditions (Al-Qadi et al. 2008). Interlayer condition is known to affect the distribution of stresses and strains in the underlying layers of the pavement structure. The bond between layers is essential as it enables the pavement to act as a composite structure, thus reducing interface stresses. The lack of interlayer bonding can significantly degrade pavement service life due to instability within the overall pavement structure and lead to pavement distresses including slippage cracking and early fatigue cracking amongst other issues. For the aforementioned reasons, the interface bonding between pavement layers has considerably occupied several researchers. The majority of these research studies have been concentrated on the bonding conditions between the individuals Hot-Mix Asphalt (HMA) layers of the pavement structure, while limited emphasis has been placed on the interface bonding between the HMA and Cement Bound Material (CBM) layers (Yildirim et al. 2005; Al-Qadi et al. 2008; Chabot et al. 2008).

The present research aims to investigate the effect of bonding conditions between HMA and CBM layers on pavement design reinforcement. In order to achieve this goal the Laboratory of Pavement Engineering of the National Technical University of Athens (NTUA) undertook a field investigation along an existing motorway that was scheduled to be repaired and upgraded in the frame of a concession system. For the investigation Non-Destructive Tests (NDT) were performed with both a Falling Weight Deflectometer (FWD) and a Ground Penetrating Radar (GPR) system. The NDT data analysis was utilized in order to evaluate the in situ condition of the existing pavement. Pavement design and strain response analysis was conducted considering two approaches: (a) full-bonding and (b) no-bonding conditions between HMA and CBM layers. The analysis was performed for different cases of road elevation, i.e. the difference between the existing vertical alignment and the new one after reinforcement. The related considerations and results are shown in the following sections.

2 Field Condition and Data Collection

The in-situ field experiment was conducted along an existing 5 km long semi-flexible pavement section of an existing motorway which is scheduled to be reinforced due to the requirements of the new vertical alignment. The pavement design is based on the requirement of two interventions during the concession period (24 years plus 10 year guarantee period) that are foreseen after the 12th and the 24th year of road operation. For the interventions two types of reinforcement are considered: (a) milling of HMA layers and replacement with high modulus HMA (EME), plus a skid wearing HMA course and (b) a new overlay with a skid wearing HMA course. Therefore, the analysis was made for three design periods: A (1–12 years), B (13–24 years) and C (25–34 years).

The proposed pavement support solution includes the initial milling to a depth of at least 3 cm of the existing HMA layers. Depending on the new elevation levels the new flexible pavement will be laid over the existing HMA after milling. On top of the unbound material layers a new HMA base, binder and wearing course will be constructed. The pavement consists of two lanes per directions. For the purposes of this research the elevation for the right lane is considered. The increase in elevation from the existing pavement levels ranges from 0.13 m to 0.68 m along the section under investigation. In order to identify the pavement structure GPR continuous measurements were performed and core extraction was conducted along the test section (COST Action TU1208 2013). Based on the available thickness data the pavement structure consists of HMA layers, base/subbase layers including CBM and unbound materials, and the subgrade. Furthermore, FWD measurements were performed along two sub-sections of the considered pavement in order to determine the structural condition of the existing pavement (NCHRP 2008). The frequency of the FWD records ranged from 10 to 100 m.

3 Data Analysis and Pavement Design

In-situ data was utilized for the determination of the existing pavement structure and the evaluation of its current condition. For both GPR and FWD data advanced statistical analysis was conducted in order to estimate a characteristic value of layers thickness and back calculated moduli (reference temperature 20 °C), based on a conservative approach. Based on the statistical test of Kolmogorov-Smirnov (Justel et al. 1996), it was found that the generalized extreme value distribution (Nadarajah 2002) provides a good fit to the considered data. The thickness and the modulus are determined for the characterization of the individual layers by using the value of the 10 % data of the distribution curve for the design as a conservative approach (Table 1). The considered thickness values were validated by using limited HMA-CBM cores. It is notable that unbound material and subgrade moduli are high. However these values are acceptable as the materials are evaluated as part of the pavement structure.

Table 1 The characteristic values of the in-situ pavement data

Layer	Thickness (cm)	Modulus (MPa)
HMA	19	3678
CBM	7	3945
Unbound material	8	852
Subgrade	–	220

The extracted cores indicated that at some locations there was lack of bonding between the HMA and the CBM layers. Therefore, for the present study two approaches will be taken into consideration in terms of pavement design. The first considers that the interface between the HMA and the CBM layers is full-bonded, while the second considers that there is no interlayer bonding for the two layers. In situ investigation showed that the existing pavement is structurally sound. For this reason the existing pavement can be used as a support for pavement reinforcement purposes. The design is based on the analytical-mechanistic pavement design method described by LCPC-SETRA (1994, 2007). The pavement is considered as a multi-layered structure built on the top of the existing pavement structure. It is a rational method, based on the computation of the stresses and strains by the elastic multi layer linear approach. Therefore it is a combination of the existing pavement upgrade followed by flexible pavement design analysis.

For the failure analysis of HMA base course material, according to the Miner law, the damage index at the end of the design life should be less than or equal to 1 (Eq. 1).

$$d_{tot} = d_1 + d_2 + d_3 \leq 1 \tag{1}$$

where d_1, d_2, d_3 are the cumulative damage index for the HMA base course during design periods 1, 2 and 3, respectively.

The three different cases of the reinforced pavement that are used for further analysis are presented below. The moduli values of the new HMA layers are considered to be equal to the relative ones of the French catalogue for temperature 20 °C and frequency 10 Hz. Since the modulus of the underlying layer is higher than the one of the new unbound material, the increase in the unbound layer thickness will cause increase of the induced strain at the bottom of the new HMA layers. For this reason, the most critical case is the case of the maximum elevation. For the case of the minimum elevation (13 cm), milling more than 3 cm is required in order to achieve a new 10 cm layer of unbound material and to meet the elevation requirements. Figure 1 shows the cross section and the damage index of each

Design period A		
Layer	h(cm)	E(MPa)
Wearing course	3	3600
Binder course	5	7450
Base course	5	7450
Unbound material	10	600
Existing HMA layers	9	3600
CBM	7	3900
Unbound material	8	600
Subgrade	600	200
	-	10000

d_A	Bond HMA-CTB	NO-Bond HMA-CTB
	0.20	0.42

Design period B		
Layer	h(cm)	E(MPa)
Wearing course	3	3600
EME	10	11000
Binder course	5	7450
Base course	5	7450
Unbound material	10	600
Existing HMA layers	9	3600
CBM	7	3900
Unbound material	8	600
Subgrade	600	200
	-	10000

d_B	Bond HMA-CTB	NO-Bond HMA-CTB
	0.10	0.22

Design period C		
Layer	h(cm)	E(MPa)
Wearing course	3	3600
Wearing course	3	3600
EME	10	11000
Binder course	5	7450
Base course	5	7450
Unbound material	10	600
Existing HMA layers	9	3600
CBM	7	3900
Unbound material	8	600
Subgrade	600	200
	-	10000

d_C	Bond HMA-CTB	NO-Bond HMA-CTB
	0.04	0.08

Fig. 1 Cross sections and damage indexes—minimum elevation

design period. It is noted that the modulus values of the existing layers have been rounded down in terms of a conservative approach.

In order to examine the effect of the elevation, the trigger value of 3 cm for the milling depth was considered. For this value the minimum elevation was estimated to be equal to 20 cm. For elevation greater or equal to 20 cm, the milling depth will be 3 cm, while for elevation less than 20 cm, the milling depth will be dependent on the new vertical alignment. Figures 2 and 3 show the pavement cross sections and the damage indexes of each period for the case of 20 cm elevation and for the case of maximum elevation (68 cm) respectively. Table 2 shows the summarized damage indexes for the two approaches of interlayer conditions.

It can be observed that as new elevation levels increase, less significant variations between the two approaches are observed.

Design period A		
Layer	h(cm)	E(MPa)
Wearing course	3	3600
Binder course	5	7450
Base course	5	7450
Unbound material	10	600
Existing HMA layers	16	3600
CBM	7	3900
Unbound material	8	600
Subgrade	600	200
	-	10000

d_A	Bond HMA-CTB	NO-Bond HMA-CTB
	0.16	0.24

Design period B		
Layer	h(cm)	E(MPa)
Wearing course	3	3600
EME	10	11000
Binder course	5	7450
Base course	5	7450
Unbound material	10	600
Existing HMA layers	16	3600
CBM	7	3900
Unbound material	8	600
Subgrade	600	200
	-	10000

d_B	Bond HMA-CTB	NO-Bond HMA-CTB
	0.07	0.12

Design period C		
Layer	h(cm)	E(MPa)
Wearing course	3	3600
Wearing course	3	3600
EME	10	11000
Binder course	5	7450
Base course	5	7450
Unbound material	10	600
Existing HMA layers	16	3600
CBM	7	3900
Unbound material	8	600
Subgrade	600	200
	-	10000

d_C	Bond HMA-CTB	NO-Bond HMA-CTB
	0.03	0.04

Fig. 2 Cross section and damage indexes—minimum elevation with milling depth 3 cm

Design period A		
Layer	h(cm)	E(MPa)
Wearing course	3	3600
Binder course	5	7450
Base course	5	7450
Unbound material	20	600
Unbound material	20	500
Unbound material	18	400
Existing HMA layers	16	3600
CBM	7	3900
Unbound material	8	600
Subgrade	600	200
	-	10000

d_A	Bond HMA-CTB	NO-Bond HMA-CTB
	0.55	0.56

Design period B		
Layer	h(cm)	E(MPa)
Wearing course	3	3600
EME	10	11000
Binder course	5	7450
Base course	5	7450
Unbound material	20	600
Unbound material	20	500
Unbound material	18	400
Existing HMA layers	16	3600
CBM	7	3900
Unbound material	8	600
Subgrade	600	200
	-	10000

d_B	Bond HMA-CTB	NO-Bond HMA-CTB
	0.24	0.25

Design period C		
Layer	h(cm)	E(MPa)
Wearing course	3	3600
Wearing course	3	3600
EME	10	11000
Binder course	5	7450
Base course	5	7450
Unbound material	20	600
Unbound material	20	500
Unbound material	18	400
Existing HMA layers	16	3600
CBM	7	3900
Unbound material	8	600
Subgrade	600	200
	-	10000

d_C	Bond HMA-CTB	NO-Bond HMA-CTB
	0.08	0.09

Fig. 3 Cross sections and damage indexes—maximum elevation

Table 2 Summarized damage index for the two approaches

Damage index	Bond	No bond
Minimum elevation	0.34	0.72
20 cm elevation	0.26	0.40
Maximum elevation	0.87	0.90

4 Conclusions

The present research focuses on the impact of interface bonding between HMA and CBM layers on the reinforcement of an existing semi-flexible pavement of a motorway. FWD and GPR data were utilized for the evaluation of pavement condition for the repair and upgrading of the existing motorway. The considered pavement is an old existing structure where a new flexible pavement will be laid over. The related analysis was performed based on two approaches: (a) full bonding conditions and (b) no-bonding conditions between the HMA and CBM layers. The analysis results indicated that the milling aspect of the existing HMA layers is a significant factor affecting the results when implementing the two approaches. The greater the milling depth, the more significant the observed differences in the damage index are. Moreover, for the same milling depth, i.e. for the same thickness of the existing HMA layers the differences between the two approaches are higher for the case of the minimum elevation. However it seems that the case of the maximum elevation is the most critical case in terms of the pavement design and as a result the consideration of the interface condition does not have a significant impact on the final proposal. Further investigation is needed.

References

Al-Qadi IL, Carpenter SH, Leng Z, Ozer H, and Trepanier JS (2008) Tack coat optimization for HMA overlays: laboratory testing. Illinois Department of Transportation, FHWA-ICT-08-023

Chabot A, Balay JM, Pouteau B, and de Larrand F (2008) FABAC accelerated loading test of bond between cement overlay and asphalt layers. Taylor & Francis Group, ISBN 978-0-415-47575-4, London

COST Action TU1208 (2013) Civil engineering applications of Ground Penetrating Radar. Proceedings, First Action's General Meeting, Pome, 22-24 July 2013

Justel A, Pena D, and Zamar R (1996) A multivariate Kolmogorov-Smirnov test of goodness of fit. Elsevier, Statistics & Probability Letters 35, p. 251-259

LCPC-SETRA (1994) Catalogue des structures types de chaussées neuves

LCPC-SETRA (2007) Guide Technique: Diagnostic et conception des renforcements de chausses, version draft du 30_07_07

Nadarajah S (2002) Reliability for extreme value distributions. Pergamon, Mathematical and Computer Modelling 37, p. 915-922

NCHPR (2008) Falling Weight Deflectometer usage. NCHRP Synthesis 381, Transportation Research Board, Washington, D.C.

Yildirim, Y, Smit, AF, and Korkmaz, A (2005) Development of a laboratory test procedure to evaluate tack coat performance. Turkish Journal of Engineering and Environmental Sciences, No. 29, p. 195-205

Part IX
Interface Debonding Behavior: Debonding Mechanisms (Bituminous Interface)

Development of a Pull-off Test to Measure the Bond Strength of Bituminous Emulsions

Lucas Leprince, Nathalie Piérard, Alexandra Destrée, Ann Vanelstraete and Joëlle De Visscher

Abstract It is now generally recognized that good adhesion between different layers of an asphalt pavement is of great importance for its lifetime. Poor interfacial bond strength leads to a higher probability of cracking, early fatigue failure, delamination, etc. Many factors have an impact on the bond strength between layers. Despite all these factors, it is important that the performance of tack coats can be evaluated independently of the asphalt complex for the purpose of comparison, optimization and selection of these products. Consequently, the development of a reproducible quantitative test method to characterize the bond strength of a bituminous tack coat is crucial. This paper describes the development and validation of a pull-off test on a bituminous tack coat film. In a first step, the test device, the breaking and curing conditions and the pull-off test conditions were optimized to ensure that failure occurs within the film and not at the interface between the film and the contacting plates of the test device. In a second step, a validation of the pull-off device and procedure was conducted by comparing the results of the pull-off tests for different emulsions with direct tensile tests carried out on multi-layer asphalt samples in which the same tack coats were applied. This paper is an overview of BRRC's work in this field.

Keywords Tack-coat · Pull-off · Bond strength · Interlayer adhesion

L. Leprince (✉) · N. Piérard · A. Destrée · A. Vanelstraete · J. De Visscher
Asphalt Pavements, Other Bituminous Applications and Chemistry Division,
Belgian Road Research Centre (BRRC), Brussels, Belgium
e-mail: l.leprince@brrc.be

N. Piérard
e-mail: n.pierard@brrc.be

A. Destrée
e-mail: a.destree@brrc.be

A. Vanelstraete
e-mail: a.vanelstraete@brrc.be

J. De Visscher
e-mail: j.devisccher@brrc.be

A. Chabot et al. (eds.), *8th RILEM International Conference on Mechanisms of Cracking and Debonding in Pavements*, RILEM Bookseries 13,
DOI 10.1007/978-94-024-0867-6_77

1 Introduction

The main goal of a tack coat is to create a durable bond between layers of a pavement. Proper layer bonding results in a pavement structure acting as a composite layer, reducing pavement stresses and ensuring the structure to properly withstand the traffic loading and environmental conditions. Inadequate bonding can result in pavement damage as a consequence of slippage of layers or poor bearing capacity.

The selection of a proper tack coat material and appropriate application parameters are thus fundamental steps in the optimization of bond strength between layers. In this regard, some guidelines and general principles have been described for the selection of tack coat type, application rate, placement and evaluation (BRRC 2012). BRRC also extensively studied the influence of these parameters and others on the bonding strength between layers (Destrée and Beaumesnil 2015, Destrée et al. 2016). Nevertheless, the selection of the bituminous tack coat emulsion is nowadays still mainly based on practical experience, fulfilment of specifications, convenience or price. None of the conventional characteristics of emulsions is uniquely related to its bonding strength, which means that the selection of the emulsion is not based on its true performance. Many studies have focused on the development of apparatus and procedures for the assessment of the performance of tack coat (Eedula and Tandon 2006; Mohammad 2012). Many parameters have been shown to influence the type of failure as well as the tension test results: apparatus (stamp design, loading rate, substrate materials and characteristics (aggregate type, surface texture, interaction between bitumen of course mixtures and bitumen of the tack coat, etc.), test temperature, etc.) and bitumen films [bitumen type, pen, $T_{R\&B}$, thickness, bitumen conditioning (curing temperature and time, humidity, etc.)]. The procedures developed were often complicated, not representative or showed high variability in their results. Furthermore these studies showed that it will never be possible to evaluate the adhesive strength of an emulsion to an asphalt layer by making use of a test on the emulsion itself, as both materials play their role in that. However, they show also the need for a new procedure to determine the cohesive strength of emulsions to ensure its cohesive performance, a simple and quick test to be carried out on emulsions only that shows a good reproducibility and allows to distinguish between emulsions.

In this project, the aim was to develop a simple apparatus and procedure to measure the cohesive resistance of a thin film of bitumen emulsion after curing, i.e. the *pull-off resistance* of an emulsion. The objectives of this research were: to identify the parameters and factors that affect the bond strength development of a bituminous emulsion; to elaborate a corresponding procedure to reproducibly prepare a thin bitumen film of cured emulsion; to develop an apparatus and test method to quantify the pull-off resistance of this film; to validate the newly developed test on emulsions by comparing the results with those of a standard tensile bond strength test on a two-layered asphalt specimen bonded by the emulsions.

2 Materials and Test Methods

The cationic bituminous emulsions tested in this study are described in Table 1.

2.1 *Description of BRRC's Pull-off Test*

The *pull-off* test developed in the Belgian Road Research Centre (BRRC) is performed on a thin layer of residual bitumen of controlled thickness from a cured emulsion on a plane metal plate. A controlled quantity of emulsion is poured on a metallic plate of a surface 120 mm × 600 mm previously stored at 50 °C in a drying oven (residual bitumen quantity of 400 g/m^2, corresponding to film thickness of ~400 µm). The emulsion is manually spread on the surface with a brush. The plate is then stored in the drying oven at a temperature of 50 °C for 24 h for curing. After 24 h, four square metal stamps of dimensions 100 mm × 100 mm (previously stored at 50 °C for 24 h) are applied on the surface of the cured emulsion with a 14 kg load applied for 3 min (see Fig. 1a). After removal of the load, the prepared plate is stored in the lab at ambient temperature until the temperature of the apparatus reaches 25 °C. The apparatus for tension is a motor-driven

Table 1 Characteristics of cationic bituminous emulsions used for this study

Sample	Type of emulsion	Characteristics of residual bitumen	
		Penetration at 25 °C (1/10 mm)	Softening point (°C)
A	C60B3	175	39.3
B	C60BP3 AA[a]	56	52.0
C	C60B3	73	47.5
D	C60BP3	102	59.5

[a]*AA* anti-adhesive, emulsion prepared with harder bitumens

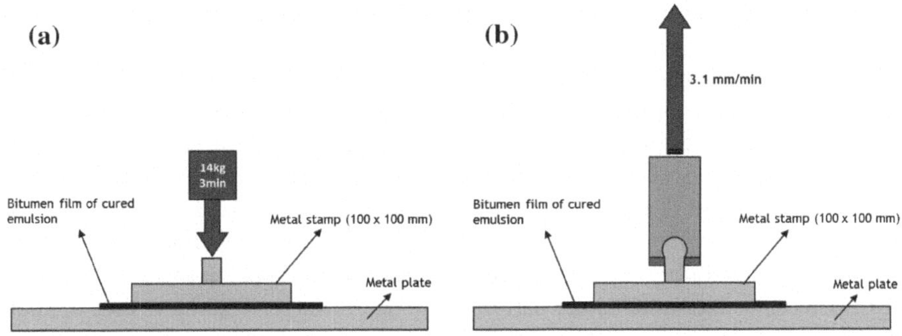

Fig. 1 Description of BRRC's pull-off test: **a** loading of a metallic stamp on the cured bitumen emulsion film, **b** pull-off resistance test by vertical tension at a constant rate of 3.1 mm/min

dynamometer commonly used for adhesion tests, It is operated at a constant displacement rate of 3.1 mm/min (see Fig. 1b). During the test, the tensile force is recorded as a function of time (or displacement). The tension is applied until the rupture of the bitumen film. After the stamps are removed from the surface, the breaking pattern is visually observed. This visual observation allows us to exclude the occurrences of adhesive failure between the metallic surfaces and the bitumen film, in order to consider only the samples which underwent cohesive failure within the film. The results are then processed to obtain a mean value for the pull-off resistance of the tested emulsion (in MPa).

During the development of the pull-off test, several parameters were evaluated to optimize both the equipment and the procedure. Residual quantity of bitumen emulsion was adapted in order to maximize the probability of cohesive rupture, in accordance to literature where it was shown that the rupture type is more of the interfacial type for films of small thickness (under 200 μm), and more of cohesive behavior for film of bigger thickness. Loading rate has also been proven to be of significant importance in pull-off tension tests. In our case, the loading rate is around 10 times lower than in the previous pull-off procedures leading to a higher probability of cohesive rupture.

2.2 Description of the Direct Tensile Test on Asphalt Specimens Used for Validation (LAMI Test)

To validate BRRC's pull-off test, the same emulsions were tested using the LC 25-010 standardized procedure which is used at the department of transportation of Quebec (Ministère des transports du Québec-MTQ) for the evaluation of the cohesive resistance of emulsions (MTQ 2014). This procedure is performed on sawed asphalt cores (plane surfaces of aggregates and bitumen) without the effects of surface texture, as described in this section.

The samples used in this LC 25-010 standardized procedure were prepared in laboratory according to the LC 25-011 standard from the MTQ (MTQ 2011). In our case, we used a slightly modified procedure. The tension tests were carried out on cores prepared from the gyratory compactor (diameter 150 mm, height 120 mm). The cores were highly compacted to avoid a rupture inside the asphalt mix (200 gyrations). The mix for these cores is a SMA 10 with a polymer modified bitumen (45/80-50). The cores were cut in two parts (upper part: 45 mm and lower part: 70 mm) to avoid any influence of the surface texture. The emulsion is spread over the sawed surface of the lower part, at a residual quantity of 0.2 l/m^2 and cured for 45 min at ambient temperature under air flow. Afterwards, the samples are prepared according to the LC 25-011 standard, by heating the upper part of the core at 115 °C and placing on the lower part of the core. A charge of 25 kg is then applied for 5 min on the core. The three prepared cores are cooled under air flow at ambient temperature for 90 to 120 min.

The tension tests were operated in laboratory conditions at ambient temperature at a constant charge rate (240 N/s ± 40 N/s) by the Layer Adhesion Measuring Instrument- LAMI (tension testing device as described in LC 25-010), to evaluate the bond strength between the pavement layers. Measures of tensile resistance are performed on cylindrical specimens of approximately 101 mm in diameter, by means of an hydraulic device controlled by an automated valve.

3 Results

The results of pull-off strength resistance and LAMI resistance tests for the bitumen films from cured emulsions can be seen in Table 2. The values for the maximum bond strength for the 4 bituminous emulsions have been measured. The LAMI maximum bond strength is calculated as the average of three cores. The pull-off bond strength is calculated as the average of four metallic stamps (or three if an interfacial rupture was observed, as described in the previous section). An analysis of variance (ANOVA) was performed with level of significance, α of 0.05, in order to analyze and compare the average maximum bond strength. For the sake of simplicity and concision this paper will not present the ANOVA analysis results of the interlayer bonding tests but they have been taken into account in our interpretations and our conclusions.

In Table 2, we can see that for the analyzed emulsions, the standard deviation is relatively low both for pull-off and standardized tests, with an error of generally less than 10 %. Pull-off resistance measurements lead to significantly different values between emulsions. It is thus possible to distinguish two emulsions based on their pull-off resistance using this procedure. The same observation can be made based on the maximum bond strengths by LAMI.

By comparing pull-off and LAMI results, we can see that the same ranking is respected in terms of maximum bond strength for emulsions A, B and C. Emulsion D shows lower results for bond strength measured by LAMI, as well as larger standard deviations. At this point, this behavior cannot be fully explained and

Table 2 Results of maximum bond strength with BRRC's pull-off test and LAMI for bitumen films from cured bituminous emulsions

Sample	Type of emulsion	Maximum pull-off resistance			Bond strength by LAMI		
		Average (MPa)	STDEV (MPa)	CV (%)	Max (MPa)	STDEV (MPa)	CV (%)
A	C60B3	0.38	0.02	5.3	0.8	0.05	6.2
B	C60BP3 AA[a]	0.62	0.03	4.8	1.27	0.02	1.6
C	C60B3	0.48	0.04	8.3	1.06	0.05	4.7
D	C60BP3	0.47	0.05	10.6	0.65	0.13	19.7

[a]*AA* anti-adhesive, emulsion prepared with harder bitumens

is possibly linked to the polymer modified nature of the bitumen. The experiment should be repeated with the same type of emulsion to confirm the trend for this particular emulsion. This experiment should be repeated with the same type of emulsion. The validation still needs to be further confirmed; the use of another test with more equivalent test conditions than these applied in the pull-off test should be considered as well.

4 Conclusion

In this paper, we presented a new method to evaluate the bond strength resistance of a bitumen film from cured emulsion based on the preparation of an emulsion film on a metal plate and the application of a square metal stamp on the surface. The results were compared to a standard method, the Layer Adhesion Measuring Instrument-LAMI used on asphalt specimens with the emulsion as tack coat. It has shown that the BRRC developed procedure leads to similar results for most of the bituminous emulsion tested, with a simpler and faster procedure and which can be carried out on the emulsion only without the need of asphalt samples. However, due to the small amount of bituminous emulsions tested, validation of the BRRC's pull-off test will need an extended research to compare emulsions regarding their characteristics. Although all in-situ parameters of bonding between pavement layers are not considered in this test, this is a promising first step in the development of a new method to quickly evaluate and compare the performance of bituminous emulsions in terms of cohesive strength.

Acknowledgements The authors also express their gratitude to the Belgian Bureau for Standardisation for the financial support (convention CC-CCN PN/NBN-808-958-14A06) as well as the cooperation received from the administration and the manufacturer of emulsion who provided us with samples and information. In addition the authors wish to thank gratefully C. Motte, O. Moens, P. Crabbé, A. Fondu from BRRC and G. Leclerc from MTQ for their technical support. The authors wish to thank generously the MTQ for the loan of the LAMI.

References

BRRC (2012) Les émulsions cationiques bitumineuses en tant que couches de collage - Recommandations pratiques de mise en œuvre Annexe au Bulletin CRR n° 90 Trimestriel: janvier – février – mars 2012 Dossier 14

Destrée A., De Visscher J., Vanelstraete A (2016) Field study to evaluate different pre-normative interlayer adhesion tests 6th Eurasphalt & Eurobitume Congress, 1– 3 June 2016 – Prague Congress Centre, Abstract accepted

Destrée A., Beaumesnil B. (2015) Evaluation du type de réparation bitumineuse le plus adéquat après l'exécution de l'essai de traction in situ avec l'AMAC – Planche d'essais sur le site du CRR à Sterrebeek, Bulletin CRR, N°103, pp 4-9, Trimestriel: avril– mai – juin

Eedula S., Tandon V. (2006) Tack coat field acceptance criterion. Texas Department of Transportation

Ministère des transports du Québec-MTQ (2011) Méthode d'essai LC 25-011 Mesure de la force de liaison avec un appareil de mesure d'adhésion des couches

Ministère des transports du Québec-MTQ (2014) Méthode d'essai LC 25-010 Mesure de la force de liaison avec un appareil de mesure d'adhésion des couches

Mohammad L.N. (2012) O ptimization of Tack Coat for HMA Placement. NCHRP Report 712, Transportation Research Board

Debonding in Airfield Pavement Hot Mix Asphalt Layers

Navneet Garg and Murphy Flynn

Abstract Pavement distresses are external indicators of pavement deterioration caused by loading, environmental factors, construction deficiencies, or combinations thereof. The Hot Mix Asphalt (HMA) design for commercial airports in United States of America (USA) is performed as per Federal Aviation Administration (FAA) Advisory Circular AC 150/5370-10G "Standards for Specifying Construction of Airports", Item P401—Plant Mix Bituminous Pavements (2014). According to FAA study "Operational Life of Airport Pavements", pavements that meet FAA standards and specification (for design, and construction) generally performed well over their design life. However, recently cases of debonding and slippage have been observed at airports in USA, and other countries. This distress was also observed in the full-scale accelerated pavement tests during flexible pavement construction cycles 1 and 3 at the FAA's National Airport Pavement Test Facility (NAPTF) and was attributed to tack coat and environmental factors. Asphalt strain gages were installed at the top of milled HMA surface and bottom of HMA overlay at Runway 4R-22L at Newark Liberty International airport (EWR) to study the debonding and slippage distress. Sensor responses were used to detect the onset of debonding and subsequent slippage in the HMA overlay. This paper summarizes the results from two case studies —HMA behavior under full-scale accelerated pavement tests at NAPTF, and the pavement instrumentation project at EWR airport.

Keywords Debonding · HMA · Strain gage · Airport

N. Garg (✉) · M. Flynn
Airport Technology R&D Branch, FAA William J. Hughes Technical Center,
Bldg. 296, Atlantic City, NJ 08405, USA
e-mail: Navneet.Garg@faa.gov

M. Flynn
e-mail: Murphy.flynn@faa.gov

A. Chabot et al. (eds.), *8th RILEM International Conference on Mechanisms of Cracking and Debonding in Pavements*, RILEM Bookseries 13,
DOI 10.1007/978-94-024-0867-6_78

1 Introduction

Pavement distresses are external indicators of pavement deterioration caused by loading, environmental factors, construction deficiencies, or combinations thereof. According to Federal Aviation Administration (FAA) study "Operational Life of Airport Pavements" (Garg et al. 2004), pavements that met FAA standards and specification (for design, and construction) generally performed well over their design life. However, recently cases of debonding and slippage in hot mix asphalt (HMA) layers have been observed at airports in USA, and some other countries. It generally occurs when there is a low-strength surface mix or a poor bond between the surface and the next layer of pavement structure. Slippage cracking (Fig. 1) occurs when braking or turning wheels cause the pavement surface to slide and deform. The slippage cracks, shown in Fig. 1b, are crescent- or half-moon-shaped cracks having two pointed ends away from the direction of traffic.

Delamination and slippage cracking was observed in full-scale accelerated pavement tests (Fig. 1a) on flexible pavement at the FAA's National Airport Pavement Test Facility (NAPTF). This paper summarizes the results from two case studies—HMA delamination under full-scale accelerated pavement tests (APT) at NAPTF, and the HMA overlay delamination at EWR international airport in USA. Asphalt strain gages were installed at interface of milled pavement and HMA overlay at Newark Liberty International Airport (EWR). Sensor responses were then used to detect the onset of debonding and subsequent slippage in the HMA overlay.

(a) (b)

Fig. 1 **a** Severe slippage cracking observed at NAPTF; **b** slippage observed at EWR airport

2 HMA Delamination Under Full-Scale APT at NAPTF

The FAA NAPTF is located at the FAA William J. Hughes Technical Center, Atlantic City, New Jersey. The primary objective of the testing performed at the NAPTF is to generate full-scale pavement performance and response data for development and verification of airport pavement design criteria. A construction cycle (CC) at the NAPTF includes test pavement construction, sensor installation, traffic tests to failure, post traffic trenching/tests, and pavement removal.

The most common distresses observed in the HMA surface layer were shoving at the longitudinal joint, fatigue (alligator) cracking, and tearing in the surface layer. Rutting in the HMA layer was minimal. Debonding/delamination between HMA layers was observed in Construction Cycles 1 (CC1) and 3 (CC3) (Fig. 1a). Inspection of cores extracted showed delamination/separation at the interfaces between lifts. Careful examination of the cores showed that all the cracks (except one core from the four-wheel traffic path) initiated from the top. Delamination in CC1 and CC3 was attributed to the tack coat. During CC1 and CC3, HMA was placed in two 63.5-mm lifts and were paved a day apart. An emulsion based tack coat was used between the two HMA layers. Since NAPTF is an indoor test facility, the tack coat did not get enough time to "break" and for water (in emulsion) to evaporate, resulting in delamination between the HMA layers. This delamination (along with high shear stresses at the location of wheels touchdown) resulted in severe tearing in the HMA surface layer and the deterioration of the longitudinal construction joint. Additional details about CC1 and CC3 tests can be found in Hayhoe et al. (2003), Hayhoe and Garg (2006). In subsequent flexible pavement construction cycles, asphalt binder was used as the tack coat between the two HMA lifts. No signs of delamination or debonding have since been observed.

3 HMA Delamination and Slippage at EWR Airport

Figure 1b shows slippage in HMA overlay at the intersection of Runway 4R-22L and high-speed taxiway exit N (HST-N) at EWR airport. Slippage cracking occurs due to high shear stresses induced by aircraft braking and turning, at high speed exits, which cause the pavement surface to slide and deform. Similar distresses at airport pavements have been reported in Netherlands and Australia. High pavement temperatures also contribute to slippage or delamination at HMA layer interface by reducing bond strength (Bognacki et al. 2007, and Collop et al. 2009).

In 2012 FAA Airport Technology R&D Branch, in partnership with Port Authority of New York and New Jersey (PANYNJ), initiated a pavement instrumentation project at EWR airport during the mill and HMA overlay construction project. EWR is one of the busiest airports in the country (serving about 33 million passengers every year) and 4R-22L is the primary landing runway. The main objective of this project was to detect the initiation of delamination and slippage in

HMA overlay. Since EWR has had a history of slippage failures in HMA overlay (Bognacki et al. 2007), several precautions were taken during the mill and overlay operation—milling without use of water to prevent slurry formation at the interface, thorough cleaning of milled surface, and increasing the application rate of tack coat. A 75 mm thick HMA overlay was paved over the milled HMA pavement. HMA mix was required to use a hard, angular aggregate to grip the milled surface.

Pavement instrumentation consisted of 32 H-bar asphalt strain gages (ASG), and 4 thermocouples to measure temperature within the asphalt overlay. Sixteen ASG's (8 longitudinal-LSG, and 8 transverse-TSG) were installed at the top of milled HMA pavement (also designated as base layer), and 16 ASG's (8 longitudinal and 8 transverse) were placed at the bottom of HMA overlay. In other words, the ASG's were installed on either side of the interface with separation of less than 25 mm between them. Figure 2 shows the ASG layout. The sensor (ASG) layout was developed considering Boeing 737-800/900 aircraft (most prevalent aircraft at EWR). The data acquisition system (DAS) at EWR provides the ability to acquire static and dynamic data from the individual sensors. This system is remotely controlled via internet with cellular communication. The temperature data is collected using a Campbell Scientific data logger. The main idea behind the development of ASG layout was that as long as the HMA overlay and existing milled pavement are fully bonded (no delamination) the magnitude and nature (compressive or tensile) of strains would be similar in ASG's installed on top of milled pavement and the bottom of HMA overlay. Once the delamination occurs at the interface, and slippage starts, significant change in strain magnitudes and nature will be observed. Figure 3 shows LSG measurements. Figure 3a shows the LSG-6 A & B strain response plots from July'2013 for pair of gages located at the same position but above and below the interface indicating strains of similar nature. No signs of delamination or slippage are observed. Strain responses from a year later

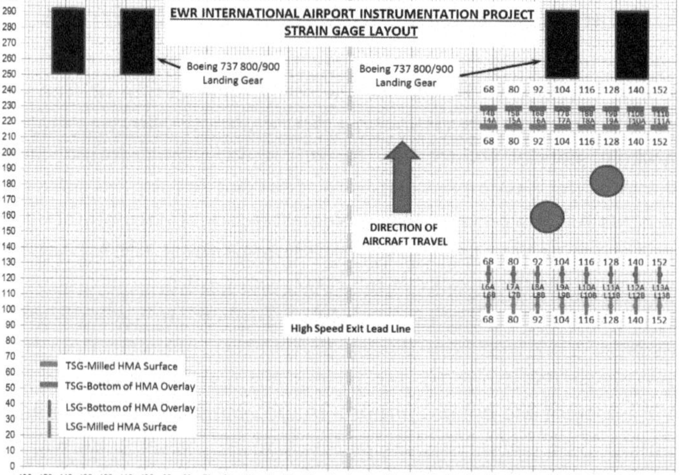

Fig. 2 ASG layout on runway 4R-22L at EWR airport

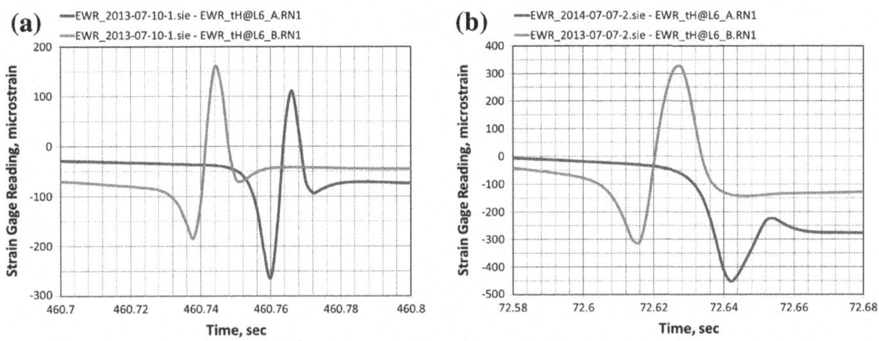

Fig. 3 LSG response measurements, **a** no delamination, **b** delamination

(July'2014) for the same pair of LSG shows that the nature of strain measurements has changed (Cook 2014). LSG-6A shows compressive strain and LSG-6B shows tensile strain which indicates delamination and slippage. Visual observations on the instrumented pavement showed slippage and is shown in Fig. 1b. Figure 4 shows correlation between strain responses from their respective pairs. In the pavement

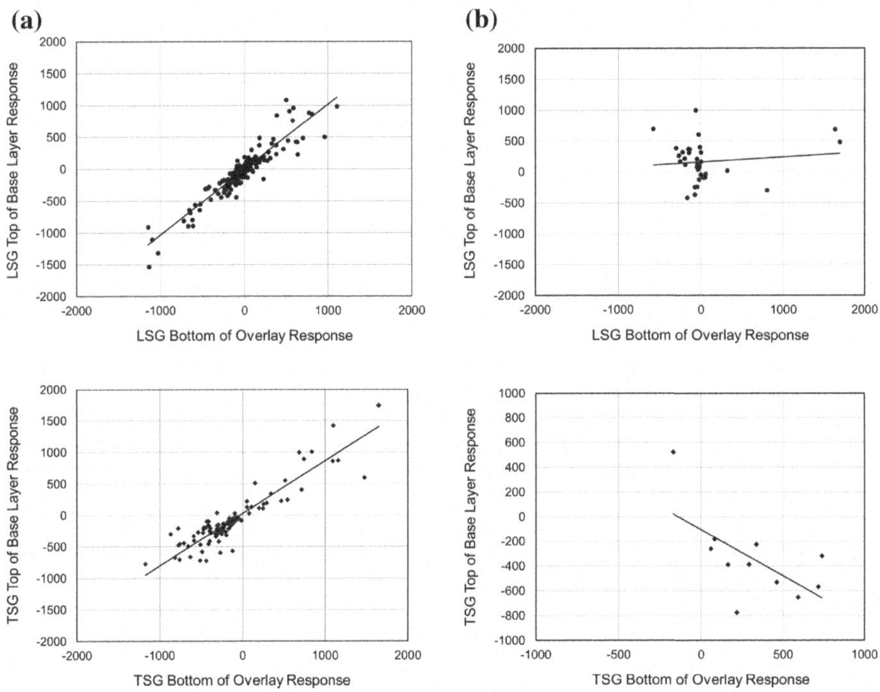

Fig. 4 Correlation between ASG response measurements (all readings in microstrain). **a** No delamination observed, **b** delamination observed

with fully bonded interface, a clear relationship is observed between the strain measurements in both layers (Fig. 4a). No correlation is observed in the case when delamination and slippage has occurred as shown in Fig. 4b. Additional details about the ASG measurements and data analyses can be found in Cook (2014).

4 Summary

Results from two case studies—HMA behavior under full-scale accelerated pavement tests at NAPTF, and the pavement instrumentation project at EWR airport are presented. Delamination can be construction/material related, or can be caused by high shear stresses induced by the loading conditions. In full-scale APT tests at NAPTF, the delamination and slippage cracks were attributed to construction/tack coat material. Asphalt strain gages were successfully used to detect the onset of delamination and slippage at EWR airport runway due to turning and braking action of the aircraft, and verified by field observations. A strong bond at HMA layer interface is necessary to prevent delamination and slippage distress.

Acknowledgements The work described in this paper was supported by the FAA Airport Technology R&D Branch. Support and coordination with the PANYNJ was integral to the successful completion of this research. The authors would like to acknowledge the support provided by Ms. Karissa Cook and Prof. Amarjit Singh in analyzing the data collected. The contents of the paper reflect the views of the authors who are responsible for the facts and accuracy of the data presented, and do not necessarily reflect the official views and policies of the FAA.

References

Karissa Cook (2014). Strain Gage Instrumentation Systems for Detecting Interlayer Delamination in Asphalt Airport Pavements. Thesis submitted to the Graduate Division of the University of Hawaii in partial fulfillment of the requirements for MS degree in Civil Engineering.

Garg, N., Guo, Edward, and McQueen, R. (2004). Operational Life of Airport Pavements. FAA Report DOT/FAA/AR-04/46.

Hayhoe, G. F., Garg, N., & Dong, M. (2003). Permanent Deformations During Traffic Tests on Flexible Pavements at the National Airport Pavement Test Facility, Proceedings, 2003 Airfield Pavement Specialty Conference "Airfield Pavements – Challenges & New Technology," Las Vegas, Nevada, USA.

Hayhoe, G. F., & Garg, N. (2006). Traffic Testing Results from CC3 Flexible Pavements at the FAA'S NAPTF, Proceedings, 10th International Conference on Asphalt Pavements, Quebec City, Quebec, Canada.

Bognacki, C. J., Frisvold, A., and Bennert, T. (2007). "Investigation of Asphalt Pavement Slippage Failures on Runway 4R-22L, Newark International Airport." Proc., 2007 FAA Worldwide Technology Transfer Conference. Atlantic City, New Jersey.

Collop, A.C., Sutanto, M.H., Airey, G.D., and Elliot, R.C. (2009). "Shear Bond Strength between Asphalt Layers for Laboratory Prepared samples and Field Cores." Constr. Build. Mater. 23(6), 2251-2258.

Impact of New High Inflation Pressure Aircraft Tyres on Asphalt Overlay Interface Debonding

Greg White

Abstract Aircraft tyre inflation pressures and wheel loads have increased over time and this trend is not expected to abate in the future. With increased tyre inflation pressures has come increased shear forces induced by braking aircraft. An increase in reports of interface debonding and delamination failures of asphalt surfaces on runways has followed in Australia, Japan and South Africa. Octahedral shear and normal stresses were calculated for extreme braking aircraft of various tyre inflation pressures and wheel loads. The historical increase in tyre inflation pressure of around 50 % was shown to have resulted in a moderate 10 % increase in the stress-strength ratio associated with the surface layer interface. The factor of safety associated with interface shear strength was shown to be negligible. New interface construction techniques or tack coat materials with increase adhesion are required to reduce the risk of further shear-related delamination of asphalt surfaces in high stress applications such as runways.

Keywords Tyre pressure · Shear stress · Interface resistance · Asphalt delamination

1 Introduction

Since their first introduction in the early 1900s, aircraft have become ever larger and heavier. Particularly since WWII, aircraft wheel loads and tyre inflation pressures have increased significantly (Roginski 2007; Fabre et al. 2009). The trend to higher aircraft tyre pressures is not expected to abate in the near future. At the same time, the pressure to minimise runway occupancy times during landing operations has led to increased use of rapid exit taxiways (White 2014). As a result, aircraft braking forces have increased and will continue to increase in the future.

G. White (✉)
University of the Sunshine Coast, Sunshine Coast, QLD, Australia
e-mail: greg.white@fultonhogan.com.au

A. Chabot et al. (eds.), *8th RILEM International Conference on Mechanisms of Cracking and Debonding in Pavements*, RILEM Bookseries 13,
DOI 10.1007/978-94-024-0867-6_79

These increases in tyre inflation pressures and the associated increase in shear stresses are not directly considered during routine pavement design. Asphalt overlay specification and construction practices for airport overlays have not changed substantially since their development in the 1960s. The link between these practices and sound performance is empirical. The increase in aircraft induced shear forces is testing this empirical approach. Current practice is for airport interfaces between asphalt overlays to incorporate:

- Texturing the existing surface to remove the top 5–10 mm of aged asphalt.
- Minimising the milling water to reduce the generation of slurry on the surface.
- Thorough cleaning and drying of the milled pavement surface.
- Application of 0.20 l/m^2 (residual bitumen) as an emulsified tack coat.
- Minimisation of trafficking on the surface to reduce pick-up of tack coat.

This practice has generally resulted in good interface bond with relatively few reported surface failures attributed to delamination. The aim of this research was to assess the theoretical impact of higher tyre inflation pressures associated with new aircraft on the risk of debonding failure of asphalt overlays. Calculated shear stresses are compared to measured shear strengths from typical overlays.

2 Research Methods

The software mePADS/GAMES (Maina and Matsui 2004) was selected for this research to calculate shear stresses under an extreme aircraft braking of differing tyre inflation pressures. The Octahedral Shear Stress (OSS) on the horizontal plan was used as the primary indicator of interface shear. OSS was calculated at a depth of 45 mm below the surface, to represent the near-interface stresses of a nominal 50 mm surface layer. OSS values were calculated under and in front of the aircraft tyre. The Octahedral Normal Stress (ONS) was also calculated and used to subsequently calculate the interface shear strength and associated stress–strength ratio. This ratio is inverse to the factor of safety for delamination.

2.1 Inputs

Table 1 summarises the tyre inflation pressures and loads adopted to represent the past, present and future aircraft. The selection of an aircraft for 'present' times is arbitrary and based on what is considered to be 'common' in Australia.

White (2015) investigated braking forces of various aircraft under normal and extreme landing conditions at a major Australian airport. The calculations were based on information gathered from experienced pilots. The resulting (single wheel) vertical and horizontal surface forces and contact stress during extreme braking by a B767 aircraft were calculated to be:

Table 1 Representative tyre inflation pressures and wheel loads

Representation	Tyre inflation pressure (MPa)	Wheel load (kN)
Past aircraft	1.10	130
Present aircraft	1.35	160
Future aircraft	1.65	195

- Vertical force. 160 kN.
- Horizontal force. 52 kN.
- Tyre contact stress. 1.35 MPa.

Note that mePADS/GAMES assumes that contract stress is uniformly equal to tyre inflation pressure. The following pavement structure was adopted as being typical of a capital city airport on a subgrade of low strength in Australia:

- Asphalt layers. 3 by 50 mm thick with modulus (E) of 3500 MPa and Poisson's ratio (ν) of 0.44.
- Fine crushed rock base. 250 mm, E = 300 MPa, ν = 0.35.
- Fine crushed rock sub-base. 1150 mm, E = 150 MPa, ν = 0.35.
- Subgrade. Design CBR (at MMDD) of 3 %, E = 30 MPa, ν = 0.35.

Typical airport asphalt overlays were cored and the interfaces tested in direct shear at various normal stresses (White 2014). This allowed Mohr-Coulomb envelopes of shear strength to be generated as shown in Fig. 1, from which the cohesion (c) and friction angle (ϕ) values for the interface shear strength were calculated. Values one standard deviation below the average were selected as being representative of the majority of any typical asphalt overlay. This resulted in c = 257 kPa and ϕ = 23° being adopted as representative, which is also shown.

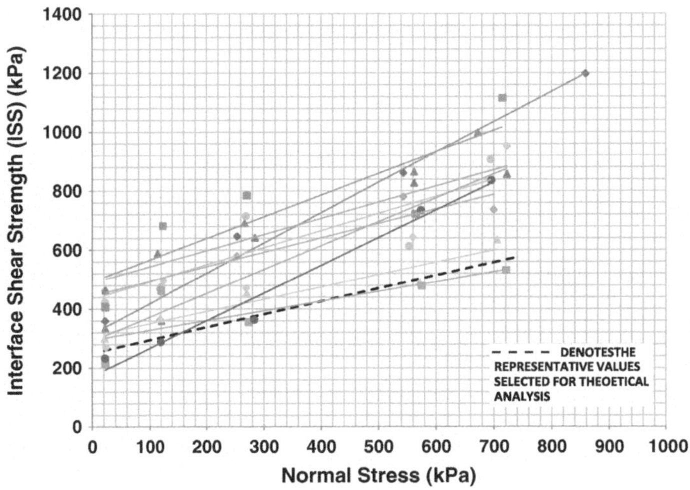

Fig. 1 Representative interface shear strength results from direct shear testing

2.2 General Stress Distributions

The general distribution of horizontal shear stress is shown in Fig. 2 for a tyre inflation pressure of 1.35 MPa. The centre of the tyre defaults to the origin (X = 0, Y = 0). X is the longitudinal direction of travel and Y is the transverse direction. The braking force was applied in the positive X to simulate aircraft braking. The depth below the surface of the pavement is represented by Z.

The effect of braking can be seen in Fig. 2 where the shear stress at the leading edge of the tyre (851 kPa) is significantly higher than the equivalent shear stress at the trailing edge (589 kPa). The zone of constant shear stress near the surface under the braking tyres is clear. Figure 3 shows the impact of tyre inflation pressure on near-surface shear stress. For an increase in tyre pressure from 1.10 to 1.65 MPa the maximum shear stress increased from 721 to 1011 kPa. The shear stress between the leading and trailing edges of the tyre also increased significantly.

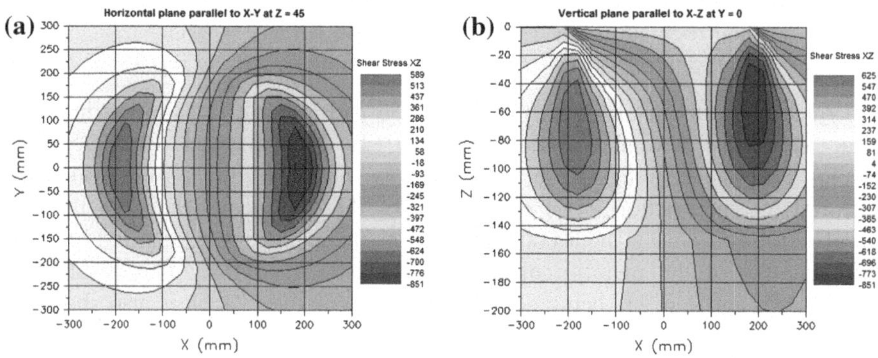

Fig. 2 Calculated shear stresses in for 1.35 MPa tyre pressure. **a** Horizontal (Z = 45) **b** vertical-longitudinal (Y = 0)

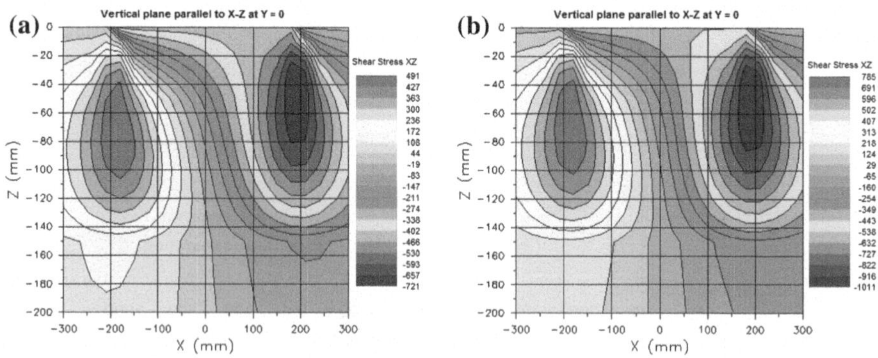

Fig. 3 Calculated shear stresses for different tyre inflation pressures. **a** Past Aircraft (1.10 MPa) **b** Future aircraft (1.65 MPa)

2.3 Calculated OSS at the Interface

Calculated OSS values at the interface (Z = 45) are presented in Table 2 and Table 3 for a 1.10 and 1.65 MPa tyre inflation pressures, representing the past and future aircraft respectively. The OSS values for 1.35 MPa tyre inflation pressures followed the same trend. The 50 % increase in tyre inflation pressure from 1.10 to 1.65 MPa resulted in a 30 % increase in maximum OSS. At this point the calculated ONS was also relatively high due to the influence of the leading edge of the tyre.

2.4 Impact of Tyre Inflation Pressure on Interface Debonding

Stress-strength ratios were calculated for the various aircraft. In all cases, the critical location was approximately 50 mm in front of the centre of the leading edge of the aircraft tyre (X = 250, Y = 0). At this location OSS remained close to its maximum

Table 2 Calculated OSS for past (1.10 MPa)

Y (Transverse) (mm)	X (longitudinal in mm)								
	0	60	120	195	210	250	300	350	400
0	184	241	263	423	484	475	359	270	211
100	156	214	268	468	471	408	315	245	196
190	263	323	394	355	338	292	240	199	166
210	317	357	374	322	308	269	225	188	159

Table 3 Calculated OSS for future (1.65 MPa)

Y (Transverse) (mm)	X (Longitudinal in mm)								
	0	60	120	195	210	250	300	350	400
0	266	315	321	548	632	615	458	343	266
100	220	276	339	621	623	533	406	314	249
190	385	460	550	481	456	389	316	258	214
210	469	514	524	438	416	359	296	245	205

Table 4 Summary of calculated stress measures for various tyre inflation pressures (Z = 45)

Measure	Aircraft representing		
	Past	Present	Future
Tyre pressure (MPa)	1.10	1.35	1.65
Maximum OSS (kPa)	484	550	632
Maximum ONS (kPa)	1270	1541	1862
Stress–Strength × 100 (%)	0.89	0.93	0.98

value and ONS was minimal in front of the leading edge of the tyre. The increase in tyre inflation pressure from 1.10 to 1.65 MPa resulted in an increase in calculated stress–strength from 0.89 to 0.98 (10 %). This was relatively modest for the 50 % increase in the tyre inflation pressure. Table 4 summarises the tyre pressures, calculated OSS, ONS and stress–strength ratios.

3 Conclusions

Aircraft tyre inflation pressures and associated braking forces will continue to increase as aircraft continue to evolve. While the maximum shear stress was calculated to occur immediately under the leading edge of the tyre, the critical location, based on stress–strength was approximately 50 mm in front of the tyre, where the normal stress was minimal. Shear stresses are consistently maximum between 30 and 80 mm below the surface and a high-quality asphalt mixture is required through this depth range to resist the high stress state.

Maximum shear stress and the calculated stress–strength ratio increased with increasing tyre inflation pressure. Surprisingly, an increase in tyre inflation pressure of 50 % resulted in a stress–strength increase of only 10 %. The calculated stress–strength ratio suggested that the factor of safety for bond failure is negligible under extreme braking events during hot weather. It should be noted that extreme braking usually only occurs in emergency situations and the factor of safety would be significantly greater during normal aircraft landing.

To reduce the risk of delamination failures in the future, either improved interface construction techniques are required or aircraft tyre inflation pressure increases must be curbed. Limiting new aircraft tyre pressures would seem unlikely to be achievable given the drive for more efficient commercial passenger aircraft design.

References

Fabre, C, Balay, J, Lerat, P & Mazars, A 2009, 'Full-scale aircraft tire pressure test', *Proceedings Eight International Conference on the Bearing Capacity of Roads, Railways and Airfields*, Urbana-Champaign, Illinois, USA, 29 June - 2 July, pp. 1405-1413.

Maina, JW & Matsui, K 2004, 'Development of software for elastic analysis of pavement structure due to vertical and horizontal surface loadings', *Proceedings 83rd Meeting of the Transport Research Board*, Washington, DC, USA, 11-15 January, Transport Research Board.

Roginski, MJ 2007, 'Effects of aircraft tire pressures on flexible pavements', *Proceedings Advanced Characterisation of Pavement and Soil Engineering Materials*, Athens, Greece, 20-22 June, Taylor and Francis, pp. 1473-1481.

White, G 2014, 'Cyclic shear deformation of asphalt at Melbourne Airport', *Proceedings 2014 Worldwide Airport Pavement Technology Transfer Conference*, Galloway, New Jersey, USA, 5-7 August, Federal Aviation Administration.

White, G 2015 'Shear Stresses in an Asphalt Surface under Various Aircraft Braking Conditions', *International Journal of Asphalt Research and Technology*, submitted but not yet published.

Localized Debonding as a Potential Mechanism for Near-Surface Cracking

David Hernando, Jeremy A. Magruder, Jian Zou and Reynaldo Roque

Abstract Research on the potential link between debonding and cracking has primarily focused on changes in location and magnitude of maximum tensile stress/strain under the load center for various interface bonding conditions. Prior modeling efforts have evaluated smeared bonding conditions, which represented the entire interface as either bonded, partially bonded or debonded. However, field evidence indicates debonding is a local phenomenon that starts at a critical location and gradually progresses through a portion of the interface. A parametric study identified a critical zone of high shear stress coupled with low confinement where the onset of debonding is likely. This critical zone is located around the mid-depth of the asphalt layer and extends for nearly 5 cm from the edge of the tire. Localized debonding should be introduced at the identified critical zone to better evaluate stress redistribution resulting from interface failure. Localized debonding was shown to potentially lead to stress states of shear-induced tension in the portion of the interface near the tire edge, which can explain the initiation of near-surface longitudinal cracking.

Keywords Debonding · Interface · Surface cracking · Stress redistribution

1 Introduction

Poor bonding conditions between asphalt layers can lead to three major distresses: slippage, delamination and cracking. Slippage is characterized by crescent or half-moon shaped cracks with two ends pointed in the direction of traffic resulting from failure of the uppermost interface located directly beneath the surface layer. Furthermore, it is associated with relatively thin surface layers (typically <4 cm) and single events that cause high horizontal shear stress, such as braking.

D. Hernando (✉) · J.A. Magruder · J. Zou · R. Roque
Department of Civil and Coastal Engineering,
University of Florida, Gainesville, USA
e-mail: dhernando@gmail.com

A. Chabot et al. (eds.), *8th RILEM International Conference on Mechanisms of Cracking and Debonding in Pavements*, RILEM Bookseries 13,
DOI 10.1007/978-94-024-0867-6_80

Delamination can result from severe slippage and disintegration of the mixture for causes not related to the interface (moisture damage and oil spills). Cracking is likely to be promoted by failure of an interface deeper in the pavement structure (no less than 4 cm). Whereas slippage may result from a single high load event, cracking involves progressive breakdown of the interface under repeated traffic loads.

In terms of cracking, most research has focused on the effect of debonding on tensile stress/strain distribution and the potential acceleration of bottom-up cracking. However, Willis and Timm (2007) showed near-surface cracks in concert with localized interface debonding at the National Center for Asphalt Technology (NCAT) test track. Studies by Muench and Moomaw (2008) in Washington State support observations at NCAT test track. Extracted cores showed interface debonding only in areas of the wheelpath with near-surface longitudinal cracking. Cores removed outside the wheelpath showed no debonding, cracking or any form of deterioration. Another finding was that average depth to the first debonded interface was 4.5 cm, which closely corresponded to the typical overlay thickness. Roque et al. (2011) also found debonding concurrent with near-surface cracking in the wheelpath on cores extracted as part of a Superpave monitoring project (Fig. 1).

Field observations associated near-surface cracking in the wheelpath with localized interface debonding. However, the question remains as to whether near-surface cracking starts first and promotes debonding or interface breakdown leads to near-surface cracking. Further research is needed to identify a potential mechanism that connects debonding and near-surface cracking.

Fig. 1 Presence of near-surface longitudinal cracking concurrent with interface debonding in I-10 interstate highway (Duval County, Florida)

2 Objectives

The objectives of this study were:

- Locate critical stress states in un-cracked pavement sections that may lead to the onset of localized interface debonding prior to near-surface cracking.
- Identify a potential mechanism connecting localized interface debonding and near-surface cracking.

3 Identification of Critical Location for Onset of Debonding

A parametric study was conducted on a three-layer pavement to locate critical stress states that may lead to the onset of debonding. Factors considered included thickness of the asphalt layer and stiffness of the asphalt and base layers. Subgrade stiffness was fixed at 100 MPa. Poisson's ratio values of 0.35, 0.40 and 0.45 were assumed for asphalt, base and subgrade layers, respectively. As shown in Fig. 2, five cases were selected: four of them defined by different stiffness ratios (SR) between the asphalt and base layer, and a fifth case in which a gradient in asphalt stiffness was introduced to simulate temperature and aging conditions in the field.

The multilayer linear elastic analysis program BISAR was employed to obtain responses at selected locations. A circular load of 40 kN with a radius of 13.5 cm was chosen. The analysis focused on locating critical stress states, defined by high shear stress combined with low confinement levels, where the onset of debonding is likely. All interfaces were assumed fully bonded. Results are shown in Fig. 3.

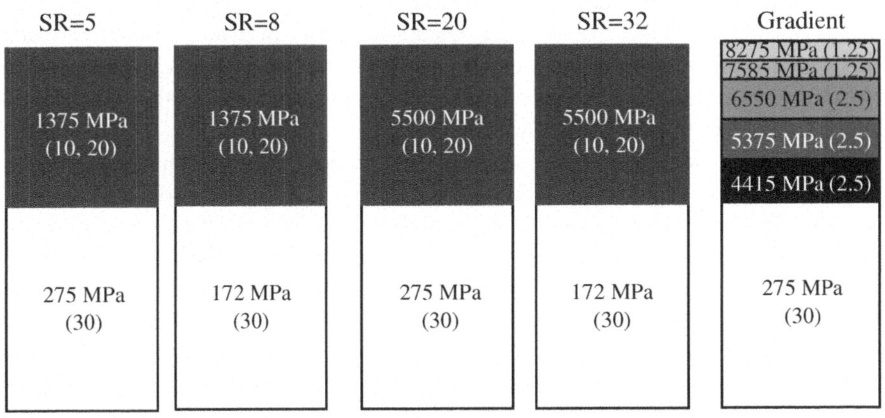

Fig. 2 Pavement sections selected for parametric study (*brackets* contain layer thickness in cm)

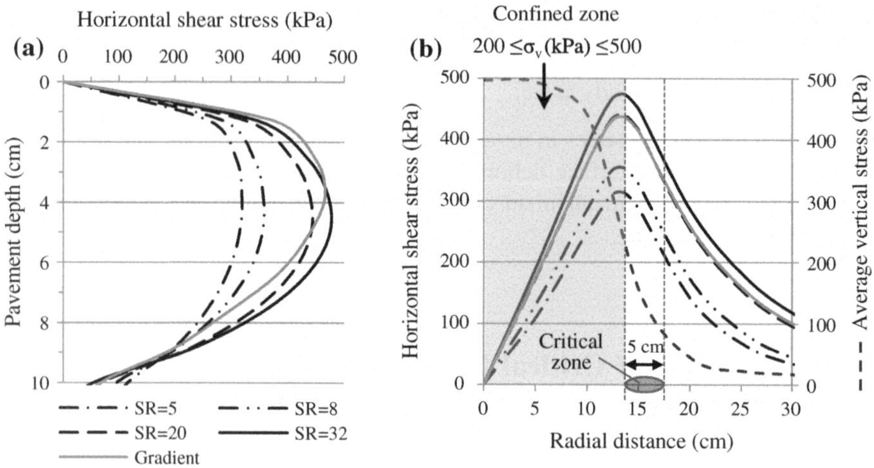

Fig. 3 Results from parametric study. **a** Horizontal shear stress through the depth of the asphalt layer under the tire edge. **b** Horizontal shear stress and average vertical stress at a depth of 5 cm

Figure 3a illustrates the horizontal shear stress through the depth of the asphalt layer under the edge of the tire (i.e., at a radial distance of 13.5 cm). High shear stress was observed for a broad range of depths around the center of the layer (2–8 cm for a 10-cm structure), with a maximum at a depth slightly less than 5 cm. This agreed with the depth at which debonding was observed in the field by Muench and Moomaw (2008) and Roque et al. (2011). Maximum shear stress occurred for SR = 32. This was expected as shear stress is a function of bending, which increases with SR.

Figure 3b shows the horizontal shear stress distribution at the mid-depth of the asphalt layer (5 cm). The maximum shear stress was under the edge of the tire for all cases. Figure 3b also illustrates the existence of an area of high confinement under the tire (confined zone), but confinement drastically decreased below 200 kPa outside the edge of the tire. Similar results in terms of stress distribution and location of critical stress states were found when AC thickness was increased to 20 cm; however, stress magnitude was lower (less critical) due to the reduction in bending resulting from a thicker section.

These results identified the existence of a critical zone of high shear stress coupled with low confinement around the mid-depth of the asphalt layer and extending for nearly 5 cm from the edge of the tire. However, it is important to be mindful of traffic wander, in addition to different tire configurations (e.g., wide base and dual tires), which can significantly increase the extent of the critical zone. Presence of an interface between asphalt layers through this critical zone could potentially lead to localized debonding. Furthermore, factors occasionally found in the field, such as uneven tack coat distribution, tack tracking and thin debonded layers left after milling, can help traffic loads to promote localized failure.

4 Evaluation of the Effect of Interface Conditions on Stress Redistribution

Most studies have evaluated the effect of interface conditions on stress redistribution through smeared bonding conditions (Uzan et al. 1978; Maina et al. 2007). These studies represent the entire interface as either bonded, debonded or partially bonded by the introduction of an interface reaction modulus (K), which is defined as the ratio of shear stress to relative horizontal displacement and represents interface compliance. Some studies have added a failure criterion to the interface response (Romanoschi and Metcalf 2001; Ozer et al. 2008). Figure 4 conceptually depicts the location and orientation of maximum tensile stress in the asphalt layer for various interface conditions.

As interface compliance increases from globally bonded (Fig. 4a) to globally debonded (Fig. 4b), shear stress transferred across the interface reduces to zero. Reduction in transferred shear stress results in tensile stress above the interface. Although this may be interpreted as an increased susceptibility to cracking, the area under the tire is subjected to high vertical stress (confined zone), which allows for the development of horizontal frictional resistance along the interface. The development of horizontal frictional resistance makes the presence of a debonded interface through the confined zone less critical.

Figure 4c shows an alternative method of evaluation based on the introduction of localized debonded areas at the critical locations identified in Sect. 3. The presence of localized debonding causes a redistribution of stress and potentially intensifies the shear stress ahead of the tip of the debonded zone (Detail 1). The increase in shear stress outside the confined zone leads to stress states where tension is present (shear-induced tension), as illustrated by the Mohr's circle in Detail 2. This scenario indicates the location of critical tension is the localized debonded area

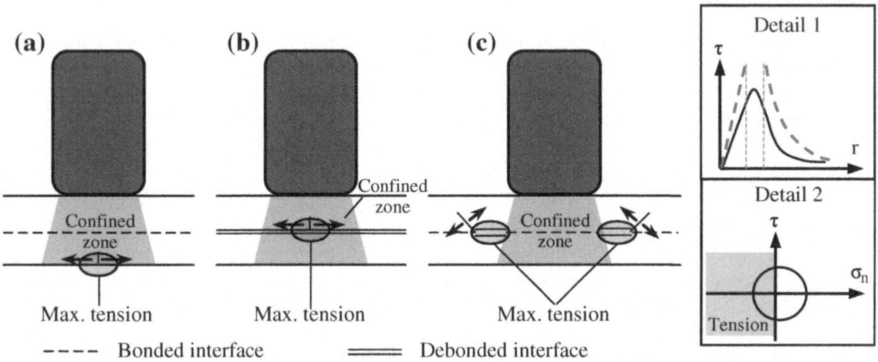

Fig. 4 Location and orientation of maximum tensile stress in the asphalt layer for various interface conditions. **a** Global bond. **b** Global debonding. **c** Localized debonding (*Detail 1* stress intensification as a result of localized debonding; *Detail 2* Mohr's circle showing shear-induced tension)

of the interface near the tire edge and not underneath the center of the tire. This finding can explain the initiation of cracking near the tire edge, which agrees with near-surface longitudinal cracks observed in the field.

5 Conclusions and Recommendations for Future Research

Field observations indicate that near-surface longitudinal cracking is associated with localized interface debonding. A parametric study identified the existence of a zone of high shear stress coupled with low confinement, where the onset of debonding is likely. This zone is located around the mid-depth of the asphalt layer and extends for nearly 5 cm from the edge of the tire, although the extent could be greater due to the effect of traffic wander and different tire configurations. Furthermore, after evaluation of previous efforts on the effect of interface conditions on stress redistribution, it was concluded that use of smeared bonding conditions is not suitable for analysis of the link between localized interface debonding and near-surface cracking. Localized debonding should be introduced at the identified critical zone to better evaluate stress redistribution resulting from interface failure. Localized debonding was shown to potentially lead to stress states of shear-induced tension in the portion of the interface near the tire edge, which can explain the initiation of near-surface longitudinal cracking.

References

Maina JW, De Beer M, Matsui K (2007) Effects of layer interface slip on the response and performance of elastic multi-layered flexible airport pavement systems. Proceedings of the 5th international conference on maintenance and rehabilitation of pavements and technological control, Utah, August 8-10: 145-150.

Muench ST, Moomaw T (2008) De-bonding of hot mix asphalt pavements in Washington State: an initial investigation. Report No. WA-RD 712.1. WSDOT, Olympia.

Ozer H, Al-Qadi IL, Leng Z (2008) Fracture-based friction model for pavement interface characterization. Transportation Research Record 2057: 54–63.

Romanoschi SA, Metcalf JB (2001) Characterization of asphalt concrete layer interfaces. Transportation Research Record 1778: 132-139.

Roque R, Chun S, Zou J, Lopp G, Villiers C (2011) Continuation on Superpave projects monitoring. Final report BDK-75-977-06. FDOT, Tallahassee.

Uzan J, Livneh M, Eshed Y (1978) Investigation of adhesion properties between asphaltic concrete layers. Proceedings of the annual meeting of the Association of Asphalt Pavement Technologists 47: 495–521.

Willis JR, Timm DH (2007) Forensic investigation of debonding in rich bottom pavement. Transportation Research Record 2040: 107-114.

Experimental Evidence of the Viscoelastic Behavior of Interfaces in Bituminous Pavements—An Explanation to Top-Down Cracking?

Damien Grellet, Guy Doré, Olivier Chupin and Jean-Michel Piau

Abstract Top-down cracking is known to be initiated near the surface of the asphalt layers. The aim of this paper is twofold: (i) to show experimental evidence of the viscoelastic behavior of interface in asphalt pavements under some temperature conditions, (ii) to show that the integration of such a behavior could provide an explanation of the mechanism involved in the initiation of bottom-up and top-down cracking. This paper documents the methodology used to investigate the behavior of the upper interface from experimental tests carried out at the IFSTTAR's track facility. The mechanical behavior of the experimental pavement is evaluated using three different models: (1) elastic model, (2) Huet-Sayegh viscoelastic model to account for the behavior of asphalt layers, and (3) same as 2, but with additional very thin viscoelastic layers to represent interfaces between the asphalt layers. Computations are performed for two load configurations using software ViscoRoute 2.0© to evaluate the stresses and strains at different depths of the pavement for the 3 models. The comparison between the test track results and the models clearly show that model 3 is that yielding the best fit. This model shows that significant tensile stresses and strains occur near the surface and at the interface between two asphalt layers. The transposition of the viscoelastic behavior of interfaces to real traffic conditions could explain top-down cracking as one of the modes of failure of asphalt pavements.

Keywords Cracking · Viscoelasticity materials · Strain measurement · Interface modeling

D. Grellet (✉) · G. Doré
Department of Civil Engineering, Laval University, Québec, QC, Canada
e-mail: damien.grellet.1@ulaval.ca

G. Doré
e-mail: guy.dore@gci.ulaval.ca

O. Chupin · J.-M. Piau
IFSTTAR, LUNAM Université, 44344 Bouguenais Cedex, France
e-mail: olivier.chupin@ifsttar.fr

J.-M. Piau
e-mail: jean-michel.piau@ifsttar.fr

A. Chabot et al. (eds.), *8th RILEM International Conference on Mechanisms of Cracking and Debonding in Pavements*, RILEM Bookseries 13,
DOI 10.1007/978-94-024-0867-6_81

1 Introduction

Two types of cracking deterioration resulting from traffic loading under various climatic conditions are commonly observed in flexible pavements. First, the bottom-up cracking is initiated at the bottom of the asphalt layer and propagates toward the surface. Then, the top-down cracking is described as initiating near the surface and propagating downward through the upper bound layer(s). Tamagny et al. (2004) recommend the use of viscoelastic properties of asphalt materials in evaluating flexible pavement response. In particular, their works suggest that the initiation of top-down cracking of asphalt pavements could be better explained by considering the viscoelastic behavior of bituminous materials. We show in this paper that interfaces must also be considered as viscous to account for the behavior of asphalt pavements under high temperature conditions. This evidence stems from the comparison between field experiments with specific monitoring and numerical simulations of the pavement response under different assumptions. This finding is used for revisiting the interpretation of top-down cracking.

2 Experimental Context

The experimental campaign sustaining this work is part of a collaborative project between Laval University (Quebec City, Canada) and IFSTTAR (Nantes, France). The field tests have been conducted at the IFSTTAR's accelerated pavement testing facility. Two retrofit techniques, which allow measuring strains in each asphalt layer under different loading and environmental conditions have been used to instrument four pavement structures. The fiber optic strain instrumentations consist of an instrumented asphalt concrete core and a thin polymeric plate instrumented and fixed inside a saw cut in the asphalt layer (Grellet et al. 2010). The use of these techniques allows measuring the strains in three directions: longitudinal (ε_{xx}), transversal (ε_{yy}) and vertical (ε_{zz}). The depths of the sensors were chosen to measure strains in the upper (15–20 mm below the surface) and lower (5 mm from the bottom) parts of the asphalt layer and at both sides of the interface between two layers (5 mm above and below the interface). The configuration of the sensors for the IFSTTAR structure is detailed in Grellet et al. (2012). Results from these studies show that optic sensors make it possible to adequately characterize the strains occurring within the layer and evaluate the effects of the load configurations (Grellet et al. 2013).

3 Pavement Modeling

3.1 Model Assumptions and Material Properties

Calculations are performed using software ViscoRoute 2.0© (Chabot et al. 2010). The 2 structures are represented by multilayered half-spaces. They have the same base and soil, but different asphalt concrete thicknesses. Structure A consists of 70 mm of HMA N°1, 60 mm of HMA N°2 over a 300 mm granular base and sandy subgrade soil. Structure B has three asphalts layers: 80 mm of HMA N°1, 90 mm of HMA N°2 and 80 mm of HMA N°2. The soil and base layer are assumed to be linear elastic. The mechanical behavior of the asphalt layers is modeled by using either linear elasticity or the Huet-Sayegh viscoelastic constitutive law. Perfect bonding is assumed between all layers (thick or thin) of the models. Three different models are used:

- Model 1, referred to as "elastic": all layers are elastic (usual approach).
- Model 2, referred to as "viscoelastic": asphalt layers are viscoelastic.
- Model 3, referred to as "interface": same as model 2, but also including thin (0.5 mm) asphalt layer(s) (tack coat) with viscoelastic properties of bitumen in order to simulate the mechanical behavior of the interface(s).

Complex moduli were measured on two of the tested bituminous materials following the French testing standards (NF EN 12697-26): HMA N°1 and HMA N° 2. The first is an asphalt concrete 0/10 with 5.48 % of 35/50 binder. The second is an asphalt concrete 0/14 with 4.49 % of 35/50. The laboratory data were used to fit the eight parameters of the Huet-Sayegh constitutive law used in models 2 and 3. A similar characterization was also carried out for the interface bitumen of model 3. The equivalent elastic moduli in model 1 are deduced using asphalt temperature, speed of the load and the master curve.

The load is assumed to move at a constant speed of 42 km/h and is discretized into eight rectangular shaped surfaces. Each rectangular load represents one of the tire tread footprints. Only a vertical load is applied over the rectangular areas. Each tridem axle is separated by 1.5 m.

3.2 Comparison Between the Strain Measurements and the Simulation Results

Figure 1 shows 15 strain basins at different depths for 3 different temperatures (and two load levels), each plotted from 55 experimental points of measurements obtained on the thin test pavement of IFSTTAR. The 3 models lead to relatively similar results close to the pavement surface (z = 15 mm, 20 mm) or close to the bottom of the base course (z = 125 mm) and provide a quite good agreement with the experimental data. The effect of the footprint (tire tread) configurations can be

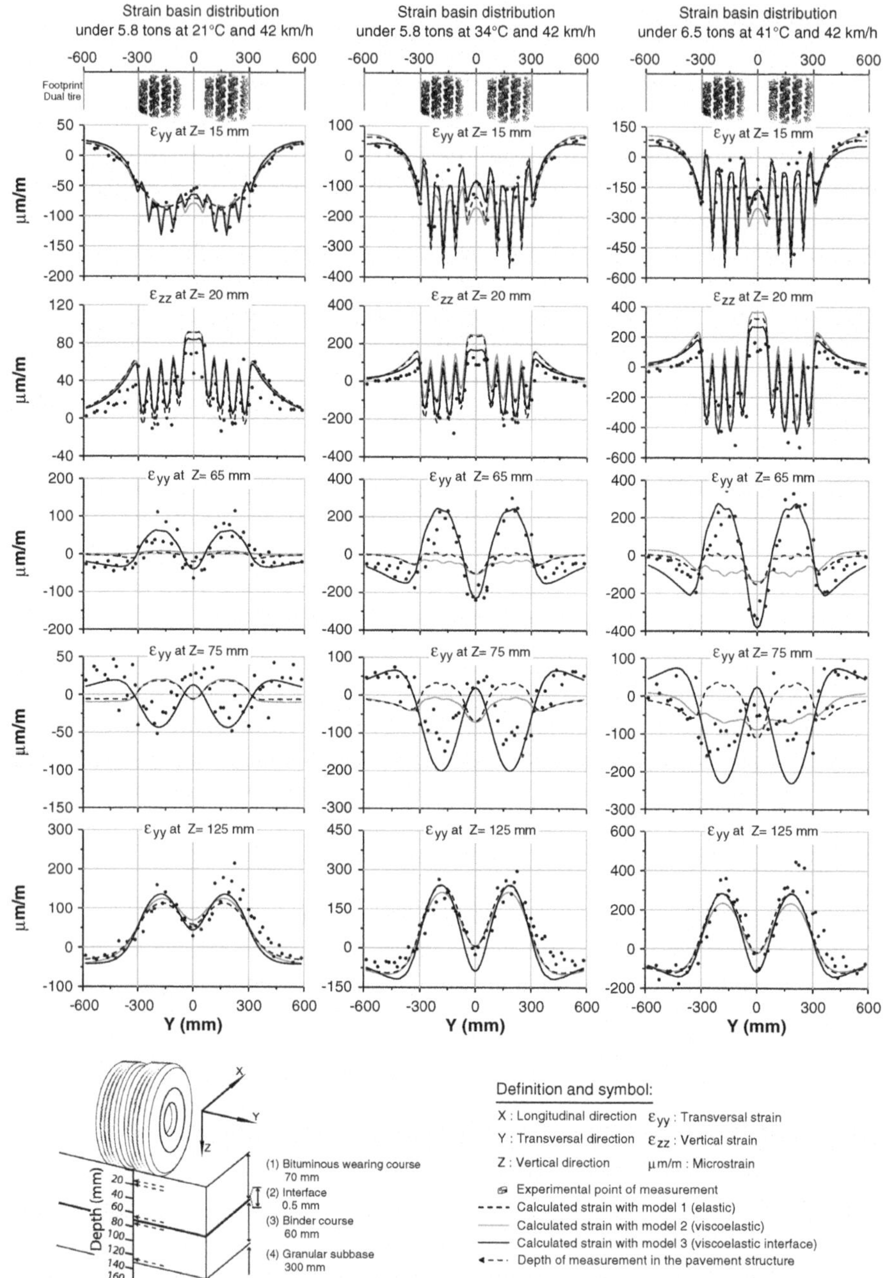

Fig. 1 Experimental and calculated strain basin for three conditions (dual tires on structure A)

observed near the surface. However models 1 and 2 are not able to match the strain measurements apart from the interface, which show high levels of extension at z = 65 mm in phase opposition with the high levels of contraction measured at z = 75 mm.

Only model 3 can reproduce these observations providing an appropriate choice for the thickness of the tack coat layer. Among the four thicknesses tested (5, 2, 1 and 0.5 mm), the best fit was obtained for 0.5 mm. The effect of temperature is well accounted for using this thickness combined with the Huet laboratory-fitted parameters for the bitumen; the higher the temperature the softer the interface. This analysis shows that attention must be paid at the behavior of interfaces when analyzing the mechanical response of asphalt pavements, especially at high temperature. The next part shows how a viscoelastic interface might provide an explanation to the so-called top-down cracking mechanism.

4 Application: Stress/Strain Calculations in the Surface Layer of Thick Pavements

Based on findings from some research projects, pavement structures with thicker asphalt layers located in hot climatic conditions are more affected by top-down cracking. Structure B is then selected to appraise this mechanism considering a uniform temperature of 30 °C and a tridem axle.

Figure 2 presents the results obtained using models 2 and 3 at two depths: at the bottom of the first layer (z = 78 mm) and at the bottom of the third layer

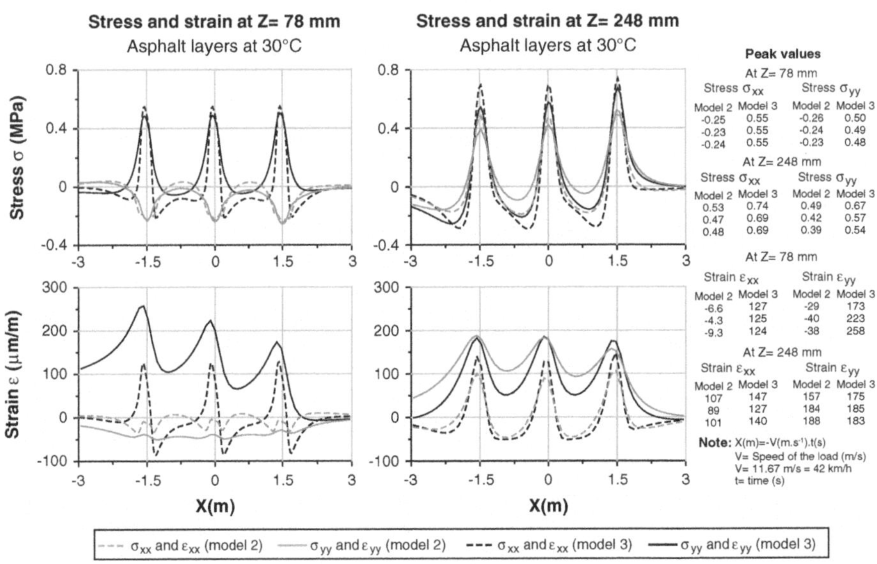

Fig. 2 Calculated stress and strain at two depths in structure B for a tridem axle

(z = 248 mm). In terms of strains, the same comments as made above for the thin structure are applicable. In particular high extensional strains (more than 250 $\mu\varepsilon$) are observed at z = 78 mm above the interface for model 3, which cumulate in the transversal direction with the passing of wheels. In parallel tensile stresses up to 0.6 MPa are obtained. Given these results, the combination of tensile stress and extension could induce cracking. Model 3 shows that in the present case cracking could initiate in the neighbouring of the first interface in layer 1, i.e. near the surface, rather than at the bottom of layer 3. If so, we suspect that in a first time the crack then propagate to the pavement surface leading to a cracking pattern which could be associated to top-down cracking. The viscoelastic behavior of interface then seems to be able to provide an explanation to the mechanism of top-down cracking. The maximum value of extensional strain in z = 78 mm is that in the transversal direction indicating that a cracking initiation at this location would develop in the longitudinal direction subsequently. Complementary studies will be undertaken so that model 3 can be adapted to several load conditions in order to evaluate the impact of each parameter on pavement damage.

5 Conclusions

Comparisons between simulations and experimental data have shown that model 3 which considers viscoelasticity for the asphalt concrete layers and their interface was the most pertinent to predict the pavement response. It was shown also that the mechanism of top-down cracking could be explained by the presence of viscous interface which tends to produce high values of tensile stress and extension in its vicinity, especially close to the pavement surface.

References

Chabot, A., Chupin, O., Deloffre, L., Duhamel, D., 2010. ViscoRoute 2.0. A Tool for the Simulation of Moving Load Effects on Asphalt Pavement. Road Materials and Pavement Design 11, pp. 227-250, 10.1080/14680629.2010.9690274

Grellet, D., Doré, G., Bilodeau, J.-P., 2010. Effect of tire type on strains occurring in asphalt concrete layers; Proc. of the 11th ICAP; Nagoya, Japan

Grellet, D., Doré, G., Bilodeau, J.-P., Gauliard, T., 2013. Wide-Base Single-Tire and Dual-Tire Assemblies. Transportation Research Record: Journal of the Transportation Research Board 2369, pp. 47-56, 10.3141/2369-06

Grellet, D., Doré, G., Kerzreho, J.-P., Piau, J.-M., Chabot, A., Hornych, P., 2012. Experimental and Theoretical Investigation of Three Dimensional Strain Occurring Near the Surface in Asphalt Concrete Layers. In: Scarpas, A., Kringos, N., Al-Qadi, I., A, L. (Eds.), 7th RILEM ICCP. Netherlands, pp. 1017-1027.

Tamagny, P., Wendling, L., Piau, J.M., 2004. A new explanation of pavement cracking from top to bottom: The visco-elasticity of asphalt materials? Proc. of the 5th International RILEM Conference; Limoges, France, pp. 3-10.

Experimental Characterization of the Interface Damage Between Bituminous Layers: From the Interphase to the Interface Properties

Ktari Rahma, Millien Anne, Fouchal Fazia, Petit Christophe, Pop Ion-Octavian and Valle Valéry

Abstract The mechanical properties of surface courses inter-layers are a key parameter for the design of repaired or reinforced multi-layered asphalt pavement structures. The insufficiency of the interlayer bonding leads to significant decrease of pavement durability and recurrent distresses (slippage failures, top-down cracking). Hence, the main objective of this paper is to identify the input parameters for interlayer modeling, using DIC techniques. On the first hand, direct tensile test (DDT) and double notched shear test (DNST) are performed on two layered bituminous concrete specimens. The full field analysis allows to characterize an interphase thickness, the constitutive laws of both interphase and interface without thickness and the critical energy release rate in mode I and II. On the other hand, the contact surface morphology is analyzed by a fringe projection technique in order to determine the roughness parameters. A strong degree of similarity seems to be obtained between the interphase thickness and the roughness.

Keywords Interphase/interface · Digital image correlation · Roughness

K. Rahma (✉) · M. Anne · F. Fazia · P. Christophe · P. Ion-Octavian
GEMH-GCD, University of Limoges, EA 3178, 19300 Egletons, France
e-mail: rahma.ktari@unilim.fr

M. Anne
e-mail: anne.millien@unilim.fr

F. Fazia
e-mail: fazia.fouchal@unilim.fr

P. Christophe
e-mail: christophe.petit@unilim.fr

P. Ion-Octavian
e-mail: ion-octavian.pop@unilim.fr

V. Valéry
Institut Pprim, PEM, University of Poitiers, CNRS UPR 3346, 86000 Poitiers, France
e-mail: valery.valle@univ-poitiers.fr

A. Chabot et al. (eds.), *8th RILEM International Conference on Mechanisms of Cracking and Debonding in Pavements*, RILEM Bookseries 13,
DOI 10.1007/978-94-024-0867-6_82

1 Introduction

The importance of the interfaces/interphases properties between pavement layers and the negative influence of the lack of bond on these infrastructures performances is well known. The interfaces behavior is highly influenced by parameters such as: the type of layers and aggregates, the surface properties (roughness, cleanness, durst, air void,...), the bitumen tack coat type and dosage, and the environmental condition (water, temperature, freeze-thaw cycles...) (Chaignon and Roffe 2001; Chabot et al. 2013; Chen et al. 2012; D'Andrea et al. 2013; Raab et al. 2012; Raposeiras et al. 2012; Santagata et al. 2008).

In this study, optical methods have been used to characterize an interphase defined as an Intermediate Transition Zone (ITZ), as well as a Zero Thickness Interface model (ZTI) and to explore their fundamental properties. The purpose is also to find a correlation between the thickness and the roughness of the interface. The first section of this paper will present materials and methods. Experimental results, will be given for two types of de-bonding mode (modes I and II). Lastly, according to these experimental data, an interpretation of results will be provided.

2 Materials and Methods

Test specimens are extracted from the experimental section of the RILEM-SIB TG4 project (Canestrari et al. 2014). They are composed of two similar AC12 layers with a nominal 70/100 penetration grade bitumen dosed at 5.5 % by aggregate weight. The maximum aggregate size is around 12.5 mm. The interlayer is reinforced by a pre-coated carbon/glass fibers grid (with a square mesh of 20 mm). Before applying the grid, a polymer-modified tack coat emulsion (C69BP3) was spread on the surface of the lower layer with a rate of 0.25 kg/m^2 of residual binder.

In order to evaluate the interface properties, monotonic direct tensile tests (DTT, Mode I) and double notch shear tests (DNST, Mode II) were performed at T ≈ 22 °C, with a displacement rate of 0.5 mm/min. Specimen geometries and boundary conditions are presented in Fig. 1.

The experimental characterization was made by using Digital Image Correlation (DIC and H-DIC) methods (Valle et al. 2015). Displacement fields were calculated into a Zone Of Interest (ZOI) and the scale factor was 0.202 mm/pixel for the DTT and 0.216 mm/pixel for the DNST. The ZOI was discretized by using several subset sizes: 8 × 8, 16 × 16, 32 × 32 and 64 × 64 pixels^2, with a longitudinal and vertical gap of 1 pixel. This analysis allowed for considering the influence of the subset sizes on the interphase geometrical properties. For both methods, the uncertainty of displacement is around 0.03 pixels.

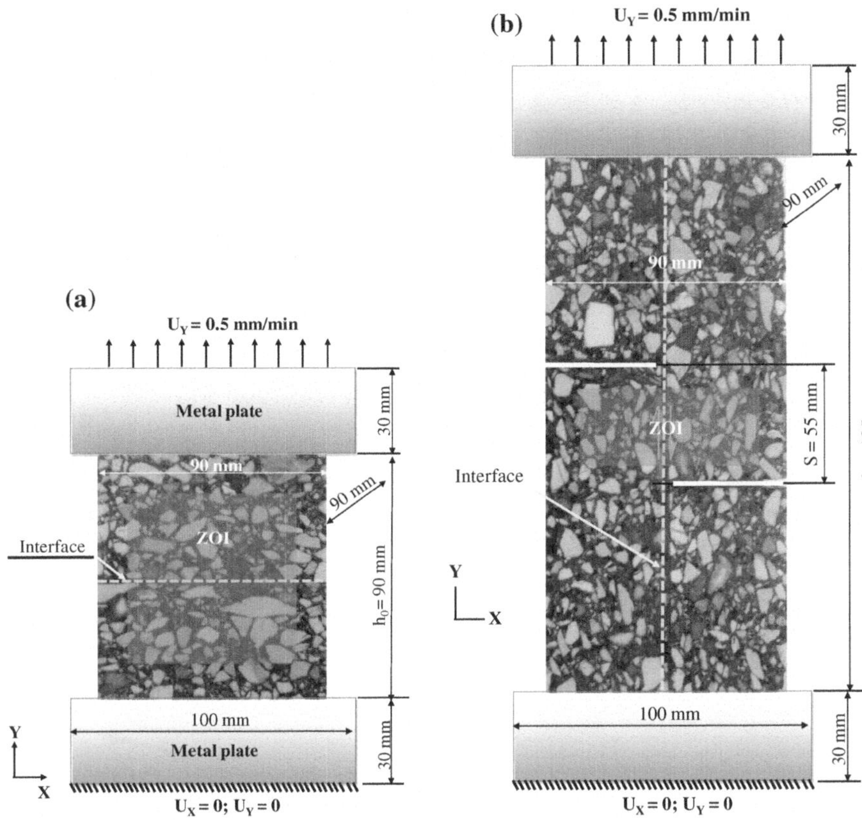

Fig. 1 Specimen geometry and boundary conditions **a** (DTT), **b** (DNST)

3 Data Analysis

Full field results, in term of the axial strain component ε_{yy} and the shear strain component ε_{xy} derived from the displacement field in the ZOI are displayed in Fig. 2a for DTT and Fig. 2b for DNST tests respectively. In this figure, the subset size is 16 × 16 pixel^2 (3.2 × 3.2 mm^2). It can be seen from H-DIC (high gradient field's detection), that the interphase localization occurs early, and before the peak load (point C). However, the global fields are highly heterogeneous at this scale.

According with these observations, the study was focused on a local analysis in order to a better understanding of the de-bonding phenomenon and the interphase-interface damage. An homogeneization process was implemented: the displacements components were calculated as an average displacement u_y in x-direction from each raw of subsets along the ZOI length and strains were derived from these values. Figure 3 shows the analysis of these mean value distributions $u_{y,}$ for five levels, until the peak load (C), for the mode I and mode II tests. Measurements of

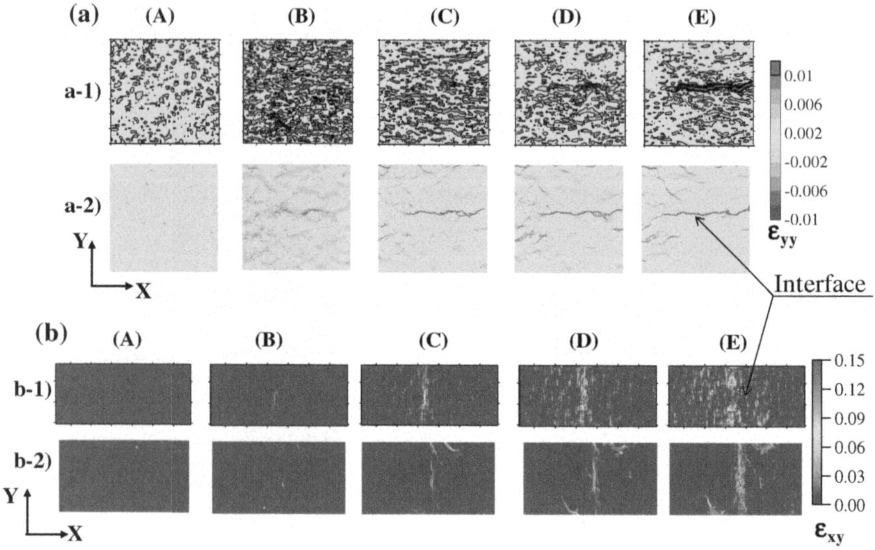

Fig. 2 Strain fields cartography: **a** DTT, **b** DNST

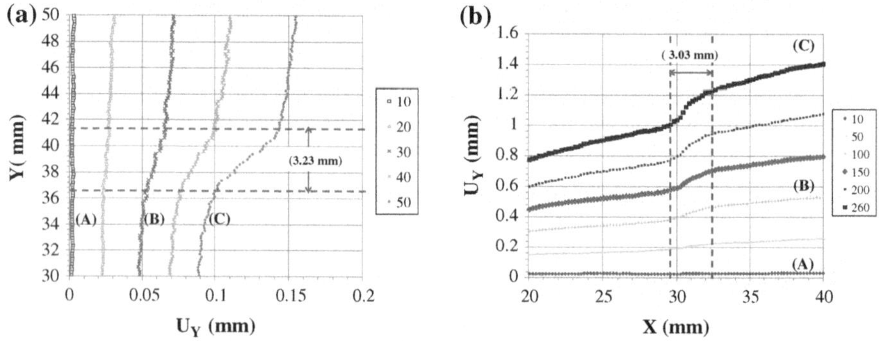

Fig. 3 Interphase localization: mean displacement versus y-position **a** DTT **b** DNST

the ITZ thickness between point A and point B, by finding the inflection points, let us conclude that the interphase thickness is about 3 mm, for these two configurations (Ktari et al. 2016).

In order to investigate correlation between the surface roughness profile and ITZ thickness, the upper layer surface texture of the slab, (assumed to be representative of the lower one), was investigated by means of a phase shifting fringe profilometry's method (Liu et al. 2003). The calculation of the surface roughness was performed using Light3D software, developed by the Pprim Institute (Leandry et al. 2012). Convergence of ITZ size has been achieved at 16×16 pixel^2 and the computed thickness seems to be highly correlated with the S_z (or R_{max}) parameter.

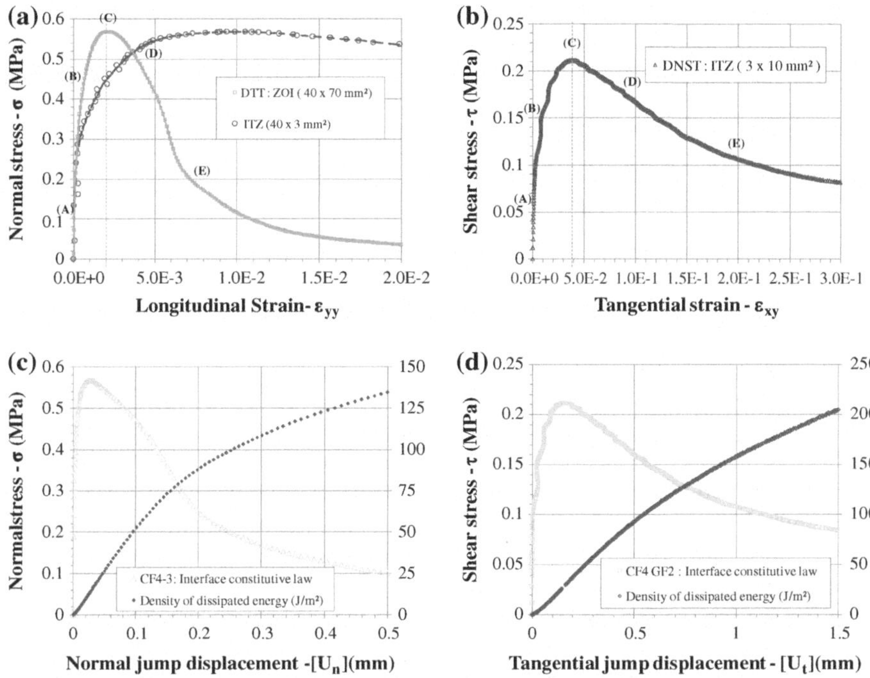

Fig. 4 Interphase and interface constitutive laws: stress versus strain at the interphase **a** DTT (global and local strain) **b** DNST (local strain), stress versus relative displacement jump at the zero thickness model interface and dissipated energy density **c** DTT **d** DNST

A comparison between the global (ZOI) and the local (ITZ) mean strain component evolution versus the nominal stress is shown in Fig. 4a, for the "homogeneous" direct tensile test. The global analysis was performed using a centered ZOI of 40 × 70 mm^2 (Fig. 1) subdivided by the subset of 16 × 16 pixels^2. For the local analysis, the size of ZOI defined near to ITZ is 40 × 3 mm^2 and the same subset parameters. The results confirm that the interphase strain localization emerges very early before the peak loading (see Fig. 4a, point A), and that the DIC method allows realistic predictions of an interphase behavior law (Fig. 4a, b). For the DNST, boundary conditions and heterogeneous fields did not allow the computation of the global mean components.

A simplified method was proposed to describe the interphase (ITZ), at the meso-scale level, using two damaging thin bituminous layers and a zero thickness interface (Ktari et al. 2016). Properties of the bituminous layers are deduced from the upper and lower layers behaviors, i.e. each bituminous concrete layer was assumed to be homogeneous until the interface, even if the air void content should be higher in the ITZ (Santagata et al. 2008). Under these simplifying assumptions, the interface relative normal and tangential jumps (deduced from that analysis) versus nominal normal and shear stresses are computed and presented in Fig. 4c, d, in parallel with the dissipated energy density.

Recall that the aim of this present study is to enable the characterization and identification of a 3D constitutive law for the interphase or of a 2D cohesive zone model for the interface, for these monotonic and decoupled tests. Parameters such as initial normal and tangential stiffness or modulus, damage and energy release rate, can be identified from the obtained experimental data.

4 Conclusions

Shear and tensile tests have been widely used to obtain strength evaluation of asphalt interlayers. In this paper, these tests are analyzed with optical methods DIC and H-DIC for determining both the interphase geometrical and mechanical properties. These methods lead to the identification of the interphase/interface constitutive laws. The described method is a first step to provide data for interlayer modeling and lifetime prediction of multilayered structures. However, there are several steps that need to be carried out in order to better understanding the interphase/interface mechanical behavior. In fact, pavements interfaces are subjected to mixed mode fatigue loadings. Then, in the first step, cyclic and coupling mode tests must be performed. In the second step, more interface roughness need to be investigated, as this work only one mixture was analyzed.

References

Canestrari F, Belogi L, Ferrotti G, Graziani A (2014) Shear and flexural characterization of grid-reinforced asphalt pavements and relation with field distress evolution. Mater. & Struct. 47 (7), 1131-1148

Chabot A., Hun M, Hammoum F. (2013). Mechanical analysis of a mixed mode debonding test for "composite" pavements. Constr. Build. Mater. 40, 1076-1087.

Chaignon F, Roffe JC (2001). Characterisation tests on bond coats: worldwide study, impact, tests, recommendations. International Bitumen Emulsion Federation (IBEF) (9) 12-19.

Chen Y, Tebaldi G, Roque R, Lopp G & Su Y (2012) Effects of interface condition characteristics on open-graded friction course top-down cracking performance. Road Mater. Pavement Des. 13(1supl) 56–75.

D'Andrea A, Tozzo C, Boschetto A & Bottini L (2013) Interface roughness parameters and shear strength. Mod. Appl. Sci. 7 (10) 1–10.

Ktari R, Millien A, Fouchal F, Pop IO & Petit C (2016) Pavement interface damage behavior in tension monotonic loading. Constr. Build. Mater. (106), 430-442

Léandry I, Brèque C & Valle V (2012) Calibration of a structured-light projection system: Development to large dimension objects. Opt. Lasers Eng. 50 (3) 373–379.

Liu H, Su WH, Reichard K, Yin S (2003) Calibration-based phase-shifting projected fringe profilometry for accurate absolute 3D surface profile measurement. Optics Communications 216 65–80

Raab C, Abd El Halim AO & Partl MN (2012) Interlayer bond testing using a model material. Constr. Build. Mater. 26 (1) 190–199.

Raposeiras AC, Vega-Zamanillo A, Calzada-Pérez MA & Castro-Fresno D (2012) Influence of surface macro-texture and binder dosage on the adhesion between bituminous pavement layers, Constr. Build. Mater., 28 (1) 187–192.

Santagata FA, Partl MN, Ferrotti G, Canestrari F & Flisch A (2008) Layer characteristics affecting interlayer shear resistance in flexible pavements. Journal of the Association of Asphalt Paving Technologists 77 222–254.

Valle V, Hedan S, Cosenza P, Fauchille AL & Berdjane M (2015) Digital image correlation development for the study of materials including multiple crossing cracks. Experimental Mechanics. 55 (2) 379-391

Moisture-Induced Debonding Mechanisms in Asphalt Mixtures

Alex K. Apeagyei, James RA. Grenfell and Gordon D. Airey

Abstract The mechanism of moisture-induced debonding in asphalt mixtures is described using data obtained from pull-off and direct tensile tests of aggregate-mastic bonds and miniature 20 mm diameter cored asphalt mixture specimens, respectively. Moisture conditioning using deionised water was conducted at multiple temperatures for periods of up to 40 days. Analyses of the results led to the development of a novel framework for describing moisture-induced debonding in asphalt mixtures in moisture-susceptible aggregates. Damage models describing debonding as cohesive damage in the bulk asphalt and/or adhesive failure at the aggregate-mastic interface resulting from monotonic tensile loading were derived based on load versus conditioning time. Debonding at the aggregate-bitumen interface was found to be the main mechanism of moisture-damage in the moisture-susceptible acidic aggregates that were tested. The trends in debonding with conditioning time was found to be neither test-mode (pull-off versus direct tension) nor conditioning temperature (20 °C versus 60 °C) dependent which suggest the presence of moisture at the aggregate-bitumen interface is a key factor influencing the mechanism of moisture-induced debonding in asphalt mixtures.

Keywords Moisture damage · Moisture-induced debonding · Pull-off test · Acidic aggregates

A.K. Apeagyei (✉) · J.RA.Grenfell · G.D. Airey
Nottingham Transportation Engineering Centre, Pavement Research Building, University of Nottingham, Nottingham NG7 2RD, UK
e-mail: alex.apeagyei@nottingham.ac.uk

J.RA.Grenfell
e-mail: james.grenfell@nottingham.ac.uk

G.D. Airey
e-mail: gordon.airey@nottingham.ac.uk

A. Chabot et al. (eds.), *8th RILEM International Conference on Mechanisms of Cracking and Debonding in Pavements*, RILEM Bookseries 13,
DOI 10.1007/978-94-024-0867-6_83

1 Introduction

Moisture-induced damage remains one of the most difficult issues in asphalt pavement durability. A common manifestation of moisture-induced damage is the loss of cohesion in the mixture and/or loss of adhesion at the bitumen-aggregate interface (Airey and Choi 2002) or more realistically, a loss of adhesion at the aggregate-bitumen mastic interface and/or cohesion within the bulk mastic (Airey et al. 2008). Major findings from previous studies suggest that the mineralogical and chemical composition of aggregates play a fundamental and more significant role in the generation of moisture-induced damage than commonly cited bitumen properties (Apeagyei et al. 2014, 2015; Horgnies et al. 2011; Zhang et al. 2015). These studies have shown that in moisture-susceptible acidic aggregates such as granites, the loss of adhesion or debonding at the aggregate-bitumen interface plays a more dominant role than the loss of cohesive within the bulk bitumen. Thus the mechanism of moisture-induced damage in asphalt mixtures can be better understood if the underlying debonding mechanisms could be better identified and quantified. This requires a multi-scale, micro-mechanical approach in which the physicochemical properties (such as moisture diffusion and absorption) and mechanical properties (adhesive strength) of the various components of asphalt mixtures (bitumen, mastic and aggregate-bitumen bonds) are captured and used as input parameters in appropriate finite element models to numerically simulate moisture damage.

This paper describes moisture-induced debonding in asphalt mixtures using data obtained from pull-off tests and direct tensile tests of aggregate-mastic bonds and cored asphalt mixture specimens, respectively. Moisture conditioning was conducted at two temperatures (20 and 60 °C) for periods of up to 40 days. Analyses of the results led to the development of a novel framework for describing damage responsible for moisture-induced debonding in asphalt mixtures containing moisture-susceptible aggregates.

2 Materials and Methods

2.1 Specimen Preparation and Moisture Conditioning

Two specimen types were used. They included a 10 mm nominal maximum size (NMAS) dense-graded macadam (DBM) mixture containing granite aggregate and 40/60 pen bitumen and aggregate-mastic butt joints made from the same aggregate and bitumen. The mastic was comprised of 25:25:50 weight ratios of bitumen, filler, and fine aggregates, respectively. Details of the mastic design including the motivation of the design have been provided elsewhere (Apeagyei et al. 2014). The DBM mixtures containing 5.5 % by weight of total mix of bitumen were compacted to 6.5 % air voids using a gyratory compactor to a size of 60 mm thick and 100 mm diameter. Direct tension specimens measuring 20 mm diameter by 60 mm were cored from the gyratory compacted mixtures. The exposed aggregates resulting from coring ensured that moisture diffused to the aggregate-bitumen interface rapidly in a similar fashion as the pull-off test described later in the paper. A key reason for using these miniature specimens was to reduce computational cost during FE simulation of moisture damage.

To fabricate the pull-off test specimens, the substrates and mastic were heated to a temperature of 140 °C. Small amounts of mastic were then poured into silicone molds to form mastic films of dimensions approximately 3 mm thick and about 26 mm diameter. The mastic films were annealed to the 23 mm diameter hot (130 ° C) aggregate substrates. A second aggregate substrate, also at 130 °C was annealed to the exposed face of the mastic to form a butt joint comprising the 3-mm thick mastic sandwiched between two aggregate substrates. The whole assembly was trimmed using a hot knife to produce the tensile butt-jointed specimens. The specimens were then kept at 70 °C for 2 h to ensure complete bonding. An aluminum-backed adhesive film was used to cover the mastic film during the two-hour period to ensure that no material leaked out of the mastic. At this stage, the specimens were either stored dry or moisture conditioned and then tested.

Two moisture conditioning processes were used for the two different specimens. The DBM specimens were kept in a water bath maintained at 60 °C, for a period of 40 days similar to that reported in a previous study (Apeagyei et al. 2014). Moisture conditioning of the butt-jointed specimens was performed at 20 °C by partially submerging the substrate in water such that only about 1–2 mm of the bottom aggregate substrate was exposed to the open air which ensured that moisture reached the aggregate/mastic bond only through the aggregate. At least three replicates of each mixture were tested.

2.2 *Pull-off Tensile and Direct Tensile Tests*

A bespoke tensile pull-off test for evaluating aggregate-bitumen bonding strength was used (Fig. 1). The system is capable of testing both butt-jointed aggregate-bitumen bonds and cored dense-graded macadam. The system consists of three major components: (1) a moisture conditioning step that limits moisture diffusion to the aggregate-mastic/bitumen interface, (2) precision bitumen/mastic thickness control, and (3) capability to precisely control loading rates. Details of the development of pull-off testing system are provided elsewhere (Apeagyei et al. 2015; Zhang et al. 2015). The specific testing conditions adopted for the current study included: mastic thickness of 3 mm, a loading rate of 20 mm per minute at 20 °C, and moisture conditioning under ambient conditions (20 °C and 50 % relative humidity). The adopted moisture conditioning enhanced the potential for moisture-induced adhesive failure at the interface and minimized moisture degradation in the bulk mastic (Kringos et al. 2008, 2011).

Direct tensile strength tests were conducted at 20 °C, using a loading rate of 20 mm/min. The exposed aggregates in the direct tension specimens enabled rapid diffusion of moisture to the aggregate-bitumen interface, therefore, it was expected

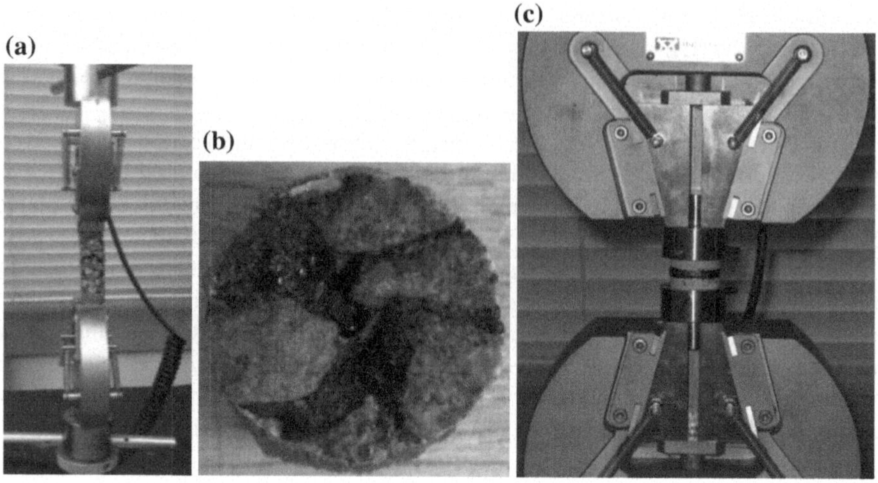

Fig. 1 Test set up used for both the pull-off test and the direct tension test: **a** direct tension 20 mm diameter DBM test, **b** cross section of 20 mm diameter DBM specimen **c** pull-off test

that moisture-induced debonding mechanisms in the direct tensile testing of the DBM will be similar to the pull-off test previously discussed. This will in turn provide credence to the use of the novel DBM geometry for numerically simulating moisture-induced damage in asphalt mixtures. In this case, the observed moisture-induced damage should be interfacial (complete separation of aggregates from bitumen film will be observed). For each combination of mixture and testing condition, three replicates were tested at three different moisture conditioning times (0, 10 and 40 days). The results were evaluated by comparing the measured adhesive strength with maximum tensile stress obtained using the Nottingham pull-off tests.

The aggregate-mastic interfacial bond strength was determined with a bespoke tensile testing rig mounted on an Instron testing machine (Fig. 1c). A constant cross-head speed of 20 mm/min was applied. All the tests were conducted at a constant temperature of 20 °C. To determine the effect of conditioning time on strength degradation of the aggregate butt joints, three conditioning times (0, 10, 40 days) were used. Three replicate specimens were tested for each combination of aggregate substrate and mastic combination.

3 Results and Discussion

3.1 Moisture Absorption, Pull-off Tensile, and Direct Tensile Tests

Results of the moisture absorption tests are presented in Fig. 2 for the aggregate substrates (in the case of the pull-off tests) and asphalt mixture in the case of the DBM. As expected the more porous (6.5 % air voids) and high temperature (60 °C)

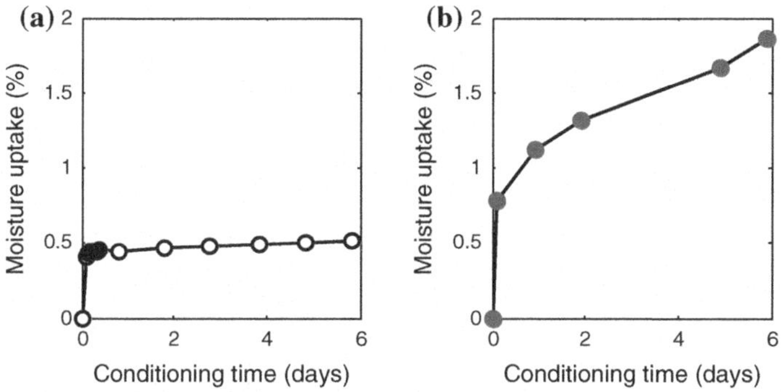

Fig. 2 Moisture absorption: **a** Aggregate substrate (pull-off test) at 20 °C, **b** DBM at 60 °C

conditioning resulted in higher overall moisture uptake in the DBM. It should be noted that by the end of the first day of conditioning, both specimen types have reached a significant percentage of their equilibrium moisture uptake (Fig. 2). The results summarized in Fig. 2 are useful for estimating the moisture absorption and diffusion properties of asphalt mixtures necessary for numerical simulation of moisture-induced damage. To characterize strength degradation in the asphalt mixtures with moisture (damage model), aggregate-mastic bond strength was computed as the ratio of the peak load divided by the cross-sectional area of the butt joint. Similarly, adhesive strength of the DBM was computed as the ratio of the peak load to the cross-sectional area of the 20 mm diameter by 60 mm tall specimens.

From Fig. 3, moisture had a significant effect on bond strength irrespective of moisture conditioning temperature (20 °C versus 60 °C) or specimen configuration (butt-joint pull-off versus DBM direct tension). In each case, significant moisture-induced debonding was observed as evidenced by the rapid degradation of bond strength (Fig. 3a, b). For a more realistic comparison and for numerical simulation purposes, damage models represented by retained bond strength were

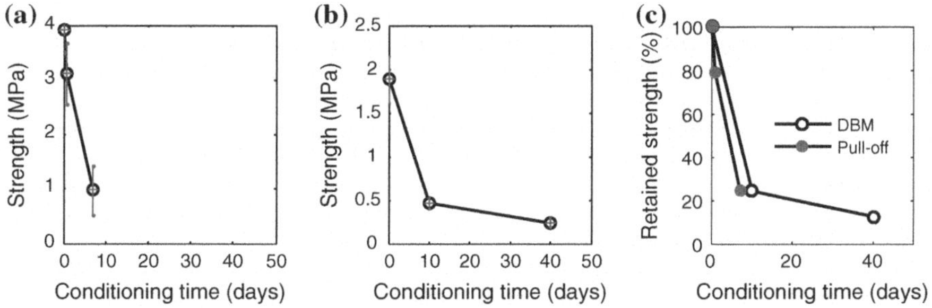

Fig. 3 Effect of moisture on strength degradation in asphalt mixtures. **a** Aggregate-mastic bond strength from pull-off test. **b** Tensile strength of dense-graded macadam. **c** Damage models describing debonding at the aggregate-mastic interface

calculated as the ratio of moisture-conditioned strength to dry condition. These later results are shown in Fig. 3c. The close similarity between the retained strength obtained for the two different specimen geometries and moisture conditioning temperatures are clear from the figure. The results suggest that in moisture susceptible aggregates, the presence of water at the interface between the aggregate and bitumen/mastic maybe a more critical factor in the debonding mechanism in asphalt mixtures than specimen configuration or the moisture conditioning temperature. The results also suggest that the miniature DBM geometry could be used for simulation of moisture-induced debonding in asphalt.

4 Conclusions

The mechanisms of moisture-induced debonding in asphalt mixtures are described using data obtained from pull-off and direct tensile tests of aggregate-mastic bonds and miniature 20 mm diameter cored asphalt mixture specimens, respectively. Moisture conditioning using deionised water was conducted at multiple temperatures for periods of up to 40 days. Damage models describing debonding as cohesive damage in the bulk asphalt and/or adhesive failure at the aggregate-mastic interface were produced. Debonding was found to be the main mechanism of moisture-damage in susceptible aggregates. The trends in debonding with conditioning time was found to be neither specimen configuration-dependent nor conditioning temperature-dependent which suggest the presence of moisture at the aggregate bitumen interface is the key parameter influencing the debonding mechanism.

References

Airey GD, Choi YK (2002). State of the art report on moisture sensitivity test methods for bituminous pavement materials. Road Materials & Pavement Design; 3(4): 355-372

Airey GD, Collop AC, Zoorob SE, et al (2008). The influence of aggregate, filler and bitumen on asphalt mixture moisture damage. Constr Build Mater, 22(9): 2015-2024

Apeagyei AK, Grenfell JRA, Airey GD (2014). Observation of reversible moisture damage in asphalt mixtures. Construction and Building Materials 60 73–80

Apeagyei AK, Grenfell JRA, Airey GD (2015). Influence of aggregate absorption and diffusion properties on moisture damage in asphalt mixtures. RMPD, 16(sup1): 404-422

Horgnies M, Darque-Ceretti E, Fezai H, Felder E (2011). Influence of the interfacial composition on the adhesion between aggregates and bitumen: investigations by EDX, XPS and peel tests. Int J Adhes Adhes, 31 (5): 238–247

Kringos N, Scarpas A, de Bondt A (2008). Determination of moisture susceptibility of mastic-stone bond strength and comparison to thermodynamical properties. Journal of the Association of asphalt Paving Technology, 77: 435–478

Kringos N, Khedoe R., Scarpas A de Bondt A (2011). A new asphalt concrete moisture susceptibility test methodology. In: 90th Annual Meeting of TRB Compendium of Papers DVD TRB

Zhang J, Apeagyei AK, Airey GD, Grenfell JRA (2015). Influence of aggregate mineralogical composition on water resistance of aggregate–bitumen adhesion. Int J Adhes Adhes, 62: 45–54

Part X
Interface Debonding Behavior: Debonding Mechanisms (Various Interfaces)

Is Debonding in Concrete Pavements Unavoidable?

Johan L. Silfwerbrand

Abstract Concrete overlays constitute one of the most frequent measures to repair and strengthen concrete bridge decks, concrete pavements, and industrial concrete floors. Good, secure, and durable bond between overlay and substrate provides the prerequisite for monolithic action and improves the durability. 30 years of Swedish research on bonded overlays thought beam tests, slab tests, and field measurements shows that a good bond can be obtained if some important demands are fulfilled. The Swedish research also includes long-term studies on repaired concrete bridge decks. These studies show that the bond is of the same magnitude ten years after concrete repair independent of structural system of the bridge, climate zone, traffic volume, or use of de-icing salt. Laboratory studies show that good bond is also able to resist fatigue loading. This paper summarizes these research activities and provides heavy arguments for answering the question in the heading negatively. Debonding is not unavoidable.

Keywords Bond · Field tests · Laboratory tests · Longterm studies

1 Introduction

Despite that solutions with unbonded overlays exist for certain purposes, the bonded concrete overlay is the usual aim of the intervention. It may be added that bonded concrete overlays are not limited to repair; contrary they are common also in new concrete structures to provide wear resistance or smoothness. One example is concrete toppings on elevated precast concrete slabs. A sufficiently good bond between base layer concrete and concrete overlay is mandatory for bonded overlays.

J.L. Silfwerbrand (✉)
Department of Civil and Architectural Engineering, Division of Concrete Structures, KTH Royal Institute of Technology, 100 44 Stockholm, Sweden
e-mail: jsilfwer@kth.se

A. Chabot et al. (eds.), *8th RILEM International Conference on Mechanisms of Cracking and Debonding in Pavements*, RILEM Bookseries 13,
DOI 10.1007/978-94-024-0867-6_84

Good bond is not only necessary for obtaining monolithic action between the two concrete layers with corresponding high load-carrying capacity and small deflections but it also (i) promotes crack control in the overlay and (ii) prevents water and detrimental substances to be transported in the interfacial zone between the two layers.

Within the field of bonded concrete overlays, there is a specific term for losing bond and it is called debonding. The use of the present participle (verb ending with —ing) indicates that it is a process; the initial bond vanishes to zero during a certain time. Several factors may contribute to debonding, e.g., fatigue, differential shrinkage, and a severe environmental event. An important question is if debonding is unavoidable. Will bond eventually disappear in all practical cases? Is it just a question of the time? Theoretically, these questions can neither be answered conclusively nor generally, but this paper provides evidence that debonding at least can be postponed significantly if all work during concrete removal, substrate preparation, casting, and curing are carried out meticulously.

2 Factors Influencing Bond

The research on factors influencing bond between the base layer concrete (substrate) and a new-cast concrete overlay is very comprehensive, see, e.g., Silfwerbrand (2009). In an attempt to rank the various factors, Silfwerbrand and Beushausen (2005) defined three classes of importance; major, medium, and minor influence. Five factors were identified as the major ones; (i) a substrate free from micro-cracking, (ii) a substrate free of laitance, (iii) a clean substrate surface at the moment of overlay placement, (iv) sufficient compaction of the overlay, and (v) good curing during a sufficiently long time after placement. Prewetting of the substrate, overlay properties, and time after overlay placement (= age of overlay) belong to the intermediate group, whereas substrate roughness, bonding agents, and early traffic are factors of minor importance.

3 Fatigue Tests

Concrete pavements are subjected to a large number of traffic loads which means that the bond will be subjected to fatigue loading. At KTH Royal Institute of Technology, Stockholm, Sweden, two test series have been carried out to investigate bond and fatigue; one on two-layer beams with various types of interfaces between the concrete layers and a second series on two-layer beams where shotcrete was used in the overlay.

In the first test series at KTH, 18 simply supported concrete beams were tested in 3-point bending (3 PB, Fig. 1) either by static or fatigue loading (Silfwerbrand 1984). This was before the water-jet technology (hydrodemolition) was introduced

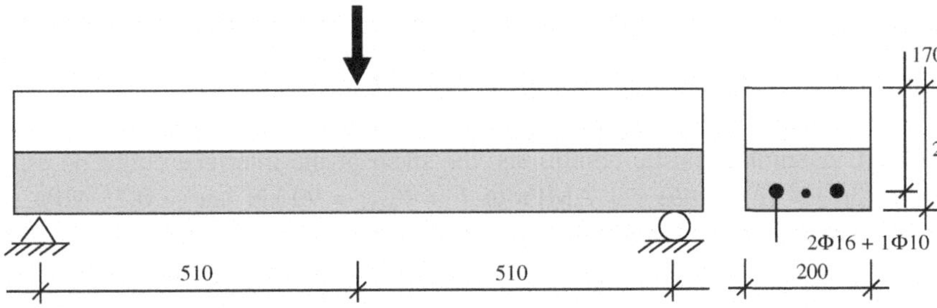

Fig. 1 The set-up for the two-layer concrete beam tests. Dimensions in mm (Silfwerbrand 1984)

in Sweden for repairing concrete bridge decks. One aim was to study the influence of the treatment of the substrate prior to overlay casting. Twelve beams were treated by pneumatic hammers. Additional measures as threaded dowel bars through the interface and epoxy bonding agent were applied to some of the beams (Table 1). In order to promote debonding, the interface on four beams was smooth as a result of steel-troweling. The tests were carried out on twin beams, one beam was statically loaded to failure and one was subjected to fatigue loading up to 1 million load cycles. After initial wet curing, the specimens were stored in room temperature and humidity in the laboratory until testing.

The most important result was that no interface failure occurred in any of the fatigue tests, whereas shear failure occurred in four cases (including the homogeneous beam, No. 14). The remaining five beams withstood the entire fatigue

Table 1 Summary of fatigue tests on two-layer beams (Silfwerbrand 1984)

No	Old	New	Inter-face	Dowels (number × diameter, mm)	Ultimate load of statically loaded twin beam (kN)	Min/max load (kN)	Number of cycles at failure	Failure mode
8	A	B2	1		130	10/90	85,266	Shear
9	A	B2	2		150	10/90	$>10^6$	
10	A	B2	1	8 × 7	200	10/90	$>10^6$	
11	A	B2	1	8 × 10	235	10/90	812,907	Shear
12	A	B2	1	8 × 12	235	10/90	$>10^6$	
13	A	C2	1		130	10/90	6681	Shear
14	B2	B2	N/A		130	10/90	109,691	Shear
16	E	F	3		70	10/40	$>10^6$	
18	E	G	3*		69	10/40	$>10^6$	

Notes. Concrete types: A = concrete cut from old bridge, B2 = concrete C50/60 cast in Jan. 1983, C2 = vacuum treated concrete of type B2, E = concrete C50/60 cast in Dec. 1983, F = G = concrete C 50/60 cast in Feb. 1983. Interfaces: 1 = treated with pneumatic hammer, 2 = as 1 but with an additional epoxy bonding agent, 3 = steel-troweled, 3* as 3 but with an additional grouting. No. 14 was a homogeneous beam program of 1 million load cycles (Table 1). The observed resistance of the interface was rather unprecedented since the maximum load applied during fatigue tests was high in relation to the failure load of the undowelled beams (90/130 ≈ 70 %, 40/70 ≈ 60 %). In design, the fatigue load will hardly be higher than 30 % of the failure load. Assuming elastic conditions, the shear at the interface could be estimated as (3/2) × $(F/2)/(bh)$ = 1,7 MPa at $F = F_{max}$ = 90 kN and = 0.75 MPa at F_{max} = 40 kN for $b = h$ = 200 mm.

In cases, where no failure occurred during the fatigue loading, the beams were subsequently loaded statically to failure. Here, a clear difference between the beams with smooth interface (Nos. 16 and 18) and the remaining ones was obtained (Fig. 2). On the latter beams, interface failure occurred, whereas shear failure occurred on all others. This fact indicates that the smooth interface resulting from steel-troweling might be lower than a threshold value for sufficient bond strength.

Aiming at investigating the possibilities to repair concrete bridge decks with shotcrete, a second test series on beams (with shotcrete overlays) were carried out at KTH (Silfwerbrand 1988, 1989). The test program consisted of the following steps: (1) casting of four reinforced concrete slabs (A, B, C and R), (2) water jetting of slabs A, B and C to different depths, (3) shotcreting (dry mix) of slabs A, B and C to original depth, (4) cutting three beams (denoted A1, A2 and A3 etc.) from each slab, (5) static loading (3 PB) of one beam from each slab (see Fig. 3, left), and (6) fatigue loading of the remaining eight beams. As indicated in steps 2 and 3, the primary test parameter was the location of the interface. It could either be outside

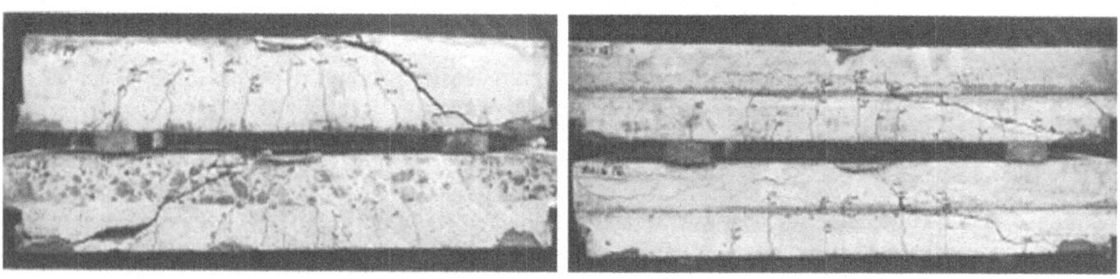

Fig. 2 Photographs of test beams Nos. 14 (*top left*), 13 (*bottom left*), 18 (*top right*) and 16 (*bottom right*) after fatigue loading and subsequent static loading to failure. Note the shear crack through the entire beam depth in beams Nos. 14 and 13 and the shear crack through the lower (reinforced) half of beams Nos. 18 and 16. Horizontal cracks, developed prior to the eventual shear failure, can be seen at the interface in beams Nos. 18 and 16

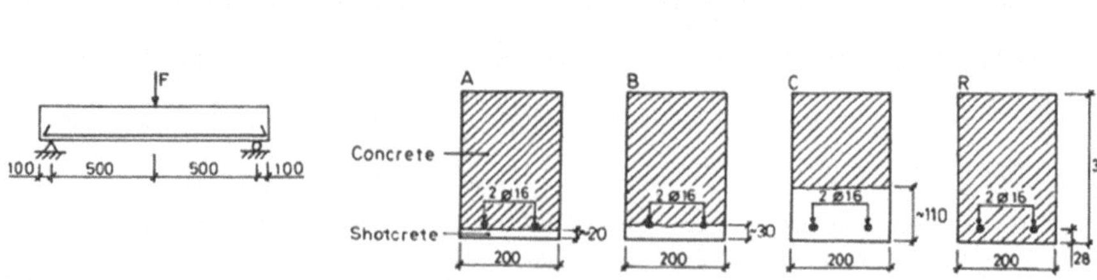

Fig. 3 The set-up (*left*) and cross-section (*right*) of the tests on beams with shotcrete overlay. Dimensions in mm (Silfwerbrand 1989)

Table 2 Summary of fatigue tests on beams with shotcrete overlay (Silfwerbrand 1989)

No	Type	Average concrete removal (mm)	Average ratio between uncovered and total circumference of rebar (%)	Ultimate load of statically loaded twin beam (kN)	Min/max load (kN)	Number of cycles at failure	Failure mode
A1	A	22	2	(A2:) 98	10/70	313,000	Flexure
B1	B	30	23	(B2:) 95	10/70	23,000	Shear + anchorage
C1	C	110	100	(C2:) 100	10/70	227,000	Flexure
R3	R	N/A	N/A	(R2:) 101	10/70	220,000	Flexure
A3	A	22	2	(A2:) 98	15/75	205,000	Flexure
B3	B	30	23	(B2:) 95	15/75	20,000	Shear + anchorage
C3	C	110	100	(C2:) 100	15/75	166,000	Flexure
R1	R	N/A	N/A	(R2:) 101	15/75	268,000	Flexure

the flexural reinforcement (A), coinciding with the reinforcement plane (B), or markedly inside the reinforcement (C). These types were compared with a homogeneous reference slab (R), see Fig. 3, right. At time of testing, the base layer concrete had a compressive strength of 31.5 MPa and the shotcrete one of 77.7 MPa.

The results of the tests were decisive; all beams of type A and C showed same behavior and obtained static failure loads and number of load cycles at fatigue loading of the same magnitude as corresponding homogeneous reference beams (R). The beams with a coinciding reinforcement and interface plane (B type) obtained a brittle failure (combined shear and anchorage) instead of a ductile flexural failure and withstood just one tenth of the number of load cycles at failure (Table 2). The recommendation drawn from these tests was that if the reinforcement is uncovered during the concrete removal, the removal has to continue behind the reinforcement to enable the new concrete to completely surround the reinforcement bars.

4 Long-Term Field Tests on Bond

In the mid-1980s, a large number of Swedish concrete bridge decks were repaired using water jetting and new-cast overlays. KTH measured bond strength both shortly after repair and 10 years later through pull-off tests on some of these bridges (Silfwerbrand and Paulsson 1998). The number of cores varied from bridge to bridge. Despite differences in structural system, climate zone, traffic, and use of de-icing salt, the average failure stress σ_f had increased slightly on all seven cases (Table 3). The probability that higher values will occur in seven out of seven areas just is a coincidence is less than 1 %.

Table 3 Long-term tests on bond strength σ_f (in MPa) in repaired concrete bridge decks (Silfwerbrand and Paulsson 1998)

Bridge	Structural system	Climate zone	Average daily traffic	Use of de-icing salt	Repair year	σ_f at time of repair	σ_f in 1995
Bjurholm	Steel girders	5	1440	Occasional	1985	1.96	1.99
Mälsund	Concrete arch	2	210	Occasional	1986	1.71	2.17
Skellefteå	Steel girders	4	25,000	Moderate	1987	1.43	1.82
Södertälje	Flat slab	2	~ 10,000	Intensive	1989	>1.5	2.05
Umeå	Steel girders	4	23,000	Moderate	1987	1.56	1.61
Vrena	Steel girders	2	170	Occasional	1987	1.49	1.56
Överboda	Concrete arch	2	7600	Intensive	1986	2.09	2.18

Notes. Climate zone 2 means an annually frost amount of 300–600 day × degree of frost, zone 4: 900–1200, and zone 5: 1200–1500. 300 day × degree = 30 days of −10 °C or 60 days of −5 °C

5 Concluding Remarks

Bond strength is crucial for obtaining monolithic action between base layer concrete and concrete overlay. Differential shrinkage, fatigue from intensive traffic, and environmental actions such as frost and de-icing salts may eventually cause horizontal cracking along the interface, debonding. Swedish research during more than 30 years show, however, that debonding is not inevitable. Field tests show that the bond strength in repaired concrete bridge decks does not diminish during 10 years and laboratory tests show that a proper interface is able to withstand fatigue loading.

References

Silfwerbrand J (1984) Interaction between partly chipped concrete slab and overlay. Beam tests. Bulletin No. 142, Dept. of Structural Mechanics & Engrg., KTH Royal Institute of Technology, Stockholm, Sweden, 72 p. (In Swedish).

Silfwerbrand J (1988) Repair of concrete damages using shotcrete. Tests on beams subjected to static and fatigue loading. Bulletin No. 153, Dept. of Structural Mechanics & Engrg., KTH Royal Institute of Technology, Stockholm, Sweden, 77 p. (In Swedish).

Silfwerbrand J (1989) Concrete Repair with Shotcrete. Proceedings, IABSE Symposium on Durability of Structures, Lisbon, September 1989, p 785-790.

Silfwerbrand J, Paulsson J (1998) The Swedish Experience: Better Bonding of Bridge Deck Overlays. Concrete International V 20, No 10, p 56-61.

Silfwerbrand J, Beushausen H (2005) Bonded Concrete Overlays – Bond Strength Issues. Proceedings, International Conference on Concrete Repair, Rehabilitation and Retrofitting. Cape Town, South Africa, 21-23 November 2005.

Silfwerbrand J (2009) Bonded Concrete Overlays for Repairing Concrete Structures. In: Delatte N (ed) Failure, Distress and Repair of Concrete Structures. Woodhead Publishing Limited, Oxford, Cambridge & New Dehli, p 208-243.

Durability of FRP to Concrete Bonded Interface Under Accelerated Ageing

Marc Quiertant, Karim Benzarti, Julien Schneider, Fabrice Landrin, Mathieu Landrin and Frédéric Boinski

Abstract Externally bonded carbon Fiber Reinforced Polymers (FRPs) are now commonly used for the strengthening and repair of Reinforced Concrete structures. However, if the effectiveness of this technique has been widely demonstrated, the durability of the adhesive bond at the concrete/composite interface is still a matter of investigation and remains a critical issue to be addressed in order to assess the long-term performance of FRP strengthening methods. The proposed paper aims at presenting the first results of an ongoing investigation on the time evolution of the concrete/composite adhesive bond strength. Such an evolution was studied by performing double lap shear tests, while changes in the mechanical properties of the polymer adhesive were investigated by means of tensile tests. Preliminary results show that shear tests are able to reveal evolutions of both the bond strength and the failure mode of concrete/composite assemblies subjected to various accelerated ageing conditions. The weakest part of the assembly, initially assigned to the concrete substrate (cohesive failure in concrete), is progressively transferred to the polymer joint (adhesive failure at the bonded interface).

Keywords CFRP · Epoxy · Hydrothermal ageing · Shear test

M. Quiertant (✉) · K. Benzarti
Département Matériaux et Structures, Université Paris-Est, IFSTTAR, Paris, France
e-mail: marc.quiertant@ifsttar.fr

K. Benzarti
e-mail: karim.benzarti@ifsttar.fr

J. Schneider · F. Landrin · M. Landrin · F. Boinski
CEREMA—DT Id/LEM/UCMA, Bron, France
e-mail: julien.schneider@cerema.fr

F. Landrin
e-mail: fabrice.landrin@cerema.fr

M. Landrin
e-mail: mathieu.landrin@cerema.fr

F. Boinski
e-mail: Frederic.BOINSKI@concrete-mail.com

A. Chabot et al. (eds.), *8th RILEM International Conference on Mechanisms of Cracking and Debonding in Pavements*, RILEM Bookseries 13,
DOI 10.1007/978-94-024-0867-6_85

1 Introduction

Externally bonded carbon Fiber Reinforced Polymers (FRPs) are commonly used for the strengthening and repair of reinforced concrete structures (Meier and Kaiser 1991; ACI 2008; AFGC 2011). However, if the effectiveness of such techniques has been widely demonstrated, the durability of the adhesive bond at the concrete/FRP interface is still a matter of investigation (Benzarti et al. 2011).

The proposed paper aims at presenting the first results of an ongoing investigation on the environmental degradation of the concrete/FRP adhesive bond. Observed evolutions of the bond strength were discussed in view of the changes in mechanical properties of the constitutive materials (bulk adhesive and concrete).

2 Experimental Program

To investigate the effects of accelerated ageing on the adhesive bond properties between concrete and FRP plates, specific double-shear test specimens were first designed. Specimens were then constructed, exposed to various environments and finally tested. Shear tests were periodically carried out to assess the bond strength of aged specimens. Evolutions of the concrete compressive strength and the tensile strength of the polymer adhesive were also monitored over time. Durability of the pultruded CFRP plate was not investigated here, as this material seems not significantly affected under environmental ageing conditions (Sen 2015).

2.1 Double Shear Test Experiments

2.1.1 Principle of the Double-Shear Test

The experimental program includes the assessment of the bond strength of concrete–FRP interfaces by means double lap-joint shear tests. Test specimens are composed of two concrete blocks attached by two parallel bonded FRP plates (Fig. 1a). A load applied (F) to the concrete blocks produces a shear stress along the adhesive joint (Fig. 1b). Each lap joint is then submitted to the average shear stress (τ_{ave}), which can be calculated according to:

$$\tau_{ave} = \frac{F}{2A_{adh}} \text{ (MPa)} \tag{1}$$

where A_{adh} is the bonded area of each lap joint (Ferrier et al. 2010).

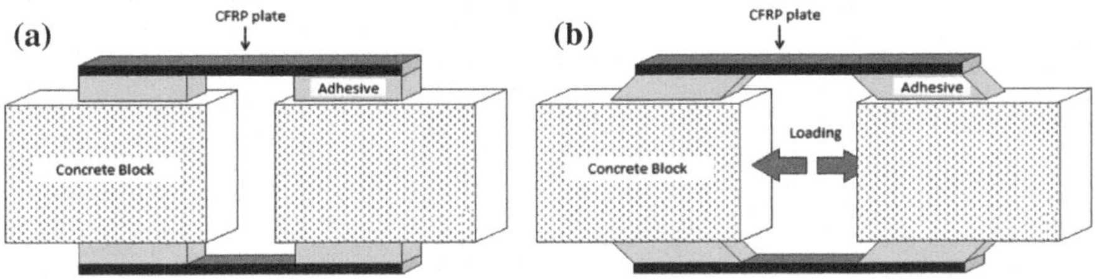

Fig. 1 Principle of the near-end supported double-shear test: **a** double lap-joint shear test specimen, **b** schematic description of the loading conditions

2.1.2 Double-Shear Test Specimens

In a preliminary step, involving the characterization of different specimen geometries, the design of specimens was optimized. The best configuration consisted of two concrete blocks of dimensions 80 × 80 × 150 mm^3 (see Fig. 2b) connected together (with a separating gap of 35 mm) by two CFRP plates symmetrically bonded (without prestressing) to the lateral faces. Commercially available pultruded carbon FRP plates (Sika® Carbodur®S, supplied by SIKA France) were used to prepare the specimens. To avoid premature failure of the block corner (Chataigner et al. 2009), plates were bonded with a gap of 30 mm starting from the edge of each block. The surface area of each bonded joints was 70 × 50 mm^2. Geometry of the FRP-strengthened concrete specimens is presented on Fig. 2b.

2.2 Characterization of the Constitutive Materials

To monitor the evolutions of concrete and polymer adhesive properties under accelerated ageing condition, specific samples were also prepared. Changes in concrete properties were determined by measuring compressive strength on concrete

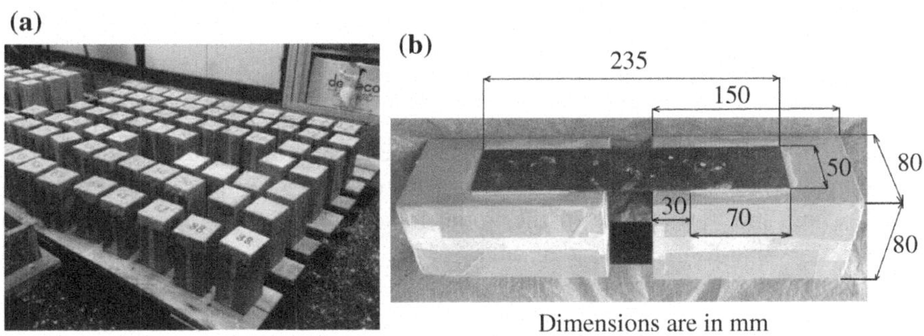

Fig. 2 FRP-strengthened concrete specimens: **a** concrete blocks, **b** geometry of the specimens

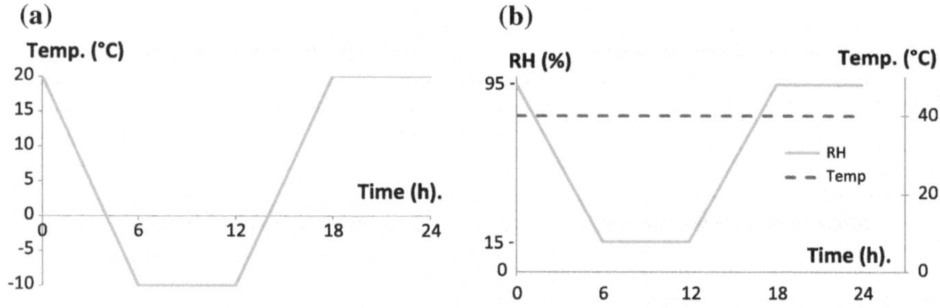

Fig. 3 Description of ageing cycles in **a** climatic chamber N°2, and **b** climatic chamber N°3

cylinders (diameter 110 mm, length 210 mm). Moreover, tensile tests were performed on dumbbell shaped samples of the polymer in order to assess possible degradation of the tensile strength of the epoxy induced by accelerated ageing.

2.3 Ageing Conditions

All specimens (FRP-strengthened concrete, concrete cylinders and polymer samples) were exposed to three different hygrothermal conditions in various climatic chambers. These environments consisted of:

- 95 % relative humidity at a temperature of 40°C (climatic chamber N°1)
- Cycles of temperature (between −10 and 20 °C, without humidity control; climatic chamber N°2; see Fig. 3a)
- Cycles of relative humidity (between 15 % RH and 95 % RH) at a constant temperature of 40 °C (see Fig. 3b; climatic chamber N°3).

Ageing conditions were chosen considering that (i) water absorbed by the epoxy reduces its bond capacity and that (ii) the chosen temperature of 40 °C accelerates water sorption kinetics, while remaining below the glass transition temperature.

3 Experimental Results and Discussions

In this section, the bond strength of the adhesive joint is analyzed as a function of ageing treatment and ageing time. Results of the double-shear tests are presented in Fig. 4, where the mean value of the initial bond strength was obtained from the average of twelve test results, and values of the bond strength of aged specimens were calculated from an average of four tests.

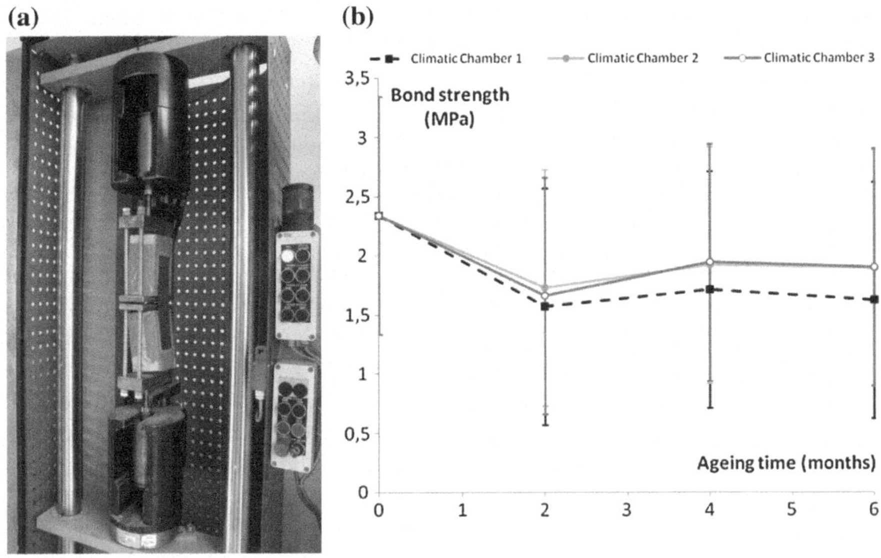

Fig. 4 Bond strength measurement: **a** test setup, **b** evolutions of the bond strength for specimens exposed to the various environments

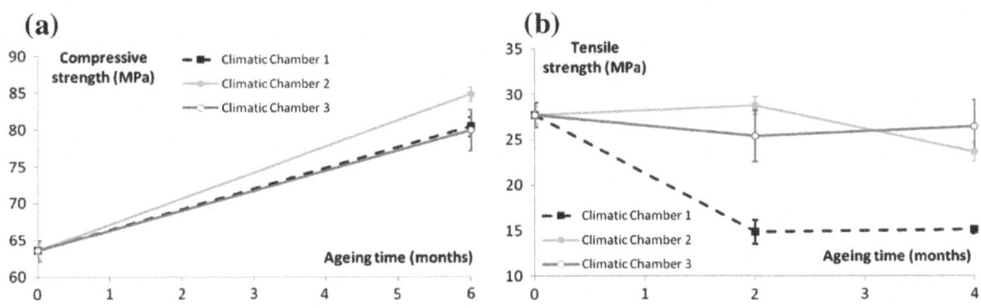

Fig. 5 Time evolution of: **a** concrete compressive strength, **b** the tensile strength of the epoxy

Changes in mechanical properties of the constitutive materials (concrete and bulk epoxy adhesive) are also presented in Fig. 5. Error bars are plotted based on calculated standard deviations.

It can be noticed from Fig. 4b that there is large dispersion on experimental bond strength values. However, it is possible to observe a slight decrease in the bond strength during the early period of exposure (first 2 months), whatever the ageing condition. After 2 months, the bond strength curves became steadier, showing even a slight increase. Whereas cyclic exposure conditions in Climatic Chamber 2 (CC2) and Climatic Chamber 3 (CC3) induce very similar effects, exposure in Climatic Chamber 1 (CC1: 40 °C and 95 % R.H.) seems to produce a more significant degradation of the bond properties. Moreover, a careful observation of fractured surfaces (not presented in this paper) indicated an evolution of the shear failure mode of specimens after ageing, whatever the type of environment. The initial failure mode was mostly within a thin concrete layer (cohesive failure in concrete) and shifted to a mixed failure after ageing (cohesive failure in concrete + cohesive failure in the adhesive layer + debonding). Such a phenomenon was also observed by Benzarti et al. (2011) and Shrestha et al. (2015).

As regards the evolution of the concrete properties over ageing (Fig. 5a), it is found that the completion hydration reaction of cement progressively leads to an increase in the concrete compressive strength. Therefore, as the bond strength of FRP/composite assemblies is decreasing while concrete strength is increasing, a significant degradation of the properties of the adhesive was expected. However, as shown in Fig. 5b, only the epoxy samples stored in CC1 exhibited a significant loss of tensile strength. It can be concluded that the degradation in bond strength of FRP reinforced specimens can be mainly attributed to alterations of bond properties rather than a degradation of the cohesive strength of the adhesive.

4 Conclusion

The first results of an ongoing experimental study on the degradation of the adhesive bond between FRP and concrete under accelerated ageing conditions are presented. As preliminary conclusions, it can be pointed out that severe environmental conditions can lead to a significant loss in the shear capacity of FRP/concrete bonded joints. However, in the light of the time evolution of the tensile strength of the adhesive, it is suggested that the main degradation mechanisms is attributed to alterations of chemical bonds between FRP and adhesive material rather than the degradation of the cohesive strength of the adhesive material itself (at least for two of the environmental conditions under study). Since experiments are still in progress, additional conclusions are expected in the forthcoming months.

References

ACI Committee 440 (2008) ACI 440.2R-08 Guide for the Design and Construction of Externally Bonded FRP Systems for Strengthening Concrete Structures

AFGC (2011) Réparation et renforcement des structures en béton au moyen des matériaux composites. Recommandations provisoires, Bulletin scientifique et technique de l'Association française de génie civil (in french)

Benzarti K, Chataigner S, Quiertant M, Marty C, Aubagnac C (2011) Accelerated ageing behaviour of the adhesive bond between concrete specimens and CFRP overlays. Construction and Building Materials. Vol. 25(2) Sp. Iss. pp. 523-538

Chataigner S, Caron JF, Benzarti K, Quiertant M, Aubagnac C (2009) Characterization of composite to concrete bonded interface. Description of the single lap shear test. Eur J Environ Civil Eng 2009;13(9):1073–82

Ferrier E.; Quiertant M, Benzarti K, Hamelin P (2010) Influence of the properties of externally bonded CFRP on the shear behavior of concrete/composite adhesive joints. Compos. Pt. B-Eng. 41(5): 354-362

Meier U, Kaiser HP (1991) Strengthening of structure with CFRP laminates. In: Iyer SL, Sen R, editors. Advanced composite materials in civil engineering structures, ASCE specialty conference, Las Vegas, NV.

Sen R, (2015) Developments in the durability of FRP-concrete bond. Constr Build Mater 78: 112–125. doi:10.1016/j.conbuildmat.2014.12.106

Shrestha J, Ueda T, Zhang D (2015) Durability of FRP Concrete Bonds and Its Constituent Properties under the Influence of Moisture Conditions. J. Mater. Civ. Eng., 2015, 27(2): A4014009

Mixed-Mode Debonding Approach to Evaluate Water Sensibility in Bi-Layer Composite Pavement

Armelle Chabot, Ferhat Hammoum and Manitou Hun

Abstract In order to evaluate water sensibility on the interface between layers of composite pavements, a four-point bending (4PB) test on bilayer structure in a water bath is proposed. Using the virtual crack closure technique, the individual strain energy release rates are calculated with a specific model. The debonding mode I should be recognized as the main failure mode. For bilayer specimens made of a cement concrete overlay on bituminous material, the specific test has shown a competition between the different failure mechanisms. A very good bond resistance between layers compared to the fracture tension resistance of the cement concrete layer is observed. In this work, first results of the effect of water on the behavior of such a material interface are presented. An aquarium is built in order to submerge under water the bi-layer specimen. The final fracture length of the specimens, curves of force-displacements and first digital image correlation results show the influence of water immersion on the debonding failure mode.

Keywords 4PB test · Water · Debonding · Composite pavements

1 Introduction

Composite pavement systems, concrete/asphalt and asphalt/concrete structure, show good potential for becoming an interesting rehabilitation option for urban pavements that exhibit structural deterioration (De Larrard et al. 2005). However,

A. Chabot (✉)
LUNAM Université, IFSTTAR, CS4, 44344 Bouguenais Cedex, France
e-mail: armelle.chabot@ifsttar.fr

M. Hun
LBL International, 34 Sothearos Boulevard, Phnom Penh BP 609, Cambodia

F. Hammoum
MAST/Laboratoire MIT, LUNAM Université, IFSTTAR, CS4,
44341 Bouguenais Cedex, France
e-mail: ferhat.hammoum@ifsttar.fr

A. Chabot et al. (eds.), *8th RILEM International Conference on Mechanisms of Cracking and Debonding in Pavements*, RILEM Bookseries 13,
DOI 10.1007/978-94-024-0867-6_86

due to shrinkage phenomenon of cement materials, the existing vertical crack combined to environmental and traffic loading lead to failures that may often develop as delamination between layers (Chabot et al. 2008). Typically, such debonding phenomenon initiates and propagates under the combined influence of normal and shear stresses (Chabot et al. 2005). The presence of water, whatever its phase is, adds to this complex phenomenon an irreversible damage that cannot be ignored (Vandenbossche et al. 2011; Raab et al. 2012). The water damage can be classified with two mechanisms: (a) loss in strength and durability of materials due to the presence of water in the pore of asphalt concrete (b) loss of mechanical behavior of bond between layers. Evaluating effect of water on the interface behavior is a complicated task. In order to approach such an effect on the interface between layers of composite pavements, a four-point bending (4PB) test on bilayer structure in a water bath is proposed. Without water, the specific test has shown significant interface debonding in mixed mode (mode I and II) condition (Hun et al. 2012; Chabot et al. 2013). In this work, first results of the effect of the water immersion on the behavior of a cement concrete/asphalt material interface are presented. Final fracture length, curves of force-displacement and digital image correlation (DIC) technics are used to investigate if the presence of water in the bi-layer specimen favors debonding failure mode during the 4PB test.

2 Theoretical Characterization of the Bond During the 4PB Test

Following results stemming from concrete beams strengthened by externally bonded plates, a 4PB test has been adapted for the pair of bituminous and cement concrete materials (Hun et al. 2012). This test has the advantage to investigate the mechanical properties of interfaces under mixed mode conditions without using any supports or applied any loads directly on the bituminous material (Fig. 1).

The multi-particle model of multilayer materials (M4) with five kinematic fields per layer is chosen to design and to investigate the debonding results of the 4PB test on bi-layer specimens. The approximations of the so-called M4-5n (n: total number of layers i) offer the advantage of defining the out-of-interface plane normal $\nu^{i,i+1}$ and shear stresses $\tau^{i,i+1}$ at interface i, i + 1, between layers i and i + 1. As opposed to other classical models, the main interest of this modeling is to yield finite stresses

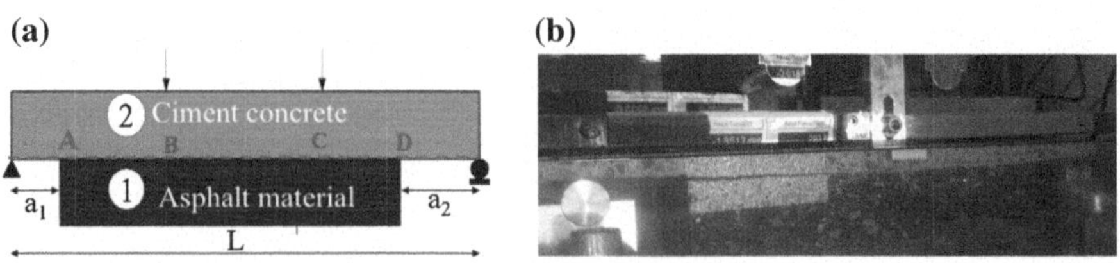

Fig. 1 **a** The four-point bending test configuration; **b** the debonding phenomenon into water

at a free edge or crack tip at the interface point location of two different layers and accelerate the equation resolution process (Chabot 1997). For crack initiation problems, two delamination criteria in the angle-ply laminates have been proposed (Chabot 1997; Caron et al. 2006). The first one is based on the maximum value of interface stresses. Using the virtual crack closure technique (VCCT) (Bui 1978; Moutou Pitti et al. 2008), the second criterion is based on an analytical calculation of the individual strain energy release rates (G_I, G_{II}, G_{III}). Applied to the 4PB test under plane strain assumptions, the M4-5n solutions allow making several parametric studies (Chabot et al. 2013). It indicates that a competition exists between tensile stress at the bottom of the cement concrete layer 2 (between points B and C) and interface normal and shear stresses (at points A and D) (Fig. 1a). In order to favor the debonding phenomenon between the two layers onto only one side, a non-symmetrical geometry of the specimens is used. A higher value of a_2 length compared to the a_1 length favors delamination on the side a_2 first. For the 4PB test with a debonding crack length "a" (Fig. 1b), the analytical expression of strain energy release rates is given in (Eq. 1) where b is the width of the beam. e^i, E^i and υ^i denote the thickness, Young modulus and Poisson ratio of each layer i respectively. Only G_I is a pure quadratic function of the normal interface stress. G_{II} takes into account the combined terms containing the shear interface stress, shear force Q_1^i of each layer i.

$$G_T(a) = G_I(a) + G_{II}(a)$$
$$with \begin{cases} G_I(a) = \frac{1}{2b} \cdot \frac{13(e^1E^2 + e^2E^1)}{35E^1E^2} \left(\nu^{1,2}(a)\right)^2 \\ G_{II}(a) = \frac{1}{2b} \cdot \frac{4(e^1(1+\upsilon^1)E^2 + e^2(1+\upsilon^2)E^1)}{15E^1E^2} \left(\tau^{1,2}(a)\right)^2 \\ \qquad - \frac{1}{2b} \cdot \left(\frac{(1+\upsilon^1)}{5E^1} Q_1^1(a) + \frac{(1+\upsilon^2)}{5E^2} Q_1^2(a) \right) . \tau^{1,2}(a) \end{cases} \quad (1)$$

3 First Test Results and Discussion

4PB tests on bi-layer specimens are performed under controlled static conditions (0.7 mm/min, 20 °C constant temperature). To improve the understanding of the fracture scenario of bi-layer materials of such a 4PB test, Digital Image Correlation (DIC) technics are used (Sutton 1983). First results are performed on a reference specimen made with an aluminum (Alu) homogeneous layer 2 sticks with an epoxy glue, on a PVC homogeneous layer 1 (L = 420 mm; b = 125 mm; a1 = 40; a2 = 71; e^1 = 30.6; e^2 = 40.6 mm; Young modulus ratio: $E^2/E^1 \approx 22.4$; υ^1 = 0.30;

Table 1 M4-5n energy release rate results compared to Dunders ones for an Alu/PVC specimen

	Maximun crack length (mm)	G_I (J/m^2)	G_{II} (J/m^2)	G_{Total} (J/m^2)
Dundurs	70 ± 1	93	35	128
M4-5n		106	21	127

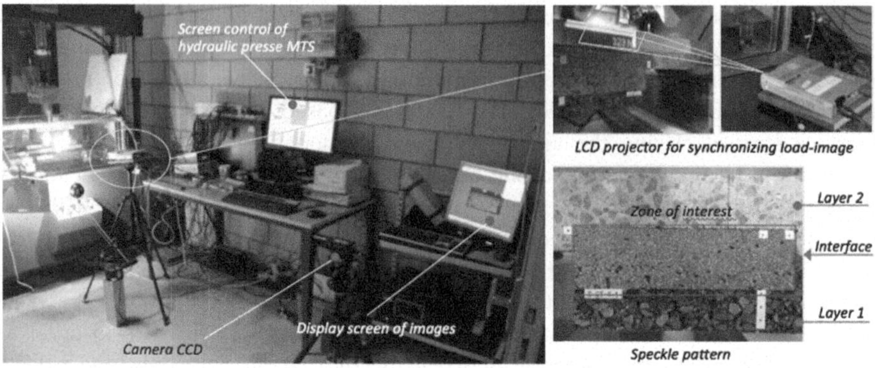

Fig. 2 General description of the 4 PB test equipped with an aquarium and a DIC system

$\upsilon^2 = 0.34$). Table 1 shows that the energy release rates of the M4-5n calculus are validated successfully compared to Dundurs's approach (Dundur 1969). Here, the mode I should be recognized as the main failure mode.

As opposed to tests conducted on Alu/PVC specimens, the determination of strain energy release rate seems more complicated to apply to a cement concrete overlay on bituminous material type of heterogeneous bi-layer materials. The competition between the poor tensile resistance of the cement concrete layer and the good bond of such an interface complicates the analysis (Chabot et al. 2013). In order to evaluate the additional effect of water during the debonding process, an aquarium is built. Before being tested, the specimens are saturated during two hours in a vacuum box filled of water. The wall glass of the aquarium, with its refraction index identical to the water one, allows using the DIC system (Fig. 2). The highest resolution of the camera is 1388 × 1038 pixels at 63 frames per second.

To control and measure the imposed displacement (U_z), a linear variable differential transducer (LVDT) sensor is placed at the mid-span (L/2) on the top of the specimen ($z = e^1 + e^2$). The force (F) is recorded directly on the hydraulic testing machines. During the test, to find the correspondence between each image and the loading conditions, the resulting force and the displacement values are displayed directly in the specimen by using a video-projector (Fig. 2).

M4-5n calculations indicate that, at a loading force of F = 4.2 kN, the tensile stress (R_t) at the bottom of the layer 2 made of the cement concrete tested air is supposed to reach its maximum tensile value (around Rt = 3.46 MPa) at point C before point B in this anti-symmetrical specimen geometry case (L = 420 mm; b = 100 mm; a1 = 40; a2 = 71; $e^1 = e^2 = 60$ mm; $E^2/E^1 \approx 17.5$; $\upsilon^1 = \upsilon^2 = 0.35$). Figure 3 resumes only the static 4PB results for specimens (I-PT-N°slab-N° specimen) instrumented with the DIC system. In this experimental work, among the specimens (4-2; 5-2; 5-3; 5-1) tested in air, only the 5-1 one has failed with an interface debonding mode between layers. The bi-layer specimens tested into water (4-3; 6-2; 6-3) failed by debonding mode. The corresponding curves of

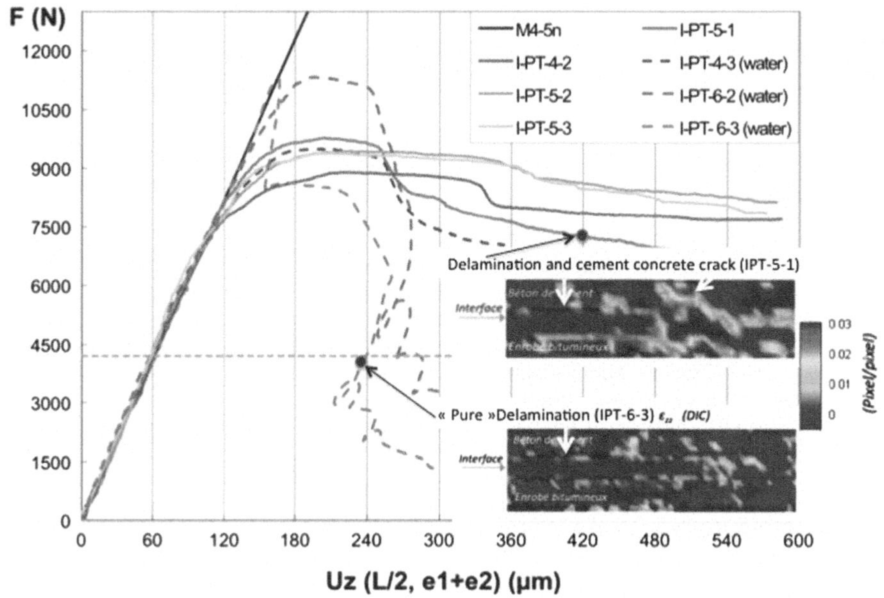

Fig. 3 Curves of force-displacement results of cement concrete/bituminous specimens tested on air and into water with the 4 PB test (0.7 mm/mn, T = 20 °C)

force-displacement show that after the maximum value of the loading force, the force decreases suddenly (Fig. 3).

The final debonding length of bi-layer specimen (Fig. 1b), obtained in this work and a previous one (Chabot et al. 2013), gives 59 mm for specimens tested in air (calculated from 4 specimens) compared to 97 mm for specimens tested into water (from 3 specimens). The image analysis confirms that even if the 5-1 failed with interface debonding mode a parasite crack is observed in the cement concrete material (Fig. 3). All this preliminary results lead to confirm that water has an effect and favors interface debonding phenomenon in such an adapted 4PB test.

4 Conclusion

In order to investigate under water the mechanical behavior of interface between layers, a 4PB test has been adapted. Using the VCCT and debonding assumptions, individual strain energy release rates are calculated with a specific model. Debonding mode I should be recognized as the main failure mode. On specimens made with a cement concrete overlay on a bituminous material layer, all the results confirms that, in air, the delamination process is in competition with micro cracks at the bottom of the cement concrete layer during the test. On the contrary, water presence seems to favor debonding phenomenon in such a 4PB test. More investigations need to be done with more tests and in comparison to other debonding tests such as those done by (Tchegg et al. 2007) and the different Rilem TCs.

References

Bui H (1978) Mécanique de la rupture fragile. Ed Masson, Paris.

Caron JF, Diaz Diaz A, Carreira RP, Chabot A, Ehrlacher A. (2006) Multi-particle modeling for the prediction of delamination in multi-layered materials. Comp. Sc. &Tech., 66(6): 755-765

Chabot A (1997) Analyse des efforts à l'interface entre les couches des matériaux composites à l'aide de Modélisations Multiparticulaires des Matériaux Multicouches (M4). ENPC PhD thesis, (https://tel.archives-ouvertes.fr/tel-00197853/)

Chabot A, Tran QD, Ehrlacher A (2005) A simplified modeling for cracked pavements. Bulletin des Laboratoires des Ponts et Chaussées, ISSN 1269-1496, (258-259): 105-120

Chabot A, Pouteau B, Balay J-M, De Larrard F (2008) FABAC Accelerated Loading Test of Bond between Cement Overlay and Asphalt layers. 6th Rilem Int. Conf. on Cracking in Pavements, June 16-18 2008, Chicago, US. Taylor & Francis Group Proceedings, 13-23

Chabot A, Hun M, Hammoum F (2013) Mechanical analysis of a mixed mode debonding test for "composite" pavements. Construction and Building Materials, 40: 1076-1087

Dundurs J (1969) Edge-bonded dissimilar orthogonal elastic wedges under normal and shear loading. Transactions of the ASME, Journal of Applied Mechanics, 650-652

Hun M, Chabot A, Hammoum F (2012) A four point bending test for the bonding evaluation of composite pavement. RILEM Bookseries, 4: 51-60, doi:10.1007/978-94-007-4566-7_6

De Larrard F, Chabot A, Sedran T, Pouteau B, Mathias V (2005) Recent developments in French concrete pavement technology. Global Construction: Ultimate concrete opportunities. Concrete for Transportation Infrastrastructure, 19-26, doi:10.1680/cfti.34020

Moutou Pitti R, Dubois F, Petit C, N Sauvat, Pop O (2008) A new M-integral parameter for mixed-mode crack growth in orthotropic viscoelastic material. EFM, 75(15): 4450–4465

Raab C, Partl MN, Abd El Halim AO (2012) Effect of moisture on interlayer bonding of asphalt pavements. 7th Int. conf. on Maintenance and Rehabilitation of Pavements and Technological Control, Auckland, New Zealand, August 28-3

Sutton M, Wolters W, Peters W, Ranson W, McNeill S (1983) Determination of displacements using an improved digital correlation method. Image and Vision Computing, 1(3): 133-139

Tschegg EK, Macht J, Jamek M, Stegenberger J., (2007) Mechanical and Fracture-Mechanical Properties of Asphalt-Concrete Interfaces. ACI Materials, 104 (5): 474-480

Vandenbossche J, Barman M, Mu F, Gatti, K (2011) Development of design guide for thin and ultra-thin concrete overlays of existing asphalt pavements. Technical report, University of Pittsburgh, http://www.lrrb.org/media/reports/201125.pdf

Effect of Incorporating Rubber Aggregates and Fiber Reinforcement on the Durability of Thin Bonded Cement-Based Overlays

S. Asad Ali Gillani, Ahmed Toumi and Anaclet Turatsinze

Abstract After certain period of time, the degradation of concrete structures is unavoidable. For large concrete areas, thin bonded cement-based overlay is a suitable rehabilitation technique. Previous research demonstrated that durability of such applications is always a problem and one of its main reason is debonding at interface. Laboratory and field researches show that fiber reinforcement in repair material can be a solution for controlling crack opening and also to enhance the durability of thin bonded cement-based overlays. In other respect, previous researches also show that by addition of rubber aggregates obtained from grinding of used tyres is also a suitable solution for improving strain capacity of cement based materials. This present research mainly focuses on synergetic effect of using rubber aggregates and fiber reinforcement in mortar as a composite for the repair work. For this study four mortar compositions to be used as overlay material were prepared: one control mortar, second with fibers at dosage of 30 kg/m^3, third containing rubber aggregates replacing 30 % sand by equivalent volume and fourth one containing fibers and rubber aggregates. Direct tension tests were conducted in order to obtain the tensile strength, strain capacity, residual post peak behaviour of the repair material and bond tensile strength of the repair-substrate interface. Results showed that although by incorporating rubber aggregates in mortar reduce

S. Asad Ali Gillani · A. Toumi · A. Turatsinze
UPS, INSA, LMDC (Laboratoire Matériaux et Durabilité des Constructions), Université de Toulouse, 135, avenue de Rangueil, 31077 Toulouse, France
e-mail: toumi@insa-toulouse.fr

A. Turatsinze
e-mail: anaclet@insa-toulouse.fr

S. Asad Ali Gillani (✉)
Civil Engineering Department, University of Engineering and Technology, Lahore 54890, Pakistan
e-mail: gillani@insa-toulouse.fr

A. Chabot et al. (eds.), *8th RILEM International Conference on Mechanisms of Cracking and Debonding in Pavements*, RILEM Bookseries 13,
DOI 10.1007/978-94-024-0867-6_87

compressive strength and modulus of elasticity but improvement in straining capacity is observed. Moreover, fiber reinforcement in repair significantly improves residual post peak tensile strength.

Keywords Overlays · Rubber aggregates · Fiber reinforcement · Bond strength

1 Introduction

Concrete has been used in construction industry from thousands of years and with the passage of time a distresses or decrement of load carrying capacity of existing structures is observed. There can be different approaches for rehabilitation of concrete among them bonded overlays found to be the most economical alternative (Bissonnette et al. 2011). This overlays repair technique is particularly suitable for the structures having larger surface areas. The main problem in overlay concrete is bond sustainability and durability of bonded overlay systems. The durability of these overlays is jeopardized if cracking cuts new repair layer, because this cracking initiates the debonding mechanism (Nguyen et al. 2010). Debonding normally occurs due to two mechanisms or their combination. First is due to applied loads on the overlaid structure and second is due to the differential length changes of the overlay and of substrate. According to some previous numerical studies, use of high tensile strength materials in repair material reduces risks of cracking (Tran et al. 2008). Moreover by using repair material having low Young's modulus is more beneficial because it improves strain capacity which in turn is helpful in delaying crack localization, particularly the cracks which normally developed due to restrained shrinkage as a result of differential length change between the substrate and the overlay (Tran et al. 2008). Previous laboratory also showed that by using metallic fiber reinforcement in overlays restrains cracking phenomena which in turn delays debonding initiation and propagation and (Nguyen et al. 2010). So in the light of previous researches, it can be said that debonding in overlays can be controlled by increasing the bond tensile strength at interface and improving repair material properties to restrain cracking. So in this study the goal is to try to achieve these conditions by a fiber-reinforcement, rubber aggregates incorporation and combination of both in overlay materials.

2 Experimental Methodology

Ordinary Portland cement (CEM I 52.5R), sand (fine river rounded aggregates 0/4 mm), super-plasticizer (BASF Glenium 27) for required workability and viscosity agent (BASF Rheomac 890F) for avoiding segregation of rubber aggregates were used in different mortar mixes. The quantity of each constituent of mortar is shown in Table 1.

Table 1 Mix design and proportioning (values of all constituent materials are in kg/m^3)

Sr. No	Mix designation	Cement	Sand	Water	Rubber aggregates	Fibers	Super-plasticizer	Viscosity agent
1	M0R0F	500	1600	250	0	0	1.2	0
2	M0R30F	500	1600	235	0	30	5	0
3	M30R0F	500	1120	235	215	0	4.5	2.5
4	M30R30F	500	1120	235	215	30	10	2.5

Using similar water to cement ratio, the superplasticizer dosage was adjusted in order to maintain almost identical workability having slump between 100 ± 20 mm. For this purpose, dosage of superplasticizer needs to be increased in rubberized and/or fiber reinforced mortars. In the light of previous research (Nguyen al. 2010), dosage of fibers used was 30 kg/m^3 and rubber aggregates was 30 % as partial replacement of natural aggregates. Also it was observed that by addition of rubber aggregates in mortar, there is a significant increase in air contents up to 65 % as expected in accordance with the previous literature (Fiore et al. 2014).

The elastic modulus tests and compressive tests were carried out on cylindrical specimens of size 110 mm in diameter and 220 mm in height. Compressive test and elasticity modulus test were conducted according to European Standard NF EN 12390-3 and RILEM recommendations (RILEM CPC8 1972) respectively.

Direct tensile tests were conducted on notched specimens as shown in Fig. 1. The test was carried out by using MTS 100 kN and was controlled by Crack Mouth Opening Displacement (CMOD) method by using clips which are called as COD attached on two faces at notched cross section as shown in Fig. 2. The loading rate was same as proposed by RILEM recommendation (RILEM TC 162-TDF 2001). Test was controlled by using this sensor called COD clip at two rates: from 0 to 0.1 mm at 5 μm/min and from 0.1 to 1 mm at 100 μm/min.

Also in order to check the bond between substrate and overlay material with different mortar compositions, direct tensile tests were carried out on notched prismatic specimens as shown in Fig. 3 in the same way as described above, the notch being created when casting overlay thanks to an adhesive tape.

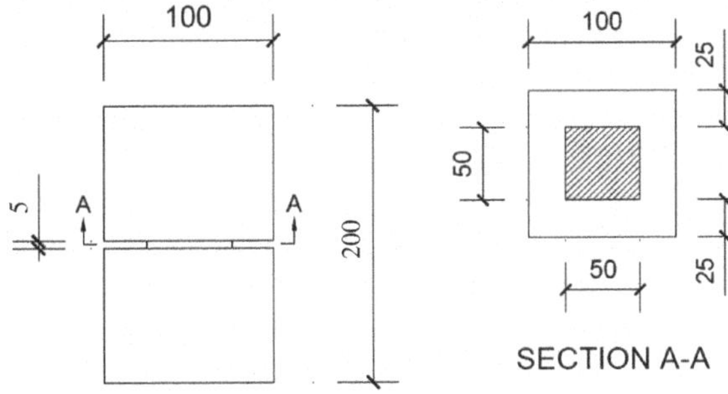

Fig. 1 Notched test specimen for direct tension tests

Fig. 2 COD attached at notch

Fig. 3 Notched specimen for bond tensile strength test (dimensions are in mm)

3 Results and Discussion

Discussion will focus on the results obtained after 28 days of curing.

3.1 Compressive Test

From the experimental results as shown in Fig. 4, it is evident that with the addition of rubber aggregates or combination of both rubber aggregates and fibers in mortar, a reduction of compressive strength is observed as compared to control mortar. This shows that results are in good agreement with the previous research of (Nguyen et al. 2010; Turatsinze et al. 2005).

Results of elasticity modulus test show that with the addition of rubber aggregates in mortar there is a significant decrement in modulus of elasticity as compared to control mortar. Also results show that presence of metallic fiber have no significant impact on modulus of elasticity as shown in Fig. 4. Similar reduction of modulus of elasticity with rubberized mortar has been observed in the research conducted by Turatsinze et al. (2005).

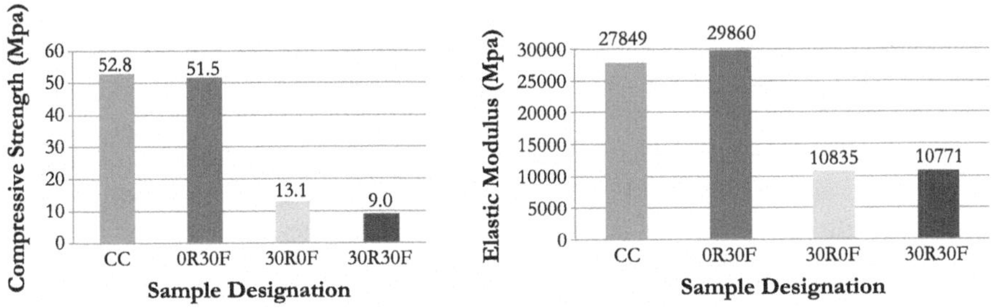

Fig. 4 Compressive strength and modulus of elasticity of different mortar mixes

3.2 *Tensile Test*

Results of direct tensile test controlled by CMOD method for all mortar mixes are presented in Fig. 5. It is evident that with the addition of 30 % rubber aggregates there is a significant reduction in tensile strength but strain capacity improved about 1.5 times with reference to control mortar as shown in Fig. 6. With the only

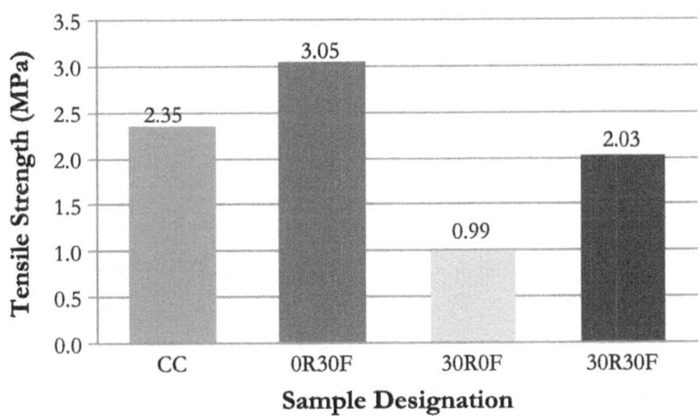

Fig. 5 Tensile strength of different repair mortar mixes

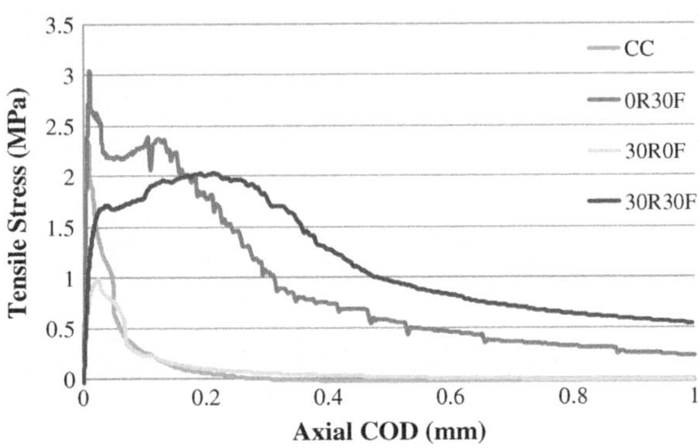

Fig. 6 Effect of rubber aggregates and fibers on strain capacity

fiber-reinforcement in mortar, a significant improvement in residual post peak strength is observed. Results also show that with the presence of both metallic fibers and of rubber aggregates not only residual post peak strength improves but also strain capacity enhances up to 3.5 times with reference to control mortar.

With regard to bond tensile test controlled by CMOD method, results for all mortar mixes are presented in Figs. 7 and 8. It indicates that improvement in bond tensile strength is observed by addition of rubber aggregates, fiber reinforcement or combination of both rubber aggregates and fibers reinforcement in repair material of overlays. This increase in bond tensile strength is because of the fact that in order to incorporate fibers and rubber aggregates in mortar certain superplasticizers and viscosity modifying agents are necessary to be used in order to make mortar workable and to prevent segregation of rubber aggregates which as a result also have impact on the material compatibility with the surface of the substrate. These repairs have a modified contact with the substrate and also cement slurry of these repair materials penetrate into porous substrate surface which as a result produce an enhanced bond between substrate and repair material. When tension test was conducted on overlay with rubber aggregates or combination of rubber aggregates and fiber reinforcement it was noticed that failure started from interface portion between repair and substrate but major part of failure occurs in repair material as shown in Fig. 3 which means that bond tensile strength is more than repair material tensile strength (Fig. 7).

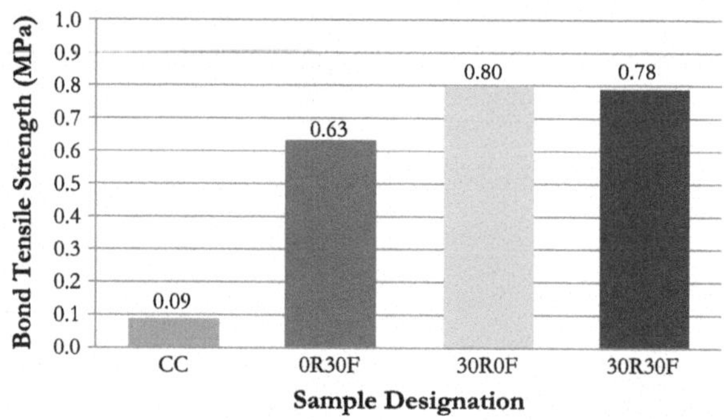

Fig. 7 Bond tensile strength variation with different repair materials

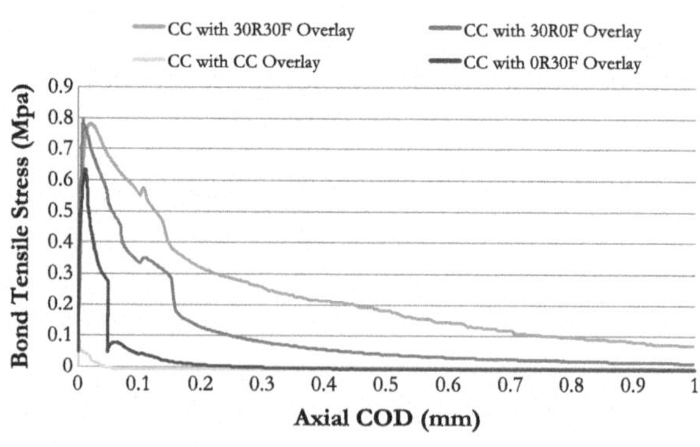

Fig. 8 Strain capacity behavior with different overlay repair materials

4 Conclusions

From the experimental results following conclusions can be drawn by incorporating rubber aggregates and metallic fibers in repair material of overlays.

1. By the addition of rubber aggregates in mortar there is a decrement in 28 days compressive strength observed as compared to control mortar.
2. Addition of rubber aggregates in repair material of overlays reduces compression modulus of elasticity significantly. This reduction of modulus of elasticity as a result improves the strain capacity of repair material.
3. It can be concluded that with the addition of rubber aggregates in repair material, tensile strength reduces significantly but along with that improvement in strain capacity of mortar is also observed. By addition of metallic fibers in mortar, residual post peak strength is significantly improved. Further it can be concluded that addition of rubber aggregates and fiber in mortar not only improves the strain capacity but also improves the residual post peak strength which is a solution to control crack opening in overlays and debonding phenomena. Also it can be said that the presence of rubber aggregates didn't change the role of fiber reinforcement. However combination of both rubber aggregates and fibers in mortar produces a positive synergetic effect.
4. Further use of rubber aggregates obtained by grinding end-of-life tyres in cement-based materials can be considered as a solution to maintain a clean environment by limiting the landfill for residual waste.

References

Bissonnette B, Courard L, Fowler DW, Granju J-L (2011) Bonded Cement-Based Material Overlays for the Repair, the Lining or the Strengthening of Slabs Pavements. State of the Art Report of RILEM Technical Committee 193-RLS. doi:10.1007/978-94-007-1239-3

Fiore A, Marano GC, Marti C, and Molfetta M (2014) On the Fresh/Hardened Properties of Cement Composites Incorporating Rubber Particles from Recycled Tires. Advances in Civil Engineering, Volume 2014, Article ID 876158. doi:10.1155/2014/876158

Nguyen T-H, Toumi A, Turatsinze A (2010) Mechanical properties of steel fibre reinforced and rubberized cement-based mortars. Mater Des 31: 641–647. doi:10.1016/j.matdes.2009.05.006

RILEM TC (1972) CPC 8 Modulus of elasticity of concrete in compression. Mater Struct 6:507-512

RILEM TC 162-TDF (2001) Test and design methods for steel fibre reinforced concrete. Mater Struct 34:3–6

Turatsinze A, Bonnet S, Granju J-L (2005) Mechanical characterisation of cement-based mortar incorporating rubber aggregates from recycled worn tyres. Build Env 40:221-226. doi:10.1016/j.buildenv.2004.05.012

Tran Q-T, Toumi A, Turatsinze A (2008) Thin bonded cement-based overlays: numerical analysis of factors influencing their debonding under fatigue loading. Mater Struct 41:951-967. doi:10.1617/s11527-007-9297-y

Debonding Mechanism of Bonded Concrete Overlay According to Horizontal Traffic Loading

Kim Young Kyu and Lee Seung Woo

Abstract Recently, bonded concrete overlay has been considered as a possible alternative material for use in pavement rehabilitation as its material properties are similar to those of the existing concrete pavements. In a bonded concrete overlaid pavement, both the overlay layer and the existing pavement should perform as one monolithic pavement and bonding between the overlay layer and the existing pavement is essential. Therefore, bonded concrete overlays are carefully constructed to achieve and maintain the bond between the overlay layer and the existing concrete pavement using milling, blasting method and bonding agent. Generally, typical debonding failure modes of bonded concrete overlay observed in the loss of bonding, failure by normal tensile stresses, and horizontal shear stresses. Previous study indicated that normal tensile stress at the interface is main cause of debonding in bonded concrete overlay. However, the bonded concrete overlay at the traffic congested section and toll booth of expressway experience due to stopping and starting of vehicle. This may lead to debonding by horizontal shear stress at the interface of bonded concrete overlay. This study aimed to investigate the debonding mechanism of bonded concrete overlay due to horizontal shear stress. A set of numerical analysis was performed in order to evaluate the horizontal shear stress at the interface of bonded concrete overlay due to vehicle stopping and starting of vehicle in the traffic congested section and toll booth of expressway.

Keywords Concrete overlay · Debonding · Tensile stress · Shear stress

K.Y. Kyu · L.S. Woo (✉)
Gangneung-Wonju National University, Gangneung, South Korea
e-mail: swl@gwnu.ac.kr

K.Y. Kyu
e-mail: kingdom1980@nate.com

A. Chabot et al. (eds.), *8th RILEM International Conference on Mechanisms of Cracking and Debonding in Pavements*, RILEM Bookseries 13,
DOI 10.1007/978-94-024-0867-6_88

1 Introduction

Bonded concrete overlays have advantages in terms of structural performance since the overlay layer and the existing pavement perform as a monolithic layer. Bonding between the two layers is essential to prevent early distress and to secure long-term performance. The bonding property ensures that the overlay and existing pavement layer perform as one pavement until that continues to carry a significant portion of the traffic and environmental loading.

Generally, three types of failure modes, the loss of bonding, bond failure by normal tensile stresses, and bond failure by horizontal shear stresses, are commonly seen in bonded concrete overlays (Chavez et al. 2007). Normal tensile stresses at the edges and corners of the overlay interface are higher than the stresses at the interior. The bond effect is the highest at the very edge and corner of the bonded interface. This is due to curling and warping stresses at the top of the overlay, as the temperature and moisture conditions change more rapidly at the top than at the bottom of the slab as shown in Fig. 1a. In addition, normal tensile stress can arise due to the influence of traffic loads. On the other hand, horizontal shear stresses occur as a result of the influences of shrinkage/expansion due to traffic loads and temperature changes, as shown in Fig. 1b. If the overlay material has a CTE (coefficient of thermal expansion) that is equal to or greater than that of the existing pavement, it will expand or shrink more than the existing pavement. This is resulted by shear stresses forming at the interface which may cause the overlay layer to crack (Semen and Rollings 2005; Granju 2001; Lange and Shin 2001).

According to Kim and Lee (2013), in addition, the normal tensile stress increased considerably with an increase in the modulus of elasticity and the coefficient of thermal expansion of overlay material in contrast to the horizontal shear stress. The normal tensile stress was dominated by the environmental loading, while the horizontal shear stress was dominated by the vertical traffic loading. It means that bonding failure occurred due to both of them, though normal tensile stress prevailed over horizontal shear stress. In an analysis of bond stresses by normal tensile stresses at the interface of bonded concrete overlays carried out to find the causes of debonding, bond-fatigue failure by cyclic loading needed to be evaluated as well as the bond strength by monotonic loading, as the bond stresses exceeded

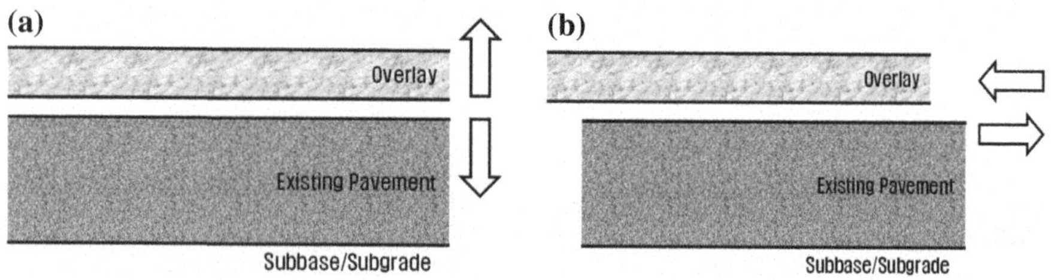

Fig. 1 Bond failure modes of the bonded concrete overlay: **a** bond failure by normal tensile stresses, and **b** bond failure by horizontal shear stresses

minimum 0.6 MPa which is close to the bond strength criterion of 1.4 MPa excluding the ultimate analysis conditions. In addition, the maximum bond stress can exceed the bond strength given the critical conditions of a high coefficient of thermal expansion of the overlay material and a thin overlay thickness (Kim and Lee 2013).

In previous study, normal tensile stress at the interface is main cause of debonding in bonded concrete overlay. However, horizontal shear stress due to stopping and starting of vehicle at the traffic congested section and toll booth in expressway may be the potential cause of debonding by horizontal loading. This study aimed to investigate the debonding mechanism of bonded concrete overlay due to horizontal shear stress. A set of numerical analysis was performed in order to evaluate the horizontal shear stress at the interface of bonded concrete overlay due to vehicle stopping and starting of vehicle in the traffic congested section and toll booth of expressway.

2 Bonded Concrete Overlay Modeling for a FE Analysis

2.1 Modeling of Bonded Concrete Overlay

This study investigated the bond behavior of bonded concrete overlay by means of a three-dimensional finite element analysis. The element type for the bonded concrete overlay modeling consists of C3D8R (an 8-node brick element with reduced integration, solid). The slab size is 3800 mm in X-direction and 6000 mm in Y-direction. The applied overlay thickness is 50 mm, which normally constructed in Korea, with the existing PCC thickness of 250 mm as shown in Table 1. This is used to estimate the bond stress, such as the normal tensile stress and the horizontal shear stress at the interface of the bonded concrete.

Table 1 Material properties and various conditions for FE analysis

	Overlay	Existing PCC	Lean	Subbase	Subgrade
Modulus of elasticity (MPa)	25,000 to 35,000	27,000	3800	206	70
Poisson's ratio	0.15	0.15	0.2	0.35	0.4
Unite weight (t/m^3)	2.4	2.4	–	–	–
Layer thickness (mm)	50	250	150	300	100
Loading cases	Corner (point A), edge (point B), interior (point C)				

2.2 Material Properties for FE Analysis

Table 1 shows the material properties and various conditions for FE analysis. The material properties of the existing pavement are fixed variables with changes in the modulus of elasticity.

2.3 Horizontal Traffic Loading Conditions

Assuming that the traffic load is 4.1 tons for a total of 8.2 tons of the equivalent single-axle load (ESAL) applied to both axles, square-type traffic loads are applied with an axle distance of 2000 mm, as shown in Fig. 2a. Furthermore, to simulate loads at various point, the traffic loads are applied on the corner point (Point A), edge point (Point B), and wheel path of the slab center (Point C), as shown in Fig. 2b. According to Tayebali et al. (2004), the horizontal force was varied from zero to 0.7 of vertical force. The maximum value of 0.68 of vertical force (Tayebali et al. 2004), described before in the shear stress analysis, is based on the surface friction coefficient representing the maximum horizontal force at approximately 50 km/h. In this study, therefore, coefficients of friction between wheel and pavement surface are assumed in the range from 0.2 to 0.8, horizontal load can be defined by multiply this coefficient with the vertical load value. The values are produced by following equation;

$$F = f \times W \tag{1}$$

where,

F Horizontal load on contact surface between wheel and pavement surface
f Coefficient of friction between wheel and pavement surface
W Vertical load from the wheel

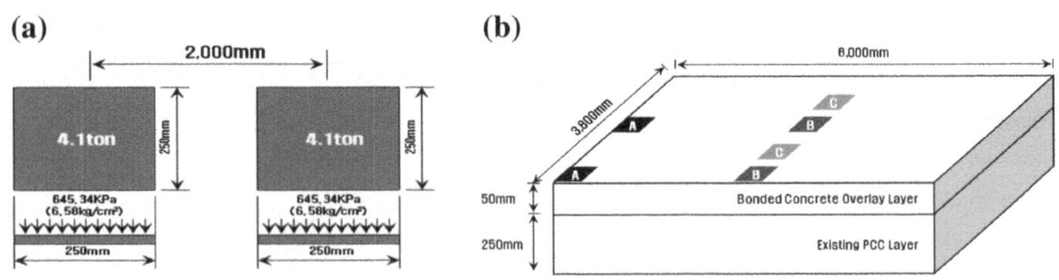

Fig. 2 Traffic loading conditions: **a** vertical traffic loads, and **b** loading points

3 Evaluation of Bond Stresses According to Horizontal Loading

3.1 Horizontal Shear Stress with Various Conditions for FE Analysis

Figure 3 shows the maximum bond stresses at the interface of bonded concrete overlay with the change of elastic modulus of the overlay material and the given elastic modulus of the existing PCC of 27 GPa. The horizontal shear stress greatly increased about 0.2 to 1.0 MPa when the horizontal loading is applied on the slab corner (Point A). However, the maximum horizontal shear stress, when the horizontal loading is applied on the slab edge and interior wheel path, was less than 0.5 MPa.

3.2 Debonding Effects by Horizontal Shear Stress

ACPA (1990) reported that a bond strength criterion of 1.4 MPa of the horizontal shear strength is generally used for quality control during construction projects. The bond stress due to horizontal shear stress, when the horizontal loading applied on the slab corner with the changes of elastic modulus from 25 to 35 GPa, did not exceed the bond strength criterion. However, cyclic stress at the interface of bonded concrete overlay may occur due to the repetition of horizontal traffic loading since the stress level is exceeded 60 %. Bond-fatigue failure by cyclic loading is needed to be evaluated because the bond stress caused by monotonic loading may exceed 0.8 MPa which is close to the bond strength criterion 1.4 MPa. In addition, the horizontal shear stress may be increased if the loading is applied by heavy axle and

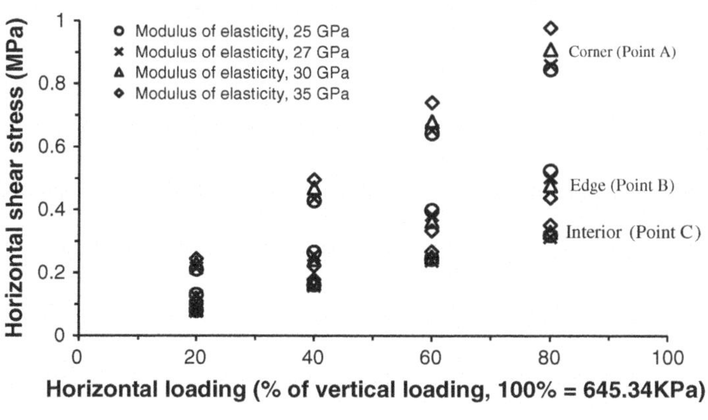

Fig. 3 Horizontal shear stresses at the interface caused by horizontal loading

thermal expansion. Therefore, it needs to be investigated the debonding failure of bonded concrete overlay due to the horizontal shear stress by various traffic and environmental conditions.

4 Conclusions

This study was performed to investigate the debonding mechanism of bonded concrete overlay due to the horizontal traffic loading as influenced by vehicle stopping and starting. Several important conclusions are summarized below:

- At the slab corner, the maximum horizontal shear stress increased considerably from 0.2 to 1.0 MPa with the increasing of elastic modulus of overlay material.
- On the other hand, the maximum horizontal shear stress was less than 0.5 MPa at the slab edge and interior.
- It means that, at the slab corner, the bond failures may not be caused by horizontal shear stress due to horizontal loading. However, the results of this study indicated that the debonding at the interface of bonded concrete overlay may arise due to potential bond failure caused by repetition of horizontal traffic loading since the horizontal shear stress exceeded 60 % of bond strength criterion.
- In addition, this study aimed to evaluate the horizontal shear stress due to traffic load as 8.2 tons equivalent single-axle load. The horizontal shear stress may be increased if the loading condition is applied by the heavy axle. Therefore, further study is needed to investigate the debonding failure of bonded concrete overlay according to the horizontal shear stress caused by various traffic and environmental conditions.

Acknowledgement This research was financially supported by the Ministry of Education (MOE) and National Research Foundation of Korea (NRF) through the Human Resource Training Project for Regional Innovation (No. 2014H1C1A1067082).

References

American Concrete Pavement Association (ACPA), 1990a. Guidelines for Bonded Concrete Overlays, *Technical Bulletin TB-007P*, Arlington Heights, IL.

Chavez M., C.I., Choi, S.C., and Won, M., 2007. Concrete Pavement Overlays and Failure Mechanisms, *Certer for Transportation Research*, Report No. 0-4893-2.

David A. Lange., and Hak-Chul Shin, 2001. Early Age Stresses and Debonding in Bonded Concrete Overlays, *Transportation Research Record 1778*, Paper No. 01-0410, pp.174 ~ 181.

Granju J. L., 2001. Debonding of Thin Cement-based Overlays, *Journal of Materials in Civil Engineering*, 13(2), pp. 114 ~ 120.

Kim, Y. K., and Lee, S. W., 2013. Performance Evaluation of Bonded Concrete Overlay, *Construction and Building Materials*, Vol. 49, pp. 464 ~ 470.

Peter M. Semen, Raymond S. Rollings, 2005. A Review and Discussion of Current Developments Involving Bonded Concrete Overlays of Airfield Pavements, *International Conference on Concrete Pavements*, pp. 900 ~ 910, Colorado, USA.

Tayebali AA, Rahman MS, Kulkarni MB and Xu Q., 2004. A Mechanistic Approach to Evaluate Contribution of Prime and Tack Coat in Composite Asphalt Pavements. Research Report 2001-04, Department of Civil Engineering, North Carolina State University, USA.

Study of Delamination and Cracking of Multilayered Systems Used as Waterproofing and Wearing Course Solutions for Orthotropic Steel Bridge Decks: 10 Years Feedback

Bertrand Pouteau and Kamal Berrada

Abstract Specific products are designed for truck traffic lanes on orthotropic steel bridge decks. These products are multilayered systems that must meet requirements for two functions: waterproofing and wearing course. This paper focuses on a specific test described in the standard NF P 98 286. This durability test reproduces mechanical behavior of the steel deck in the vicinity of a stiffener. A typical test specimen is comprised of a 10–14 mm thickness steel plate covered by a multi-layered waterproofing system including asphalt layer from 50 to 80 mm thickness. After surface preparation of the steel plate in a blowing chamber, the test sample is manufactured in laboratory. A waterproofing layer is applied on the steel plate. An asphalt mix is compacted over substrate elements with a wheel track compactor. After sawing, strain sensors are glued on the asphalt mix sides of the specimen near the stiffener. The specimen is subjected to fatigue flexural load at control temperature that produces two kinds of distress: delamination and cracking. The authors present ten years of feedback on the use of this test at Eurovia Research Center in Bordeaux. This paper focuses on the study of the slab calibration and proposes as a conclusion to extend the standard specifications to any slab thickness.

Keywords Fatigue damage · Orthotropic steel-deck · Cracking · Debonding

---

B. Pouteau (✉) · K. Berrada
Eurovia Research Center, 22 Rue Thierry Sabine, 33290 Mérignac, France
e-mail: bertrand.pouteau@eurovia.com

K. Berrada
e-mail: kamal.berrada@eurovia.com

A. Chabot et al. (eds.), *8th RILEM International Conference on Mechanisms of Cracking and Debonding in Pavements*, RILEM Bookseries 13,
DOI 10.1007/978-94-024-0867-6_89

1 Introduction

Specific products are designed for the pavement on orthotropic steel bridge decks. These products are multilayered systems that must meet requirements for two functions: waterproofing and wearing course. Their durability regarding truck traffic is essential. In case of cracks appearance, the waterproofing layer can be damaged and results in deck corrosion.

A fatigue test, named five points bending test, has first been designed in the early 70s by former LCPC and SETRA (Hameau et al. 1981) to reproduce the most severe conditions of the waterproofing products. This test method resulted into the standard NF P 98 286 in 2006 and was supported by Houel works (Arnaud and Houel 2010; Houel 2008). EUROVIA team took part in the standardization and has been conducting since then. During this 10 years of experimental works, finite element modeling 2D and 3D were developed to support the calibration and for the cracking and debonding analysis.

After a brief description of the test, this paper focuses on the findings: generalization of test to any slab thickness, impact of an edge debonding on the test severity and ability of the sensors set to detect cracks.

2 Test Description

For each waterproofing product the experimental campaign consist in 2 tests of 1 million cycles at −10 °C and 2 million cycles at +10 °C. The vertical load cycles are applied at a frequency of 4 Hz as shown on the Fig. 1. Resistance to cracking is required to pass the test. The criteria of success is the non appearance of any cracks upon the stiffener.

The intensity of the vertical load is determined after a calibration test on a virgin steel plate following the standard specification. These values are provided for thicknesses of 10, 12 and 14 mm only.

The following failures modes can be observed on the specimen:

1. Surface cracking upon the stiffener
2. Debonding at the sample edges
3. Permanent deformation of the asphalt concrete under the loading plates.

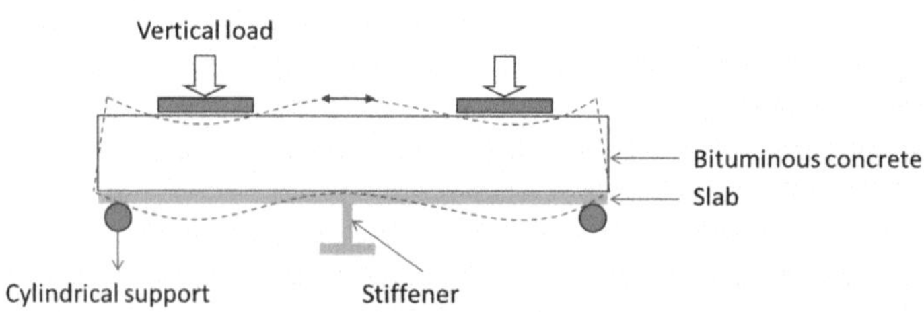

Fig. 1 Test description

Table 1 Maximum stress to be applied and corresponding deformation based on slab thickness

Slab thickness (mm)	Maximum stress upon the stiffener (MPa)	Maximum deformation admitted upon the stiffener (udef)
10	200	1000
12	125	625
14	90	450
>14	To be determined	To be determined

3 Study Issue

The table below is taken from the NF P 98-286 standard (NF P 98 286 Détermination de la résistance à la fatigue d'une étanchéité/roulement sur tôle métallique 2006). It specifies the maximum stress value that can be undergone by the sample and the associated admitted deformation according to the slab thickness (Table 1).

As part of the study of the Jacques Chaban Delmas bridge surface layer (Bordeaux, FRANCE), Eurovia had to determine the tests conditions for three slab thicknesses: 16, 19 and 20 mm which corresponded to the steel plate thickness at different locations of the bridge deck.

4 Test Modeling

Finite element modeling was carried out using the software abaqus (Dassault Systèmes Simulia Corp). Quarter of the sample is represented due the symmetric geometry. Figure 2 shows the sample geometry for the calibration stage. The modeling parameters are shown in Table 2. The elastic modulus for the steel material is 200,000 MPa and the Poisson's ratio is 0.3.

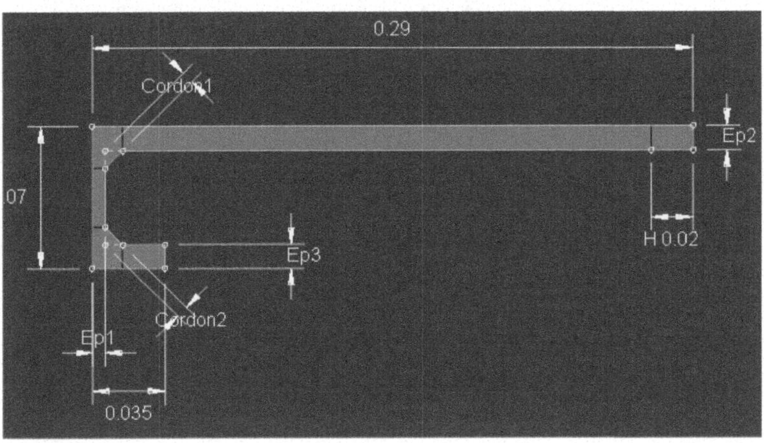

Fig. 2 Geometry description

Table 2 Model parameters

Name	Value (m)	Expression
e	12e-03	0.012
a	6e-03	0.006
Ep1	6e-03	e/2
Ep2	12e-03	e
Ep3	12e-03	e
Cordon1	6e-03	a
Cordon2	6e-03	a

For the durability test stage, this model has been completed by adding the multilayered waterproofing system on top of the slab composed of a membrane (thickness: 2 mm, Elastic modulus 100 MPa) and an asphalt concrete (Elastic modulus = 10000 MPa (arbitrary choice), Poisson ratio = 0.35 Thickness = 70 mm). All modeling are done under elastic linearity assumption.

Computations were carried out in order to:

1. Determine a relation between the loss of durability and an edge debonding between the waterproofing layer and the slab.
2. Determine the ability for the sensors to detect cracking (initiation and propagation) regarding the crack size and the product thickness.

5 Modeling Results and Analysis

Table 3 shows the modeling results of the maximum stress upon the stiffener corresponding to a vertical load of **40 kN**. This value is an arbitrary choice but is representative of the common loading level during laboratory tests.

By crossing Table 1 and Table 3, the numerical values of the vertical load corresponding to the standard values can be calculated. The corresponding mechanical pressure is also calculated using the contact surface Table 4.

These pressures are very close to 0.662 MPa which is the mechanical pressure reference in the French pavement structure model. The disparity between the values and the reference may result from rounded values.

Table 3 Modeling results for a vertical load of 40 kN

Slab thickness (mm)	Maximum stress upon the stiffener (MPa)
10	216
12	146
14	104.1

Table 4 Theoretical vertical loads and corresponding pressure for standard thicknesses

Slab thickness (mm)	Maximum stress upon the stiffener (MPa)	Corresponding vertical load (kN)	Corresponding mechanical pressure (MPa)
10	200	37.040	0.712
12	125	34.257	0.659
14	90	34.583	0.665

Table 5 Maximum stress to be applied and corresponding deformation based on slab thickness

Slab thickness (mm)	Maximum stress upon the stiffener (MPa)	Corresponding mechanical pressure (MPa)
10	186	0.662
12	126	0.662
14	90	0.662
16	67	0.662
19	45	0.662
20	40	0.662

Table 5 proposes new values for the vertical loads for all the slab thicknesses using a mechanical pressure of 0.662 MPa.

For each new test, the pressure applied to the specimen will be increased until obtaining the maximum value that correspond the value provided on the table above. The associated vertical load will be used to conduct the test.

6 Conclusion

The finite element modeling permitted to determine the link between the slab thickness and the maximum stress upon the stiffener for a given bituminous concrete stiffness and thickness. Thanks to this relation, it appears that the mechanical pressure used for the specifications is close to the standard mechanical pressure used in the French pavement structure model standard.

In the light of this result, we propose to take those new values to determine the vertical load for the test whatever the steel plates thickness.

The severity of the test is increased by a debonding at the sample extremities.

The next study will focus on surface cracking detection methods for very thin waterproofing product (<1 cm).

References

G Hameau, C Puch, A-M Ajour (1981) Comportement à la fatigue en flexion sous moment négatif. Bull LPC n°111,p 29-38

L Arnaud, A Houel (2010) Modelling of pavement materials on steel decks using the five-point bending test: thermo mechanical evolution and fatigue damage. doi:10.1088/1757-899X/10/1/012060

A Houel (2008) Endommagement à la fatigue et fissuration mécanique des enrobes bitumineux sur dalle orthotrope, Ecole doctorale MEGA-ENTPE, France, 2004

NF P 98 286 (2006) Détermination de la résistance à la fatigue d'une étanchéité/roulement sur tôle métallique - Méthode d'essai sur banc de fatigue en flexion sous moment négatif

Dassault Systèmes Simulia Corp The Abaqus Software is a product of Dassault Systèmes Simulia Corp, Providence, RI, USA © Dassault Systèmes, 2011.

Effect of Hydrodemolition and Cutting Diameter of Pull-off Test on the Bond Strength of LMC Bridge Deck Pavement

Kyong-Ku Yun, Kyeo-Re Lee, Seung-Yeon Han and Kyong Namkung

Abstract Latex modified concrete (LMC) has many advantages against an ordinary Portland concrete. Hundreds of projects have been completed using LMC for overlay in Korea. A sound bond is an essential requirements of bridge deck pavements, and the bond property of a bonded overlay to its substrate concrete during the life-time is one of the most important performance requirements. This paper compare and evaluate the effect of the hydrodemolition and cutting diameter of the pull-off test into the bond strength of latex-modified concrete overlay by a series of experimental variables. The concrete surface preparation methods included hydrodemolition, cold milling and jack hammer. The core cutting diameters of pull-off test included four levels: 50, 75, 100 and 150 mm.

Keywords LMC · Bond strength · Pull-off · Hydrodemolition

1 Introduction

Latex-modified concrete (LMC) is made by using a composite binder of cements and latexes. LMC has many advantages against an ordinary Portland concrete Ramakrishnan (1992). It has a higher strength, an improved durability, a good

K.-K. Yun (✉)
Department of Civil Engineering, Center for Regional Construction Technology, Kangwon National University, 1 Kangdae-Gil, Chuncheon 200-701, South Korea
e-mail: kkyun@kangwon.ac.kr

K.-R. Lee · S.-Y. Han · K. Namkung
Department of Civil Engineering, Kangwon National University, 1 Kangdae-Gil, Chuncheon 200-701, South Korea
e-mail: ds2vqk@nate.com

S.-Y. Han
e-mail: seungyeon0420@nate.com

K. Namkung
e-mail: doadsa@naver.com

A. Chabot et al. (eds.), *8th RILEM International Conference on Mechanisms of Cracking and Debonding in Pavements*, RILEM Bookseries 13,
DOI 10.1007/978-94-024-0867-6_90

resistance to corrosion, a reduced water permeability and a greater resistance to damage from freeze-thaw cycles (Sprinkel 1980). Common applications of LMC are overlays of new bridge decks. Hundreds of projects have been completed using LMC for overlay in Korea. A sound bond is an essential requirements of bridge deck pavements, and the bond property of a bonded overlay to its substrate concrete during the lifetime is one of the most important performance requirements.

Hydrodemolition is a method of surface preparation using high-pressure water jets. A hydrodemolition system consists of high-pressure water pump; a robotic cutting head; and a trailer. Hydrodemolition cutting depth is dependent on the length of time that the water jet is directed at the concrete surface that determines the depth of removal, and contact time is controlled at the robotic cutter. During hydrodemolition concrete removal, the cement matrix is removed from the aggregate, leaving bulk debris of sand and aggregate to be collected and disposed.

A standard bond strength measurement method is required at field for screening, selecting materials and quality control for overlay or repair materials (Kuhlmann 1990). The bonding capacity of the prepared surface can be randomly checked using a direct tensile pull-off test, as described in ASTM C1583, Standard Test Method for Tensile Strength of Concrete Surfaces and the Bond Strength or Tensile Strength of Concrete Repair and Overlay Materials by Direct Tension (Pull-off method) (ACI Committee 503 1992).

The pull-off test method may measure the bond strength, in situ, between a repair overlay and the substrate. The pull-off test involves applying a direct tensile load to a partial-depth core. The test is used for a variety of purposes, from monitoring strength development and quality control of newly cast concrete, to the condition assessment of older concretes.

This study was performed by a series of experimental program in order to investigate the effect of the hydrodemolition and cutting diameter of the pull-off test into the bond strength of latex-modified concrete overlay (Yun et al. 2001, 2002).

2 Experimental Program

The experimental variables were concrete surface preparation methods of hydrodemolition and cutting diameters of pull-off test. The concrete surface preparation methods included hydrodemolition, cold milling and jack hammer. The core cutting diameters of pull-off test included four levels: 50, 75, 100 and 150 mm.

2.1 Concrete Mixtures

The concrete used for the substrate had a water-cement ratio of 0.49 and a maximum aggregate size of 25 mm, which targeted high strength concrete of 40.0 MPa. Three cylinders, 100 × 200 mm, to determine compressive strength and three beams, 100 × 100 × 450 mm, to determine flexural strength, were also cast from each mix. The measured compressive strength and flexural strength at 28 days of base concrete were 44.1 and 6.9 MPa, respectively.

The concrete used for overlay was latex-modified concrete (LMC). The LMC with very-early strength cement had a 0.38 water-cement ratio, 0.15 latex solids/cement, 58 % fine aggregate, 13 mm maximum aggregate size, and 390 kg/m^3 of cement. The measured compressive strength and flexural strength at 28 days of LMC were 50.0 and 8.2 MPa, respectively.

2.2 Concrete Slabs

Six slabs of 1100 × 1100 × 200 mm with OPC were cast as a substrate; the height of the slabs was determined considering real bridge deck slab. After casting, all slabs were cured in moisture for 7 days with wet burlap, followed by air curing in the laboratory of 20 °C and 50 % RH for an additional 21 days.

All substrate slabs were treated by a surface preparation method (hydrodemolition, cold milling and jack hammer) to remove concrete laitance and extraneous material that could inhibit bond. Loose particles and dust were thoroughly removed from all slab surfaces using high pressure air. All of the slabs were wetted and kept moist for 24 h before casting the overlay. After the overlay casting the slabs were moist cured for 48 h using wet burlap, followed by air curing until pull-off test. The overlay of latex-modified concrete was 50 mm thick, similar to LMC overlays in the field.

2.3 Hydrodemolition and Other Surface Preparations

A sound bond of LMC overlay to its substrate concrete during the lifetime is one of the most important performance requirements which should be quantified. The Guide of Korea Highway Corporation describe the required bond strength to be 1.4 MPa for the bonded concrete overlay.

A hydrodemolition is one method to achieve the sound bond using high-pressure water jets. The removal of concrete surface would be dependent on the water pump, robot types, water pressure, water flow, nozzle number, nozzle diameter, shooting angle, and duration times. Two water system were adopted in this study: One water robot has 2 nozzles with 110 MPa and 195 LPM (liters per minute) water flow. The other one has 4 nozzles with 250 MPa and 82 LPM water flow. Jack hammer and cold milling are another concrete surface preparation methods.

2.4 Pull-off Test

A pull-off test method was adopted to measure the bond strength, in situ, between the LMC overlay and the substrate slab. The pull-off test involves applying a direct tensile load to a partial-depth core. Cores with 3 diameters were drilled from top surface to the designated depths. To minimize damage from coring, cores were cut by fixing the equipment with anchor bolts. Steel plates were glued to the top surface of the core and a pure axial tensile load was applied at a constant rate.

3 Experimental Results and Discussions

The failure mode of pull-off test could be grouped into 4 types (steel disk interface, overlay material, concrete interface, and base concrete material). This study, however, only had 3 types because failure never occurred solely in the latex-modified overlay due to its high tensile strength. Thus, the three failure modes for this study were:

Failure Mode I: steel disc interface or mixed with overlay concrete near at disc
Failure Mode II: substrate concrete
Failure Mode III: concrete interface.

3.1 Effect of Hydrodemolition

Table 1 compares the bond strength results with concrete surface preparation methods: jack hammer, cold milling, and hydrodemolition. The bond strength was the highest at hydrodemolition as to be over 1.85 MPa, followed by cold milling as to be over 0.71 MPa and, and by hand jack hammer as to be over 0.60 MPa. The reason for the "over 1.85 MPa" at hydrodemolition is that all three samples were failed at steel disk interface (Mode I). The bod strength was the biggest at the hydrodemolition as anticipated. This would be due to the sound surface after removal of concrete.

The failure modes were at the steel disk interface (Mode I) for hydrodemolition, at the concrete interface (Mode III) for cold milling, and at the substrate (Mode II) for jack hammer. This is, also, due to the sound surface after removal of concrete.

Table 1 Bond strength Results with concrete surface methods

Preparation method	Jack hammer				Cold milling				Hydrodemolition			
Core	1	2	3	mean	1	2	3	mean	1	2	3	mean
Bond strength (MPa)	0.36	0.89	0.55	0.6	0.86	0.70	0.56	0.71	Over 1.51	Over 2.14	Over 1.91	Over 1.85
Failed area	substrate				Concrete interface				Steel disk interface (epoxy bonding)			

3.2 *Effect of Core Cutting Diameter*

The experimental effects of cutting diameter on failure mode and failure strength at each core diameter (D = 50, 75, 100 and 150 mm) are shown at Fig. 1. The failures at D50 and D75 were totally governed by mode II, while the failures at D100 and D150 were mostly governed by mode III. These trends shows higher strength from base concrete and lower strength from concrete interface.

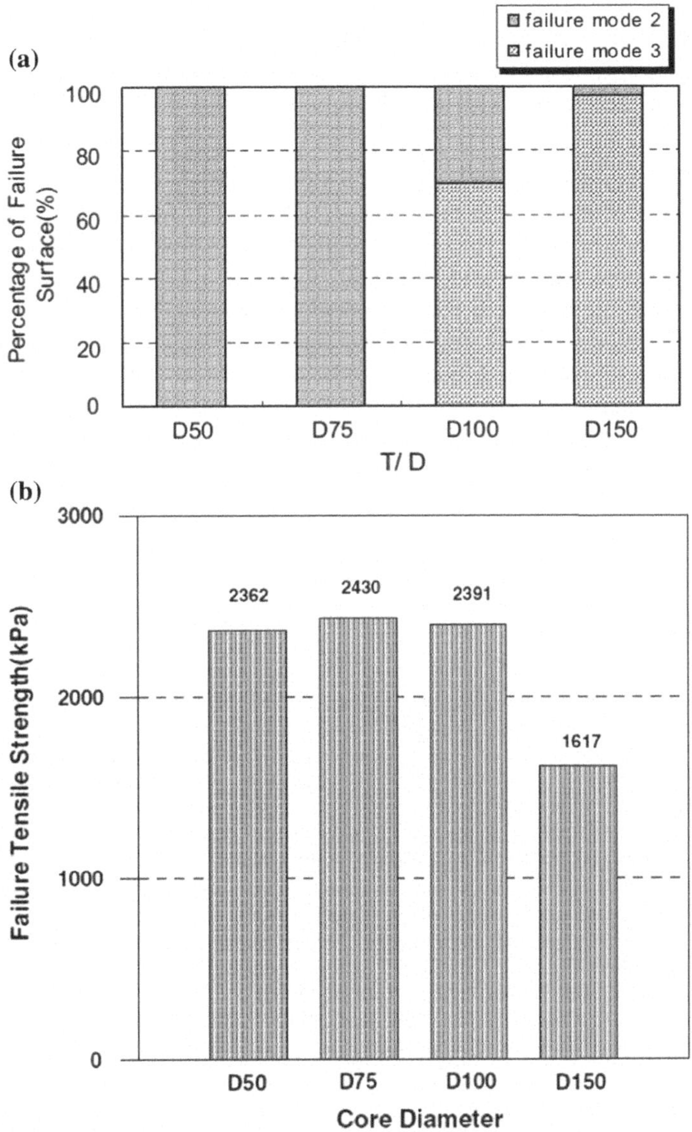

Fig. 1 Effect of cutting diameter of pull-off test **a** on failure mode **b** on tensile strength

For theoretical purposes, in order to measure the adhesive strength at concrete interface the core diameter should be larger than 100 mm. For practical purposes, to reduce the damage from coring the smaller core diameter may be better. Thus, 100 mm of core diameter could be the optimum size based on these theoretical and experimental results.

4 Conclusions

The purpose of this study was to evaluate the effect of the hydrodemolition and cutting diameter of the pull-off test into the bond strength of latex-modified concrete overlay by a series of experimental variables. The experimental results are following:

1. The bond strength was the highest at hydrodemolition as to be over 1.85 MPa, followed by cold milling as to be over 0.71 MPa and, and by hand jack hammer as to be over 0.60 MPa. The reason for the "over 1.85 MPa" at hydrodemolition is that all three samples were failed at steel disk interface (Mode I). The bod strength was the biggest at the hydrodemolition as anticipated.
2. The failure modes were at the steel disk interface (Mode I) for hydrodemolition, at the concrete interface (Mode III) for cold milling, and at the substrate (Mode II) for jack hammer.
3. The failures at D50 and D75 were totally governed by mode II, while the failures at D100 and D150 were mostly governed by mode III. For theoretical and practical purposes, 100 mm core diameter would be the optimum.

References

Ramakrishnan, V.(1992), Latex Modified Concrete and Mortars, NCHRP Synthesis 179, Transportation Research Board, National Research Council, Washington, D. C.

Sprinkel, M. M.(1980) Twenty Year Performance of Latex Modified Concrete Overlays, Transportation Research Record 1335, TRB, National Research Council, Washington, D.C., pp. 27-35.

Kuhlmann, L. A. (1990),Test method for Measuring the Bond Strength of Latex Modified Concrete and Mortar, ACI Materials Journal, Vol.87, NO.4

ACI Committee 503 (1992), Guide for the Selection and use of Structural Adhesives with Concrete, ACI Materials Journal, Vol.89, Issue 1, 1992.

Yun, K.K., Hong, C.W., Lee, J.H. and Choi, S.L.,(2002) "Strength Development and Permeability of Latex-Modified Concrete with Rapid-Setting Cement (in Korean)," Journal of Korea Concrete Institute, Vol.14, No.3, June, 2002, pp299-306.

Yun, K.K., Lee, J.H., Choi, S.L. and Kim, K.H.(2001), " Bond Strength Properties of Latex Modified Concrete (in Korean)", Journal of Korea Concrete Institute, Vol.13, No.5, October, 2001, pp507-517.

Part XI
Advanced Measurement Systems for Crack Characterization: Advance Measurements

Instrumentation of Large Scale Direct Shear Test to Study the Progressive Failure of Concrete/Rock Interface

Hussein Mouzannar, M. Bost, P. Joffrin, C. Pruvost, F. Rojat, J. Blache, A. Houel, M. Valade, A. Khadour, S. Chataigner, L. Gaillet, J.F. David, Y. Falaise and M. Quiertant

Abstract The shear strength of concrete/rock interface must be clearly assessed to design the rock foundations. However, the concrete/rock interface has different failure modes which depend on the mechanical characteristics of the materials and the morphology of the rock surface. Consequently, to study thoroughly the shear behavior of concrete/rock interface and to define the cracking mechanisms and debonding propagation along the interface, different instrumentation devices were applied during a large scale direct shear test. A large sample formed by pouring concrete over a large rock surface (shear surface of 1.5 m^2), was instrumented using the following techniques: strain gauges, acoustic emission and distributed optical fiber sensor. The different observations made during the test were combined to determine the successive stages of interface failure. The strain gauges, which measure local strains within the materials, showed their ability to detect the crack propagation in the materials and the debonding at the interface. The acoustic emission technique was able to detect the damages in the materials and at their interface. The optical fiber sensor measured strains evolution of the concrete near the interface showing the crack propagation during the test and the progressive debonding at the interface.

H. Mouzannar (✉) · M. Bost · P. Joffrin · C. Pruvost
IFSTTAR, GERS, RRO, 69500 Bron, France
e-mail: hussein.mouzannar@ifsttar.fr

F. Rojat · J. Blache · A. Houel · M. Valade
CEREMA, DLL, Bron, France
e-mail: fabrice.rojat@cerema.fr

A. Khadour
IFSTTAR, COSYS, LISIS, Marne-la-Vallée, France
e-mail: aghiad.khadour@ifsttar.fr

S. Chataigner · L. Gaillet · J.F. David · Y. Falaise
IFSTTAR, MAST, SMC, Nantes, France
e-mail: sylvain.chataigner@ifsttar.fr

M. Quiertant
IFSTTAR, MAST, EMMS, Marne-la-Vallée, France

A. Chabot et al. (eds.), *8th RILEM International Conference on Mechanisms of Cracking and Debonding in Pavements*, RILEM Bookseries 13,
DOI 10.1007/978-94-024-0867-6_91

Keywords Concrete/rock interface · Strain gauges · Acoustic emission · Optical fiber

1 Introduction

Rock foundation is usually designed to avoid sliding at the concrete/rock interface. The shear strength of the concrete/rock interface has thus to be evaluated.

The shear strength of the concrete-rock interface is mostly obtained by performing laboratory shear tests on specimens of smaller dimensions than those of the rock foundation. According to CFBR (2013), for dam foundations, the values obtained in laboratory are smaller than those evaluated by back-analysis on the real scale.

In this work, a large concrete/rock interface was tested in a large direct shear box. To fully understand the behavior of the tested rock/concrete interface, this sample was instrumented using the following technologies: strain gauge, acoustic emission and distributed optical fiber sensor. These different instrumentation devices aimed at identifying cracking mechanisms and debonding propagation along the interface during the large scale direct shear test.

In the following, the large scale direct shear box is first introduced. Then, the different instrumentations used are described, and finally, the main results of the shear test are presented and analyzed.

2 The Large-Scale Shear Test Apparatus

The large-scale shear test apparatus (Fig. 1) is a world-unique giant shear box owned by the soil and rock mechanics Laboratory of CEREMA (Center for Expertise and Engineering on Risks, Urban and Country Planning, Environment and Mobility) in Lyon, France.

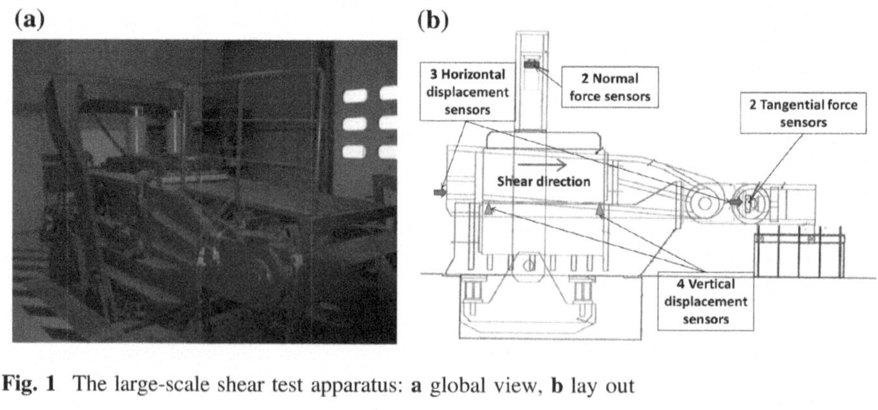

Fig. 1 The large-scale shear test apparatus: **a** global view, **b** lay out

The machine consists of a static lower box and a mobile upper box, of $1 \times 1.5 \times 0.625\ m^3$ each. The global behavior of the sample is measured during the test through force sensors positioned on each jack and through sensors that monitor horizontal (3 sensors) and vertical (4 sensors) displacements. This giant experimental device can be used for various purposes. Some examples of large scale testing programs conducted with this machine on rock bolts can be found in Maiolino and Pellet (2015) or in Bidaut et al. (2006) for instance.

In the research program presented here, the machine allowed to obtain the shear failure of intact rock/concrete contacts at a metric scale. To fully apprehend the behavior of the tested rock/concrete interface, a complete instrumentation program was undertaken on the samples, as shown in the following sections.

3 Large Scale Direct Shear Test

3.1 Sample Description

The sample was formed by an upper concrete half-sample poured directly on a lower half-sample formed of a granite rock block with a natural rough surface. The rock block was cleaned with a water high-pressure cleaner before concrete casting. The contact between the two half-samples was the object of this study (Fig. 2a). Each half-sample had a volume of 1.5 (L) × 1 (W) × 0.625 (H) m^3. The lower half-sample represents the rock foundation and the upper one the civil engineering structure. A concrete layer below the rock part was poured in order to adjust the best-fit plane of the natural rock surface at the level of the shear plane of the large shear apparatus.

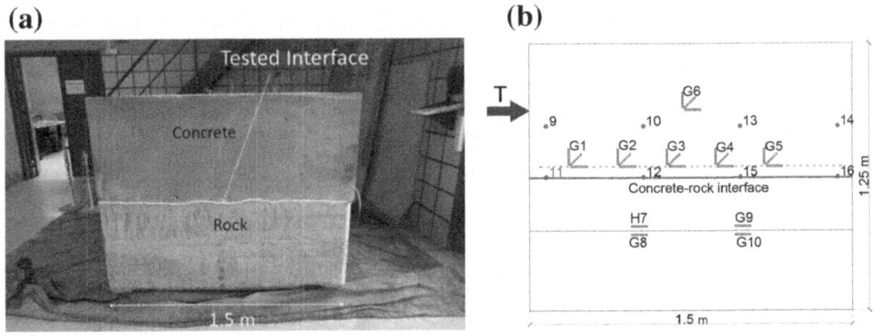

Fig. 2 The direct shear test: **a** the large sample, **b** instrumentations set up on the sample

3.2 Instrumentations

Different instrumentation devices were set up on the specimen as:

- Strain gauges: the principle of bonded strain gauges is to convert their electrical resistance variation to local strain: expansion or contraction of material along the strain gauge axis causes a proportional change in the strain gauge resistance. In the elastic range, recorded strains make it possible to evaluate the corresponding local stresses in the material. Here, the strain gauges are positioned on the concrete along the interface (Fig. 2b—yellow rectangles). During the shear test, the strain gauges were monitored by PEGASE WiFi connected sensors (wireless platforms developed by Ifsttar and marketed by the company A3IP). The supervising software SYSADYP, developed by the Cerema's Mediterranean France Territorial Directorate using Labview, allows instrument data acquisition at a frequency of 10 Hz.
- Optical fiber: optical fiber measurement is based on the monitoring of a transmitted light into a fiber bonded to material. The light characteristics along the fiber are modified as a function of the strain distribution. These changes are detected in the back-scattered light which is later analyzed and then converted into strain data. The principle of Rayleigh back-scattering used here allows quasi-continuous measurement of strain profiles over a distance up to 70 m. Here, a multi-cable optical fiber was implemented in the concrete block with a serpentine form in a horizontal plane, 30 mm above the best-fit plane of the rock surface (Fig. 2b—blue dotted line).
- Acoustic emission: The acoustic emission technique consists in using specific sensors to detect and record acoustic waves induced by a sudden release of energy such as the one caused by crack initiation or growth for instance (Grosse and Ohtsu 2008). The form of these waves is studied through the determination of specific parameters such as the amplitude, the rise time, the duration, the energy or the counts. This monitoring technique has been used in several civil engineering applications on concrete substrate (ElBatanouny et al. 2014). Parameters and data process allow to identify damage mechanisms and locate them in the tested volume. Here 16 sensors were positioned on the two lateral concrete faces on both sides of the interface (Fig. 2b—green circles).

3.3 Test Procedure

The direct shear test was performed with a constant normal stress applied to the concrete/rock interface.

After applying the normal stress of 0.6 MPa to the sample, a constant shear displacement rate of 0.1 mm/min was applied to the upper shear box during the test.

4 Test Results and Analysis

Figure 3a shows the evolution of the shear stress applied to the concrete/rock interface against the shear displacement of the mobile upper part of the box. There are two distinct stress peaks on the curve. Peak 1 corresponds to a shear stress of 2.2 MPa and a shear displacement of 5.3 mm while peak 2 corresponds to a shear stress of 2.05 MPa and a shear displacement of 8.4 mm. The two peaks suggest two successive brittle failure mechanisms.

Figure 3b shows the form of the rock surface after the test. The red arrow indicates the shear direction. We can observe that the failure occurred mainly at the contact between concrete and rock. However, there is a crack in the rock on the opposite side to the push movement.

In the following, only the main results are presented and analyzed.

At peak 1 the vertical strain gauge of rosette G5, bonded near the interface on the opposite side to the push (Fig. 2b), broke indicating a crack opening in concrete. Then the following vertical strain gauge (part of G4), broke at peak 2. These observations agree with crack propagation through rock and then concrete before peak 1.

The data process with the acoustic emission results enabled the 2D localization of the acoustic events in a plane parallel to the best-fit plane of the rock surface. Each point on Fig. 4a, b indicates that an acoustic event was detected and localized at these coordinates. The color of each point indicates the energy level of the event. The results show that before peak 1 (Fig. 4a) the most energetic acoustic events are localized on the side opposite to the push. These observations are consistent with the failure of the vertical strain gauge of G5. Then (Fig. 4b) the events are distributed over the whole monitored surface with few significant acoustic events detected in the central area. Before peak 2 the acoustic events are more concentrated in the central area and in the push side. So it means that the failure occurred in the center of the interface at peak 2.

Finally, the 2D strain distributions in the concrete close to the interface extrapolated from the measurement of the optical fiber at different steps of the shear

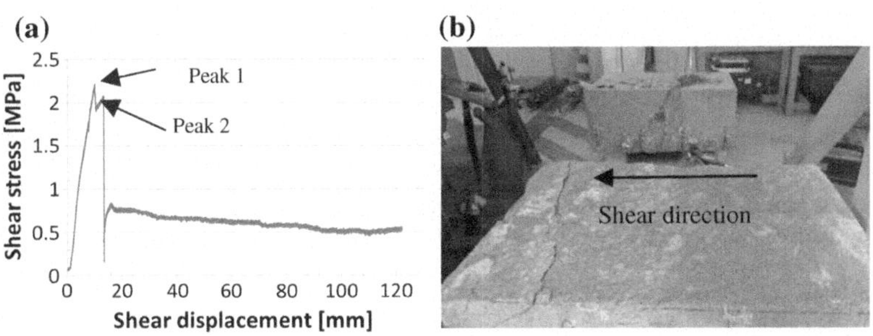

Fig. 3 Results: **a** shear stress versus shear displacement of the upper part of the box, **b** rock surface after the test

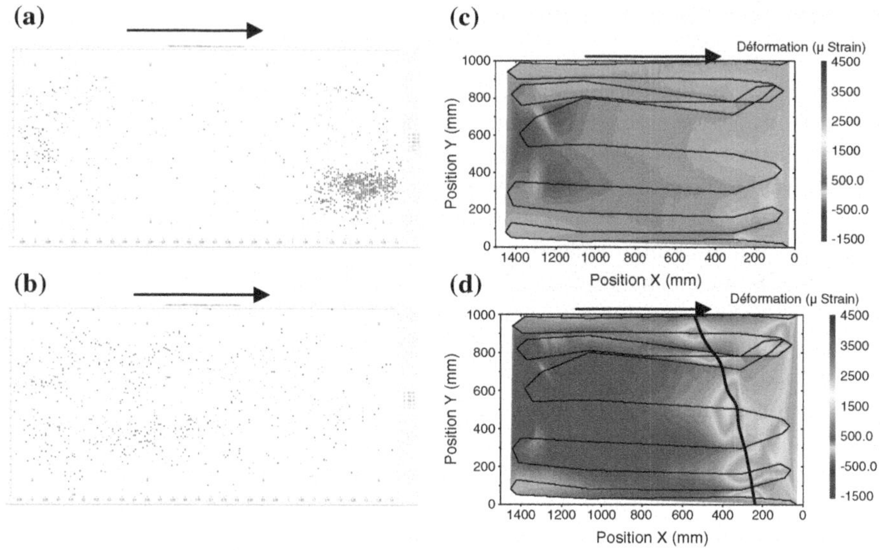

Fig. 4 The 2D acoustic emission events localization and the strain distribution in the concrete at different moments: **a**, **c** before the peak 1, **b**, **d** after the peak 1. The arrow shows shear direction

test show that before peak 1 (Fig. 4c) the interface, except the side opposite to the push, was in compression. After peak 1 (Fig. 4d) a significant increase in compressive strain is observed and a crack opening in the concrete in the side opposite to the push is detected (black line). After peak 2 the tensile zones become compressive zones and there is a global strain release.

It can thus be concluded that peak 1 corresponds to local failures on both sides of the tested concrete/rock interface and that peak 2 corresponds to the failure of the central part of the concrete/rock interface.

5 Conclusion

In this study of the concrete/rock interface, the measurements given by the various instrumentations were consistent and well suited. The combination between the different instrumentations allowed us to figure out the different steps of the failure mechanism along the interface. These instrumentations allow to study the dependency between failure mechanism and rock surface roughness.

The authors wish to acknowledge EDF for supporting this study.

References

Bidaut P, Durville JL, Guillemin P, Richard JC, Viktorovitch M (2006) Essais de cisaillement sur discontinuités armées par ancrages passifs : utilisation d'une boîte de cisaillement de grandes dimensions. Bulletin de Liaison des Ponts et Chaussées, 263-264: 7-18

CFBR Comité Français des Barrages et Réservoirs (2013) Recommendations for the justification of the stability of gravity dams. Justification of gravity dams work group, France

ElBatanouny MK, Ziehl PH, Larosche A, Mangual J, Matta F, Nanny A (2014) Acoustic emission monitoring for assessment of prestressed concrete beams. Construction and Building Materials, 58: 46–53. doi:10.1016/j.conbuildmat.2014.01.100

Grosse C, Ohtsu M (2008) Acoustic emission testing - Basics for research – Applications in civil engineering. Springer,Verlag Berlin Heidelberg

Maiolino S, Pellet F (2015) Full-scale lab testing for the determination of rock bolt contribution to reinforced joint shear strength. Paper presented at the International Symposium on Rock Mechanics (ISRM), Montreal, Canada, 10-13 May 2015

Assessment of Low-Temperature Cracking in Asphalt Concrete Pavements Using an Acoustic Emission Approach

Zhe Sun, Behzad Behnia, William G. Buttlar and Henrique Reis

Abstract An acoustic emission (AE) approach to evaluate low-temperature cracking susceptibility of asphalt binders and asphalt concrete mixtures is presented. In the approach thin films of asphalt binders were bonded to granite substrates and exposed to temperatures ranging from 15 °C to −50 °C. Differential thermal contraction between granite substrates and asphalt binders induces progressive higher thermal stresses in the binders resulting in the thermal crack formation, which is accompanied by a release of elastic energy in the form of transient stress waves (AE). These AE-based T_{cr} predictions showed excellent correlations with predictions based upon AASHTO TP1 and with AASHTO MP1A protocols. Similar results were also obtained when asphalt concrete mixtures samples were exposed to temperatures ranging from 15 °C to −50 °C. The AE-based approach for low-temperature characterization of binders and asphalt concrete mixtures is faster and has a lower coefficient of variance than the traditionally used methods based upon the binder rheological properties.

Keywords Asphalt concrete · Embrittlement temperatures · Acoustic emission · Asphalt binder

Z. Sun · H. Reis (✉)
Department of Industrial and Enterprise Systems Engineering, University of Illinois at Urbana-Champaign, 104 S. Mathews Ave, Urbana, IL 61801, USA
e-mail: h-reis@illinois.edu

Z. Sun
e-mail: zhesun3@illinois.edu

B. Behnia · W.G. Buttlar
Department of Civil and Environmental Engineering, University of Illinois at Urbana-Champaign, 104 S. Mathews Ave, Urbana, IL 61801, USA
e-mail: bbehnia@illinois.edu

W.G. Buttlar
e-mail: buttlar@illinois.edu

A. Chabot et al. (eds.), *8th RILEM International Conference on Mechanisms of Cracking and Debonding in Pavements*, RILEM Bookseries 13,
DOI 10.1007/978-94-024-0867-6_92

1 Introduction

In the United States, there is an effort to develop more efficient maintenance of asphalt pavements. It has been estimated that about 96 % of the approximately 2.4 million miles of paved roads in the country are surfaced with asphalt. The popularity of asphalt concrete derives from the fact that it delivers a smooth, quiet surface, and can be rapidly constructed, particularly during rehabilitation, i.e., resurfacing, operations. Immediately after construction, asphalt is a remarkably tough and resilient material. This is mainly due to the fact that asphalt concrete is comprised of a highly ductile and healable matrix, i.e. asphalt binder, combined with hard aggregate particles, i.e., crushed stone, which provides stiffness and strength.

However, with time, asphalt binder ages, particularly near the pavement surface, which causes the binder to lose its ductility and resilience. Oxidative hardening leads to stiffness and embrittlement of asphalt binders, see Fig. 1, which reduces healing capacity, and increases the rate of micro-crack propagation. Furthermore, the pavement system is more prone to coalesced micro-crack formation, and may begin to develop surface-initiated fatigue cracking. In addition, the brittle pavement surface will be prone to channeling cracks, such as thermal and block cracks, see Fig. 1. Fatigue, thermal, and block cracking lead to an exponential decline in pavement serviceability and a resulting exponential increase in maintenance costs to restore pavement condition.

Traditionally, determination of cracking temperature is based on results from the bending beam rheometer (BBR) test in accordance with the standard test methods. These methods, which are based upon the rheological properties of the binder, are very time consuming and expensive. For a review of these traditional methods, the reader is referred to the following AASHTO standards, i.e., AASHTO TP1 and AASHTO MP1a, and to the following references (Boundin et al. 2000; Dongre et al. 1999; Kim 2000; Kim et al. 2006; Mirza et al. 1995; Roy and Hesp 2001; Shenoy 2000).

Recently, an acoustic emission (AE) based test was developed by the authors that has been shown to be highly effective as a tool to rapidly determine the

Fig. 1 Thermal cracking (*left*) and block cracking (*right*) on asphalt concrete pavements in Illinois

low-temperature cracking threshold (i.e., embrittlement temperature) of asphalt binders and mixtures. The test, developed under a recent NCHRP IDEA project, involves the use of an acoustic emission sensor to listen for microcrack development in binder or mixture samples as they are cooled to relatively low temperatures (Apeagyei et al. 2009; Dave et al. 2011; Behnia et al. 2011a, b; Hill et al. 2012; Behnia et al. 2014). The system has a number of important applications, such as providing an alternate tool for the "rapid" characterization of the low temperature 'grade' of existing and newly developed binders, e.g., bio-binders, for evaluating trial asphalt mixtures such as mixtures involving recycled materials (RAP) (Behnia et al. 2011a, recycled asphalt shingles (RAS) (Arnold et al. 2013), and warm mix asphalt (Behnia et al. 2011b; Hill et al. 2012). The approach allows a rapid estimation of the embrittlement temperatures of the binders and mixtures with a lower variability of results when compared with currently used testing methods, e.g., bending beam rheometer (BBR). The AE-based T_{cr} predictions showed strong correlations ($R^2 = 0.9$) with AASHTO TP1 and AASHTO MP1A protocols and have less variability (Apeagyei et al. 2009; Behnia et al. 2011a, b).

2 Experimentation

Below is a brief description of on-going research work at the NDT Laboratory regarding the assessment of embrittlement temperatures of asphalt concrete. Results from both asphalt binders and asphalt mixtures are discussed.

2.1 Acoustic Emission Testing Setup

Acoustic Emission (AE) tests were conducted with the test samples using a cooler capable of reaching −70 °C. To binder samples were constrained with a granite substrate. The temperature was recorded using a K-type thermocouple, which was placed adjacent to the asphalt sample at the interface between the granite substrate and asphalt binder specimen. Wideband AE sensors with a nominal frequency range of 50–1.5 MHz were used, and high-vacuum grease was used to couple the sensors to the granite substrate. For AE testing of asphalt concrete mixture, specimens of 150 mm diameter semicircular shape with 50 mm thickness were prepared and cooled down from room temperature to −50 °C, while the specimen's temperature and AE activities were recorded. When testing asphalt concrete mixtures, the thermocouple and the AE wideband sensors were coupled to the AC test sample. During cooling, the asphalt mastic contracts more than the aggregates causing thermal stresses to develop. When temperatures drop below the point where the asphalt mastic becomes brittle, thermal cracking develops in the mixture, which is detected using acoustic emission. Additional information regarding the experimental set-up is provided in (Behnia et al. 2011b).

Evaluation of AE activity during thermal loading of asphalt binder samples is performed on recorded event signals and their associated test temperatures. An AE event is hereafter referred to as a local material change giving rise to an acoustic emission with energy equal to or greater than 4 V^2 μs, where energy is defined as the integral of the voltage squared over the duration of the waveform. Please see Eq. 1, where *t* and *V(t)* are the time and the recorded voltage, respectively.

$$Energy = \int V(t)dt \tag{1}$$

The temperature corresponding to the event with the first peak energy level equal to or larger than 4 V^2 μs has been termed the "Embrittlement temperature." It is hypothesized that the embrittlement temperature represents a fundamental material state which is independent of material constraint, sample size (as long as a statistically representative volume or larger is used.

3 Experimental Results

Some acoustic emission based results of the embrittlement temperatures of binders and asphalt concrete mixtures are discussed below.

3.1 Acoustic Emission Test Results of Asphalt Binders

The embrittlement temperature (T_{EMB}) of asphalt binders evaluated through AE testing was compared with the critical cracking temperature (T_{CR}) obtained from the Superpave asphalt binder test procedures as described by Marasteanu et al. (2007) and using the procedure described in NCHRP Report 452, see Table 1.

3.2 Acoustic Emission Results of Asphalt Concrete Mixtures

Microcracks result primarily from a combination of asphalt mastic brittleness (at lower temperatures) and from the thermally-induced tensile stresses within the material. Progressively higher stresses in the specimen eventually causes thermal microcracks to develop in the asphalt mastic as well as at the interface between asphalt mastic and aggregates. AE tests were conducted on several lab compacted

Table 1 AE-based and BBR-based cracking temperatures of SHRP core asphalt binders

Sample ID	AE-based embrittlement temperature		BBR-based cracking temperature (°C)
	T_{EMB} (°C)	CoV (%)	
TANK-AAA1	−36.3	1.7	−34.5
RTFO-AAA1	−35.4	2.0	−32.3
PAV-AAA1	−30.9	1.9	−29.7
TANK-AAD1	−37.2	1.3	−34.4
RTFO-AAD1	−35.9	1.9	−33.1
PAV-AAD1	−30.6	1.5	−30.1
TANK-AAF1	−24.6	1.8	−25.4
RTFO-AAF1	−21.1	1.3	−24.7
PAV-AAF1	−18.7	2.6	−20.4
TANK-AAG1	−20.7	2.9	−20.0
RTFO-AAG1	−19.2	2.1	−19.1
PAV-AAG1	−17.2	1.5	−17.0

Table 2 Acoustic emission test results of asphalt institute field cores

Sample ID	T_{Emb} (°C)	CoV (%)	$T_{Max\ Energy\ Event}$ (°C)	CoV (%)
Conchas lake airport, NM	−12.65	4.92	−27.64	1.77
Clayton airport, NM	−21.57	10.10	−29.72	1.38
Roundup top, MT	−13.02	17.33	−26.06	4.91
Roundup bottom, MT	−12.17	0.17	−23.27	2.52

asphalt mixtures field cores using three replicas. In Montana, cores were taken from one airport—identified as Roundup—that had recently received an overlay of an older, cracked pavement. The estimated age of the pavement is 12–15 years old. Cores were split into layers with the upper layer representing the new pavement (Roundup Top) and the lower layer representing the older pavement (Roundup Bottom).

In New Mexico, cores were taken from the Clayton airport and the Conchas Lake airport. Site reports for the Clayton airport indicate that it was paved in 2004. Obtained AE test results of provided field cores are presented in Table 2. The embrittlement temperature of field cores is consistent with the reported cracking condition of the airfields. Both Conchas and Clayton airfields are paved using Western Canadian Crude asphalt binder. It is seen that T_{EMB} of the Conchas samples is much higher than those obtained on the Clayton samples. This can be attributed to the fact that Conchas pavement encountered more severe age hardening, as the pavement was observed to be more oxidized than the Clayton pavement. In Table 2, T_{Max} represents the temperature at which the AE event with maximum energy develops, which occurs because of the indeterminate nature of asphalt concrete.

4 Conclusions

Application of an acoustic emission approach to evaluate embrittlement temperatures of asphalt binders and asphalt concrete mixtures is presented. It is observed that the acoustic emission approach is faster, more reliable, and has a lower coefficient of variance than the traditionally used methods, which are based upon the binder rheological properties.

Acknowledgements The authors acknowledge those who contributed to the success of NCHRP IDEA project no. 170, including the program manager, Dr. I. Jawed. Any opinions, findings and conclusions or recommendations expressed in this publication are those of the authors and do not necessarily reflect the views of the sponsoring agency.

References

Arnold J Behnia B Buttlar WG Reis H (2013) Evaluation of low-temperature properties of HMA mixtures containing recycled asphalt shingls (RAS) using an acoustic emission approach. Journal of Construction and Building Materials 58C:1-8.

Apeagyei AK Buttlar WG Reis H (2009) Assessment of low-temperature embrittlement of asphlat binders using an acoustic emission approach. Insight 51(3): 129-136.

Behnia, B Dave EV Ahmed S Buttlar WG Reis H (2011a) "Effects of Recycled Asphalt Pavement Amounts on Low Temperature Cracking Performance of Asphalt Mixtures using Acoustic Emissions." Tranportation Research Record 2208:64-71.

Behnia B Buttlar WG Reis H (2011b) An Acoustic-Emission Based Test to Determine Asphalt Binder and Mixture Embrittlement Temperature. NCHRP-IDEA Project 144 Final Report, NCHRP IDEA PROGRAM, University of Illinois at Urbana-Champaign, Illinois.

Behnia B Buttlar WG, Reis H (2014) Cooling Cycle Effects on Low Temperature Cracking Characteristics of Asphalt Concrete Mixtures. Materials and Structures Journal - Special Issue on Bituminous Materials. RILEM, 47(8):1359-1371.

Bouldin MG Dongre R Rowe GM Sharrock MJ Anderson DA (2000) Predicting thermal cracking of pavements from binder properties: theoretical basis and field validation." *AAPT* 69 (2000):455-496.

Dave EV Behnia B Ahmed S Buttlar G Reis H (2011) Low Temperature Fracture Evaluation of Asphalt Mixtures using Mechanical Testing and Acoustic Emission Techniques. Association of Asphalt Paving Technologists Journal 80: 193-226.

Dongre R Bouldin M.G Anderson D.A Reinke G D'Angelo J Kluttz RO Zanzotto L (1999) Overview of the development of the new low-temperature binder specification. Technical Report. Federal Highway Administration Binder Expert Task Group. Washington, D.C.

Hill B Behnia B Hakimzadeh S Buttlar WG Reis H (2012) "Evaluation of the Low Temperature Cracking Performance of WMA Mixtures." Transportation Research Record, 2294: 81-89.

Kim SS (2000) Direct measurement of asphalt binder thermal cracking. Journal of Materials in Civil Engineering 17(6): 632-639.

Kim SS Wysong ZD Kovach J (2006) Low-temperature thermal cracking of asphalt binder by asphalt binder cracking device. *Transportation Research Record* 1962:28-35.

Marasteanu, M., Zofka, A., Turos, M., Li, X., Velasquez, R., Li, X., Williams, C., Bausano, J., Buttlar, W., Paulino, G., Braham, A., Dave, E., Ojo, J., Bahia, H., Gallistel, A., and McGraw, J., "Investigation of Low Temperature Cracking in Asphalt Pavements", Report No. 776, Minnesota Depart. of Transportation, Research Services MS 330, St. Paul, MN 55155, 2007.

Mirza MW Witczak MW (1995) Development of a global aging system for short term and long term aging of asphalt cements. *J. AAPT* 64: 393-430.

Roy D Hesp SAM (2001) Low-temperature binder specification development: Thermal stress restrained specimen testing of asphalt binders and mixtures. Transportation Research Record 1766 9:7–14.

Shenoy A (2000) Single-event cracking temperature of asphalt pavements directly from bending beam rheometer data. *J. of Transportation Engineering* 128(5):465-471.

Damage Detection in Pavement Structures Using Self-powered Sensors

Karim Chatti, Amir H. Alavi, Hassene Hasni, Nizar Lajnef and Fred Faridazar

Abstract This paper presents a new approach for the continuous health monitoring of asphalt concrete pavements using self-powered wireless sensors. Numerical and experimental studies were carried out to evaluate the damage detection performance of the proposed self-sustained sensing system. A three-dimensional finite element analysis was performed to obtain the pavement responses under moving tire loading. Damage was introduced as bottom-up fatigue cracks at the bottom of the asphalt layer. Thereafter, features extracted from the simulated dynamic strain data for a number of sensing nodes were used to detect damage. Laboratory tests were carried out on an asphalt concrete specimen in three point bending mode to verify the sensor response. The results indicate that the proposed method is effective in detecting different damage states including crack propagation.

Keywords Pavement health monitoring · Energy harvesting · Fatigue cracking · Damage

K. Chatti (✉) · A.H. Alavi · H. Hasni · N. Lajnef
Department of Civil and Environmental Engineering,
Michigan State University, East Lansing, MI 48823, USA
e-mail: chatti@egr.msu.edu

A.H. Alavi
e-mail: alavi@msu.edu

H. Hasni
e-mail: hasniha1@msu.edu

N. Lajnef
e-mail: lajnefni@egr.msu.edu

F. Faridazar
Federal Highway Administration, Turner-Fairbank Highway Research Center,
McLean, VA, USA
e-mail: Fred.Faridazar@dot.gov

A. Chabot et al. (eds.), *8th RILEM International Conference on Mechanisms of Cracking and Debonding in Pavements*, RILEM Bookseries 13,
DOI 10.1007/978-94-024-0867-6_93

1 Introduction

A major limitation of the traditional wired sensors for pavement health monitoring pertains to their deployment and maintenance. To cope with these limitations, wireless sensor networks (WSNs) are increasingly utilized as alternatives to traditional monitoring systems (Lynch and Loh 2006). However, a significant concern for the application of wireless sensors is about their power supply. Periodic replacement of batteries for embedded sensors or use of solar power technology would be cost-prohibitive and in some cases impractical. Harvesting ambient energy seems to be an attractive solution for tackling this problem. Energy harvesting devices can convert mechanical energy into electrical energy (Elvin et al. 2003). For pavement health monitoring, piezoelectric transducers can be used for the self-powering of wireless sensors by harvesting energy from the mechanical loading experienced by the pavement (Rhimi et al. 2012; Lajnef et al. 2013). Recently, the authors have developed a new class of self-powered wireless sensors (SWS) (Huang et al. 2010; Lajnef et al. 2013). The developed SWS is a small size battery-less sensor. The communication between the sensor and a service vehicle is done using a Radio Frequency Identification (RFID) scanner. Based on an experimental study done by our team, the scanner can read the data stored on-board the sensor from a distance of about 25 cm within the asphalt concrete layer. This study presents a new system for the continuous long-term health monitoring of pavement structures based on the SWS data. The proposed approach uses features extracted from the cumulative time strain distributions at preselected discrete levels. A finite element model of an asphalt concrete pavement layer was used to simulate pavement response. The main goal was to detect fatigue cracking due to tensile strains at the bottom of the asphalt concrete. In order to verify the response of sensors embedded within the asphalt concrete, a series of laboratory tests were conducted on an asphalt concrete beam under a three point bending configuration.

2 The Proposed Pavement Health Monitoring System

The new smart SWS has a series of memory cells that cumulatively store the duration of strain events, at a preselected level discretization. They measure the duration of events when the amplitude of the input signal, coming from the piezoelectric transducer exceeds different thresholds. At constant loading frequency, the sensor output is the cumulative histogram of the loading strain distribution. Figure 1 presents a schematic representation of the level crossing cumulative time counting implemented by the developed SWS (Huang et al. 2010; Lajnef et al. 2013; Alavi et al. 2016).

Since the summation of Gaussian distributions can be assumed to be a Gaussian distribution, this study rationally assumes that the sensor output can be characterized by the following cumulative density function (CDF) (Alavi et al. 2016):

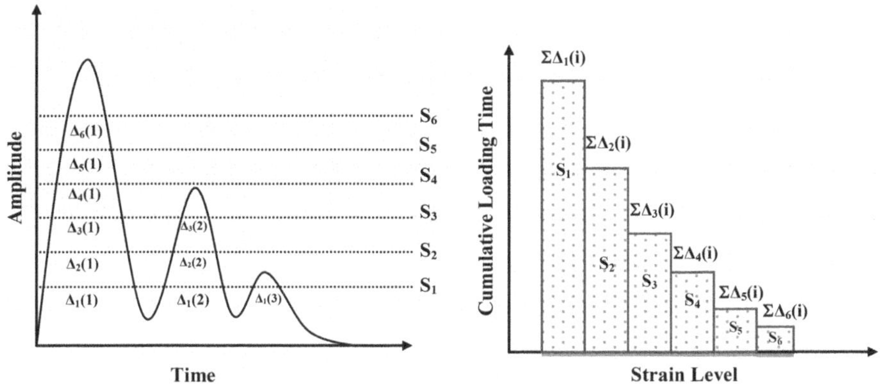

Fig. 1 The level crossing cumulative time counting implemented by the SWS

$$F(\varepsilon) = \frac{\alpha}{2}\left[1 - erf\left((\varepsilon - \mu)/\sigma\sqrt{2}\right)\right] \quad (1)$$

where

μ Mean of the strain distribution

σ Standard deviation that account for the load and frequency variability

α Total cumulative time of the applied strain.

The statistics μ and σ of the strain distribution can be regarded as indicators of the damage progression. In fact, μ and σ are the only viable tools to define the SWS output data. These parameters are obtained by a curve fitting of the sensor output distribution collected from the memory cells. Then, more in-depth analyses are conducted to find the relationship between the PDF parameters (μ and σ) and damage state.

2.1 Finite Element Modeling and Laboratory Testing

A 3D FE model was developed to analyze the dynamic response of the pavement under a moving truck tire loading. The ABAQUS^{TM} FE software was used for the modeling and post-processing of the results. A series of acquisition nodes were considered at the bottom of the asphalt layer as the potential sensors. The pavement model is composed of 3 layers: asphalt (E = 2757 MPa; ν = 0.35; Thickness = 152.4 mm), base (E = 344 MPa; ν = 0.35; Thickness = 127 mm) and subgrade (E = 35 MPa; ν = 0.48; Thickness = 4826 mm) layers. The model has a dimension of 6200 mm along the direction of traffic and 4170 mm along the transverse direction (the width of one lane). To simulate the movement of the load at the desired speed, a quasi-static analysis was adopted (Yoo and Al-Qadi 2007; Al-Qadi and Wang 2010). The tire pressure was taken as 0.69 MPa. The sensors

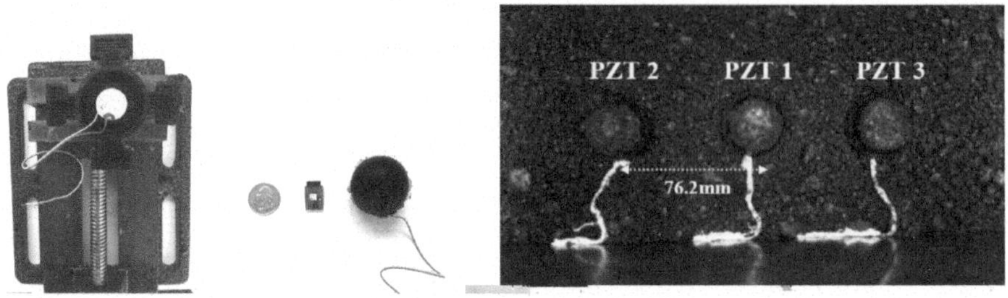

Fig. 2 Manufacturing process and layout of spherical packaging

were located at a distance of 50.8 mm from the bottom of the asphalt layer. The distance between two consecutive sensors is 304.8 mm. Also, the distance of Sensors 1 and 9 from vertical boundaries was equal to 1880 mm. Out of the considered 55 sensors, only those along the loading path which experienced strains higher than 20 $\mu\varepsilon$ were kept for the analysis. Later, a new FE model was developed with a damage zone at the bottom of the asphalt layer. The length of the crack zone was 12.7 mm from the bottom of the asphalt layer.

The performance of the sensors was tested through an experimental study on an asphalt concrete specimen. The sample was tested under a three point bending configuration (Marasteanu et al. 2012). In this study, the fixture was developed for an asphalt concrete sample with a span length of 381 mm, thickness of 165.1 mm and a width of 152.4 mm. The slab was built using the HMA, 4E1 mixture type provided by the Michigan Department of Transportation (MDOT). The temperature was around 150 °C. The weight of the hot mix asphalt (HMA) was 25 kg and the length of the slab was equal to 450 mm. All tests were conducted under constant displacement rate control. The tests were done at 2 and 5 Hz loading frequency for 0.1, 0.15 and 0.2 mm amplitudes. Damage was introduced by making a notch at the bottom of the asphalt layer using a circular saw. The damage states were defined by increasing the notch size (a) as follows: Intact: Intact beam (a = 0 mm); Damage 1: a = 22.2 mm; Damage 2: a = 31.75 mm. After a number of cyclic loadings, a crack propagation phenomenon was observed. The crack length was 12.7 mm. This new damage phase was considered as Damage 3. Accordingly, the total length of crack for Damage 3 was 44.45 mm. Figure 2 presents the manufacturing process and layout of the packaging system with embedded piezoelectric transducers. Among different epoxies tested (e.g. Araldite® GY-6010 and EPIC S7514), Conathane® TU-981 epoxy was found to perform better. The elastic modulus of Epoxy TU-981 was found to be 258.8 MPa.

3 Damage Detection Using Smart Sensing Technology

Using the μ and σ values from the FE simulations, the PDF plots corresponding to sensors 4, 5 and 6 were obtained and shown in Fig. 3. As can be seen in Fig. 3a–e, μ decreases and σ increases by transitioning from intact to damaged mode. Accordingly, the PDFs shift to the left, their width increases and their height decreases due to damage progression.

It is evident that for Sensor 5, which is located right above the crack, there is a distinct change in the PDF shape. It is worth mentioning that even for sensors far from the damage, the PDFs still shift to the left and expand due to damage progression. It was also tried to find the relationship between the PDF parameters of a group of sensors and damage progression. Figure 4 presents the variation of the standard deviation (STD) of μ and σ of all of the sensors. As it is seen, the STD of μ and σ of group of sensors increases with damage progression.

For the experimental study, the damage detection results are provided for amplitude of 0.2 mm and loading frequency of 5 Hz. The μ and σ values were used to plot the PDFs corresponding to each sensor (Fig. 5). Also, since there were only 2 sensors installed, performing the group effect analysis was not done. As can be

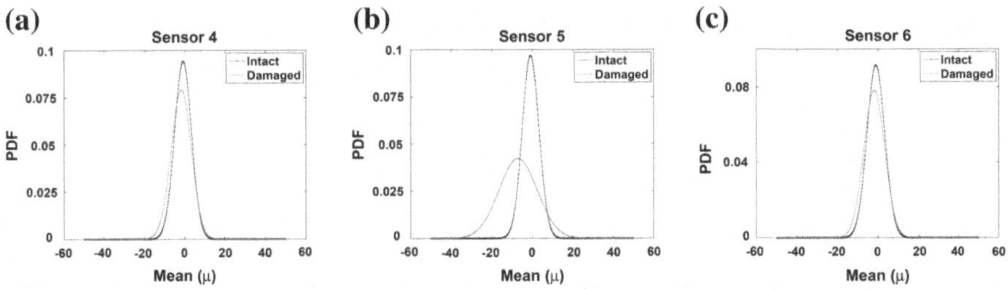

Fig. 3 Changes of PDFs due to damage for the FE simulations **a** sensor 4 **b** sensor 5 **c** sensor 6

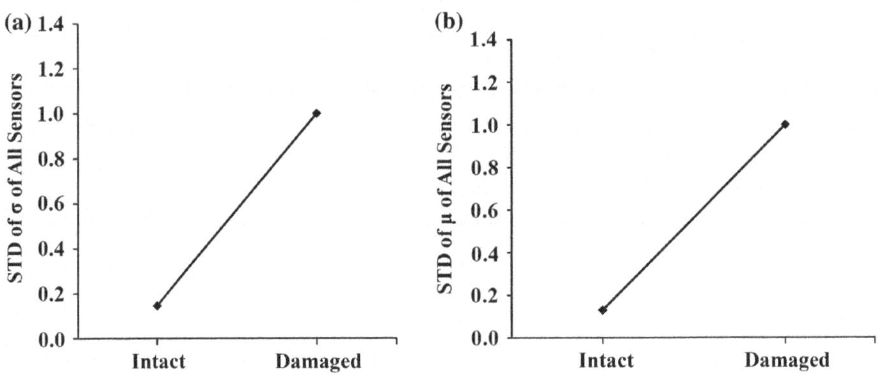

Fig. 4 The STD of **a** σ and **b** μ of group of sensors for the FE simulations

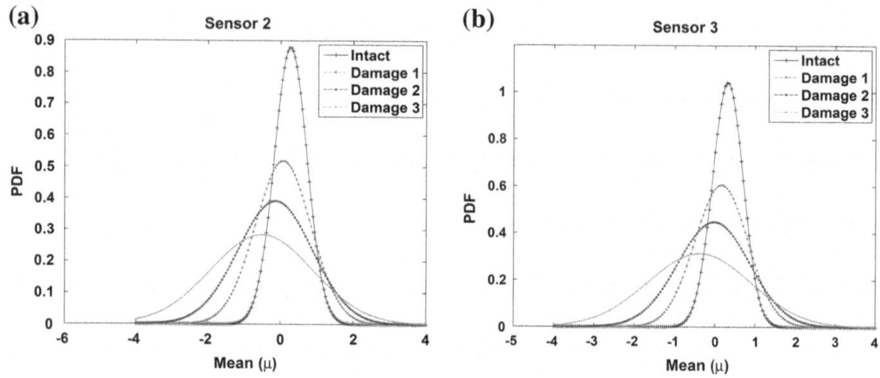

Fig. 5 Changes of PDFs due to damage progression for the experimental study **a** sensor 2 **b** sensor 3

seen in Fig. 5, the values of μ and σ decrease and increase, respectively, due to damage progression. This is in agreement with what was observed from the FE model. Interestingly, the PDFs change notably by transiting from Intact to Damage 1 compared to other stages. This is because the first introduced notch was deeper than those considered for Damage states 2 and 3. Additionally, the crack propagation process was accurately detected by the sensors.

4 Conclusions

This paper presents the performance analysis of a newly developed self-powered wireless sensor for the continuous health monitoring of asphalt concrete pavements. The uniqueness of this sensor pertains to the fact that there is no need to directly measure the absolute value of strain in order to estimate damage. The rate of variation of strain distributions is related to the rate of damage. The FE simulation results indicate that the proposed system is capable of detecting bottom-up fatigue cracks introduced at the bottom of the asphalt layer. Similar results were observed from laboratory beam testing. In this case, the sensors could detect different damage states including a crack propagation phase. The important observation from the individual sensor analysis was that the PDFs shift to the left (μ decreases) and their width increases (σ increases) due to the damage progression. Based on the results, variations of σ can be regarded as a better indicator of damage. On this basis, it was shown that the STD of μ and σ of a group of sensors has a sound relationship with the damage progression.

References

Alavi A.H., H. Hasni, N. Lajnef, K. Chatti, and F. Faridazar (2016) An Intelligent Structural Damage Detection Approach Based on Self-Powered Wireless Sensor Data. Aut Const 62: 24–44.

Al-Qadi IL, Wang H (2010) Pavement damage due to different tire and loading configurations on secondary roads. NEXTRANS University Transportation Center, West Lafayette, Indiana

Elvin N, Elvin A, Choi DH (2003) A self-powered damage detection sensor. J Strain Anal 38 (2):115 -124

Huang C, Lajnef N, Chakrabartty S (2010) Self-calibration and characterization of self-powered floating-gate usage monitors with single electron per second operational limit. IEEE Trans Biomed Circuits Syst 57:556-567

Lajnef N, Chatti K, Chakrabartty S, Rhimi M, Sarkar P (2013) Smart pavement monitoring system. Report: FHWA-HRT-12-072, Federal Highway Administration, Washington, DC.

Lynch JP, KJ Loh (2006) A summary review of wireless sensors and sensor networks for structural health monitoring. Shock and Vibration Digest 38:91-128

Marasteanu MO, Buttlar W, Bahia H, Williams C, Moon KH, Teshale EZ, Falchetto AC, Turos M. (2012) Investigation of low temperature cracking in asphalt pavements. National Pooled Fund Study -Phase II. Final Report 2012-232012, University of Minnesota

Rhimi M, Lajnef N, Chatti K, Faridazar F (2012) A self-powered sensing system for continuous fatigue monitoring of in-service pavements. Int J Pavement Res Technol 5(5):303-310

Yoo PJ, Al-Qadi IL (2007) Effect of transient dynamic loading on flexible pavements, J Transportation Research Board TRR 1990:129-140, TRB, National Academies, Washington, DC

Detection and Survey of Interface Defects Within a Pavement Structure with Ultrasonic Pulse Echo

Jean-Michel Simonin and Géraldine Villain

Abstract A pavement structure, which contains artificial interface defects, has been built on the full scale accelerated pavement testing facility of IFSTTAR in Nantes. This test section is made of two bituminous layers (8 cm thick base layer, and 6 cm thick wearing course), over a granular subbase. Several types of defects have been included at the interface between the two asphalt layers. Rectangular debonded areas of different size (of longitudinal or transversal direction) have been created artificially, using different techniques (sand, textile, absence of tack coat). The construction has been carried out by a road construction company, using standard road works equipment. Then, the pavement fatigue testing facility has been used to apply traffic loading on this pavement, to study the effect of such sliding interfaces on the mechanical behaviour of the pavement, and the evolution of the defects with traffic. The pavement structure has been monitored with several non-destructive methods. The paper presents the survey of the test section with an Ultrasonic Pulse Echo method. This Non Destructive Technique used high frequency wave propagation. It has been able to detect and locate interface defects. The method can also be used to evaluate material properties.

Keywords Interface defects · NDT · Fatigue testing · UPE

1 Introduction

French roads consist mainly of old bituminous pavements often more than 30 years old. Usually, they have been maintained several times with thin overlays (less than 8 cm thick). In recent years, potholes and alligator cracking has been frequently observed, in particular after periods of heavy rain or freeze/thaw. Frequently, this

J.-M. Simonin (✉) · G. Villain
IFSTTAR, LUNAM Université, Route de Bouaye, CS4, 44344 Bouguenais Cedex, France
e-mail: jean-michel.simonin@ifsttar.fr

G. Villain
e-mail: geraldine.villain@ifsttar.fr

A. Chabot et al. (eds.), *8th RILEM International Conference on Mechanisms of Cracking and Debonding in Pavements*, RILEM Bookseries 13,
DOI 10.1007/978-94-024-0867-6_94

type of damage is associated with moisture effects linked to interface debonding between the overlays and the old pavement. Such debonding mechanisms reduce the residual life of the pavement, and thus their early detection is a very important issue for pavement maintenance. To detect such interface damage, non-destructive techniques (NDT), such as seismic wave propagation methods, appear as promising approaches.

This paper presented tests performed on a pavement structure with Ultrasonic Pulse Echo (UPE). Shear waves have been used to detect and locate debonding during an experiment carried out on the large pavement fatigue carrousel of IFSTTAR in Nantes. Additional tests, based on compression waves propagation, have been performed to characterize the mechanical behavior of the wearing course.

2 Description of the Full Scale Experiment

The test site is located on the pavement fatigue carrousel of IFSTTAR (Fig. 1). A 25 m long test section has been built by a road construction company, using standard equipment. It consists of two bituminous layers (8 cm thick base layer, and 6 cm thick wearing course), over a granular subbase (30 cm thick). Several types of defects were intentionally incorporated or at the interface between the two asphalt layers. Figure 2 presents a photograph of the debonded areas before the wearing course construction (Fig. 2a) and a map of the experimental section, with the location of the different defects (Fig. 2b).

Three techniques were used to simulate different level of sliding interface: without tack coat, textile and sand. Defect I10 is narrower than the wheel-path (0.5 m versus 1 m) for evaluating the transversal extension of the defect. Defects I1, I2, I3 and I11, I12, I13 are wider than the wheel-path to study a longitudinal extension.

Fig. 1 The pavement fatigue carrousel of IFSTTAR

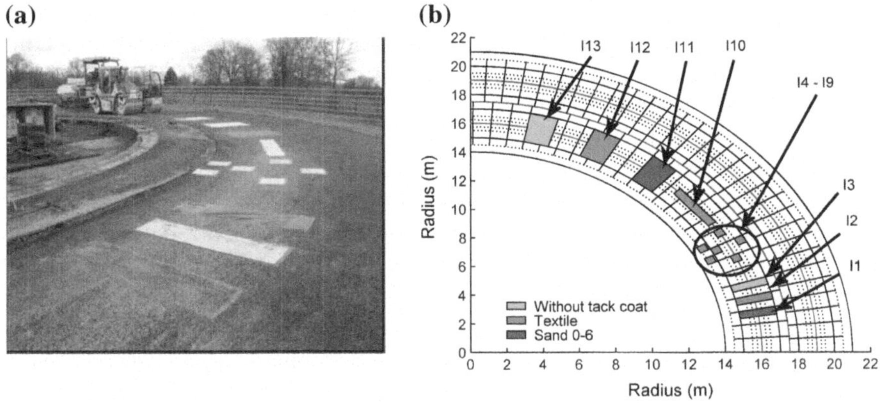

Fig. 2 **a** Interface defects before wearing course construction; **b** Map of the different debonded areas

The pavement structure built on the existing subgrade of the test track (E ≈ 70 MPa) includes the following layers:

- A granular subbase consisting of unbound granular material (UGM);
- A bituminous base layer (8 cm), consisting of Road Base Asphalt (RBA);
- A bituminous wearing course (6 cm), consisting of bituminous concrete (BC).

Complex modulus tests on trapezoidal specimens were performed on the two bituminous mixes. The standard elastic moduli obtained for the 2 materials at 15 °C and 10 Hz were respectively 11,320 MPa (RBA) and 10,700 MPa (BC). Using the Huet-Sayegh visco-elastic model (Chailleux et al. 2006), infinite moduli (at very high frequency) can be extrapolated and have been estimated to 23 GPa (RBA) and 29 GPa (BC).

3 Loading and Monotoring

The main objective of the experiment was to compare different NDT techniques based on mechanic (deflection measurement, modal testing, US wave propagation) or electromagnetic (Radar) phenomena for detecting different geometrical characteristics of artificial defects. Some results obtained with the other NDT methods are presented in papers (Simonin et al. 2013). Other objectives were to survey the defects extension during loading. This will be helpful in optimizing pavement monitoring with the different NDT methods. Investigations have been done at different stages of the experiment:

- At the beginning of the experiment (initial state);
- After 10,000 loads when the structure is consolidated;
- After 50,000, 100,000, 200,000, and 300,000 loads (structure survey).

The experiment started in the spring 2012 with 100,000 loads applied in April and May. At the beginning of July, 100,000 loads have been applied on the debonding sector. First results show any extension of the debonding zones. So, it has been decided to apply 100,000 additional loads before the end of July. Thus 300,000 loads have been apply to the road. After this experiment, the fatigue machine has been transferred to another test site. However the test site with debonding area is always available to test NDT methods.

This paper presented only the results obtained with the Ultrasonic Pulse Echo (UPE) system. Tests have been performed using an antenna array (ACSYS system) which includes 24 dry-point-contact transducers. Half of them emit waves, with a central frequency equal to 55 kHz, into the structure while the others record the reflected waves. At each measurement point the antenna is apply and maintain at the road surface during few seconds. Compared with the practice on hydraulic concrete (Taffe and Wiggenhauser 2006), the data analysis is more difficult due to attenuation of wave in the bituminous material phenomena which is enhanced at high temperature. Two antenna arrays are available and can be used one after the other, either equipped with shear S-wave transducers or compression P-wave transducers.

Concerning the structural survey to monitor the defects extensions, the UPE S-wave device was used at different test times. The UPE S-wave device was used to investigate a longitudinal profile at the 16 m radius above the large debonded areas (I1, I2, I3, I11, I12 and I13). These profiles began 0.20–0.5 m before the defect and ended 0.2–0.5 m after. Measurements were made at points spaced at a distance of 0.05 m far from the theoretical limits of the defect and 0.02 m close to the limits. Moreover, three transverse profiles 1.5 m long has also been performed with a measurement spacing of 0.02 m to survey a lateral extension of the I10 textile debonded zone. Two profiles were recorded above the I10 zone and one reference profile was recorded between I10 and I11. Recorded signals show a high amplitude wave with a saturation of the signal before 50 μs. Then the signal is highly attenuated, and interface echoes are difficult to detect. However, the differences are significant between the signals corresponding to different defects and the one of the well bonded zone.

Regarding the mechanical characterization and the evaluation of the dynamic modulus, tests were carried out with both S-wave and P-wave devices only after 300,000 cycles in July 2015. Longitudinal profiles were investigated at the 16 m radius above the wider debonded areas (I1, I11 and I12) with a measurement spacing of 0.05 m. Acquisition parameters were identical in the 2 cases. The surface temperature was 28 °C. Compare to S-wave signals, P-wave signals have lower amplitude. This is probably due to a lower energy emits with the P-wave antenna. However a significant difference also exists between signals recorded above the different zones.

4 Application of the UPE

UPE measurement can be used to detect and locate debonding zones. Measurements along each profile are presented as a B-Scan where:

- The X-coordinate is the abscissa along the road section;
- The Y-coordinate is time (UPE);
- Colors (or level of gray) represent the signal amplitude.

Figure 3 compares B-scans obtained with S-wave antenna measurements recorded on the zone I11 to I13. The colors represent the amplitude level from amplitude −100 (blue) to 100 (red). Above the debonded zone, an interface echo can be detected at around 60 μs. This echo is easier to locate for the debonded interfaces with sand and textile than for the interface without tack coat. The length of the debonded zone is consistent with the design and with measurement done with other NDT methods such as GPR or FWD. S-wave UPE mappings are able to detect and locate the debonded areas and to make a distinction between them.

Tests performed with the UPE S-wave device during the loading phase 2012 allow detecting the different defects. B-scans were used to estimate the size of each defect. A small increase of the longitudinal defect size can be perceptible (≈5 % eq. 5 cm) between the 10,000 cycle time and the 300,000 cycle time, but it remains not significant compare with the spacing with consecutive measurement point (2 or 5 cm). The increase of the transversal profile is more significant (10 and 20 %). The site has been preserved and it is planned to add loading (at least 500,000 loads) in conjunction with other accelerated pavement testing as soon as possible.

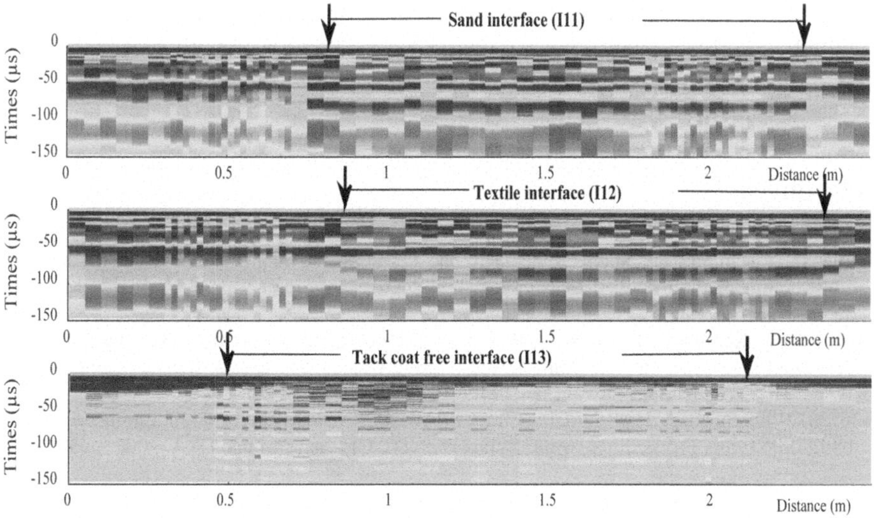

Fig. 3 B-scans of UPE (S-wave) measurements above the I11 (*upper*), I12 (*middle*) and I13 (*lower*) defects after 50,000 loads (May 2012)

A second application consists to estimate the mechanical parameter of the wearing course. To do so, surface and interface echo are picked on the raw signals for both S and P wave antennas. B-scans can be helpful to locate the interface echo. Assuming the layer thickness, S-wave and P-wave velocities can be estimated from travel time picking. Using the density of the material controlled after construction (2440 kg/m^3) the dynamic Young's modulus (E_{dyn}) and Poisson's ratio (ν) can be evaluated at 55 kHz and in the testing conditions at a temperature of 28 °C. These mechanical parameters have been evaluated for each signal. The Poisson's ratio is quite constant about 0.30. The elastic modulus is about 25 GPa outside defected zones and slightly lower above them. This value is consistent with the one of the infinite modulus around 28.9 GPa, calculated by the Huet-Sayegh model.

5 Conclusions and Outlooks

Tests performed with UPE antennas have been performed on the IFSTTAR Accelerated Pavement Testing facility which includes different artificial interface defects. B-scans base on S-wave show very good sensitivity to debonded areas. The method could provide to the road engineer a picture of the internal damage of the road base and wearing course which is helpful to detect and locate damages. Signals analysis can deliver other information such as depth of defect or mechanical properties (E, ν) of the wearing course.

After 300,000 loads, small evolution of the interfaces has been observed with the different NDT methods including UPE. The test section will continue to be loaded in conjunction other programs on the carousel to perform at least 500,000 loads. NDT tests will be continued on the pavement at regular intervals.

Future experiments will improve the data interpretation in particular taking into account the temperature and frequency effects by using the master-curve.

References

Chailleux E., Ramond G., Such C., de la Roche C., 2006, A mathematical-based mastercurve construction method applied to complex modulus of bituminous materials. Roads Materials and pavement Design 7 (EATA Special Issue): 75-92

Simonin J-M., Kerzreho J-P., Hornych P., Gouy T., 2013, Comparison of NDT method to detect debonded interfaces, 9th Int. Conf. on the Bearing Capacity of Roads, Railways and Airfiels, June 2013, Trondheim, Norway, pp. 171-180

Taffe A., Wiggenhauser H., Validation for Thickness Measurement in Civil Engineering with Ultrasonic Echo, Proceedings of Int. Symp. on NDT-CE, Saint-Louis, USA, 2006, pp506-512.

Investigation of Energy-Based Crack Initiation Threshold from Meso-Scale Asphalt Concrete Response

Ibrahim Onifade and Björn Birgisson

Abstract The existence of a fundamental energy threshold for meso-scale crack initiation is investigated using micromechanical modeling techniques. X-ray Computed Tomography (CT) is used to acquire the internal structure of an asphalt concrete mixture while Digital Image Processing (DIP) techniques is used to segment and analyze the different phases present in the mixture. Finite Element (FE) modeling is used to simulate a tensile loading condition to establish a critical micromechanical criterion for meso-scale crack initiation. The meso-scale asphalt concrete mixture is subjected to different loading rates to obtain the global strain energy density at the instance when the critical micromechanical crack-initiation criterion threshold is attained at different deformation rates. The result from the study shows that there exists a fundamental global strain energy density threshold that is invariant of the rate of loading at the instance of meso-scale crack initiation. The result of this study also shows the potential of the use of X-Ray computed tomography in understanding the cracking phenomenon in asphalt mixtures.

Keywords Micromechanical damage · Micro-crack initiation · Asphalt concrete · X-ray tomography

I. Onifade (✉) · B. Birgisson
Department of Civil and Architectural Engineering,
KTH—Royal Institute of Technology, Stockholm, Sweden
e-mail: onifade@kth.se

B. Birgisson
e-mail: bjorn.birgisson@aston.ac.uk

B. Birgisson
School of Engineering and Applied Science, Aston University,
Aston Triangle, Birmingham B4 7ET, UK

A. Chabot et al. (eds.), *8th RILEM International Conference on Mechanisms of Cracking and Debonding in Pavements*, RILEM Bookseries 13,
DOI 10.1007/978-94-024-0867-6_95

1 Introduction

The mechanisms of damage and fracture in asphalt concrete are complex phenomena that cannot be fully described by traditional continuum damage and linear elastic fracture mechanics theories. These complexities are due to the viscoelastic properties of the asphalt material and the interactions between the different constituents in the mixture. Asphalt mixtures consist of aggregates, bitumen-based binder and air-voids. The aggregates interact to transfer load while the integrity and durability of the mixture is dependent on the condition of the binder. Loss of integrity in the binder usually occurs at mastic regions between adjacent interacting aggregate particles and at the mastic-air voids interface. The air-voids are usually of numerous sizes, spatially distributed inside the mixture, and are potential locations for stress concentrations and damage initiation.

Quite a number of research efforts in the area of damage and fracture of asphalt mixtures are based on energy considerations for damage formulation, e.g., (Roque et al. 1999; Zhang et al. 2001) identified the existence of fundamental energy thresholds, the Dissipated Creep Strain Energy (DCSE) and the Fracture Energy (FE) density that distinguishes between healable and non-healable macro-crack formation under cyclic and fast or instantaneous load application conditions respectively. The viscoelastic continuum damage which is based on energy considerations and the principle of work potential have also been used to characterize damage in asphalt mixtures by various researchers. In (Luo et al. 2015), an energy balance equation is introduced which is a function of the Dissipated Pseudo Strain Energy (DPSE), Recovered Pseudo Strain Energy (RPSE) and the surface energy to determine the energy threshold for micro-crack initiation in asphalt mixtures. (Onifade et al. 2015a, 2016) developed a viscoelastic damage model and proposed a critical energy threshold for micro-crack initiation which is a function of the strain energy density at the instance of damage initiation.

X-ray computed tomography and image processing techniques have been used by researchers to characterize the internal structure of asphalt mixtures, e.g, Kutay et al. (2010), Masad et al. (2002), Onifade et al. (2015b, 2014, 2013), Tashman et al. (2002), You et al. (2012). Micromechanical modeling provides an insight into the interaction of the constituent phases in the material and their effect on the overall strength and deformation mechanism. It provides the opportunity to consider the geometric and spatial distribution of the constituent phases to capture the meso-scale response and investigate its effect on the global response on the macroscale level.

In the present study, micromechanical modeling is used to establish critical criteria for meso-scale crack initiation and to investigate the effect of varying strain rate on the global strain energy density threshold at the instance of meso-crack initiation. X-ray Computed Tomography (CT) and image processing techniques are used to capture, identify, and segment the different phases in asphalt concrete mixtures using the established procedures in Onifade et al. (2015b, 2014, 2013).

2 X-ray Computed Tomography and Image Processing

X-ray computed tomography is used to capture the internal structure of the asphalt concrete core sample used in this study. The asphalt concrete core sample has a diameter of 100 mm and a height of 80 mm. The sample is scanned at an energy intensity of 225 kV which produced images with a voxel size of 59 μm. Digital image processing techniques which include beam hardening correction, image contrast enhancement, filtering, segmentation were all carried out using image processing tools in Avizo Fire®. Figure 1 shows the image of the asphalt concrete core used in this study, a sample slice from the X-ray CT scan and the phase segmented image. Details of the scanning procedure and the image processing techniques used for the phase identification and segmentation can be found in Onifade et al. (2015b, 2013).

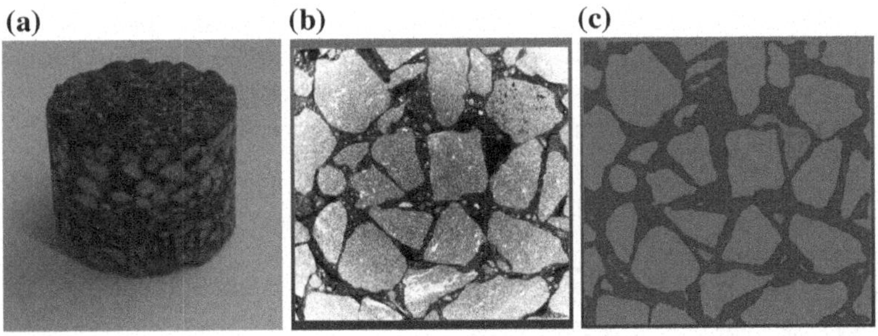

Fig. 1 **a** Asphalt concrete core sample, **b** sample slice from X-ray CT scan, **c** phase segmented asphalt concrete image

3 Numerical Simulation

Micromechanical analysis is carried out to establish critical conditions for meso-crack initiation and to investigate the effect of varying strain rate on the global strain energy density threshold at the instance of meso-scale crack initiation using the Finite Element (FE) method. The numerical simulation is performed using COMSOL Multiphysics®. A Slice from the stack of images obtained from the X-ray scan is processed and used for the analysis. The segmented asphalt concrete image consists of 59 % aggregate, 37.5 % mastic and 3.5 % air voids. The aggregates are modeled as isotropic linear elastic materials with a Young's modulus of 25 GPa and a Poisson's ratio of 0.2. The mastic is considered as a viscoelastic material and modeled using the generalized Maxwell's model. The WLF shift factors C1 and C2 are 27.83 and 250.1 K respectively, at a WLF reference temperature of −20 °C. The mastic material parameters and the WLF shift function parameters can be obtained from Dai et al. (2006), Onifade et al. (2015b). The numerical simulation is carried out at a temperature of −10 °C with a mastic Poisson's ratio of 0.3.

The FE model geometry is subjected to a tensile loading condition with a reference deformation rate of 0.009 mm/s prescribed at the top of the geometry while symmetry boundary conditions are prescribed at the left and bottom. The COMSOL mesh utility is used to mesh the geometry of the FE model. The final mesh consists of 118,321 triangular elements with very fine mesh size at the interface between aggregates and the mastic, and at the air voids boundaries. Figure 2a shows the mesh of asphalt concrete model used for the numerical simulation.

3.1 Damage and Crack Initiation Criteria

Meso-scale damage initiates on the meso-scale level when the local strain energy density (G) is greater than a critical local energy value (G_c). Due to the concentration and accumulation of the damage in front of the crack tip, a certain critical damage density threshold is reached and a meso-scale crack initiates in from of the

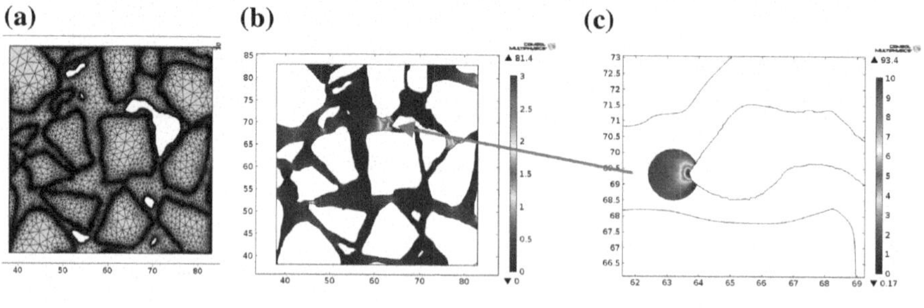

Fig. 2 **a** FE mesh, **b** plot of meso-scale strain energy density (kJ/m^3) showing the critical damage region, **c** plot of the local strain energy (kJ/m^3) in the contour around the critical crack tip with 1 % damage density

crack tip. The following procedure can be used to establish critical conditions for meso-scale crack initiation from the micromechanical model response at the reference deformation rate of 0.009 mm/s. The local stress and strain in the meso-scale are denoted using the normal convention as $\boldsymbol{\sigma}$ and $\boldsymbol{\varepsilon}$ respectively. The corresponding stress and strain on the macroscopic level are denoted as $\boldsymbol{\Sigma}$ and $\mathbf{E}$ respectively.

In the study on the development of an energy-based damage model by (Onifade et al. 2015a), it was observed that the global tensile stress (Σ_t) at the instance of initiation of a meso-scale crack at −10 °C is generally below 1.95 MPa for the unmodified asphalt mixtures used in the study. In this study, a global stress value of 1.6 MPa is selected and used as the global stress value at the instance crack initiation to establish damage and crack initiation criteria at the reference deformation rate. The local energy (G) at the crack tip at the instance of meso-scale crack initiation is obtained as a product of the local first principal stress (σ^1) and strain (ε^1) and denoted as G_c. To quantify the extent of damage needed to initiate a meso-scale crack, a contour (Γ) is defined in front of the crack tip and used to establish a critical damage density criterion for meso-scale crack initiation. The critical meso-scale damage energy (G_c) is obtained as 10 kJ/m^3 and the corresponding critical contour damage density is 1 % of the total contour area (A_Γ). Figure 2b shows a plot of the local energy density (G) in the mastic and Fig. 2c shows a plot of crack tip contour with a damage density of 1 %. The meso-scale damage and crack initiation criteria are summarized below:

Damage criterion: $G = 0.5\sigma^1\varepsilon^1 > G_c$; where $G_c = 10$ kJ/m^3
Crack-initiation criterion: $A_{\Gamma d}/A_\Gamma = 1\ \%$; where $A_{\Gamma d} = A_\Gamma(G > G_c)$

3.2 *Effect of Varying Deformation Rates on Crack Initiation*

A parametric study is carried out to investigate the effect of varying deformation rates on the global strain energy density at the instance of meso-scale crack initiation. The established damage and crack initiation criteria are used to obtain the global strain energy density at the instance of meso-scale crack initiation for the corresponding deformation rate. Figure 3a shows the evolution of the damage density in the crack tip contour with respect to the global strain. It can be seen that the damage density is higher at faster deformation rates and the damage density approaches an asymptotic value at a strain value of about 600 με. Figure 3b shows a closer view of the damage density evolution and the critical damage density threshold.

Figure 4a shows the evolution of the global strain energy density for the different deformation rates. The global strain energy threshold at the instance of meso-scale crack initiation is obtained and shown in Fig. 4b. The global energy threshold at the reference deformation rate of 0.009 mm/s is 0.128 kJ/m^3. It was observed that the changes in the global energy threshold as a result of varying deformation rates are not significant as a 90 % change in the deformation rate from the reference value resulted in only 2.4 % change in the global critical energy threshold.

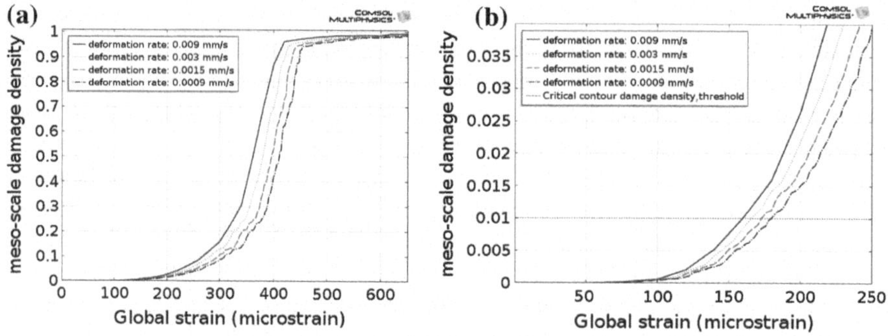

Fig. 3 **a** Evolution of the damage density in the crack-tip contour as a function of global strain, **b** detailed plot of crack-tip contour damage density versus global strain

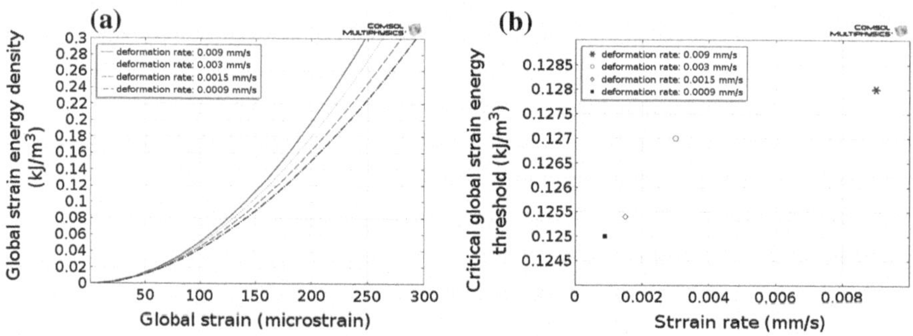

Fig. 4 **a** Evolution of the global strain energy (kJ/m^3) density at different deformation rates, **b** plot of critical global strain energy (kJ/m^3) threshold at the instance of meso-scale crack initiation

4 Conclusions

In this study, meso-scale damage and crack initiation criteria are established and the existence of a fundamental global energy threshold for meso-scale crack initiation is investigated. The results obtained from this study show that there are no significant differences in the values of the global strain energy density at the instance of meso-scale crack initiation obtained at different deformation rates. The findings show that there exists a fundamental global energy-threshold that is invariant of the rate of loading for meso-scale crack initiation.

References

Dai, Q., Sadd, M.H., You, Z., 2006. A micromechanical finite element model for linear and damage-coupled viscoelastic behaviour of asphalt mixture. Int. J. Numer. Anal. Methods Geomech. 30, 1135–1158. doi:10.1002/nag.520

Kutay, M.E., Arambula, E., Gibson, N., Youtcheff, J., 2010. Three-dimensional image processing methods to identify and characterise aggregates in compacted asphalt mixtures. Int. J. Pavement Eng. 11, 511–528. doi:10.1080/10298431003749725

Luo, X., Luo, R., Lytton, R.L., 2015. Energy-Based Crack Initiation Criterion for Viscoelastoplastic Materials with Distributed Cracks. J. Eng. Mech. 141, 04014114. doi:10.1061/(ASCE)EM.1943-7889.0000830

Masad, E., Jandhyala, V., Dasgupta, N., Somadevan, N., Shashidhar, N., 2002. Characterization of Air Void Distribution in Asphalt Mixes using X-ray Computed Tomography. J. Mater. Civ. Eng. 14, 122–129. doi:10.1061/(ASCE)0899-1561(2002)14:2(122)

Onifade, I., Balieu, R., Birgisson, B., 2016. Interpretation of the Superpave IDT strength test using a viscoelastic-damage constitutive model. Mech. Time-Depend. Mater. 1–19. doi:10.1007/s11043-016-9297-9

Onifade, I., Birgisson, B., Balieu, R., 2015a. Energy-based damage and fracture framework for viscoelastic asphalt concrete. Eng. Fract. Mech. 145, 67–85. doi:10.1016/j.engfracmech.2015.07.003

Onifade, I., Jelagin, D., Birgisson, B., Kringos, N., 2015b. Towards Asphalt Mixture Morphology Evaluation with the Virtual Specimen Approach. Road Mater. Pavement Des. doi:10.1080/14680629.2015.1098561

Onifade, I., Jelagin, D., Guarin, A., Birgisson, B., Kringos, N., 2014. Effect of micro-scale morphological parameters on meso-scale response of Asphalt Concrete, in: Asphalt Pavements. CRC Press, North Carolina, pp. 1775–1784.

Onifade, I., Jelagin, D., Guarin, A., Birgisson, B., Kringos, N., 2013. Asphalt Internal Structure Characterization with X-Ray Computed Tomography and Digital Image Processing, in: Kringos, N., Birgisson, B., Frost, D., Wang, L. (Eds.), Multi-Scale Modeling and Characterization of Infrastructure Materials, RILEM Bookseries. Springer Netherlands, Stockholm, Sweden, pp. 139–158.

Roque, R., Zhang, Z., Sankar, B., 1999. Determination of crack growth rate parameters of asphalt mixtures using the superpave IDT. Proc. Assoc. Asph. Paving Technol. 68, 404–433.

Tashman, L., Masad, E., D'Angelo, J., Bukowski, J., Harman, T., 2002. X-ray Tomography to Characterize Air Void Distribution in Superpave Gyratory Compacted Specimens. Int. J. Pavement Eng. 3, 19–28. doi:10.1080/10298430290029902a

You, T., Abu Al-Rub, R.K., Darabi, M.K., Masad, E.A., Little, D.N., 2012. Three-dimensional microstructural modeling of asphalt concrete using a unified viscoelastic–viscoplastic–viscodamage model. Constr. Build. Mater. 28, 531–548. doi:10.1016/j.conbuildmat.2011.08.061

Zhang, Z., Roque, R., Birgisson, B., Sangpetngam, B., 2001. Identification and Verification of a Suitable Crack Growth Law (with Discussion). J. Assoc. Asph. Paving Technol. 70.

Mechanical Behavior of Asphalt Mixture Based on X-ray Computed Tomography Images and Random Generation of Aggregates Skeleton

J. Absi, F. Fakhari Tehrani, F. Courreges and C. Petit

Abstract The microstructure of an asphalt mix is influenced by the aggregate content, orientation and contacts. In several previous studies, the asphalt concrete is treated as homogeneous material and its microstructures are ignored. However, it has been reported that the deformation and strength of asphalt concrete are not only influenced by volume fraction of its components but also affected by the spatial distribution of its microstructures. Thus, to investigate the mechanical behaviors of asphalt concrete, it is imperative to notice more accurately a micromechanical model containing information of components and microstructures. The objective of this study is to investigate the influence of the microstructure characteristics on the mechanical behaviour of asphalt mixes. The details of the microstructure are obtained by X-ray computed tomography (CT). A comparison with results issues from digital models of random aggregate generation will be considered. The dynamic modulus of the matrix and elastic properties of aggregates were introduced

J. Absi (✉)
SPCTS, CNRS, CEC, Université de Limoges, 12 Rue Atlantis,
87068 Limoges Cedex, France
e-mail: josept.absi@runilim.fr

F. Fakhari Tehrani
Groupe d'Étude des Matériaux Hétérogènes—Équipe Génie Civil et Durabilité,
Université de Limoges, Boulevard Jacques Derche, 19300 Egletons, France
e-mail: fatehftehrani@yahoo.com

F. Fakhari Tehrani · F. Courreges
Conservatoire national des Arts et métiers, Paris, France
e-mail: fabien.courreges@unilim.fr

C. Petit
XLIM—UMR CNRS, No 7252, 123 Avenue Albert Thomas,
87060 Limoges Cedex, France
e-mail: christophe.petit@unilim.fr

A. Chabot et al. (eds.), *8th RILEM International Conference on Mechanisms of Cracking and Debonding in Pavements*, RILEM Bookseries 13,
DOI 10.1007/978-94-024-0867-6_96

as input parameters into numerical models qualified by heterogeneous microstructure. Finally, the influence of the microstructure on the mechanical response of material at local level is also highlighted.

Keywords Microstructure characteristics · X-ray tomography · Random generation · Complex modulus

1 Introduction

An asphalt mixture is a composite material that comprises aggregates, binder and air voids. Aggregates usually take up about 85 % of the volume of asphalt concrete with different sizes, irregular shapes and random locations. In the most previous studies, the asphalt concrete is treated as homogeneous material and its microstructures are ignored. However, it has been reported that the deformation and strength of asphalt concrete are not only influenced by volume fraction of its components but also affected by their spatial distribution. Thus, to investigate the mechanical behaviors of asphalt concrete, it is imperative to notice more accurately a micromechanical model containing information of components and microstructures. Many researchers used micromechanical models and numerical tools that implement these models such as finite element method (FEM) and discrete element method to predict fundamental asphalt properties based on the properties of the mastic and aggregates (Abbas et al. 2005; Sadd and Dai 2005; You et al. 2009).

Generally, the recent micromechanical modeling methods can be divided into two main categories. The first one is the random aggregates generation. A number of researchers used this method to describe the heterogeneous microstructure of asphalt concrete. According to aggregates generation curve, the aggregates are generated and randomly located within matrix phases. This method can reflect, for aggregates of an asphalt sample, the size, shape and distribution in a statistical sense. Since there are various technical algorithms to generate the virtual microstructure of asphalt mixture, and each method has its advantages and weaknesses, so this approach is developed in function of own needs.

Another method is the numerical image processing technique such as X-ray computed tomography (CT) imaging techniques. In this method the internal microstructure of asphalt mixture is captured by computed tomography scanner and then 2D or 3D digital samples are created with the reconfiguration of the scanned surface images. This approach has been paid significant attention in the past 10 years to investigate and identify the spatial microstructure distribution of bituminous materials and also to investigate the influence of several parameters such as binder content, fine aggregate and coarse aggregate gradation, air void distribution on the mechanical behavior of this media (Massad et al. 2002; Bhasin et al. 2011; You et al. 2008; Coleri et al. 2012).

In this study, X-ray computed technology and random aggregate generation have been respectively applied for acquisition of the internal microstructure of the considered material and generation of virtual microstructure. This later was generated from a custom software (called M.O.A., French acronym for Random Object Modeler) developed in our laboratory. On the numerical simulations, asphalt mixture is considered as a heterogeneous material containing irregular aggregate skeleton bonded with asphalt mastic. By this approach we are able to take into account the capacity of energy dissipation of the heterogeneous viscoelastic matrix either by X-ray tomography and random generation method. Results can be used to investigate the influence of shape and filling rate of the granular skeleton on the dissipated energy due to crack formation and propagation in the heterogeneous medium.

2 Material Description and Characteristics

In this study, three components are used to construct the coming numerical models.

A 50/70 penetration grade bitumen where its mean penetrability is 60 tenths of mm at 25 °C, 5 s, 100 g and the ring and ball temperature is 48.5 °C.

A mastic formulated with a well—graded Limestone filler. The volume concentration of the filler is 40 % (V_{filler}/V_{mastic} = 40 %) (Delaporte et al. 2005).

The HMA concerned by this study is a 50/70 asphalt mixture. The air void content in this asphalt mixture is equal to 7.4 % and bitumen content is 5.7 ppc.

The complex modulus tests of bitumen, asphalt mastic are carried out by Delaporte et al. (2005) experimental studies.

3 Microstructure Acquisitions and Digital Sample Generation

3.1 X-ray Tomography and Particle Acquisition

X-ray computed tomography (CT) offers a non-destructive technique for acquiring internal microstructure of heterogeneous materials. In the case of asphalt concrete, the aggregates used in the asphalt mixture can be visualized inside an asphalt concrete cylinder. The initial dimensions of the laboratory prepared specimen are: diameter d = 10 mm and height h = 150. In order to capture the fines particles, the

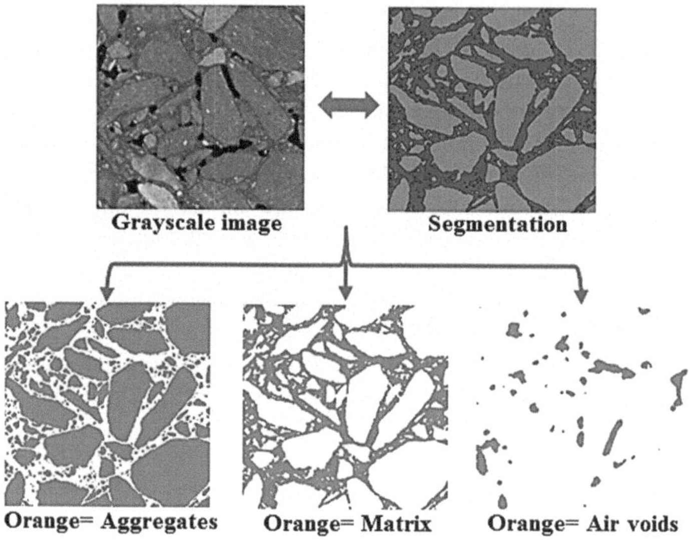

Fig. 1 Illustration of *gray image* acquired by X-ray CT, segmentation for aggregates, matrix and air voids distribution in 2D

cubic samples are seen from center of cylindrical initial sample with a diameter of 50 mm and a height of 50 mm. The sample is scanned at an energy intensity of 80 kV. The scanning resolution is 783 × 782 pixels with a total of 1270 slices. In the X-ray CT images the aggregates, matrix and air voids have different grayscale densities ranging from 0 to 255. The dense materials like aggregates are revealed with brighter pixels (close to 255); in contrast the air voids with negligible density are seen with dark pixels (close to 0).

Following being recognized the thresholds for each domain, masks with different colors were assigned to visualize the boundaries between each domain. This process which is so-called segmentation is a very important step in CT procedure to separate the constituents of the asphalt mixture. The quality of segmentation essentially depends on the resolution height and gray level of each image. The procedure of segmentation to separate aggregate, matrix and air voids is shown in Fig. 1.

3.2 Random Generation of Aggregate Skeleton

The commercial finite element codes ABAQUS is unable to construct random microstructures in heterogeneous media. Therefore, it is necessary to use dedicated software. The M.O.A. program, developed in our laboratory in C++ language, is an

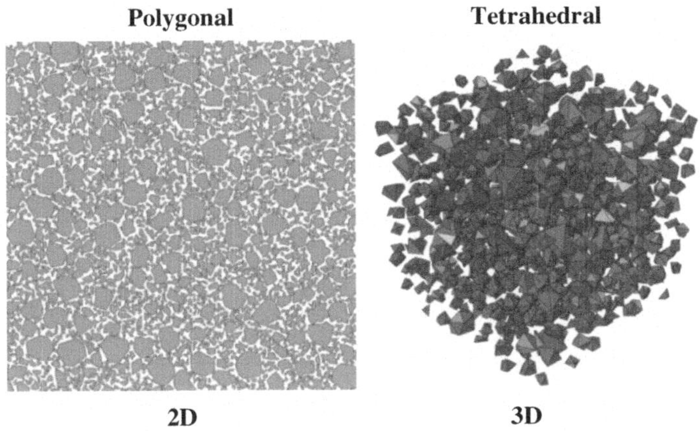

Fig. 2 Generation of aggregates in 2D with polygonal and in 3D with tetrahedral shapes

original software able to generate automatically samples derived from a matrix with a given percentage of air voids and containing a known number of inclusions. Aggregates present different shapes, extending from spherical to polyhedral and randomly distributed in the volume defined by the matrix in the form of a rectangular prism or cylinder. Examples of patterns generated by M.O.A are shown in Fig. 2. The generated files can be written in various formats (IGES, STEP) and then imported into the ABAQUS software.

4 Numerical Results for Asphalt Mixture at Global and Local Sale

Figure 3a, b present the master curves of the complex modulus and phase angle of asphalt mixture obtained respectively by the random aggregate generation and X-ray computed tomography methods. It can be observed that the results obtained by both of these approaches agree reasonably. The average difference between simulation results at global scale is less than 4 % for complex modulus and 3 % for phase angle.

In order to evaluate the influence of microstructure on the mechanical response of material at local scale, the intensity of strain at different zones of numerical models were obtained and compared to global imposed strain ($\varepsilon = 1E^{-4}$).

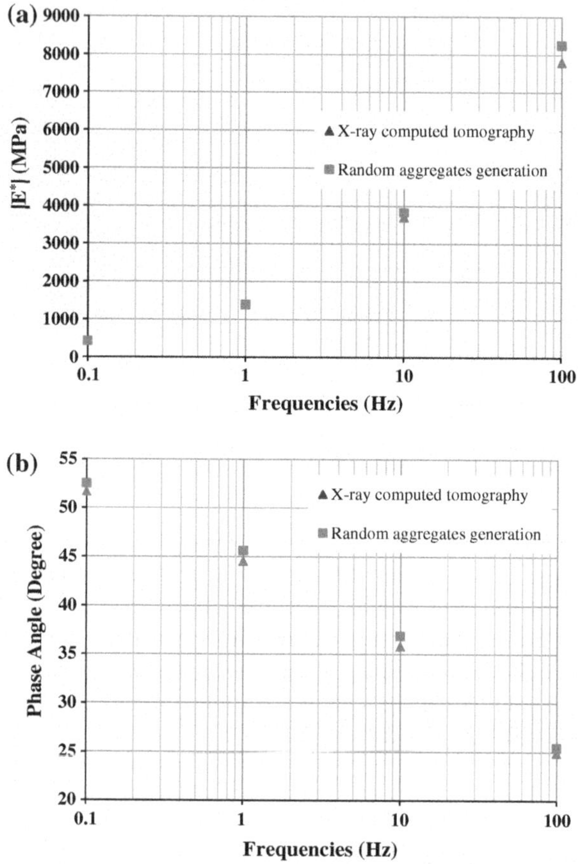

Fig. 3 Comparison between random generation and X-ray tomography methods, **a** complex modulus, **b** phase angle

Figure 4a, b show the distributions of local vertical strain obtained respectively by X-ray computed tomography and random aggregates generation methods. It can be observed that the deformation intensity ratio depends generally on the position and the thickness of the thin viscoelastic layer of matrix. As it shown, the intensity of strain at local scale can be reached up to 10 times greater than imposed global strain.

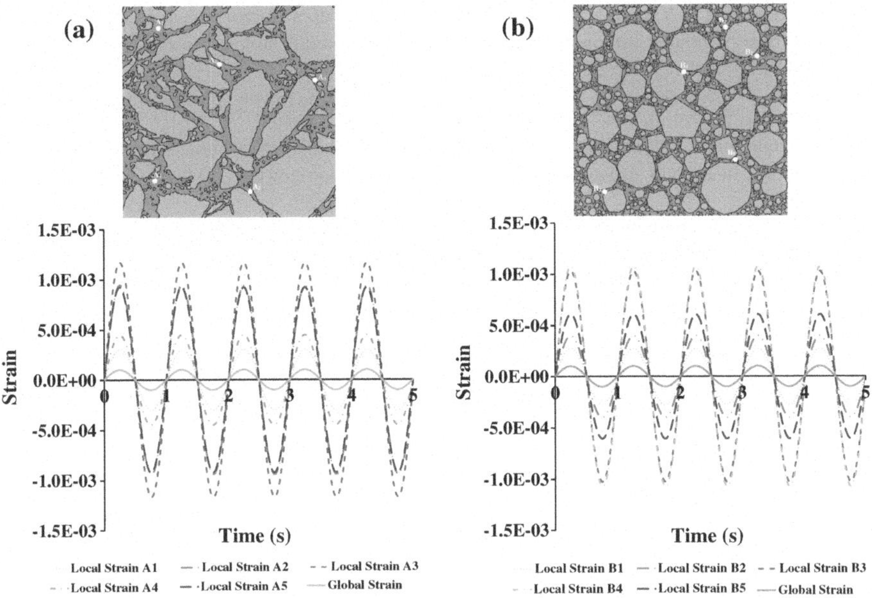

Fig. 4 Strain intensity obtained at different zones of **a** X-ray tomography, **b** random aggregate generation models

5 Conclusion

In this paper, we have offered a heterogeneous numerical approach able to calculate the mechanical properties of hot mix asphalt from the properties of its components. The microstructures of the material were obtained by two methods: random aggregates generation and X-ray computed tomography. In order to investigate the influence of microstructure into mechanical response of asphalt mixes, the complex modulus and phase angle issue from the two methods under study, were compared. Moreover, the intensity of deformation within the viscoelastic matrix at local scale was investigated. In summary, the proposed heterogeneous numerical approach, yields very acceptable results for highly heterogeneous media of asphalt mixes at global and local scales in comparison to those issues from X-ray method.

References

Abbas A, Masad E, Papagiannakis T, Shenoy A. Modelling asphalt mastic stiffness using discrete element analysis and micromechanics-based models. Inter J Pave Eng 2005; 6: 137.

Bhasin A, Izadi A, Bedgaker S. Three dimensional distribution of the mastic in asphalt composites. Constr Build Mater 2011; 25: 4079.

Coleri E, Harvay JT, Yang K, Boone JM. Developement of a micromechanical finite element model from computed tomography images for shear modulus simulation of asphalt mixtures. Constr Build Mater 2012; 30: 783.

Delaporte B, Di Benedetto H, Sauzéat C, Chaverot P(2005). Linear viscoelastic properties of mastics: results from a new annular shearing rheometer, Trondheim.

Massad E, Jandhyala VK, Dasgupta N, Somadevan N, Shashidhar N. Characterization of air void distribution in asphalt mixes using X-ray Computed Tomography. J Mater Civil Eng 2002; 14:122.

Sadd M, Dai Q. A comparison of micro-mechanical modeling of asphalt materials using finite elements and doublet mechanics. Mech Mater 2005; 37: 641.

You Z, Adhikari S, Kutay ME. Dynamic modulus simulation of asphalt concrete using the X-ray computed tomography images. Mater Struct 2009; 42: 617.

Part XII
Advanced Measurement Systems for Crack Characterization: Optical Measurements

Determination of Asphalt Pavements Fracture Parameters Using Optical Approaches

Amine Jamaaoui, Rahma Ktari, Octavian Pop, Anne Millien, Valéry Valle, Christophe Petit and Frédéric Dubois

Abstract This paper deals with the characterization of mode I fracture parameters using a kinematic approach integrating the experimental displacement measured by optical techniques. Tests are carried out using a wedge splitting sample made in asphalt concrete. The WST-cube specimen is a bi-layered asphalt concrete AC12 with Carbon Fiber grid at the interface. The tested material was provided in the RILEM-SIB. During the test the displacement field evolution close to the crack tip was recorded by using the optical techniques. An adjustment procedure was also used to improve the displacement fields and avoid experimental noise. Based on the experimental optical measurements, the energy release rate was performed by means the Crack Relative Displacement Factor and Stress In-tensity Factor. The Mark Tracking method is employed in order to measure the Crack Opening Displacement. This approach allows to consider the assessment of fracture parameters for the real structures.

A. Jamaaoui · R. Ktari · O. Pop (✉) · A. Millien · V. Valle · C. Petit · F. Dubois
GEMH, EA 3178, University of Limoges, 19300 Egletons, France
e-mail: ion-octavian.pop@unilim.fr

A. Jamaaoui
e-mail: amine.jamaaoui@etu.unilim.fr

R. Ktari
e-mail: Rahma.ktari@unilim.fr

A. Millien
e-mail: anne.millien@unilim.fr

V. Valle
e-mail: valery.valle@univ-poitiers.fr

C. Petit
e-mail: chirstophe.petit@unilim.fr

F. Dubois
e-mail: frederic.dubois@unilim.fr

A. Jamaaoui · R. Ktari · O. Pop · A. Millien · V. Valle · C. Petit · F. Dubois
PEM, CNRS UPR 3346, University of Poitiers, Institut Pprim, 86000 Poitiers, France

A. Chabot et al. (eds.), *8th RILEM International Conference on Mechanisms of Cracking and Debonding in Pavements*, RILEM Bookseries 13,
DOI 10.1007/978-94-024-0867-6_97

Keywords Opening mode · Mark tracking · Integral invariant · Energy

1 Introduction

Cracking is one of the recurring pathological consequences in asphalt concrete. The crack presence reduces the durability of the structure or even affects its behavior, by facilitating the migration of aggressive agents within the material. In these circumstances and considering the relatively high cost of repairs, the structural health assessment of asphalt concrete structures must include a strengthening in terms of monitoring of the cracking risks.

In the literature, several experimental devices and testing procedures are proposed in order to analyze the fracture process. Among these experimental techniques Wedge-Splitting Test (WST) (Tschegg et al. 1995), Louisiana Tack Coat Quality Tester (LTCQT) (Mohammad et al. 2009), ATacker Test (InstroTek Inc 2003) allow to perform the direct tensile tests (i.e. mode I). Since some years, optical techniques were increasingly applied to the fracture behavior analysis. Among these experimental techniques Digital Image Correlation (DIC) (Sutton et al. 1983; Buttlar et al. 2014) seem to be most adapted to analyze the fracture process in asphalt concrete. Associated with full-fields techniques the DIC can be easily used to measure the displacement and strain fields at different scales.

This paper presents an experimental Wedge Splitting Test completed by an analytical approach in order to evaluate fracture parameters in opening mode (mode I). In order to characterize local or global fracture parameters, several methods have been developed. Among these, the local approach based on Stress Intensity Factor (Sanford 1982) and Crack Relative Displacement Factor (Méité et al. 2013) evaluation or the energetic approaches based on energy release rate evaluation via integral invariants (Lanczos 1970) are the most used ones in fracture mechanics. Based on the possibility to associate the Crack Relative Displacement Factor with optical full-field methods in the present study, the fracture phenomenon is characterized from DIC measurements (Méité et al. 2013). The advantage of these approaches is the possibility to use the mechanical fields' remote from the crack tip not affected by the crack tip singularity.

1.1 Crack Relative Displacement Factor

Based on the Dubois developments, the kinematic state in the crack tip vicinity can be defined using CRDF (Dubois and Petit 2005; Méité et al. 2013). When considering two opposite points on crack lips identified by their polar coordinates $(\xi, -\pi)$ and $(\xi, +\pi)$ the CRDF associated with the opening mode can be expressed in terms of relative displacements in x_2 direction:

$$K_1^{(\varepsilon)} = |u_2(\xi, -\pi) - u_2(\xi, +\pi)| \cdot \sqrt{\frac{2 \cdot \pi}{\xi}} \tag{1}$$

Moreover, by applying the superposition principle, Dubois et al. shows that by separating the relative displacement description from the stress distribution the energy release rate may be expressed in the fallowing form (Méité et al. 2013):

$$G_1 = \frac{1}{C_1} \cdot \left(K_1^{(\varepsilon)}\right)^2 = \frac{K_1^{(\varepsilon)} \cdot K_1^{(\sigma)}}{8} \tag{2}$$

where C_1 is the reduced elastic compliance, $K_1^{(\varepsilon)}$ is the crack relative displacement factor and $K_1^{(\sigma)}$ is the stress intensity factor.

The recent works (Méité et al. 2013) show also that the Crack Relative Displacement Factor can be related with the first weighting coefficient A_1^1 of the Muskhelishvili series (Muskhelishvili 1933; Williams 1957) and DIC measurements.

$$K_1^{(\varepsilon)} = A_1^1 \cdot (\kappa + 1) \cdot \sqrt{8 \cdot \pi} \tag{3}$$

where κ is the Kolossov constant defined according with plane stress configuration.

The calculation of Muskhelishvili series weighting coefficients is performed from an adjustment procedure. Based on the Newton-Raphson iterative method, the adjustment procedure requires substituting experimental displacement fields measured by Digital Image Correlation by a Muskhelishvili series solution fitted to the experimental data (Méité et al. 2013). It should be noted that the adjustment procedure allows the calculation of the rigid body movements.

In the present study the $K_1^{(\sigma)}$ was calculated according with Guinea et al. approach taking into account a short crack (Guinea et al. 1996).

2 Materials and Methods

The tested material was provided from the experimental section of the RILEM-SIB project. The WST-cube specimen is a bi-layered asphalt concrete AC12 with Carbon Fiber grid at the interface. The grid is a square mesh with 20 by 20 mm (Canestrari et al. 2015). The AC12 material is a typical Italian formulation, with 12.5 mm maximum aggregate dimension and 50/70 penetration bitumen dosed at above 5 % by aggregate weight.

The fractures parameters were evaluated from Wedge Splitting test (Brühwiler and Wittmann 1990; Harmuth et al. 1996). The sample geometry and the experimental were presented schematically in Fig. 1. The testing was carried out using an

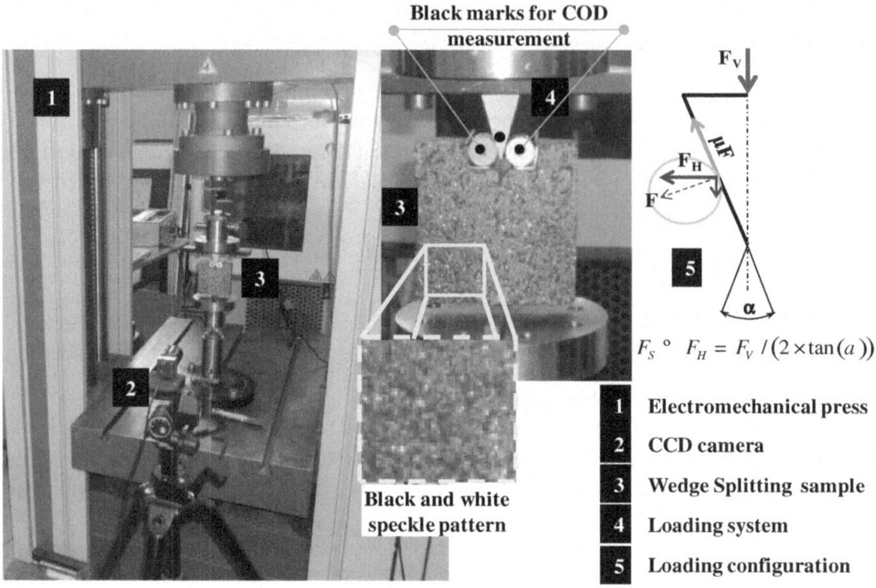

Fig. 1 Wedge splitting test setup

electromechanical press under displacement control. All the tests were performed in the ambient conditions of the test room.

As illustrated in Fig. 1, during the test, the applied vertical load component F_V and the crack opening displacement (COD) at load line are recorded. The (COD) and loading were measured by means of Mark Tracking method (Bretagne et al. 2005) and a load cell, respectively. The capacity of the load cell is 50 kN with 0.5 % of accuracy. The basic principle of Mark Tracking method is based on comparison of two images acquired during the test, one before deformation and another one after deformation (Bretagne et al. 2005). The displacement of each mark is in fact the translation vector (u_1, u_2) in x_1 and x_2 directions of the centre of gravity.

In addition with the measurement devices of testing machine, the sample deformation was performed from the measurements by Digital Image Correlation. These measurements allowed for determining displacement and strain fields. The Digital Image Correlation configuration used consisted of a CCD camera and a LED light source. Deftac and Correla software's, developed by PEM team of Pprim Institute of Poitiers, were used to perform the mark tracking and digital image correlation. The estimating uncertainty of displacement is 0.026 pixels.

According to DIC principle prior to testing, a very thin black and white speckle pattern was sprayed on the specimen surface (as the paint is spread in a very thin layer the structural behavior of sample is not modified). As described in Méité et al. (2013) displacement fields were calculated into Zone Of Interest (ZOI) discretized by small group of pixels called subsets. It should be noted that the displacement is

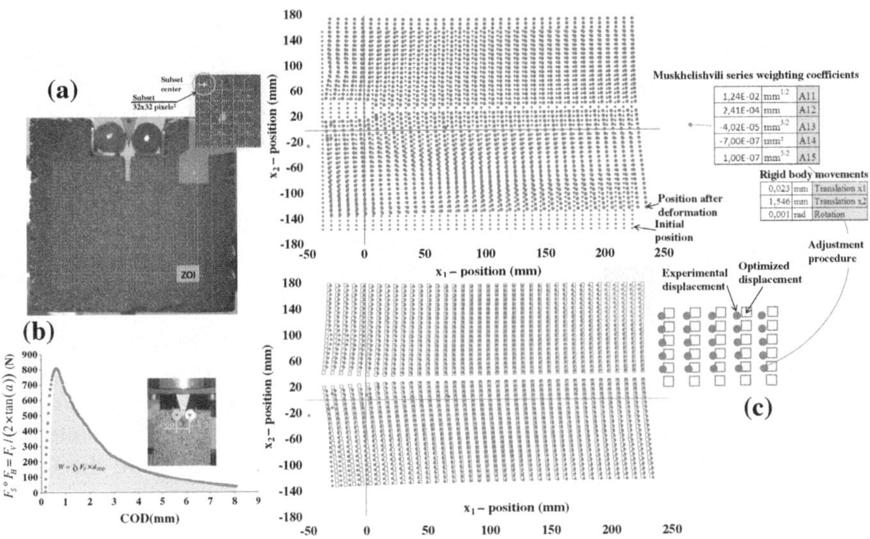

Fig. 2 Experimental results

supposed homogeneous and bilinear inside the subset. Supposing that the light intensity distribution during the test does not change, the displacement fields are estimated by searching the subset distortions in terms of translations, rotations and rigid body motions (Sutton et al. 1983).

3 Results

In the present study the analysis is performed for a splitting load of 575 N.

The Zone Of Interest (ZOI) defined in DIC analysis is illustrated in Fig. 2a. Their size is 586 × 1448 pixels^2. The ZOI was subdivided by the subsets of 32 by 32 pixels^2 with vertical and horizontal gaps of 1 pixel. The factor scale is 0.220338 mm/pixel.

The splitting load versus the Crack Opening Displacement (COD) was plotted in Fig. 2c. In fact, the plotted COD corresponds to the rollers displacement measured by Mark Tracking method. Knowing that the crack ligament area is about 0.00341 m^2, the fracture energy calculated from graph evolution is evaluated to 558 J/m^2.

As explained above the Crack Relative Displacement Factor calculation was performed by using the first weighting coefficient A_1^1. As described in Fig. 2c the weighting coefficients of the Muskhelishvili series and the rigid body movements was performed from an adjustment procedure. By using the weighting coefficient A_1^1 calculated for a splitting load of 575 N (see Fig. 2b) the CRDF is equal to

0.0056 $m^{1/2}$ (the Poisson's ratio is about 0.4). In the same time, by using (Guinea et al. 1996) the Stress Intensity Factor calculated for F_S equal to 575 N is 0.32 MPa $m^{1/2}$.

By replacing the CRDF and SIF value in Eq. (1) the energy release rate corresponding a splitting loading of 575 N is 223 J/m^2.

4 Conclusion

In this study, we have proposed a coupling between experimental and analytical approaches for the purpose of characterizing fracture parameters in opening mode. Based on the experimental optical measurements, the energy release rate was performed by means the Crack Relative Displacement Factor and Stress Intensity Factor. The Mark Tracking method is employed in order to measure the Crack Opening Displacement.

The present study allows also concluding that the Crack Relative Displacement factor coupled with Stress Intensity Factor approach provides a good estimation of energy release rate without the knowledge of material elastic properties. This approach allows considering the assessment of fracture parameters for the real structures.

References

Bretagne N, Valle V, Dupre JC (2005) Development of the marks tracking technique for strain field and volume variation measurements. NDT&E Int 38 : 290–298.

Brühwiler E, Wittmann FH (1990) The wedge splitting test, a new method of performing stable fracture mechanics tests. Eng Fract Mech 35:117–25.

Buttlar WG, Hill BC, Kim YR, Kutay ME, Millien A, Montepara A, Paulino GH, Petit C, Pop IO, Romeo E, Roncella R, Safavizadeh SA, Tebaldi G, Wargo A (2014) Digital image correlation techniques to investigate strain fields and cracking phenomena in asphalt materials. Mater Struct 47:1373–1390.

Canestrari F, Belogi L, Ferrotti G, Graziani A (2015) Shear and flexural characterization of grid-reinforced asphalt pavements and relation with field distress evolution. Mater Struct 48:959-975.

Dubois F, Petit C (2005) Modelling of the crack growth initiation in viscoelastic media by the Gtheta integral. Engng Fract Mech 36:72-2821.

Guinea GV, Elices M, Planas J (1996) Stress intensity Factor for wedge-splitting geometry. Int J of Fracture 81:113-124.

Harmuth H, Rieder K, Krobath M, Tschegg E (1996) Investigation of the nonlinear fracture behaviour of ordinary ceramic refractory materials. Mater Sci Eng A 214:53–61.

InstroTek Inc (2003) ATACKERTMA Tack Coat Testing Device. USA.

Lanczos C (1970) The variational principles of mechanics. 4th edn. Dover Publications, New York.

Méité M, Pop O, Dubois F, Absi J (2013) Characterization of mixed-mode fracture based on a complementary analysis by means of full-field optical and finite element approaches. International Journal of Fracture 180:41-52.

Mohammad LN, Bae A, Elseifi MA, Button J, Scherocman JA (2009) Evaluation of Bond Strength of Tack Coat Materials in Field. Transp Res Rec J Transp Res Board 1:1–11.

Muskhelishvili IN (1933) Some basic problem of mathematical theory of elasticity. English translation, Noordhoff.

Sanford RJ (1982) Fracture mechanics. ASTM STP 833.

Sutton M, Wolters W, Peters W, Ranson W, McNeill S (1983) Determination of displacements using an improved digital correlation method. Image Vis Comput 1:133–139.

Tschegg EK, Kroyer G, Tan DM, Stanzl-Tschegg SE, Litzka J (1995) Investigation of bonding between asphalt layers on road construction. J Transp Eng 309–316.

Williams M (1957) On the stress distribution at the base of a stationary crack. ASME Journal Applied Mechanics 24:109-114.

Applying a Full-Field Measurement Technique to Study the Mechanical Behavior of Asphalt Mixtures

M.-C. Teguedi, B. Blaysat, M. Grédiac, S. Liandrat, S. Moreira and E. Toussaint

Abstract This study is an experimental investigation of the mechanical response of asphalt mixtures using a full-field measurement technique: the grid method. This measuring technique is able to detect small strain amplitudes (up to hundreds of microstrains) with a spatial resolution which is often compatible with the high strain gradients which generally occur in thin binder layers between aggregates. Compression tests were carried out on cylindrical specimens. Displacement and strain fields were measured during these tests. The main conclusion is that the mixtures exhibit highly concentrated strain peaks in the binder, while the aggregates remain almost undeformed. In addition the analysis of the local temporal response of the binder shows that it exhibits a nonlinear local response.

Keywords Grid method · Asphalt · Displacement measurement · Strain measurement

M.-C. Teguedi (✉) · B. Blaysat · M. Grédiac · E. Toussaint
Institut Pascal, Clermont Université, Université Blaise Pascal,
BP 10448, 630002 Aubière, France
e-mail: mohamed_cheikh.teguedi@univ-bpclermont.fr

B. Blaysat
e-mail: Benoit.BLAYSAT@univ-bpclermont.fr

M. Grédiac
e-mail: michel.grediac@univ-bpclermont.fr

E. Toussaint
e-mail: evelyne.toussaint@univ-bpclermont.fr

S. Liandrat · S. Moreira
Département Laboratoire de Clermont-Ferrand,
CEREMA, DterCE, Clermont-Ferrand, France
e-mail: sebastien.liandrat@cerema.fr

S. Moreira
e-mail: Sylvain.Moreira@cerema.fr

A. Chabot et al. (eds.), *8th RILEM International Conference on Mechanisms of Cracking and Debonding in Pavements*, RILEM Bookseries 13,
DOI 10.1007/978-94-024-0867-6_98

1 Introduction

The characterization of the mechanical behavior of asphalt materials is an issue that concerns directly the sustainability of pavement infrastructures. The highly heterogeneous nature of these materials induces a complex mechanical response at different scales. Moreover, their behavior at the microscale result from the interaction between the components (aggregates and binder) whose physical and mechanical nature is completely different. It is therefore clear that the investigation of the mechanical response of asphalt materials requires the use of suitable measurement tools which provide information at the scale of the asphalt constituents. In recent years, full-field measurement techniques, particularly the digital image correlation (DIC) technique, have become popular for observing and characterizing the heterogeneities in composite materials (Grédiac 2004). This technique is nowadays widely accepted and increasingly used by the asphalt pavement community for the characterization and analysis of local mechanical response of asphalt materials. Studies found in the literature addressing the application of the DIC technique to study the mechanical behavior of asphalt mixture can be categorized as follows: the characterization of strain properties of asphalt microstructures (Masad et al. 2001; Montepara et al. 2010; Yi-qiu et al. 2012); the investigation of cracks and damage distribution (Aragao and Kim 2011; Chehab et al. 2007; Moriyoshi et al. 2013) and the validation of models (Chehab et al. 2007).

However, the use of full-field measurement techniques for studying the mechanical behavior of asphalt materials is still limited compared to their application to other fields of experimental mechanics. Moreover, the obtained results allow generally only qualitative analysis of phenomena, without addressing the quantitative parameters characterizing them. This is related to the fact that DIC is not really suitable for detecting significant strain gradients which take place in narrow binder layers between aggregates. Besides, most of the aforementioned studies were conducted using commercial packages. These programmes have the disadvantage not to provide sufficient flexibility to suit the specific asphalt testing requirements. Recently, the grid method has been used to characterize the mechanical behavior of asphalt mixture in compression (Grédiac et al. 2014; Grédiac and Toussaint 2013). The obtained results show that this technique is able to detect small strain amplitudes (up to hundreds of microstrain), with a spatial resolution often compatible with the strain gradients, which that take place in narrow bands of binder.

This study aims at investigating the mechanical behavior of asphalt mixtures in compression using the grid method. The materials and methods are firstly described. The strain fields obtained with the grid method are finally presented and analyzed.

2 Tested Specimens and Measurement Technique

Three cylindrical specimens made of the same recycled hot-mixed asphalt supplied by Cerema Direction Territoriale Centre-Est, France, were tested in this study. The specimens were made from EB 14 wearing course on the one hand, and with a recycled asphalt mixture (RAP) content of 15 % on the other hand. The latter was fabricated according to the EN 13108-1 standard (EN 13108-1 2007). The matrix is a bitumen (grade: 30/50). The void content in this asphalt mixture is equal to 9 %. The bitumen content is 5.3 %, including 0.6 % of RAP bitumen. The tests were carried on cylindrical specimens manufactured according to the procedure proposed by Duriez (EN 12697-26 2012) (cylindrical-shaped specimens with 80 mm diameter). The front face under study is flat. This was obtained by cutting each cylindrical specimen along a plane parallel to its axis. The dimension of surface under investigation for all specimens is 6.5 cm × 6.5 cm. The tests carried out in this study are compression tests. They were performed in a temperature controlled room at 21 °C, with a displacement rate equal to 0.01 mm/s. The displacement/strain measurement technique is the grid method. Its principle is to analyze the displacements of a grid. This grid is transferred onto the surface of the specimen beforehand (Badulescu et al. 2009). Thus, the grid is assumed to perfectly follow the deformation of the specimen surface. It is worth noting that the procedure of grid transferring described in (Piro and Grédiac 2004) involves a curing phase of 40 h at 37 °C. To prevent heat deterioration of the sample, the glue polymerization is carried out at room temperature (20 °C) during a polymerization time of one week.

The next step consists in acquiring images of this grid by a camera at different stages of the load. These images are finally processed after testing by using an in-house algorithm. This algorithm uses a specific grid image processing procedure which relies on the calculation of the phase change of this pseudo-periodic surface marking between the reference and the current stage (Badulescu et al. 2009). Typically, a resolution in strain of some hundreds of microstrains with a spatial resolution equal to 30 pixels are achieved (Grédiac and Sur 2014). The camera used here to capture the grid images was a PCO 2000 camera featuring a 14-bit/2048 × 2048 pixel CCD sensor (CCD = charge-coupled device). The number of pixels used to digitize one grid period is equal to 5 pixels/period.

3 Results

As the method used in this study is a 2D measurement technique and since the movement of the specimen surface does not always remain plane, especially the aggregates. The latter are likely to exhibit an out-of-plane motion. The use of the measuring method shall be limited here to the study of the early stages of the tests. A preliminary compression test was performed on one specimen to determine the force from which this out-of-plane motion occurs. It has been verified that for a force less than 11kN, the movement of the specimen surface could reasonably be considered as plane.

3.1 Global Response

The global load-strain curves for three typical specimens are presented in Fig. 1a. In this case the macroscopic strain (or mean strain ε_{yy}^{mean}) is calculated directly from the displacement field provided by the grid method. This was done by measuring the displacement difference between two lines of pixels located 100 pixels under the top face and 100 pixels above the bottom face of the specimens, and dividing the obtained result by the initial distance between these two lines of pixels. This procedure enables us to remove the effect of parasitic displacements that generally occur during compression tests. The global temporal response of the specimens is presented in Fig. 1b. This figure presents the evolution of the mean vertical strain versus time. The two main stages of the loading phase are clearly visible: a loading phase followed by an unloading one. It has been checked that the response of the specimens was almost stabilized after 6000 s (for the sake of clarity only the first 4000 s of the test are presented in Fig. 1b). At the end of the test, residual strains remained for all specimens. They are due to the combination of the bitumen viscoplastic behavior and the irreversible degradation in the mastic (mix of bitumen and fine aggregates that binds coarse aggregates) during the test. The value of these residual strains is proportional to the peak strain reached during the loading phase.

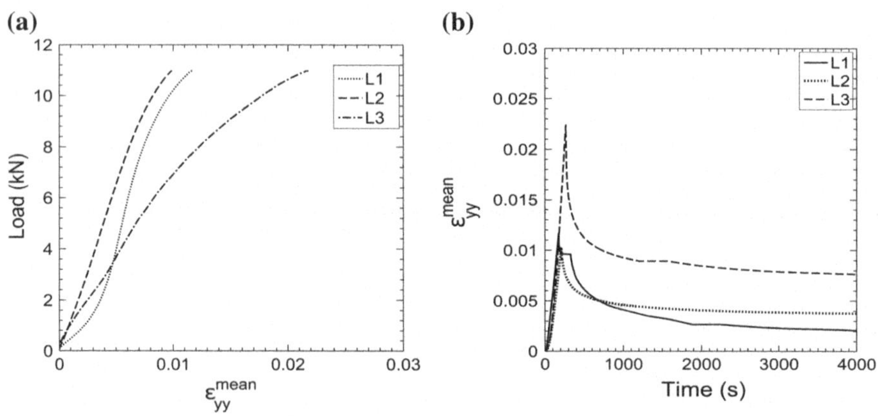

Fig. 1 **a** Load strain curves, **b** ε_{yy}^{mean} versus time

3.2 Local Response

Only the results related to specimen L2 are presented in the following. The other specimens have a similar response. The front face of the specimen and the corresponding strain fields obtained at the end of loading phase (F = 11 kN) are presented in Fig. 2. The horizontal longitudinal strain maps (ε_{xx}) emphasize the global expansion of the specimen along this direction, while a compressive longitudinal deformation appears along vertical direction (ε_{yy}). The strain maps feature some localized peaks, which can be superposed with the voids that are clearly visible in Fig. 2a. This is also the case with the other specimens tested within the framework of this study. It is worth noting that the strain is sustained by the binder while the aggregates remain almost undeformed. This is due to the important difference in rigidity between bitumen matrix and aggregates. Hence, it is possible to distinguish the aggregates beyond a certain size because a quasi-null strain takes place in the corresponding zones. This difference in behavior assigned to each of the three phases (aggregate, mastic and voids) highlights the complexity of the mechanism governing asphalt deformation. Each of these maps that have been are available for each image (600 per test).

To visualize the local response of the specimen, some points (see crosses in Fig. 2a) were selected in the mastic. The ε_{yy} at these points is plotted vs. time in Fig. 3a. The magnification between local and global strain is clearly visible especially for point 1 whose location is close to a void. The residual strain remaining at the end of the test are clearly visible. The small fluctuation which occur during the unloading phase are presumably due to random strain measurement errors. They are negligible compared to strain values.

It is interesting to compare local strains in the mastic and the average one. As in (Grédiac and Toussaint 2013), this is done by plotting at each point the ratio between local and global strains *vs* time. The corresponding curves are shown in Fig. 3b. Note that fluctuations are noticeable in Fig. 3b. These fluctuations are of two kinds: the small fluctuations (less than 1 %) correspond to measurement errors and do not modify the trend, the others which can be considered as trend fluctuation

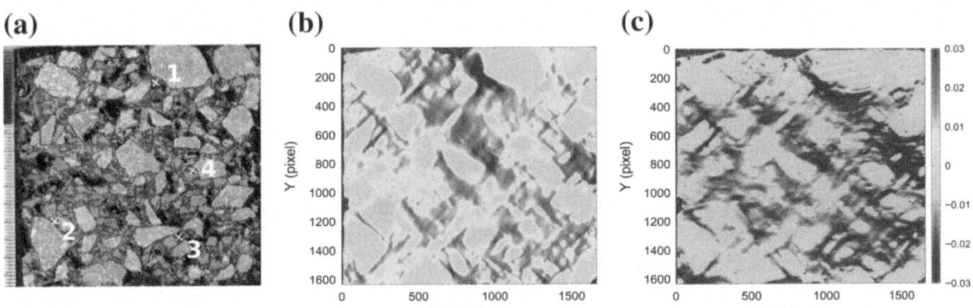

Fig. 2 **a** Front face of the specimens under test before depositing the grid, **b** vertical (ε_{yy}) and **c** horizontal longitudinal strain components (ε_{xx}) at the end of the loading phase

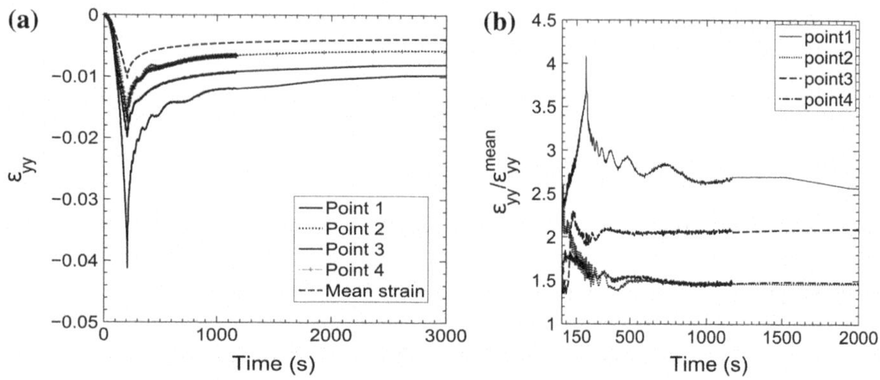

Fig. 3 **a** ε_{yy} and **b** ratio between local and global strain versus time, specimen L2

are related to the nonlinear *local* asphalt behavior of asphalt. During the unloading phase, these fluctuations become smaller and vanish progressively during specimen recovery. Except for point 1, for which behavior is affected by void crushing.

4 Conclusion and Perspectives

Full-field measurements obtained by the grid method are highly promising to investigate the local response of heterogeneous materials like asphalt concrete. The mechanical response of the material was observed at length scale ranging from binder scales to the mixture scale. Results show that the mixture exhibits highly concentrated strain that is developed in the binder around aggregates while the aggregates remain almost undeformed. The local response of the binder was also analyzed and compared to macroscopic strain. For some particular points, the local stain magnification between bulk and local strain is up to 4.

Acknowledgments The Auvergne Regional Council, is gratefully acknowledged for providing the financial support of this study.

References

Aragao, F.T.S., Kim, Y.-R., 2011. Characterization of fracture properties of asphalt mixtures based on cohesive zone modeling and digital image correlation technique. presented at the Transportation Research Board 90th Annual Meeting.

Badulescu, C., Grédiac, M., Mathias, J.D., 2009. Investigation of the grid method for accurate in-plane strain measurement. Meas. Sci. Technol. 20, 095102.

Chehab, G., Seo, Y., Kim, Y., 2007. Viscoelastoplastic damage characterization of asphalt–aggregate mixtures using Digital Image Correlation. Int. J. Geomech. 7, 111–118.

EN 12697-26, 2012. Mélanges bitumineux - Méthodes d'essai pour mélange hydrocarboné à chaud - Partie 26: module de rigidité.

EN 13108-1, 2007. Mélanges bitumineux - spécifications des matériaux - Partie 1: enrobés bitumineux.

Grédiac, M., 2004. The use of full-field measurement methods in composite material characterization: interest and limitations. Compos. Part Appl. Sci. Manuf. 35, 751–761.

Grédiac, M., Sur, F., 2014. Effect of sensor noise on the resolution and spatial resolution of displacement and strain maps estimated with the Grid Method. Strain 50, 1–27.

Grédiac, M., Toussaint, E., 2013. Studying the mechanical behaviour of asphalt mixture with the grid method. Strain 49, 1–15.

Grédiac, M., Toussaint, E., Petit, C., Millien, A., Ngyuen, D.C., 2014. A comparative study of the heterogeneous local mechanical response of two types of asphalt mixes. Mater. Struct. 47, 1513–1529.

Masad, E., Somadevan, N., Bahia, H., Kose, S., 2001. Modeling and experimental measurements of strain distribution in asphalt mixes. J. Transp. Eng. 127, 477–485.

Montepara, A., Romeo, E., Birgisson, B., Tebaldi, G., 2010. Strain Localization and Damage Distribution in SBS Polymer Modified Asphalt Mixtures. Road Mater. Pavement Des. 11, 899–915.

Moriyoshi, A., Takahashi, N., Ikeda, O., Kawashima, M., Akabane, T., 2013. Strain distribution in asphalt mixtures during the wheel tracking test at high temperatures. Constr. Build. Mater. 40, 1128–1135.

Yi-qiu, T., Lei, Z., Meng, G., Li-yan, S., 2012. Investigation of the deformation properties of asphalt mixtures with DIC technique. Constr. Build. Mater. 37, 581–590.

Application of a Coupled Digital Image Correlation and Discrete Element Method Approach to Model Low Temperature Asphalt Concrete Fracture

Brian Hill, Oliver Giraldo-Londono, William G. Buttlar and Glaucio Paulino

Abstract The transportation industry has become increasingly focused on thermal cracking in asphalt concrete. As such, fracture tests such as a single-edge notched beam test have become the norm. The most important output from this type of test is the crack mouth opening displacement (CMOD) based fracture energy, but it only takes into account the applied load and displacement at a discrete location on the test specimen. Full field displacement evaluation such as digital image correlation (DIC) and advanced material models used in discrete element modeling (DEM) provide further avenues to characterize the complex fracture process associated with asphalt mixtures. An ongoing research study couples DEM with DIC displacement fields to study the Mode I response of asphalt concrete in the beam fracture test. Previous research studies defined local properties to match the global load-CMOD behavior using elastic contact bonds. However, improved property calibration is possible through the optimized use of DIC displacement fields and viscoelastic DEM traction-separation laws. Research findings in the proposed work demonstrate the applicability of the DIC-DEM optimization scheme.

Keywords Fracture · DIC · DEM · Optimization

B. Hill (✉) · O. Giraldo-Londono · W.G. Buttlar
University of Illinois at Urbana-Champaign, Champaign, USA
e-mail: bchill2@illinois.edu

O. Giraldo-Londono
e-mail: grldlnd2@illinois.edu

W.G. Buttlar
e-mail: buttlar@illinois.edu

G. Paulino
Georgia Institute of Technology, Atlanta, USA
e-mail: glaucio.paulino@ce.gatech.edu

A. Chabot et al. (eds.), *8th RILEM International Conference on Mechanisms of Cracking and Debonding in Pavements*, RILEM Bookseries 13,
DOI 10.1007/978-94-024-0867-6_99

1 Introduction

The discrete element method (DEM) has shown to be a powerful tool for the analysis of fracture in hot mix asphalt (HMA). This material can be assumed to be composed of three material phases each with unique bulk and fracture properties. A typical fracture analysis of HMA in the context of DEM requires input parameters defining the bulk material behavior as well as the fracture properties according to Kim et al. (2008).

Determining the bulk properties is relatively simple; however, determining the fracture properties (e.g., fracture energy and cohesive strength) is a highly nontrivial task. The results obtained are often far from resembling the actual behavior of the tested specimen. Hence, further calibration of these properties is required to obtain representative parameters to be used in the simulations. Kim et al. (2005), Kim (2007) have performed this calibration by comparing the global response of the specimen with the results of heterogeneous DEM simulations (e.g., calibration based of load-CMOD curve). This approach, although approximated, is not enough to guarantee that the results obtained from the simulations will predict accurate local behavior of the tested specimen. One way of overcoming this issue is obtaining the calibrated fracture parameters by comparing the full displacement field obtained experimentally with the displacement fields obtained from the DEM simulations. Shen and Paulino (2011) used a hybrid inverse technique to estimate the cohesive fracture properties of a fiber-reinforced cementitious composite, by comparing the DIC measurements of the displacement fields with the results of FEM simulations.

In the current research paper, a nonlinear optimization method for obtaining the cohesive fracture properties of asphalt mixtures is presented. The method consists of comparing the DIC displacement field measurements, performed on: a single-edge notched beam SE(B) under three-point bending, with the results from heterogeneous DEM simulations. Different mastic and interface fracture properties are varied within predefined values, creating a discrete grid in the fracture parameters space. Additionally, a newly developed viscoelastic contact model for mastic and interface components is employed. The difference between the displacement fields obtained from DIC and the DEM simulations is computed at each point in the fracture parameters grid to identify an optimal solution.

2 Experimental and Optimization Procedure

The experimental portion of the study evaluated an virgin aggregate asphalt surface mixtures. The mixture was a 9.5 mm nominal maximum aggregate size mixture containing PG 64-22 asphalt binder and used central Illinois dolomitic limestone aggregate. The test samples were compacted such that they contained approximately 7.0 ± 0.5 % air voids following ASTM D7313-07 volumetric protocols.

The beam test geometry and loading rate were chosen in this study based on the experimental observations from previous work. The geometry has a thickness of

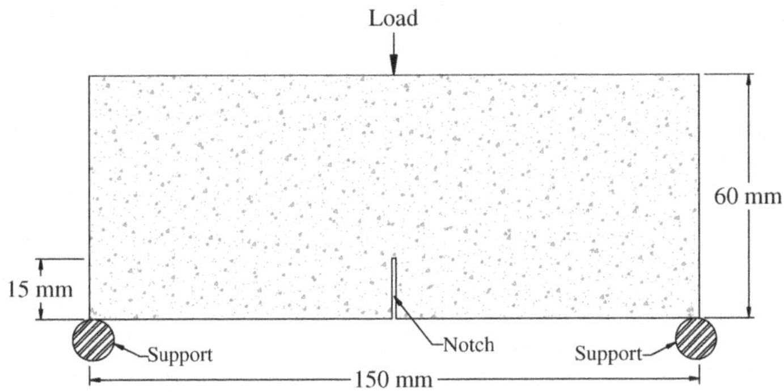

Fig. 1 Asphalt concrete beam geometry

40.0 mm, a schematic is provided in Fig. 1, and has a shorter ligament of 45.0 mm as compared to the 82.5 mm found with the DC(T) configuration. However, in both cases, the notch to depth (a/W) ratio remains constant with a value of 0.25. Furthermore, the loading rate was chosen as 0.5 mm/min in order for the peak load to occur at approximately 5 s meeting the asphalt concrete strength criteria discussed by Wagoner et al. (2005). Specimens were cooled to the −12 °C test temperature for approximately 2 h prior to testing to ensure uniform material temperature.

Stiffness properties required for the DEM applications were found using the indirect tension (IDT) test. This test, completed according to AASHTO T-322 at −12 °C, provided relaxation modulus ($E(t)$) properties for both mixtures and mastics (phase of an asphalt mixture with particle sizes less than 2.36 mm). The results are shown in Hill et al. (2016). This study used digital image correlation (DIC) as a tool to evaluate the deformation fields produced during the fracture tests. Due to the presence of a propagating crack, the subset splitting DIC approach developed by Poissant and Barthelat (2010) was used in the current study. The DIC set-up used in this study consisted of a Stingray F-201C CCD camera, a Tamron 50 mm lens, and multiple LED light sources. The region of interest was 45 mm in length by 45 mm in height.

A total of 16 nodal locations on the fracture specimen within the DIC region of interest were considered in the optimization process. These locations were spaced 5 and 10 mm on either side of the notch tip. This spacing was used in the horizontal direction for the beam geometry and vertical direction for the DC(T) geometry. The vertical grid spacing was 1.5, 3.0, and 4.5 mm, respectively beginning at the notch tip and moving in the vertical direction. A total of 10 analysis times were considered during the softening portion of the fracture tests.

The discrete element method (DEM) provides a useful tool to model the particulate behavior of granular systems. A complete description of the DEM implementation of bilinear cohesive softening in asphalt mixtures is provided in Kim et al. (2008) and Hill et al. (2016). Previous research only used the DEM technique with elastic cohesive contacts between particles. The current study implemented viscoelastic mastic contacts between mastic and interface particles to better model the behavior of the time-dependent asphalt components. A description of these contacts is also available in

Hill et al. (2016). Particle size in the DEM simulations was chosen to be 0.6 mm in diameter. In order to complete the optimization process, only the local fracture strength and fracture energy were optimized for the mastic and interface components. All other properties such as stiffness and aggregate fracture strength remained unchanged.

In this work, the authors use a nonlinear optimization method to find the cohesive fracture properties of an asphalt sample. The method consists of minimizing the error between the displacement field obtained from DIC and heterogeneous DEM simulations. To quantify this error, the DEM and DIC displacement field vectors are measured at a given number of grid points, n_n, near the notch tip. For each grid point, the displacement field vector is measured for n_p different values of CMOD on the load-CMOD curve. Then, the square of the 2-norm of the difference in the displacement field vectors obtained for all the grid points is added. This leads to the following measure of error:

$$\Phi(\boldsymbol{\lambda}) = \sum_{j=1}^{n_p} \sum_{i=1}^{n_n} \left\| \mathbf{u}_{DIC}^{(i,j)} - \mathbf{u}_{DEM}^{(i,j)}(\boldsymbol{\lambda}) \right\|_2^2 \tag{1}$$

where $\Phi(\boldsymbol{\lambda})$ is the error function or objective function; n_n is the number of grid nodes used to measure the displacements in the DEM; n_p is the number of points selected on the load-CMOD curve; $\boldsymbol{\lambda}$ is the set of fracture parameters (e.g., cohesive strength and maximum crack opening) used for each one of the DEM simulations; $\mathbf{u}_{DIC}^{(i,j)}$ and $\mathbf{u}_{DEM}^{(i,j)}$ are the displacement field vectors at the grid point i and for the jth value of CMOD on the load-CMOD curve. Further details can be found in Hill et al. (2016).

3 Results

The average experimental load-CMOD curve for the virgin aggregate mixture is provided in Fig. 2. As shown, this test yielded a peak load of approximately 4.0 kN and a total fracture energy of 494.2 J/m^2. A snapshot of the DIC measurement of

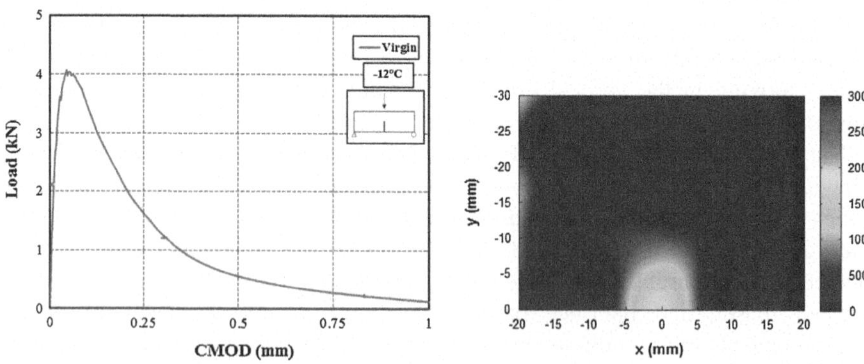

Fig. 2 *Top* Load-CMOD curve for SE(B) geometry. *Bottom* DIC strain field (ε_{xx}) virgin mixture

tensile micro-strain the horizontal direction is provided in Fig. 2 as well. As shown, the DIC field at the peak load of the material has a localized field of strain that encompasses the entire imaging ROI. Consequently, in the case of the beam, the entire remaining ligament in the specimen is encompassed by the size of the fracture process zone. The numerical optimization results for the elastic and viscoelastic simulations are provided below in Fig. 3. The plots are in three dimensions with the error function presented vertically and the parameter grid for the cohesive strength and maximum cohesive bond displacement shown as f_{max} and δ_{max}, respectively. As

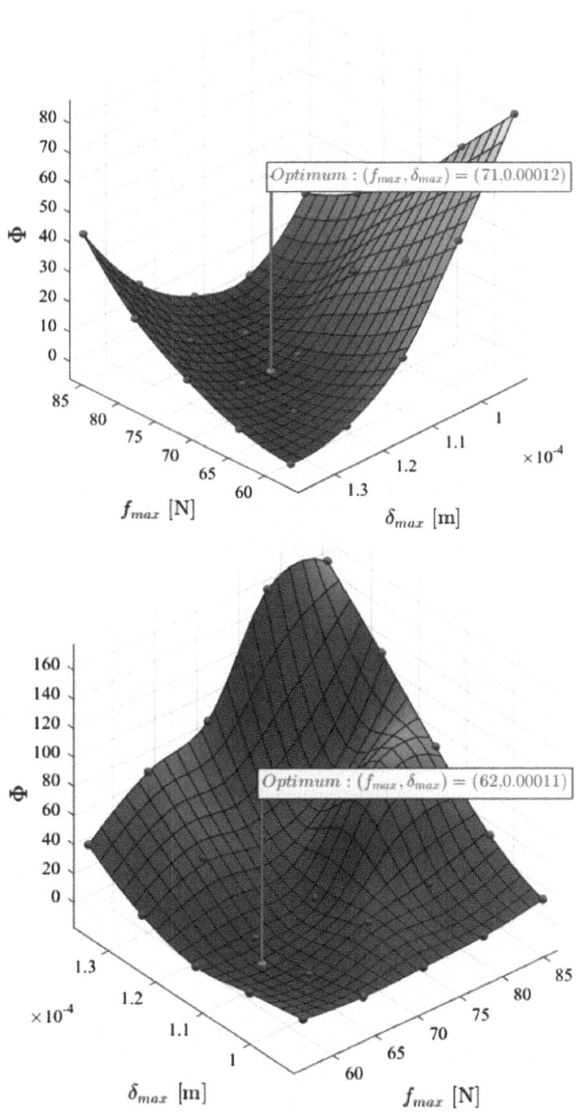

Fig. 3 *Top* Elastic contact error function surface. *Bottom* Viscoelastic contact error function surface

shown, a minimum value for the error function is found for both cases with approximately similar strengths and maximum bond displacements in the elastic and viscoelastic cases. The reduction in optimized strength and maximum displacement in the viscoelastic case is a function of the amount of energy dissipated by the viscoelasticity present in the contacts between mastic and interface particles. The amount of difference between the optimized parameter set for elastic and viscoelastic properties will likely change as a function of the material evaluated. In the case of recycled materials, it is hypothesized that the difference may be reduced as the recycled material may reduce the relaxation ability of the mixture.

4 Concluding Remarks

The current study demonstrated the ability of a hybrid DIC-DEM optimization approach to asphalt mixture fracture in a beam geometry. Fracture testing using the SE(B) geometry was completed and DIC methods were applied to identify experimental displacements to compare with those found using DEM. This study found that minimized error values could be found within a grid of local cohesive strength and maximum displacement in both viscoelastic and elastic models. The difference in optima between the viscoelastic and elastic models was determined to be a function of the dissipative effects of viscoelastic mastic and interface contacts.

References

Hill, B.C., Londono, O., Paulino, G., Buttlar, W.G. (2016) Inverse Estimation of Cohesive Fracture Properties of Asphalt Mixtures using Nonlinear Optimization, To be Submitted to Journal of Experimental Mechanics.

Kim, H., Buttlar, W.G. (2005) Micromechanical fracture modeling of asphalt mixture using the discrete element method. Proc., GeoFrontier 2005, ASCE, Reston, Va.

Kim, H (2007) Investigation of Toughening Mechanisms in the Fracture of Asphalt Concrete using the Clustered Discrete Element Method. Ph.D. Dissertation, University of Illinois at Urbana-Champaign.

Kim, H., Wagoner, M.P., Buttlar, W.G. (2008) Simulation of Fracture Behavior in Asphalt Concrete using a Heterogeneous Cohesive Zone Discrete Element Model, Journal of Materials in Civil Engineering 20, pp. 552-563.

Shen, B., Paulino, G. (2011) Identification of Cohesive Zone Model and Elastic Parameters of Fiber-Reinforced Cementitious Composites using Digital Image Correlation and a Hybrid Inverse Technique, Journal of Cement and Concrete Composites, 33 pp. 572-585.

Poissant, J., Barthelat, F. (2010) A Novel Subset Splitting Procedure for Digital Image Correlation on Discontinuous Displacement Fields, Experimental Mechanics 50 (3): pp. 353-364.

Automatic Crack Detection on Pavement Images for Monitoring Road Surface Conditions—Some Results from the Collaborative FP7 TRIMM Project

V. Baltazart, J.-M. Moliard, R. Amhaz, L.-M. Cottineau, A. Wright, D. Wright and M. Jethwa

Abstract This paper presents the two image processing techniques that have been developed within the scope of the TRIMM project to automatically detect pavement cracking from images. The first technique is a heuristic approach (HA) which was originally developed to process pavement images from the French imaging device. The Minimal Path Selection (MPS) method is a new technique which provides the accurate segmentation of the crack pattern along with the estimation of the crack width. HA has been assessed against the field data collection over UK roads provided by Yotta and TRL using the Tempest 2 device. MPS has been assessed against Aigle-RN pavement images over the French network. The benchmarking of five automatic segmentation techniques has been provided at both the pixel and the grid levels. Among others, MPS reached the best performance at the pixel level while it is matched to the FFA method at the grid level.

V. Baltazart (✉) · J.-M. Moliard · R. Amhaz · L.-M. Cottineau
IFSTTAR, LUNAM University, Nantes, France
e-mail: vincent.baltazart@ifsttar.fr

J.-M. Moliard
e-mail: jean-marc.moliard@ifsttar.fr

R. Amhaz
e-mail: rabih.amhaz9@gmail.com

L.-M. Cottineau
e-mail: louis-marie.cottineau@ifsttar.fr

R. Amhaz
IRIT, Paul Sabatier University, Toulouse, France

A. Wright · D. Wright
TRL, Wokingham, UK
e-mail: dean.wright@yotta.co.uk

A. Wright · M. Jethwa
Yotta, Leamington Spa, UK
e-mail: manish.jethwa@yotta.co.uk

A. Chabot et al. (eds.), *8th RILEM International Conference on Mechanisms of Cracking and Debonding in Pavements*, RILEM Bookseries 13,
DOI 10.1007/978-94-024-0867-6_100

Keywords Cracking · Automatic segmentation · Pavement · Surface distress

1 Introduction

Monitoring road surface conditions is an important issue in many industrial countries. The objective is to detect surface distresses, like cracking, in order to plan effective road maintenance and to afford a better sustainability of the pavement structure (Wang and Smadi 2011; Vavrick et al. 2013). Several projects have looked into this issue in recent years, e.g., the European FP7 project TRIMM (Karlsson 2014; Wright et al. 2014), the joint research action (Ifsttar-Cerema 2012).

For monitoring road surface conditions, in situ human visual inspection has been gradually replaced by automatic image data collection at traffic speed. Off-line image processing techniques have been developed in support of human visual control. This paper briefly presents the two image processing techniques that have been developed within the scope of the TRIMM project. They assume the cracks are composed of contiguous dark pixels embedded within the image background. The first technique is a heuristic approach (HA) based on the search for gradient within the image (Wright et al. 2014). The second technique, the Minimal Path Selection (MPS) method is detailed in (Amhaz et al. 2015) and has been developed in collaboration with IRCCyN (www.irccyn.ec-nantes.fr) and IRIT (http://www.enseeiht.fr/fr/recherche/irit.html).

HA and MPS have been assessed against the field data collection provided by Yotta DCL and TRL with the imaging device Tempest 2 over UK roads, and against 36 Aigle-RN (Cerema 2004) pavement images on the French network, respectively. The benchmarking includes five automatic segmentation techniques.

2 Developments

In monitoring surface condition, a major breakthrough would be the provision of an efficient automatic processing of the pavement images. As the algorithms included in commercial dedicated software are proprietary, the review was based on the literature.

Assuming the crack pattern are dark connected pixels embedded within a textured image background, the most powerful approaches rely on both photometric and geometric characteristics of the cracks, e.g., see (Amhaz et al. 2015) for a review. Among the existing processing approaches, two main different strategies can be distinguished. The first strategy is based on the detection of thin and rare defects within regions of interest (ROI), and then, on the data aggregation of the latter segmented results over the full size image and beyond. New statistical tools have allowed the second strategy to emerge. This is illustrated for example by the block-based approach, where some ROIs are selected within the image on the basis of classification tools, e.g., see (Correia and Oliveira 2014) and the references therein. If required, the second step would refine the image analysis at the pixel level to detect the crack pattern. Compared to the first strategy, this one proposes a progressive search into the details of images at two scales. The work carried out at IFSTTAR over the last 15 years fits within the first strategy, and as such the two following developments have been proposed.

The heuristic approach (HA) which is described in (Wright et al. 2014; Sect. 3.2.1) uses a combination of different processing tools (contrast enhancement, morphological, etc.). The leading principle for detecting cracks relies on the detection of significant gradient magnitude within the image. Some pre- and post-processing are required to make the results more robust to the small contrast between the crack and the image background.

By contrast, the Minimal Path Selection (MPS) in (Amhaz et al. 2015) is based on a single mathematical principle and hypothesis. Assuming the cracks are connected dark pixels, the average grey level along the cracks, namely the path cost, is much lower than anywhere else. The use of the latter path cost enables a more reliable way to distinguish between the two categories of pixels, i.e., the crack and the background pixels, than the grey level distribution can afford. Besides, the cost path is used for different purposes at the five steps of the algorithm.

3 Testing Methodology

Two field data collections were used to evaluate the performance of the automatic crack detection algorithms introduced in Sect. 2. Conventional criteria were introduced for assessing the performance of the latter algorithms at both the pixel and the grid levels. To this aim, the pseudo ground truth (PGT) has been established on the selected data sets. At the pixel level, the PGT provides the skeleton of the crack pattern and the crack thickness. Table 1 synthesizes the assessment protocols used for the evaluation purpose.

Table 1 The assessment protocols associated to the 3 image data sets

Data collection	Pavement images	Assessment protocol	Assessment scale	Evaluation criteria
Reference data set (tempest)	2660	Statistic	Over 50 m	Density of defective grids per square meters
Hectometric data set (tempest)	119	Similarity	Grid-based	DSC and SM
Image samples (aigle-RN)	36	Amhaz et al. (2015)	Pixel-based	DSC

DSC and SM stand for the Dice similarity and the simple matching coefficients, respectively

4 Results

- MPS results at the pixel and the grid levels

The performance of MPS was compared with three other processing families: the HA method introduced in Sect. 2, Markov-based processing, labelled M2 (Chambon and Moliard 2011), and two methods based on computing minimal path, namely, GC-POI (Chambon and Moliard 2011) and FFA (Nguyen et al. 2011).

As an illustration, Fig. 1 compares the segmented results which are obtained by the 5 methods on a pavement image at both the pixel and the grid levels. Figure 2 provides the performance assessment on the Aigle-RN database in terms of Dice similarity index (DSC) at both scales (Amhaz et al. 2015; Wright et al. 2014).

At the pixel level, Markov-based and HA methods provide discontinuous crack segmentation and detects a lot of small false positive (FP) pixels within the image

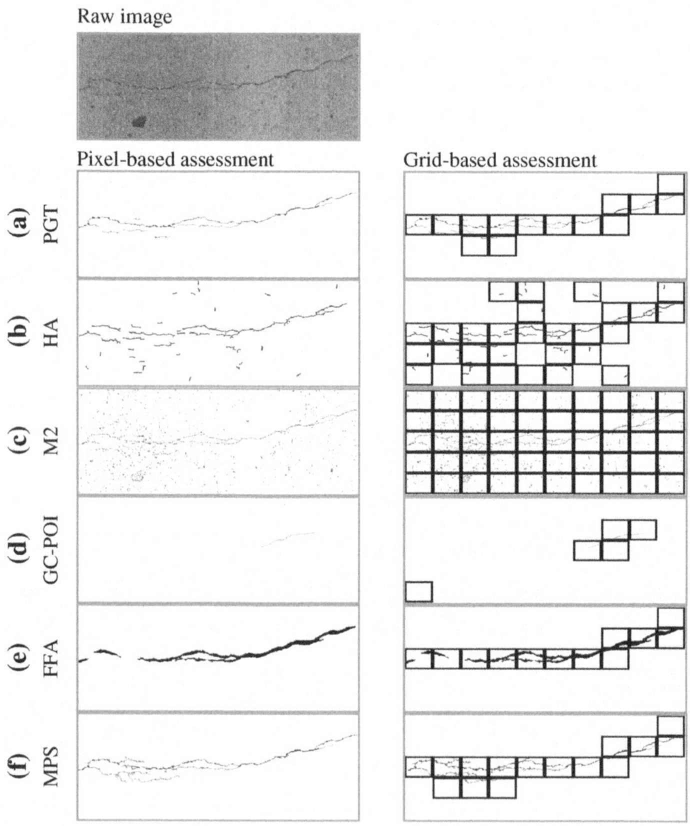

Fig. 1 Pixel-based (*left*) and grid-based (*right*) segmentations by the five methods included in the benchmarking; the original raw image at the *top*, **a** the pseudo-ground truth, **b–f** represent the segmentation results by the algorithms HA, Markov-2, GC-POI, FFA and MPS, respectively

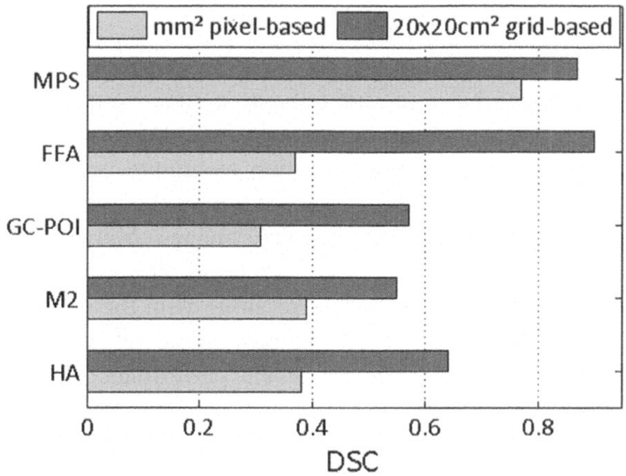

Fig. 2 Pixel-based (*light grey*) and grid-based (*darker grey*) Dice Similarity coefficient (DSC) obtained by the five segmentation methods over 36 Aigle-RN pavement images

background. GC-POI gives the worst performance, because only a small part of the crack has been detected. The segmented result by FFA appears as a thick line, implying a lot of FPs within the vicinity of the crack pattern. By contrast, MPS reveals the thin and chaotic structure of the whole crack pattern, as well as the crack width variations along the crack skeleton. For the grid-based segmentation, M2 shows the worst result because of scattered FP pixels all over the image. By contrast, MPS and FFA are put forward by the grid-based benchmarking.

On Fig. 2, grid-based performance (in dark grey) shows a higher DSC rate for all methods compared to the pixel-based performance (in light grey). The ranking of methods has changed between the two scale levels: FFA becomes the best method, followed by MPS; HA is the third best method. It is worth noticing that the two methods which have shown the best ability to detect the crack pattern in a more contiguous manner, namely FFA and MPS, outperform the other methods, i.e., the DSC rate is above 0.8.

- HA results at hectometric scale

The HA method presented in Sect. 2 was tested on the 133 km Tempest 2 data set. As proposed in Table 1, the comparison between the segmented results and the grid-based PGT has been performed over 50 m long pavement subsections and expressed in terms of the density of defective grids. It was observed that HA provides a larger density of defective grids (at least twice larger than the one in the PGT) except for a few rare cases. Further processing provides a similarity-like index between the segmented results and the PGT thanks to the binarization of the latter results with a specific threshold. However, HA still detects a large false alarm rate in the segmented results.

5 Conclusion and Perspectives

This paper has presented the two image processing techniques that have been developed within the scope of the TRIMM project to automatically detect pavement cracking from images. The first technique is a heuristic approach (HA) which is based on the search for gradient within the image; it has been successfully used for selecting the distressed pavement images (Jouin et al. 2015). Minimal Path Selection (MPS) technique is a new technique which provides a refined and accurate segmentation of the crack pattern along with the estimation of the crack width.

HA has been assessed against the field data collection provided on UK roads with the imaging device Tempest 2. The performance of MPS has been assessed against the Aigle-RN database on the French network, and on some Tempest 2 image samples after the end of the project, e.g., see (Amhaz et al. 2015). MPS reached the best performance at the pixel level while it is matched to the FFA method at the grid level.

Owing to its accuracy, MPS may serve as a reference benchmark for other methods to provide the automatic segmentation of pavement images. The ongoing work aims at reducing the computing time of MPS.

References

Amhaz R, Chambon S, Idier J, Baltazart V (2015), Automatic crack detection on 2D pavement images : An algorithm based on minimal path selection, accepted to IEEE Trans. Intell. Transp. Syst.

Cerema (2004), Aigle-RN imaging device, DTer Ouest, http://www.ouest.cerema.fr/IMG/pdf/_route-AigleRN_cle78acd3.pdf, last accessed on 18th January 2016

Chambon S, Moliard J-M (2011) Automatic Road Pavement Assessment with Image Processing: Review and Comparison, Int. Journal of Geophysics, vol. 2011, pp. 1-20

Correia P.L., Oliveira H. (2014) Special session 19 "Image Processing for the Detection of Road-Surface Degradations", in IEEE Int Conf. on Image Processing (ICIP), Paris, France

Ifsttar-Cerema (2012), Research action "Fissures", http://actions-incitatives.ifsttar.fr/orsi-infrastructures/encours/fissures/, last accessed on 20th January 2016

Jouin D, Baltazart V, Moliard J-M, Cord A (2015) Evaluation de deux algorithmes de classification automatique d'images de chaussées pour l'exploitation des campagnes de mesures Visiodec, paper presented at the national workshop Club CMOI, Lannion, France

Karlsson R (2014) Implementation of advanced monitoring techniques in road asset manage-ment results from the TRIMM project, paper presented in Transport Research Arena 2014, Paris

Nguyen TS, Begot S, Duculty F, Avila M (2011) Free-form anisotropy: A new method for crack detection on pavement surface images, in Proc. IEEE Int. Conf. Image Process., 2011, pp. 1069–1072

Vavrick W., L. Evans, J. Stefanski and S. Sargand (2013) PCR Evaluation –Considering Transition from Manual to Semi-Automated Pavement Distress Collection and Analysis, Ohio University

Wang K. and O. Smadi (2011) Automated imaging technologies for pavement distress surveys, Transportation Research Circular, E-C156, Transportation Research Board

Wright D, Baltazart V, Elsworth N, Hamrouche R, Krarup J, Lurdes Antunes M, McRobbie S, Merecos V, Saarenketo T (2014) TRIMM D4.3: Monitoring structural and surface conditions, European Commission DG Research, Tech. Rep. FP7-285119, accessed from http://trimm.fehrl.org/?m=3&id_directory=7539.

Digital Image Correlation to Monitor Cracking and Induction Healing of Asphalt Roads

Moisés Bueno, Josep Andrés, Andreas Treuholz, Martin Arraigada and Manfred N. Partl

Abstract The Road Engineering/Sealing Components Laboratory at Empa has been working during the last four years in an innovative experimental approach aiming at developing a new kind of healable asphalt road via induction heating. The last step of this research focused on the analysis of 1.8 m long test slabs damaged by the Model Mobile Load Simulator MMLS3 in order to prove the feasibility of the healing concept at larger scale. It is known that visible cracks could not be completely healed by this technique and therefore, the recovering of the mechanical performance was not significant. In this context, it seems clear that the healing treatment must be applied not later than the initiation of microcracks to avoid their propagation. For this reason, a digital image correlation system has been used successfully to monitor the damage level of a number of test slabs during the loading phase. This system employs two digital cameras which record the deformation process by means of high resolution images. Then, these images are analysed with correlation algorithms in order to calculate the 3D displacements and strain components for every object point. This optical method allowed to visualize the accumulated damage as well as to select the right moment for initiating the healing process. In addition, this method was useful to confirm that the strength is partially recovered after the healing process resulting in an increase of life of the road. Finally, it was shown that the healing procedure by induction heating can be a feasible alternative for maintenance purposes by acting before irreversible damage of the pavement occurs.

M. Bueno (✉) · J. Andrés · A. Treuholz · M. Arraigada · M.N. Partl
Empa, Swiss Federal Laboratories for Materials Science and Technology, Ueberlandstr. 129, 8600 Duebendorf, Switzerland
e-mail: moises.bueno@empa.ch

J. Andrés
e-mail: Josep.Andres@empa.ch

A. Treuholz
e-mail: Andreas.Treuholz@empa.ch

M. Arraigada
e-mail: Martin.Arraigada@empa.ch

M.N. Partl
e-mail: Manfred.Partl@empa.ch

A. Chabot et al. (eds.), *8th RILEM International Conference on Mechanisms of Cracking and Debonding in Pavements*, RILEM Bookseries 13,
DOI 10.1007/978-94-024-0867-6_101

Keywords Induction healing · Crack formation · DIC · MMLS3

1 Introduction

The self-healing capability of asphalt materials has been documented since several decades (Bazin and Saunier 1967). For example, Little et al. (2001) showed that a rest period of 24 h applied in traditional flexural bending beam experiments resulted in an increase of fatigue life. Nevertheless, in practice, it is often problematic to stop traffic circulation in order to give the road time to heal at ambient temperature. In this context, it is reported that the rate of natural healing increases with an increase in temperature due to the decrease of the viscosity of the bitumen which allows the binder to flow and close the cracks and to recover the adhesive bonds with the mineral aggregates. As a consequence, the use of induction heating was proposed as an artificial healing technique to increase the lifespan of an asphalt concrete pavement.

Previous studies accomplished at Empa concluded that visible cracks could not be completely healed by this innovative technique and therefore, the recovering of the mechanical performance was not significant. For this purpose, the control of the damage accumulation and microcrack initiation becomes a key factor in order to define when the healing treatment must be applied. In this way, microcracks will be promptly closed avoiding their propagation.

Nowadays, full-field non-contact optical measurement techniques based on digital image correlation (DIC) have been applied to asphalt materials in order to investigate cracking phenomena (Romeo 2013). The basic concept of this method is to measure the displacement of a material under testing by tracking and correlating the deformation of a random speckle pattern applied to the surface of the material in digital images acquired during loading. In this work, a full-field optical technique was used for monitoring the damage level of a number of test slabs during the mechanical loading with the Model Mobile Load Simulator (MMLS3). This experimental approach was useful to select the right moment to accomplish the induction heating as well as to confirm the performance recovery after the healing process.

2 Materials and Experimental Setup

Different slabs (1800 mm × 435 mm × 40 mm) were roller compacted from a reference dense asphalt concrete mixture with 5.9 wt% of bitumen 50/70 and a maximum mineral aggregate size of 8 mm. Moreover, cast iron particles of 0.6–1 mm diameter were added to the mixture in order to ease the heating process via induced electromagnetic fields. In addition, two 10 mm deep and 5 mm wide rectangular grooves were cut along the bottom surface of each slab. These grooves were used as stress concentration points for crack initiation during the test and in order to control the propagation path of the cracks.

One of the aims of this study was to develop an appropriate experimental method to induce damage in asphalt slabs without generating unexpected cracking phenomena. The healing capacity of the asphalt slabs was evaluated following a multiple-phase test procedure, which involved the mechanical loading of the slabs and the healing phases (García et al. 2013) followed by new loading phases until total failure. The loading phases simulated the effect of real traffic on the deterioration of a pavement by inducing damage followed by formation and propagation of cracks in the slabs due to repetitive loading of rolling tyres. This step represented an upscaling of the laboratory tests carried out on small specimens. On the one hand, the fabrication of slabs is closer to reality than the production of specimens with Marshall procedures. On the other hand, the effect of rolling tyres on the slabs produces stresses and strains that are similar to those generated in real pavements. In order to upscale the healing concept to a real scenario, the model mobile load simulator MMLS3 was used as a so-called third-scale load accelerated pavement testing system (Arraigada et al. 2014).

During the experimental process, a 1.8 m long asphalt slabs was placed such that the loading direction was perpendicular to the grooves. The slab was positioned under the MMLS3 on two transversal supports (14 mm thickness), so that the rolling tyres could produce bending stresses on the bottom. The distance between the supports was 1450 mm. In order to simulate the supportive effect of the subgrade produced on the asphalt layers and to buffer the total bending of the slabs when loaded, a 14 mm thick rubber mat was placed bellow the slab. In this way, any creep deformation in the slab due to its own weight was avoided.

The experimental setup of the DIC system involved two digital cameras mounted on a rigid bar to avoid relative motion (Fig. 1).

The cameras recorded the deformation processes with an accuracy of 20 μm and a measurement rate of 20 Hz. The stereo correlation between the images of left and right cameras allows the measurement of the 3D position for each point taking into account the rigid body movements. The speckle pattern was randomly applied around the notched areas on the lateral surface of the test slabs by using white paint and air-brush technique. Regarding the dynamic loading conditions of the damage phase with the MMLS3, sequences of 124 digital images were captured every 1000

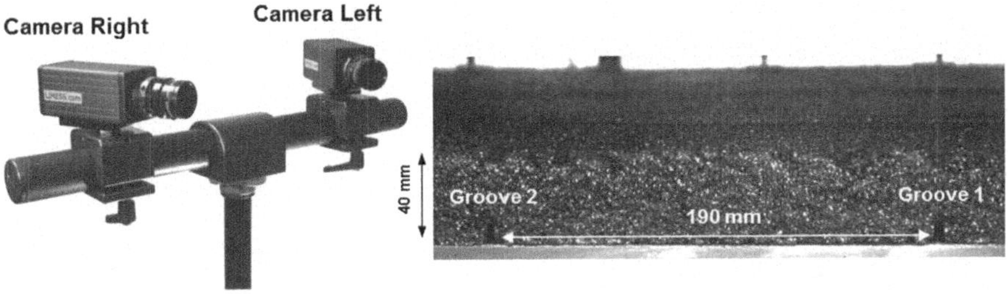

Fig. 1 Setup for digital image correlation system (*left*). Real image of the lateral surface of a test slab (*right*)

loading cycles. After processing, those images when a loading tyre was just above the notch area were taken as reference for characterizing the damage evolution.

In addition, it was expected that the strength of the slabs would decrease due to progressing damage (micro-cracking) caused by the MMLS3 loading. This strength reduction usually leads to an increase of the vertical bending deformation. Therefore, along with the DIC system, a number of deflection sensors were also installed at different positions on the surface monitoring the development of the damaging.

3 Results and Discussion

The optical measurement system allowed to detect accumulated damage as well as to select the proper moment to stop the damaging process in order to initiate the healing process. As an example, Fig. 2a shows the horizontal displacement on the lateral surface of a test slab at different times (or loading cycles) during the damage phase. It shows how the damage of an unloaded test slab starts and is accumulated under the load until crack initiation. It is important to note that at this level of deformation, no visible crack is perceived. Thus, this technique allows stopping the fatigue test before irreversible damage is reached. As expected and due to the experimental loading configuration, the horizontal displacement at the right area of groove number 1 grows following a fracture mode-I path up until reaching total failure. At this point, damage can also be observed around groove number 2.

Analysis of the displacement evolution at different heights (Fig. 2b) allows an estimation of the crack pattern and the velocity of crack propagation. Note, that the deformations around the groove at the bottom of the slab during loading procedure are increasing.

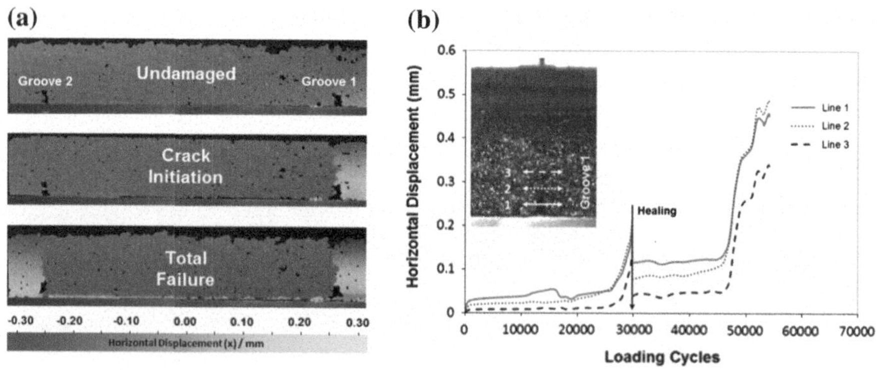

Fig. 2 Full-field displacement maps during the different test phases (**a**) and horizontal displacement evolution obtained from DIC system at different height positions (**b**)

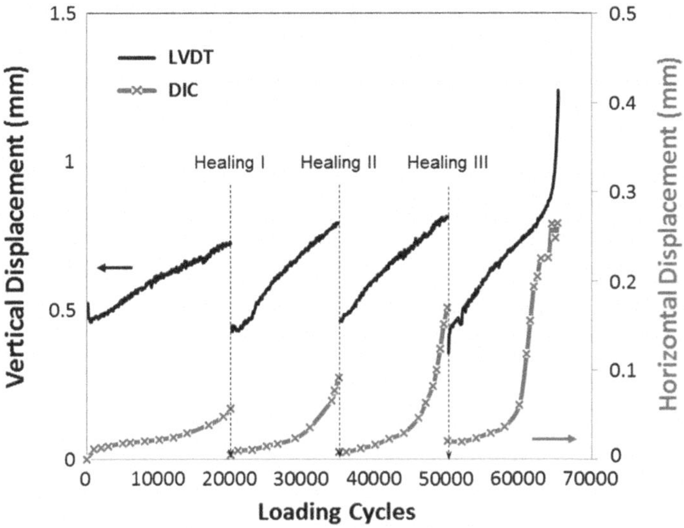

Fig. 3 Evolution of the vertical and horizontal displacement obtained from deflection sensors and DIC system during the loading cycles before and after healing processes

The variation of the horizontal displacement around the grooves monitored by the DIC system was taken as a sign of damage while running the test. This allowed defining the effect of the artificial healing. Figure 3 shows an example of the evolution of the vertical and horizontal deformations during a multiple healing process. It is remarkable that the range of horizontal displacement obtained by DIC was smaller than the vertical deformation measured by the deflection sensors; however, both loading curves follow the same pattern during the test cycles. Taking into account the anisotropy of the asphalt material, this difference could be related to the fact that the optical method analysed the lateral surface of the slab whereas the sensors measured the deflection on the rolling surface.

The analysis of the effect of the healing process in terms of performance of the test slab shows that the deformation after the heating treatment was similar to that obtained at the beginning of the test, when the asphalt slab was still new and unloaded. This demonstrates the initial stiffness of the slab was recovered. However, it is possible that temperature increase, due to the continuous friction induced by the wheel over the surface, influences deformation history due to softening of the bitumen at higher temperature. Hence, healing effect should also be assessed by considering the slopes of the displacement-loading cycle curves during the multiple healing phases. The slopes define the change in damage rate during the test. Following this criteria for the qualitative analysis of the curves plotted in Fig. 3, different phases can be clearly distinguished. During a first phase, the vertical amplitude and horizontal displacement increase constantly with the number of loading cycles until the accumulated damage leads to crack initiation which is linked to a sudden increase of the slope. After the induction healing process, the slope decreased compared to the previous phase and showed a value similar to the first damage phase. This performance recovery can be taken as a clear indication of the healing effect. After additional loading cycles, the slope increased again leading to a new phase where the cracks were initiated and propagated until total failure occurred.

In this particular case, the life-span of the asphalt slab was increased up to 65,000 loading cycles through a multiple healing procedure. Crack initiation only started after 64,000 cycles when the deformation on the groove reached 0.2 mm. This analysis confirms that healing by induction heating can be a feasible method for extending the life-span of the asphalt pavements when it is applied as maintenance technique before the structure has been irreversibly damaged.

4 Conclusions

This work shows the development and validation of a new method for the evaluation of the artificial healing of asphalt slabs (up-scaled) by induction heating. The conditions of the traffic over a road surface were reproduced by inducing loading damage with the MMLS3 on roller compacted asphalt slabs. The deformation during the loading process was successfully monitored with the DIC system. This optical approach allowed us to detect the crack initiation which is defined as the critical limit for still obtaining a significant performance recovery. Finally, this study demonstrated in a large scale, that the enhancement of the pavement life by induction healing process was feasible.

References

Arraigada M, Pugliessi A, Partl MN, Martínez F (2014) Effect of full-size and downscale accelerated traffic loading on pavement behavior. Mater Struct 47, 1049–1424

Bazin P, Saunier J (1967) Deformability, fatigue and healing properties of asphalt mixes. Paper presented at the 2nd international conference on the structural design of asphalt pavements, Ann Arbor, Michigan, 1967

García A, Bueno M, Norambuena-Contreras J, Partl MN (2013) induction healing of dense asphalt concrete. Constr Build Mater, 48, 1-7

Little DN, Lytton RL, Williams Ad, Chen CW (2001) Microdamage healing in asphalt and asphalt concrete, Volume 1: Microdamage and microdamage healing. Project summary report, Texas Transportation Institution, College Station

Romeo E (2013) Two-dimensional digital image correlation for asphalt mixture characterization: interest and limitations. Road Mater Pavement, 14, 747-763

Digitized Measurement of the Cracking Index on the Facings of Concrete Structures

J.-M. Moliard, V. Baltazart, B. Bérenger, T. Perrin and C. Tessier

Abstract In civil engineering, the survey of reinforced concrete structures is performed mostly from visual indicators by specialized operators. This paper focuses on the internal swelling of concrete, which is revealed by the cracking at the material surface. The manual survey method consists in estimating the average elongation of the material from the cumulative measurement of cracks thicknesses in 4 directions. To overcome some drawbacks of the manual method, this paper proposes to develop a semi-automated measurement of the cracking index from digital images and image processing tools. This allows us to obtain a more reliable crack index and refined crack characterization. This method has been first tested in laboratory and on a bridge near Nantes in 2012. Finally, the authors have been involved in 2015 on a long-term survey of the stacks of a larger bridge in Nantes. The results of the first-year survey will be reported by the time of the congress in 2016.

Keywords Cracking · Concrete · Image processing · Features characterization

J.-M. Moliard · V. Baltazart (✉) · C. Tessier
LUNAM Université, IFSTTAR, CS4, 44344 Bouguenais cedex, France
e-mail: vincent.baltazart@ifsttar.fr

J.-M. Moliard
e-mail: jean-marc.moliard@ifsttar.fr

C. Tessier
e-mail: christian.tessier@ifsttar.fr

B. Bérenger
Cerema, DTer-Ouest, BP 69, 49136 Les Ponts de Cé, France
e-mail: bruno.berenger@cerema.fr

T. Perrin
Cerema, DTer-Est, 11 rue Jean Mentelin, 67035 Strasbourg, France
e-mail: thibaut.perrin@cerema.fr

A. Chabot et al. (eds.), *8th RILEM International Conference on Mechanisms of Cracking and Debonding in Pavements*, RILEM Bookseries 13,
DOI 10.1007/978-94-024-0867-6_102

1 Introduction

In civil engineering, the survey of reinforced concrete structures is performed from visual indicators by specialized operators (Thauvin 2012). Recent trends have emerged to develop aerial robotic and unnamed aerial vehicles to perform image data collection in difficult access areas, e.g., (Chrobocinski 2015).

Among the pathologies under scope, this paper focuses on the internal swelling of concrete, which is revealed by the cracking at the material surface. Some image processing methods have been already proposed in the literature to afford the cracking pattern (Ammouche et al. 1998; Glinicki 2006; Gunkel et al. 2012; Moliard 2012; Prasanna et al. 2012; Jahanshahi and Masri 2013; Ifsttar-Cerema 2012).

The manual survey method that was defined in France to measure the cracking index, namely, ME-47 in (Fasseu and Michel 1997), consists in estimating the average elongation of the material from the cumulative measurement of crack widths in 4 specific directions. The measurement is generally carried out in some vertical facings of the structure with limited areas, typically 1 m^2 frame.

To overcome some drawbacks of the manual method (subjective measurement, low statistical representativeness, difficult to assess the reproductibility), this paper proposes to develop a semi-automated measurement of the cracking index from image processing tools. This paper first introduces the principle of the French manual method for measuring the cracking index, and then the digitized version from CCD images and image processing. A generalization of the latter is then proposed to exploit the large volume of information that represents the images. This allows us to obtain a more reliable crack index and refined crack characterization. This method has been first tested in laboratory and on a bridge near Nantes in 2012.

2 Basis on the Manual Measurement of the Cracking Index

The manual measurement of the cracking index (CI) on concrete structures is subject to the referenced French method LPC-47 (Fasseu and Michel 1997). CI is obtained by collecting the thickness of the cracks which intersect 4 segments (2 sides and 2 diagonals) on a 1 m × 1 m reference square frame, as shown on Fig. 1. For each segment orientation (0°, 90°, 45° and 135°), a cracking index is obtained by summing the crack thickness per length unit. Only the cracks beyond 5/100 mm in thickness are taken into account for the calculation. The final CI is the average of the indices along the 4 latter directions as follows:

$$\mathrm{CI}_{\mathrm{ME\_47}} = [CI(0) + CI(45) + CI(90) + CI(135)]/4 \quad (1)$$

CI provides the isotropic deformation of the material. The internal swelling phenomena can be neglected for a CI < 0.5 % roughly. This is considered

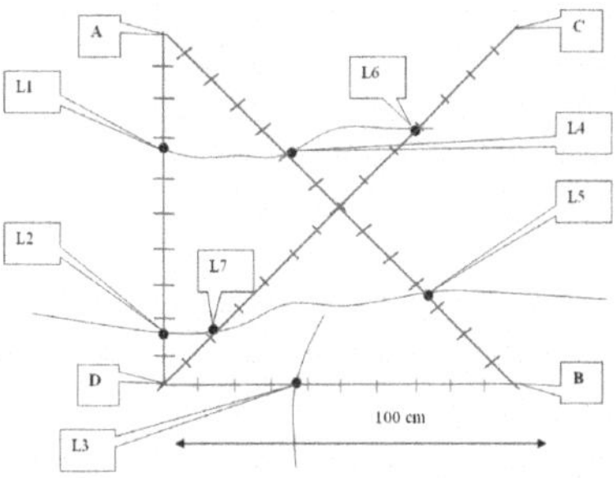

	CI vs. segment m/m	CI (ME_47), m/m
AB	CI (135°) = (L4+L5)/1.4	
CD	CI (45°) = (L6+L7)/1.4	eqn. (1)
AD	CI (90°)= (L1+L8+L2)	
DB	CI (0°) = L3	

Fig. 1 Basis of the manual measurement of the cracking index: *left* Referenced frame superimposed on the concrete surface; the operator determines the thickness of the crack with a crack-width gauge at each point intersecting the crack pattern with the segment of the referenced frame (shown by *black dots*); *right* Table showing the calculation of the cracking index associated to the scheme on the *left*

important above 2 % and some specific repair actions can be planned (Fasseu and Mahut 2003). The thickness of the cracking is measured with a pocket crack-width gauge, made of a transparent polycarbonate ruler holding various calibrated thickness graduations from 0.05 to 2 mm.

3 Digitized Measurement of the Cracking Index

This section presents the four steps to provide the measurement of the CI from the image processing of CCD images. For now, the processing is performed on desk on the images collected in situ on the area under evaluation (Moliard 2012).

The numerical method implies the geometrical calibration of the imaging system as the first step. To this aim, some markers are placed in the corners of the 1×1 m^2 reference frame as shown on Fig. 2. Back to the laboratory, the operator clicks on the markers on the screen. The program determines the centroid pixel of the markers which serves for the reference distance. The graphical interface then converts pixels into millimeters from the distance between centroid points. When a side view is required, further processing allows converting the side view to front view.

The existing CCD cameras cannot afford the measurement of crack width below 5/100 mm from the 1 m^2 reference image. A mixed solution has been used, reducing the field of view to 0.25 m^2 by coming closer at first, and second, matching the image resolution to the most representative crack width. Then, the 1 m^2 reference frame is covered by 4 images of 0.25 m^2 each (see Fig. 2). Within each 0.25 m^2 image, the standard 4 MPx CCD camera (resp. 12 MPx) allows measuring the crack width beyond 2.5/10 mm (resp. 1.5/10 mm).

The detection of the cracking pattern is based on a partial automatic process. First, automatic image segmentation is performed within the image reference frame thanks to the processing tools that have been successfully used for crack detection

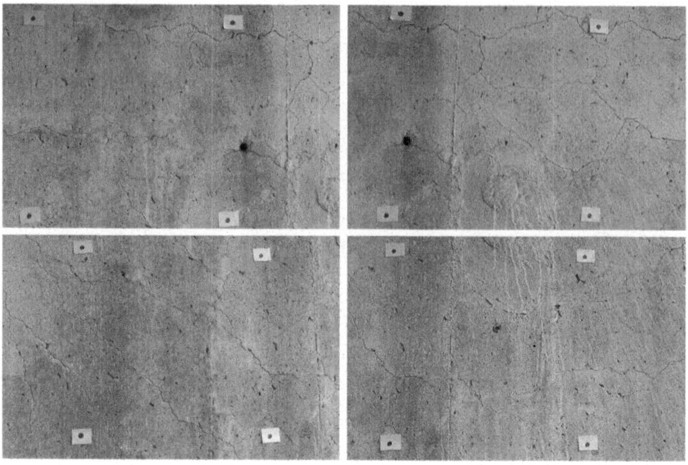

Fig. 2 1 m^2 image frame to be analysed and splitted into four subimages of 0.25 m^2 each, in order to verify the required pixel resolution introduced in this paper; the four corners of the 0.5 × 0.5 m^2 square frame are spotted by post-it markers (with *black dot*) and serve for image calibration

on pavement surfaces, e.g., (Amhaz et al. 2015; Baltazart et al. 2016). The list of the various detected cracks is eventually completed by the operator in a semi-automatic manner: the operator manually selects the two endpoints of the crack and a minimal path selection algorithm determines the path of the crack between them (Charbonnier and Moliard 2003).

The thickness of the cracks is estimated at each intersection of the crack pattern with the four oriented segments of the reference frame. The thickness is calculated along the profile perpendicular to the main direction of the crack. To make the result more reliable, the calculation is performed on two other adjacent profiles and the crack width is taken as the average of the three latter calculations. According to Fig. 1, the estimated crack openings are cumulated along each oriented segment to obtain the elongation of the material into four specific directions, namely, 0°, 45°, 90° and 135°. The conventional CI is taken as the average of the latter values according to Eq. (1).

4 First Results

First tests have been performed in laboratory using a camcorder and a CCD reflex camera. These devices provide 1920 × 1080 image pixels (2.65 Megapixels) and 4896 × 3672 pixels (18 Megapixels), respectively.

The tests were performed on targets composed of black straight lines with a calibrated width, namely 0.5 and 0.75 mm, from which the theoretical CI value can be calculated. The relative error on the thickness was within about 5 % at worst. But it degrades when the crack width tends to the pixel size. Then, interpolation techniques mitigate the latter discrepancy to a limited extent.

Table 1 Comparison between the CI values which are calculated by the conventional manual ME-47 method and the digitized version of the latter on the four subimages displayed in Fig. 2

	Top left	Top right	Bottom right	Bottom left	Average
Manual method	2.9	2.1	2.4	2.7	2.5
Digitized method	2.8	2.1	2.0	2.6	2.4

A first experimental assessment has been realized on a bridge structure located nearby the Nantes' beltway (France) according to the method proposed in Sect. 3 (Moliard 2012). Table 1 shows the fair agreement between the results by the conventional manual ME-47 method and the digitized version of the latter. A more systematic survey with Cerema has been recently initiated on another bridge in Nantes and will be lasting 5 years.

5 Full Surface Scanning Method

The digitized method in Sect. 3 is extended to the scanning of the full surface of the image frame by a dense gridding of oriented parallel segments at angle θ. Each oriented segment has a varying length L(θ), according to its position within the frame. The crack width is estimated at each intersection point between the cracking pattern and the oriented segments. The crack index CI(θ) estimates the material elongation for a given orientation θ as the sum of the latter crack widths along the segments divided by the cumulative length of the oriented segments.

In practice, the CI measurement is performed by image processing using the oriented secant method in (Ammouche et al. 1997) as shown in Fig. 3 (left). The new index CI(θ) allows characterizing the angular distribution of the material elongation, and thus, to detect any anisotropic behavior of the concrete material as

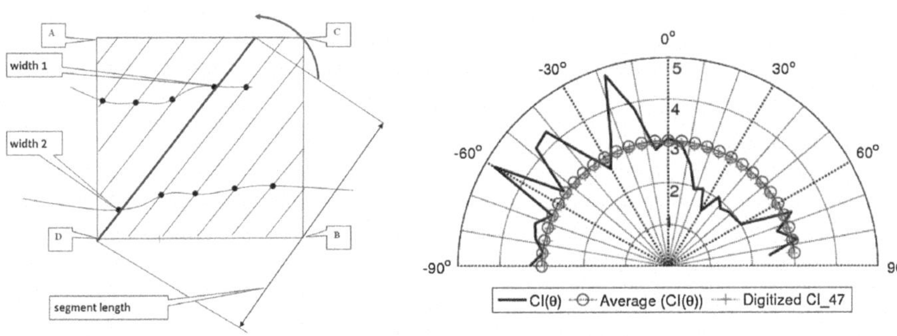

Fig. 3 *left* Illustration of the oriented secant method (the final step of the angular scanning is oriented to 85° on the picture); *right* Angular distribution of the CI associated to the *top-left* image in Fig. 2, showing the anisotropic material elongation toward −60° to −20° (*filled lines*); the average CI over 180° ('o') is superimposed on the digitized CI_47 measurement ('+')

illustrated in Fig. 3 (right). The conventional CI measurement in Eq. (1) is obtained by averaging the latter CI(θ) over the four specific angles 0°, 90°, 45° and 135°. Finally, a more reliable isotropic CI measurement can be computed as the average of the CIs over the whole set of scanning angles. As shown on Fig. 3 (right), both CI values are very close to each other.

6 Conclusion and Perspectives

This paper has proposed a first assessment of the digitized French method ME-47 for measuring the cracking index on the facings of concrete structures for the survey of concrete swelling. Tests in laboratory on specific targets and in situ have shown that the digitized method and the conventional manual method get close results. A more rigorous and intensive experiment has been recently undergone in Nantes and will last 5 years and some first results will be available by the time of the congress. Finally, the new CI which was introduced in Sect. 4 is expected to afford a better accuracy and sensitivity to perform the survey of the concrete swelling over a few years.

References

Amhaz R, Chambon S, Idier J, Baltazart V (2015), Automatic crack detection on 2D pavement images: An algorithm based on minimal path selection, accepted to IEEE TITS

Ammouche A, Breysse D, Hormain H, Dirdry O, Marchand J (1998) A new analysis technique for the quantitative assessment of microcracks in cement-based materials, Cement and Concrete Research

Baltazart V., Moliard J-M., Amhaz R., Cottineau L-M., Wright A., Jethwa M. (2016), Automatic crack detection on pavement images for monitoring road surface conditions – Some results from the collaborative FP7 TRIMM project, Rilem Conference, Nantes, France, July

Charbonnier P, Moliard J-M (2003), Calculs de chemins minimaux, suivi de fissures et autres applications, paper presented at « Journées scientifiques du LCPC » , Dourdan

Chrobocinski P. (2015), AErial RObotic System for In-Depth Bridge Inspection by Contact, kick-off meeting of the Aerobi project (H2020 european project in robotics)

Fasseu P., Mahut B. (2003), Aide à la gestion des ouvrages atteints de réactions de gonflement interne, ISSN 1151-1516, IFSTTAR

Fasseu P., Michel M. (1997) Méthode d'essai des Laboratoires des Ponts et Chaussées n°47, Détermination de l'Indice de Fissuration d'un parement de béton, ISBN 2-7208-3700-8

Glinicki M.A., Litorowicz A. (2006) Crack system evaluation in concrete elements at mesoscale, Bulletin of the Polish Academy of Sciences, vol 54, 6, pp371-379

Gunkel C, Stepper A, Müller AC, Müller CH (2012) Micro crack detection with Dijkstra's shortest path algorithm, Machine Vision and Applications, Volume 23, Issue 3, pp 589-601

Ifsttar-Cerema's research action "Fissures" (2012), http://actions-incitatives.ifsttar.fr/orsi-infrastructures/encours/fissures/, last accessed on 20th Jan. 2016

Jahanshahi MR, Masri SF (2013) A new methodology for non-contact accurate crack width measurement through photogrammetry for automated structural safety evaluation, Smart Materials and Structures, 22

Moliard J-M (2012) ANDROA : Analyse Numérique des Défauts Répertoriés sur Ouvrages d'Art, internal report, IFSTTAR

Prasanna P, Dana K, Gucunski N, Basily B (2012) Computer Vision Based Crack Detection and Analysis, Proc. SPIE 8345, Sensors and Smart Structures Technologies for Civil, Mechanical, and Aerospace Systems

Thauvin B. (2012), Indicateurs d'état visuels pour l'évaluation des structures en béton armé, internal communication to IFSTTAR meeting « Gestion des risques structuraux » , IFSTTAR.

Monitoring of Debonding or Cracking in Bending Tests by Virtual Image Correlation

Marc L.M. François

Abstract Three or four point bending tests are able to initiate controlled mixed mode cracking between two layers of different materials (Hun in Influence de l'eau sur le décollement d'une chaussée urbaine, 2012). The monitoring of such phenomena is a difficult challenge as if global force-displacement measurements are able to detect from the apparition of a non linearity the existence of a defect, they cannot determine its nature or location. More recent Digital Image Correlation (DIC) techniques can help however their application is limited because they require an image of a speckled side of the beam. Due to the form factor of the plate or beam, this image cannot cover all the side of the test except if one owns a ultra-high definition camera and lens. On the contrary the Virtual Image Correlation (VIC) method (François et al. in Eur Phys J App Phys 56:1–10, 2011) allows the measurement of the plate shape without any speckle thus does not require ultra high definition imaging. This global image is correlated to a virtual image whose parameterized shape is simply issued from the beam or plate theory. The delamination or the cracking is respectively introduced in the computation from simple reduction of the beam stiffness or a plastic joint. It will be shown that both their magnitude and location can be measured by using the VIC.

Keywords Bending test · Displacement measurement · Delamination · VIC

1 The Delamination Between Concrete and Asphalt

The work of Hun (2012) and the work presented in this conference by Chabot and Hammoum (2015) is related to the measurement of the interface strength between concrete and asphalt. Although the mechanisms of degradation can be complex, we retain in this study the simplest case of a crack propagation in mode 1 at the

M.L.M. François (✉)
Lab. GeM, Faculté des Sciences de Nantes, 2 rue de la Houssinière, 44322 Nantes Cedex 3, France
e-mail: marc.francois@univ-nantes.fr

A. Chabot et al. (eds.), *8th RILEM International Conference on Mechanisms of Cracking and Debonding in Pavements*, RILEM Bookseries 13,
DOI 10.1007/978-94-024-0867-6_103

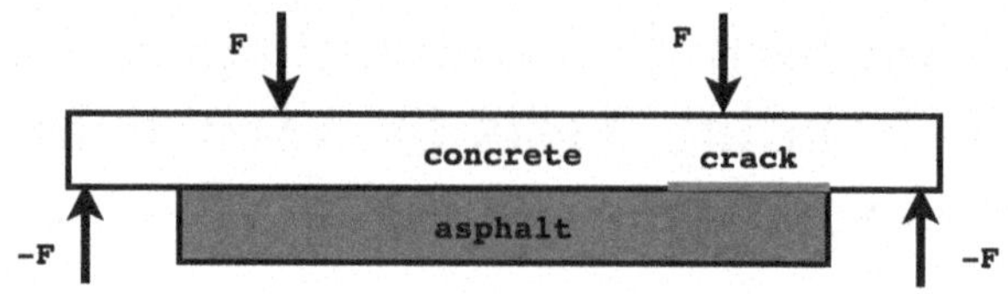

Fig. 1 Four point bending test used for delamination study (Chabot and Hammoum 2015; Hun 2012)

concrete and asphalt interface. The authors retained astute 4-point bending test in which the induced shear at the end of the layer induces stable crack propagation (Fig. 1).

The monitoring of such test is not obvious. The force-displacement curve does not indicate the crack location (neither the crack nature). The direct image observation indicates the crack tip but only where its opening exceeds the picture resolution. The Digital Image Correlation techniques as used by authors returns, via the error map, an exploitable crack tip location but with a weak precision, at the surface of the specimen, and requires the picture to be taken close enough thus to know a priori the approximative crack tip location. As soon as a four point bending test provides a constant bending moment along the beam, according to the beam theory the curvature is constant and its value is $\rho = M/EI$ where M is the bending moment, ρ the curvature, EI the Young modulus and inertia of the beam. This term can be easily computed from the beam or plates theory on both the sides where the assemblage is sound and where the crack has occurred (in this case, EI represents the rigidity of the sole concrete). As a consequence, in case of mode 1 crack delamination, the shape of the central part of the assembly is constituted of two arcs of circles of different radii. We propose to measure the curvature of these circles and the crack tip will be located at the junction point between the two arcs.

2 The Virtual Image Correlation Technique

The virtual image correlation (VIC) technique (François and Semin 2010; François et al. 2011; Réthoré and François 2014) applies for contour or silhouette measurement. On the contrary of filtering techniques, its objective is not the detection (furthermore one needs to identify roughly the contour or silhouette at the beginning) but the precision which can reach 10^{-3} pixels. The technique consists in finding the mapping $\mathbf{x}(\mathbf{X})$ which transforms a so-called virtual image G* in an image G as close as possible to the physical image F (see Fig. 2). In its proper frame the virtual image consists in a simple linear variation from white to black. The reverse mapping $\mathbf{X}(\mathbf{x})$ transforms it in an image G which represents a band in the neighborhood of a mathematical curve $\mathscr{C}$. The curve is parameterized by a set of shape parameters λ_k. The goal is to find the set of parameters for which the images F and G are in best correspondence. This is achieved when the black part of the image G lies in the black part of the image F and similarly for the white part, thus the edge in image F can be assimilated to the equation of the curve $\mathscr{C}$ which passes

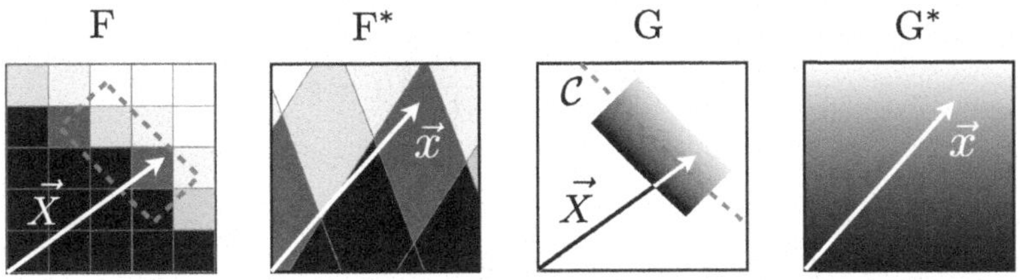

Fig. 2 VIC scheme. F is the physical image which includes a boundary of interest and G^* is the virtual image in its proper frame. F^* and G correspond to the previous images deformed by a mapping in order to fit them at the best

by the 50 % grey line of the image G. One may note that the definition domain of image G is only the vicinity of the curve $\mathscr{C}$ thus the number of pixel of image F to be computed for the correlation is very small and the method is perfectly insensible to far artifacts. The mapping is based on the curve $\mathscr{C}$ equation (the choice of the curve family depends upon the addressed problem and will be discussed further). For instance we denote as $\mathbf{X}^c$ the position of a current point of $\mathscr{C}$. Naming $(\mathbf{e}_r, \mathbf{e}_s)$ the normal and tangent vectors, the reverse mapping $\mathbf{X}(\mathbf{x})$ is:

$$\mathbf{X} = \mathbf{X}^c(x_1, \lambda_1 \ldots \lambda_n) + Rx_2\, \mathbf{e}_r(x_1, \lambda_1 \ldots \lambda_n), \tag{1}$$

in which x_1 is at the same time the curve $\mathscr{C}$ parameter (which defines the curvilinear abscissa) and the abscissa of $\mathbf{x}$ in image G*. The term R corresponds to the half width of the virtual image G. The correlation is computed in the frame $\mathbf{x}$ i.e. between F* and G* (instead of between F and G) because this allows one to use of the reverse mapping (Eq. 1) which is straightforward. As in the DIC (Hild and Roux 2006) method the correlation is measured by the mean square distance:

$$2\Psi(\lambda_1, \ldots, \lambda_n) = \int_{-1}^{1} \int_{0}^{1} (f^*(\mathbf{x}) - g^*(\mathbf{x}))^2 dx_1 dx_2 \tag{2}$$

in which f^* and g^* denotes the normalized gray level of corresponding images. Considering that the input data is $f(\mathbf{X})$ and that the gray level of the virtual image depends only upon x_2 (with respect to a linear function) gives:

$$2\Psi(\lambda_1, \ldots, \lambda_n) = \int_{-1}^{1} \int_{0}^{1} (f(\mathbf{X}(\mathbf{x})) - \bar{g}(x_2))^2 dx_1 dx_2 \tag{3}$$

The goal is to find the parameters λ_p which minimize this distance. This is helped by the knowledge of the gradient and Hessian of Ψ. By using differential geometry one obtains

$$\frac{\partial \Psi}{\partial \lambda_p} = \int_{-1}^{1}\int_{0}^{1}\left(\text{grad}(f)\cdot\frac{\partial \mathbf{X}}{\partial \lambda_p}\right)(f-\bar{g})dx_1 dx_2 \tag{4}$$

$$\begin{aligned}\frac{\partial^2 \Psi}{\partial \lambda_p \partial \lambda_q} = &\int_{-1}^{1}\int_{0}^{1}\left(\frac{\partial \mathbf{X}}{\partial \lambda_p}\cdot \mathrm{H}(f)\cdot\frac{\partial \mathbf{X}}{\partial \lambda_q} + \text{grad}(f)\cdot\frac{\partial^2 \mathbf{X}}{\partial \lambda_p \partial \lambda_q}\right)(f-\bar{g})dx_1 dx_2 \\ &+\int_{-1}^{1}\int_{0}^{1}\left(\text{grad}(f)\cdot\frac{\partial \mathbf{X}}{\partial \lambda_p}\right)\left(\text{grad}(f)\cdot\frac{\partial \mathbf{X}}{\partial \lambda_q}\right)dx_1 dx_2\end{aligned} \tag{5}$$

It can be shown that the derivatives of $\mathbf{X}$ can be expressed with respect to the derivatives of the curve $\mathscr{C}$ (i.e. $\partial \mathbf{X}^c/\partial \lambda_p$). These derivatives can be computed either numerically or analytically depending upon the retained family of curve $\mathscr{C}$. Finally the minimum of Ψ is obtained iteratively thanks to a Newton integration scheme, the above equations leading to a linear $n \times n$ system. The choice of the half width of the virtual image R depends upon the sharpness of the image. One may notice that even a perfect image (such as the simulation in Fig. 2 leads to some blur due to the pixel interpolation. In this case R can be set to $1/\sqrt{2}$ pixel. It has been tested that the identification precision (the location of the curve $\mathscr{C}$) was quite insensible to R when ranging up to 2 or 3 pixels (which is often necessary in case of real images). The choice of the curve family $\mathscr{C}$ depends upon the addressed problem. In the general case one can use generic curves such as B-Splines (Réthoré and François 2014). In the case of use for metrology measurement geometrical curves can be used. For example one may use ellipses in case of circular defect measurement. In the present case the shape of the object is given by a mechanical expression (according to the beam theory, see Sect. 1). This will be shown in Sect. 3. Many tests (François et al. 2011) have shown that the contour was identified with a theoretical precision up to 10^{-3} pixel. This precision can be surprising. The reason lies in the filtering effect of the VIC: the function Ψ, a surface integral, can be seen as an optimal filter between F and G. Of course this precision is only obtained if the curve family is able to fit the shape of interest. In this case the precision is due to the spatial richness of the information. On the opposite case, if one chooses a too large number of shape parameters (comparable to the—quite large—number of pixels which cross $\mathscr{C}$), this advantage vanishes.

3 The VIC Applied to the Delamination Measurement

As recalled in Sect. 1, the four point bending test applied onto a set of partially delaminated beams leads to a set of two arcs of circles (considering the—here—simpler large transformation beam theory). The curve family of $\mathscr{C}$ is then depicted as shown in Fig. 3, by a set of 8 parameters which are (λ_1, λ_2) the two components of the point K, λ_3 the curvilinear length L between A and B, λ_4 the angle θ at point K, (λ_5, λ_6) (R_1, R_2) the two radii (proportionals to M/EI with respect to the figure scale), λ_7 the abscissa $\xi(K)L$ (where ξ is the normalized curve parameter) of the point K and λ_8 the beam thickness. Whenever the radii can be estimated if one knows the material properties and the scale, we decided to let them as unknown for the present test. The parameter of interest is mainly the seventh, which defines the position of K i.e. the location of the fracture tip. Since we do not dispose at this time of usable images of the test (which requires a low distortion objective, the painting of a black on white stripe on the specimen's side and a camera precisely perpendicular to the device) we shall show the ability of the method onto a synthesized image and shall compare initial and final shape parameters. The image in Fig. 4 is 1734×440 pixels. The "beam" edges are blurred as if the image would have been obtained thanks to a real camera (with a—quite large—4 pixel blur). The mathematical curve $\mathscr{C}$ is obtained from the beam (circles) equation thanks to an Euler-Bernoulli kinematics (the points are located at the distance along the normal of the curve). The curve $\mathscr{C}$ describes the beam boundary counterclockwise. The measurement has been initialized by hand clicking and large value for the virtual image width R. After about 100 iterations, the program converges to the state shown Fig. 5. During the computation the virtual image width R has been gradually reduced from 50 to 10 to 2 pixels. The final state is shown by Fig. 5 and the parameter value in Table 1. One may pay attention to the images F* which are a good indicator of the quality of the identification at it represent the physical image seen in the "unwrapped" frame $\mathbf{x}$. Once the correlation is perfect this image should

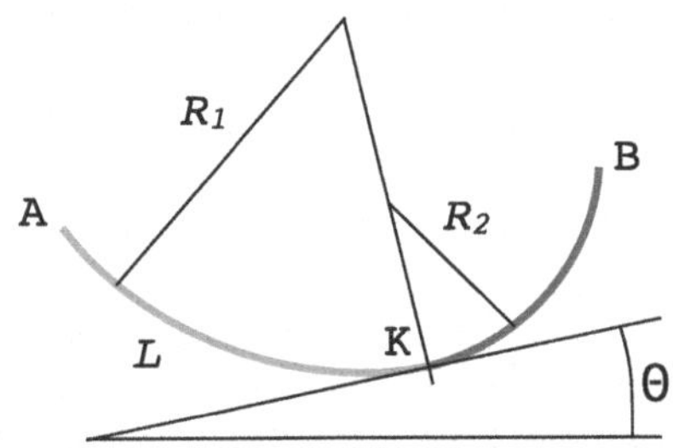

Fig. 3 The retained curve parameters

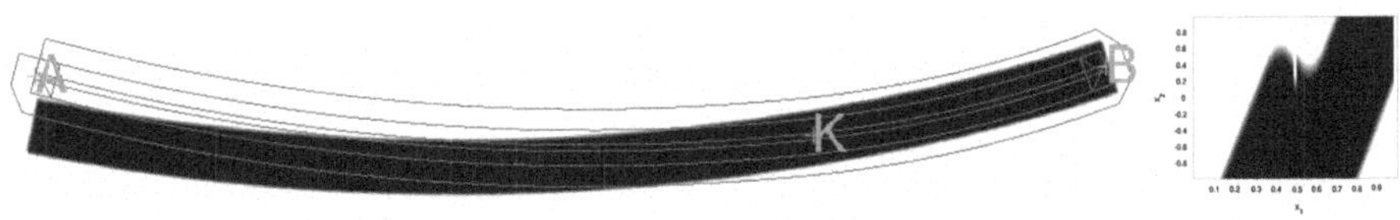

Fig. 4 Initial state. *Left* the image F and the virtual image position in red. *Right* corresponding image F* with $R = 50$ (half height) pixel

Fig. 5 Final state. *Left* the image F and the virtual image position in red. *Right* corresponding image F* with $R = 2$ (half height) pixel

Table 1 Key beam characteristics

Value	R_1/L	R_2/L	xi_K
Prescribed	2	1.8	0.8
Measured	2.0028	1.8000	0.7985

be very similar to the virtual image G*. This is achieved despite the existence of corners in the curve $\mathscr{C}$. The results are shown in both Fig. 5 and Table 1. The key information, the abscissa of K—the abscissa of the crack tip—which was set at 80 % of the length of the beam is recovered with a relative precision of 210^{-3}. The program has been launched several times varying the initial conditions and it converges any time to similar values within the announced precision.

4 Conclusion

The VIC is yet used in some laboratories (of course mine but also the PRISME in Orleans, the Groupo dos Medios Porosos in Argentina) for the boundary measurements. This benchmark tends to show its utility in the case of beam or plate shaped tests in pavement science. Of course, the announced precision will certainly not apply in case of real images and some points remain to be studied such as the mean line which varies in ordinate from the bonded section to the de-bonded section. Further straightforward extension will consist in enriching the mechanical model. The time-dependent viscoelastic behavior of asphalt based materials can be taken into account either with a simple beam model as presented here either with a more sophisticated plate model such as the M4 N plate model (Naciri et al. 1998; Chabot et al. 2013) used at the IFSTTAR.

References

Naciri, T. and Ehrlacher, A. and Chabot, A.: Interlaminar stress analysis with a new multiparticle modelization of multilayered materials (M4). Comp. Sci and Tech. **58**, 337–343 (1998).

Chabot A., Hun M., Hammoum, F.: Mechanical analysis of a mixed mode debonding test for "composite" pavements. Construction and Building Materials, **40**, 1076–1087, 2013.

Chabot, A. and Hammoum, F.: Mixed-mode debonding approach to evaluate water sensibility in bi-layer composite pavement. In proceeding of 8th RILEM int. Conf., 2015.
François, M. and Semin, B.: Identification of the shape of curvilinear beams and fibers. Applied Mechanics and Materials **24–25**, 359–364 (2010).
François, M. and Semin, B. and Auradou, H.: Accurate measurement of curvilinear shapes by Virtual Image Correlation. Eur. Phys. J. App. Phys. **56**, 1–10 (2011).
Hild, F. and Roux, S.: Measuring stress intensity factors with a camera: integrated digital image correlation (I-DIC). Comptes Rendus de MÃ©canique **334**, 8–12 (2006).
Hun, M.: Influence de l'eau sur le décollement d'une chaussée urbaine. PhD thesis, École Centrale de Nantes, 2012.
Réthoré, J. and François, M.: Curve and boundaries measurement using B-splines and virtual images. Optics and Lasers in Engineering **52**, 145–155 (2014).

Part XIII
Advanced Measurement Systems for Crack Characterization: Field Measurements and Back Calculations

Progress in Monitoring the Debonding Within Pavement Structures During Accelerated Pavement Testing on the Fatigue Carousel

J.-M. Simonin, V. Baltazart, C. Le Bastard and X. Dérobert

Abstract The paper gives an overview of the ongoing experiment to survey debonding areas within pavement structure during accelerated pavement testing on the Ifsttar's fatigue carrousel. Several defects have been embedded during the construction phase and have been probed by different NDT techniques. Among them, the paper focuses on radar NDT&E techniques which have been used to detect and locate the artificial defects at the early stage of the experiment and then to perform the survey at different loading cycles. Besides, the test-site has been used to test different GPR materials, including the 3D GPR technology. The experiment has also motivated some related studies in the field of GPR data modelling and data processing, which are summarized in the paper. The experiment is expected to be pursued beyond 300 loading kcycles to reach larger pavement degradations.

Keywords Debonding · Reflexive cracks · GPR · Data processing

1 Introduction

Monitoring structural conditions of road is an important issue in many industrial countries. The objective is to detect structural distresses, like debonding and reflexive cracking, in order to plan effective road maintenance and to afford a better sustainability of the pavement structure, e.g., Pajewski et al. (2013), Picoux et al.

J.-M. Simonin · V. Baltazart (✉) · X. Dérobert
LUNAM Université, IFSTTAR, Nantes, France
e-mail: vincent.baltazard@ifsttar.fr

J.-M. Simonin
e-mail: jean-michel.simonin@ifsttar.fr

X. Dérobert
e-mail: xavier.derobert@ifsttar.fr

C. Le Bastard
Cerema, Dter-Ouest, Les Ponts de Cé, France
e-mail: cedric.lebastard@cerema.fr

A. Chabot et al. (eds.), *8th RILEM International Conference on Mechanisms of Cracking and Debonding in Pavements*, RILEM Bookseries 13,
DOI 10.1007/978-94-024-0867-6_104

(2011). Within this scope, the paper gives an overview of the ongoing experiment to monitor artificial debonding areas within pavement structure by different NDT&E techniques during accelerated pavement testing on the Ifsttar's fatigue carrousel.

The paper especially focuses on the Radar NDT techniques which use electromagnetic waves to probe the pavement structure, e.g., Dérobert (2004). Besides, the test-site has been used to test different GPR materials, including the 3D GPR technology.

The paper first presents a short overview of the past data collection which have been collected so far at the early stage of the experiment and then at different loading cycles. The ongoing experiment has prompted since 4 years a lot of related studies (including Ph.D. topics, published and unpublished materials, research projects and collaborations) which are briefly recalled in the paper.

2 Test-Site

The Ifsttar's pavement fatigue carrousel is a one-off circular outdoor test track which is characterized by a 120 m long test track and allows various loading capabilities and speed. It can be divided into different test sections, which can be simultaneously loaded. The width of the test track (6 m) allows applying the traffic loads on the same track at two different radii (Fig. 1).

A quarter of the test track, namely 25 m, has been dedicated to the experiment. The pavement structure consists of two bituminous layers (8 cm thick base layer, and 6 cm thick wearing course), over a granular subbase. Rectangular patches of materials were incorporated at the interface between the two asphalt layers to create embedded debonding areas. Patches differ by size, location, and materials, namely, sand, geotextile, tack-coat free interface.

(a) **(b)**

Fig. 1 Fatigue carousel at Ifsttar in Nantes: *left* overview, *right* 25 m long section with embedded artificial defects made of (*a*) sandy interface, (*b*) geotextile and (*c*) interface free of tack coat

3 Radar NDT&E Techniques

Among the different NDT&E techniques which have been used to probe the pavement structure, this section focuses on radar techniques. The measurements based on mechanical waves are subject to other papers in the same Rilem Conference.

Electromagnetic NDT&E techniques take advantage of the penetration capability of the waves to detect any dielectric contrast within the material. They usually allow contactless and fast measurements. For the tests, air-coupled radar afforded larger central frequency and bandwidth compared to the ground-coupled radar, resulting in a refined timely signal analysis (see Fig. 2). For probing the pavement, zero-offset radar configuration has been mostly used. The radar is dragged along the scanning direction to provide 2D Bscan images.

Radar Bscans were collected at the early stage of the experiment and then at the different loading cycles, namely 50, 100, 150, 200, and 300 kcycles. The signal features at the different loading cycles were compared to each other and between the signal features over flawless areas. Besides, the opportunity has been found to perform the benchmarking of two other advanced GPR imaging devices at 300 kcycles. The contribution of 3D GPR is clearly demonstrated in Simonin et al. (2014), especially to detect the spatial extension of the debonded areas.

As an illustration, Fig. 3a shows the radar Bscans over the three largest defects, namely, A, B and C on Fig. 1b. Figure 3b depicts the time slice of Cscan data over the whole tested area of the fatigue carousel. For both results, the most detectable interface is the geotextile; by contrast, the signature over the interface free of tack coat provides the smallest echo because of small dielectric contrast.

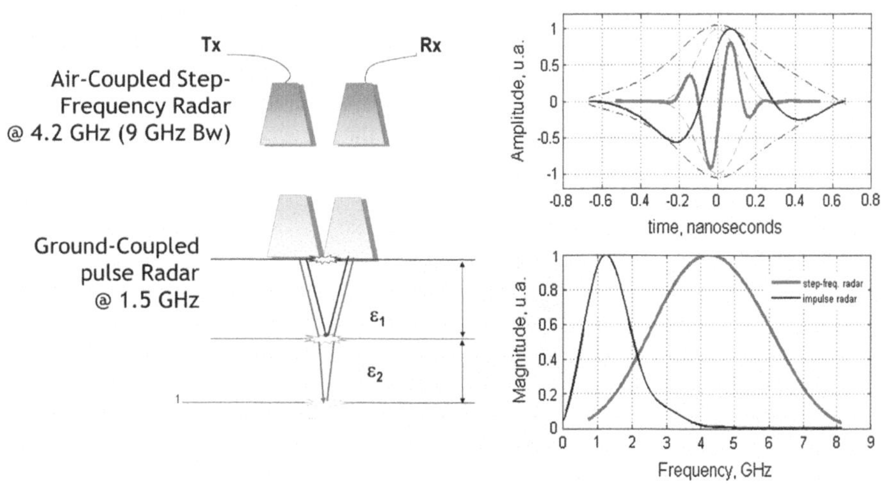

Fig. 2 The two antenna configurations, radar technologies and the associated radar wavelets which have been used so far to probe the pavement subsurface of the test-site: *left* air-coupled (AC) step frequency radar (SFR) and the ground-coupled (GC) pulse radar; both uses quasi-monostatic configuration; *right* radar wavelets and spectrum for both radar configurations

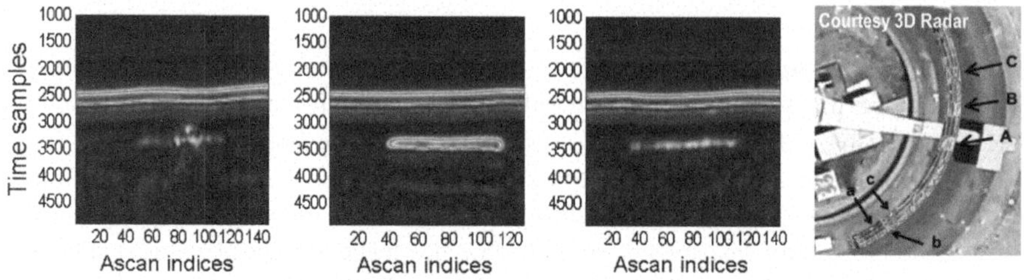

Fig. 3 Radar signatures over the defects shown in Fig. 1b: **a–c** Bscans collected by the Ifsttar's air-coupled step-frequency radar over one meter long **a** interface free of tack coat (*A*), **b** geotextile (*B*) and **c** sandy interface (*C*), **d** Cscan time slice measured by 3D Radar technology

4 GPR Data Modelling and Processing Techniques

The experiment has motivated related studies, which are summarized in this section. As ground-coupled radar configuration was used at the beginning, some ground-clutter removal techniques have been tested in addition to the ones in Tebchrany et al. (2014). Figure 4 shows the results of a time-frequency technique when the clutter and the echo from the first interface do not fully overlap in the time-frequency domain.

The simplest modelling of the radar signal backscattered from the debonding interface relies on the overlapping of two primary echoes. The magnitude of the resulting echo may be reinforced compared to the one from a single interface. Three processing have been planned to allow a refined waveform analysis.

The conventional technique for monitoring the interface has been illustrated in Simonin et al. (2012a, b). The larger bandwidth of the step-frequency radar shown on Fig. 2 (right) allowed enhancing the magnitude of the echoes over the debonding interface and eased the detection of the latter.

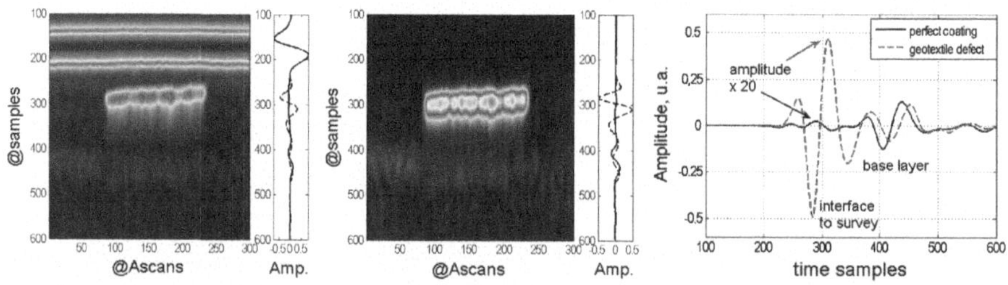

Fig. 4 Testing time-frequency clutter removal technique over the embedded 1 m long geotextile interface (labeled B on Fig. 1a): *left* ground-coupled radar Bscan; *middle* clutter-free Bscan; *right* clutter-free time signature in *red* compared to the signature over the healthy interface (with tack coat) in *black*; the echo over the geotextile is likely reinforced by some water seeping

Specific signal processing techniques have been applied to separate overlapping echoes at some extent, namely, model-based processing in Le Bastard et al. (2007) and supervised SVR technique in Le Bastard et al. (2014). Model-based processing techniques can separate overlapping echoes which the minimum B × Δτ product is 0.1 at medium SNR (20 dB). The latter time resolution capability matches to the detection of debonding larger than a few mm in thickness for the UWB step-frequency radar. This may be not sufficient enough within the scope of the application.

Time-frequency analysis of the echoes has been then explored in Ma (2014) and was included in a machine learning-based signal processing technique. The Kolmogorov distance between the debonding and the debonding-free Time Frequency Signatures (TFS) was found sensitive enough to allow classifying the simulated data by SVM. As an illustration, Fig. 5 depicts the path that follows the peak location of the short-term Fourier transform (STFT) in the time-frequency domain with increasing debonding thicknesses.

Besides, rigorous data simulation in Pinel et al. (2011) have shown that inter-layer roughness modifies the shape of the backscattered echoes and then limit the performance of model-based time delay estimation techniques. Then, 2D signal processing have been proposed in Sun et al. (2015) to recover full time resolution.

Finally, Bourlier et al. (2015) have shown that the interlayer roughness within a 10 mm thick debonding interface further modifies the shape of the backscattered

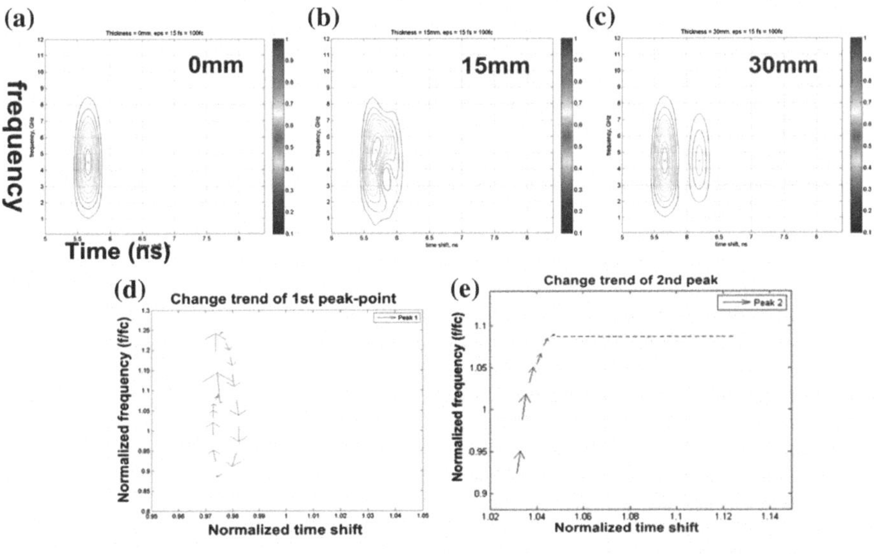

Fig. 5 Time-Frequency analysis for a refined waveform analysis of simulated radar data; *top* magnitude of the STFT with three debonding thicknesses; *bottom* path of the peak location in the time-frequency domain with increasing debonding thickness on the first peak on the left and for the second peak on the right

echo and favors the rise of multiple scattering. Multiple overlapping echoes are then intended to greatly limit the resolution capability of the signal processing techniques which were tested previously.

5 Conclusion and Perspectives

The paper has given an overview of the ongoing experiment for monitoring artificial debonding areas within pavement structure with a special emphasis on EM radar techniques and related studies. The experiment is expected to be pursued beyond 300 loading kcycles to reach larger pavement degradations.

In the near future, data which have been collected at the different loading cycles and beyond may be made available to the GPR community, e.g., the COST TU 1208 (Pajewski et al. 2013). Data processing will be further developed to help the user for data interpretation.

References

Bourlier C, Le Bastard C, Baltazart V (2015) Generalization of PILE method to the EM scattering from stratified subsurface with rough interlayers - Application to the detection of debondings within pavement structure, IEEE Trans. Geosc. Remote Sensing, Vol. 53 (7)

Dérobert, X (2004) Step-frequency radar technique applied on very-thin layer pavements. In: Surface-penetrating radar, Daniels D.J. (ed), Inst. Electrical Engineers. London, pp. 386-394.

Le Bastard C, Baltazart V, Wang Y, Saillard J (2007) Thin pavement thickness estimation using GPR with high and super resolution methods, IEEE Trans. on Geosc. Rem. Sensing, Vol. 45, 8, pp 2511-2519

Le Bastard C, Wang Y, Baltazart V, Dérobert X (2014) Time Delay and Permittivity Estimation by Ground Penetrating Radar with Support Vector Regression", IEEE Geosc. Remote Sensing Letters, Vol. 11(4)

Pinel N, Le Bastard C, Baltazart V, Bourlier C, Wang Y (2011) Influence of surface roughness on thin layer thickness evaluation: Application to the theoretical study of road layers, IET Sonar and Navigation, Vol. 5(6), pp 650-656

Ma S (2014) GPR signal classification with SVM for the detection of interface debonding within pavement structure, Master Aria, LUNAM University, Ecole Centrale de Nantes

Pajewski L, Benedetto A, Dérobert X, Giannopoulos A, Loizos A, Manacorda G, Marciniak M, Plati C, Schettini G, Trinks I (2013) Applications of Ground Penetrating Radar in Civil Engineering – COST Action TU1208, in Proc. IWAGPR, 2-5 July, Nantes, France.

Picoux, B., Takarli, M., Petit, C. (2011). Non-destructive testing of an experimental pavement. Paper presented at the Proceedings of the 13th International Conference on Civil, Structural and Environmental Engineering Computing

Simonin J-M, Hornych P, Baltazart V, Dérobert X (2014), Case study of detection of artificial defects in an experimental pavement structure using 3D GPR systems, paper presented in International conference on GPR, Brussels

Simonin J-M, Baltazart V, Hornych P, Kerzrého J-P, Dérobert X, Trichet S, Durand O, Alexandre J, Joubert A (2012a) Detection of debonding and vertical cracks with ND techniques during accelerated tests, paper presented in 4th International conference on Accelerated Pavement Testing (APT), Davis, California, USA

Simonin J-M, Fauchard C, Hornych P, Guilbert V, Kerzrého, J-P, Trichet S. (2012b) Detecting unbounded interface with non destructive techniques, paper presented at Rilem Conference on Craking in pavement, Delft, Netherlands

Sun M, Pinel N, Le Bastard C, Baltazart V, Ihamouten A, Wang Y (2015) Time delay and interface roughness estimations by GPR for pavement survey, Near Surface Geophysics, EAGE, Vol. 13 (3), pp.279 – 287

Tebchrany E, Sagnard F, Baltazart V, Tarel J-P, Dérobert X (2014) Assessment of ICA and PCA - based time domain Clutter Reduction Techniques on ground-coupled GPR data, paper presented at GPR 2014, Brussels, Belgium

Relative Near Surface Pavement Performance for Dual and Wide-Base Tyre Assemblies Using a Finite Element Method

Dermot B. Casey, Gordon D. Airey and James R. Grenfell

Abstract There is a need to assess the realistic tyre contact pressure created by a tyre in contact with a pavement. A traditional representation of tyre-pavement contact pressure is a circular uniform contact patch. This is overly simplistic, the contact pressure is a function of tyre type, axle loading and tyre inflation pressure (TiP). The research carried out considered dual tyre and wide-based tyre assemblies across a range of axle loading and TiP. These contact pressures were incorporated into a finite element package (CAPA-3D) and modelled on a thin pavement structure. The strains from this modelling were sorted to produce key shear strains associated with the key mechanisms of near surface pavement distress. The main distress mechanisms being top down cracking and asphalt cracking/rutting. This gave a method to fairly compare the dual and wide-based tyre assemblies with the same axle loading and TiP. The analysis gave interesting results for the different distress mechanisms of the pavement. The wide-based tyre gives consistently higher shear strains for all the areas of distress investigated. There is great variation in shear strains due to the different combinations of axle loading and TiP. It is clear that the wide-based tyre is a more damaging tyre for all combinations of TiP and axle loading. It is also apparent that how these factors interact has a great influence on the damaging potential of a tyre.

Keywords Tyre-pavement · Contact pressure · Shear strain · FEM

D.B. Casey (✉)
AECOM, Nottingham NG9 6RZ, UK
e-mail: Dermot.Casey@aecom.com

G.D. Airey · J.R. Grenfell
Nottingham Transportation Engineering Centre (NTEC), University of Nottingham, University Park, Nottingham NG7 2RD, UK
e-mail: Gordon.Airey@nottingham.ac.uk

J.R. Grenfell
e-mail: James.Grenfell@nottingham.ac.uk

A. Chabot et al. (eds.), *8th RILEM International Conference on Mechanisms of Cracking and Debonding in Pavements*, RILEM Bookseries 13,
DOI 10.1007/978-94-024-0867-6_105

1 Introduction

The increase in traffic is leading to increased deterioration of pavements. There is also the issue of how different tyre assemblies transfer a load into the pavement for the same TiP and axle loading. This paper endeavours to highlight this issue for the near surface of a pavement.

The magnitude of the 3-D tyre contact pressures has been shown to be highly influenced by the tyre type, tyre loading and TiP (De Beer et al. 1997; Blab 1999; Novak et al. 2003; Fernando et al. 2006; Wang and Roque 2010). It has been shown that there are significant differences to surface cracking potential caused by non-uniform tyre contact pressure (Casey et al. 2014). An individual tyre can have a large range of vertical, transverse and longitudinal tyre contact pressure depending on the TiP and tyre loading (Prozzi and Luo 2005; Fernando et al. 2006).

2 Objectives

This paper uses 3-D tyre contact pressure representations of dual and wide-based tyre assemblies. This is to understand the effects these tyres have on the pavement if they have the same axle loading and TiP. The key objectives of the paper to assess the effects are:

- To assess the shear strains generated on the surface of the pavement for the two tyre assemblies
- To assess the shear strains generated between 50-80 mm from the surface of the pavement for the two tyre assemblies.

These measures act as a method of comparing the two tyres for key areas for the performance of a pavement for the major forms of near surface pavement distress.

3 Method

The model in this paper utilises a static loading, a linear-elastic material model, a simple pavement structure and an intricate 3-D tyre contact pressure loading technique using a finite element method. There are 2 tyres used, one a wide-based tyre and the other a dual tyre assembly. Each of these tyres is loaded with 3 TiPs and 3 axle loadings for a total of 5 loading combinations between both of them. It gives a reasonable spread of inflation and tyre loading conditions to compare the two tyre types against each other.

The dual tyre assembly consists of two 295/75R22.5 tyres and the wide-base radial tyre consists of a 425/65R22.5 tyre. These parameters give a basis to conduct a comparison of the relative aggression of each axle to a specific mode of distress.

The loads used are from a software package called Tireview from the Texas Transportation Institute. This is a library of tyre contact pressures from the VRSPTA (Vehicle-Road Surface Pressure Transducer Array) stress in motion sensor for different tyres, axle loads and TiPs (De Beer et al. 1997).

The pavement is modelled as a three-layer system with an asphalt layer, a sub-base layer and a subgrade layer. The asphalt layer is 150 mm in thickness, the sub-base is 300 mm and the subgrade layer is assumed to behave as a semi-infinite layer but for the purposes of the model is 2 m deep. This is a simple pavement structure that is of reasonable composition for a regional pavement and could be classed as a thin pavement structure.

The material properties for the pavement are isotropic linear elastic and represent values that would be typical for such layers and for the type of pavement. The asphalt modulus is 7500 MPa with a Poisson's ratio of 0.35. These values represent good quality asphalt at lower temperatures. The sub-base layer has an elastic modulus value of 690 MPa and Poisson's ratio of 0.35. The subgrade layer is modelled as a semi-infinite layer with a modulus of 100 MPa and a Poisson's ratio of 0.35.

The finite element analysis utilises a 4 m × 4 m and 2.45 m deep finite element mesh consisting of 20-noded quadratic elements using full integration. The mesh is fixed in all directions at the base and restrained in the lateral direction at the sides. The tyre contact area is set to match the discretisation of the raw tyre contact pressure data, which are 15 mm in the longitudinal direction and 17 mm in the transverse direction (Fig. 1).

The 3-D tyre contact pressure loading method works by matching the mesh dimensions to the dimensions of the discretisation in the loading area of the mesh. This was checked for each of the tyre contact pressures used to assure that this reduced data remains representative of the raw data. The stress and strain outputs for the model for a simple circular contact pressure representation were compared against BISAR, they compared excellently against each other (De Jong and Peutz

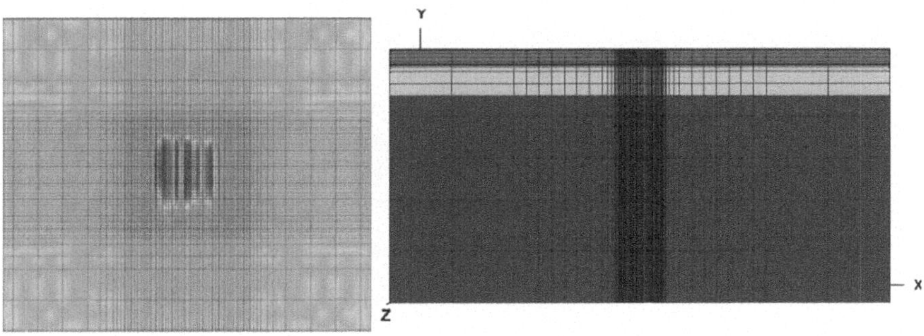

Fig. 1 Refinement of the top of the mesh and cross-section of the pavement

1973). This showed that the model behaves well and the mesh density and boundary conditions are stable.

There were over 45,000 strain output values recorded for each run of the model. The areas of interest were on the surface of the asphalt layer and to a depth of 50–80 mm in the asphalt layer. The procedures and the model layout presented here represent a good basis for the goals of this part of the research.

4 Results

In this section a direct comparison between a dual 295 tyre and a wide-based 425 tyre is presented. This was undertaken to show under the same conditions which would result in higher maximum shear strains for the identified key locations for near surface pavement life. The values used for the comparison were restricted due to the values available in the Tireview software for both tyres. However, this does at least allow for a comparison for equal TiP and axle loading to be undertaken. There were 5 loading combinations for each tyre.

The distribution of the differences between the two tyres is interesting. The half axle load of 65 kN results in the two values being rather close for the maximum shear strain on the pavement surface and the three tyre inflation pressures (Fig. 2). However, at the two extremes of axle load there is a large variation between the two tyres. Therefore, in these two cases the wide-based tyre would cause much greater shear strain on the surface. The shear strain in the asphalt between 50 and 80 mm from the surface shows considerable differences between the two tyres. The wide-based tyre has created higher values for all the loading combinations. The greatest difference was for the 82 kN half axle load which was 47 microstrain (Fig. 3).

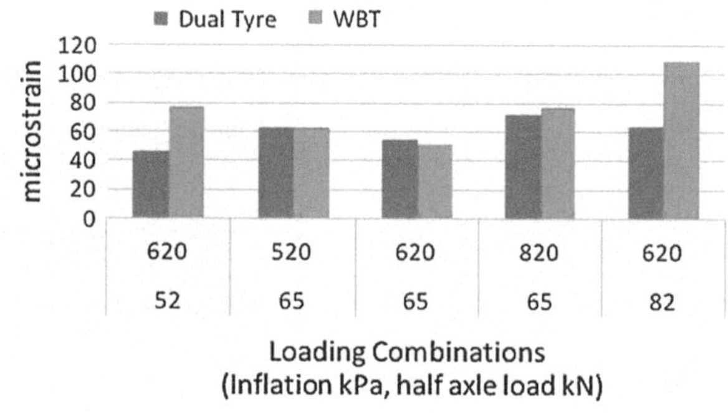

Fig. 2 Comparison of the maximum surface shear for a dual and wide-base tyre

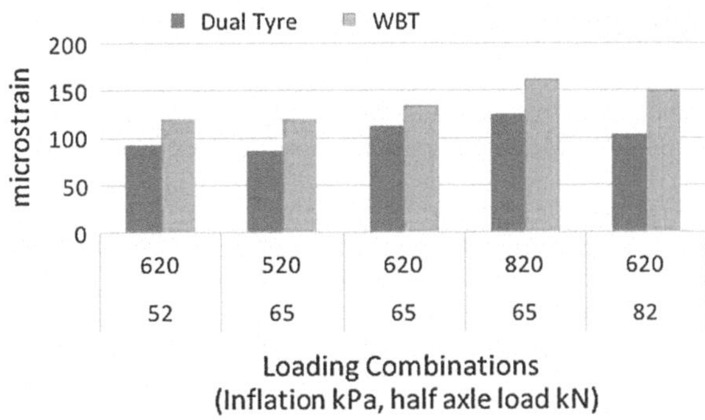

Fig. 3 Comparison of the maximum asphalt shear for a dual and wide-base tyre

5 Discussion

The two tyres with the same loads and same TiPs were tested against each other to see if one could replace the other without undue change to the shear strain in the pavement. This was most definitely not the case as the wide-based tyre created higher strains for every location and strain measured. The shear strain on the surface was similar over the 65 kN axle load but for the lower 52 kN and higher 82 kN there was a large increase for the wide-based tyre (Fig. 2). The situation for the shear strain in the asphalt gave consistently higher shear strains for the wide-based tyre across all the loading scenarios (Fig. 3). This would no doubt increase the propensity of the pavement to crack or rut.

6 Conclusion

These results point to a clear outcome that the wide-based tyre causes varying degrees of increased shear strain on a pavement compared with a dual tyre for the same axle load and tyre inflation pressure.

- A wide-based tyre for the same axle loadings and TiPs creates higher shear strain across all presented measurements than a dual tyre with the same loading combinations.
- The difference between the wide-based tyre and the dual tyre were greatest at the extremes of axle loading for the presented shear strain measurements for the direct comparison.

This means by consequence the pavement become damaged quicker than if dual tyres were used. This could lead to more maintenance interventions in the future.

Acknowledgements The authors would like to acknowledge the support of the European Commission under the Marie Curie Intra-European Fellowship Programme. The authors would like to thank the CAPA-3D group at TU Delft under the leadership of Prof. Tom Scarpas for their help.

References

Blab, R. (1999). Introducing Improved Loading Assumptions into Analytical Pavement Models Based on Measured Contact Stresses of Tires. *Paper presented at the International Conference on Accelerated Pavement Testing, Reno, NV, USA.*

Casey, D. B., Grenfell, J. R., & Airey, G. D. (2014, April). 3-D longitudinal and transverse cracking and the influence of non-uniform contact pressure on the stress intensity factors of these Cracks. In *Transport Research Arena (TRA) 5th Conference: Transport Solutions from Research to Deployment.*

De Beer, M., Fisher, C., & Jooste, F. J. (1997). Determination of pneumatic tyre/pavement interface contact stresses under moving loads and some effects on pavements with thin asphalt surfacing layers. In *Proceedings of the 8th International Conference on Asphalt Pavements, Seattle, Washington, USA* (Vol. 1, pp. 10-14).

De Jong, P. D., & Peutz, M. (1973). Korswagen. *A R. Computer program BISAR layered systems under normal and tangential surface loads. Koninklijke/Shell Laboratorium, Amsterdam, External Report AMSR, 6.*

Fernando, E. G., Musani, D., Park, D. W., & Liu, W. (2006). *Evaluation of effects of tire size and inflation pressure on tire contact stresses and pavement response* (No. FHWA/TX-06/0-4361-1).

Novak, M., Birgisson, B., & Roque, R. (2003). Near-surface stress states in flexible pavements using measured radial tire contact stresses and ADINA. *Computers & Structures, 81*(8-11), 859-870.

Prozzi, J. A., & Luo, R. (2005). Quantification of the joint effect of wheel load and tire inflation pressure on pavement response. *Transportation Research Record.*(1919), 134-141.

Wang, G., & Roque, R. (2010). Three-Dimensional Finite Element Modeling of Static Tire-Pavement Interaction. *Transportation Research Record: Journal of the Transportation Research Board, 2155*(-1), 158-169.

Assessment of HWD Ability to Detect Debonding of Pavement Layer Interfaces

Sadoun Amir, Broutin Michaël and Simonin Jean-Michel

Abstract Usual Falling or Heavy Weight Deflectometer (F/HWD) backcalculation methods assume that pavement layers are fully bonded. In order to assess the ability of HWD testing to detect debonding of the interfaces between pavement layers, the French civil Aviation technical center (STAC) performed an HWD campaign on the circular Accelerated Pavement Testing (APT) facility of IFSTTAR in Nantes, on an experimental pavement with artificially created local sliding interface between bituminous surface and base layers. Both the sound and the defective tested pavement sections present a common structure in terms of subgrade, materials and layer thicknesses. HWD tests have been performed along a large circular arc covering the two sections. A fine spatial measure discretization (10 cm) was used. The experimentation shows that the HWD is a valuable tool for the detection of extended interface defects detection, the central deflection being significantly affected by the interface quality, the outer deflections being less sensitive to this parameter. This paper presents raw deflection results, allowing to precisely locate the underlying defects in the test section. Then a numerical study is detailed, which aims at assessing the effect of the bonding conditions over deflections. The theoretical analysis of this HWD survey is performed in two steps: firstly, on the basis of the finite elements method (FEM) dynamical modeling of the HWD test developed by the STAC, a backcalculation of material moduli is performed using the deflections measured in the sound area. Secondly the backcalculated moduli are introduced in a forward calculation, taking into account the debonded interface. The numerical results are compared with experimental deflections measured over the defects. They are consistent with in situ results.

S. Amir · B. Michaël (✉)
French Civil Aviation Technical Center, 31 av. du Maréchal Leclerc, 94381 Bonneuil, France
e-mail: michael.broutin@aviation-civile.gouv.fr

S. Amir
e-mail: sadoun.amir@gmail.com

S. Jean-Michel
French Institute of Science and Technology for Transport Development and Network, IFSTTAR, Route de Bouaye, CS4, 44344 Bouguenais, France
e-mail: jean-michel.simonin@ifsttar.fr

A. Chabot et al. (eds.), *8th RILEM International Conference on Mechanisms of Cracking and Debonding in Pavements*, RILEM Bookseries 13,
DOI 10.1007/978-94-024-0867-6_106

Keywords Debonded interface · APT · HWD testing · Dynamical backcalculation

1 Experimentation Presentation

A full-scale experimentation was launched on the IFSTTAR's APT facility (Kerzrého 2014) in 2012 to assess the ability of non destructive testing (NDT) devices to detect interface defects. It included debonded areas artificially created in the pavement during the construction works. The STAC performed HWD tests on both the sound pavement and the areas presenting localized interface defects.

The pavement structure is presented in (Simonin et al. 2012). It is composed of a 6 cm thick wearing course of Bituminous Concrete (BC) and a 8 cm thick base layer of Road Base Asphalt material (RBA) over a 30 cm thick unbound granular material (UGM). These material properties meet the specifications of the NF EN 13,108 series of standards. The depth to bedrock (corresponding to the APT concrete structure) is 3 m. The modulus of the subgrade between UGM and concrete was estimated using static plate tests. A 70 MPa value was obtained, corresponding to a value in the 140–210 MPa range for dynamic tests. Three 2 m × 1.5 m rectangular defects, 2 m apart from each other, were created at the interface between the bituminous layers by means of a thin layer of sand, a geotextile or the absence of tack coat.

HWD tests were performed with a 10 cm step, starting from 0.90 m before the first defect boundary, to 0.90 m after the end of the third one. Tests were performed in March 2012, when the pavement had not yet been circulated by the APT device, so that the areas without defects can be considered as sound. Mean temperature in the bituminous layer during the HWD was in the 15–20 °C range.

2 Qualitative Approach

Figure 1 presents raw results relative to peak deflection values recorded on the central geophone (positioned at plate center) as a function of the distance between the loading plate and the defects boundary D_X. HWD peak load was kept constant for the whole survey (equal to 82 kN ± 2 kN).

It appears that the central deflection increases significantly (on average plus 53 %) when moving from sound pavement to defect area. This underlines that HWD testing is relevant for the detection of such size of localized defects. Same analysis on outer geophone (2.40 m offset from load center) shows that measured peak deflection is almost constant ($\sim$30 ± 1 µm) regardless of D_X. This was predictable considering that the defect is relative to the upper layers. It is thus confirmed that outer deflections are almost exclusively influenced by deep layers deformations, whereas the center deflection is impacted by the whole structure.

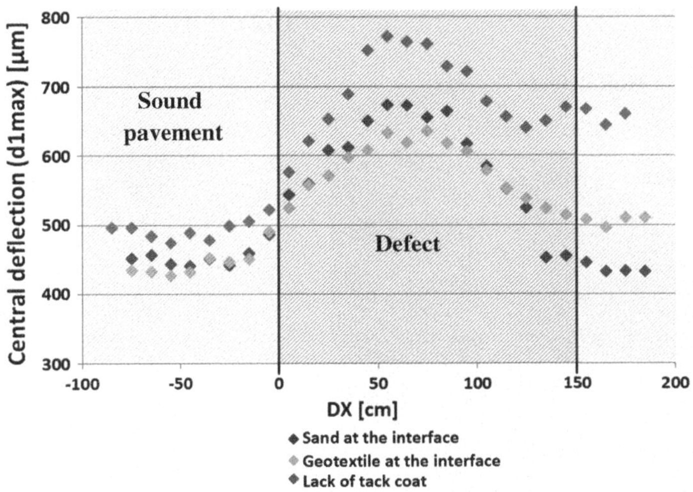

Fig. 1 Central deflection peak values—in situ measurements

3 Numerical Analysis

The analysis methodology proposed in this paper is as follows: a dynamical 3D-backcalculation on the "sound" pavement is first performed, in order to determine the layer material moduli. Then several forward calculation using the afore-backcalculated moduli are performed, which correspond to each experimental load plate position. These forward calculations include the defect in the modeling. This paragraph describes first the modeling used for back and forward calculations. The backcalculation results are then displayed. The synthesis of the forward calculations is finally presented, and compared to in situ data.

3.1 Modeling

The modeling is inspired by the 2D-axisymmetric dynamical method developed by the STAC for sound flexible pavement analysis (Broutin 2010; STAC 2014). The FE calculation, performed using the CESAR-LCPC software (LCPC 2005) has been adapted for the 3D-case under study. The 3 m deep bedrock is taken into account in the modeling. The defect was modeled using double-nodes. For each pair of nodes, linear relationships enable fixing at zero the vertical relative displacements of the two nodes, whereas relative horizontal displacements remain free. Therefore the two conditions of complete debonding and non-interpenetration between layers are verified.

Using symmetry considerations, half a plate and half a defect are taken into account in the model. Figure 2a presents the mesh used for the "sound" pavement

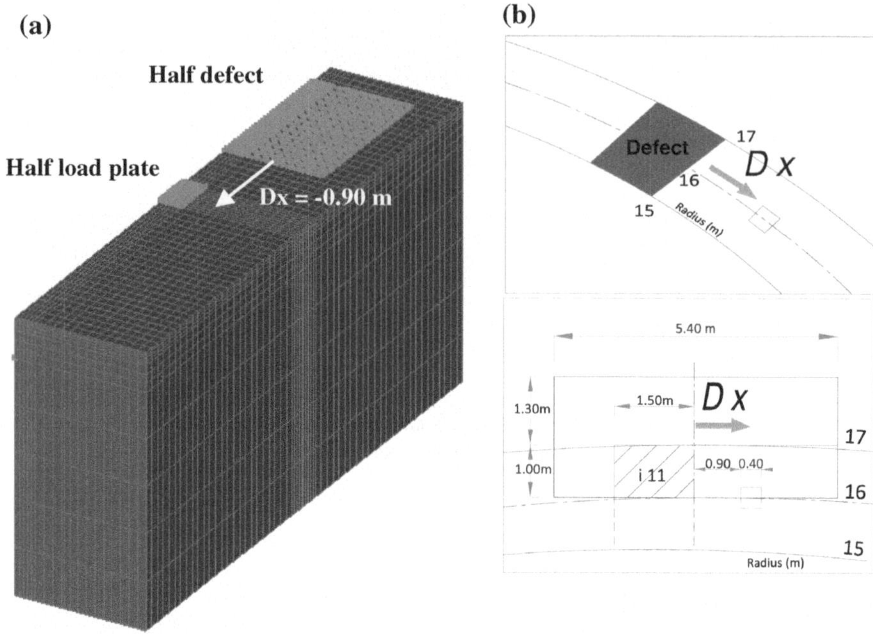

Fig. 2 Back- and forward calculation **a** FE mesh used for backcalculation, $D_X = -0.90$ m; **b** defect positions, $D_X = -0.90$ to $D_X = 2.40$ m, positioned between the 15 and 17 m APT radius

configuration, i.e. with a load plate outside the defect area (in practice at 0.90 m left from boundary). Figure 2b displays in a schematic view the different meshes used for forward calculations. These meshes are similar to the previous one, but moving the defect along the X-axis. Let again D_X be the distance between load plate center and defect boundary. In both cases the research version of the PREDIWARE software (Broutin 2010) developed by the STAC for flexible pavement HWD dynamical back-and forward calculation automation was adapted.

3.2 *Backcalculation*

Figure 3 presents the backcalculation results, obtained with PREDIWARE. In situ measurements are displayed in solid lines whereas numerical deflection signals appear in dotted lines. For readability purposes only five deflections are displayed, corresponding to geophones 1, 3, 7, 10 and 12, respectively positioned at 0, 0.4, 0.9, 1.5 and 2.1 m from load center.

Note that no structural damping has been considered in the current calculation. This may explain that numerical curves present slight slope changes between 15 and 20 ms which are not encountered for the smoother curves from in situ results.

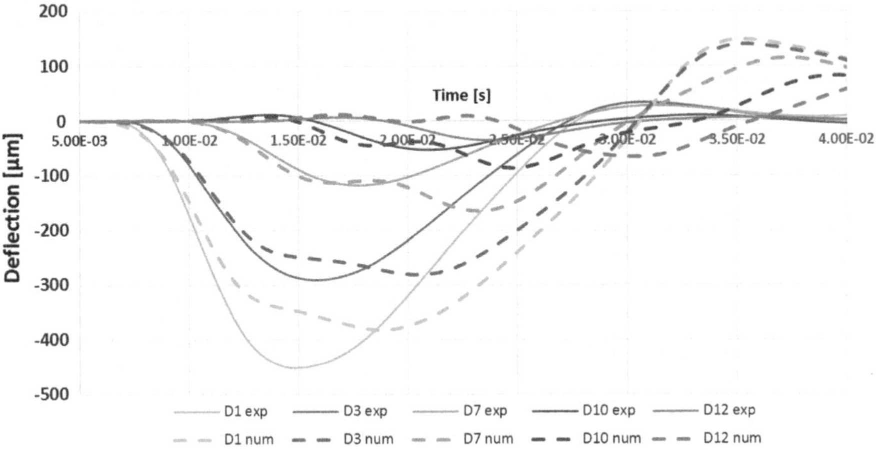

Fig. 3 Backcalculation results—distance between load plate and defect $D_X = 0.90$ m

Table 1 Backcalculated moduli

	BC	RBA	UGM	Subgrade
Backcalculated moduli (MPa)	7000	7000	240	180
Expected values (MPa)	4500–6500	8000–11,000	210–315	140–210

Backcalculated moduli corresponding to Fig. 3 fitting are given in Table 1, and compared with typical values at temperature between 15 and 20 °C and a frequency of 30 Hz, representative of the HWD impulse loading.

Subgrade and UGM moduli are consistent with expectations. Moduli of bituminous material fall between the expected ranges, with the BC value slightly higher and the RBA value slightly lower than expected; this is in accordance with general observations made in Broutin (2010) showing that bituminous materials moduli are generally misestimated due to the fact that their viscoelastic behavior is neglected. Work is in progress in STAC to better account for this effect.

3.3 Forward Calculation Results

Figure 4 depicts (blue points) central deflection peak values extracted from forward calculations results for $D_X = -0.90$ to 2.40 m, using the backcalculated moduli in Table 1.

It shows that numerical resolution enables bringing to light the defect. The deflection evolution shape and order of magnitude are consistent with in situ results. Difference observed may be explained by the fact that the backcalculation fitting is not perfect. Actually numerical peak value of the central deflection (geophone d1)

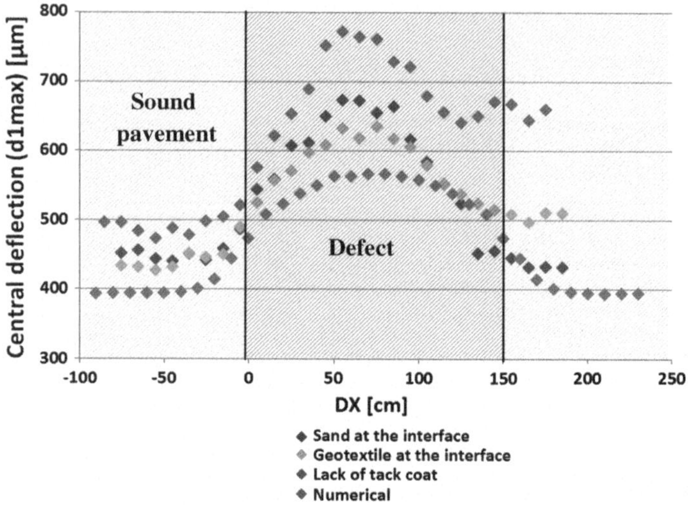

Fig. 4 Central deflection peak values—comparison between numerical results and in situ measurements

under study in this paper is slightly lower than the experimental one according to Fig. 3. This explains partly the underestimation of the deflections on this geophone in the forward calculations performed from backcalculated moduli. Furthermore defects have been considered as independent in the calculation, even if they are relatively close to each other (2 m).

4 Conclusion and Future Work

It has been shown in this paper that HWD is a valuable tool for detecting large-size (1.50 m × 2 m in this study) defects. Qualitative analysis using raw central and outer deflections enables identifying a surface defect. Visual inspection or complementary tests should enable differentiating between interface defect and bulk problems (material damage leading to a fall in modulus). The possibility of integrating localized interface defects in the modeling has also been proven for both back-and forward calculations. These results will be completed in the future using on the one hand complementary calculations taking into account bituminous materials viscoelastic behavior, and on the other hand the analysis of additional in situ data. To achieve this, 2012 test survey presented in this paper was reiterated in 2013 at pavement mid-life. STAC plans a last HWD test survey in 2016, after completion of the full scale APT life cycle test.

References

Broutin M. (2010) Assessment of flexible airfield pavements using Heavy Weight Deflectometers; Development of a FEM dynamical time-domain analysis for the backcalculation of structural properties, PhD thesis, http://www.stac.aviation-civile.gouv.fr/chaussee/ausc_hwd.php Accessed 10 oct 2014

Kerzreho J.-P. (2014) Setup and use of the IFSTTAR's fatigue test track, RGRA ERR review 23-24, Fall 2013- Spring 2014.

LCPC (2005) Numéro spécial CESAR-LCPC, Bulletin des Laboratoires des Ponts et Chaussées, N° 256-257

STAC (2014) Guidance for flexible pavement testing using Heavy Weight Deflectometer (HWD)

Simonin J.-M., Fauchard C., Hornych P., Guilbert V., Kerzrého J-P., Trichet S., Detecting unbounded interface with non destructive techniques. 7th Rilem International Conference on Cracking in Pavements, June 20-22 2012, Delft, The Netherlands. RILEM Bookseries, 4: 179-190, DOI:10.1007/978-94-007-4566-7_18, ISBN 978-94-007-4565-0,

Author Index

A. Chabot et al. (eds.), *8th RILEM International Conference on Mechanisms of Cracking and Debonding in Pavements*, RILEM Bookseries 13,
DOI 10.1007/978-94-024-0867-6

The manufacturer's authorised representative in the EU is Springer Nature Customer Service Centre GmbH, Europaplatz 3, 69115 Heidelberg, Germany. If you have any concerns regarding our products, please contact ProductSafety@springernature.com

Printed and bound by CPI Group (UK) Ltd, Croydon, CR0 4YY
25/09/2026
02222884-0020